Environmental Physiology of Fishes

NATO ADVANCED STUDY INSTITUTES SERIES

A series of edited volumes comprising multifaceted studies of contemporary scientific issues by some of the best scientific minds in the world, assembled in cooperation with NATO Scientific Affairs Division.

Series A: Life Sciences

Recent Volumes in this Series

Volume 25 – Synchrotron Radiation Applied to Biophysical and Biochemical Research
edited by A. Castellani and I. F. Quercia

Volume 26 – Nucleoside Analogues: Chemistry, Biology, and Medical Applications
edited by Richard T. Walker, Erik De Clercq, and Fritz Eckstein

Volume 27 – Developmental Neurobiology of Vision
edited by Ralph D. Freeman

Volume 28 – Animal Sonar Systems
edited by René-Guy Busnel and James F. Fish

Volume 29 – Genome Organization and Expression in Plants
edited by C. J. Leaver

Volume 30 – Human Physical Growth and Maturation
edited by Francis E. Johnston, Alex F. Roche, and Charles Susanne

Volume 31 – Transfer of Cell Constituents into Eukaryotic Cells
edited by J. E. Celis, A. Graessmann, and A. Loyter

Volume 32 – The Blood–Retinal Barriers
edited by Jose G. Cunha-Vaz

Volume 33 – Photoreception and Sensory Transduction in Aneural Organisms
edited by Francesco Lenci and Giuliano Colombetti

Volume 34 – Lasers in Biology and Medicine
edited by F. Hillenkamp, R. Pratesi, and C. A. Sacchi

Volume 35 – Environmental Physiology of Fishes
edited by M. A. Ali

This series is published by an international board of publishers in conjunction with NATO Scientific Affairs Division

A Life Sciences
B Physics
Plenum Publishing Corporation
London and New York

C Mathematical and Physical Sciences
D. Reidel Publishing Company
Dordrecht, Boston and London

D Behavioral and Social Sciences
E Applied Sciences
Sijthoff & Noordhoff International Publishers
Alphen aan den Rijn, The Netherlands, and Germantown U.S.A.

Environmental Physiology of Fishes

Edited by

M. A. Ali

Département de Biologie
Université de Montréal
Montréal, Quebec, Canada

PLENUM PRESS • NEW YORK AND LONDON
Published in cooperation with NATO Scientific Affairs Division

Library of Congress Cataloging in Publication Data

Nato Advanced Study Institute on Environmental Physiology of Fishes, Bishop's University, 1979.
Environmental physiology of fishes.

(NATO advanced study institutes series: Series A, Life sciences; v. 35)
"Lectures presented at the 1979 NATO Advanced Study Institute on Environmental Physiology of Fishes, held at Bishop's University, Lennoxville, Québec, Canada, August 12–25, 1979."
"Co-sponsored by The Université de Montréal"
Includes indexes.
1. Fishes–Physiology–Congresses. 2. Adaptation (Physiology)–Congresses. I. Ali, Mohamed Ather, 1932- II. Université de Montréal. III. Title. IV. Series.
QL639.1.N35 1979 597'.01 80-22156
ISBN 0-306-40574-1

Lectures presented at the 1979 NATO Advanced Study Institute on Environmental Physiology of Fishes, held at Bishop's University, Lennoxville, Quebec, Canada, August 12–25, 1979.

Co-sponsored by the Université de Montréal

Director: M. A. Ali, Montréal, Quebec, Canada

Advisory Committee: J.H.S. Blaxter, Oban, Scotland
F.E.J. Fry, Toronto, Ontario, Canada
K. F. Liem, Cambridge, Massachusetts

A Division of Plenum Publishing Corporation
227 West 17th Street, New York, N.Y. 10011

Printed in the United States of America

PREFACE

A very good piece of work, I assure you, and a merry.
-Now, good Peter Quince, call forth your actors by the scroll.
-Masters, spread yourselves.

A Midsummer Night's Dream. Act 1, Sc. 2

This volume is the outcome of a NATO Advanced Study Institute held in August 1979 at Bishop's University, Lennoxville, Québec, Canada. About 130 participants from all the countries of the alliance as well as India and Japan attended this event which lasted two weeks. Seventeen of these participants had been invited to present reviews of chosen topics, usually in their specialty. This book is constituted mainly of these presentations, which were prepared as chapters. In addition, six of the participants, whose seminars were found to complement the main chapters, were coopted by the invited lectures/authors to provide additional chapters. Although a lecture was given on electric fields, a chapter on this matter is unfortunately absent due to the lack of preparation time.

One may say that Environmental Physiology of Fishes as a discipline originated in Canada. Having been involved as a teacher and worker in this field since 1954, it was but natural that I was tempted to organise an ASI and get a volume out on the matter. I was encouraged by discussions with colleagues and the acceptance on the part of a large number of eminent colleagues to attend the ASI, deliver lectures and write chapters. It was felt that participants made up of a heterogenous group varying from pre-doctorate students to workers nearing retirement, on the one hand and, workers interested in the entire animal, e.g. organs, tissues, even cells, chemicals factors, physical factors, behaviour, rhythms etc., on the other would have nothing but to gain much from the company of one another, were they to spend two weeks working, eating and playing together. I am glad to say that this proved to be correct judging by the constant discussions which took place at all times, the friendships which were made and the post-ASI contacts which were established. I feel that more than the impact the volume may have on the field, the interaction between younger and older workers and among workers from so many different fields will prove to be more advantageous for the advancement of this field of endeavour in the next ten or even twenty years.

Fishes form the largest class of vertebrates and occupy a very wide variety of habitats, some even non-aquatic. In every aspect of their physiology they demonstrate a larger repertoire of adaptive features, more than any other class of animal with the possible exception of the crustaceans. My hope is to have this volume reflect these numerous physiological adaptations in them, with the different environmental factors and variations. I asked the authors to prepare critical, or general reviews, and encouraged them to be as provocative and speculative as they wanted. This book was planned to be reasonable in length and yet cover as much of the field as possible. Obviously not all fields could be covered due to the paucity of space, time or lecturers/authors. In a volume of this sort a certain amount of overlap or repetition is unavoidable. I have tried to keep it to a minimum but may not have entirely succeeded since I am certainly not expert in all aspects of the physiology of fishes. As the director of the ASI, I attended every lecture and seminar and, as the editor I have read every chapter in this volume carefully. This has considerably broadened my knowledge and I am grateful for having had this opportunity and hope that the reader will find the effort worthwhile.

It appears also desirable to explain briefly the process by which the lecturers/authors are selected for the ASI and the ensuing volume. Being a NATO-ASI, lecturers have to be drawn from as many member countries of the alliance as possible. In most cases, travel costs have also to be taken into consideration. Other factors which are not insignificant are lecturing and writing ability of the person, his (or her) ability to deal with a heterogenous group scientifically and socially over a two-week period. Taking all these points into consideration a list is drawn up with the help of the advisory committee and correspondence begins - often as early as two years before the ASI is to take place and chapters are to be submitted. In spite of such early arrangements some persons found it unavoidable to withdraw their presence a few months before the event. In this case they have usually been replaced but if the withdrawel was just a few weeks before the ASI it was virtually impossible to ask someone else to take their place. The presentations at the ASI are followed by discussions and towards the end a meeting of the authors and the editor takes place. At this meeting matters are discussed openly and suggestions are made to fill the lacunes in the volume by inviting one or two or more of the other participants, generally those whose seminars were found to fill such a need, to contribute chapters on subjects and lines established by the consensus. After this, of course, matters are between the editor and the individual author.

I am grateful to Dr. Mario di Lullo of the NATO-Scientific Affairs Division and his predecessor Dr. Tilo Kester for the advice and encouragement they gave during the planning and organising of this ASI. I thank Dr. René J.A. Lévesque, vice-rector for research of my University and Prof. Jean-Guy Pilon, chairman of my department, for the material and moral support they provided, in the organisation of the ASI and the preparation of this volume.

It would have been impossible to conduct the ASI without the kind support and cooperation of Monsieur J.L. Grégoire, vice-principal and his assistant Mrs. Lillian Garrard and Mr. Ivan Saunders, directors of buildings and grounds of Bishop's University. I am grateful to them for everything they did to make my task less arduous and our stay enjoyable. I thank also the members of my advisory committee for their help in the choice of lecturers and participants. I am grateful to my colleague Dr. Mary Ann Klyne for all the assistance she so willingly gave in the organising and running of the ASI as well as the editing of the volume. My librarian, Miss Margaret Pertwee returned from her second retirement to help in the preparation of the volume, particularly the bibliographies and indices and I wish to record my thanks to her for that.

I thank Mesdemoiselles Marielle Chevrefils and Jocelyne Trudeau for preparing the camera ready manuscript with a Xerox word processing system.

Montréal
March 1980

Contents

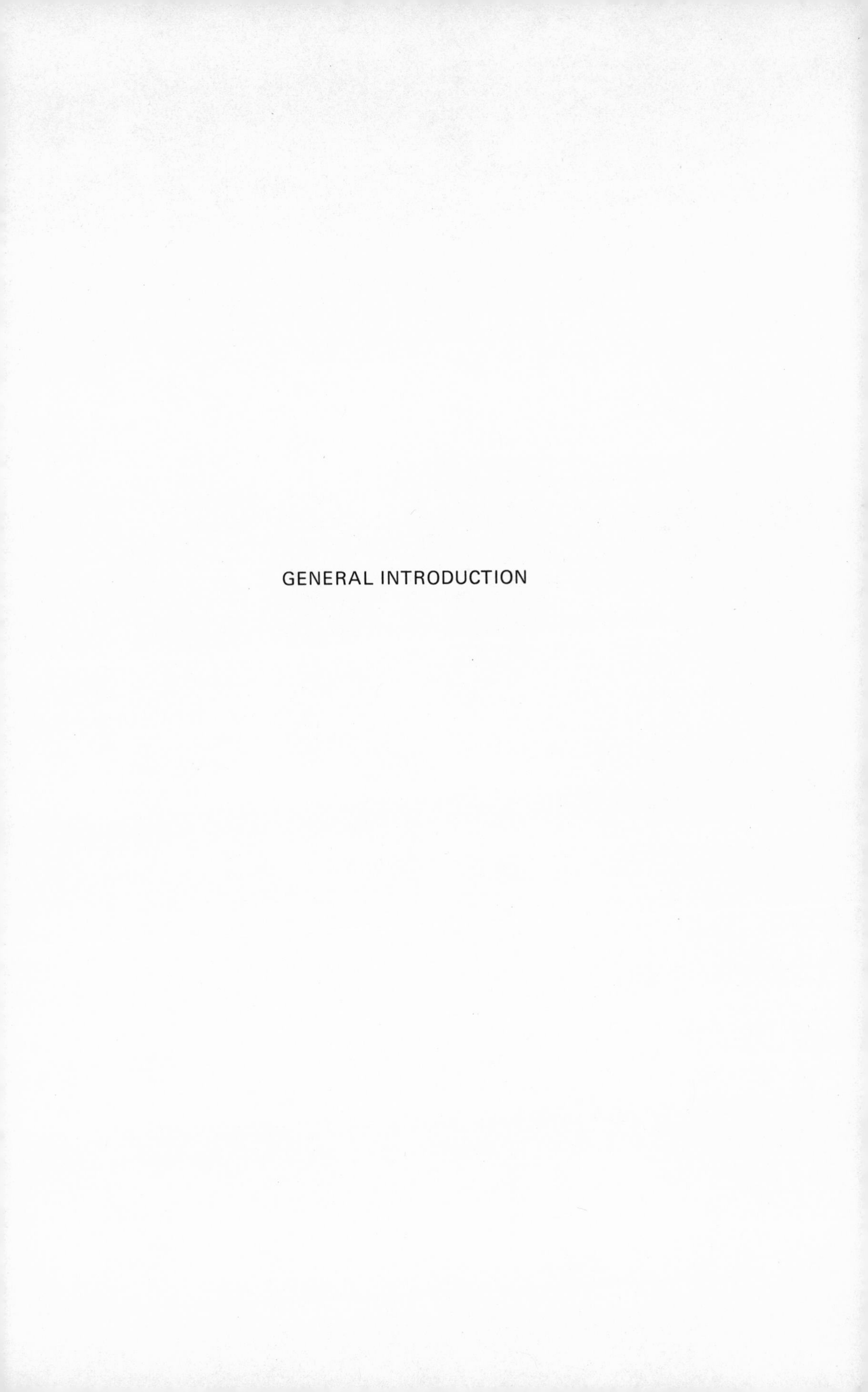

GENERAL INTRODUCTION

GENERAL INTRODUCTION

M.A. Ali

Département de Biologie, Université de Montréal

Montréal, P.Q. H3C 3J7 Canada

Bait the hook well; this fish will bite.
Much Ado about Nothing, Act II, Sc. III.

Almost half of the vertebrate species are fishes. These are estimated to be about 18,800 living species as compared to 21,100 living tetrapods. However, estimates vary greatly, ranging from 17,000 to 30,000. The eventual number of piscine species may well be 50% greater than presently known, going perhaps as high as 28,000![1] This vast group is also very heterogenous with tremendous diversity in morphology, habitats and physiology. Despite this diversity, members of the super class Pisces can be defined as aquatic poikilothermic vertebrates that have gills throughout life and, if present, fins as appendages. Thus, forms varying from lamprey, rays, eels, flatfishes and lungfishes are distantly related.

Fishes range in size from a 12 mm goby from the Philippines (*Pandaka pygmaea*) to the 15 m whale shark (*Rhincodon typus*). Some have stringlike bodies while some others are ball shaped. Some, such as the reef living forms are highly coloured while benthic forms are usually drab. Some, as the salmon and tuna are sleek, graceful and fast while some others, as the chimaeras and stonefishes are grotesque and slow. Fins, when present, may be modified as holdfast organs or as a lure for attracting prey. Some Cyprinids, Silurids, Amblyopsids, Ophidiids and Gobiids lack eyes, mainly as an adaptation to cave dwelling while in a number of forms inhabiting the great ocean depths, which are also devoid of light, eyes may either get to be very large or, greatly reduced. In fishes inhabiting turbid waters also, the eyes may either be reduced or, be of normal size but with reflecting layers to facilitate the capture of light quanta. In many species, bodies are inflatable or encased in inflexible armour. Scales may be present and indeed be very conspicuous as in the carp or, may be totally absent as in eels and catfishes. Internal anatomy too is very diversified.

Fishes inhabit every conceivable aquatic habitat. They are found in lake Titicaca, the world's highest (3,812 m), and in lake Baikal, the world's lowest (7,000 m below sea level) and deepest (1,000 m). Some are found in almost distilled water containing only 0.01 parts per thousand (ppt) dissolved solids, while some live in very salty lakes with 100 ppt dissolved solids. Most lakes have between 0.05 to 1.0 ppt dissolved solids while the value for average seawater is about 35 ppt. One finds fishes such as the African _Tilapia_ in hot soda lakes (44°C) while some, such as _Trematomus_ occur under the Antarctic ice sheet at temperatures of - 2.0°C. Many have acquired the ability to breath air in order to sustain life in stagnant tropical swamps. Some of these, such as the Dipnoi (lungfishes) have become almost amphibious being capable of living outside water for long periods. On the other hand, fishes such as the Salmonids need fast flowing, well oxygenated waters to survive. Similarly, considerable variation is also manifested as far as the ability to withstand differences in salinity and temperature is concerned. Those able to cope with wide variations are termed euryhaline or eurythermal while those restricted to a narrow range are called stenohaline or stenothermal. Dissolved matter has direct and indirect effects upon fishes. Dissolved oxygen is affected by temperature and pressure and is often a determining factor in the occupation of a habitat. Let us recall that water is about 800 times denser than air. Air contains about 21% oxygen by volume but freshwater can dissolve very little, with about 10.23 ml/l at 0°C. This becomes even lesser as the temperature increases. Seawater of 35 ppt salinity has about 8.8 ml/l oxygen at 0°C. Some fishes need saturated conditions (about 6.5 ml/l at 20°C) but most can exist at about 50% of this value. Dissolved carbon dioxide (in the free state) can affect the ability of the blood of fishes to take up oxygen.

Fishes show a remarkable ability to overcome barriers and one often observes some kind of a continuity in their distribution. Exceptions are some marine genera which occur in the temperate and polar regions of both hemispheres but not in the tropics (anti-tropicality). But the vast majority of genera are tropical and most of the others are found either in the Northern or the Southern hemisphere.

The majority of species live entirely either in freshwaters or in the oceans. A few are diadromous, living part of their lives in freshwater and part in the sea. Among these, most are anadromous which spawn in freshwater but spend most of their lives in the oceans. Catadromous forms do the opposite. However, in the case of some species it is not possible to classify as marine, freshwater, catadromous or anadromous since variations may be found between populations and in some cases even individuals may differ. Of course, one often finds many marine and freshwater species in brackish water whose salinity undergoes remarkable variations even during a 24 h period. It is also interesting to note that freshwater, which covers only about 1% of the earth's surface and accounts for a little less than 0.01% of its water has about 6,850 species of fishes while saltwater which covers about 70% of the globe with 97% of its water has only about 11,650 species! Most of the freshwater species are found in Southeast Asia and in the Amazon basin. As far as the marine forms are concerned, the Indo-Pacific is the richest region, followed by the Caribbean area. Fish fauna is sparse in areas recently exposed from the last ice age such as Western Europe, the West African marine region, Arctic and Antarctic.

The habits of fishes vary as much as their morphology, physiology and ecology do. Some are territorial, others form schools. Some take care of their young, some others just lay millions of eggs and leave their survival to chance. Some feed on plants, some on corals and some on plancton. Many are voracious predators, even capable of swallowing an entire large prey. A few are parasitic, some even on their own females! Quite a few are capable of producing sound, light, electricity or, venom. Some are hermaphrodites, others show sex reversal. Some migrate far, some not so far and some not at all. Some migrate vertically during a 24 h period. In many species the habitat of the juveniles is different from that of the adults. For example, tide pools and marshy areas may serve as nursery grounds for the young of many species.

Economically, the interest in, and the importance of fishes is considerable. The quantity of fish caught in 1978 exceeded 45,000 metric tons. Recreational fisheries yielded millions of dollars to a number of countries. The aquarium trade is worth about a billion dollars per year in North America alone and over 25 million persons all over the world indulge in this hobby. Fishes are used for industrial products and food; for such things as fish meal (for agriculture), food of trout, catfish, poultry, pigs, etc. Fish (cod) liver oil is a supplement to diet in many areas and oil from a number of other speciesis another important product of reduction plants and is used in many manufacturing processes and in foodstuffs. Fishes are used for making glue, are a source of leather, and a source of guanine for some paints. Nearly 30% of the fishes caught are used for industrial purposes. Fish culture is getting to be more and more important every year, not only in the warmer but also in the temperate countries. Much greater use could be made of fishes as biological control agents (algal, insect and fish control). Knowledge of poisonous and dangerous fishes (those that attack humans, vectors of parasites and diseases) would also enable us to better control them.

Our purpose is to examine the physiology of fishes in relation to variations in environmental factors so that we could understand the manner in which they adapt themselves to various habitats. This has two points of interest. First, we would be able to comprehend how different species adapt to different habitats; how the same family, genus or species adapts to different environments and, how different species, genera, or families adapt to the same environmental conditions. Second, environmental pollution is being studied more methodically and frequently and in view of their importance, as outlined above, the effect of pollution on fishes is receiving greater attention. Since pollution is essentially an artificial modification of the environment, its effect on the physiology of a fish may be considered to be the study of an aspect or of aspects of its environmental physiology. Also, in order to understand what a pollutant does to a fish, one must first know what the physiological responses of a fish to its "normal" environmental factors are.

Environmental physiology of fishes has not existed as a discipline or sub-discipline for long although papers may be found sporadically in the literature even in the latter part of the 19th century. It is probably correct to say that a concerted effort to study the environmental physiology of fishes commenced in Canada, particularly in Professor F.E.J. Fry's laboratory. Persons who were trained there and some of us who were trained elsewhere in the country have spread out literally almost all over the globe, establishing groups studying various aspects of the environmental physiology of fishes. It is thus a signal honour that Professor Fry was a member of the advisory committee of the Advanced Study Institute which led to this volume. This volume was planned so that it would be as complete as possible yet not put too much emphasis on matters that have been dealt with thoroughly in the recent past, such as Vision in Fishes (Vol. A1 in this series, 1975), Electroreception (Handbook of Sensory Physiology, Vol. III/3, 1974, Springer-Verlag, N.Y.). The chapters dealing with light for example have been written with the theme of this volume in mind and try to bring a fresher perspective. On the other hand, considerable space has been given to gases, gas exchange organs, ions, pressure, sound, temperature, photoperiods, rhythms and reproduction. It is hoped that it will give an overall view of the effect of various environmental factors and a combination of them on fishes, both at the organ's and individual's level. It will be evident from the chapters that the gaps in our knowledge of the environmental physiology of fishes are still much too large to permit generalisations and it is hoped that this volume would succeed at least to a small extent in encouraging further concerted work in this field.

ACKNOWLEDGEMENTS

I thank Michel Anctil, Naercio Menezes and Mary Ann Klyne for comments on an earlier draft of this essay. However, I alone am responsible for its shortcomings. In preparing this essay I drew heavily from many books dealing with the biology, ecology, physiology and taxonomy of fishes and it does not appear feasible to list them in a bibliography.

OXYGEN AS AN ENVIRONMENTAL FACTOR OF FISHES

G.F. HOLETON*
Department of Zoology, University of Toronto
Toronto, Ontario, M5S 1A1, Canada

*Dr. Holeton passed away while this volume was in press

I - ABUNDANCE OF OXYGEN IN NATURAL WATERS

A) Sources of Oxygen

Ultimately the source of oxygen in the aquatic environment is photosynthesis. Bodies of water may be net exporters or importers of oxygen. In circumstances where plant life is abundant and light is available water bodies release a considerable amount of oxygen to the atmosphere.

B) Solubility of Oxygen

The solubility of oxygen in water is low. A liter of air-saturated fresh water at 10 C will contain 11.3 mg of oxygen which corresponds to a concentration of 0.35 millimoles per liter. In contrast a liter of air at the same temperature and at atmospheric pressure will contain about 296 mg or 9.25 millimoles of oxygen.

A number of factors affect oxygen solubility. As water temperature rises oxygen solubility decreases such that the oxygen content of air-saturated fresh water at O C is approximately twice what it is at 30 C. The solubility of oxygen in water declines with increasing salinity. At normal temperatures air-saturated sea water will hold 20-23% less oxygen than fresh water. A number of useful tables on oxygen solubility have been reproduced in a recent review by Davis (1975).

It is important to remember that CO_2 is 25 to 30 times more soluble in water than is oxygen. This means that water is a much better medium in which to carry off metabolic CO_2 than it is as a source of oxygen.

C) Variability of Oxygen in Aquatic Environments

Perhaps the most striking feature about oxygen as an environmental factor of fishes relates to its variability in abundance. In contrast with the atmosphere, in which the pressure and concentration of oxygen (P_{O_2} & $[O_2]$) are relatively constant, the P_{O_2} and $[O_2]$ of water can vary dramatically. Not only can the P_{O_2} of water drop below air-saturation levels or even disappear, it can also rise well above atmospheric levels.

The variations of oxygen in natural waters arise from a number of sources, and the variations may be spatial or temporal. Some of the most spectacular fluctuations are due to the diurnal changes in the balance between photosynthetic oxygen production and overall community oxygen demand from respiration and chemical processes. Such fluctuations are particularly common in shallow freshwater systems (Davis, 1975; Garey and Rahn, 1970; Jones, 1961; Kramer et al. 1978), where P_{O_2} may drop to 20 to 30 torr during the night and rise to 200 to 400 torr during the middle of the day.

For the majority of cases it is only photosynthetic activity which can produce elevations in $[O_2]$ above air-saturation levels. However sudden warming of water such as occurs in industrial and thermal power station effluents can increase P_{O_2} in the absence of an increase in $[O_2]$.

Even large bodies of water may have oxygen levels higher than air-saturation. Fairbridge (1966) reports that oxygen levels in many oceans may be as high as 130% of air saturation in the photic layers. Certainly this extends to the antarctic oceans where productivity is high and summer photoperiod is prolonged (Holeton, 1970).

Variations in oxygen concentration also occur on a seasonal basis and are pronounced in situations where there is density dependent stratification of water bodies. Such effects are particularly pronounced in the temperate fresh water lakes where in summer the formation of a warm surface layer effectively seals off a colder hypolimnion from the atmosphere. Often the hypolimnetic oxygen demand exceeds the supply and $[O_2]$ drops or even disappears as the warm season progresses. In the temperate or high latitudes the formation of winter ice cover effectively seals off water bodies from access to the atmosphere. Under these circumstances the oxygen levels are very dependent upon the balance between photosynthetic activity and community oxygen demand. Often the balance swings in favor of oxygen demand due to the influence of short daylengths and light attenuation due to snow cover. In shallow biologically productive waters this can result in nearly total depletion of oxygen which is lethal to fish thereby producing a "winter kill".

In some circumstances, particularly in shallow arctic lakes, the downwards growth of the ice layer throughout the winter results in a "freezing out" or exclusion of oxygen from the ice. This can have a concentrating effect on the oxygen levels in the remaining liquid part of the lake.

In tropical regions the reduced oxygen solubility which occurs at the prevailing high temperatures coupled with higher metabolic rates results in oxygen depletion in most water bodies of any substantial depth (Kramer et al., 1978).

Pollution is another factor which can lead to a change in oxygen levels in natural waters. For example, it is common for progressive addition of nutrients or eutrophication to lead to greater fluctuations in oxygen concentration and in particular, oxygen depletion. On the other hand poisonous pollutants such as heavy metals or extreme acidification such as is becoming common in Scandinavia and north-eastern North America can reduce aquatic life to the point where oxygen levels can become relatively stable.

II - OXYGEN DEMANDS OF FISH

A) General Metabolic Strategy

The low oxygen content of water coupled with its high viscosity and density precludes very high rates of oxygen uptake by most fishes. Brett (1979) has calculated the maximum oxygen uptake rate of 2 kg salmon just approaches the basal rate of a rabbit. The evolutionary response of fish to such difficult access to oxygen has been to develop a high capacity for sprinting. The main axial musculature of fish which may exceed 50% of the body mass is comprised mainly of white muscle fibers. White muscle is poorly vascularized and is specialized for burst activity utilizing the fermentation of glycogen to produce ATP while accumulating lactic acid as a temporary end-product. The glycogen is present in the muscle fibers so the burst of activity can be carried out without a supply of oxygen. The lactic acid built up over a very short period of strenuous activity is retained by the fish and metabolized slowly over a period of several hours.

Although fish can consume oxygen at fairly high rates compared to other aquatic organisms their maximum sustained rate of oxygen uptake falls short of that of comparable sized birds and mammals by a factor of from 10 to 100 (Brett, 1972). This does not mean that fish are weak. On the contrary fish are very muscular animals which rely on glycolytic fermentation or so called "anaerobic" metabolism to cover their short-term energy requirements. For periods of about 20 seconds fish may expend energy at a rate about 100 times the basal rate (Brett, 1979). Brett (1972) has calculated that a 100 gram salmon can have short-term oxygen demand during burst exercise which is comparable to the aerobic capacity of a similar sized mammal, that is about 8000 mg $O_2 . kg^{-1} . hr^{-1}$.

In summary the majority of fish are well developed sprinters capables of very high levels of activity for short periods. The resulting "oxygen debts" based on lactic acid accumulation are usually paid back over several hours or even a full day.

B) General Range of Oxygen Demands

(i) Measurement of Oxygen Uptake

Although it is fairly easy to make a determination of oxygen consumption of a fish in the laboratory, it is extremely difficult to obtain a precise estimate of the oxygen demand of a fish in its natural environment. Virtually all our knowledge of oxygen demands of fish is based on laboratory studies. Furthermore this laboratory based information often requires careful interpretation in order to be useful.

Since fish can vary their oxygen demands widely, some of the most useful forms of laboratory study involve determination of the upper and lower limits of oxygen uptake rate at which a fish can function over an extended period.

The lower boundary of oxygen uptake is called the standard metabolic rate. This is the rate of oxygen uptake of a resting post-absorptive fish which has been fully acclimated to the experimental conditions (Krogh, 1914; Fry, 1947). Use of activity measuring devices (see Beamish and Mookherji, 1964) or exercise chambers (Blazka et al., 1960; Brett, 1964; Farmer and Beamish, 1969) has permitted the experimenter to eliminate the effects of activity on oxygen uptake. This is done by extrapolating the relationship between activity and oxygen uptake back to zero activity which yields an estimate of standard metabolic rate. With many fish this method produces estimates of standard metabolism which are lower than would ever be measured directly since it eliminates even the efforts that a fish would make to remain upright. Brett (1972) has pointed out that standard metabolism measured in the way outlined yields a theoretical value which in all cases is quite comparable to the basal metabolism applied to other vertebrates. Although standard metabolic rate is more difficult to measure than resting or routine metabolism (Fry, 1957) and is usually a theoretical value, it is a most desireable laboratory measurement because of its consistency and because it permits more valid intra and interspecific comparisons.

The upper limit of sustained oxygen uptake is called the active metabolic rate. Its determination requires the use of a swimming chamber or water tunnel respirometer (Blazka et al., 1960; Brett, 1964; Farmer and Beamish, 1969). The active metabolic rate of fish is far more sensitive to the availability of environmental oxygen than is standard metabolism (Job, 1955; Brett, 1964).

(ii) Factors Affecting Oxygen Demands

A number of factors affect the oxygen demand of fish. Aside from the effects of methodology applied to make the measurements, the foremost factors affecting oxygen demand are; activity, temperature, size, and the availability of oxygen.

The degree to which activity can change oxygen uptake is largely the difference between the active and standard rates of oxygen uptake. This difference is known as the scope for activity (Fry, 1947). Fry (1971) has pointed out that the definition of activity should include not only locomotor activity but all other supplementary processes such as digestion, assimilation and deposition of new materials. Many fish have a large scope for activity. The active rate of oxygen uptake of most fish ranges from 2.5 to 10 times the standard rate and a considerable body of numerical data is presented in reviews by Brett (1972) and by Brett and Groves (1979).

Of the fish species so far examined, salmonids in particular have a large scope for activity. It seems quite possible that some marine pelagic species such the scombrids may be able to surpass the performance of the salmonids, although at present there are few data available except for those of Stevens (1972) on small tuna.

In general most measurements of active metabolic rates of fish have been made under normoxic conditions. Job (1955) demonstrated that even slight reductions in oxygen levels below air saturation reduced the active oxygen uptake rate of brook trout *Salvelinus fontinalis*. More strikingly, Brett (1964) demonstrated that at warmer temperatures, sockeye salmon *Onchorhynchus nerka* could increase their active rate of oxygen uptake under hyperoxic conditions. It is possible that many species of fish, particularly pelagic fish which inhabit photic zones where hyperoxia is a normal event for part of the year, can only realize their full aerobic potential in water in which the oxygen pressure is above air saturation. This is an area that deserves further research.

With only a few exceptions most fish have body temperatures which are within less than one degree of the ambient water (Stevens & Sutterlin, 1976) and hence their metabolic rates will reflect changes in water temperature. There are some spectacular exceptions to this generalization involving large lamnid sharks (Carey and Teal, 1969a) and active pelagic teleosts of the family Scombridae (Carey and Teal,

1969b; Stevens and Fry, 1971). These fish maintain elevated temperatures in some regions of the body, particularly the muscles, by employing counter current heat exchangers or retia in the blood vessels supplying the tissues concerned (Carey et al., 1971, Graham, 1973; Stevens et al., 1974).

Because of the close relationship between body temperature and ambient temperature, temperature is a major factor affecting oxygen demands of fish. The effects of temperature on fish have been extensively reviewed (Fry and Hochachka, 1970; Fry, 1971; Brett and Groves, 1979). Furthermore the change in oxygen uptake rate in fish which occurs when temperature is changed follows a complex pattern and varies depending upon the rate at which the temperature is changed and upon the previous thermal history of the fish. Although there is considerable interspecific variation, the common pattern is for the change in oxygen consumption to be more profound shortly after abrupt temperature change than it is after acclimation to that temperature for a few days Bullock (1955). The process of acclimation is widely acknowledged but seldom studied. Brett (1970) reviews most of the early work on the subject.

Some of the acclimatory process following temperature change and reflected by oxygen uptake may be traced to acid-base adjustments. The time course for the establishment of new acid-base steady states in fish following a temperature change is similar to those observed for oxygen consumption adjustments (Heisler, 1978; Randall and Cameron, 1973; Cameron, 1976).

Very roughly the rate of oxygen uptake of fish increases by a factor of 2.0 to 2.3 fold for every 10 C rise in temperature (Fry and Hochachka, 1970; Evans, 1977; Brett and Groves, 1979). Exceptions to this rule are legion indicating the limitations of the generalization. Much of the problem arises because there are few species of fish which can maintain "normal" metabolic activity over a wide temperature range. Comparisons, therefore, are mainly interspecific rather than intraspecific and this contributes greatly to the scatter in the data reported.

One widely held hypothesis relating to the effect of temperature upon oxygen demand which does not hold up to critical scrutiny is the concept of metabolic cold adaptation of polar fish. This hypothesis was originally put forward by Scholander et al., (1953) and was subsequently elaborated and extended by Wohlschlag (1964). The hypothesis states that, although arctic and antarctic fish have lower metabolic rates than temperate or tropical fish, their metabolic rates are higher than expected of temperate or tropical fish when the oxygen uptake versus temperature relationships are extrapolated to permit such a comparison. This hypothesis was challenged (Holeton, 1973, 1974) on the basis of its logic, interpretation of the existing data and upon the collection of more extensive data on arctic fish. Virtually all of this phenomenon can be explained as arising from a series of artefacts related to the methodology, use, and interpretations applied to the data and to other published data upon which the theory was based. Recently Brett and Groves (1979) have suggested that compensatory cold adaptation may still exist at a much reduced level compared to that suggested in earlier reports (Brett, 1970; Wohlschlag, 1964). However to arrive at this conclusion Brett and Groves (1979) are forced to concentrate completely on only one of the 12 species of arctic fish reported on by Holeton (1973, 1974). Brett and Groves (1979) have also apparently overlooked my cautionary statement (Holeton, 1974) that this one species Boreogadus saida was the one species in which the determination of routine metabolism was most likely to be well above the standard metabolic rate due to the high level of spontaneous activity exhibited by this fish under experimental conditions. The hypothesis of metabolic cold adaptation has been unable to withstand a crucial test. Its ultimate acceptance or rejection must await much more extensive testing based upon a wider range of species under standardized conditions.

One aspect of the effect of temperature upon oxygen consumption of fish is the complicating relationship between temperature and oxygen solubility in water. At cold temperatures the oxygen demand of fish is very low while oxygen solubility in water (and in blood plasma) is higher. This explains to a great extent why such biological oddities as the hemoglobinless fish of the family Chaenichthyidae are found only in the cold well oxygenated waters of the antarctic. High metabolic demand and low solubility of oxygen at high temperatures is also at least a partial explanations as to why there is such an abundance of airbreathing fish in the tropics.

There is a widespread size-related effect on oxygen demand of fish in which large fish have lower weight-specific metabolic rates than small fish. The power function:

$$Y = aX^b$$

where Y = oxygen uptake
X = body weight

is readily applied to oxygen consumption rate versus weight relationships of fish. The exponent b for most fish has an average value of 0.86 (Brett and Groves, 1979) although variations among species ranges from 0.5 to 1.05. Brett and Groves (1979) conclude that the interspecific variation are extensive enough to warrant determination of species specific relationships.

The usual technique of applying a linear regression analysis to logarithmically transformed size versus metabolic rate data has been criticized as potentially inaccurate by Glass (1969) who suggests use of an iterative regression process. However Ricker (1973, 1979) suggests that the process suggested by Glass (1969) is not applicable to most data and recommends use of functional regression analysis which gives metabolism and weight the same mathematical weighting in the relationship. Where a small size range is involved and correlation is not so good, it is particularly important to follow Ricker's (1973) method in order to avoid a bias in the b value.

Many other factors affect oxygen demands of fish, for example; season, feeding, osmotic loading, reproductive activity. These are numerous and an adequate catalogue of their effects is beyond the scope of this chapter and the reader is referred to reviews by Brett and Groves (1979) and Brett (1970, 1979).

III - THE EFFECT OF LOW OXYGEN ON FISH SURVIVAL

A) Effects related to life history

Severe oxygen depletion can be lethal to fish but the actual lethal level may change at different stages in life history. Eggs, being non-motile are especially vulnerable to localized oxygen depletions. This vulnerability is offset by a number of adaptations such as choice of nest sites, parental care such as mouthbrooding, fanning and bubble nests. Davis (1975) points out that, in general, oxygen requirements of fish eggs increase with ongoing development, and with this increase requirement a greater sensitivity to environmental oxygen supply.

After hatching, the acquisition of motility means that fish can at least escape local depletions of oxygen. In this regard size makes a difference since smaller fish can exploit smaller microhabitats such as oxygen-rich surface films and shallow water. On the other hand large fish can move further and faster and can resist the effects of potentially lethal oxygen concentrations for longer (Shepard, 1955). In part, the longer resistance times of larger fish to potentially lethal oxygen concentrations may be due to lower weight-specific oxygen consumption. At present there is insufficient evidence to assume that this generalization of resistance time of larger fish is applicable to all species.

B) Resistance and Anaerobiosis

In cases where fish are subjected to chronic very low oxygen levels where continued oxygen uptake is not physiologically possible their continued survival depends upon resistance strategies. Most of the strategies involve processes such as exploiting body oxygen stores, using anaerobic metabolism or slowing down of metabolism (oxygen dependence). The two most extensive studies on this subject are those of Shepard (1955) on brook trout *Salvelinus fontinalis* and of Anderson (1975) on goldfish (*Carassius auratus* L.).

Anderson (1975) found that *Carassius auratus* could survive for extended periods in water containing less than 0.01 mg $O_2 . l^{-1}$, particularly at lower temperatures. The resistance of these fish to anoxia was quite sensitive to seasonal effects, being 11 hours in autumn at 5 C and 75 hours in spring at 5 C.

Both Shepart (1955) and Anderson (1975) report that the rate at which hypoxic or anoxic conditions occur has an effect on the ultimate resistance of the fish. Shepard (1955) also reported that the lethal threshold for hypoxia of Salvelinus fontinalis was lowered considerably when the fish were acclimated to an intermediate level of hypoxia.

C) Pollution Effects on Oxygen Supply

Oxygen lack severe enough to be lethal can arise from pollution as well as from natural sources. Some pollutants such as sewage and agricultural wastes enrich water leading to biological and chemical oxygen demand leading to a reduction in oxygen supply. Other classes of pollutants act by impairing the ability of fish to extract oxygen from the water. In this category would be classes of materials which precipitate and smother eggs as well as classes of materials, such as heavy metals, which disrupt the physiological and structural integrity of the gills (Skidmore, 1970).

IV - ENERGETICS OF BREATHING

The high cost of breathing a dense viscous medium in which oxygen is only sparingly soluble must restrict the maximum oxygen uptake of fish. To put the problem in perspective, a fish in the 100 gram weight range extracting approximately 30% of the available oxygen from air saturated water must breath the equivalent of one to two thirds of its own body weight per minute just to stay alive. For a fish with higher metabolic rates this breathing requirement is much greater. For example Stevens (1972) reported that it is necessary for the skipjack tuna, Katsuwonus pelamis, to breath about twice its own body weight in water every minute merely to sustain life. Furthermore, when a fish is hypoxic or is physically active the amount of water which must be respired can rise by just under an order of magnitude.

Another factor affecting breathing energetics arises from the effect of temperature on oxygen solubility and metabolic rate. These opposing trends are such that breathing requirements must increase considerably with increasing temperature. A more detailed analysis involving salmonid fish is presented by Hughes and Roberts (1970). Because of the potentially large proportion of a fish's energy budget which can be spent on breathing, the study of the metabolics of breathing has received considerable attention. The pioneering work of van Dam (1938) showed clearly that the concerted actions of the buccal and opercular pumps could provide a nearly continuous flow over the gills. This work was extended with the application of sensitive pressure measurements which started with Hughes and Shelton (1958). Much of the subsequent work based on pressure studies has been reviewed by Shelton (1970) and Ballantijn (1972). The general physiology of the respiratory pumps of fish are well understood although there are major uncertainties as to the precise nature of water flow over the respiratory lamellae. These uncertainties can only be cleared by the application of direct measurement techniques. Such techniques await the development of small rapid responding flow sensors which can be located on the gills.

A recent study on the nature of water movement in the buccal chamber of large carp *Cyprinus carpio* has yielded new insights on respiratory dynamics of fish (Holeton and Jones, 1975). This study revealed that water moving in the buccal chamber during quiet respiration reached velocities as high as 40 cm per second. These studies also indicated that dynamic components were just as important as hydrostatic pressures and that the assumption, long-held, that hydrostatic pressures within the buccal chamber are uniform is probably incorrect.

An assessment of the costs of breathing poses an extremely difficult problem because it involves the measurement of unsteady flow of a dense viscous fluid through a non-uniform system which is, at best, ill-defined. The compliance of the respiratory tract varies temporally and spatially and the major resistive components, the gills, are mobile - both activity and passively. The problem is further compounded because many of the structures involved in breathing are greatly elaborated as part of the feeding metabolism. Perhaps the best testimony to the difficulties involved is given by the wide variations in estimated breathing costs, ranging from 0.5% of the total metabolism (Alexander, 1967) to 70% (Schumann and Piiper, 1966).

There have been at least three different approaches applied in estimating the cost of breathing in fish. Hydrostatic pressure measurements, made in the buccal and opercular chambers, analyzed along with either measured or estimated respiratory flow rates have provided the data base for one major approach (Alexander, 1967; Davis and Randall, 1973; Jones and Schwarzfeld, 1974). In this approach an estimate is made of the work done on the fluid pumped but requires the following simplifying assumptions: steady states of pressure and flow, dimensional and resistive uniformity, negligible inertial and kinetic factors, and an estimate must be made of respiratory muscle efficiency. Virtually none of these assumptions is satisfied by the real system, and most are obviously wrong (Cameron & Cech, 1969; Holeton and Jones, 1975). In view of the mobility of the structures involved, the pulsatility of the flow into and out of the system, the fluctuating hydrostatic pressures, and the fairly high water velocities involved, it would seem that it is only a matter of chance whether the estimates of cost of breathing obtained by this approach are near the actual value.

A second major approach to estimating cost of breathing involves simultaneous measurement of oxygen consumption and respiratory water flow. By manipulation of respiratory water flow any observed changes in oxygen consumption are attributed solely to the respiratory pumps (Van Dam, 1938; Schumann and Piiper, 1966; Edwards, 1971; Jones and Schwarzfeld, 1974). This assumption of cause and effect has been justly criticized by Cameron and Cech (1969), and certainly there has been no experimental validation to warrant its acceptance.

A third approach has been to examine all of the muscles which take part in actuating the respiratory pumps. By measuring their mass and capillary density it is possible to set some practical upper limits on the probable cost of breathing (Cameron and Cech, 1969). This approach like the second will tend to yield an overestimate of the cost of quiet breathing inasmuch as the muscles involved will not be maximally active. Also some of the muscles are part of specialized feeding mechanisms and may never become maximally active in a respiratory role. For example the estimated cost of breathing for the mullet *Mugil cephalus*, a fish with only minor elaboration of respiratory muscles for feeding, is probably between 5% and 15% of total resting metabolism (Cameron and Cech, 1969). This cost, as a proportion of the total can be expected to decrease when the fish becomes active and raises its metabolic rate (Cameron and Cech, 1969).

On the comparatively sound basis of muscle mass and potential of that muscle for activity, it appears that the cost of breathing as a proportion of the total energy budget will generally be under 10% for most fish under normoxic conditions. The proportion of the total can be expected to rise sharply under hypoxic conditions, although the absolute energy consumption will not be great because of the small mass of muscles involved. However it should be remembered that the respiratory muscles may limit oxygen uptake by the fish simply because of the limited amount of pumping work they can do.

Up to this point the discussion of breathing costs has centered upon fish which are stationary with respect to the ambient water. In cases where the water moves past the fish either due to locomotion of exogenous factors the fish can, merely by opening its mouth, obtain a respiratory water flow. The respiratory pumps can stop working entirely if the velocity is sufficient. This process is called ram ventilation, and a good theoretical treatment of its energetics is given by Brown and Muir (1970). Roberts (1975) reviews the subject and points out that above a critical velocity most fish rely on ram ventilation for their respiratory water flow. Recently Freadman (1979) has presented evidence that the shift to ram ventilation by swimming fish results in substantial metabolic savings.

It is generally believed that some fish, particularly the scombrids are so committed to continuous pelagic swimming that they cannot breath by means other than by ram ventilation (Hall, 1930; Alexander, 1967). Recently my coworkers and I had an opportunity to study the respiration of atlantic mackerel Scomber scomber and found that this particular species could breath and oxygenate its blood even when stationary (Holeton, Pawson, and Shelton, in preparation). This finding is in direct contradiction to that of Hall (1930) who suggested that this same species was dependent upon ram ventilation. The difference between our findings and Hall's may be due to a secondary factor; the extreme fragility of mackerel skin. Handling these fish damages their skin leading to ionoregulatory breakdown and death usually within 24 hours. It is apparent that the reason the confined mackerel used by Hall were unable to oxygenate their blood was because they were dying from ionoregulatory dysfunction.

V - RESPONSES OF FISH TO CHANGES IN ENVIRONMENTAL OXYGEN CONCENTRATION

A) Behavioral Responses

There appears to be a wide variety of behavioral responses of fish to low oxygen and to oxygen gradients (Davis, 1975). A common response of fish to very low oxygen near lethal levels is an increase in general activity and a movement towards the surface where the oxygen-rich surface film can be exploited (Shepard, 1955; Dandy, 1970; Gee et al., 1978). A number of fish show definite avoidance responses to low oxygen but there is considerable variation in the threshold oxygen level required between species and within a species with respect to temperature and season (Jones, 1952; Whitmore et al., 1960; Hoglund, 1961). Some of the variation undoubtedly arises from differing experimental approaches (Davis, 1975). It is possible also for fish to acquire or "learn" behavioral responses with manipulation of environmental oxygen as the primary conditioning stimulus (Van Sommer, 1962) although the oxygen levels involved were all-or-none in nature.

Many fish which can breath air facultatively, switch over to predominantly air breathing from water breathing in response to hypoxic conditions (Hughes and Singh, 1970; Stevens and Holeton, 1978).

In summary, it is clear that responses such as surfacing or a switch to air breathing can offset hypoxic conditions although the value of such responses is difficult to quantify. It is even more difficult to interpret avoidance observations and extrapolate these to natural conditions (Whitmore et al., 1960).

B) Physiological Responses

In preface I wish to point out that the definition of normoxic conditions in an aquatic milieu must be considerably less precise than when applied to the atmosphere. This is because oxygen levels vary so widely in water and, out of necessity, many fish show little reaction to modest changes in environmental oxygen to either side of atmospheric P_{O_2}. In this sense the responses of fish to hyperoxia and hypoxia should be viewed as parts of a continuum where "normoxia" is essentially a midpoint.

It is also possible to generalize on the nature of the responses of fish to oxygen variations. What varies between species is not so much the nature of the response but rather the degree of the response and the specific oxygen levels at which the responses occur. I will deal with the responses by use of the following categories:

Responses of the respiratory pumps
The blood to water diffusion barrier
Responses of the circulatory system
The tissue to blood diffusion barrier(s)
Oxygen receptors and respiratory control.

(i) Responses of the respiratory pumps

This topic has been reviewed by Shelton (1970) but there have been further advances in the past decade. In general there is a close inverse relationship between the amount of water respired by fish and the oxygen content of the water (Saunders, 1962; Holeton and Randall, 1967b; Davis and Cameron, 1970; Shelton, 1970; Davis and Randall, 1973; Eddy, 1974; Dejours et al., 1977; Itazawa and Takeda, 1978). The respiratory water flow rate is altered by changes in both the rate and amplitude of respiratory movements with amplitude usually being the more important variable.

This generalization does not appear to extend to elasmobranchs which show little or no breathing response to hypoxia (Shelton, 1970; Butler and Taylor, 1971; Piiper et al., 1970) although Dejours et al. (1977) have demonstrated that two species of dogfish do reduce their respiratory water flow considerably when hyperoxic.

Most fish remove about 40 to 80% of the available oxygen in the water they breath under normoxic conditions (Shelton, 1970). This efficiency of extraction tends to remain constant under hyperoxic conditions while respiratory flow rate drops commensurately (Holeton, 1972; Dejours, 1973; Randall and Jones, 1973; Dejours et al., 1977). However, under hypoxic conditions fish appear to be unable to maintain a high oxygen extraction efficiency, especially when hypoxia is severe (Saunders, 1962; Holeton and Randall, 1967b; Shelton, 1970; Randall and Jones, 1973; Itazawa and Takeda, 1978). Under severe hypoxia the amount of water respired increases in greater proportion than the proportionate drop in environmental oxygen. It is not unusual for fish to increase respiratory water flow rate by a factor of 10 to 20 fold in response to severe hypoxia (Shelton, 1970; Itazawa and Takeda, 1978).

The reduction in oxygen extraction efficiency under hypoxic conditions has been attributed to a variety of deadspace phenomena, all of which relate to the passage of water through the branchial chambers without coming into intimate diffusional contact with the respiratory surfaces (Hughes, 1966; Randall, 1970b).

(ii) The blood to water diffusion barrier

It has been demonstrated that a major response to hypoxia is an apparent increase in the oxygen diffusing capacity or transfer factor (T_{O_2}) of the gills (Randall et al., 1967; Eddy, 1974; Itazawa and Takeda, 1978). Observations of this capacity of hypoxic fish to change the apparent oxygen permeability (T_{O_2}) of the gills is based upon empirical analysis of the mean P_{O_2} gradient from blood to water divided into the weight specific oxygen uptake rate. The mean P_{O_2} gradient is usually averaged from the P_{O_2} of blood and water entering and leaving the gills (Hughes and Shelton, 1962; Randall et al., 1967). An improved method of computing T_{O_2} is given by Scheid and Piiper (1976) which corrects for the nonlinearity effect of the oxygen dissociation curve.

The physiological basis of changes in T_{O_2} has been the subject of considerable attention but is still not fully understood. Most attention has focussed on circulatory patterns within the gills. Steen and Kruysse (1964) postulated that there was large scale non-respiratory shunting of blood through the gills. However Cameron (1974a) presented logical arguments why such a strategy is unlikely and potentially counter-productive for the fish. The shunting hypothesis of Steen and Kruysse (1964) is not supported by more recent anatomical evidence of blood pathways through the gills (Laurent and Dunel, 1976).

Recently one of my students, John Booth, has demonstrated that some of the changes in gill T_{O_2} can be explained in terms of changes in the proportion of gills secondary lamellae which are activity perfused with blood at any given instant (Booth, 1978, 1979). In the rainbow trout <u>Salmo</u> <u>gairdneri</u> the proportion of secondary lamellae perfused increased from a mean of 59% during normoxia to 71% during fairly severe hypoxia. What is most striking about Booth's findings is the observation that lamellar recruitment can only be responsible for a minor part of the changes in T_{O_2} which occur during hypoxia. The question remains, but with one variable accounted for, what is the basis of the observed change in T_{O_2} in fish gills?

One aspect relating to T_{O_2} which has not been examined closely is the nature of the fluid layers immediately adjacent to the respiratory surfaces. It is likely that along with changes in the flow of water over the gills there are also changes in the size of the relatively static boundary layers of water in contact with the gills. Certainly a thinning of the relatively unstirred boundary layers should produce an increase in T_{O_2}, the real question is how much?

The blood can also have an effect on T_{O_2}. Reductions in blood oxygen capacity or blood viscosity can reduce T_{O_2}. Inactivation of hemoglobin function with carbon monoxide results in a large drop in T_{O_2} of trout even though blood and water flows through the gills are increased (Holeton, 1971a). Severe anaemia causes a small drop in T_{O_2} as well, even though blood flow rate is increased (Cameron and Davis, 1970).

(iii) Responses of the circulatory system

This topic has been covered in reviews by Randall (1970a, 1970b), Johansen (1971), and Satchell (1971, 1978). The generalized circulatory response to hypoxia is an increase in blood pressure and heart stroke volume which is largely offset by a decrease in heart rate or bradycardia (Holeton and Randall, 1967b; Butler and Taylor, 1971, 1975; Eddy, 1974; Piiper et al., 1970).

The circulatory response of fish to hyperoxia is less well documented. The hemoglobinless antarctic icefish _Chaenocephalus aceratus_ reduces its cardiac output in response to hyperoxia either by reducing its heart rate (Holeton, 1972) or stroke volume (Hemmingsen et al., 1972).

The ratio of flow of blood to water through the gills ranges from 1:15 to 1:19 under normoxic conditions (Satchell, 1978). This ratio increases greatly in favor of water flow during hypoxia. There is insufficient information to say what happens to the flow ratio during hyperoxia, but in view of a diminished respiratory water flow (Dejours et al., 1977) it may be that the ratio declines.

It has been repeatedly suggested that during rest under normoxic conditions there may be a partial shutdown of blood flow through secondary lamellae and this has been demonstrated in the case of rainbow trout (Booth, 1978). The reason put forward previously for such a partial shutdown, or for comparable changes in gill circulation which would shunt blood past the respiratory surfaces, is to reduce osmotic or ionic loading on the fish (Satchell, 1971; Randall et al., 1972; Satchell, 1978). However this does not appear to be the case. Booth has clearly shown that blood is still present in secondary lamellae in which blood flow has stopped. Thus there must still be standing osmotic and ionic gradients from blood to water or _vice versa_ even in those lamellae in which perfusion of blood is temporarily suspended. It seems unlikely that changes in gill microcirculatory blood flow in response to hypoxia are directly involved with iono- of osmoregulatory loading (Booth, 1979). The increased ionic and osmotic loads alluded to by Randall et al., (1972) seem more likely to stem from some common causal factor such as blood pressure or changes in levels of circulating catecholamines. Increased levels of circulating catecholamines have been observed in hypoxic dogfish _Scyliorhinus canicula_ (Butler et al., 1978).

(iv) The nature of the blood to tissue diffusion barrier

It is in the realm of oxygen delivery to the tissues that information is particularly scarce. Garey and Rahn (1970) and Rahn et al., (1973) have demonstrated that tissue P_{O_2} estimated from gas pockets in trout underwent substantial changes in response to changes in environmental P_{O_2}. Also it has been demonstrated that rainbow trout are able to maintain their oxygen consumption in the face of hypoxia severe eough to depress their blood P_{O_2} considerably (Holeton and Randall, 1967a). Both these examples illustrate the point that there must be changes in peripheral circulation or diffusion characteristics within the tissues in response to environmental oxygen changes. The nature of these changes is not clear. Cameron (1975) found no major changes in blood flow to various tissues of grayling Thymallus arcticus subjected to hypoxia.

(v) Oxygen receptors and respiratory control

Respiratory control metabolisms of fish have been reviewed by Shelton (1970) and oxygen receptors by Laurent (1977). On the basis of logical deduction it appears that oxygen supply must be the major cue for respiratory control in fish. This is because fish must respire large volumes of water in order to obtain sufficient oxygen. On the other hand CO_2, the other gas upon which respiratory control might be based, is 25 to 30 times as soluble as oxygen in water. Therefore the capacity to eliminate CO_2 greatly exceeds the ability to take up oxygen. Randall and Cameron (1973) point out that fish cannot compromise oxygen transfer in order to regulate arterial P_{CO_2} and pH. The conclusion that oxygen supply must dominate respiratory control of fish is supported by evidence that fish are more responsive to changes in oxygen than they are to CO_2 (Dejours, 1973; Dejours et al., 1977).

The active and coordinated responses of fishes to changes in environmental oxygen implicate the involvement of oxygen-sensitive chemoreceptors. Of the responses, the bradycardia response to hypoxia is one of the best understood except, paradoxically, for one crucial aspect. It is still not clear what its function does for fish. The bradycardia is mediated by cardiac branches of the vagus nerve exerting inhibitory action on the heart (Satchell, 1971; Randall, 1970b; Butler and Taylor, 1971). Usually the bradycardia appears at oxygen levels substantially below atmospheric P_{O_2} and slows heart rate to 1/4 to 1/2 of its normal rate.

A striking feature of this response is that the 'on' response tends to be slower than the 'off' response although both can be extremely rapid (Daxboeck and Holeton, 1978, 1980; Smith and Jones, 1978). The rapidity of the 'off' response (for example see Daxboeck and Holeton, 1980) has led to a search for external or superficial oxygen receptors particularly since Randall and Smith (1967) demonstrated that the 'off' response could occur in the absence of blood flow through the gills.

Studies with carbon monoxide have shown that the bradycardia response is a readily separable component of the overall response of rainbow trout to hypoxia (Holeton, 1971a, 1971b, 1977). Furthermore in developing rainbow trout the bradycardia response to hypoxia becomes operational later than the other responses (Holeton, 1971b).

Recently sites responsible for initiation of the hypoxic bradycardia have been localized in rainbow trout. The oxygen sensitive sites are located at the dorsal portion of the first functional gill arch (Daxboeck and Holeton, 1978; Smith and Jones, 1978) and are not present on the posterior hemibranch of that same arch (Daxboeck and Holeton, 1978). The pseudobranch, suggested as a potential site for oxygen reception (Laurent, 1977), was found to have little effect on the bradycardia response to hypoxia (Randall and Jones, 1973; Daxboeck and Holeton, 1978; Smith and Jones, 1978).

Elasmobranchs have superficial oxygen receptor sites controlling bradycardia but these must be more widely dispersed since abolition of the response requires sectioning of portions of four cranial nerves (Butler and Taylor, 1971, 1975; Butler et al., 1977).

The function of the bradycardia of fish is not clear. Although there have been suggestions (See Satchell, 1971) there is little conclusive evidence to verify the suggestions. One crude but informative experiment performed by Ms G. Jacobs in my laboratory involved an examination of the P_{O_2} at which rainbow trout lost equilibrium. Two groups of fish were subjected to slowly intensifying normocapnic hypoxia at 5 C. The rate of change of P_{O_2} was slow and an experiment required 40 to 50 minutes. The peridardial cavities were intubated with a fine bore (PE 20) cannula 24 hours prior to the experiment. Half of the experiments were conducted on fish given a dose of 0.1 ml of $8.7 \cdot 10^{-4}$ M atropine to abolish the bradycardia. The results are tabulated below.

Group	Control	Atropinized
Critical P_{O_2} $\pm$ S.E.	23.3 ± 0.97	24.4 ± 1.14
Number of expts.	(9)	(9)
Hypoxic heart rate	17.7	33.4

t = 0. 769 No significant difference

In this case the bradycardia did not affect the P_{O_2} at which the trout were unable to cope with hypoxia and lost equilibrium. Close examination of the gills revealed no hemmorhage which suggests that the bradycardia does not appear to be a vital protective measure to prevent pressure damage in the vasculature of the gills.

In the elasmobranch Scyliorhinus canicula (Butler and Taylor, 1971; Taylor et al., 1977) it was found that abolition of bradycardia during severe hypoxia reduced arterial oxygen saturation from 20% down to 8%. These authors suggest that the bradycardia may serve to maintain P_{O_2} and perhaps reduce the cardiac work load.

As yet the bradycardia in teleost fishes seems to have had little demonstrable effect although blood P_{O_2} maintenance and cardiac function are implicated. One new possibility that would be worth investigating relates to cardiac function. A large portion (usually 65% or more) of the ventricle of fish consists of a spongy trabeculated muscle which must obtain its oxygen supply from the venous blood it pumps (Poupa et al., 1974; Cameron, 1975). A photograph of a casting of the blood space of the fish ventricle given by Cameron (1975) is particularly interesting in this context. When venous blood P_{O_2} declines, as it does in hypoxic conditions, a reduction in heart rate permits two potentially beneficial changes:

1. an increased distension of the myocardium through greater venous filling permits greater infiltration of blood into the trabeculated spaces.

2. a longer time between contractions during which oxygen can diffuse from the blood into the muscle down the reduced diffusion gradients. Both of these changes may be important in improving the oxygen supply to the inner portion of the ventricular myocardium. I suggest that the bradycardia response of fish may in part, be a strategy to improve oxygen supply to the ventricle and improve its aerobic efficiency.

Many of the other responses of fish to changes in oxygen can be implemented fairly rapidly, taking from a little over a second (Eclancher, 1972) to several minutes (Randall and Jones, 1973). Much of the temporal variation observed may be due to experimental design rather than being indicative of the performance of the fish. Virtually nothing is known of location or nature of the other oxygen receptors which are responsible for activation of the respiratory responses of fish. DeKock (1963) described structures in the first epibranchial arteries which appears to be sensory structures. Certainly the first arch appear to be more important than the others to respiratory control in teleost fish (Davis, 1971; Daxboeck and Holeton, 1978; Smith and Jones, 1978; Booth, 1979). In view of this the structures would be worth investigating further.

C) Responses at the Biochemical Level

It is beyond the scope of this chapter to discuss in depth the biochemical aspects of the responses to oxygen, however hemoglobin function does merit attention. Examination of hemoglobins from a wide variety of species reveals that there is a tendency of those species which inhabit environments in which hypoxia is common to have high affinity hemoglobins. The absolute amount of hemoglobin present does not seem to show the same correlation to oxygen availablity. Thus at an evolutionary level it appears that fish respond to chronic hypoxia with left-shifted or high oxygen affinity hemoglobin but do not necessarily have any more hemoglobin than fish with low affinity pigments.

This strategy is a logical one since only a high affinity pigment is capable of binding oxygen if the environmental P_{O_2} is very low. Increasing the quantity of hemoglobin without adjusting the affinity would not help in the case of severe hypoxia.

An increased oxygen binding affinity of hemoglobin to offset chronic hypoxia has been demonstrated in eels (Wood and Johansen, 1972). These same workers have also demonstrated that plaice *Pleuronectes platessa* exposed to oxygen levels higher than they normally were accustomed to reacted by lowering the oxygen affinity of their hemoglobin (Wood et al., 1975). This suggests the adjustment can work both ways. The alternations are linked to changes in organic phosphate concentrations such as ATP within the red blood cells.

Shepard (1955) demonstrated that brook trout *Salvelinus fontinalis* could improve their tolerance to severe hypoxia through prolonged acclimation to hypoxia. It seems possible that a goodly portion of this acclimatory process may be due to a leftward shift towards higher oxygen affinity of their hemoglobin oxygen dissociation curves.

REFERENCES

Alexander, R. Mc. (1967). Functional design in fishes, London, Hutchinson.

Anderson, J.R. (1975). The anaerobic resistance of *Carassius auratus* (L.). Ph. D. Thesis, Australian Nat. Univ., 424 p.

Ballantijn, C.M. (1972). Efficiency, mecanics and motor control of fish respiration. Respir. Physiol. 14: 125-141.

Beamish, F.W.H. and Mookherji, P.S. (1964). Respiration of fishes with special emphasis on standard oxygen consumption. I. Influence of weight and temperature on respiration of goldfish. Can. J. Zool. 42: 161-175.

Blazka, P., Volt, M. and Cepela, M. (1960). A new type of respirometer for determination of the metabolism of fish in an active state. Physiol. Bohemoslov. 9: 553-558.

Booth, J. (1978). The distribution of blood flow in the gills of fish: application of a new technique to rainbow trout (*Salmo gairdneri*). J. Exp. Biol. 73: 119-129.

Booth, J. (1979). The effects of oxygen supply, epinephrine, and acetylcholine on the distribution of blood flow in trout gills. J. Exp. Biol. 83: 31-39.

Brett, J.R. (1964). The respiratory metabolism and swimming performance of young sockeye salmon. J. Fish. Res. Board Can. 21: 1183-1226.

Brett, J.R. (1970). 3. Temperature. 3.3 Animals. 3.3.2 Fishes. In Marine Ecology Vol I, Environmental Factors, O. Kinne, ed. Part I. New York, Wiley (Interscience), p. 515-560.

Brett, J.R. (1972). The metabolic demand for oxygen in fish, particularly salmonids, and a comparison with other vertebrates. Respir. Physiol. 14: 151-170.

Brett, J.R. (1979). Environmental factors and growth. In "Fish Physiology", W.S. Hoar, D.J. Randall and J.R. Brett eds., N.Y., Academic Press, p. 599-675.

Brett, J.R. and Groves, T.D.D. (1979). Physiological energetics. In "Fish Physiology" Vol VIII, W.S. Hoar, D.J. Randall and J.R. Brett eds. N.Y., Academic Press, p. 279-352.

Brown, C.E. and Muir, B.S. (1970). Analysis of ram ventilation of fish gills with application to skipjack tuna (Katsuwonus pelamis). J. Fish. Res. Board Can. 27: 1637-1652.

Bullock, T.H. (1955). Compensation for temperature in the metabolism and activity of poikilotherms. Biol. Rev. 30: 311-342.

Butler, P.J. and Taylor, E.W. (1971). Response of the dogfish (Scyliorhinus canicula L.) to slowly induced and rapidly induces hypoxia. Comp. Biochem. Physiol. 39A: 307-323.

Butler, P.J. and Taylor, E.W. (1975). The effect of progressive hypoxia on respiration in the dogfish (Scyliorhinus canicula) at different seasonal temperatures. J. Exp. Biol. 63: 117-130.

Butler, P.J., Taylor, E.W. and Short, S. (1977). The effect of sectioning cranial nerves V, VII, IX and X on the cardiac response of the dogfish Scyliorhinus canicula to environmental hypoxia. J. Exp. Biol. 69: 233-245.

Butler, P.J., Taylor, E.W., Capra, M.F. and Davison, W. (1978). The effect of hypoxia on the levels of circulating catecholamines in the dogfish Scyliorhinus canicula. J. Comp. Physiol. 127: 325-330.

Cameron, J.N. (1974a). Evidence for the lack of by-pass shunting in teleost gills. J. Fish. Res. Board Can. 31: 211-213.

Cameron, J.N. (1974b). Blood flow distribution as indicated by tracer microspheres in resting and hypoxic arctic grayling (Thymallus arcticus). Comp. Biochem. Physiol. 52A: 441-444.

Cameron, J.N. (1975). Morphometric and flow indicator studies of the teleost heart. Can. J. Zool. 53: 691-698.

Cameron, J.N. (1976). Branchial ion uptake in arctic grayling: resting values and effects of acid-base disturbance. J. Exp. Biol. 64: 711-725.

Cameron, J.N. and Cech, J.J. (1970). Notes on the energy cost of gill ventilation in teleost. Comp. Biochem. Physiol. 34: 447-455.

Cameron, J.N. and Davis, J.C. (1970). Gas exchange in rainbow trout (Salmo gairdneri) with varying blood oxygen capacity. J. Fish. Res. Board Can. 27: 1069-1085.

Carey, F.G. and Teal, J.M. (1969a). Mako and porbeagle; warm bodied sharks. Comp. Biochem. Physiol. 28: 199-204.

Carey, F.G. and Teal, J.M. (1969b). Regulation of body temperature by the bluefin tuna. Comp. Biochem. Physiol. 28: 205-213.

Carey, F.G., Teal, J.M., Kanwisher, J.W., Lawson, K.D. and Beckett, J.S. (1971). Warm-bodied fish. Am. Zool. 11: 137-145.

Dam, L. van (1938). On the utilization of oxygen and regulation of breathing in some aquatic animals. Dissertation, Univ. Groningen, Netherlands.

Dandy, J.W.T. (1970). Activity response to oxygen in the brook trout Salvelinus fontinalis (Mitchill). Can. J. Zool. 48: 1067-1072.

Davis, J.C. (1975). Minimal dissolved oxygen requirements of aquatic life with emphasis on Canadian species: a review. J. Fish. Res. Board Can. 32: 2295-2332.

Davis, J.C. and Cameron, J.N. (1970). Water flow and gas exchange at the gills of the rainbow trout Salmo gairdneri. J. Exp. Biol. 54: 1-18.

Davis, J.C. and Randall, D.J. (1973). Gill irrigation and pressure relationships in rainbow trout, Salmo gairdneri. J. Fish. Res. Board Can. 30: 99-104.

Daxboeck, C. and Holeton, G.F. (1978). Oxygen receptors in the rainbow trout, Salmo gairdneri. Can. J. Zool. 56: 1254-1259.

Daxboeck, C. and Holeton, G.F. (1980). The effect of MS-222 on the hypoxic response of the rainbow trout (Salmo gairdneri). Comp. Biochem. Physiol. In Press.

Dejours, P. (1973). Problems of control of breathing in fishes. In "Comparative Physiology" L. Bolis, K. Schmidt-Nielsen and S.H.P. Maddrel eds. North Holland Publ. p. 117-133.

Dejours, P., Toulmond, A. and Truchot, J.P. (1977). The effect of hyperoxia on the breathing of marine fishes. Comp. Biochem. Physiol. 58A: 409-411.

DeKock, L.L. (1963). A histological study of the head region of two salmonids with special reference to pressue and chemoreceptors. Acta. Anat. 55: 39-50.

Eclancher, B.(1972). Action des changements de P_{O_2} de l'eau sur la ventilation de la truite et de la tanche. J. Physiol. Paris 65: 397A.

Eddy, F.B. (1974). Blood gases of the tench (Tinca tinca) in well aerated and oxygen-deficient waters. J. Exp. Biol. 6071-83.

Edwards, R.R.C. (1971). An assessment of the energy cost of gill ventilation in the plaice (Pleuronectes platessa L.). Comp. Biochem. Physiol. 40A: 391-398.

Evans, D.O. (1977). Seasonal changes in standard metabolism, upper and lower thermal tolerance and thermoregulatory behavior of the pumpkinseed, Lepomis gibbosus (Linnaeus). Ph.D. Thesis, Univ. Toronto, 429 p.

Fairbridge, R.M. (1966). The encyclopaedia of Oceanography. N.Y. Reinhold, 1021 p.

Farmer, G.J. and Beamish, F.W.H. (1969). Oxygen consumption of Tilapia nilotica in relation to swimming speed and salinity. J. Fish. Res. Board Can. 26: 2807-2821.

Freadman, M.A. (1979). Swimming energetics of striped bass (Morone saxatilis) and bluefish (Pomatomus saltatrix): gill ventilation and swimming metabolism. J. Exp. Biol. 83: 217-230.

Fry, F.E.J. (1947). Effects of the environment on animal activity. Univ. Toronto Studies, Biol. Ser. 55; Publ. Ontario Fish. Res. Lab. 68: 1-62.

Fry, F.E.J. (1957). The aquatic respiration of fish. In "The Physiology of Fishes" M.E. Brown ed., Vol. I, pp. 1-63. N.Y., Academic Press.

Fry, F.E.J. (1971). Effect of environmental factors. In "Fish Physiology" W.S. Hoar and D.J. Randall Eds. Vol. VI, p. 1-98. N.Y., Academic Press.

Fry, F.E.J. and Hochachka, P.W. (1970). Fish. In "Comparative Physiology of Thermoregulation" G.C. Whittow ed. Vol. I, pp. 79-134. N.Y., Academic Press.

Garey, W.F. and Rahn, H. (1970). Gas tensions in tissues of trout and carp exposed to diurnal changes in oxygen tension of the water. J. Exp. Biol. 52: 575-582.

Gee, J.H., Tallman, R.F. and Smart, H.J. (1978). Reactions of some great plains fishes to progressive hypoxia. Can. J. Zool. 56: 1962-1966.

Glass, N.R. (1969). Discussions of calculation of power function with special reference to respiratory metabolism in fish. J. Fish. Res. Board Can. 26: 2643-2650.

Graham, J.B. (1973). Heat exchange in the black skipjack and the blood gas relationship of warm-bodied fishes. Proc. Nat. Acad. Sci. 70: 1964-1967.

Hall, F.G. (1930). The ability of the common mackerel to remove dissolved oxygen from sea water. Am. J. Physiol. 93: 417-421.

Heisler, N. (1978). Bicarbonate exchange between body compartments after changes of temperature in the larger spotted dogfish (Scyliorhinus stellaris). Respir. Physiol. 33: 145-160.

Hemmingsen, E.A., Douglas, E.L., Johansen, K. and Millard, R.W. (1972). Aortic blood flow and cardiac output in the hemoglobin-free fish Chaenocephalus aceratus. Comp. Biochem. Physiol. 43A: 1045-1051.

Hoglund, L.B. (1961). The reactions of fish in concentration gradients. Rep. Inst. Freshwater Res. Drottningholm 43: 1-147.

Holeton, G.F. (1970). Oxygen uptake and circulation by a hemoglobinless antarctic fish (Chaenocephalus aceratus Lonnberg) compared with three red-blooded antarctic fish. Comp. Biochem. Physiol. 34: 457-471.

Holeton, G.F. (1971a). Oxygen uptake and transport by the rainbow trout during exposure to carbon monoxide. J. Exp. Biol. 54: 239-254.

Holeton, G.F. (1971b). Respiratory and circulatory responses of rainbow trout larvae to carbon monoxide and to hypoxia. J. Exp. Biol. 55: 683-694.

Holeton, G.F. (1972). Gas exchange in fish with and without hemoglobin. Respir. Physiol. 14: 142-150.

Holeton, G.F. (1973). Respiration of arctic char (Salvelinus alpinus) from a high arctic lake. J. Fish. Res. Board Can. 30: 717-723.

Holeton, G.F. (1974). Metabolic cold adaptation of polar fish: fact or artefact? Physiol. Zool. 47: 137-152.

Holeton, G.F. (1977). Constancy of arterial blood pH during CO-induced hypoxia in the rainbow trout. Can J. Zool. 55: 1010-1013.

Holeton, G.F. and Jones, D.R. (1975). Water flow dynamics in the respiratory tract of the carp (Cyprinus carpio L.). J. Exp. Biol. 63: 537-549.

Holeton, G.F. and Randall, D.J. (1967a). Changes in blood pressure in the rainbow trout during hypoxia. J. Exp. Biol. 46: 297-305.

Holeton, G.F. and Randall, D.J. (1967b). The effect of hypoxia upon the partial pressure of gases in the blood and water afferent and efferent to the gills of rainbow trout. J. Exp. Biol. 46: 317-327.

Hughes, G.M. (1966). The dimensions of fish gills in relation to their function. J. Exp. Biol. 45: 177-195.

Hughes, G.M. and Roberts, J.L. (1970). A study on the effects of temperature changes on the respiratory pumps of the rainbow trout. J. Exp. Biol. 52: 177-192.

Hughes, G.M. and Shelton, G. (1958). The mechanism of gill ventilation in three freshwater teleosts. J. Exp. Biol. 35: 807-823.

Hughes, G.M. and Shelton, G. (1962). Respiratory mechanisms and their nervous control in fish. In "Advances in Comparative Physiology and Biochemistry.", O. Lowenstein ed., New York, Academy Press, p. 274-364.

Hughes, G.M. and Singh, B.N. (1970). Respiration in an air breathing fish, the climbing perch Anabas testudineus (Bloch). I. Oxygen uptake and carbon dioxide release into air and water. J. Exp. Biol. 53: 265-280.

Itazawa, Y. and Takeda, T. (1978). Gas exchange in the carp gills in normoxic and hypoxic conditions. Respir. Physiol. 35: 263-269.

Job, S.V. (1955). The oxygen consumption of Salvelinus fontinalis. Univ. Toronto Biol. Ser. 61; Publ. Ontario Fisheries Res. Lab. 73: 1-39.

Johansen, K. (1971). Comparative physiology; gas exchange and circulation in fishes. A. Rev. Physiol. 33: 596-612.

Jones, D.R. and Schwarzfeld, T. (1974). The oxygen cost to the metabolism and efficiency of breathing in trout (Salmo gairdneri). Respir. Physiol. 21: 241-254.

Jones, J.D. (1961). Aspects of respiration in Planorbis corneus L. and Lymnaea stagnalis L. (Gastropoda: Plumonata). Comp. Biochem. Physiol. 4: 1-29.

Jones, J.R.E. (1952). The reactions of fish to water of low oxygen concentration. J. Exp. Biol. 29: 403-415.

Kramer, D.L.; Lindsey, C.C., Moodie, G.E.E. and Stevens, E.D. (1978). The fishes and aquatic environment of the central amazon basin, with particular reference to respiratory patterns. Can. J. Zool. 56: 717-729.

Krogh, A. (1914). The quantitative relation between temperature and standard metabolism in animals. Intern. Z. physik. chem. Biol. 1: 491-508.

Laurent, P. (1977). Arterial chemoreceptive structures in fish. In "Morphology and Mechanics of Chemoreceptors." A.S. Paintal ed., Pub. by Vallabhbhai Patel Chest Institute, Univ. Delhi, Delhi - 110 007. p. 275-281.

Laurent, P. and Dunel, S. (1976). Functional organization of the teleost gill. I. Blood pathways. Acta. Zool. (Stockh.) 57: 189-209.

Piiper, J., Baumgarten, D. and Meyer, M. (1970). Effects of hypoxia upon respiration and circulation in the dogfish Scyliorhinus stellaris. Comp. Biochem. Physiol. 36: 513-552.

Poupa, O., Gesser, H., Johnsson, H. and Sullivan, L. (1974). Coronary supplied compact shell of ventricular myocardium in salmonids: growth and enzyme pattern. Comp. Biochem. Physiol. 48A: 85-95.

Rahn, H., Wangensteen, O.D. and Crowley, G.J. (1973). Tissue O_2 and CO_2 tensions of trout in high altitude lakes. Trans. Am. Fish. Soc. 102: 132-134.

Randall, D.J. (1970a). The circulatory system. In "Fish Physiology" Vol. IV, W.S. Hoar and D.J. Randall eds., N.Y., Academic Press. p. 133-172.

Randall, D.J. (1970b). Gas exchange in fish. In "Fish Physiology." Vol. IV, W.S. Hoar and D.J. Randall eds., N.Y., Academic Press. p. 253-293.

Randall, D.J., Baumgarten, D. and Malyusz, M. (1972). The relationship between gas and ion transfer across the gills of fishes. Comp. Biochem. Physiol. 41A: 629-637.

Randall, D.J. and Cameron, J.N. (1973). Respiratory control of arterial pH as temperature changes in rainbow trout Salmo gairdneri. Am. J. Physiol. 225: 997-1002.

Randall, D.J., Holeton, G.F. and Stevens, E. Don (1967). The exchange of oxygen and carbon dioxide across the gills of rainbow trout. J. Exp. Biol. 6: 339-348.

Randall, D.J. and Jones, D.R. (1973). The effect of deafferentation of the pseudobranch on the respiratory response to hypoxia and hyperoxia in the trout (Salmo gairdneri). Respir. Physiol. 17: 291-301.

Randall, D.J. and Smith, J.C. (1967). The regulation of cardiac activity in fish in a hypoxic environment. Physiol. Zool. 40: 104-113.

Ricker, W.E. (1973). Linear regressions in fishery research. J. Fish. Res. Board Can. 30: 409-434.

Ricker, W.E. (1979). Growth rates and models. In: Fish Physiology, Vol. VIII, Eds. W.S. Hoar, D.J. Randall and J.R. Brett. New York, Academic Press, p. 677-743.

Roberts, J.L. (1975). Active branchial and ram gill ventilation in fishes. Biol. Bull. 148: 85-105.

Satchell, G.H. (1971). Circulation in fishes. Cambridge Univ. Press. 131 p.

Satchell, G.H. (1978). Microcirculation in fishes. In "Microcirculation", Vol. 2, ed. G. Kaley and B.M. Altura, 756 p. Baltimore, Univ. Park Press.

Saunders, R.L. (1962). The irrigation of the gills in fishes. II. Efficiency of oxygen uptake in relation to respiratory flow activity and concentration of oxygen and carbon dioxide. Can. J. Zool. 40: 817-862.

Scheid, P. and Piiper, J. (1976). Quantitative functional analysis of branchial gas transfer: theory and application to Scyliorhinus stellaris (Elasmobranchii). In "Respiration of Amphibious Vertebrates." G.M. Hughes ed. London, Academic Press. p. 17-38.

Scholander, P.F., Flagg, W., Walters, V. and Irving, L. (1953). Climatic adaptations in arctic and tropical poikilotherms. Physiol. Zool. 26: 67-92.

Schumann, D. and Piiper, J. (1966). Der Sauerstoffbedarf der Atmung bei Fischen nach Messungen an der Narkotisierten Schleie (Tinca tinca). Pflugers Arch. ges. Physiol. 288: 15-26.

Shelton, G. (1970). The regulation of breathing. In "Fish Physiology" Vol. IV. Hoar, W.S. and Randall, D.J. eds. N.Y., Academic Press, p. 293-359.

Shepard, M.P. (1955). Resistance and tolerance of young speckled trout (Salvelinus fontinalis) to oxygen lack, with special reference to low oxygen acclimation. J. Fish. Res. Board Can. 12: 387-434.

Skidmore, J. (1970). respiration and osmoregulation in rainbow trout with gills damaged by zinc sulphate. J. Exp. Biol. 52: 484-494.

Smith, F.M. and Jones, D.R. (1978). Localization of receptors causing hypoxic bradycardia in trout (Salmo gairdneri). Can. J. Zool. 56: 1260-1265.

Sommers, P. van (1962). Oxygen-motivated behavior in the goldfish. Science 137: 678-679.

Steen, J.B. and Kruysse, E. (1964). The respiratory function of teleostean gills. Comp. Biochem. Physiol. 12: 127-142.

Stevens, E. Don (1972). Some aspects of gas exchange in tuna. J. Exp. Biol. 56: 809-823.

Stevens, E.D. and Fry, F.E.J. (1971). Brain and muscle temperatures in ocean-caught and captive skipjack tuna. Comp. Biochem. Physiol. 38A: 203-211.

Stevens, E. Don and Holeton, G.F. (1978). The partitioning of oxygen uptake from air and from water by erythrinids. Can. J. Zool. 56: 965-969.

Stevens, E.D., Lam, H.M. and Kendall, J. (1974). Vascular anatomy of the counter current heat exchanger of skipjack tuna. J. Exp. Biol. 61: 145-153.

Stevens, E. Don and Sutterlin, A. (1976). Heat transfer between fish and ambient water. J. Exp. Biol. 65: 131-145.

Taylor, E.W., Short, S. and Butler, P.J. (1977). The role of the cardiac vagus in the response of the dogfish Scyliorhinus canicula to hypoxia. J. Exp. Biol. 70: 57-75.

Whitmore, C.M., Warren, C.E. and Doudoroff, P. (1960). Avoidance reactions of salmonids and centrarchid fishes to low oxygen concentrations. Trans. Am. Fish. Soc. 89: 17-26.

Wohlschlag, D.E. (1964). Respiratory metabolism and ecological characteristics of some fishes in McMurdo sound, Antarctica. In "Biology of the antarctic seas" M.O. Lee ed. Am. Geophys. Union. Washington D.C., Antarct. Res. Ser. i: 33-62.

Wood, S.C. and Johansen, K. (1972). Adaptation to hypoxia by increased Hb_{O2} affinity and decreased red cell ATP concentration. Nature, New Biol. 237: 278-279.

Wood, S.C., Johansen, K. and Weber, R.E. (1975). Effects of ambient P_{O2} on hemoglobin-oxygen affinity and red cell ATP concentration in a benthic fish, Pleuronectes platessa. Respir. Physiol. 25: 259-267.

MORPHOMETRY OF FISH GAS EXCHANGE ORGANS IN RELATION TO THEIR RESPIRATORY FUNCTION

G.M. Hughes
Research Unit for Comparative Animal Respiration
University of Bristol
Woodland Road, Bristol, England, BS8 1UG

INTRODUCTION

The respiration of fish living in water is severely limited by the physico-chemical properties of water (Hughes, 1961) which are largely overcome by some fine adaptations of the ventilatory pumping mechanism and structure of the gills. These organs are the site of important exchanges with the environment, not only of the respiratory gases, but also of inorganic ions and temperature. The double pumping mechanism produces a fairly continuous flow of water across the gills and is operated by muscles whose activity and coordination may vary, but always ensure a pressure differential across the gills as a result of a number of coupling mechanisms (Ballintijn & Hughes, 1965; Osse, 1969). The pumping mechanism and gill structure have evolved in relation to the respiratory needs of different species and in some cases may include "coughing" movements as a regular part of their rhythm (Kuiper, 1907; Hughes & Shelton, 1958). Such " coughs " have well defined patterns of muscular activity (Ballintijn, 1969; Young, 1970; Hughes, 1975) and provide a useful index for the study of some environmental stresses, for example they often become reduced in frequency during hypoxia but are increased when the fish breathes water containing pollutants (Hughes & Adeney, 1977a). The precise pattern of coughing movements may vary (Hughes & Adeney, 1977b), but the most common types involve a reversal of water flow across the gills. Thus these organs form a central feature of the fish respiratory system and a quantitative study has led to the greater understanding of their function. Modification of these structures has also occured in relation to air breathing.

This paper reviews some of the methods available for a morphometric study of respiratory surfaces, and a discussion is given of results which show the variety of environmental circumstances to which they have recently been applied. No account is given of the skin and cutaneous respiration, but the same principles and methods are applicable.

It is important to note, however, that a number of recent studies have drawn attention to the importance of oxygen uptake across the skin in fish where this had previously been thought to be absent (Kirsch & Nonnotte, 1977). The amount of oxygen entering is small relative to that taken up at the gills and much of it is utilised by the skin itself or tissues very close to the body surface.

NATURE OF GAS EXCHANGE SURFACES

The study of gas exchange in any animal must centre upon the nature and extent of the barrier which separates the external and internal media of the animal. Among vertebrates this is usually between air or water and the circulating blood. The conditions at such a surface are indicated diagrammatically in Fig. 1 which shows that the physical processes involved are those of diffusion and convection. The diffusion takes place across a surface of a given area (A) and a distance (t). Analysis using classical methods in such a situation (Weibel, 1970; Hughes, 1972a; Scheid & Piiper, 1976) involve a consideration of the conditions for diffusion and convection in each of the layers into which the overall distance for gas transfer can be subdivided. Such an analysis requires accurate measurements not only of the area and thicknesses of the different layers but also the permeability coefficients, i.e. the solubility (or capacitance, β) and diffusion coefficient (D) for the respiratory gases in each of these layers. Unfortunately, many of these features are unknown for the gas exchange surfaces of fish or any other organism, although approximations have been made. An alternative view is that taken by engineers in which they consider an overall gas transfer coefficient between the two media (Hills & Hughes, 1970; Hills, 1974). They then are able to give variations for this coefficient in relation to the dimensions and flow conditions on the two sides of the barrier.

Both types of approach have been adopted by different authors and fortunately the general nature of the deductions obtained are in substantial agreement.

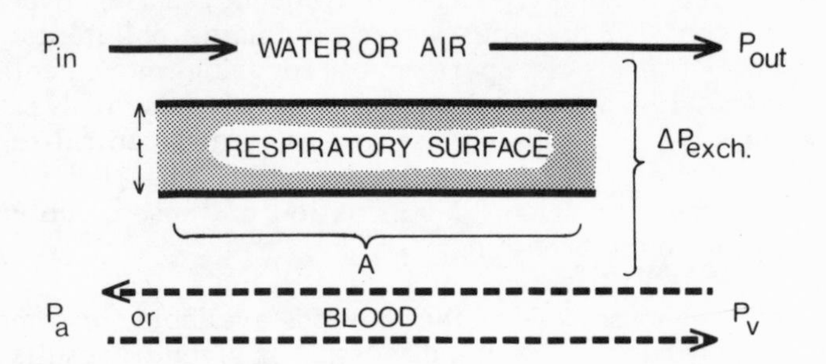

FIG. 1. Diagram to show basic parameters of importance in oxygen exchange at a respiratory surface (from Hughes, 1977).

SITES OF GAS EXCHANGE AND THEIR MORPHOLOGY

Among fish the main regions of the body surface where gas exchange takes place are the skin, gills, lungs and accessory air breathing organs. Uptake of oxygen and release of carbon dioxide usually occur at all these places but the relative amounts can vary not only at a given time, but also in relation to the season and stage of development. Consequently the gas exchange ratio (R) can vary quite markedly. In air breathing fish the ratio is often greater than one for the gills but less than unity for the air breathing organs - indicating a greater uptake of oxygen from the air and greater release of CO_2 to the water (Hughes, 1966a). Gas exchanges across the skin can occur with either water or air, whereas those of the gills are mainly with water, except in some exceptional cases. The lung is an air breathing organ as are the accessory organs, but some recent evidence indicates that ventilation of the accessory organs of some fish includes a phase when the organ is ventilated with water between phases in which it contains air (Hughes, 1978a; Peters, 1978). Analysis of pressure recordings from the buccal, suprabranchial and opercular cavities made it possible to recognise four phases during the ventilatory cycle of Osphronemus (Hughes & Peters, 1980). From the dimensional point of view, the gas exchange area is identical regardless of the external medium, the main differences being related to the physical properties of air and water with respect to diffusion and convection. It is with respect to the flow conditions that there are important differences between respiratory surfaces, notably that the gills tend to be external branchings of the body surface whereas lungs and the other accessory organs are generally contained within expanded pouches of the alimentary canal. These differences necessitate a generally tidal flow of the air or water during ventilation of the lung or accessory organs whereas the gills are characterised by ventilation occuring as a uni-directional and more or less continuous process. Ventilation of the skin obviously depends upon the degree of movement shown by the fish and there are few cases in which special mechanisms are to be found. Ventilation of the air breathing organs is not always a purely tidal mechanism insofar as the air may enter through one opening and leave through another so that there is a certain degree of flushing out of the respiratory cavity. This also seems to occur in Lampreys and is therefore not a truly tidal mechanism (Hughes & Shelton, 1962; Randall, 1971).

The basic morphology of the organs specifically concerned with gas exchange, i.e. excluding the skin, is one in which there is a great increase in surface of contact (Fig. 2b). The gills are a fine example of the way in which a very large surface may be accomodated within a small space. The design of these organs seems particularly adapted for this purpose and ensures that the resistance to water flow is minimized while increasing the facility for gas exchange (Hughes, 1966b). The lungs and other air breathing organs also have greatly increased surfaces but as they are not normally supported by an aqueous medium they tend to be less finely divided and consequently a given surface area needs a greater space within the animal. Among fish there are remarkable examples of the way in which parts of the

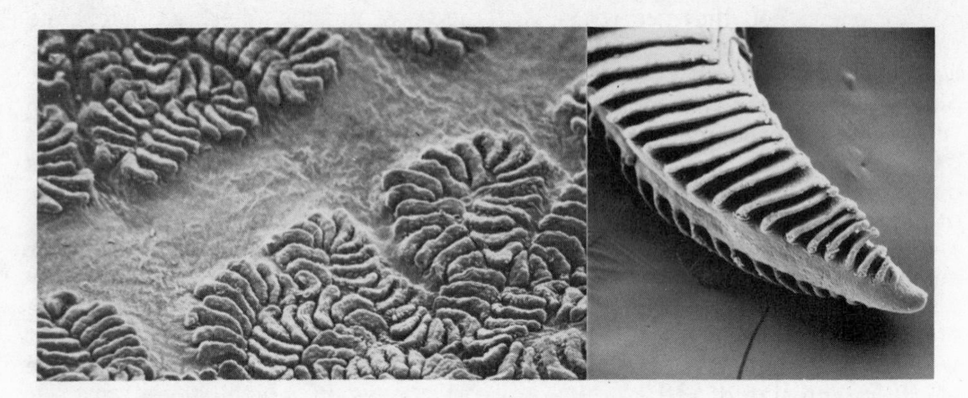

FIG. 2. Scanning electron-micrographs showing (a) the surface of the air sac of Heteropneustes fossilis (x 257) (from Hughes & Munshi, 1978), and (b) the shape and arrangement of the secondary lamellae of the giant sturgeon, Huso huso (x 81).

normal gill structure have become modified as air breathing organs. Perhaps an extreme case being shown by Heteropneustes where modification of the gill has led to the evolution of a pair of sac-like elongations which are very reminiscent of a lung. Morphological investigations, both with the light microscope (Munshi, 1968) and more recently with the scanning (Fig. 2a) and transmission electron microscopes (Hughes & Munshi, 1968, 1978) have shown that they are based upon the structure of a gill.

Apart from these sites of external gas exchange, there are important exchanges between the blood and tissues where both the physiological and morphological relationships are important. There have been very few morphometric studies of fish tissues from this point of view. It is clearly important to known the degrees of vascularisation at the tissue level which would give information regarding the diffusion distances and surface areas across which gas exchange occurs. Differences in the density of mitochondria and distances from the cell surfaces are also relevant parameters which could be quantified by suitable morphometric investigations.

MORPHOMETRY

1. Methods

The methods most appropriate for a morphometric study depend upon the nature of the material and the purpose for which the measurements are made. Many investigations are not concerned with functional aspects but are of value from a taxonomic point of view. In this account, I wish to emphasise the value of choosing measurements which are relevant for the interpretation of the physiology of a given organ system. In some cases simple linear measurements are appropriate either with or without a microscope. For the most important features of respiratory organs some sampling method is necessary because of their large and complex form. Gross morphometry of the gills is more readily achieved than that of lungs and other air breathing organs, mainly because of their external position. The sampling problem remains, however, and for area measurements it is usual to measure every tenth filament on one side of the fish. As a consequence, measurements of total filament length (L) are probably the most reliable. When discrepancies are found in the area measurements of different authors, figures for total filament length are often in the agreement. Measurements of the frequency of the secondary lamellae (1/d) along the filaments are also fairly reliable, but a area of a secondary lamella (bl) which forms the third component is much more difficult and clearly can have a very marked effect on the final result ($Ag = L.bl/d$). Variations in the form and size of secondary lamellae are well known throughout the gill system and it is for that reason that sampling and weighting methods have been introduced (Muir & Hughes, 1969).

For more detailed morphometry of the surface and volume components of the secondary lamellae and diffusion distances, fixed material is studied either in the light or the electron microscope. These treatments can alter some of the dimensions, and fixation followed by paraffin embedding produces the greatest shrinkage (20%). Sampling is necessary and light microscopy has the advantage of utilising a larger portion of the total respiratory system, whereas electron microscopy of necessity is restricted to more localised regions. Analysis of this material using stereological methods can be very useful (Hughes, 1972a; Hughes & Perry, 1976) and involves the superposition of a suitable grid. Where the material is sectioned at random a rectilinear grid is appropriate, but in studies with such oriented structures it is possible to ensure a fairly uniform plane of sectioning and with such material a more randomising grid such as the Merz is used. By means of these grids and point and intersection counting, it is possible to estimate the relative surfaces and relative volumes within a structure which are respectively proportional to the intersection and point counts.

With air-breathing organs the problem of area measurement is much more difficult, but for some simple sac-like structures it is possible to approximate to a cylinder and thence by determining areas and surfaces for sections chosen at measured distances along the cylinder, estimates can be made for the total volume (Fig. 3a & b) and surface area. Such a study was carried out for the lung of Lepidosiren and is applicable to other structures of this kind (Hughes & Weibel, 1976). Very often the surface area of air breathing organs has been estimated by opening up the organ structure and using an overlying piece of paper. These measurements give values for the total internal surfaces but do not take account of irregularities and usually assume that the total surface is respiratory. Recent scanning electron micrographs (Fig. 2a) reveal the extent of the errors involved and could be used to estimate appropriate correction factors.

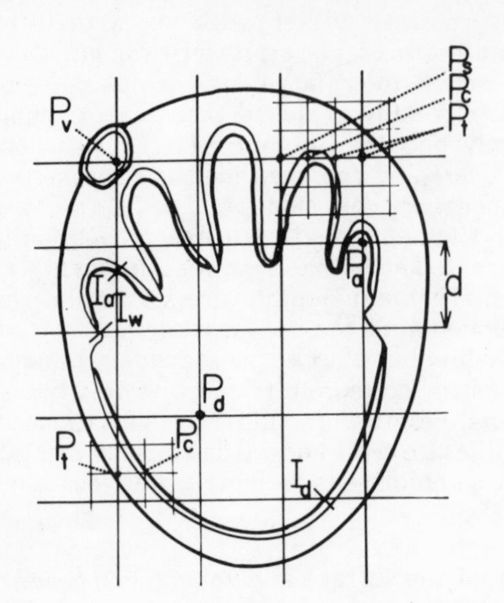

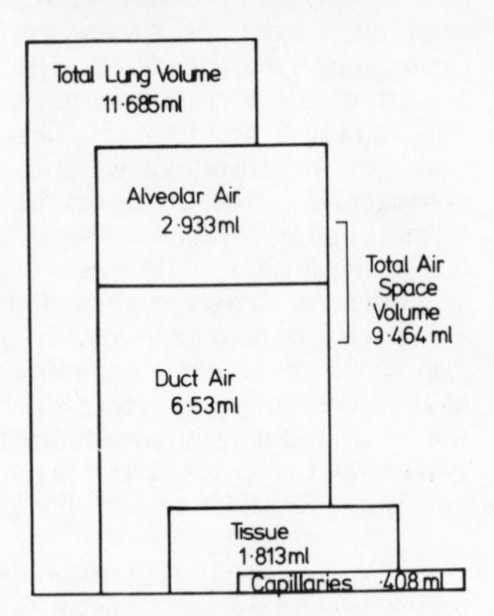

FIG. 3a. Diagrammatic section through the lung of Lepidosiren with superimposed rectilinear grid showing the regions used for point and intersection counting.

FIG. 3b. Distribution of volumes of air and tissues in a lung of Lepidosiren (500 gms) based upon morphometric study.

(from Hughes & Weibel, 1976).

2. Model

The model which tends to become accepted in studies of this kind is based upon the Fick equation: $\dot{V}_{O_2} = KA\Delta P_{O_2}/t$ and at the outset it is important to appreciate some of the limitations attached to the components of this relationship.

a) Oxygen uptake ($\dot{V}o_2$)

In all animals but especially fish the oxygen consumption varies markedly in relation to activity (Fry, 1957). It is therefore important to define as far as possible the conditions under which the measurements of oxygen uptake are made. Another aspect which is of importance is that fish are also capable of much anaerobic metabolism and it is clearly relevant to know the amount of energy being provided for the animal's activity which is not directly related to the oxygen uptake.

b) The gill area (Ag or Sg)

Measurements of the gill surface usually try to estimate the total surface of the secondary lamellae exposed to the water. It is clear from a number of studies on different species of fish that by no means all of this area is functional at any given time. These variations are due in part to the limitations of the surface which is ventilated by the water current but also differences in the vascular beds which are perfused will also limit gas exchange. Inequalities between ventilation and perfusion of different regions of the gill surface can lead to an inefficient overall gas transfer. This was emphasised in an early analysis which showed that the effectiveness of gas transfer was greatest when the capacity rates of the two media were equal (Hughes & Shelton, 1962). Use of capacity rates (the product of volume flow/unit time and gas content/unit volume/unit partial pressure) emphasised that ventilation and perfusion alone give an incomplete description of the situation. Such a view has become generally accepted by respiratory physiologists when discussing comparative aspects of gas exchange. For example, in *Scyliorhinus stellaris* the volume of water ventilated ($\dot{V}_g$) is nearly 10 times greater than that of the blood perfusing ($\dot{Q}$) the gills in unit time, whereas the ratio of the capacity rates (G_{vent}/G_{perf}) of water and blood is 0.42 (Scheid & Piiper, 1976). To avoid confusion it should be appreciated that the term " conductance" (G_{vent}, G_{perf}) is now often used in such discussions but the basic relevance of taking into account the ability of the media to transport gases (or heat) remains. Thus any environmental factor which alters the ability of water or blood to transport oxygen will change the condition for gas transfer just as would any factor which interferes with the flow of blood, water or air at the respiratory surface.

c) Barrier thickness (t, or τ)

The distance separating the oxygen molecules in the external medium from the haemoglobin molecules in the red blood cells is the most important dimension from the point of view of oxygen or O_2 transfer. Even this dimension can vary in different regions of a given respiratory organ so that it is important when making detailed estimates of the barrier thickness, that samples should be taken from different parts of the organ. In most respiratory organs there are channels below the outer epithelial surface which contain blood cells; the blood channels are separated by cells the best known being the pillar cells of the gill secondary lamella (Hughes & Grimstone, 1965; Newstead, 1967). Analogous supporting cells are also found in the accessory organs and lungs (Munshi, 1976; Hughes, 1978b). The outermost surface is thus a mosaic of regions which lie immediately over the blood channels and those which lie above the supporting cells. The diffusion distances in the latter regions are greater than those above the blood channels. In consequence it is important to define which thickness measurements are used in any calculations and to ensure that they refer to the same fraction of the total surface as is used in the area measurement. For example, if the total secondary lamellar area is used then the most appropriate thickness measurement is the harmonic mean thickness which takes into account both the long distances for diffusion into the blood channels across the pillar cells as well as the short distances directly across the barrier. Very often in calculations made in the literature the total respiratory area is used together with a mean value for the thickness separating the water from the blood in the barrier region alone.

When sections are made of respiratory organs it is often very difficult to ensure that the plane of section is perpendicular to the gas exchange surface. Consequently measurements of oblique sections will give much higher values for thickness than those that are more perpendicular. In a number of studies this fact accounts for the wide range in values. In some studies the harmonic mean of such measurements is taken and compensates for this problem to some extent. In other studies, especially with gills, it is possible to orientate the gill filaments so that the plane of section is very close to 90° to the surface of the secondary lamellae. In light microscopic studies it is possible to check this plane quite accurately (Hughes & Perry, 1976). In other cases (Hughes, 1972a) the plane of section is deliberately randomised so that harmonic mean values can be obtained.

There is also variation in the measurements made - in some cases the tissue barrier is measured, assuming that in most cases the red cells come very close to the wall of the blood channel. In other studies (e.g. Hughes & Perry, 1976) using light microscopy, measurements were made to the nearest red blood cell.

Another feature of the exchange surface which has recently been investigated is the detailed sculpturing of the outermost epithelial surface in contact with the air or water. Scanning electron microscopy has emphasised regional differences in such surfaces insofar as the microvilli or microridges may be developed to different extents. The inner surfaces of air breathing organs are not smooth and this would increase the surface area. However, many measurements have also ignored the non-vascular "lanes" and these two features probably compensate for one another. In fish the outer surface is almost certainly covered with mucus so that for the thickness measurements the most raised parts of the villi or ridges should be used to define the limit of the tissue barrier (Hughes, 1979a).

In addition to these complications there is also the probability that the thickness of the different parts of a respiratory organ may vary both during a respiratory cycle and also according to environmental conditions.

TABLE 1. Summary of results of regression analysis for total filament length against body mass of 9 species of fish studied by Gray (1954). Values for the intercept (a) and slope (b) are given together with the correlation coefficient (r). The average of figures obtained from these nine separate analyses are given and also the result of a regression analysis for all 88 specimens.

FISH SPECIES	n	a	b	r
Sea Trout, Cynoscion regalis	7	172.3	0.403	0.437
Sheepshead, Archosargus probatocephalus	6	187.1	0.364	0.984
Scup, Stenotomus chrysops	10	717.9	0.372	0.871
Striped Mullet, Mugil cephalus	9	160.1	0.423	0.641
Butterfish, Poronotus triacanthus	14	152.9	0.357	0.824
Tautog, Tautoga onitus	9	92.9	0.430	0.876
Sea Robin, Prionotus strigatus	11	144.1	0.375	0.931
Fluke, Paralichthys dentatus	6	124.5	0.358	0.819
Puffer, Spheroides maculatus	16	69.4	0.423	0.814
All Fish	88	102.5	0.429	0.617
Average of results of 9 species	88	141.7	0.389	0.800

TABLE 2. Table showing diffusing capacity of the tissue barrier (Dt) and component measurements for respiratory surfaces of Channa, Anabas, Heteropneustes and Monopterus (from Hakim et al., 1978).

FISH SPECIES	Respiratory surface for 1g fish (mm^2)	Thickness of tissue barrier t (μm)	Area/g wt. 100g fish (mm^2)	Diffusing capacity 100g fish (ml/min/mmHg/kg)	Reference
Channa punctata					
(i) Total gills	470.39	2.0333	71.8229	0.0530	Hakim et al (1978)
(ii) Suprabranchial chamber	159.08	0.7800	39.1705	0.0753	Hakim et al (1978)
Anabas testudineus					
(i) All gill arches	278.00	10.0000	47.2000	0.0071	Hughes, Dube & Munshi (1973)
(ii) Suprabranchial chamber	55.40	0.2100	7.6500	0.0539	Hughes, Dube & Munshi (1973)
(iii) Labyrinthine organ	80.70	0.2100	32.0000	0.2286	Hughes, Dube & Munshi (1973)
Heteropneustes fossilis					
(i) All gills	186.10	3.5800	57.7000	0.0242	Hughes et al (1974)
(ii) Air sac	145.90	1.6000	30.7000	0.0288	Hughes et al (1974)
(iii) Skin	851.10	98.000	200.0000	0.0031	Hughes et al (1974)
Monopterus cuchia					
(i) Air sac	12.90	0.44	4.8400	0.0165	Hughes et al (1974)

d) Mean Oxygen Pressure Gradient (ΔPO_2)

Many factors affect the driving force responsible for the passage of oxygen across a respiratory membrane. The blood coming to the respiratory organ is low in oxygen and the level of this venous partial pressure is related to the activity of the animal. Following its passage through the respiratory organ the arterialised blood is usually at a partial pressure close to that of the water or air entering the respiratory organ. During exchange processes between the respiratory medium and the blood there are, however, many variations in the size of the driving force which are closely related to the flow conditions on both sides of the membrane.

The magnitude of the oxygen tension difference across the gill surface depends upon the ventilation volume ($\dot{V}g$) and cardiac output ($\dot{Q}$) and there have been a number of ways of expressing the overall mean value. A common method is to use ΔPg but this has a number of disadvantages which have been discussed by Piiper & Baumgarten-Schumann (1968) and Hughes (1972a).

Returning to the Fick equation, it can be arranged as follows:

$$\frac{\dot{V}O_2}{\Delta \overline{PO}_2} = \frac{K\,A}{t}$$

This ratio of O_2 transfer to mean difference in oxygen tension is called the " diffusing capacity " by physiologists. Consequently we can also refer to K A/t as the morphometric diffusing capacity. This concept has been applied to many fish respiratory surfaces and has the important advantage of emphasising that both the area and thickness measurements are important. In some cases (Table 2) respiratory organs may have an extensive surface but because of the thickness of the barrier they are relatively unimportant in gas exchange, as is indicated by their low diffusing capacity.

3. Modes of analysis

A common way of analysing morphometric data is the use of log/log plots of the particular measurement against body mass. This depends on the relationship $Y = aW^b$ or the logarithmic transformation $\log Y = \log a + b \log W$. From such studies it has become clear that it is important to specify the body size when giving a measurement per unit body weight. In most cases the slope (b) is less than one and consequently larger fish have smaller weight specific measurements. In an early study of this kind (Muir & Hughes, 1969) it was shown that when the three primary components of the gill area of tunas were plotted logarithmically then the sum of the slopes of the individual lines was very close to that for the total area. The intercept value (a) corresponds to the measurement for a 1 gm specimen and in general indicates the position of the regression line on the Y axis. Although the intercept value is important, perhaps of greatest interest is a value about the mean normal weight distribution for a particular species. One method of comparing species is to use values for 10, 100 or 1000 gm, etc. More recently a suggestion has been made (de Jager and Dekkers, 1975) that a standard fish of 200 gm should be taken. Clearly this would be appropriate in some cases but quite inappropriate in others.

The validity of the logarithmic transformation analysis has now been demonstrated for a large number of species in which the sum of the exponents is close to the overall slope for gill area. It has become clear that the slopes of not only the individual components but also for the total gill area vary quite significantly between different species. Some suggestions have been made that the higher slopes (about 1.0) are found in fish with greater maximum weights (Pauly, 1979). Such findings have emphasised the importance of making intraspecific studies of this kind. For many groups of animals interspecific plots are carried out for small numbers of specimens of a wide range of species of different sizes. The resulting slope of the logarithmic plot is often about 0.8 both for respiratory surface area and for oxygen uptake. While such studies are of value in making general comparisons, it is important to appreciate that there may be wide divergences from this mean value. Although de Jager and Dekkers (1975) emphasised this point, nevertheless they obtained a value (0.811) for the mean regression line between gill surface and body weight by averaging slopes ranging from 0.639 to 0.961 for 11 different species and also averaged the correlation coefficient. A similar procedure has been carried out for data obtained by Gray (1954) using nine species of marine fish (Table 1). The values for total filament lenght when averaged in this way gave a value of 0.389, which is significantly different however from the value (0.429) obtained by plotting all the points separately and obtaining the regression line for all these points. Clearly this latter procedure is much more extensive than the usual interspecific plot but indicates how different results can be obtained using the two methods. A further way of illustrating the point is shown in Fig. 4, where species, each having quite different slopes, are plotted and then a single specimen from each species is used to plot the overall interspecific relationship which has a slope different from most of its component species.

Oxygen consumption of an animal is usually specified in relation to unit bidy weight. In many ways this is not ideal as it varies according to body size. Thus in most cases with increasing size the oxygen consumption per unit body weight falls. In fact the slope of the log/log regression line may range between 1.0 and 0.6, but is most usually about 0.8. The figure of 0.85 has been used as an average figure for many fish (Winberg, 1956) and seems to indicate a dependence halfway between the weight and surface area. It is perhaps important to recognise the differences which may occur between species in this respect as there are now a number of well authenticated cases where the exponent is about one (e.g. icefish, Holeton, 1976; Torpedo, Hughes, 1978c). The well-known relationship for mammals with a slope of 0.8 applies to interspecific plots in which small numbers of many different species are used.

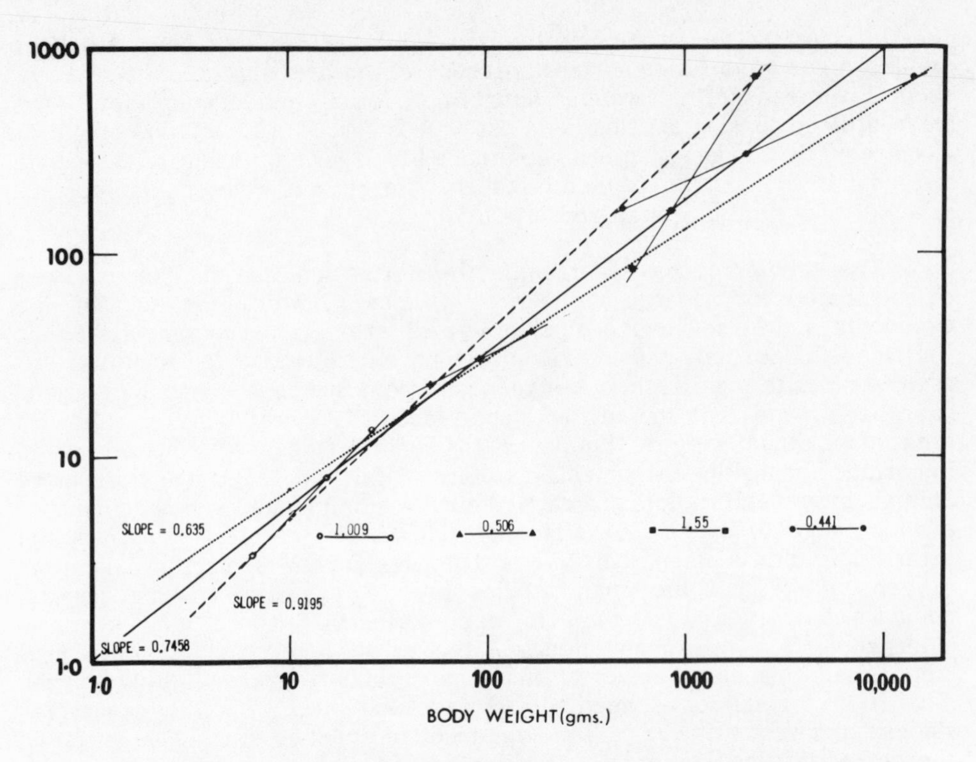

Fig. 4 Theoretical plots of oxygen consumption against body mass on logarithmic coordinates for four different species. The slopes of the lines for each species are 1.009, 0.506, 1.55 and 0.441. A combined overall plot using the middle size specimen from each species is plotted as a full line and has a slope of 0.746. Lines (dotted, dashed) using other specimens of these species have slopes of 0.6358 and 0.9195.

In relation to the morphometry of the respiratory surfaces similar differences in slope of the log plot have been obtained. Once more these vary from 0.5 to 1.0. These differences in slope are especially important if one wishes to compare the slope of the lines for respiratory surface and oxygen uptake. In our earliest studies of this kind it was thought that the two slopes were almost identical (Muir and Hughes, 1969). More recent work, although confirming this in a number of cases, has also shown (Hughes, 1972a, 1977) that for some fish the slopes diverge in such a way that the oxygen transferred per unit surface decreases with body size and consequently the fish has much greater scope for activity (Fry, 1957) as the gas exchange surfaces are not limiting and the oxygen uptake can be proportional to body weight. In other cases the two slopes converged and theoretically a body size is reached above which the gas exchange surface is no longer sufficient even to transfer the oxygen required for the resting metabolism. The dependence of metabolic scope on body size has been demonstrated experimentally at least for Salmonids (Brett, 1972; Brett & Glass, 1973).

There are also several examples among the air breathing fish where changes in the degree of dependence upon water and air for the supply of oxygen is correlated with the changes in slope and the relative slopes for the different respiratory surfaces.

SOME APPLICATIONS OF MORPHOMETRIC METHODS

The value of morphometric studies of respiratory organs in relation to environmental conditions is illustrated by the following example chosen to show the variety of fields in which this method can provide useful information:

1. Development

a) In some studies on the development of the gills and accessory respiratory organs of Indian air breathing fishes, it was shown that these different organs develop at different rates. Thus in *Channa punctata* the slope of the gill surface/B.wt regression line is 0.592, whereas it is 0.696 for the suprabranchial chamber (Hakim, Munshi & Hughes, 1978). Furthermore, during the development of some species (e.g. *Amphipnous*, Hughes, Singh, Thakur & Munshi, 1974) a significant change occured in slope of the regression line relating surface area to body mass.

These observations proved of great interest as they correlated very well with the relative dependence of these fish on water or air for their oxygen supply at different stages of the life cycle. For example, in *Anabas* and *Channa* the fish can survive if it is prevented from surfacing up to a certain size but above that size access to the air is essential.

b) Among water breathing fish, studies of the development of the gills in plaice and herring have shown a significant decrease in growth rate of the gill surface following metamorphosis (de Silva, 1974). At metamorphosis the cutaneous contribution to respiration is markedly reduced by the acquisition of scales, and perhaps the change in slope is related to a switch from skin to gill respiration.

c) It seems likely that different fish species will vary in their respiratory requirements during development, and evidence that the size of gas exchange surfaces may be linked to some limiting factor has been obtained from morphometric analysis of sections through the gill region of developmental stages of the Black Sea Bream (Iwai & Hughes, 1978). Measurements were made of the surfaces and volume of the developing gill using a rectilinear grid. The surface/volume ratio for the gill arch and filament regions were determined separately. For the arch region, this ratio (I_a/P_a) showed a gradual decline with increasing

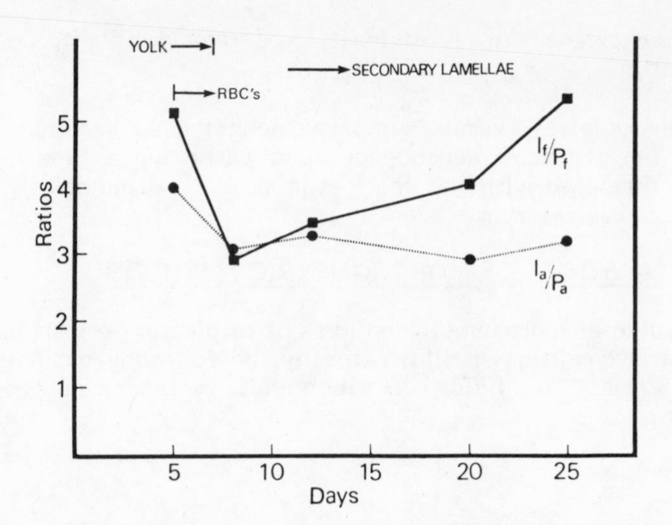

Fig. 5 Plots showing the change in the surface/volume ratio for the gill arch and gill filament regions of the black sea bream. The relative timing of several other events is also indicated (from Iwai & Hughes, 1977).

size, whereas for the filaments the S/V ratio (I_f/P_f) reached a minimum about the time when the secondary lamellae began to develop, after which the ratio gradually increased (fig.5). It was of special interest to find that the minimum S/V ratio of the gill filaments coincided with the sensitive period during which the fish larvae are liable to die. Previously most authors had considered a change in nutrition as the most likely factor involved, but this morphometric analysis focuses attention on the importance of maintaining a good supply of ambient oxygen during this critical stage of the fish's development.

2. Taxonomy

a) Some of the earliest electron microscopic studies of air-breathing fish were made of the climbing perch (Hughes & Munshi: 1968, 1973) and showed that the air breathing organs were not modified gill structures as had been thought in some earlier studies (Rauther, 1910; Munshi, 1968). Measurements of the thickness of the water/blood barrier of the gills gave high values (10 - 15 μm) which contrasted markedly with the very thin (1 μm) barrier between the air and blood in the labyrinthine organs. Later studies of *Anabas* gave values for the water/blood distances which were much less than the earlier measurements and this drew attention to the fact that the two studies were made using two quite distinct varieties of climbing perch, the so-called broad and narrow trunked forms (Dube & Munshi, 1974).

TABLE 3. Exponent (b) and correlation coefficients obtained from regression analysis of measurements of total filament length against body mass for different populations of the Dolphin fish (Coryphaena hippurus). Estimated values for a 1 kg. fish are given and also for a 313 gm fish being the same size as the specimen measured at Malta.

n		b	r	1000g	313g
12	Coryphaena hippurus (Hughes & Gray 1972)	0.431	0.819	36,892	22,362
9	Gulf of Mexico	0.515	0.920	29,526	16,234
4	Carolina	0.511	0.946	38,331	21,178
9	Beaufort, 1977	0.482	0.984	36,552	20,882
1	Measured (313g Lampuki at Malta)				21,263
22	Carolina, G of M, Beaufort	0.429	0.911	38,183	23,198

b) Investigations of the respiratory morphology and physiology of icefish are of general biological interest and it was early found that there were no significant differences between the general structure of the gill (Hughes, 1966b; Hughes & Shelton, 1962; Steen & Berg, 1966). The actual areas were not particularly small, although the water/blood barrier was thicker than in most fish gills. These studies were made with single specimens which had been preserved in formalin for some time and it was only at a later stage that a systematic study was made of the gills of specimens obtained in the Antarctic and of different body weights (Holeton, 1976). This study showed that the slope of the regression line relating gill area to body mass had a slope of about 1. It also showed that the earlier measurements were rather different from those for these well identified specimens of Chaenocephalus aceratus. Small numbers of other specimens of icefish made it seem likely that the earlier specimens were not, in fact, of the same species and most likely Pseudochaenichthys georgianus (Hughes, 1972b; Holeton, 1976). Here again, gill morphometry gave a valuable clue to the taxonomic differences.

c) The dolphin fish (Coryphaena hippurus) has a very wide distribution throughout the oceans and seas of the world. Details of its life, cycle and migrations are by no means clear. In an earlier analysis of measurements for this fish collected by Dr. I.E. Gray, I analysed the components of the surface area in relation to body mass. In the original data, distinction was made between specimens collected in the Gulf of Mexico and those collected off the Carolinas. Later it was noticed that perhaps these two populations might be considered separate especially as, for example, the regression lines relating total filament length to body mass have different "a" values. These observations became of interest when some preliminary studies were made of the gills of this species in the Mediterranean, where in the waters around Malta it forms a popular fish locally called "Lampuki". Unfortunately very few specimens were available in the particular area at the time the study was begun. However, the total filament lenght of the specimen measured (313 gms) fitted the same regression line as obtained for the Carolina fish (Table 3) although not significantly different from the Gulf of Mexico line as it lay within its wide 95% confidence limits at this size. It was therefore decided to try to obtain further specimens of Coryphaena from American waters, but unfortunately specimens from the Gulf have not become available. At Beaufort (North Carolina) fishing for small specimens was attempted but none below 2 kg in weight was obtained. Nevertheless, the gills of these specimens had a similar relationship to body size as the relatively small number of Carolina fish in the original data (Table 3).

It is impossible with the limited data available to conclude that the Lampuki is more related to the fish from the Carolina waters than from the populations in the Gulf of Mexico, but at least this method has pinpointed ways in which such questions might be approached in the future.

3. Effects of Environment

a) As the respiratory organs are the sites where oxygen enters the fish, it is evident that any environmental effects which interfere with the extent and thickness of these surfaces will have a significant effect on the activity and life of the animal. Such effects have been demonstrated for a long time in fish exposed to pollutants in natural waters (Herbert & Merkens, 1961; Brown, Mitrovic & Stark, 1968; Skidmore & Tovell, 1972; Hughes, 1976). Some of these so reduce the gas exchange surfaces that the growth rate of the fish is severely limited and death frequently occurs, especially when the oxygen levels in the water fall (Lloyd, 1961). The question as to whether such effects can be quantified especially at sub-lethal concentrations has recently been tested using morphometric methods (Hughes & Perry, 1976; Hughes, Perry & Brown, 1979). These studies have shown that the gill morphometry of fish exposed to heavy metals and controls may be compared using the concept of relative diffusing capacity of the experimental fish and those not exposed to the pollutant. In the case of nickel at concentrations of 2 & 3.2 mg/l, reductions in this index of the gas exchange capability of the gill were quite clear and the fish recovered over a period of 19 days in normal waters. Two components of the morphometric diffusing capacity (thickness and area) are plotted in Fig. 6 which shows significant changes for nickel but not for chromium.

Indications of a slight improvement in the gas exchange capability of the gill were found in Rainbow trout exposed to very low concentrations of cadmium (0.002 mg/l) and it is suspected that this may be due to the "disinfectant" effect of such a low concentration of this heavy metal. These methods were carried out using light microscopy and

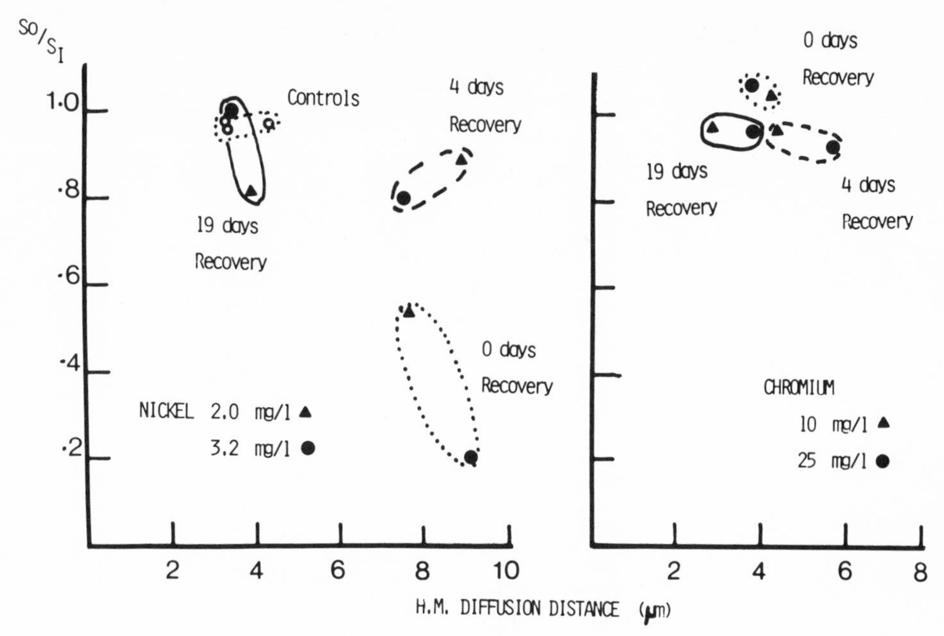

Fig. 6 Plots of harmonic mean diffusion distances against an index of surface area (So/Si, ratio of the outer surface of the gill secondary lamella and the internal surface as defined by the basement membrane).

could be applied to many other situations where it is desired to test environmental effects on the gas exchange apparatus.

b) The chance observation of some Rainbow trout with damaged gill filaments was also investigated morphometrically and comparisons made with fish in which the filaments were less damaged (Hughes & Nyholm, 1979). In this way the extent of the damage was quantified and correlated with changes in respiratory behaviour of the fish as those specimens with most damage (46%) exhibited a greater tendency for ram jet ventilation than a more normal specimen (11%).

c) Fish from a completely different environment also proved suitable material for morphometric studies, as for example fish from deep waters in the ocean where the general observation of a poorly developed gill system has been known for many years (Marshall, 1960). Quantification of this feature has been carried out, initially on some specimens collected during an expedition to the Comores when two specimens of Latimeria were also obtained (Hughes, 1972c). A simple way of quantifying gill size which was scarcely been used, is to weigh the whole gill apparatus including the skeletal arches. Such data is plotted out to Fig. 7 and compared with corresponding measurements

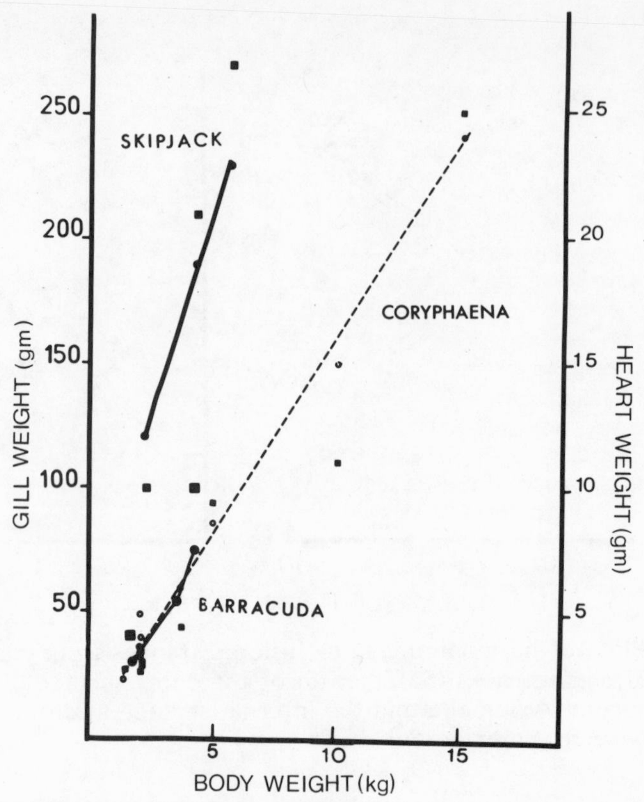

FIG. 7. Plot of the weight of gills (squares) and heart (circles) for specimens of Skipjack and Barracuda collected at the Comores (Hughes, 1980), and for Coryphaena collected at Beaufort. Data is plotted against body weight and indicates some marked differences in gill weight which is related to their surface areas. Lines are drawn for heart weights.

of heart weight. It is clear that a very active fish such as skipjack tuna has a much greater development of both these organs when compared with fish such as Barracuda. Measurements of rare specimens such as Latimeria are sometimes restricted to studies of a single gill arch and in comparing measurements obtained for Latimeria and other fish, it is apparent that a study of the filament lenght of the second gill arch gives a good indication of the total filament length and enables comparisons to be made with comparable data from other fish (Hughes, 1980). In this case it is shown that the poor development of the gills of Latimeria is related to its deep water habitat, and in those specimens where complete measurements were possible, the gill area was found to be extremely small (19 mm^2/gm for a 10 g fish).

This very low surface area indicates that the oxygen consumption of this fish will also be very small (Hughes, 1976) and correlates with the view that this interesting species is extremely inactive during most of its life.

d) Estimation of the oxygen uptake of Latimeria involved consideration of the relationship between gill area and oxygen consumption and it seems reasonable to suggest that the total respiratory area is indicative of maximum rates at which oxygen may be transferred across these surfaces during active swimming. Normally the relationship between oxygen consumption and body mass is plotted with reference to the resting or standard metabolism of the fish and consequently a comparison of the slopes of these two lines can give suggestive information regarding the change in scope for activity with increasing body mass (Hughes, 1977). In some species the slopes of the two lines diverge and consequently suggest an increase in scope for larger fish and this confirms the measurements for Salmonids by Brett (1972) and Brett & Glass (1973).

4. Dimensional Analysis

a) Using a standard model of gas exchange across the gill (Fig. 1) it is possible to test this model by means of the exponent values for different gill dimensions obtained from morphometric analysis of fish of different body mass. Such an analysis was carried out by Hills & Hughes (1970) using the classical data for Micropterus (Price, 1931). A similar analysis for the Rainbow trout has also led to similar conclusions (Hughes, 1977), although the exponent values were quite different. Of particular interest were the conclusions that water flow is probably laminar and counter current to that in the blood. The resistance to overall oxygen transfer due to the water film is 5-10 times that due to the tissue barrier. A conclusion borne out by subsequent analyses using more classical methods (Scheid & Piiper, 1976).

b) Other analyses using such methods were carried out on data obtained for the ventilation of Rainbow trout gills and suggested that recruitment of water channels is an important factor during increased gill ventilation (Hughes, 1972a).

5. Nature of Blood Pathways in Secondary Lamellae

a) Morphometric studies of material fixed during normoxia and hypoxia and during exposure to pollutants has given interesting information regarding the proportion of the total secondary lamellar volume occupied by different structures, e.g. pillar cells (20%), blood (30%) and epithelium (50%). It also provides data for the relative volumes of the blood (Vb) and red blood cells (Ve) and consequently the secondary lamellar haematocrit. Comparison of these values with those obtained for normal blood samples gave values that are about 15% greater than those normally obtained. This observation suggests that during the flow of blood through the secondary lamellae there are alternative pathways along which plasma alone or containing very few red blood cells may be circulated (Hughes, 1979b). Variations during hypoxia and anaesthesia also indicate that some regulation of the relative volumes occupied by these different pathways can be achieved. Morphometry also showed, for example, that in the tench red blood cells, the nucleus occupies about 17% of the total red cell volume (Hughes, 1979b).

6. Other parts of fish respiratory systems can also be treated by morphometric methods in order to establish the proportion of the surface or volume of an organ that they occupy. One example concerns the distribution of mucous cells on the gill filaments. Counts were made of mucous cells stained in strips of epithelium removed from the inner and outer surfaces of the filaments (Hughes & Mittal, 1980). Estimates of the numbers of mucous cells/unit area were obtained, together with the proportion of the total surface occupied by these cells. Measurements confirmed the view that the number of cells on the inner or trailing surface was much greater than on the outer. However, the differences in area were not so great because the area of individual mucous cells was greater on the outer or leading surface. These methods can therefore give quantitative information about aspects of the morphology which may be of functional importance and draw attention to the more detailed nature of differences which may not be the same as appears at first sight.

CONCLUSIONS

The morphometric study of respiratory structures can provide a valuable method for investigations into relationships between environment and the structure and function of fishes. Because of their importance in many basic physiological functions, gill morphometry can provide a valuable guide to the demand placed upon these organs during adaptation to adverse conditions. Consequently, measurements which indicate any impairment with their normal function can give clues as to the degree of environmental stress to which the fish is subjected.

REFERENCES

Ballintijn, C.M. (1969) Muscle co-ordination of the respiratory pump of the carp (Cyprinus carpio L.). J. exp. Biol. 50: 569-91.

Ballintijn, C.M. & Hughes, G.M. (1965) The muscular basis of the respiratory pumps in the trout. J. exp. Biol. 43: 349-62.

Brett, J.R. (1972) The metabolic demand for oxygen in fish, particularly salmonids, and a comparison with other vertebrates. Respir. Physiol. 14: 151-170.

Brett, J.R. & Glass, N.R. (1973) Metabolic rates and critical swimming speeds of sockeye salmon (Oncorhynchus nerka) in relation to size and temperature. J. Fish. Res. Bd. Can. 30: 379-387.

Brown, V.M., Mitrovic, V.V. & Stark, G.T.C. (1968) Effect of chronic exposure to zinc on toxicity of a mixture of detergent and zinc. Water Res. 2: 255-63.

Dube,S.C. & Munshi, J.S.D. (1974) Studies on the blood-water diffusion barrier of secondary gill lamellae of an air-breathing fish, Anabas testudineus (Bloch.). Zool. Anz., Jena. 193: 35-41.

Fry, F.E.J. (1957) The aquatic respiration of fish. Physiology of Fishes, Vol. 1, Ed. M.E. Brown, London & New York, Academic Press, p. 1-63.

Gray, I.E. (1954) Comparative studies of the gill area of marine fishes. Biol. Bull. Mar. Biol. Lab. Woods Hole 107: 219-225.

Hakim, A., Munshi, J.S.D. & Hughes, G.M. (1978) Morphometrics of the respiratory organs of the Indian green snake-headed fish, Channa punctata J. Zool. (Lond.) 184: 519-543.

Herbert, D.W.M. & Merkens, J.C. (1961) The effect of suspended mineral solids on the survival of trout. Int. J. Air. Wat. Poll. 5: 46-55.

Hills, B.A. (1974) Gas transfer in lungs and gills. Cambridge. Cambridge Univ. Press, 169p.

Hills, B.A. & Hughes, G.M. (1970) A dimensional analysis of oxygen transfer in the fish gill. Respir. Physiol. 9: 126-40.

Holeton, G.F. (1976) Respiratory morphometrics of white and red blooded antarctic fish. Comp. Biochem. Physiol. 54A: 215-220.

Hughes, G.M. (1961) How a fish extracts oxygen from water. New Sci. 11: 346-48.

Hughes, G.M. (1966a) Evolution between air and water. In: Ciba Foundation Symposium on Development of the Lung. Churchill, Ed. A.V.S. de Reuck and R. Porter p. 64-80.

Hughes, G.M. (1966b) The dimensions of fish gills in relation to their function. J. exp. Biol. 45: 177-95.

Hughes, G.M. (1972a) Morphometrics of fish gills. Respir. Physiol. 14: 1-25.

Hughes, G.M. (1972b) Distribution of oxygen tension in the blood and water along the secondary lamella of the icefish gill. J. exp. Biol. 56: 481-92.

Hughes, G.M. (1972c) Gills of a living coelacanth. Latimeria chalumnae. Experientia 28: 1301-1302.

Hughes, G.M. (1975) Coughing in the rainbow trout (Salmo gairdneri) and the influence of pollutants. Rev. Suisse Zool. 82: 47-64.

Hughes, G.M. (1976) On the respiration of Latimeria chalumnae. J. Linn. Soc. (Zool.) 59: 195-208.

Hughes, G.M. (1977) Dimensions and the respiration of lower vertebrates. In: Scale effects in animal locomotion. Ed. T.J. Pedley, London Academic Press, p. 57-81.

Hughes, G.M. (1978a) Some features of gas transfer in fish. Bull. Inst. Maths. and its applications 14: 39-43.

Hughes, G.M. (1978b) A morphological and ultrastructural comparison of some vertebrate lungs. In: XIX Congressus Morphologicus Symposia, Ed. E. Klika, Prague, Charles Univ. Press, p. 393-405.

Hughes, G.M. (1978c) On the respiration of Torpedo marmorata. J. exp. Biol. 73: 85-105.

Hughes, G.M. (1979a) Scanning electron microscopy of the respiratory surface of trout gills. J. Zool. (Lond.) 188: 443-453.

Hughes, G.M. (1979b) The paths of blood flow through the gills of fishes - some morphometric observations. Proc. Bulg. Acad. Sci. (in Press).

Hughes, G.M. (1980) Ultrastructure and morphometry of the gills of Latimeria chalumnae - and a comparison with the gills of associated fishes. Proc. Roy. Soc. B. (in Press).

Hughes, G.M. & Adeney, R.J. (1977a) The effects of zinc on the cardiac and ventilatory rhythms of rainbow trout (Salmo gairdneri, Richardson) and their responses to environmental hypoxia. Water Res. 11: 1069-1077.

Hughes, G.M. & Adeney, R.J. (1977b) Variations in the pattern of coughing in rainbow trout. J. exp. Biol. 68: 109-122.

Hughes, G.M., Dube, S.C. & Munshi, J.S.D. (1973) Surface area of the respiratory organs of the climbing perch, Anabas testudineus. (Pisces: Anabantidae). J. Zool. (Lond.) 170: 227-243.

Hughes, G.M. & Grimstone, A.V. (1965) The fine structure of the secondary lamellae of the gills of Gadus pollachius. Quart. J. Micr. Sci. 106: 343-353.

Hughes, G.M. & Hills, B.A. (1971) Oxygen tension distribution in water and blood at the secondary lamella of the dogfish gill. J. exp. Biol. 55: 399-408.

Hughes, G.M. & Mittal, A.K. (1980) Comparison of mucous glands from the gills and opercular epithelium of rainbow trout. (in preparation).

Hughes, G.M. & Munshi, J.S.D. (1968) Fine structure of the respiratory surfaces of an air-breathing fish, the climbing perch Anabas testudineus (Bloch.). Nature (Lond.) 219: 1382-1384.

Hughes, G.M. & Munshi, J.S.D. (1978) Scanning electron microcospy of the respiratory surfaces of Saccobranchus (= Heteropneustes) fossilis (Bloch.). Cell.Tiss. Resp. 195: 99-109.

Hughes, G.M. & Nyholm, K. (1979) Ventilation in rainbow trout (Salmo gairdneri, Richardson) with damaged gills. J. Fish Biol. 14: 285-288.

Hughes, G.M. & Perry, S.F. (1976) Morphometric study of trout gills: a light microcospic method suitable for the evaluation of pollutant action. J. exp. Biol. 64: 447-460.

Hughes, G.M., Perry, S.F. & Brown, V.M. (1979) A morphometric study of effects on nickel, chromium and cadmium on the secondary lamellae of rainbow trout gills. Water Res. 13: 665-679.

Hughes, G.M. & Peters, H.M. (1980) Pressure changes in the respiratory cavities of Osphronemus goramy during air and water ventilation (in preparation).

Hughes, G.M. & Shelton, G. (1958) The mechanism of gill ventilation in three fresh water teleosts. J. exp. Biol. 35: 807-823.

Hughes, G.M. & Shelton, G. (1962) Respiratory mechanisms and their nervous control of fish. In: Advances in Comparative Physiology and Biochemistry, Vol. 1. Ed. O. Lowenstein, London & New York, Academic Press, p. 275-364.

Hughes, G.M., Singh, B.R., Thakur, R.N. & Munshi, J.S.D. (1974) Areas of the air-breathing surfaces of Amphipnous cuchia (Ham.). Proc. Indian Nat. Sci. Acad. 40: 379-392.

Hughes, G.M. & Weibel, E.R. (1976) Morphometry of fish lungs. In: Respiration of Amphibious Vertebrates. Ed. G.M. Hughes, London & New York, Academic Press, p. 213-232.

Iwai, T. & Hughes, G.M. (1977) Preliminary morphometric study on gill development in Black Sea bream (Acanthopagrus schlegeli). Bull. Jap. Soc. Sci. Fish. 43: 929-934.

Jager, S. de & Dekkers, W.J. (1975) Relations between gill structure and activity in fish. Neth J. Zool. 25: 276-308.

Kirsch, R. & Nonnotte, G. (1977) Cutaneous respiration in three freshwater teleosts. Respir. Physiol. 29: 339-354.

Kuiper, T. (1907) Untersuchungen uber die Atmung der Teleostei.. Pflug. Arch. ges. Physiol. 177: 1-107.

Lloyd, R. (1961) Effects of dissolved oxygen concentrations on the toxicity of several poisons to rainbow trout (Salmo gairdneri, Richardson). J. exp. Biol. 38: 447-455.

Marshall, N.B. (1960) Swimbladder structure of deepsea fishes in relation to their systematics and biology. Discovery Rep. No. 31, p. 1-122.

Muir, B.S. & Hughes, G.M. (1969) Gill dimensions for three species of tunny. J. exp. Biol. 51: 271-85.

Munshi, J.S.D. (1968) The accessory respiratory organs of Anabas testudineus (Bloch.), (Anabantidae, Pisces). Proc. Linn. Soc. Lond. 170: 107-126.

Munshi, J.S.D. (1976) Gross and fine structure of the respiratory organs of air-breathing fishes. In: Respiration of Amphibious Vertebrates. Ed. G.M. Hughes, London & New York, Academic Press, p. 73-104.

Newstead, J.D. (1967) Fine structure of respiratory lamellae of teleostan gills. Z. Zellforsch-mikrosk. Anat. 79: 396-428.

Osse, J.W.M. (1969) Functional morphology of the head of the perch (Perca fluviatilis) - an electromyographic study. Neth. J. Zool. 19: 290-392.

Pauly, D. (1979) Gill size and temperature as governing factors in fish growth: a generalization of von Bertalanffy's growth formula. Beritch. Inst. Meereskunde Kiel, No. 63, 156p.

Peters, H.M. (1978) On the mechanism of air ventilation in anabantoids (Pisces: Teleostei). Zoomorphology. 89: 93-124.

Piiper, J. & Baumgarten-Schumann, D. (1968) Effectiveness of O_2 and CO_2 exchange in the gills of the dogfish (Scyliorhinus stellaris). Resp. Physiol. 5: 338-349.

Price, J.W. (1931) Growth and gill development in the small-mouthed black bass, Micropterus dolomieu, Lacépède, Ohio State Univ. Stud., 4: 1-46.

Randall, D.J. (1971) Respiration. In: Biology of Lampreys, Vol. 2, Ed. M.W. Hardisty & I.C. Potter, London Academic Press, p. 287-306.

Rauther, M. (1910) Die akzessorischen Atmungsorgane der Knochenfische. Ergb. Zool. 2: 517-75.

Scheid, P. & Piiper, J. (1976) Quantitative functional analysis of branchial gas transfer: theory and application to Scyliorhinus stellaris (Elasmobranchii). In: Respiration of Amphibious Vertebrates, Ed. G.M.

Hughes, London & New York, Academic Press, p. 17-38.

Silva, C. de (1974) Development of the respiratory system in herring and plaice larvae. In: The Early Life History of Fish, Ed. J.H.S. Blaxter, Springer-Verlag, New York, p. 465-485.

Skidmore, J.F. & Tovell, P.W.A. (1972) Toxic effects of zinc sulphate on the gills of rainbow trout. Water Res. 6: 217-230.

Steen, J.B. & Berg, T. (1966) The gills of two species of haemoglobin-free fishes compared to those of other teleosts, with a note on severe anemia in the eel. Comp. Biochem. Physiol. 18: 517-26.

Weibel, E.R. (1970) Morphometric estimation of pulmonary diffusion capacity. I. Model and Method. Respir. Physiol. 11: 54-75

Winberg, G.G. (1950) (1) Rate of metabolism and food requirements of fishes. (2) New information on metabolic rate in fishes. Fish Res. Board Can. (translation series 194 & 362).

Young, S. (1970) EMG activity in tench (Tinca tinca L.) gill lamellae and its association with coughing. J. Physiol. (Lond.) 215: 37-38.

AIR VENTILATION IN ADVANCED TELEOSTS:
BIOMECHANICAL AND EVOLUTIONARY ASPECTS.

Karel F. Liem

Museum of Comparative Zoology, Harvard University

Cambridge, Massachusetts 02138, USA

INTRODUCTION

Physiologists, ecologists, ethologists and evolutionary biologists have long been interested in the phenomenon of air breathing in fishes. Aside from the intrinsic significance of understanding the numerous and intriguing multiple adaptations that have evolved in air breathing fishes, biologists hope to find the biological solutions to the environmental problems associated with the transition of aquatic to terrestrial life in vertebrates. Since some recent air breathing fishes possess fairly primitive gas exchange systems, and exhibit distinct behavioral and ecological responses to the fluctuating environmental factors characterizing their habitats, they offer extraordinary opportunities to gain understanding of the early evolution of air breathing in the vertebrates.

In the past decade we are witnessing an ever increasing research effort in the biology of air breathing fishes. The extraordinary vitality of this kind of research is reflected in several major reviews and symposia: Johansen (1970) has dealt with the nature of the structural and physiological adaptation for air breathing; Gans (1970) has focussed on the strategy and sequence in the evolution of external gas exchangers in poikilothermic vertebrates; Hughes (1976) has edited a major monograph on the respiration of amphibious vertebrates, in which morphological, physiological, and ecological aspects have been reviewed exhaustively; Randall and Hochachka (1978) organized an in-depth and eloquent, special volume of the Canadian Journal of Zoology on air breathing vertebrates of the Amazon; finally, Riggs (1979) edited a special issue of the journal Comparative Biochemistry and Physiology on the findings of the Alpha Helix expedition to the Amazon for the study of fish bloods and hemoglobins.

The common thread in this wealth of anatomical, physiological and biochemical papers on air breathing fishes is a search for how evolution has optimized respiration, which represents a clearly defined fitness component of the organism. Most investigators have adopted the extreme adaptationist program in which air breathing is assumed to have been molded by natural selection as an optimal solution to a problem posed by a preexisting hostile environment. To illustrate this approach it is best to start with the equation for the rate of diffusion $R = D.A\Delta p/d$. D is diffusion constant, A the respiratory surface, p the difference in partial pressures of the medium and the cell, and d the distance along which diffusion occurs. Morphometrics has been used to show, how A is maximized in aquatic as well as air breathers, and it has been demonstrated by elegant ultrastructural studies that the air-blood pathway (d) is reduced. Numerous physiological and biochemical features (Randall and Hochachka, 1978; Riggs, 1979) of air breathing fishes are thought to increase Δp under various physical and chemical conditions. Thus, the notion that the rate of diffusion (R) is optimized is supported by an impressive body of data. Adaptations in the respiratory apparatus of air breathing fishes are in general agreement with the theory of natural selection as a process of differential survival in hostile environments and therefore of increasing frequency of the more fit.

In sharp contrast to the impressive progress made in ultrastructure, physiology and biochemistry of the respiratory apparatus of air breathing fishes, knowledge on the mechanism of air ventilation has been virtually stagnant. The neuromuscular and biomechanical mechanisms of intake and expulsion of air into respiratory swim bladders and air chambers have not been investigated experimentally. Recently, Peters (1978) and Kramer (1978) have applied cineradiography to the analysis of the mechanics of air ventilation of the air chambers and respiratory gas bladder in some teleosts with remarkable results. It is generally accepted (Gans, 1970; Peters, 1978; Kramer, 1978) that the air pump is a modification of the gill perfusion pump, in which the buccal cavity plays a key functional role. Because the parameters of the buccal apparatus are strongly affected by the powerful selective pressures of the feeding system, the mechanism of ventilation in air breathing fishes may represent a functional and selective compromise rather than optimization toward the demands of most effective ventilation.

In this paper, I will not furnish a historical general review of the mechanism of air ventilation. Instead, I will limit myself to the presentation of new experimental data and to the discussion of the evolution of the ventilative apparatus.

APPROACHES

Anatomical: Because all previous studies on the mechanisms of air ventilation have been based almost entirely on anatomy, it is necessary to present a structural description of the major features underlying the function of the air pump. The necessity to deal with in-depth morphology is amplified by the fact that many pervasive inaccuracies exist in the literature, hampering further progress in our attempts to explain the evolutionary patterns of air breathing teleosts. Once a more precise body of anatomical data has been produced, truly meaningful statements can be made on homology, which in turn will enable us to evaluate the nature and origin of evolutionary novelties that have emerged in association with air breathing habits.

Experimental: The pitfalls of extrapolating from structures to function have become evident from recent contributions in functional morphology, in which modern analytical tools have been used. In this study air ventilation has been analyzed by means of high speed light movies (up to 200 frames $sec.^{-1}$), high speed cineradiography (up to 150 frames $sec.^{-1}$) and electromyography. As far as possible, the data of the three sources have been integrated to obtain accurate biomechanical and electromyographic profiles of the act of air ventilation. It is hoped that by the use of these tools, the speculative element of the analysis has been minimized.

Phylogenetic and Evolutionary: Many of the structural elements involved in air ventilation also play an important role in the feeding apparatus. Thus, the selective forces acting on the feeding apparatus may exert constraining influences on the optimization of air breathing or vice versa. The basic approach I have followed here is that the study of constraints-on-optimization is as informative as the study of optimization itself. An attempt will be made to trace the physiological origin of the mechanisms underlying air ventilation in air breathing fishes of different phylogenetic lineages.

MECHANISM OF VENTILATION IN SOME ANABANTOIDEI

The Anabantoidei are well known as specialized air breathers, utilizing a highly modified first epibranchial (the "labyrinth organ") and a suprabranchial chamber as the air breathing organs (Bader, 1937; Liem, 1963; Munshi, 1965, 1968; Hughes and Munshi, 1973; and Peters, 1978). Their respiratory physiology has been studied by Hughes and Singh (1970). The mode of air ventilation in several anabantoids has been analyzed by both light movies and cineradiography in an eloquent study by Peters (1978). However, no electromyographic analysis has been attempted to-date. Thus, most statements on the mechanism of air ventilation (e.g., Munshi, 1965, 1968; Liem, 1967) remain speculative. The elegant hypotheses formulated by Peters (1978) on the basis of kinematic profiles must be tested by electromyography. Here, I present an electromyographic and cineradiographic analysis of Helostoma temmincki, Anabas testudineus, Trichogaster trichopterus, and Osphronemus goramy.

The "Constrictor Suprabranchialis" Muscle and the Suprabranchial Chamber: The salient branchial muscles of Anabas testudineus are described briefly as a basis for the subsequent biomechanical analysis. The levators externi and interni muscles are only slightly modified (Fig. 1), if compared with those of more generalized percoids (e.g., Pristolepis, Liem, 1974). Topographically, the origins of the levatores from the neurocranium have been shifted posteriorly to the pterotic region (Fig. 1). However, the insertions of all four external and two internal levatores conform (Fig. 2) to the pattern in generalized percoids. Thus, each of the four levatores externi inserts on the dorsolateral aspect of the epibranchial of the corresponding arch. Misra and Munshi (1958) and Munshi (1968) have stated that: "The dorsal part of the second branchial muscle does not become differentiated as its levator, but probably becomes subordinated to the function of raising and lowering the posterior part of the respiratory membranes. Thus it serves as a constrictor suprabranchialis". Referring to Trichogaster fasciatus, Munshi (1965) states: "The second levator muscle completely encircles the posterior sacular part of the suprabranchial chamber and becomes part of its wall. This muscle plays the role of constrictor suprabranchialis". In all anabantoids that I have studied, the second levator externus is present as a parallel-fibered straplike muscle with a tendinous insertion on the second epibranchial (Figs. 1, 2: le). It is not modified into a specialized "constrictor suprabranchialis" as described by Munshi.

Morphologically, the only specialization exhibited by the second levator externus is the posterior shift in its origin, a feature characterizing all other levatores externi. In all other aspects, the second levator externus of the anabantoids resembles that of other generalized perciforms. Because the designation constrictor suprabranchialis implies a drastic anatomical change, which, in fact has not occured, it is not warranted and should be considered a misnomer.

The suprabranchial chamber of Osphronemus has been described exhaustively and accurately by Peters (1978). In all major features, Peters' description also applies to other anabantoids, although relatively minor differences occur between anabantoid taxa.

The suprabranchial chamber extends from the orbit to the shoulder girdle, except in Helostoma, in which it reaches posteriorly as far as the eighth vertebra (Figs. 3, 4). The floor of the chamber is also the roof of the caudal part of the buccal cavity. Posteriorly the floor is attached along the medial border of the second gill arch. As emphasized by Peters (1978) the chamber has three openings. The

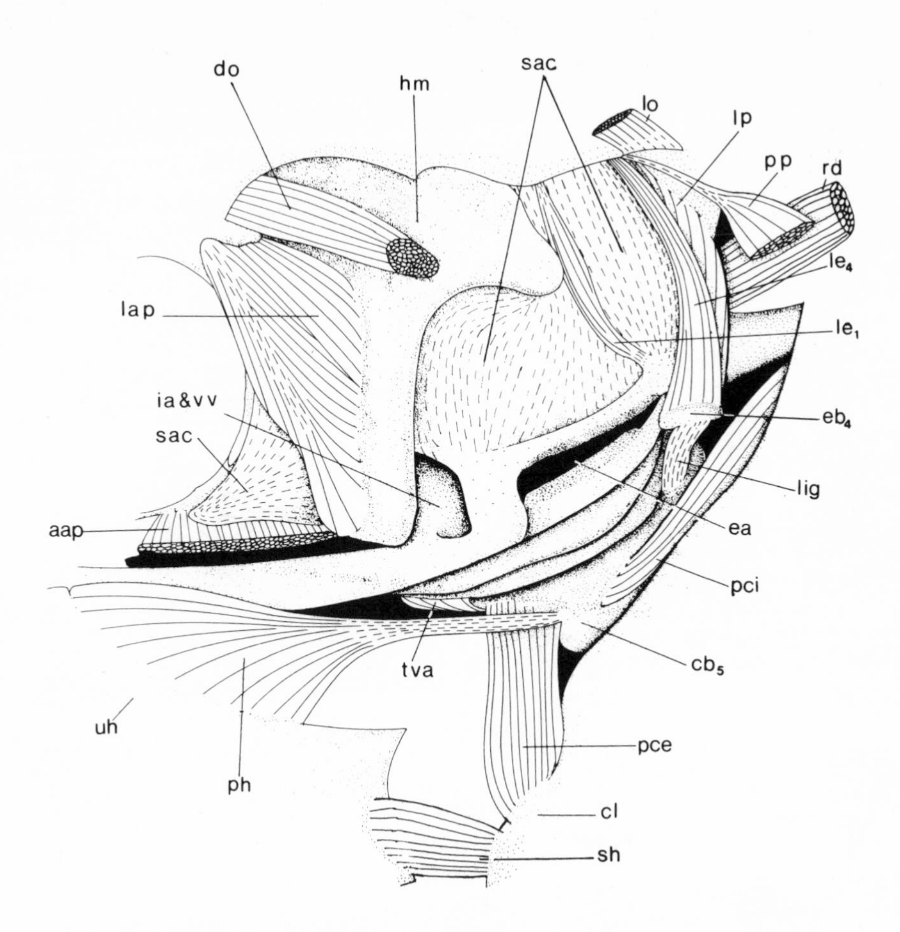

Fig. 1. Lateral view of the branchial apparatus and muscles and the suprabranchial chamber (sac) of Anabas testudineus. Suspensory apparatus (except hyomandibulae, hm), operculum, gills and gill rakers and mucous membrane have been removed. Abbreviations: aap, adductor arcus palatini; cb_5, fifth ceratobranchial (lower pharyngeal jaw); cl, cleithrum; do, dilatator operculi; ea, branchial opening; eb_4, fourth epibranchial; hm, hyomandibulae; ia, pharyngeal opening; lap, levator arcus palatini; le_1, le_4, first and fourth levator externus; lig, ligament; lo, levator operculi; lp, levator posterior; pce, pharyngocleithralis externus; pci, pharyngocleithralis internus; ph, pharyngohyoideus; pp, protractor pectoralis; rd, retractor dorsalis; sac, suprabranchial air chamber; sh, sternohyoideus; tva, transversus ventralis anterior; uh, urohyal; vv, valve and shutter.

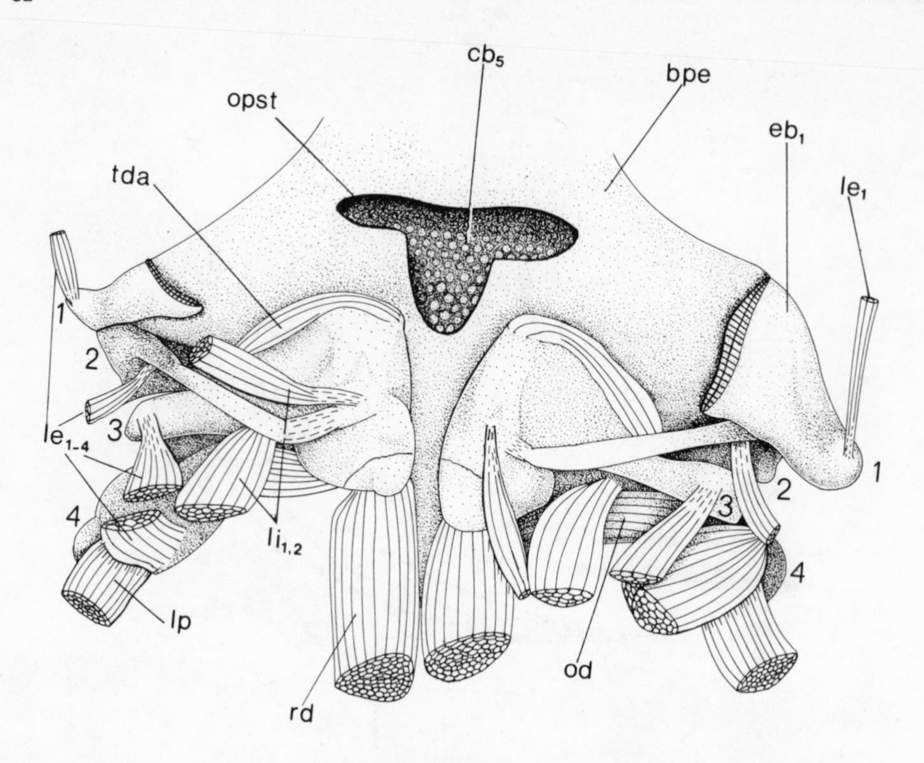

Fig. 2. Dorsal view of dissected and isolated branchial apparatus, muscles and intact floor of the suprabranchial air chamber. As we look down on this preparation, a T-shaped opening (opst) in the mucous membrane outlines the pattern of the parasphenoid teeth. The teeth of the lower pharyngeal jaw (cb_5) are visible through this opening. The highly enlarged and folded first epibranchial (eb_1) ("labyrinth organ") has been removed.
Abbreviations: bpe, buccopharyngeal epithelium; cb_5, fifth ceratobranchial (lower pharyngeal jaw); eb_1, first epibranchial; le_{1-4}, levator externus 1-4; $li_{1,2}$, levator internus 1-2; lp, levator posterior; od, obliquus dorsalis; opst, opening outlining the patch of the parasphenoid teeth; rd, retractor dorsalis; tda, transversus dorsalis anterior.

pharyngeal opening (inhalent aperture of Munshi, 1968) is a semicircular recess in the floor (Fig. 3: po) just anterior to the ceratobranchial-epibranchial joint of the first arch. It can be closed by a valve or shutter (Figs. 1, 8: vv), which is derived from modified gill rakers. The term pharyngeal opening, (Peters, 1978) is preferred over the older term inhalent aperture (Munshi, 1968), because, as we will see later, air can and does travel in two directions through this opening.

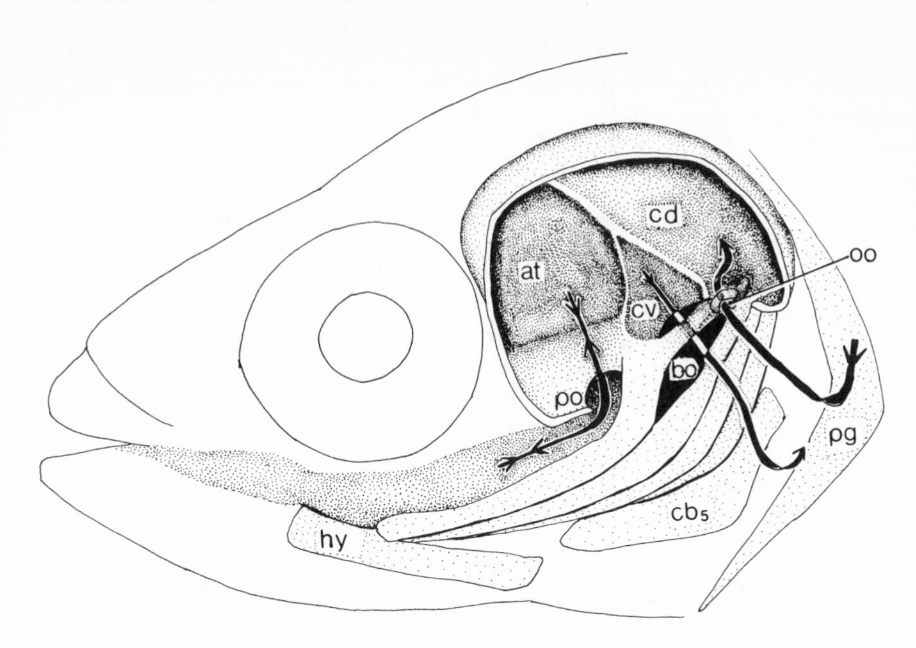

Fig. 3. Lateral view of the suprabranchial chamber of an anabantoid fish opened by incision through the wall. Gills and "labyrinth organ" have been removed. Arrows indicate pathways for air and water. Structure over oo is a thickening of the operculum, shown here separately as a C-shaped sausagelike bulge. This bulge can be pressed tautly against the muscular process of the first epibranchial on which it lies, closing the opercular opening (oo).
Abbreviations: at, atrium; bo, branchial opening; cb_5, fifth ceratobranchial; cd, caudodorsal compartment; cv, caudoventral compartment; hy, hyoid; oo, opercular opening; pg, pectoral girdle; po, pharyngeal opening in floor of the suprabranchial cavity.

The second opening, the branchial opening, is located between the first and the second arches (Fig. 3: bo). It represents a somewhat specialized gill slit, and has been named exhalent aperture by Munshi (1968). In Anabas, the large opening is unguarded (Fig. 1: ea), but in Helostoma (Fig. 8: ea, vv) the branchial opening is reduced and guarded by a distinct valve or shutter. The third opening, the opercular opening (Peters, 1978), is situated between the first gill arch and the operculum (Fig. 3: 00). It is essentially a modification of the slit between the inner surface of the operculum and the first gill arch. The inner surface of the operculum is differentiated into a prominent bulge, which can be pressed tightly on the muscular process of the first epibranchial (Fig. 3:

00), forming an effective seal. As noted by Peters (1978), there is a free pathway between the opercular cavity and suprabranchial chamber when the bulge of the operculum is separated from the muscular process of the first epibranchial by abduction of the operculum. The pathway is closed by adduction of the operculum.

The suprabranchial chamber can be subdivided into 3 interconnected compartments (Fig. 3), each associated with a particular opening. The three subdivisions have been recognized, defined and named by Peters (1978) as the atrium, ventrocaudal and dorsocaudal compartments. The atrium is the anterior chamber communicating with the buccal cavity through the pharyngeal opening. In the ventral region, the atrium passes into the ventrocaudal compartment (Fig. 3), which is connected with the opercular cavity by the branchial opening. Finally the caudodorsal compartment extends to the posterior wall of the chamber and communicates with the opercular cavity by means of the opercular opening.

Biomechanical Patterns of Air Ventilation: Peters (1978) has discovered two modes of air ventilation, which he termed "monophasic" and "diphasic". All anabantoids can ventilate both "diphasically" and "monophasically", except for adult Anabas, which uses the "monophasic" mode exclusively.

My cineradiographic observations confirm those of Peters'. Most anabantoids rely predominantly on the "diphasic" pattern, although all seem to breathe "monophasically" on occasion. My analysis is based on Helostoma, which has a dual strategy of air ventilation (Peters, 1978).

"Monophasic" (sensu Peters, 1978) pattern in Helostoma. One of the fundamental characteristics of this pattern is that the suprabranchial air sac (Fig. 4: sac) remains completely filled and does not change in either volume or configuration (Fig. 4) throughout the ventilative cycle. As the fish rises and reaches the surface of the water, it opens its mouth and widens the sidewalls of the buccal cavity. By this action air fills the buccopharyngeal cavity (Fig. 4, photos 2, 3). The closed mouth sinks below the water surface and compression of the buccopharyngeal cavity begins (Fig. 4, photos 4-7). As a result "new" air is forced into the suprabranchial chamber, while old air is forced out of the chamber into the opercular cavity (Fig. 4, photo 4) and thence out into the water (see arrows in Fig. 4, photos 5-7). Compression proceeds, until all air in the buccal cavity has been forced into the suprabranchial chamber (Fig. 4, photos 8, 9). The most important features of this "monophasic" pattern are: (1) inhalation preceeds exhalation; (2) the suprabranchial chamber retains both its volume and form throughout the ventilating cycle; (3) air invariably is expelled from behind the operculum (opercular expulsion).

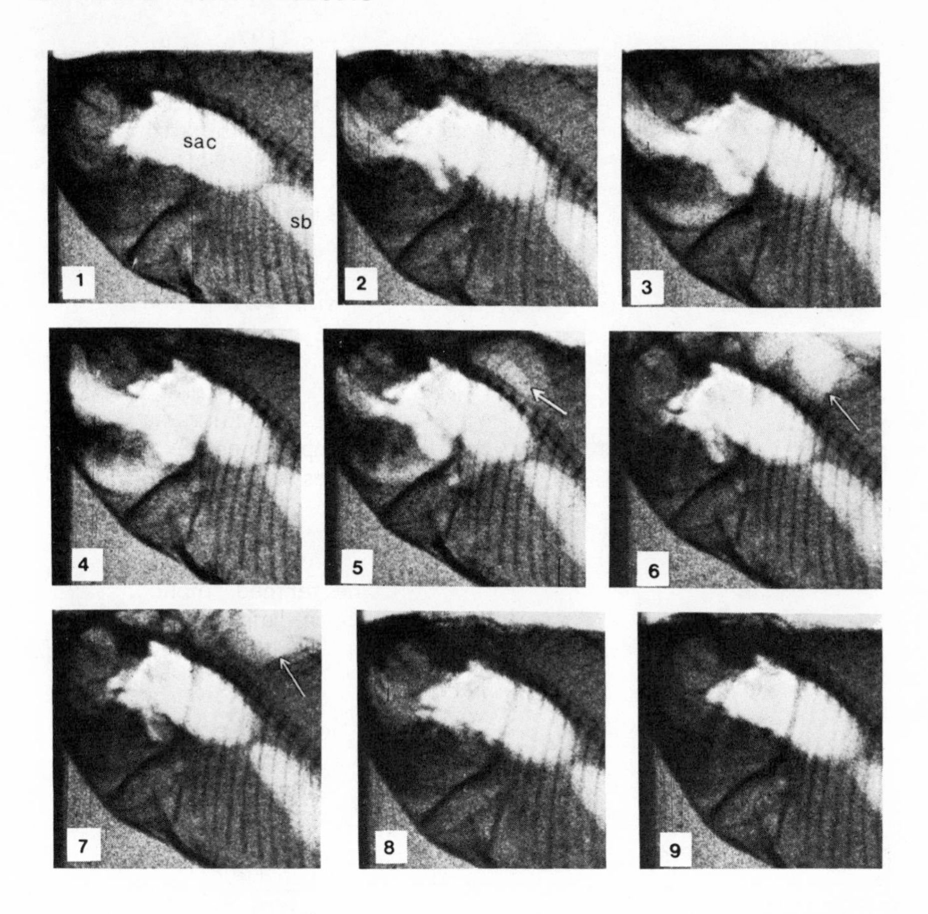

Fig. 4. Helostoma temmincki during triphasic ("monophasic" of Peters, 1978) air ventilation. Total length 15 cm, X-ray film 150 frames/sec. Note that the suprabranchial air sac (sac) maintains its volume and shape throughout the cycle. The arrows point to air bubbles that have escaped from under the operculum. Photos are selected from a sequence and do not represent successive frames from the high speed film. Abbreviations: sac, suprabranchial air chamber; sb, swim bladder.

"Diphasic" pattern of air ventilation in Helostoma. Ventilation starts with the emptying of the air from the suprabranchial chamber (Fig. 5: 2-5). First the air is emptied from the dorsocaudal region and followed by the expulsion of air from the atrial region and finally the ventrocaudal compartment (Fig. 5: 3-4) until all air has moved out from the suprabranchial chamber, although, occasionally a small air bubble remains. This process proceeds while the fish is rising to the surface and protruding the mouth out of water. Once the mouth is protruded out of the water it opens to let air escape.

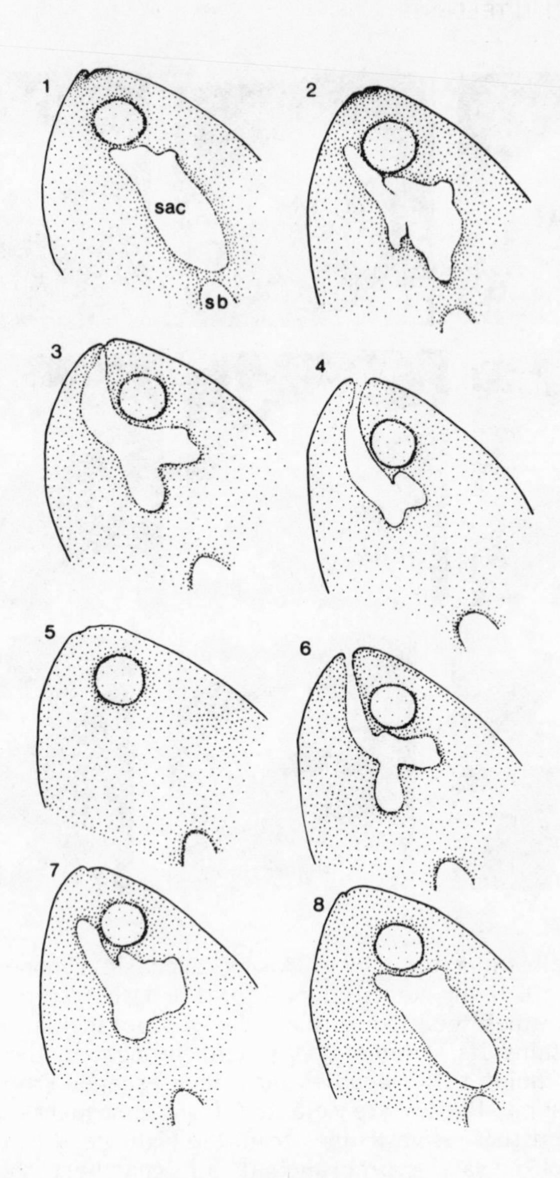

Fig. 5. Helostoma temmincki during quadriphasic ("diphasic" of Peters, 1978) air ventilation. Total length 10 cm. Diagram based on X-ray film, taken at 150 frames/sec. Note that suprabranchial chamber (sac) is emptied first from the dorsocaudal compartment, then the caudoventral compartment and finally from the atrium (compare with Fig. 3). The entire chamber is completely emptied, and air invariably escapes from the mouth. Filling takes place in exactly the opposite sequence.
Abbreviations: sac, suprabranchial air chamber; sb, swim bladder.

Immediately after all air has been expelled through the mouth, the buccal cavity is filled while the mouth is still above the water. The fish sinks below the water surface with the mouth closed. Air in the buccal cavity is then compressed into the suprabranchial chamber, which becomes filled in the reversed order of the sequence seen during emptying (Fig. 5: 6-8). The important features of the "diphasic" (sensu Peters, 1978) pattern are: (1) exhalation preceeds inhalation; (2) there is a total or nearly total evacuation of air from the suprabranchial chamber; (3) air is invariably expelled from the mouth, mostly when the fish has its mouth above water level; (4) inhalation proceeds by filling the buccal cavity first and then compressing the air into the suprabranchial chamber.

Electromyographic Profiles During Air Ventilation: I have used electromyography as an experimental tool to test the hypotheses proposed on purely anatomical grounds (Munshi, 1968) or by a combined approach of morphology and kinematics (Peters, 1978).

The "constrictor suprabranchialis" hypothesis of Munshi. Munshi (1968), following Bader (1937) and Misra and Munshi (1958), ascribed a key functional role to the constrictor suprabranchialis during exhalation: "The suprabranchial chamber contracts due to the action of the constrictor suprabranchialis and the adductor arcus palatini muscles; at the same time the operculum is lifted up by the levator and the dilator operculi muscles and the gas escapes through the exhalent opening into the opercular cavity and thence through the opercular opening. The other branchial muscles take part in the contraction and expansion of the pharynx". According to this hypothesis, inhalation occurs as follows: "Relaxation of the muscles leads to the elastic recoil of the skeleton, which brings about a fall in pressure within the supra-branchial cavity". As shown in the kinematic studies using cineradiography, whenever air is expelled from behind the operculum, the pattern is "monophasic", during which the suprabranchial chamber maintains the same volume and form and remains filled with air throughout the ventilative cycle (Fig. 4: 1-9 and Peters, 1978). Actually the species studied by Munshi (1968), Anabas testudineus, is the only anabantoid known to employ exclusively the "monophasic" pattern, inhalation invariably preceeds exhalation. Thus, the constrictor suprabranchialis hypothesis must be rejected on the basis of cineradiographic evidence alone.

However, since the hypothesis has been extended to other anabantoids (e.g., Trichogaster, Munshi, 1965) with "diphasic" patterns during which all air is expelled from the suprabranchial chamber prior to inhalation, the constrictor suprabranchialis hypothesis may still be valid. All branchial muscles that have been recorded (levatores externi 1-4, levator posterior, retractor dorsalis, pharyngohyoideus and pharyngocleithralis) remained silent during both "monophasic" (Fig. 6) and "diphasic" (Fig. 7) patterns of air ventilation. It is only during feeding that the branchial muscles are active and showing distinct

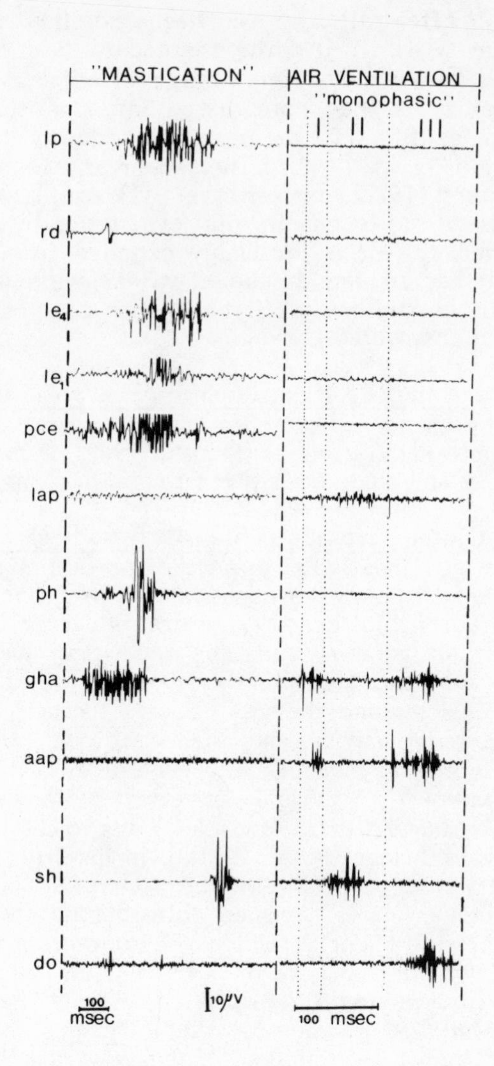

Fig. 6. Electromyographic profile of representative branchial and cephalic muscles of Anabas testudineus during mastication and triphasic ("monophasic" of Peters 1978) air ventilation. The three phases during air ventilation are: I, the preparatory; II, expansive, and III, the compressive phase. Abbreviations: aap, adductor arcus palatini; do, dilatator operculi; gha, geniohyoideus anterior; lap, levator arcus palatini; $le_{1,4}$, levator externus 1 and 4; lp, levator posterior; pce, pharyngocleithralis externus, ph, pharyngohyoideus; rd, retractor dorsalis; sh, sternohyoideus.

electromyographic profiles (Fig. 6). During air ventilation these muscles are not active, and do not function to empty the suprabranchial air chamber. Therefore, the constrictor suprabranchialis hypothesis must be rejected.

Electromyographic profile of the "monophasic" pattern. In Anabas and Helostoma all branchial muscles are silent during this mode of air ventilation (Figs. 6, 7, 8). The ventilative cycle begins with a short burst of activity in the adductor arcus palatini and geniohyoideus muscles (Figs. 6, 8: aap). Actually, the adductor mandibulae complex also shows such early activity (Figs. 7, 8: AM_3). Immediately following this essentially compressive phase, the sternohyoideus and levator arcus palatini become active (Fig. 6: SH, LAP). Finally, the adductor arcus palatini, geniohyoideus and adductor mandibulae and dilatator operculi muscles fire for a prolonged time with a high amplitude (Fig. 6: PH, AM, DO). The electromyographic profile for Helostoma (Fig. 8) during "monophasic" ventilation is very similar to that in Anabas. During my recordings of the adductor mandibulae and geniohyoideus in Helostoma, the first burst of activity is not a pure myogram but, unfortunately, it is always mixed with a motion artefact (Fig. 7: AM_3) caused by the turbulence and disturbance of the electrode wires at the air-water interface.

The information gathered from the kinematic and electromyographic profiles may be interpreted as follows: As the fish rises to the surface it compresses its buccal cavity by action of the adductor arcus palatini, geniohyoideus and adductor mandibulae complex, to minimize the volume of the buccopharyngeal cavity and to expel as much water as possible from the buccal chamber in preparation for air intake. Air is sucked in by the sudden expansion of the buccal cavity, with a slightly opened mouth. The sudden increase in volume of the buccal cavity is caused by action of the levator arcus palatini pulling the sidewalls laterally and the sternohyoideus lowering the buccal floor. It is possible that the levator arcus palatini is aided by an elastic recoil, because the excursion of the sidewalls is extreme, yet the degree of activity as measured by the amplitude of the myogram is consistently very low (Figs. 6, 7: lap). It is during this stage that air fills the entire buccal cavity (Fig. 4, photos 2, 3). With the mouth closed, the air in the buccal cavity is compressed by actions of the adductor arcus palatini, geniohyoideus and adductor mandibulae complex. Because of this compression, a pressure gradient is formed, forcing old air out of and fresh air into the suprabranchial chamber. This pressure gradient is amplified by the delayed action of the dilatator operculi muscle (Figs. 6, 7: do).

The "monophasic" pattern (sensu Peters, 1978) is actually triphasic. The term "monophasic" applied to this pattern of ventilation (Peters, 1978) is both inaccurate and misleading. It is true that on the basis of cineradiography this pattern appears to take place in a single step. However, the electromyographic profiles and a more refined and

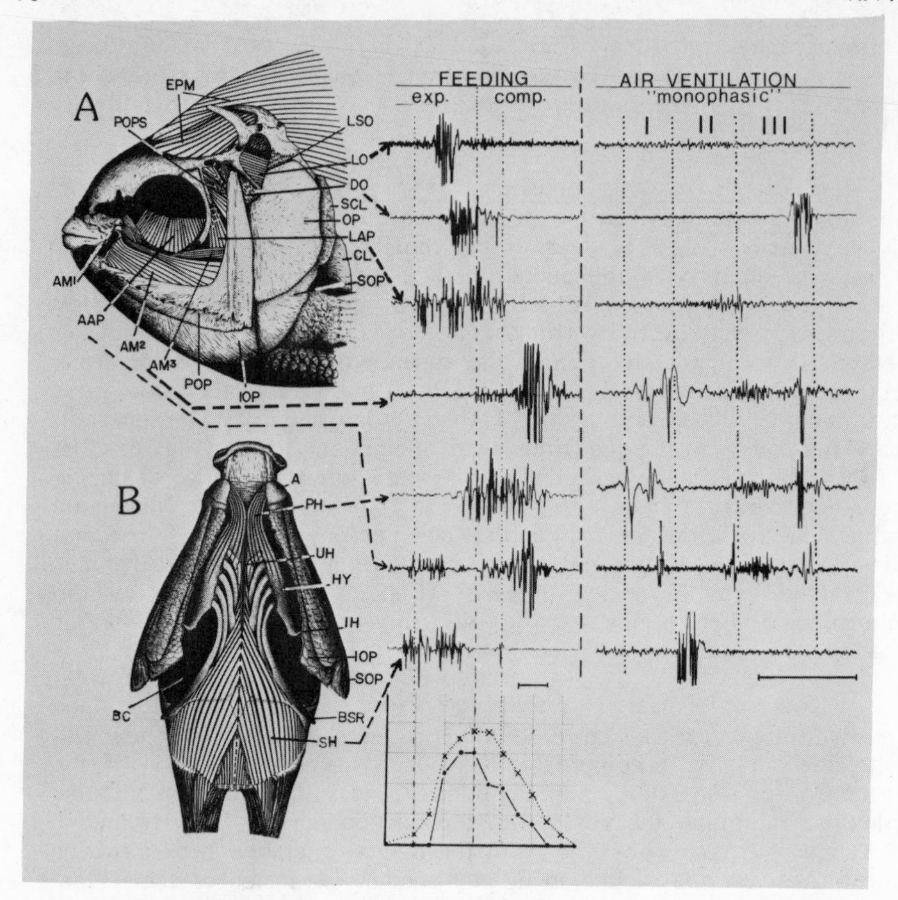

Fig. 7. Lateral (A) and ventral (B) aspect of the cephalic muscles of Helostoma temmincki (from Liem, 1967), electromyographic profiles during feeding and triphasic ("monophasic" of Peters, 1978) air ventilation. Kinematic profile expressed in a graph, the vertical axis expressing deviations from the fully adducted (o) position; horizontal axis indicate the time (frame numbers of the film). Solid line distance between the opercula (suspensory abduction). In the electromyographic profiles the horizontal lines indicates 100 msec. The three phases are: I, preparatory; II, expansive; and III, compressive.

Abbreviations: A, articular; AAP, adductor arcus palatini; AM_1, A_1 portion of adductor mandibulae; AM_2, A_2 portion of adductor mandibulae; AM_3, A_3 portion of adductor mandibulae; BC, branchial cavity; BSR, branchiostegal ray; CL, cleithrum; comp, compressive phase; DO, dilatator operculi; EPM, epaxial muscles; exp., expansive phase; HY, hyoid; IH, interhyal; IOP, interoperculum; LAP, levator arcus palatini; LO, levator operculi; LSO, ligament, OP, operculum;

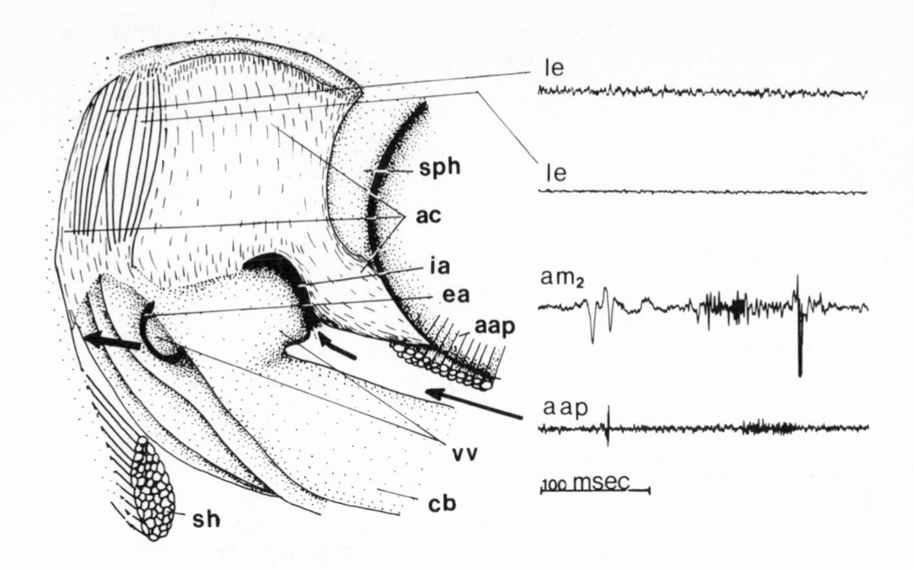

Fig. 8. Lateral view of right side of suprabranchial air chamber (ac), muscles and gill arches of Helostoma temmincki. Electromyographic profile on the right during triphasic ("monophasic" of Peters, 1978) air ventilation. Arrows indicate pathway and direction of air bubble.
Abbreviations: aap, adductor arcus palatini; ac, suprabranchial air chamber; am_2, A_2 portion of adductor mandibulae; cb, ceratobranchial; ea, branchial opening; ia, pharyngeal opening in floor of suprabranchial chamber; le, levatores externi; sh, sternohyoideus; sph, sphenotic; vv, valve or shutter.

closer kinematic analysis have revealed that the pattern is distinctly triphasic (Figs. 6, 7, 8). The components of the triphasic pattern can be identified as: (1) Preparatory phase, representing an initial compressive action in order to start the ventilative cycle from a small volume and to allow a subsequent large increase in volume; both the electromyographic and kinematic profiles are reminiscent of the preparatory phase in the feeding cycle of some predatory percoids (Osse, 1969; Elshoud-Oldenhave and Osse, 1976; Liem, 1978); (2)

PH, geniohyoideus; POP, preoperculum; POPS, postorbital process; SCL, supracleithrum; SH, sternohyoideus; SOP, suboperculum; UH, urohyal.

Expansive phase, during which the sternohyoideus and levator arcus palatini are active, enlarging the buccopharyngeal cavity to suck in air. This expansive phase differs from that of the feeding cycle in having fewer expansive muscles fire (Fig. 7: exp); (3) Compressive phase, resembling that of the feeding cycle rather closely (Fig. 7: comp), except for the delayed action of the dilatator operculi muscle (Fig. 7: DO).

Thus, the triphasic pattern of air ventilation in Anabantoidei may represent a modification of the basic feeding cycle. The greatest difference is in the expansive phase. I must emphasize that in some recordings of the feeding cycle, a preparatory phase is present and that the expansive phase resembles that of air ventilation much more closely than depicted in Fig. 7.

Electromyographic profile of the "diphasic" pattern. Peters (1978) has formulated the theory of a double-pumping mechanism for the "diphasic" pattern of air ventilation in anabantoids. According to this theory, exhalation is brought about by a coughing mechanism (as described by Hughes, 1975) and inhalation by a gill ventilation mechanism (as described by Hughes and Shelton, 1962). A typical electromyographic profile of "diphasic" ventilation of *Helostoma* is depicted in Fig. 9. The salient features are the double bursts of the adductor mandibulae complex and the adductor arcus palatini. Identical double bursts have been recorded during coughs in the perch (Fig. 10, Osse, 1969). "Diphasic" air ventilation begins with activity of the adductor mandibulae complex and the adductor arcus palatini (Fig. 9: $am_{2,3}$, aap). When activity in these two muscles stops, the levator arcus palatini, dilatator operculi and sternohyoideus (lap, do, sh) fire synchronously. A very brief pause follows, after which synchronous activity of the levator arcus palatini, levator operculi and sternohyoideus take place. The dilatator operculi (do) becomes active late in this essentially expansive phase. Air ventilation ends with high amplitude activity of the adductor mandibulae complex ($am_{2,3}$), adductor arcus palatini (aap) and geniohyoideus (gh) muscles.

The term "diphasic" for this pattern of ventilation (Peters, 1978) implies too simple a mechanism and is rather misleading.

The "diphasic" pattern (sensu Peters, 1978) is actually quadriphasic. The electromyographic profile (Fig. 9) reveals that this pattern is not "diphasic" but clearly quadriphasic. The four phases can be characterized as follows: (1) Preparatory phase is a compressive one, minimizing the volume of the buccal cavity by action of the adductor arcus palatini and adductor mandibulae muscles (Fig. 9: aap, $am_{2,3}$); although this phase is variable in its occurrence and amplitude, the preparatory phase is an integral part of the quadriphasic pattern of air ventilation; (2) Reversal phase that begins with a sudden and dramatic increase in volume of the buccal cavity by the high amplitude firings of the levator arcus palatini (lap) and sternohyoideus

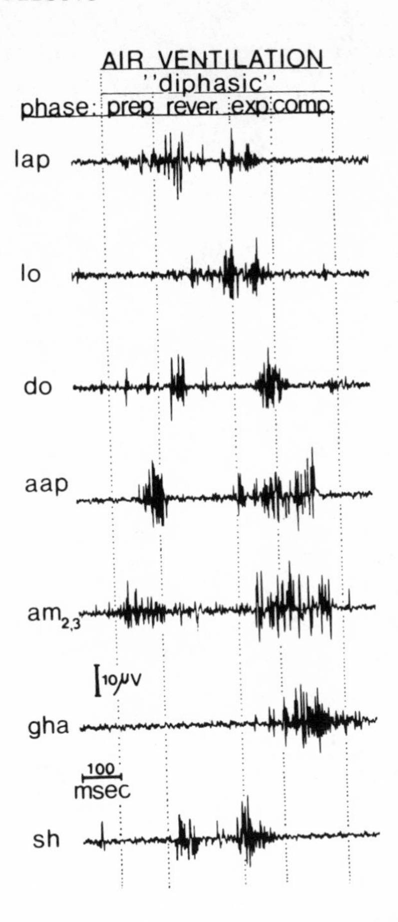

Fig. 9. Electromyographic profile of key cephalic muscles in Helostoma temmincki during quadriphasic ("diphasic" of Peters, 1978) air ventilation.
Abbreviations: aap, adductor arcus palatini; $am_{2,3}$, A_2 portions of the adductor mandibulae; comp, compressive; do, dilatator operculi; exp., expansive; gha, geniohyoideus anterior; lap, levator arcus palatini; lo, levator operculi; prep., preparatory; sh, sternohyoideus.

(sh) muscle, while action of the dilatator operculi muscle (do) opens (abducts) the operculum. The sudden negative pressure that presumably results from the expansion of the buccal cavity sucks air from the suprabranchial chamber into the buccal cavity. As the operculum is opened, water enters into the opercular cavity and the suprabranchial cavity because of the reversed pressure gradient. As the reversal phase proceeds the mouth is opened by activity of the levator operculi muscle, and air is expelled through the mouth while the suprabranchial chamber becomes filled with water; (3) Expansive phase, which resembles that of the feeding cycle (Fig. 7: exp) rather closely. The levator arcus palatini (lap), sternohyoideus (sh) and levator operculi (lo) muscles fire synchronously to suck in air through the opened mouth. Action of the dilatator operculi creates a low pressure in the opercular cavity, causing the exit of water from the suprabranchial chamber. The pattern of muscle activity during this phase can be considered an exaggerated inspiration (Osse, 1969); (4) Compressive phase, which is virtually identical to that encountered in the feeding cycle (Fig. 7). Rapid closure of the mouth by action of the adductor mandibulae complex ($am_{2,3}$), traps the air bubble in the buccal cavity (Fig. 5: 7). The air is immediately forced into the suprabranchial chamber by action of the adductor arcus palatini (aap) and geniohyoideus (gh) muscles. Toward the end of the compressive phase, the dilatator operculi often shows low level activity. The pattern of muscle activity of this phase may be compared to that of an exaggerated expiration (Osse, 1969).

The quadriphasic electromyographic profile of air ventilation in Helostoma is almost indentical to the electromyographic pattern during the cough of the perch (Fig. 10, Osse, 1969). The only major difference is that the reversal phase in Helostoma has become extended temporally.

Conclusions and Problems: In view of the electromyographic evidence presented here, we must reject, reinterpret or refine and modify previous hypotheses and theories explaining the mechanism of air ventilation in anabantoids.

The branchial muscles (levatores externi, levator posterior, retractor dorsalis, pharyngocleithralis externus, and pharyngohyoideus) do not participate in air ventilation (Figs. 6, 8). They are only active during mastication and swallowing. Thus, the key role ascribed to the "constrictor suprabranchialis" (Munshi, 1968) in air ventilation must be rejected.

As described by Peters (1978) air ventilation proceeds in two modes in anabantoids. Electromyographic evidence has shown that the two modes of ventilation are more complex than originally thought. The two patterns are triphasic and quadriphasic, which can be derived from, respectively, feeding and coughing cycles. Although the similarities in the electromyographic profiles between feeding and triphasic ventilative cycles are indeed important, differences can also be discerned. Anabantoids can feed freely without ever loosing any air from the suprabranchial chambers. Thus, "protective" valves must play a key role in regulating the air flow. Unfortunately the precise functional mechanisms of the valves at the pharyngeal and branchial openings remain unknown. Of course, anabantoids feed by inertial suction by creating a water current. Since no air is sucked in by the

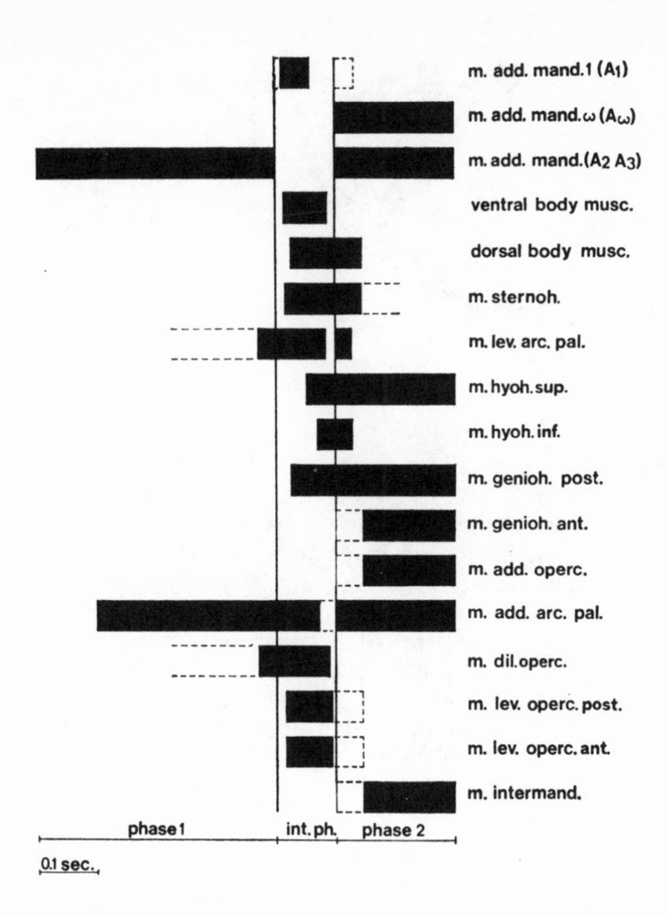

Fig. 10. Diagram of the active periods of the muscles during coughs in the perch Perca fluviatilis (from Osse, 1969). The dashed lines indicate frequently observed variations of the periods. The division into phases is based on the activity of the adductor mandibulae.
Abbreviations: int. ph., interphase; m. add. mand., adductor mandibulae; m. add. arc. pal., adductor arcus palatini; m. add. operc., adductor operculi; m. genioh. post., geniohyoideus posterior; m. hyoh. sup., hyohyoideus superior; m. intermand., intermandibularis; m. lev. arc. pal., levator arcus palatini; m. lev. operc. ant., levator operculi posterior; m. sternoh, sternohyoideus.

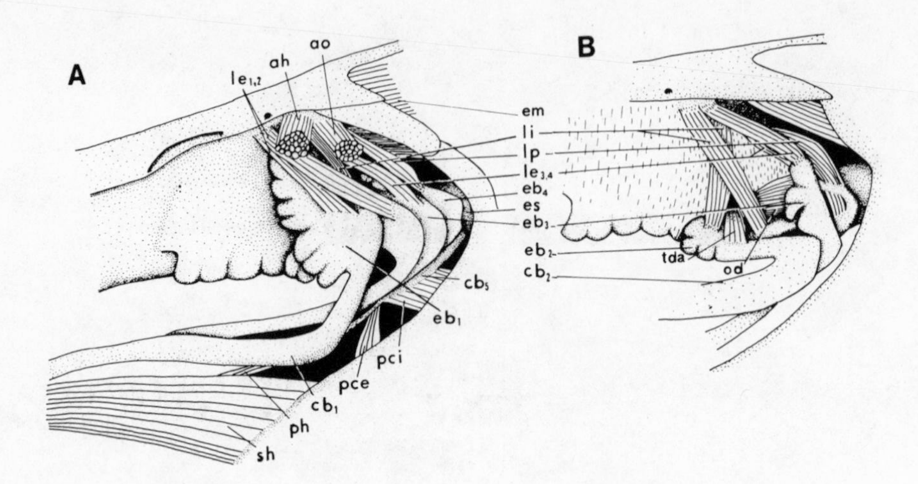

Fig. 11. A. Lateral aspect of the branchial muscles, "labyrinth organ", and suprabranchial chamber of Channa micropelts after removal of the gills and mucous membrane. B. The same dissection as in A, after removal of the entire first gill arch, associated muscles and labyrinth organ.
Abbreviations: ah, adductor hyomandibulae; ao, adductor operculi; cb_1-cb_5, first to fifth ceratobranchial; eb_{1-4}, first to fourth epibranchial; em, epaxial muscles; le_{1-4}, first to fourth levator externus; li, levator internus; lp, levator posterior; od, obliquus dorsalis; pce, pharyngohyoideus; sh, sternohyoideus, tda, transversus dorsalis anterior.

activation of the expansive and compressive actions of the feeding cycle, the air stored away in a hydrostatically strategic suprabranchial chamber may simply remain undisturbed. Preliminary behavioral studies in progress seem to indicate that if anabantoids feed from the surface they often release air bubbles from behind the operculum. This supports the hypothesis that the triphasic pattern of ventilation is only a moderately modified profile of the basic feeding cycle. If air is sucked in, the mechanism causes air ventilation, but if water is sucked in, it creates a water current into the buccal cavity and thence into the opercular cavity.

The electromyographic profile of the quadriphasic pattern of air ventilation supports the theory of Peters (1978) that, during exhalation, the air in the suprabranchial chamber is replaced by water flowing in a reversed direction, i.e. from the opercular to the suprabranchial chamber and thence to the buccal cavity. The electromyographic profile during quadriphasic air ventilation in anabantoids is almost identical to the electromyographic profile during coughing in the perch (Osse, 1969). It is not known whether anabantoids always loose air from the suprabranchial chamber when coughing.

Although the mechanisms of ventilation may superficially appear as complex, evolutionary novelties, they are in fact only minor or at most moderate modifications of the feeding and coughing mechanisms, both representing preexisting and fundamentally primitive physiological activities.

AIR VENTILATION IN THE CHANNIFORMES

All members of the order Channiformes (Ophicephaliformes) are facultative air breathers inhabiting the Old World tropics. They have been the subject of many anatomical studies. Although electromyographic data are not yet available, anatomical data combined with cineradiographic analysis enable us to evaluate previous theories on air ventilation in the channiforms.

Anatomical Specializations: The small suprabranchial chamber extends from the hyomandibula to the anterodorsal tip of the pectoral girdle (Fig. 13, photos 1, 12). It communicates freely with the buccal cavity via three large pharyngeal openings located respectively in front of the dorsal extremity of the first gill arch, between first and second gill arches, and between second and third gill arches. Munshi (1962) named these openings inhalent apertures. But as we will see below, air passes in two directions through these openings. Thus the term inhalent is inappropriate. Communication between the suprabranchial chamber and the opercular cavity is via the first two unmodified gill slits.

The small suprabranchial chamber is incompletely subdivided in compartments in a varying way depending on the species. A distinct anterodorsal recess (Fig. 13, photo 7) remains filled with air throughout the ventilative cycle.

Considerable interspecific variation occurs in respect to the presence of respiratory dendritic nodules. In Channa micropeltes dendritic nodules are present on the hyomandibula, parasphenoid (Figs. 11, 12: lao) first, second and third epibranchials (Fig. 11: eb).

The topography and morphology of the four levatores externi remain relatively unspecialized in Channa micropeltes. Each levator externus inserting on the dorsal aspect of the epibranchial of the corresponding arch (Figs. 11, 12: le_{1-4}). The levator posterior (Fig. 12: lp) is also unmodified. The levatores interni, inserting on the second and third pharyngobranchials, and the retractor dorsalis, inserting on the third pharyngobranchialis also retain a generalized pattern (Fig. 12:

li, rd). Thus the branchial myology in channiforms has remained surprisingly generalized during the evolution of the air breathing apparatus. In Channa punctatus the second and third levatores externi are reduced, as is often the case in many teleosts. In C. striatus the reduction of the second and third levatores externi is less pronounced than in punctatus. Thus morphologically there seems little justification for the term "constrictor suprabranchialis" (Munshi, 1962, 1976) to designate the reduced levatores externi two and three. The term constrictor supra-branchialis implies an important morphological specialization with a definite functional role. Yet we are actually dealing with muscles, which have undergone evolutionary reduction and, which, do not constrict the suprabranchial chamber.

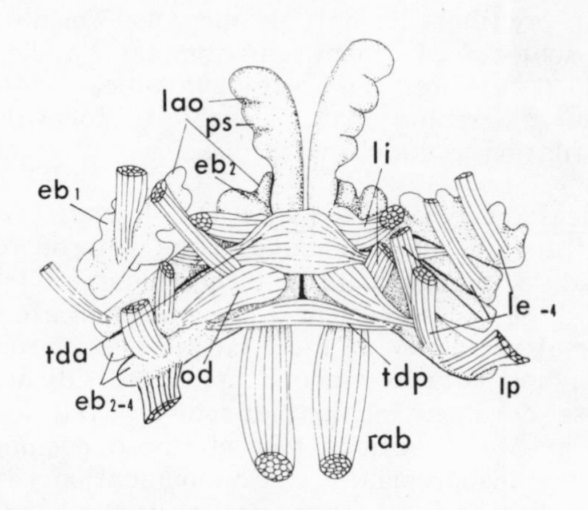

Fig. 12. Dorsal view of dissected and isolated branchial apparatus and muscles of Channa micropeltes.
Abbreviations: eb_{1-4}, first, second, third, fourth epibranchial; lao, "labyrinth epibranchials and parasphenoid (ps); le_{1-4}, first through fourth levator externus; li, levator internus; lp, levator posterior; od, obliquus dorsalis; ps, respiratory nodules associated with the parasphenoid bone; rab, retractor dorsalis; tda, transversus dorsalis anterior.

Cineradiographic Analysis of the Pattern of Air Ventilation: Adult Channiforms exhibit only one pattern of air ventilation. First it is important to note that in addition to the suprabranchial chamber, the entire buccopharyngeal cavity is filled with air (Fig. 13, photos 1, 12). This fact has never been recorded or discussed. All previous studies have ignored the fact that both the suprabranchial and buccopharyngeal cavities are filled with air, carrying on air breathing functions. Thus gas exchange by air is not restricted to the suprabranchial chamber as

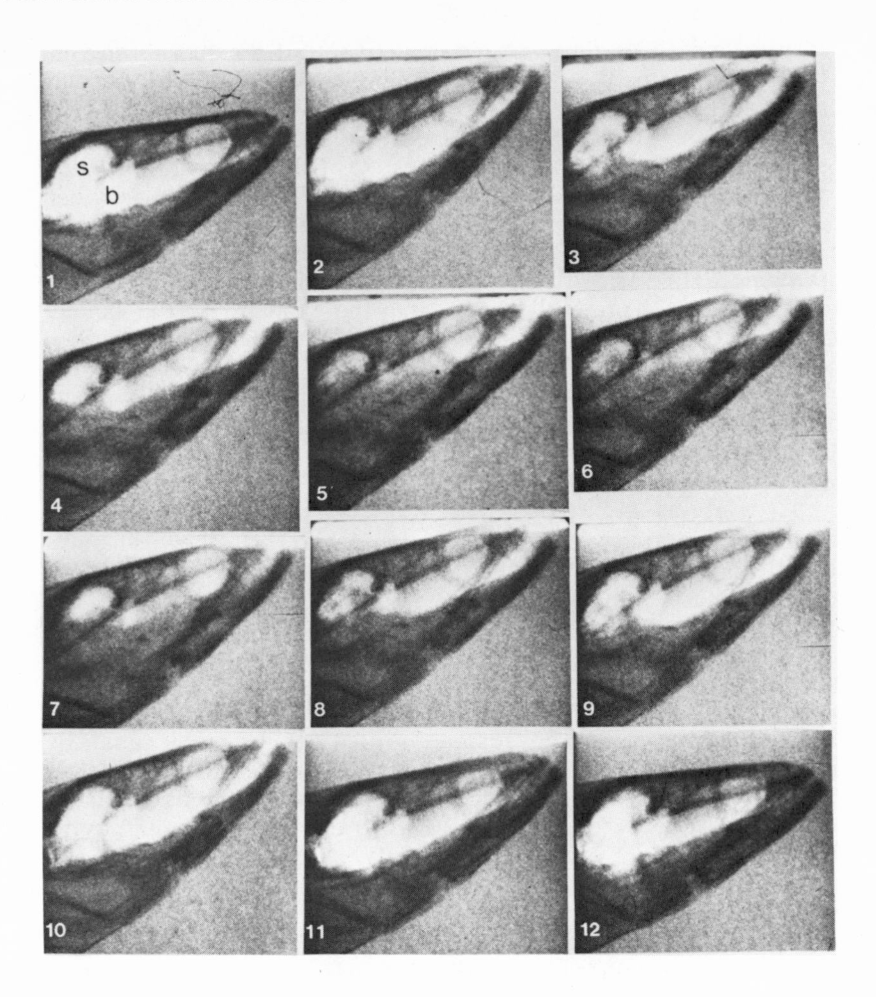

Fig. 13. Channa punctatus, total length 30 cm, air ventilation, X-ray film at 150 frames/sec. Selected frames from a sequence to show the kinematic events of air ventilation. Air from the suprabranchial cavity is partially replaced by water (3-7). Water enters posteroventrally. Consequently, a dorsal air pocket invariably remains as a residual volume (6). Inhalation and exhalation occur through the mouth. Air fills the entire buccopharyngeal cavity (6) in addition to the suprabranchial chamber (5).

previously assumed. As the fish rises to the surface the buccal cavity expands by a lowering of its floor and by moving the sidewalls apart (Fig. 13, photo 2) but the mouth is still closed. At the same time the operculum is opened or abducted. As a result water enters the opercular cavity and flows forward and upward to fill the suprabranchial chamber (Fig. 13, photo 3). But in sharp contrast to the process in anabantoids, water enters the suprabranchial chamber ventrocaudally (Fig. 13, photos 3, 4). The mouth is opened (Fig. 13, photos 4, 5, 6) and air leaves the mouth as water continues to fill the suprabranchial chamber and buccal cavity. In Fig. 13, photo 7, most of the air has been expelled and exhalation has been completed. However, one can clearly see that a pocket of air in the anterodorsal recess of the suprabranchial chamber remains as a residual. Exhalation is immediately followed by very fast inhalation. The mouth opens and the volume of the buccal cavity increases greatly, so that air is sucked into the buccal cavity and suprabranchial chamber (Fig. 13, photos 8, 9, 10). It appears as if the flow of air into the suprabranchial cavity is facilitated not only by the second abduction of the operculum but also by a passive mechanism by which air bubbles tend to rise in a water column. Once the suprabranchial chamber is completely filled (Fig. 13, photo 10), the entire buccal cavity is also filled with air (Fig. 13, photo 11). Upon completion of the ventilative cycle (Fig. 13, photo 12), the fish sinks with air in both the suprabranchial and buccal cavities.

A New Hypothesis on Air Ventilation in Channiformes: Based on the kinematic profile obtained through cineradiography and behavioral observations, a hypothesis on the mechanism of air ventilation in channiforms can now be formulated with reasonable precision. The hypothesis will be tested by electromyographic analysis now in progress. Exhalation invariably preceeds inhalation and takes place through the mouth and not from behind the operculum as reported by Munshi (1962). A reversed flow of water from the opercular cavity into the suprabranchial chamber and buccal cavity accounts for the expulsion of air. In the X-ray movie one can see the considerable turbulence of the inflowing water, pushing the air forward to the opened mouth protruding above the water level. This mechanism resembles the reversal phase of the quadriphasic pattern of anabantoids. *Channa* always has a residual air pocket at the end of exhalation, making ventilation incomplete. Inhalation is accomplished by expansion of the buccal cavity, and abduction of the operculum. Distribution of air in the buccal cavity and suprabranchial chamber may well be a totally passive process with hydrostatic pressure being the driving force.

Concluding Remarks: It is important to emphasize that more air is present in the buccal cavity than in the relatively small suprabranchial chamber of the channiforms. The presence of air in the buccal cavity and dendritic nodules on the parasphenoid in *Channa micropeltes* indicates a respiratory function of the buccopharyngeal mucosa, a fact that has been ignored by all previous workers, even those who studied the morphometry of the total respiratory surface area of *Channa* (Hakim et al., 1976).

Because both the suprabranchial chamber and the buccopharyngeal cavity are completely filled with air, it is not necessary to have the communicative openings between the two chambers guarded by valves or shutters. Although the presence of a shutter has been reported by Munshi (1962) I have not observed such a structure, and judging from the distribution of air (Fig. 13, photos 1, 12) on X-ray films, no shutters are involved in the compartmentalization of air in channiforms.

Finally, it has been observed that Channa invariably looses much of its air during prey capture. Channa, as a pursuit hunter, captures prey by a combination of overtaking the prey and inertial suction. Inertial suction causes air to be expelled mainly from behind the operculum. Shortly after prey capture Channa invariably surfaces to replenish the lost air. Thus in the channiforms, feeding and air respiration are not separated functionally and morphologically. Conflicting selection forces may act on this closely coupled system, resulting in compromise rather than optimal solutions.

AIR VENTILATION IN THE SYNBRANCHIFORMES

The Synbranchiformes (swamp eels) is a highly specialized acanthopterygian order of which many members have become remarkably well adapted to truly amphibious life (Johansen, 1968, 1970; Lomholt and Johansen, 1974, 1976; Munshi and Singh, 1968). Adult synbranchiforms are known to use the skin (Liem, 1967), buccopharyngeal mucosa (Wu and Liu, 1940), esophagus (Liem, 1967), specialized saclike extensions of the buccopharynx (Das, 1927; Munshi and Singh, 1968) and gills (Carter, 1957; Johansen, 1968, 1970) as air breathing organs.

The mode of air ventilation of synbranchids has never been analyzed experimentally. Singh and Munshi (1965) claim that in Monopterus (Amphipnous) cuchia "air is thus expelled from the air-sac through exhalent aperture into the opercular cavity thence to the outside through a single opercular opening situated on the mid ventral line". But the same authors working on the same species state (Munshi and Singh, 1968): "It has been observed that the cuchia releases its air contained in the air-sacs, through the mouth in the form of large air-bubbles". "While keeping its head above the water level, it exhales and inhales atmospheric air through the mouth at intervals of one minute to three minutes...". In this study, air ventilation in Synbranchus marmoratus is analyzed by cineradiography and electromyography.

Structural Basis of Air Ventilation: In order to understand the mechanics of air ventilation it is important to appreciate the principal specializations of the muscles in synbranchids, which have been described by Conner (1966) and Munshi and Singh (1968). In Synbranchus marmoratus the key muscles (Fig. 14) involved in respiration are: (1) The greatly enlarged adductor mandibulae complex

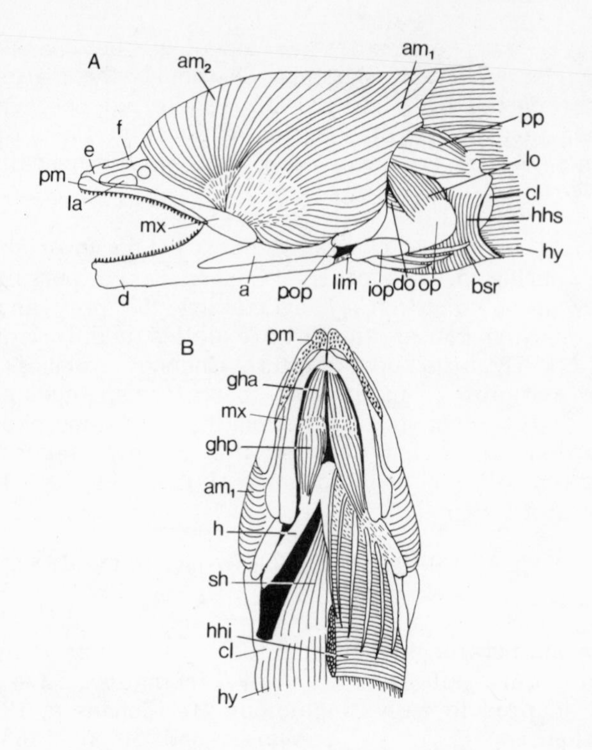

Fig. 14. A. Lateral view of head musculature of Synbranchus marmoratus. B. Ventral view of head musculature; the hyohyoideus and branchiostegal rays from the right side have been removed.

Abbreviations: a, articular; am_1, am_2, adductor mandibulae parts A_1 and A_2; bsr, branchiostegal ray; cl, cleithrum; d, dentary; do, dilatator operculi; e, ethmoid; f, frontal; gha, geniohyoideus anterior; ghp, geniohyoideus posterior; h, hyoid; hhi, hyohyoideus inferior; hhs, hyohyoideus superior; hy, hypaxial muscles; iop, interoperculum; la, lachrymal; lim, interoperculomandibular ligament; lo, levator operculi; mx, maxilla; op, operculum; pm, premaxilla; pop, preoperculum; pp, protractor pectoralis; sh, sternohyoideus.

(am_{1-3}). One part, the adductor mandibulae A_1 has its fleshy origin on the roof of the cranial vault, the posterodorsal rim of the neurocranium, and along the anterolateral edge of the preoperculum. The fibers run anteriorly and anteroventrally to a broad tendinous insertion on the dorsal edge of the mandible. Thus, the insertion of the adductor mandibulae A_1 is specialized, since, in more generalized acanthopterygians, it is invariably on the maxilla. The adductor mandibulae A_2 has a fleshy origin with that of the first and along the length of the dorsal and lateral surfaces of the frontal. The fibers run ventrally and insert by a broad tendon on the medial side of the ascending process of the dentary. The deeper adductor mandibulae A_3 has a fleshy origin on the anterior face of the sphenotic. The fibers run anteroventrally to a tendinous insertion that is continuous with that of the preceding segment. (2) The levator operculi is the largest muscle associated with the opercular apparatus. It originates from the posterolateral surface of the pterotic and inserts on both lateral and medial aspects of the dorsal part of the operculum. (3) The dilatator operculi is located immediately anterior to the levator operculi and inserts on the outer aspects of the head and shaft of the operculum (Fig. 14: do). (4) The geniohyoideus anterior is a large spindle shaped muscle, which is separated from the equally well developed geniohyoideus posterior by a transverse tendinous septum. The geniohyoideus anterior attaches to the inner surface of the anterior part of the dentary, while the geniohyoideus posterior connects to the ceratohyal near the point where the third branchiostegal ray articulates. (5) The most specialized muscle (Munshi and Singh, 1968) in synbranchiforms is probably the hyohyoideus. It has become a large and thick muscle mass occupying the entire ventrolateral surface of the branchial region of the fish. The hyohyoideus inferior runs between the branchiostegal rays and the fibers from either side fuse completely in the midline (Fig. 14B). The dorsal portion, hyohyoideus superior, is extremely well developed and runs between the first and second branchiostegals to the inner aspect of the operculum, the pectoral girdle (supracleithrum and cleithrum), a tendon associated with the anterior trunk and protractor pectoralis muscles. (6) Both the sternohyoideus and hypaxial muscles, although enlarged, seem to retain the typical acanthopterygian morphology. Both muscles play a dominant role in inhalation.

Behavioral and Cineradiographic Analysis of Air Ventilation in Synbranchus: *Synbranchus marmoratus* always exhales prior to inhalation. Two patterns of air ventilation can be seen: exhalation can proceed either passively or actively.

During the "passive exhalation pattern of ventilation" air always escapes through the mouth. As the fish swims up to the surface or lifts the anterior one-third to one-half of the body, the snout is protruded out of the water and the mouth is opened slightly allowing air to escape from the mouth (Luling, 1958). The body is raised upward at an angle of at least 45 degrees. Although in many instances all air escapes, at times some air pockets remain in the buccopharyngeal and branchial regions. This mode of expiration seems to proceed passively, without muscular effort, the driving force being hydrostatic pressure on the air bubbles. After exhalation the fish may return to the bottom without gulping air or it may first fill the buccopharyngeal and opercular cavities with fresh air. Inhalation is accomplished by a sudden lowering of the buccopharyngeal floor with the opened mouth above the water. The "passive exhalation pattern" of ventilation invariably takes place at or near the surface of the water.

The "active exhalation pattern" of air ventilation is the dominant mode, which can proceed while the fish is on the bottom, in the water column or at the surface. It is independent of the position of the body and occurs invariably prior to prey capture. Air is forced out of the mouth by muscular effort. Some air may also escape from the median external gill opening so characteristic of synbranchids. In most instances, exhalation is complete, leaving no air behind. Inhalation proceeds in exactly the same way as during the "passive exhalation pattern".

Cineradiography reveals that in Synbranchus marmoratus air fills both the buccopharyngeal and opercular cavities (Fig. 15: e). When the cavities are filled with air, the hyoid is greatly depressed at an angle of nearly 90 degrees with the neurocranial base in a lateral X-ray view (Fig. 15: e). From a dorso-ventral X-ray profile, the angle between the hyoid rami also reaches its highest value (Fig. 15: f). During expiration, either passively or actively, both hyoid-neurocranial base and interhyoid rami angle are reduced (Fig. 15: a, b). Furthermore, exhalation is also accompanied by a shortening of the distance between the posterior extremities of the mandible in a dorso-ventral X-ray profile (Fig. 15: b). The only difference between "passive and active" exhalation patterns is in the velocity by which the hyoid moves. In "active exhalation" the hyoid moves at higher velocities.

Inspiration (Fig. 15: a-f) takes place with the snout out of the water, the mouth slightly opened (Fig. 15: c, d), and a sudden and extreme movement of the hyoid (Fig. 15: c-f), lowering the floor of the buccopharynx dramatically. Air is sucked in very rapidly by this action, and the mouth is closed at high velocity (Fig. 15: e).

Electromyographic Profile of Air Ventilation: Inspiration is always preceded by extended but variable bursts of activity in the geniohyoideus posterior and anterior muscles (Fig. 15: GHA, GHP). The former is the more active one. Early action of the geniohyoideus muscle ensures a minimal volume of the buccopharyngeal cavity from which the great inspiratory expansion can start. It is reminiscent of the preparatory phase seen during air ventilation in anabantoids. After the preparatory phase, there is synchronous high amplitude activity in the hypaxial and sternohyoideus muscles lasting as long as 0.3 seconds. Very low amplitude activity of the levator operculi (Fig. 15: LO) accompanies the firings of the hypaxial (HYP) and sternohyoideus (SH) muscles. There is no doubt that the hypaxial and sternohyoideus are the dominant inspiratory muscles. The dilatator operculi remains silent. Thus, the expansion of the buccopharyngeal cavity is caused mainly by the combined action of the hypaxial and sternohyoideus muscles.

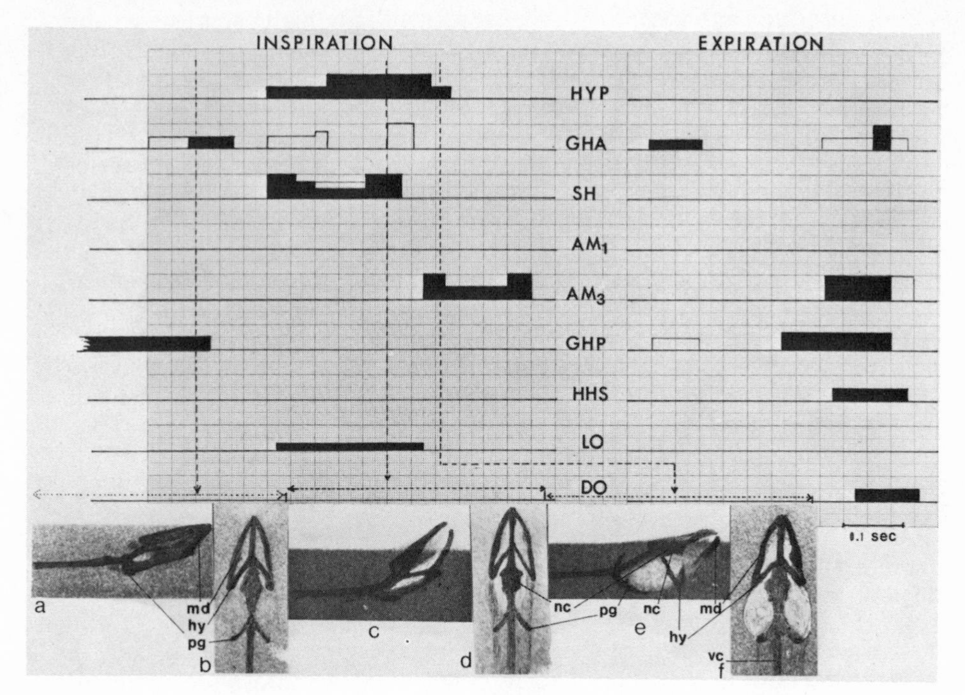

Fig. 15. Representative diagram summarizing activities of principal cephalic muscles of Synbranchus marmoratus during inspiration and expiration of air. Open blocks indicate frequently observed variations of the active periods. a-f are representative frames of an X-ray film taken at 60 frames per second of a specimen of 24 cm in total length. The frames have been slightly retouched to enhance the contrast of the bony structures. The sequence is that of an inspiration. The dashed vertical lines with arrows indicate the approximate correspondence between electromyogram and the kinematics.
Abbreviations: AM_1, AM_3, adductor mandibulae parts A_1 and A_3; DO dilatator operculi; GHA, geniohyoideus anterior; GHP, geniohyoideus posterior; HHS, hyohyoideus superior; hy, hyoid; HYP, hypaxial muscles; LO, levator operculi; md, mandible; nc, neurocranium; pg, pectoral girdle; SH, sternohyoideus; vc, vertebral column.

Toward the end of inhalation, the adductor mandibulae A_3 becomes active closing the mouth. In some recordings the geniohyoideus anterior (GHA) shows some intermittent bursts during the extended activity of the sternohyoideus muscle (SH).

Active exhalation is characterized by high amplitude and extended firings of the geniohyoideus posterior, adductor mandibulae A_3, and hyohyoideus superior muscles in precisely that sequence (Fig. 15: GHP, AM_3, HHS). Often the geniohyoideus anterior (GHA) fires synchronously with the adductor mandibulae A_3 muscle. Toward the end, the dilatator operculi fires together with the hyohyoideus superior muscles. By this action most air is expelled from the mouth although occasionally some air may escape from the median external gill slit. The dominant expiratory muscles seem to be the geniohyoideus posterior and adductor mandibulae A_3 muscles. Combined action of these two muscles must generate the compressive force to expel the air.

The electromyographic profile of a passive exhalation is a totally silent one. None of the muscles (hypaxial, geniohyoideus anterior and posterior, sternohyoideus, adductor mandibulae A_1, A_3, hyohyoideus superior, levator operculi and dilatator operculi) show any activity, yet air escapes from the mouth and the buccopharyngeal cavity is reduced in volume at the end of passive exhalation. It is possible that the electromyographic profile is not sufficiently comprehensive to justify the conclusion that this type of exhalation indeed proceeds without any cephalic muscle actions.

Some Functional and Evolutionary Implications: The Synbranchiformes is a rather diverse order containing a wide array of purely aquatic species, specialized facultative air breathing forms, and highly specialized obligatory air breathers. Thus it is premature to make any general conclusion and theories on the mode of ventilation, since our studies have been limited to one species. However, attention should be called to the fact that air is not restricted to the gill cavity in *Synbranchus*, but fills the entire buccopharyngeal cavity, suggesting that air respiration is not only performed by the gills (Carter, 1957; Johansen, 1968) but also by the entire buccopharynge mucosa (Bicudo and Johansen, 1979). The general response to life out of water for prolonged periods (more than 30 minutes) by *Synbranchus marmoratus* (Johansen, 1968) and *Monopterus albus* is to invariably open the mouth widely and continue air breathing with intermittent movements of the mandible and hyoid. There is little doubt that during such extended terrestrial excursions the buccopharyngeal mucosa of *Synbranchus marmoratus* becomes very vascularized and provides favorable conditions for prolonged air breathing gas exchange.

Because all air breathing Synbranchiformes hold air in the buccopharyngeal cavity, air must always be expelled prior to feeding. When in water, synbranchids rely on inertial suction to capture prey. Before striking at prey, air is invariably exhaled actively. Thus in synbranchids, the air breathing mechanism and feeding apparatus are closely coupled, exerting constraining mutual influences.

The neuromuscular apparatus underlying air ventilation in the Synbranchiformes is known to be the most highly specialized among teleosts. The electromyographic profiles of inspiration and expiration of air (Fig. 15) cannot be derived from those of aquatic respiration, coughing or feeding. Thus the neuromuscular and biomechanical process by which air ventilation is accomplished are truly independent and quite complex. The dominant muscles during air ventilation are confined to the ventral group: geniohyoideus, sternohyoideus, and hypaxial muscles. It is probable that such an innovative mode of ventilation has evolved under strong selection forces favoring the ability to exchange gas by air breathing.

GENERAL CONCLUSIONS

Although the comparison has been limited to just three lineages of advanced acanthopterygian fishes, several generalizations can be made concerning the meaning of physiological approaches as applied to the problems of adaptive mechanisms and evolution of air-breathing fishes.

The nature of the electromyographic profiles of air ventilation in the Anabantoidei has given us an insight on the origin of the neuromuscular and kinematic mechanisms underlying the mode of air ventilation. The recognition that we are actually dealing with triphasic and quadriphasic patterns add appreciably to an understanding of the physiological nature of air ventilation. For example, the triphasic and quadriphasic modes of air ventilation are virtually indistinguishable from those for respectively the feeding cycle and the cough. Therefore, the two modes of air ventilation have originated from ever so slightly modified basic physiological mechanisms present in almost every teleost. In the triphasic pattern the basic function of feeding is put in a dramatically different context by simply substituting an air bubble for the prey item. Similarly, the quadriphasic pattern with its original gill cleaning functions, assumes a drastically novel function: In the beginning of the cycle the air bubble is being flushed out instead of debris on the gills, and toward the end of the cycle, a new air bubble rather than water is sucked in. Thus, what appears to be a drastic change in function, i.e. air ventilation, is actually the assumption of a new function by an existing functional complex without interference with the original functions, i.e. feeding and coughing. The assumption of a new function occurred without much change in the neuronal circuitry, basic morphology and kinematics. We are merely dealing with the execution of multiple functions by one functional complex using identical neuromuscular and kinematic pathways. Theoretically most teleosts possess the building blocks and potential to evolve the triphasic and quadriphasic modes of air ventilation. It is therefore impossible to determine which of the two patterns, i.e., triphasic or quadriphasic, is, evolutionarily, the most primitive. The reason that adult Anabas relies solely on triphasic ("monophasic" sensu Peters, 1978) ventilation may well be correlated with the primitive predaceous feeding mechanism, which is not interfered with by the triphasic pattern. Because the triphasic pattern is derived from the feeding cycle, it may have been reduced to a secondary role in favor of the quadriphasic pattern in other anabantoids, which have evolved diverse feeding specializations. Many of the feeding adaptations are associated with surface feeding, and it is clear that the quadriphasic pattern, which has been derived from the cough mechanism, possesses a selective advantage since it interferes the least with the feeding apparatus.

Thus the emergence of mechanisms for air ventilation in anabantoids is due to the acquisition of a new function by existing structures. The resulting new function is just a variation of a preceding function. The shift in function exposes the new functional complex to the new selection pressure. The selection pressure in favor of the functional modification is greatly increased by a shift into new habitats made possible by the acquisition of the air breathing habits. The simultaneous functioning of the feeding cycle--triphasic air ventilation, and the cough mechanism--quadriphasic air ventilation gives us a physiological answer to a major evolutionary problem: The gradual change and improvement of a machine while it is running (Frazzetta, 1975, p. 20). It demonstrates the physiological mechanism underlying transitions from one type to the next involving a great continuity because not only must the end product be feasible, but so must be all intermediates.

When comparing the Anabantoidei with the Channiformes and Synbranchiformes, one of the most striking differences is the fact that the latter two groups invariably loose air when feeding. In sharp contrast, the Anabantoidei retain the air within the respiratory chamber throughout the feeding activities. In both the Channiformes and Synbranchiformes the parameters of the buccopharyngeal cavity are strongly affected by the selective pressures by both the feeding and respiratory systems. The constraints posed by the two conflicting selection pressures may have rendered the Channiformes and Synbranchiformes less well adapted than they could conceivably be. In respect to one fitness component, i.e. air breathing, the Channiformes and Synbranchiformes have evolved highly specialized mechanisms for air ventilation, giving them the optimum level of habitat selectivity. The other fitness component, i.e. feeding, remains unspecialized making the Channiformes and Synbranchiformes highly tolerant, unselective, predatory generalists. The complex coupling between the trophic and respiratory functional complexes may have prevented optimization of both fitness components in the Channiformes and Synbranchiformes.

In the Anabantoidei, the mutual constraints between the functional complexes underlying the trophic and air ventilation mechanisms have been uncoupled. The trophic and air breathing mechanisms have been separated with the emergence of the quadriphasic mode of air ventilation. We may consider the uncoupling of the constraints a major adaptive breakthrough, because it has allowed the Anabantoidei to adopt a mixed strategy in the exploitation of trophic resources, while simultaneously being able to occupy hypoxic environments.

ACKNOWLEDGEMENTS

This study owes much to verbal and written discussions with my colleagues, students and friends, most especially Donald Kramer, David Rand, William L. Fink, Gordon Sze, George Lauder, John Roberts, Warren Burggren, Jeffrey Graham and George Hughes. I am much indebted to them. I like to extend many thanks to those who, through their generosity and very special skills, have made it possible for me to complete this study. They include David Kraus, Christine Fox, Al Coleman, Lisa Mueller, Karsten Hartel, and Lawrence Nobrega. As always Lou Garibaldi of the New England Aquarium has been especially cooperative in dealing with many practical problems. This research has been supported in part by a fellowship of the John Guggenheim Memorial Foundation and by grants from the National Science Foundation.

REFERENCES

Bader, R. (1937). Bau, Entwicklung und Funktion des azessorischen atmungsorgans der Labyrinth fische. Z. Wiss. Zool. 149: 323-401.

Bicudo, J.E.P.W. and Johansen, K. (1979). Respiratory gas exchange in the airbreathing fish, Synbranchus marmoratus. Env. Biol. Fish. 4: 55-64.

Carter, G.S. (1957). Air-breathing. In: The physiology of fishes. Vol. 1, edited by Brown, M.E., N.Y. Academic Press: 65-79.

Conner, J.V. (1966). Morphology of the neotropical swamp eel, Synbranchus marmoratus (Pisces: Teleostei), with emphasis on adaptive features. Master of Science Thesis, Texas A and M University, College Station, Texas: 1-71.

Das, D.K. (1927). The bionomics of certain air breathing fishes of India, together with an account of the development of their air-breathing organs. Phil. Trans. Roy. Soc., Lond., Ser. B. 216: 182-219.

Elshoud-Oldenhave, J.J.W. and Osse, J.W.M. (1976). Functional morphology of the feeding system in the ruff (Gymnocephalus cernua L. 1758) (Teleostei, Percidae). J. Morph. 150: 399-422.

Frazzetta, T.H. (1975). Complex Adaptations in Evolving Populations. Sinauer Publishers, Sunderland, Mass.: 1-267.

Gans, C. (1970). Strategy and sequence in the evolution of the external gas exchangers of ectothermal vertebrates. Forma et Functio 3: 61-104.

Hakim, A., Munshi, J.S.D. and Hughes, G.M. (1976). Morphometrics of the respiratory organs of the Indian green snake-headed fish, Channa punctatus (Bloch) J. Zool., Lond., 184: 519-543.

Hughes, G.M. (1975). Coughing in the rainbow trout (Salmo gairdneri) and the influences of pollutants. Rev. Suisse Zool. 83: 47-64.

Hughes, G.M. (1976). Respiration of amphibious vertebrates, London, Academic Press: 1-402.

Hughes, G.M. and Munshi, J.S.D. (1973). Fine structure of the respiratory organs of the climbing perch, Anabas testudineus (Pisces, Anabantoidea). J. Zool., Lond., 170: 201-225.

Hughes, G.M. and Shelton, G. (1962). Respiratory mechanisms and their nervous control in fish. In: Advances in Comparative Physiology and Biochemistry edited by Lowenstein, O., Vol. 1. N.Y., Academic Press: 275-364.

Hughes, G.M. and Singh, B.N. (1970). Respiration in an air-breathing fish, the climbing perch Anabas testudineus Bloch. I. Oxygen uptake and carbon dioxide release into air and water. J. Exp. Biol. 53: 265-280.

Johansen, K. (1968). Air-breathing fishes. Sci. Am. 219: (4) 102-111.

Johansen, K. (1970). Air breathing in fishes. In: Fish Physiology, Vol. 4, edited by Hoar, W.S., and Randall, D.J. N.Y. Academic Press: 361-411.

Kramer, D.L. (1978). Ventilation of the respiratory gas bladder in Hoplerythrinus unitaeniatus (Pisces, Characoidei, Erythrinidae). Can. J. Zool. 56: 921-938.

Liem, K.F. (1963). The comparative osteology and phylogeny of the Anabantoidei (Teleostei, Pisces). Illinois Biol. Monogr. 30: 1-149.

Liem, K.F. (1967). Functional morphology of the head of the Anabantoid teleost fish Helostoma temmincki. J. Morph. 121: 135-158.

Liem, K.F. (1967). Functional morphology of the integumentary, respiratory and digestive systems of the synbranchoid fish, Monopterus albus. Copeia 1967: 375-388.

Liem, K.F. (1974). Evolutionary strategies and morphological innovations: Cichlid pharyngeal jaws. Syst. Zool. 22: 425-441.

Liem, K.F. (1978). Modulatory multiplicity in the functional repertoire of the feeding mechanism in cichlid fishes. J. Morph. 158: 323-360.

Lomholt, J.P. and Johansen, K. (1974). Control of breathing in Amphipnous cuchia, an amphibious fish. Respir. Physiol. 21: 235-340.

Lomholt, J.P. and Johansen, K. (1976). Gas exchange in the amphibious fish, Amphipnous cuchia. J. Comp. Physiol. 107: 141-156.

Luling, K.H. (1958). Uber die Atmung, amphibische Lebensweise und Futteraufnahme von Synbranchus marmoratus. Bonn. Zool. Beitr. 9: 68-94.

Misra, B.V. and Munshi, J.S.D. (1958). On the accessory respiratory organs of Anabas scandens. Fifteenth International Congress of Zoology, London, (Abstract: Title-32).

Munshi, J.S.D. (1962). On the accessory respiratory organs of Ophiocephalus punctatus (Bl.) and O. striatus, (Bl.). J. Linn. Soc. Lond. (Zoo) 44: 616-626.

Munshi, J.S.D. (1965). On the accessory respiratory organs of Trichogaster fasciatus. Proc. Zool. Seminar (Vikram Univ., Ujjain, India): 119-124.

Munshi, J.S.D. (1968). The accessory respiratory organs of Anabas testudineus (Bloch) (Anabantidae, Pisces). Proc. Linn. Soc., Lond. 179: 107-126.

Munshi, J.S.D. (1976). Gross and fine structure of the respiratory organs of air-breathing fishes. In: Respiration of Amphibious Vertebrates edited by Hughes, G.M. N.Y. Academic Press: 73-104.

Munshi, J.S.D. and Singh, D.N. (1968). On the respiratory organ of Amphipnous cuchia (Ham. Buch.) J. Morph. 124: 423-444.

Osse, J.W.M. (1969). Functional anatomy of the head of the perch (Perca fluviatilis L.): An electromyographic study. Netherlands J. Zool. 19: 289-392.

Peters, H.M. (1978). On the mechanism of air ventilation in anabantoids (Pisces: Teleostei). Zoomorphologie 89: 93-123.

Randall, D.J. and Hochachka, P.W. (1978). Water-air breathing transition in vertebrates of the Amazon. Can. J. Zool. 56: 713-1016.

Riggs, A. (1979). The Alpha Helix expedition to the Amazon for the study of fish bloods and hemoglobins. Comp. Biochem. Physiol. 62A: 1-271.

Singh, B.N. and Munshi, J.S.D. (1965). Accessory respiratory organs in Amphipnous cuchia. Proc. Zool. Seminar, Vikram Univ. Ujjain, India, 125-133.

Wu, H.W. and Liu, C.K. (1940). The buccopharyngeal epithelium as the principal respiratory organ in Monopterus javanesis Sinensia 11: 221-239.

OSMOTIC AND IONIC REGULATION BY FRESHWATER AND MARINE FISHES

David H. Evans

Department of Biology, University of Miami

Coral Gables, FL 33124, USA

INTRODUCTION

The cellular metabolism which defines life is apparently only possible in a milieu whose ionic and organic solute concentrations are relatively consistent since all organisms which have been examined regulate both the osmotic and ionic concentrations of their cells rather carefully. The constraints of this regulation result in a limited tolerance to salinity variation by most aquatic animals and plants. Because of their evolutionary history, the vertebrates in general, and the fishes in particular, show some nearly unique patterns of osmotic and ionic regulation. Moreover, some species of fishes are able to adjust their patterns to allow tolerance of rather wide salinity variation. Thus, investigation of osmotic and ionic regulation in fishes has been rather extensive; however, as will become obvious, we are still far from knowing the underlying mechanisms of this regulation.

Like all multicellular animals and plants, the osmotic and ionic concentration of fish cells is actually relatively well buffered against changes in the external salinity because of the presence of an extracellular fluid whose solute concentrations can be maintained rather stable due to epithelial transport mechanisms (kidney, gut, gill, and possibly skin). However, the epithelial mechanisms are not perfect so that extracellular - fluid solute concentrations may change when salinities vary, with the resulting need for cellular regulation by the plasma membrane. Fish cellular volume regulation has been relatively neglected but it appears that, like the invertebrates (Lange, 1972), fishes actually maintain cell volume consistency in the face of changing extracellular fluid concentration by variations in the cellular free amino acid levels (Schmidt-Nielsen, 1977). However, Cala (1977) has shown recently that flounder erythrocytes exposed to variation in the tonicity of the surrounding culture medium regulate their volume by alterations in relative ionic permeabilities of the plasma membrane. The permeability changes appear to be the result of changes in the electrical potential across the membrane and lead to net movements of ions and water down the prevailing electrochemical and osmotic gradients.

Most of this discussion of fish osmotic and ionic regulation will center around epithelial regulation of extracellular ionic concentrations in the face of rather substantial osmotic and ionic gradients across the permeable surfaces of the fish epithelium. Most evidence indicates that the major site of osmotic and ionic permeability is the branchial epithelium. Motais et al. (1966) found that the efflux of tritiated water measured from intact eels, Anguilla anguilla, could be accounted for by branchial clearance alone, so it appears that the skin is rather water impermeable. In addition, recent perfused head preparations of rainbow trout, Salmo gairdneri, in fresh water (Payan et al., 1975; Payan, 1978), and gill irrigation of the same species in sea water (Greenwald et al., 1974) indicate that most of the isotopically measured Na fluxes can be accounted for by the branchial epithelium. However, it should be noted that no extensive, systematic study of the relative roles of the gill and skin in fish osmoregulation has been published. Since they are the most prevalent and certainly best studied ions, we will be most concerned with the regulation of the Na^+ and Cl^- content of the body fluids of fish.

With the exception of the myxinid hagfishes, all fish (in fact all vertebrates) maintain their body fluid Na^+ and Cl^- concentration equivalent to approximately 30% sea water (Holmes and Donaldson, 1969; Evans, 1979a). Thus, all freshwater fish face an imbalance caused by diffusional loss of Na^+ and Cl^- and osmotic gain of water, while jawed, marine species must balance a net diffusional gain of NaCl and (except the chondrichthyan fish) an osmotic loss of water. The sharks, skates, rays and chimeras are an exception to the latter since they have avoided the osmotic problems of marine life by retaining exceptionally high concentrations of urea (and trimethylamine oxide) in their body fluids. Extra-cellular osmotic and ionic regulation by fish is therefore the use of various epithelial-based physiological mechanisms to offset or balance these net passive osmotic and ionic movements.

While the vast majority of fish species are able to regulate their body fluids (and secondarily cell volumes) in a relatively restricted salinity range (stenohaline), a few species are able to make substantial salinity transitions (euryhaline) and actually alter their regulatory mechanism to meet the reversed net movements of water and NaCl.

Osmotic and ionic regulation in fish has been reviewed rather extensively in the past few years and the reader should consult Maetz (1974a) and Evans (1979a) for general reviews and Maetz and Bornancin (1975), Maetz et al. (1976), Potts (1977), Kirschner (1977, 1979b) and Evans (1979b) for more specific discussions.

FRESHWATER TELEOSTS

The osmotic and ionic regulatory patterns employed by freshwater teleosts were first proposed by Smith (1930) and Krogh (1939). To balance the osmotic influx of water the fish produces relatively large volumes of urine secondary to a substantial glomerular filtration rate and minimal tubular water reabsorption (Evans, 1979a). The urine is rather dilute, but not distilled, so that renal NaCl loss adds to the diffusional loss of salt. Krogh (1939) demonstrated that this renal and diffusional salt loss was balanced by the active extraction of NaCl from the environment. The effector organs of osmotic and ionic regulation by freshwater teleosts are therefore the renal complex (kidney tubules plus urinary bladder) and the branchial epithelium.

Effector organs

a. Renal Complex: The renal complex of the freshwater teleosts usually is comprised of many thousands of nephrons with a substantial glomerulus, a ciliated neck region of variable length, proximal segments I and II, a ciliated intermediate segment, a distal segment, a collecting tubule and a collecting duct which, sometimes, drain into a urinary bladder of mesodermal origin (Hickman and Trump, 1969). The pathways for renal salt and water movement have been discussed and diagrammed by Hickman and Trump (1969) and Evans (1979a) and are proposed based upon mammalian homologies and studies of the urine composition of non-teleostean groups which lack various segments (see below). Briefly, it is supposed that the urine which has been filtered across the glomerulus is moved into the first proximal segment via ciliary action of the neck segment. (The neck segment is lost in the amniotes, which have a relatively higher blood pressure, Hickman and Trump, 1969). The first proximal segment is the site of iso-osmotic Na^+ and Cl^- removal and possible divalent ion secretion. The second proximal segment is probably the site of the majority of divalent ion secretion while the distal segment, collecting tubule, collecting duct and bladder are the site of net reabsorption of Na^+ and Cl^- with little concomitant, reabsorption of water. Contrary to what might be expected, the distal segment is apparently not a prerequisite for freshwater survival. Some euryhaline species such as *Fundulus heteroclitus* and *Gasterosteus aculeatus* and even the aglomerular *Opsanus tau* and *Microphis boaja* are able to invade the freshwater environment without possessing a distal tubule. One must therefore suppose that the collecting tubule and duct (which are present in these species) are major sites for final urine dilution.

The importance of the teleost urinary bladder in osmotic and ionic regulation has only been appreciated in the last few years. When it is present, it is the site of active reabsorption of both Na^+ and Cl^- with water following osmotically (Hirano et al., 1973, Renfro, 1975, Fossat and Lahlou, 1977). In freshwater teleosts it appears that the bladder reabsorbs a urine-hyperosmotic fluid due to a rather low osmotic permeability (Hirano et al., 1973). Since, in isolated bladders, both Na^+ and Cl^- are absorbed in the absence of any electrochemical gradient (under short-circuited conditions), it appears that both ions are actively absorbed (Renfro, 1975). Addition of ouabain or removal of K^+ from the serosal (blood) side of isolated bladders results in the inhibition of net Na uptake (Renfro, 1975). Thus, Na^+-K^+-activated ATPase seems to mediate a Na^+/K^+ exchange system for Na^+ uptake. Interestingly, Cl^- uptake was inhibited during the same treatments so it appears that Cl^- uptake is linked to a Na^+/K^+ exchange by some presently unknown means.

b. Branchial epithelium: Krogh (1939) was the first to demonstrate that fish (and other freshwater organisms) can extract Na^+ and Cl^- independently from the freshwater environment and recent experiments which have functionally separated the branchial region from the rest of the body (Chester Jones et al., 1965; Kerstetter et al., 1970; Payan et al., 1975; Payan, 1978) indicate clearly that the branchial epithelium is the site of active ionic uptake.

Krogh (1939) suggested that, despite independent mechanisms for Na^+ and Cl^- uptake, approximate electroneutrality was maintained across the epithelium by the exchange of NH_4^+ for Na^+ and HCO_3^- for Cl^-. Clear evidence supporting this proposition came from the studies of Maetz and Garcia Romeu (1964) who demonstrated, using radioactive isotopes, that injection of either NH_4^+ or HCO_3^- into the peritoneal fluids of the goldfish, *Carassius auratus*, stimulated Na^+ and Cl^- influx, respectively, while addition of these ions to the external medium inhibited the uptake of Na^+ and Cl^- independently. The proposal for a Cl^-/HCO_3^- ionic exchange system was corroborated by studies of Kerstetter and Kirschner (1972) on *S. gairdneri* and it has recently been shown (DeRenzis and Bornancin, 1977) that the branchial epithelium of *C. auratus* contains a Cl^- and HCO_3^- activated ATPase. There are some indications that Na^+ uptake may be coupled to proton (H^+) efflux rather than NH_4^+ efflux. This conclusion was prompted by the finding that the ammonia efflux from *Cyprinus carpio* (deVooys, 1968) and *S. gairdneri* (Kerstetter et al., 1970) was unaffected by substantial changes in the external Na^+ concentration while there seemed to be a correlation between the rate of Na^+ uptake and rate of acidification of the medium surrounding *S. gairdneri* (Kerstetter et al.,1970). In addition, other studies (Kirschner et al., 1973) showed that the compound Amiloride (which is known to inhibit Na^+ movement in a wide variety of tissues (Cuthbert et al., 1979)) substantially inhibited Na^+ uptake and H^+ efflux in *S. gairdneri* with a rather smaller effect on NH_4^+ efflux. However, more recently, Payan (1978) has demonstrated clearly (utilizing an isolated, double-perfused trout head preparation) that Na^+ uptake is associated with NH_4^+ efflux. Maetz (1973) calculated that Na uptake could best be correlated with the sum of NH_4^+ and H^+ efflux and this is probably the most realistic conclusion from these rather disparate studies.

The extraction of Na^+ and Cl^- from the medium by freshwater teleosts is indeed active transport since it is against electrochemical gradients. The blood concentration of Na^+ and Cl^- is distinctly above that of the environment (Holmes and Donaldson, 1969) and Maetz and Garcia Romeu

(1964) calculated that it would require transepithelial electrical potentials (TEPs) of nearly -200mV (inside relative to outside) for Na^+ and +200mV for Cl^- to maintain these two ions at such high concentrations in the blood relative to the environment. The few determinations of the T.E.P. across f.w. fish (Kerstetter et al., 1970; Maetz, 1974b; Eddy, 1975) indicate that it is of the order of -30 to -10mV in calcium-free freshwater (approximately 1mM NaCl) but becomes inside positive (approximately +10mV) when calcium is added to the freshwater. The electrical potential appears to be the result of both differential (and Ca^{++}-sensitive) ionic permeability (with P_{cation} greater than P_{anion}) and electrogenic Na^+ uptake (Maetz, 1974b).

The ionic uptake mechanisms are generally thought to be on the apical surface of the transporting cell since fish are able to extract both Na^+ and Cl^- from solutions below 1mM/l, well below any proposed cellular NaCl concentrations. The involvement of carbonic anhydrase to promote the production of both H^+ and HCO_3^- is implied by experiments which show that injection of the carbonic anhydrase inhibitor, acetazolamide (Diamox), is followed by a fall in both Na^+ and Cl^- influx in some (Maetz and Garcia Romeu, 1964; Kerstetter et al.,1970; Payan et al., 1975) but, importantly, not all (Kerstetter and Kirschner, 1972) teleost species studied to date. The mechanisms for the movement of Na^+ and Cl^- across the basal membrane into the blood are relatively unstudied. Na^+/K^+ exchange is suggested because Richards and Fromm (1970) found that ouabain inhibited Na uptake by isolated, perfused gills of S. gairdneri and Shuttleworth and Freeman (1974) showed that removal of K^+ from the perfused solution was followed by a decline in the Na^+ influx into isolated, perfused gills of A. anguilla.

The actual cell type involved in ionic uptake is unknown. Keys and Willmer (1932) described a specialized, silver-nitrate staining cell (termed the "chloride cell") but its distribution in various salinities and histochemistry indicate that it is probably involved with ion extrusion in marine teleosts rather than ionic uptake by freshwater teleosts (see below). However, recent work (Kikuchi, 1977) indicates that, at least in some freshwater species, the chloride cells may be quite active.

c. The gut: Smith (1930) found that the eel, A. rostrata, did not ingest the medium when in fresh water which makes osmotic sense for the organism. However, more recent studies (see Evans, 1979a for relevant literature) have shown that teleosts do ingest freshwater, but the drinking rates are generally far below that described for marine teleosts. Whether this osmotically inappropriate action is a consequence of the irrigation of the gills for respiratory gas exchange is unknown. The animals used in these experiments were not fed so it is unlikely that intake of water was secondary to feeding movement. While drinking is an obvious osmotic detriment in freshwater one must be reminded that it can be offset by an increase in water clearance by the renal complex. This is an example of the complex feedback interrelationships of the effector organs of the teleost's osmotic and ionic regulatory system; an overshoot or inappropriate response by one can be offset by another effector system.

MARINE TELEOSTS

The physiological patterns utilized by marine teleosts to offset the osmotic loss of water and diffusional gain of NaCl were first proposed by Smith (1930). Smith showed that the seawater adapted A. rostrata ingested sea water and absorbed much of the gut NaCl and water across the enteric epithelium. Most of the divalent ions remained in the gut fluids and were excreted via the anus; those absorbed were excreted via the kidney. The monovalent ions were almost completely absorbed. While this oral ingestion is obviously adaptive relative to osmotic regulation, it presents a NaCl load to the animal since the sea water is of the order of three times as salty as the body fluids. Smith showed that the marine teleost kidney is not capable of producing a urine that is hyperosmotic to the blood so he proposed that a net secretion of salt must take place through some extrarenal pathway. Subsequent experiments have determined that the site of extrarenal salt extrusion is the branchial epithelium (Keys, 1931; Kirschner et al., 1974; Shuttleworth, 1972, 1978). Thus, the effector organs of osmotic and ionic regulation by marine teleosts are the renal complex, gut and branchial epithelium.

Effector organs

a. Renal Complex: The marine teleosts renal complex is unable to produce a urine that is hyper-osmotic to the blood because it lacks the Loop of Henle which is present in those vertebrates (birds and mammals) which are able to reabsorb more water than salt under conditions of dehydration. The renal complex of the marine teleost usually consists of the segments already described for the freshwater teleost with the exception that the distal segment is usually missing and the glomeruli are sometimes reduced and are actually missing in the 23 or so species which are termed "aglomerular" (Hickman and Trump, 1969). The sea horse (Hippocampus), the toadfish (Opsanus) and the goosefish (Lophius) are good examples of aglomerular fish. Hickman and Trump (1969) and Evans (1979a) have diagrammed the supposed pathway for salt and water movement across the marine renal tubules, and these involve proximal and collecting tubular reabsorption of NaCl and water (isotonically) with proximal tubular secretion of divalent ions such as Mg^{++}, $SO_4^{=}$ and Ca^{++}, with water following osmotically. The final urine concentration can be adjusted by the urinary bladder (when present) which has a relatively higher osmotic permeability that in freshwater teleosts (Hirano, et al., 1973) so that volume reduction but little solute reduction takes place. In aglomerular fish, urine formation is secondary to secretion of monovalent and divalent ions across the proximal tubule with water following osmotically.

The relatively low glomerular filtration rate of marine teleosts is presumably secondary to either a reduced renal blood flow or to a reduction in functioning glomeruli. The latter may be a result of the evolutionary reduction of the number of glomeruli or to physiological variation of the number of perfused glomeruli (glomerular intermittancy) (Lahlou, 1967). Unfortunately, there is no recent systematic study of these two alternatives and their role in maintenance of a relatively low urine flow in marine teleost species.

There is also no systematic study of the role of the urinary bladder in determining the final volume and concentration of the urine of various marine teleosts. However, recent studies have shown that euryhaline species that possess a bladder actually increase the osmotic permeability of their urinary bladder epithelium when in sea water. The result is removal of an approximately blood iso-osmotic NaCl solution with a concomittant reduction in urine volume without major changes in the urinary solute concentration (Hirano, 1975). The importance of the urinary bladder is demonstrated by the recent work of Howe and Gutknecht (1978) who showed that urinary bladder reabsorption of needed water by O. tau amounts to 10% of the osmotic loss of water in this species. In other words, without urinary bladder reabsorption of water this species would have to increase its oral ingestion of water (and therefore, secondarily the branchial extrusion of salt) by 10%.

b. The Gut: The function of the gut of the marine teleost fish is analogous to the function of the freshwater teleosts renal complex: to balance the osmotic movement of water across branchial membranes. Evans (1979a) has presented representative rates of oral ingestion and it is clear that, while rates are variable when species are compared, oral ingestion may be substantial and may present a significant NaCl load to the organism which must be excreted (see below). It is interesting to note that the extremely low drinking rates of the intertidal blennioid fish Xiphister and Pholis are correlated with rather low water permeabilities (Evans, 1967, 1969a). This is to be expected if the oral ingestion is the response to osmotic loss of water. Hickman (1968) has presented the most complete data on the fate of ingested ions and water. His study of the Southern Flounder, Paralichthys lethostigma, supports the earlier conclusions of Smith (1930) that the vast majority of Na^+, Cl^- and water is reabsorbed by the gut epithelium with Mg^{++} and $SO_4^=$ remaining to be excreted via the anus. The Na^+ and Cl^- are then excreted principally via extra-renal mechanisms. Ca^{++} seems to be intermediate with some reabsorption from the gut and some excretion via the anus. Reabsorbed Ca^{++} is excreted via renal and, apparently, extrarenal mechanisms. The NaCl load presented by gut reabsorption is not trivial; it may, in fact, represent the dominant Na^+ load presented to the marine teleost fish (see below).

While the majority of NaCl and water is absorbed by the intestinal epithelium (Shehadeh and Gordon, 1969; Hickman, 1968) it has recently been discovered that the esophagus may also participate in salt removal from the ingested sea water (Kirsch and Laurent, 1975; Hirano and Mayer-Gostan, 1976). However, esophageal salt absorption is apparently down diffusion gradients.

That water absorption from the gut is secondary to monovalent ion absorption is shown clearly by the fact that a chemical inhibitor of Na^+ transport (ouabain) also inhibited water uptake from the lumen (Huang and Chen, 1971). In addition, net water movement into the lumen took place in isolated guts perfused with a blood-iso-osmotic sucrose solution in the lumen (Lotan and Skadhauge, 1972).

The actual mechanisms of active NaCl uptake are unknown at present. The ionic uptake mechanisms seem to be linked since, in perfused gut preparations, removal of either Na^+ or Cl^- from the perfusion fluid resulted in a decline in the rate of uptake of the other ion (Huang and Chen, 1971; Ando and Kobayashi, 1978; Field et al., 1978). The ionic transport system(s) are apparently coupled to Na^+-K^+-activated ATPase since addition of ouabain inhibited Na^+ and water uptake by the eel intestine (Oide, 1967) as well as the trans-epithelial potential and net Cl^- flux across the isolated intestine of the flounder, *Pseudopleuronectes americanus* (Field et al., 1978). The transepithelial potential (TEP) across intestines is usually of the order of a few millivolts, serosal (or blood) side negative (Ando et al., 1975; Ando and Kobayashi, 1978; Field et al., 1978). Field et al. (1978) have recently proposed that NaCl uptake is via a coupled transport mechanism on the mucosal (lumen) membrane followed by transport of Na^+ into the lateral, intercellular spaces which are continuous with the serosal extracellular fluids. Cl^- follows electrically, and moves into the serosal fluids, while Na^+ is recycled into the mucosal fluids through Na^+-selective, leaky, "tight-junctions". Thus, under short-circuited conditions, net Cl^- flow predominates over net Na^+ flow (Ando et al., 1975; Field et al., 1978). However, under open-circuited conditions, the serosal negativity draws Na^+ into the serosal fluids electrically.

The fate of ingested Ca^{++} is especially interesting. Rectal loss cannot account for more than 31% of that ingested (Hickman, 1968) so the rest must be absorbed and excreted renally and branchially. However, renal excretion can only account for 11% of that absorbed from the gut. The mechanisms for the branchial excretion of the remaining 89% of the absorbed Ca^{++} are unknown; however, most marine teleosts are blood positive to sea water (see below) so it may be possible that they face a net diffusional loss of Ca^{++} which balances this Ca^{++} load presented by gut absorption.

c. Branchial Epithelium: Recent data on gill-irrigated fish (Kirschner et al., 1974), isolated heads (Girard, 1976), isolated gills (Shuttleworth, 1978), and isolated opercular skin (Karnaky, et al.,1977) demonstrate that extrarenal salt extrusion by marine teleosts is, indeed, across the branchial epithelium. However, there is also evidence that, at least in the goby, *Gillichthys mirabilis*, jaw-skin may also be involved (Marshall, 1977). Despite a rather extensive literature, the mechanisms of Na^+ and Cl^- extrusion are still relatively unclear. Recent reviews include those of Maetz and Bornancin (1975), Kirschner (1977, 1979), Potts (1977) and Evans (1979a,b).

While it is generally agreed that Cl^- is extruded by an active transport mechanism, there is some controversy over whether Na^+ is actively extruded or actually maintained in electrochemical equilibrium passively. Both ions are maintained at levels below 200mM in the fish blood compared with approximately 500mM in the sea water, but when discussing distribution of ions, one must consider electrochemical rather than merely chemical gradients. The electrical potential (termed equilibrium potential) which can account for a chemically unequal distribution of a given ionic species across a membrane is defined by the Nernst Equation:

$$V = RT/zF \ln\frac{Cm}{Cb}$$

Where V is the electrical potential in volts (transepithelial potential or TEP), RT/zF have their usual values which reduce to 0.026 volts (26mV) at 23°C, and C_m/C_b is the ratio of the concentration of the ion (more properly, activity) in the medium and the blood, respectively. The sign of the TEP will be blood relative to medium (i.e., a positive sign indicates that the blood of the fish is positive to the medium). Given the normal marine teleost blood Na^+ and Cl^- concentrations vs. their respective concentrations in sea water the equilibrium potential for Na^+ is of order of +26mV while that for Cl^- is approximately -30mV. Kirschner (1977, 1979) and Evans (1979a,b) have recently compiled the published data on the TEP across marine teleosts and it is clear that all marine teleosts studied to date maintain TEPs far different from the equilibrium potential for Cl^-. However, some 11 species maintain TEPs near to the Na^+ equilibrium potential while 7 other species have TEPs much below the Na^+ equilibrium potential. It, therefore, appears that one must invoke active Cl extrusion mechanisms in marine teleosts, but active extrusion of Na^+ is still debatable (compare Evans, 1979a,b and Kirschner, 1977, 1979).

Despite the apparent near equilibrium of Na^+, one must remember that marine teleosts drink sea water, which constantly presents a Na^+ load to the animal. In fact, the oral ingestion rate may approach 25-50% of the isotopically measured Na^+ influx (Evans, 1979a). To extrude this added Na^+ load passively the fish blood would have to be distinctly more positive to sea water than the equilibrium TEP and this has been found only in the euryhaline, freshwater species *Tilapia mossambica* (Dharmamba et al., 1975).

As one might expect, most evidence supports the conclusion that Na^+ extrusion is coupled to K^+ influx via Na^+-K^+-activated ATPase. The enzyme is present in branchial tissue and usually increases in specific activity during seawater acclimation (Jampol and Epstein, 1970, Maetz and Bornancin, 1975). In addition, Maetz (1969) demonstrated that Na^+ efflux was stimulated by external K^+ and this was corroborated by various other studies (see Evans, 1979a,b for reviews). However, Potts et al. (1973), House and Maetz (1974) and Kirschner et al. (1974) suggested that the K^+ stimulation of Na^+ efflux was secondary to alterations of the TEP. Other evidence indicates that at least some of the K^+-stimulated Na^+ efflux is exclusive of changes in the TEP (Evans et al., 1974; Maetz and Pic, 1975; Evans and Cooper, 1976; Fletcher, 1978). Finally, externally applied or injected ouabain inhibits Na^+ efflux and results in an increased blood Na^+ concentration (Motais and Isaia, 1972, Silva et al., 1977a).

However, as pleasing as the model for Na^+-K^+-activated ATPase-driven Na^+ extrusion is, there are sufficient negative data extant to preclude a positive statement about its viability at the present time. For instance, the model would propose that the enzyme is located on the apical border of the transporting cell. The chloride cells first described by Keys and Wilmer (1932) are generally accepted as the transporting cell because of histological changes during the acclimation to sea water (Maetz and Bornancin, 1975). Recent elegant histochemical localization of Na^+-K^+-activated ATPase in these cells (Karnaky et al., 1976; Hootman and Philpott, 1979) shows clearly that the enzyme is predominantly on the borders of the basal-lateral tubular invaginations of the chloride cells rather than the apical membrane. In addition, ouabain injection was much more effective in inhibiting gill Na^+-K^+-activated ATPase than external application of the drug (Silva et al., 1977a). Thus, the enzyme is apparently in the wrong position in the cell to extrude Na^+ in exchange for seawater K^+. Unfortunately, at the present time this rather clear cytological evidence cannot be reconciled with the physiological evidence for Na extrusion (see above).

While it is generally accepted that Cl extrusion by marine teleosts is active, the mechanisms involved are relatively unknown. However, there is good evidence that the extrusion mechanism is somehow functionally connected to Na^+-K^+-activated ATPase. Epstein et al. (1973) and Maetz and Pic (1975) showed that at least *A. anguilla* and *Mugil cephalus* possess a K^+-stimulated Cl^- as well as Na^+ efflux and that the inhibition of Cl^-efflux by thiocyanate (SCN) also inhibited K^+ stimulation of Na^+ efflux. In addition, Silva et al. (1977a) showed that ouabain injected into *A. rostrata* inhibited both Na^+ and Cl^- efflux as well as Na^+-K^+-activated ATPase. Silva et al. (1977a) propose that basal, gill Na^+-K^+-activated ATPase maintains favorable electrochemical gradients into the cell for Na^+. The passive movement of Na^+ from the blood into the cell somehow draws Cl^- up its electrochemical gradient. The Cl^- is then extruded passively across the apical surface of the Cl^- cell down its electrochemical gradient. Thus, the final extrusion of Cl^- is indirectly linked to basal Na^+/K^+ exchange. This model also proposes that Na^+ is in electrochemical equilibrium between the blood and sea water-a controversial assumption as discussed above. However, it does involve a basal rather than apical Na^+-K^+-activated ATPase, which is supported by the histochemistry (see above).

Kormanik and Evans (1979) have recently shown that in the gulf toadfish, *Opsanus beta*, Cl^- extrusion is apparently linked to the uptake of HCO_3^-. They found that external HCO_3^- could stimulate Cl^- efflux exclusive of changes in the T.E.P. A recently described *in vitro* system may help to clear the rather cloudy picture of Na^+&Cl^--extrusion by marine teleosts. Karnaky and Kinter (1977) have found that the opercular lining of F. heteroclitus contains chloride cells (50%-70% of the total cell population) and can be easily isolated and mounted as a sheet in the classical method first described for frog skin by Ussing's laboratory (Koefoed-Johnson and Ussing, 1958). Karnaky et al. (1977) and Degnan et al. (1977) found that this tissue actively extruded Cl^- from the blood into the outside medium. Further, Cl^- extrusion was inhibited by ouabain. Na^+ was maintained in electrochemical equilibrium in this preparation. Surprisingly, tissue from freshwater-acclimated F. heteroclitus also extruded Cl^- and, once again Na^+ was not moved actively. The data from f.w. fish are especially interesting since in vitro studies showed clearly that F. heteroclitus actively extracts both Na^+ and Cl^- from freshwater (Potts and Evans, 1967). The efflux of Cl^- by the opercular epithelial sheet isolated from 100% seawater-acclimated fish was stimulated by increasing the HCO_3^- concentration of the bathing media but it was unaffected by the addition of the carbonic anhydrase inhibitor acetazolamide (Diamox). Thus, it appears that, if Cl^-/HCO_3^- does take place in this system, it is not limited by the availability of HCO_3^- produced by the hydration of CO_2. This isolated opercular epithelium preparation is especially important for future research on fish ion transport since it enables the manipulation of both the blood and medium side of the tissue without the problems of hormonal or cardiovascular changes found in whole animal studies. In addition, it indicates that the gills themselves may not be the only site for ion regulation in teleosts. In fact, Marshall (1977) has shown that the isolated, lower jaw skin of the mudsucker, Gillichthys mirabilis is also able to actively extrude Cl^-. The extrusion is inhibited by ouabain which again supports a link between Cl^- extrusion and Na^+-K^+-activated ATPase.

Besides the readily apparent diffusive and active transfer pathways across the branchial epithelium there appear to exist at least two other pathways for ionic movement. Based upon the fact that the efflux of either Na^+ of Cl^- was grossly affected by the absence of the same ionic species in sea water, Motais et al. (1966) proposed that some of the isotopically measured Na^+ and Cl^- efflux from marine teleosts is coupled with Na^+ and Cl^- from the sea water in a Na^+/Na^+ and Cl^-/Cl^- ionic exchange-diffusion system. Since the coupling is believed to be 1:1 it would not result in a net movement of either ion across the epithelium - only a net movement of isotope. Such exchange-diffusion (or self-exchange) has been described for other tissues (Glynn and Karlish, 1975) and is generally thought to be a non-energy demanding artifact of an imperfect ionic exchange system such as Na^+/K^+ or Cl^-/HCO_3^-. For instance, while the affinity of the transport moiety for K^+ is great on the seawater side of a fish gill, there is so much Na^+ in the surrounding medium that it sometimes outcompetes the K^+ for the transport site and Na^+/Na^+ exchange diffusion results. The presence of at least Na^+/Na^+ exchange diffusion is debated because in S. gairdneri changes in the TEP can account for much of the fall in Na efflux after the removal of external Na (Kirschner et al., 1974). However, other studies have shown that there exists a Na^+/Na^+ component exclusive of TEP changes (Maetz and Pic, 1975; Evans and Cooper, 1976.

Rather surprisingly, there is now substantial evidence that the freshwater Na^+ uptake carrier system (Na^+/NH_4^+) is also functioning in the teleost branchial epithelium when the fish is acclimated to sea water. Evans (1973, 1975a) found that Na influx into the euryhaline molly, *Poecilia latipinna*, acclimated to sea water, displayed saturation kinetics and was inhibited by increasing the external NH_4^+ concentration. Similar data have now been published for the marine pinfish, *Lagodon rhomboides* (Carrier and Evans, 1976) and, most recently, Evans (1977) has shown that NH_4^+-inhibited Na^+ uptake is carried out by 4 species of marine teleosts and the ammonia efflux from 3 of these species is partially dependent upon external Na^+. In addition, in *O. beta*, injection of NH_4^+ stimulated both Na^+ uptake and an external-Na^+-dependent efflux of NH_4^+. Finally, in the same species, a drastic increase in the external NH_4^+ concentration could actually reverse the ionic exchange mechanism so that the Na^+ efflux (measured radio-isotopically) was increased significantly. Thus, more evidence has been published supporting Na^+/NH_4^+ exchange in marine teleosts than in freshwater teleosts'. Evans (1975b) has proposed that, despite the added Na^+ load presented by this ionic exchange system to the marine teleost, it has been retained because of its importance in nitrogenous waste excretion and acid /base regulation. The magnitude of the Na^+ load presented by Na^+/NH_4^+ exchange in marine teleosts will depend upon the stoichiometry of exchange, as well as the oral and diffusional gain of Na^+ by a particular species. Evans (1979c) has computed that the Na load presented to *O. beta* by Na^+/NH_4^+ exchange is only of the order of 6% that produced by electrochemical gradients and oral ingestion of sea water.

What is of particular interest in this finding of the Na^+/NH_4^+ exchange in marine teleosts is that it now appears that they possess the ionic regulatory system for at least Na^+ balance in freshwater. Since movement into freshwater involves quantitative changes in renal function and permeability and a single qualitative change (i.e., initiation of active ionic uptake) one might argue that something other than the lack of the latter precludes euryhalinity in most teleosts. Carrier and Evans (1976) have shown that at least one of the factors may be external Ca^{++} concentrations. They found that *L. rhomboides* carries on Na^+/NH_4^+ exchange when in sea water but can only acclimate to salinities below 25mM NaCl/l if the external Ca^{++} concentration is maintained at 5-10mM/l. This finding is ecologically relevant since it has been known for many years that some normally stenohaline marine teleosts can tolerate freshwater if sufficient Ca^{++} is present (Breder, 1934).

ELASMOBRANCHS

We know much less about the mechanisms of osmotic and ionic regulation utilized by the elasmobranchs, and the other subclass of the Chondrichthyes, the holocephalans, are nearly unstudied. This is primarily due to the difficulty of laboratory maintenance of sharks, skates and rays, but secondarily, to their extremely low permeability to ions, which precludes many of the rapid-change isotopic experiments which have enabled preliminary delineation of various branchial effector systems in the teleosts. In fact, this extremely low ionic permeability is one of the hallmarks of elasmobranchian ionic regulation. Various isotopic studies have shown that the epithelial membranes of the typical elasmobranch are of the order of 1% as permeable to Na^+ as those of the marine teleost.

The vast majority of elasmobranchs are marine with only a very few species able to enter brackish or freshwater. A single genus of ray, Potamotrygon, is the only member of the Elasmobranchii actually a permanent resident of freshwater. In fact, Potamotrygon, cannot tolerate salinities greater than 58% S.W. (Griffith, et al., 1973). Like the teleosts, the elasmobranchs characteristically possess a blood NaCl concentration far below that of sea water. This logically argues for a freshwater origin for this group but the fossil record is unclear. However, unlike the teleosts, the elasmobranchs are iso- or slightly hyperosmotic to their seawater environment due to their retention of the organic solutes, urea and trimethylamine oxide. This retention of urea and TMAO is a characteristic of chondrichthyan osmoregulation since it is also found in the Holocephali. Smith (1931b) was the first to propose that it was used as an osmoregulatory device to reduce osmotic loss of water from the animal to zero.

Maintenance of high blood urea (and presumably TMAO) levels by elasmobranchs is a combination of extremely low branchial epithelial permeability and nearly complete reabsorption in the renal apparatus (95% vs 40% for mammals) (Forster, 1967). It appears that renal reabsorption of urea is secondary to Na reabsorption (Schmidt-Nielsen, et al., 1972).

The consequence of urea retention and the concomittant iso- or even hyperosmolarity to sea water are fundamental to the typical elasmobranch pattern of osmoregulation, which are distinctly different from those already described for the teleosts. Since the fish are in osmotic equilibrium with their environment (or may even be gaining some water osmotically) there is no need to ingest the medium as the teleosts do. Unfortunately, only a single determination has been published, but Payan and Maetz (1971) found that the cat shark, Scyliorhinus canicula, only ingested something of the order of 0.01% of the body weight per hour, 1-10% that found for most teleosts. In addition, since osmotic water loss is not a problem, renal water loss is not as restricted as it is in the teleosts. Elasmobranchs have well developed glomeruli and both glomerular filtration and urinary flow rates are in the range of those found in freshwater rather than marine teleosts. Lastly, elasmobranchs do not maintain a restricted epithelial water permeability as teleosts do (Evans, 1979a). The structural modifications which maintain this rather unique high water permeability and low urea and Na^+ permeability are unknown.

Despite their neat solution of the osmotic problems of maintenance of a reduced blood salt concentration in sea water, marine elasmobranchs still face a theoretical influx of both Na^+ and Cl^- down their respective concentration gradients. The data from Bentley et al. (1976) on Scyliorhinus canicula and from our laboratory (Evans, 1979b) indicate that the electrical potential across the epithelium of elasmobranchs is only of the order of 1-4mV (blood side negative), so it appears that neither Na^+ and Cl^- are in electrochemical equilibrium. Thus, despite their extremely low epithelial ionic permeability, marine elasmobranchs do face a net influx of Na^+ and Cl^-. Interestingly enough, it is unclear what role various effector organ pathways play in excretion of these excess ions.

Effector organs

a. Renal Complex: The renal complex of elasmobranchs consists of well developed glomeruli, a long neck segment, proximal tubules I and II, distal tubule and a collecting tubule and duct. (Hickman and Trump, 1969). A urinary bladder has not been described. Like all lower vertebrates, the elasmobranchs lack a loop of Henle in their renal complex and so, therefore, are unable to produce a urine more concentrated than their body fluids; thus, excretion does not provide a balance for the net diffusional influx of salts. The proposed sites for renal solute and water reabsorption and secretion have been diagrammed by Hickman and Trump (1969) and are basically those described above for the freshwater teleost, with the addition that urea and TMAO are probably reabsorbed in the proximal segment II.

b. Rectal Gland: Burger (1962) cannulated the rectal gland of the dogfish Squalus acanthias and determined that the fluid secretions were iso-osmotic to the blood but contained concentrations of Na^+, Cl^- and K^+ that were distinctly above those found in the body fluids. Thus, it is obvious that the production of a fluid by this specialized structure can represent the site of extrarenal salt extrusion which is necessary for ion regulation in marine elasmobranchs. However, the rectal gland is apparently not the sole (or possibly even major) site of salt extrusion. Burger (1965) found that S. acanthias could maintain near normal blood Cl^- levels after rectal gland extirpation and Chan et al. (1967) showed that the lip shark, Hemiscyllium plagiosum, maintained normal Na^+ blood concentrations and Na^+ efflux after removal of the rectal gland.

The mechanisms for ionic extrusion by the rectal gland are still poorly understood. The gland tissue is extremely high in Na^+-K^+-activated ATPase (Bonting, 1966) so one might propose that Na^+/K^+ ionic exchange is involved but this has not been proven experimentally. Recently, investigators at the Mt. Desert Island Biological Laboratory in Maine have developed an isolated, perfused rectal gland preparation which has enabled preliminary delineation of the physiological steps in the elaboration of the Na^+ and Cl^--concentrated fluid (Silva et al., 1977b). They showed that both Na^+ and Cl^- were necessary for the fluid formation, and that the fluid secretory rate and fluid Na^+ and Cl^- concentrations fell after the addition of ouabain to the perfusion fluid. In addition, thiocyanate (SCN) and furosemide (both known, Cl^- transport inhibitors) also inhibited Cl^- secretion by the isolated gland. Since the lumen of the gland was negative to the blood, the authors propose that, like their model for salt secretion by the teleost branchial epithelium (Silva et al., 1977a), Cl^- enters the cell up its electrochemical gradient coupled through a membrane carrier to the downhill entry of Na^+. The Na^+

electrochemical gradient is maintained by basal Na^+-K^+-activated ATPase. Cellular Cl^- is then extruded across the apical membrane down its electrochemical gradient since the cellular contents are negative to the lumen. Na^+ is presumably concentrated in the lumen because of the latter's negative TEP relative to the blood.

c. Branchial epithelium: It appears that the branchial epithelium may represent a significant site for the net extrusion of Na^+ and Cl^- from marine elasmobranchs. There is no direct evidence to support this conclusion but various studies have shown that the sum of rectal and urinary loss of radioactive Na^+ can accound for something less than 50% of the total efflux (Burger and Tosteson, 1966; Maetz and Lahlou, 1966; Horowicz and Burger, 1968). No studies of the mechanisms for Na^+ and Cl^- extrusion by the branchial epithelium have been published but recent data from our laboratory indicate that, at least in the nurse shark, *Ginglymostoma cirratum*, the yellow stingray, *Urolophus jamaicensis*, and dogfish "pups", *Squalus acanthias*, the efflux of Na^+ is not coupled to uptake of any potential counter ion from sea water. This conclusion is supported by the finding that the efflux of radioactive Na^+ actually increased when these animals were transferred to Na^+ and K^+-free seawater (Evans and Kormanik, unpublished), rather than declining significantly as has been found for many species of teleosts (see above). Bentley et al. (1976) found that the Cl^- efflux from *S. canicula* in various saline solutions was always greater than could be accounted for by the electrochemical gradient and, therefore, postulated an active Cl^- efflux component. They could find no evidence for active Na^+ efflux. Thus, it is unclear what role the branchial epithelium plays in salt extrusion and even less clear what transport mechanisms are involved.

Surprisingly, there is some evidence that marine elasmobranchs, like marine teleosts (see above), actually extract Na from sea water in exchange for blood NH_4^+ and/or H^+ in order to excrete these unwanted ions. Payan and Maetz (1973) found that injection of ammonia into *S. canicula* stimulated Na^+ influx while injection of acetazolamide, known to inhibit production of H^+ and, secondarily, NH_4^+, led to a decline in Na^+ influx. Further, Bentley et al. (1976) showed that reduction of the external pH led to a fall in Na^+ influx in *S. canicula*. More recently, we have found that the efflux of titratable acid from the skate, *Raja erinacea*, is stopped when this species is placed into Na^+- and K^+-free sea water; however, ammonia efflux is unaffected (Evans et al., 1979). Thus, it appears that there may be species variation in the actual ionic exchange system involved, i.e., Na^+/NH_4^+ vs Na^+/H^+. It is important to note that, contrary to the situation in marine teleosts, this obligatory coupling between acid/base regulation, nitrogen excretion and Na^+ uptake may represent a significant Na^+ load to the organism because of low ionic permeability, characteristic of elasmobranchs. This obviously deserves further investigation.

Euryhaline and Freshwater Elasmobranchs

A few species of chondrichthyan fish are able to tolerate reduced salinities and Potamotrygon is actually a freshwater fish. Smith (1931a) found that species of elasmobranchs such as Pristis microdon (sawfish) and Carcharhinus melamopterus (Ganges Shark) in southeast Asia respond to dilute salinities by lowering their body NaCl and urea concentrations and producing larger volumes of a more dilute urine than their marine counterparts.

Goldstein and Forster (1971a) showed that the blood NaCl decline is probably at least in part secondary to the increased urine flow (despite reduced urinary NaCl concentrations). They found that in the ray, Raja erinacea, urinary Cl^- loss increased from 3.4 to $14 \mu M.100g^{-1}.hr^{-1}$ when this species was transferred to 50% sea water. One must also suppose that decreased diffusional gain of both ions could also result in a reduced blood ionic concentration.

The decline of blood urea in euryhaline elasmobranchs acclimated to lowered salinities appears to be chiefly secondary to a reduced rate of synthesis (Goldstein and Forster, 1971a; Payan et al., 1973; Wong and Chan, 1977). However, increased urea loss may also play a role since Goldstein et al. (1968) found that the ^{14}C-labelled urea clearance from the lemon shark, Negaprion brevirostris, trebled when this species was acclimated to 50% sea water.

This turned toward reduced urea retention is completed in the freshwater stingray, Potamotrygon, Thorson et al. (1967) showed that the urea concentration of the blood of this species is only of the order of 1mM/1. The ability to retain urea has apparently been lost by Potamotrygon since it is unable to increase its blood urea levels when it is placed in brackish water salinities (maximum 90% sea water) for periods up to 6.4 days (Thorson, 1970). This inability to maintain high urea levels in the blood is apparently secondary to a relatively low rate of synthesis coupled with high renal and branchial effluxes (Goldstein and Forster, 1971b).

Since Potamotrygon is distinctly hyperosmotic to its freshwater environment one might suppose that, like freshwater teleosts, it would maintain rather low branchial (skin?) salt and water permeabilities. In fact, Carrier and Evans (1973) did find that, like marine elasmobranchs, Potamotrygon, has an extremely low rate of Na^+ and Cl^- efflux and, therefore, presumably ion permeability. However, they also found that, unexpectedly, Potamotrygon, has retained the extremely high water permeability characteristic of its marine elasmobranch relatives. Potamotrygon has a water turnover rate of 96% per hour which means, simply, that nearly every water molecule in the fish's body will exchange with a water molecule of the medium in any given hour. A typical freshwater teleost may have a turnover rate of 10-20% per hour (Evans, 1969a) so it is plain that Potamotrygon has not evolved a low water permeability as an adaptation to a dilute medium. Despite this rather high water permeability, Carrier and Evans (1973) have calculated that, given the measured urine flow rates of other elasmobranchs in dilute salinities, Potamotrygon can theoretically excrete enough urine to balance the osmotic influx of water. The apparent inability of Potamotrygon to lower its water permeability is especially interesting since it has been shown that two euryhaline skates (Raja erinacea and Raja radiata) decrease their water permeability when placed into 50% sea water (Payan et al., 1973).

Unfortunately, we have no direct information on the ionic uptake mechanisms which are presumably acting to maintain ionic balance in Potamotrygon in freshwater. Pang et al. (1972) found that there was an influx of ^{22}Na from radioactively labelled freshwater solutions, which increased with increased salinities, but this may be diffusional rather than active uptake. It is likely that ionic uptake mechanisms for Na^+ and Cl^- extraction exist, as in all the other freshwater organisms which have been studied, since recent data indicate that marine elasmobranchs are extracting Na^+ from sea water via an Na^+/NH_4^+-H^+ ionic exchange mechanism (see above).

Finally, as might be expected, euryhaline sharks in freshwater have atrophied rectal glands (Oguri, 1964), as does Potamotrygon (Thorson et al., 1978).

CHONDROSTEAN FISH

The modern chondrostean fish (sturgeons, paddlefish and bichirs) are relics of the paleoniscoid grades of bony fish evolution which dominated the fossil record some 250-350 million years ago. While the sturgeons are euryhaline and are found in both marine and freshwater environments, the paddlefish and bichirs are stenohaline, freshwater species. Holmes and Donaldson (1969) have compiled the few data on the blood analyses of chondrostean species and it is apparent that these fish maintain ionic levels similar to freshwater or marine species of teleosts. However, blood calcium levels are usually lower than those described for teleosts and Urist (1966) (1967) have proposed that this may be secondary to the progressive reduction of bone during their evolution.

Chondrostean fish in fresh and seawater, therefore, face the same sort of osmotic and ionic problems as teleostean groups. Potts and Rudy (1972) have published the only study which compares some of the parameters of Chondrostean regulation with those found in teleosts in either fresh or sea water. They investigated two species of sturgeon, Acipenser medirostris and Acipenser transmontanus, in both fresh and sea water. In fresh water the urine contained only 12.5 mM Na/l and the isotopically determined efflux of Na^+ averaged 2.1% of the body Na^+ per hour, both in the same range as a typical freshwater teleost. The rate of water turnover was also near to that described for freshwater teleosts (Evans, 1979a). In seawater-acclimated individuals both the blood and urine Na^+ concentrations rose but urine levels were still distinctly below blood levels (5-40 mM/l). The turnover of Na^+ increased in sea water, while that of water decreased, again, in the typical teleostean pattern. Determinations of the transepithelial potential were not made so one cannot be certain that the sturgeons were maintaining Na^+ out of equilibrium with sea water.

So preliminary data indicate that the osmotic and ionic patterns of the chondrostean fish are basically similar to those more carefully defined for the teleostean fishes.

HOLOSTEAN FISH

The extant garfish and bowfin (Amia) are the only living relatives of the holostean grade of osteichthyan evolution which was prevalent some 250 million years ago. While the bowfin is apparently only found in freshwater there have been various descriptions of gar in brackish or sea water (Sulya et al., 1960) and Zawodny (1975), has maintained Lepisosteus platyrinchus in 75% sea water for up to three weeks. The few published determinations of the ionic concentrations in holostean blood indicate that this group maintains concentrations similar to those found in teleost blood (Holmes and Donaldson, 1969).

Zawodny (1975) has investigated the patterns of osmotic and ionic regulation of L. platyrinchus in both freshwater and 75% sea water. The rates of Na^+, Cl^- and water efflux (determined radioisotopically) were in the same range of those described for freshwater teleosts but, interestingly, urine levels of Na^+ and Cl^- were rather high (30mM Na^+/l, 24mM Cl^-/l) so that renal loss of these ions represents a rather more significant pathway (60% of the total) than commonly described for teleosts (Evans, 1979a). Transfer of this species to 75% seawater for approximately 3 weeks resulted in a substantial increase in blood Na^+ and Cl^- levels (34% and 82%, respectively) and a rather astounding fall in the total body water (51%). The animal apparently maintains an extremely low permeability to Na^+ so the efflux of Na^+ in 75% sea water was only $287 \mu M.100g^{-1}.hr^{-1}$, far below that described for marine teleosts (Evans, 1979a). The fish did ingest the medium in response to osmotic loss of water but, while it was measured to be in the same range as that described for marine teleosts (0.26% of the body weight per hour), it is obviously far below that required for osmoregulation since 50% of the total body water was lost. L. platyrinchus had a water turnover rate of 25% (of body water) per hour in freshwater and 14% in 75% sea water. These values are similar to those described for teleosts (Evans, 1979a). The increased water permeability in freshwater may possibly be accounted for by a fall in environmental calcium rather than intrinsic changes in the water permeable (presumably branchial) epithelial membranes. The same phenomenon has been described for teleosts (Evans, 1969b).

Hanson et al. (1976) found that the single representative of the other extant holostean group, Amia calva, was very stenohaline and could only survive in salinities that were hypo-osmotic to the blood (i.e., below 25% sea water). This species appears to be quite Na^+ impermeable with isotopically measured effluxes ranging from 0.03% to 0.1% per hour, far below that described for either freshwater teleosts or elasmobranchs (Evans, 1979a).

It should be obvious that much more needs to be learned about the patterns of osmoregulation utilized by holostean fish.

CROSSOPTERYGII AND DIPNOI

These fish represent the modern descendents of the line of fish evolution which gave rise to the Amphibia some 300 million years ago. The only living crossopterygian fish is the coelacanth, Latimeria chalumnae, which is apparently restricted to the marine environment in the neighborhood of the Comoro Islands off the coast of East Africa. The rather more specialized Dipnoi are represented by lungfish restricted to freshwater environments in Africa, South America and Australia. It is interesting to note that, while they are found in habitats presumably similar to those of the fish-amphibian transition, lungfish are off the main line of crossopterygian-amphibian evolution and must, therefore, be regarded as specialized cousins.

The few data available on the blood ionic levels of lungfish and the coelacanth indicate that, while the former display the typical teleostean pattern (Evans, 1979a), the coelacanth, like marine chondrichthyan fish, has evolved the retention of high blood urea levels to reduce the osmotic loss of water in the marine environment. However, unlike the marine chondrichthyes, it appears that the coelacanth blood is still slightly hypo-osmotic to sea water so that some osmotic loss of water must take place. Nothing is known about the mechanisms of osmotic and ionic regulation by the coelacanth since no studies have been performed on living individuals. It is interesting to note that a post-anal gland is present (Millot and Anthony, 1972) and recent studies indicate that, like the shark rectal gland, it possesses high concentrations of Na^+-K^+-activated ATPase (Griffith and Burdick, 1976).

Studies of the osmotic and ionic regulation by lungfish have centered on renal function (Sawyer, 1970) and indicate that this species produces, as expected, a rather large volume of dilute urine. No studies of presumed mechanisms of ionic uptake by lungfish have been published.

AGNATHA

This largely fossil group is represented by two, rather dissimilar, groups: the lampreys and hagfish. Despite many morphological similarities it is thought by many that these two lines of fish evolution have been separate for nearly 500 million years and, therefore, represent a rather unnatural combination (Stensio, 1967). This separation is very apparent in the mechanisms of osmotic and ionic regulation in the two groups. Analyses of their respective blood constituents (Evans, 1979a) indicate that the wholly marine hagfish display the vertebrate-unique characteristic of electrolyte iso-osmolarity to sea water while the marine and freshwater lampreys display a pattern similar to that described for the teleost fish.

Despite similar total osmotic concentrations, it is apparent that hagfish maintain rather more Na^+ and rather less K^+, Mg^{++}, Ca^{++} and $SO_4^=$ in their body fluids than those found in sea water. Urine flow rates of Myxine glutinosa are quite low, of the order of $20 \mu l.100g^{-1}.hr^{-1}$ (Morris, 1965). Urine: serum ratios of Eptatretus stouti are 0.97, 1.57, 0.80, 1.23, 8.59 and 3.87 for Na^+, K^+, Ca^{++}, Mg^{++}, $SO_4^=$ and $PO_4^{\equiv}$, respectively, so it appears that renal reabsorption and secretion may play a major role in the maintenance of some ionic imbalances between hagfish blood and their marine environment (Munz and McFarland, 1964). The renal complex of the hagfish consists of glomerulae which empty via short neck segments into an ureter (archinephric duct) which superficially resembles the proximal tubule of the other vertebrates. Riegel (1978) has recently found that the colloidal osmotic pressure of the hagfish blood is greater than the blood pressure and has, therefore, questioned the ability of the hagfish to filter urine across the glomerulus. In addition, ouabain and dinitrophenol markedly reduced the GFR so it appears that active processes may be involved in the formation of urine by the hagfish glomerulus. The U/B ratios different than 1.0 found in hagfish urine (Munz and McFarland, 1964) indicate that the archinephric duct is capable of both secretion and reabsorption of various ions. The copiously-secreted slime has been shown to contain rather high concentrations of K^+, Ca^{++} and Mg^{++} (Munz and McFarland, 1964), so this may also be a site of ionic regulation. It is unknown if branchial transport mechanisms may also play a role since the electrochemical gradients have not been determined.

It is interesting to note that, like marine elasmobranchs, the hagfish are extremely permeable to water (turnover rate of 381% of its body water per hour, Rudy and Wagner, 1970) and either do not drink the medium (Morris, 1965) or do not absorb any ingested salt and water from the gut contents after some slight oral ingestion (McFarland and Munz, 1965).

McFarland and Munz (1965) examined the responses of E. stouti to salinity changes and found that, while dilution of the medium was followed by weight gain and subsequent return to control weights (presumably secondary to increased urinary loss of water), transfer to 120% sea water was followed by weight loss which was not compensated for after one week in the elevated salinity. This was due presumably to an inability to excrete the excess ions or initiate gut water uptake in the hyper-osmotic salinities. Despite the hagfish's ability to tolerate some dilution of the medium, it is still extremely stenohaline and has been found to be unable to tolerate salinities below approximately 60% sea water (Strahan, 1962; McFarland and Munz, 1965).

All lampreys, on the other hand, spawn in freshwater and the ammocoete larval form develops into a postmetamorphic lamprey which may either remain in freshwater or enter the marine environment, depending on the species.

Since freshwater lampreys are distinctly hyper-osmotic to their environment they face the same osmotic and ionic problems faced by freshwater teleosts (and indeed all freshwater animals). It appears that the mechanisms employed to offset these problems are basically the same in the two groups. The few determinations of the ionic turnover indicate a rate in the same range as that described for freshwater teleosts (Evans, 1979a). Lamprey urine flows are relatively large ($100\text{-}800\mu l.100g^{-1}.hr^{-1}$; Morris, 1972) so it appears that freshwater lampreys may be more permeable to water than freshwater teleosts. No direct measurements have been made but it would be of great interest to measure the flux of tritiated water across freshwater lampreys to determine if they have retained the relatively high water permeability found in their totally marine and iso-osmotic hagfish relatives, much in the same way that the freshwater elasmobranch, Potamotrygon, has retained the extremely high water permeability characteristic of marine elasmobranchs (Carrier and Evans, 1973). The urine concentration of Na^+ is quite low (12mM/l) (Morris, 1972) so it appears that lampreys are able to reabsorb significant quantities of Na^+ in their renal tubules (Morris, 1972). Lampreys do possess a convoluted segment distal to the proximal segment but it is not known if this is the site of ionic reabsorption.

Renal and diffusional loss of ions must be balanced by active uptake in freshwater and Morris and Bull (1970) have shown that Lampetra planeri can extract Na^+ from the environment via a saturable carrier system with a K_m of 0.25mM/l which is in the same range as that described for both the goldfish (Maetz, 1972) and rainbow trout (Kerstetter et al., 1970). One might suppose that Na^+ and Cl^- uptake are, like in teleosts, coupled to NH_4^+, H^+ and HCO_3^- extrusion but no data have been published to support this proposition.

It appears that those anadromous species of lampreys which migrate to the marine environment after metamorphosis and then return to freshwater to spawn have evolved osmoregulatory mechanisms similar to those described for marine teleosts (Morris, 1972). Unfortunately, very few data have been gathered to support this conclusion, and, in fact, there is only a single published analysis of the blood ionic levels of a marine lamprey (Robertson, in Morris, 1972) and this indicates a concentration similar to that described for marine teleosts (Na^+ and Cl^-, each approximately 160mM, total osmolarity of 333mOsm/l). What little we do know about marine osmotic and ionic regulation has come from studies of individuals caught just after entering freshwater rivers on their spawning runs ("fresh run") and then returned to various salinities experimentally in the laboratory.

Mature (freshwater) or "fresh run" Lampetra fluviatilus can maintain their body fluids hypo-osmotic to salinities up to 70% sea water but the animals lose weight at salinities above 33% sea water (Morris, 1956). The urine flow from animals in 33% sea water was significantly below that described for individuals in freshwater but interestingly the urine Cl^- concentration remained at freshwater levels (Morris, 1965). Transfer of either L. fluviatilus or Petromyzon marinus to 50% sea water after a sojourn in 33% sea water was followed by a reduction in urine flow to from zero to 258µl.100g^{-1}.hr^{-1} (Morris, 1958; Pickering and Morris, 1970), far below that described for lampreys in freshwater (see above). The animals did ingest the medium but the drinking rate was extremely low (0.012%-0.41%/hr) and may not be sufficient to balance the osmotic loss of water. Morris (1958) and Pickering and Morris (1970) have suggested that the low drinking rate may be the cause of osmoregulatory failure in higher salinities but one must know the osmotic permeability and consequent loss of water before a definitive statement can be made. Approximately 80% of the ingested water was absorbed by the gut epithelium, and it appears that this was secondary to uptake of NaCl since the concentration of both ions did not change as the ingested sea water travelled down the gut. However, like the teleosts, Ca^{++} and Mg^{++} concentrations rose significantly in the gut fluids, presumably secondary to osmotic withdrawal of water (Pickering and Morris, 1970).

Since the urine NaCl concentrations are below that of the blood (Pickering and Morris, 1970) it appears that extrarenal mechanisms must exist for the extrusion of NaCl absorbed from the gut as well as gained diffusionally from the sea water. However, nothing is known about the actual extrusion mechanisms involved and one must assume that they are relatively inefficient in "fresh run" individuals since at least L. fluviatilus is not able to osmoregulate in salinities above 33% sea water. Of course, this is not surprising since these animals are caught during their spawning migration into freshwater when one would expect that ionic extrusion mechanisms to be non-adaptive and, therefore, not functioning. However, the degree of euryhalinity may be species specific since Beamish et al. (1978) have recently found that adults of P. marinus caught after entry into freshwater can maintain blood Na^+ and Cl^- concentrations below 150mM/l over a salinity range from freshwater to over 32 o/oo. Further study of this species acclimated to sea water is certainly warranted.

Morris (1958) described branchial cells which are morphologically similar to the "chloride cells" of the teleosts and showed that their numbers increased after transfer of L. fluviatilus from 33% to 50% sea water. In addition, they were more prevalent in "fresh run" individuals than freshwater specimens. One might assume that these cells are the site of NaCl extrusion (as in the teleosts) but no direct evidence exists to support this assumption.

It should be obvious that osmoregulation by marine lampreys is a relatively untouched area for future research.

REFERENCES

Ando,M. and Kobayashi, M. (1978). Effects of stripping of the outer layers of the eel intestine on water and salt transport. Comp. Biochem. Physiol. 61A: 479-501.

Ando,M., Utida, S. and Nagahama, H. (1975). Active transport of chloride in eel intestine with special reference to sea water adaptation. Comp. Biochem. Physiol. 51A: 27-32.

Beamish, F.W.H., Strachan, P.D. and Thomas, E. (1978). Osmotic and ionic performance of the anadromous sea lamprey, Petromyzon marinus. Comp. Biochem. Physiol. 60A: 435-443.

Bentley, P.G., Maetz, J. and Payan, T. (1976). A study of the unidirectional fluxes of Na and Cl across the gills of the dogfish Scyliorhinus canicula (Chondrichthyes). J. Exp. Biol. 64: 629-637.

Bonting, S. (1966). Studies on sodium-potassium-activated adenosine triphosphatase XV. The rectal gland of the elasmobranch. Comp. Biochem. Physiol. 17: 953-966.

Breder, C.M., Jr. (1934). Ecology of an oceanic freshwater lake, Andros Island, Bahamas, with special reference to its fishes. Zoologica 18: 57-80.

Burger, J.W. (1962). Further studies on the function of the rectal gland in the spiny dogfish. Physiol. Zool. 35: 205-217.

Burger, J.W. (1965). Roles of the rectal gland and the kidneys in salt and water excretion in the spiny dogfish. Physiol. Zool. 38: 191-196.

Burger, J.W. and Tosteson, D.C. (1966). Sodium influx and efflux in the spiny dogfish (Squalus acanthias). Comp. Biochem. Physiol. 19: 649-653.

Cala, P.M. (1977). Volume regulation by flounder red blood cells in an anisotonic media. J. Gen. Physiol. 69: 537-552.

Carrier, J.C. and Evans, D.H. (1973). Ion and water turnover in the freshwater elasmobranch, Potamotrygon sp. Comp. Biochem. Physiol. 45A: 667-670.

Carrier, J.C. and Evans, D.H. (1976). The role of environmental calcium in freshwater survival of the marine teleost, Lagodon rhomboides. J. Exp. Biol. 65: 529-538.

Chan,D.K.O., Phillips, J.G. and Chester Jones, I. (1967). Studies on electrolyte changes in the lip-shark Hemiscyllium plagiosum (Bennett), with special reference to hormonal influence on the rectal gland. Comp. Biochem. Physiol. 23: 185-198.

Chester Jones, I., Henderson, I.W. and Butler, D.G. (1965). Water and electrolyte flux in the European eel (Anguilla anguilla L.). Arch. Anat. Micro. Morph. Exp. 54: 453-469.

Cuthbert, A.W., Fanelli, G.M., Jr., and Scriabine, A. (1979). Amiloride and Epithelial Sodium Transport. Urban and Schwarzenberg, Baltimore and Munich.

Degnan, K.J., Karnaky, K.J., Jr., and Zadunaisky, J.A. (1977). Active chloride transport in the in vitro skin of a teleost (Fundulus Heteroclitus) a gill-like epithelium rich in chloride cells. J. Physiol. (Lond.) 271: 155-191.

DeRenzis, G. and Bornancin, N. (1977). A Cl^-/HCO_3^--ATPase in the gills of Carassius auratus, its inhibition by thiocyanate. Biochim. Biophys. Acta. 467: 192-207.

Dharmamba, M., Bornancin, M. and Maetz, J. (1975). Environmental salinity and sodium chloride exchanges across the gill of Tilapia mossambica. J. Physiol. (Paris) 70: 627-636.

Eddy, F.B. (1975). The effect of calcium on gill potentials and on sodium and chloride fluxes in the goldfish, Carassius auratus. J. Comp. Physiol. 96: 131-142.

Epstein, F.H., Maetz, J. and deRenzis, G. (1973). On the active transport of chloride by the teleost gill. Inhibition by thiocyanate. Am. J. Physiol. 224: 1195-1199.

Evans, D.H. (1967). Sodium, chloride and water balance of the intertidal teleost, Xiphister atropurpureus. II. The role of the kidney and the gut. J. Exp. Biol. 47: 519-542.

Evans, D.H. (1969a). Sodium, chloride and water balance of the intertidal teleost, Pholis gunnelus. J. Exp. Biol. 50: 179-190.

Evans, D.H. (1969b). Studies on the permeability to water of selected marine, freshwater and euryhaline teleosts. J. Exp. Biol. 50: 689-703.

Evans, D.H. (1973). Sodium uptake by the sailfin molly, Poecilia latipinna: Kinetic analysis of a carrier system present in both freshwater-acclimated and sea-water-acclimated individuals. Comp. Biochem. Physiol. 45A: 843-850.

Evans, D.H. (1975a). The effects of various external cations and sodium transport inhibitors on sodium uptake by the sailfin molly, Poecilia latipinna, acclimated to sea water. J. Comp. Physiol. 96: 111-115.

Evans, D.H. (1975b). Ionic exchange mechanisms in fish gills. Comp. Biochem. Physiol. 51A: 491-495.

Evans, D.H. (1977). Further evidence for Na/NH_4 exchange in marine teleost fish. J. Exp. Biol. 70: 213-220.

Evans, D.H. (1979a). Fish. In: Comparative Physiology of Osmoregulation in Animals. Ed. G.M.O. Maloiy. New York and London, Academic Press, p. 305-390.

Evans, D.H. (1979b). Kinetic studies of ion transport by fish gill epithelium. Am. J. Physiol. (in press).

Evans, D.H. (1979c). Na/NH_4 exchange in the marine teleost, Opsanus beta: stoichiometry and role in Na balance. In: Epithelial Transport in the Lower Vertebrates. Ed. B. Lahlou. Cambridge, Cambridge University Press, in press.

Evans, D.H., Carrier, J.C. and Bogan, M.B. (1974). The effect of external potassium ions on the electrical potential measured across the gills of the teleost, Dormitator maculatus. J. Exp. Biol. 61: 277-283.

Evans, D.H. and Cooper, K. (1976). The presence of Na/Na and Na/K exchange in sodium extrusion by three species of fish. Nature (Lond.) 259: 241-242.

Evans, D.H., Kormanik, G.A. and Krasny, E.J., Jr. (1979). Mechanisms of ammonia and acid extrusion by the little skate, Raja erinacea. J. Exp. Zool. 208: 431-437.

Field,M., Karnaky, K.J. Jr., Smith, P.L., Bolton, J.I. and Kitner, W.B. (1978). Ion transport across the isolated intestinal mucosa of the winter flounder, Pseudopleuronectes americanus. I. Functional and structural properties of cellular and paracellular pathways for Na and Cl. J. Membr. Biol. 41: 265-293.

Fletcher, C.R. (1978). Osmotic and ionic regulation in the cod (Gadus callarias L.). II. Salt balance. J. Comp. Physiol. 124B: 157-168.

Forster, R.P. (1967). Osmoregulatory role of the kidney in cartilagenous fishes (Chondrichthyes). In: Sharks, Skates and Rays. Eds. P.W. Gilbert, R.F. Mathewson and D.P. Rall. Baltimore, Johns Hopkins Press, p. 187-195.

Fossat, B. and Lahlou, B. (1977). Osmotic and solute permeabilities of isolated urinary bladder of the trout. Am. J. Physiol. 233: F525-F531.

Girard, J.P. (1976). Salt excretion by the perfused head of trout adapted to sea water and its inhibition by adrenaline. J. Comp. Physiol. 111: 77-99.

Glynn, I.M. and Karlish, S.J.D. (1975). The sodium pump. Ann. Rev. Physiol. 37: 13-55.

Goldstein, L. and Forster, R.P. (1971a). Osmoregulation and urea metabolism in the little skate Raja erinacea. Am. J. Physiol. 220: 742-746.

Goldstein, L. and Forster, R.P. (1971b). Urea biosynthesis and excretion in freshwater and marine elasmobranchs. Comp. Biochem. Physiol. 39B: 415-421.

Goldstein, L., Oppelt, W.W. and Maren, T.H. (1968). Osmotic regulation and urea metabolism in the lemon shark Negaprion brevirostris. Am. J. Physiol. 215: 1493-1497.

Greenwald, L., Kirschner, L.B. and Sanders, M. (1974). Sodium efflux and potential differences across the irrigated gill of sea water-adapted rainbow ttout (Salmo gairdneri). J. Gen. Physiol. 64: 135-147.

Griffith, R.W. and Burdick, C.J. (1976). Sodium-potassium activated adenosine triphosphatase in coelacanth tissues: high activity in rectal gland. Comp. Biochem. Physiol. 54B: 557-559.

Griffith, R.W., Pang, P.K.T., Srivastava, A.K. and Pickford, G.E. (1973). Serum composition of freshwater stingrays (Potamotrygonidae) adapted to fresh and dilute sea water. Biol. Bull. 144: 304-320.

Hanson, R.C., Duff, D., Brehe, J. and Fleming, W.R. (1976). The effect of various salinities, hypophysectomy, and hormone treatments on the survival and sodium and potassium content of juvenilie bowfin, Amia calva. Physiol. Zool. 49: 376-385.

Hickman, C.P. Jr. (1968). Ingestion, intestinal absorption and elimination of sea water and salts in the southern flounder, Paralichthys lethostigma. Can. J. Zool. 46: 457-466.

Hickman, C.P., Jr. and Trump, B.F. (1969). The kidney. In: Fish Physiology, Vol. 1. Eds. W.S. Hoar and D.J. Randall. New York and London, Academic Press, p. 91-239.

Hirano, T. (1975). Effects of prolactin on osmotic and diffusion permeability of the urinary bladder of the flounder, Plathichthys flesus. Gen. Comp. Endo. 27: 88-94.

Hirano, T., Johnson, D.W., Bern, H.A. and Utida, S. (1973). Studies on water and ion movements in the isolated urinary bladder of selected freshwater, marine and euryhaline teleosts. Comp. Biochem. Physiol. 45A: 529-540.

Hirano, T. and Mayer-Gostan, N. (1976). Eel esophagus as an osmoregulatory organ. Proc. Nat. Acad. Sci. USA 73: 1348-1350.

Holmes, W.M. and Donaldson, E.M. (1969). The body compartments and the distribution of electrolytes. In: Fish Physiology, Vol. 1. Eds. W.S. Hoar, D.J. Randall. New York and London, Academic Press, p. 1-89.

Hootman, S.R. and C.W. Philpott. (1979). Ultracytochemical localization of Na^+-K^+-activated ATPase in chloride cells from the gills of a euryhaline teleost. Anat. Rec. 193: 99-130.

Horowicz, P. and Burger, J.W. (1968). Unidirectional fluxes of Na in the spiny dogfish, Squalus acanthias. Am. J. Physiol. 214: 635-642.

House, C.R. and Maetz, J. (1974). On the electrical gradient across the gill of the sea water-adapted eel. Comp. Biochem. Physiol. 47A: 917-924.

Howe, D. and Gutknecht, J. (1978). Role of urinary bladder in osmoregulation in marine teleost, Opsanus tau. Am. J. Physiol. 235: R48-R54.

Huang, K.C. and Chen, T.S.T. (1971). Ion transport across intestinal mucosa of winter flounder, Pseudopleuronectes americanus. Am. J. Physiol. 220: 1734-1738.

Jampol, L.M. and Epstein, F.H. (1970). Sodium-potassium-activated adenosine triphosphatase and osmotic regulation by fishes. Am. J. Physiol. 218: 607-611.

Karnaky, K.L. Jr., Degnan, K.J. and Zadunaisky, J.A. (1977). Chloride transport across isolated opercular epithelium of killifish: a membrane rich in chloride cells. Science (N.Y.) 195: 230-235.

Karnaky, K.J. and Kinter, W.B. (1977). Killifish opercular skin: a flat epithelium with a high density of chloride cells. J. Exp. Zool. 199: 355-364.

Karnaky, K.L., Kinter, W.B. and Stirling, C.E. (1976). Teleost chloride cells. II. Autoradiographic localization of gill Na, K-ATPase in killifish Fundulus heteroclitus adapted to low and high salinity environments. J. Cell. Biol. 70: 157-177.

Kerstetter, T.H. and Kirschner, L.B. (1972). Active chloride transport by the gills of rainbow trout (Salmo gairdneri). J. Exp. Biol. 56: 263-272.

Kerstetter, T.H., Kirschner, L.B. and Rafuse, D.D. (1970). On the mechanisms of sodium ion transport by the irrigated gills of rainbow trout (Salmo gairdneri). J. Gen. Physiol. 56: 342-350.

Keys, A.B. (1931). The heart-gill preparation of the eel and its perfusion for the study of a natural membrane in situ. Z. Vergl. Physiol. 15: 353-363.

Keys, A.B. and Willmer, E.N. (1932). "Chloride-secreting cells" in the gills of fishes with special reference to the common eel. J. Physiol. (Lond.) 76: 368-378.

Kikuchi, S. (1977). Mitochondria-rich (chloride) cells in the gill epithelia from four species of stenohaline freshwater teleosts. Cell Tiss. Res. 180: 87-98.

Kirsch, R. and Laurent, P. (1975). L'esophage, organe effecteur de l'osmorégulation chez un téléosteen euryhalin l'Anguille (Anguilla anguilla L.). Comp. Rend. Acad. Sci. Paris 280: 2013-2015.

Kirschner, L.B. (1977). The sodium chloride excreting cells in marine vertebrates. In: Transport of Ions and Water in Animals. Eds. B.L. Gupta, R.B. Moreton, J.L. Oschman and B.W. Wall. London and New York, Academic Press, p. 427-452.

Kirschner, L.B. (1979). Control mechanisms in crustaceans and fishes. In: Osmoregulation in Animals. Ed. R. Gilles. New York, Wiley Interscience, p. 157-222.

Kirschner, L.B., Greenwald, L. and Kerstetter, T.H. (1973). Effect of amiloride on sodium transport across body surfaces of freshwater animals. Am. J. Physiol. 224: 832-837.

Kirschner, L.B., Greenwald, L. and Sanders, M. (1974). On the mechanisms of sodium extrusion across the irrigated gill of sea water-adapted rainbow trout (Salmo gairdneri). J. Gen. Physiol. 64: 148-165.

Koefoed-Johsen, V. and Ussing, H.H. (1958). The nature of the frog skin potential. Acta Physiol. Scand. 42: 298-308.

Kormanik, G.A. and Evans, D.H. (1979). HCO_3-stimulated Cl efflux in the gulf toadfish acclimated to sea water. J. Exp. Zool. 208: 13-16.

Krogh, A. (1939). Osmotic Regulation in Aquatic Animals. Cambridge, Cambridge University Press.

Lahlou, B. (1967). Excretion rénale chez un poisson euryhalin, le flet (Platichthys flesus L.): caractéristiques de l'urine normale en eau douce et en eau de mer et effets des changements de milieu. Comp. Biochem. Physiol. 20: 925-938.

Lange, R. (1972). Some recent work on osmotic, ionic and volume regulation in marine animals. Oceanogr. Mar. Biol. Ann. Rev. 10: 97-136.

Lotan, R. and Skadhauge, E. (1972). Intestinal salt and water transport in a euryhaline teleost, Aphanius dispar (Cyprinodontidae). Comp. Biochem. Physiol. 42A: 303-310.

McFarland, W.N. and Munz, F.W. (1965). Regulation of body weight and serum composition by hagfish in various media. Comp. Biochem. Physiol. 14: 393-398.

Maetz, J. (1969). Sea water teleosts: evidence for a sodium-potassium exchange in the branchial sodium-excreting pump. Science 166: 613-615.

Maetz, J. (1972). Branchial sodium exchange and ammonia excretion in the goldfish Carassius auratus. Effects of ammonia-loading and temperature changes. J. Exp. Biol. 56: 601-620.

Maetz, J. (1973). Na^+/NH_4^+, Na^+/H^+ exchanges and NH_3 movement across the gill of Carassius auratus. J. Exp. Biol. 58: 255-275.

Maetz, J. (1974). Aspects of adaptation to hypo-osmotic and hyper-osmotic environments. In: Biochemical and Biophysical Perspectives in Marine Biology. Vol. 1. Eds. D.C. Malins and J.R. Sargent. London and New York, Academic Press, p. 1-167.

Maetz, J. (1974b). Origine de la différence de potentiel électique transbranchiale chez le poisson rouge Carassius auratus. Importance de l'ion Ca^{++}. Comp. Rend. Acad. Sci. Paris 279: 1277-1280.

Maetz, J. and Bornancin, M. (1975). Biochemical and biophysical aspects of salt secretion by chloride cells in teleosts. International Symposium on Excretion. Forts. der Zool. 23: Heft 2/3. 322-362.

Maetz, J. and Garcia Romeu, F. (1964). The mechanism of sodium and chloride uptake by the gills of a freshwater fish, Carassius auratus. II. Evidence for NH_4^+/Na^+ and HCO_3^-/Cl^- exchanges. J. Gen. Physiol. 47: 1209-1227.

Maetz, J., Payan, P. and deRenzis, G. (1976). Controversial aspects of ionic uptake in freshwater animals. In: Perspectives in Experimental Biology, Zoology, Vol. 1. Eds. P. Spencer Davies. Oxford and New York, Pergamon Press, p. 77-92.

Maetz, J. and Pic, P. (1975). New evidence for a Na/K and Na/Na exchange carrier linked with the Cl pump in the gill of Mugil capito in sea water. J. Comp. Physiol. 102: 85-100.

Maetz, J. and Lahlou, B. (1966). Les échanges de sodium et de chlore chez un Elasmobranche, Scyliorhinus, mesurés à l'aide des isotopes ^{24}Na et ^{36}Cl. J. Physiol. Paris 58: 249 (abstract).

Marshall, W.S. (1977). Transepithelial potential and short-circuit current across the isolated skin of Gillichthys mirabilis (Teleostei: Gobiidae), acclimated to 5% and 100% sea water. J. Comp. Physiol. 114: 157-165.

Millot, J. and Anthony, J. (1972). Le glande post-anale de Latimeria. Ann. Sci. Natur. Zool. Paris 12^{e} Ser. 14: 305-318.

Morris, R. (1956). The osmoregulatory ability of the lamprey (Lampetra fluviatilis L.) in sea water during the course of its spawning migration. J. Exp. Biol. 33: 235-248.

Morris, R. (1958). The mechanisms of marine osmoregulation in the lamprey (Lampetra fluviatilis L.) and the causes of its breakdown during the spawning migration. J. Exp. Biol. 35: 649-665.

Morris, R. (1965). Studies on salt and water balance in Myxine glutinosa (L.). J. Exp. Biol. 42: 359-371.

Morris, R. (1972). Osmoregulation. In: The Biology of Lampreys. Vol. 2. Eds. M.W. Hardisty and I.C. Potter. London and New York, Academic Press, p. 193-239.

Morris, R. and Bull, J.M. (1970). Studies on freshwater osmoregulation in the ammocoete larva of Lampetra planeri (Block). 3. The effect of external and internal sodium concentration on sodium transport. J. Exp. Biol. 52: 272-290.

Motais, R., Garcia Romeu, F. and Maetz, J. (1966). Exchange diffusion effect and euryhalinity in Teleosts. J. Gen. Physiol. 50: 391-422.

Motais, R. and Isaia, J. (1972). Evidence for an effect of ouabain on the branchial sodium-excreting pump of marine teleosts: interaction between the inhibitor and external Na and K. J. Exp. Biol. 57: 367-373.

Munz, F.W. and McFarland, W.N. (1964). Regulatory function of a primitive vertebrate kidney. Comp. Biochem. Physiol. 13: 381-400.

Oguri, M. (1964). Rectal glands of marine and freshwater sharks: comparative histology. Science 144: 1151-1152.

Oide, M. (1967). Effects of inhibitors on transport of water and ion in isolated intestine and Na^{+}-K^{+}-ATPase in intestinal mucosa of the eel. Annotnes. Zool. Jap. 40: 130-135.

Pang, P.K.T., Griffith, R.W. and Kahn, N. (1972). Electrolyte regulation in freshwater stingrays (Potamotrygonidae). Fed. Proc. 31: 344 (abstract).

Payan, P. (1978). A study of the Na^{+}/NH_4^{+} exchange across the gill of the perfused head of the trout (Salmo gairdneri). J. Comp. Physiol. 124: 181-188.

Payan, P., Goldstein, L. and Forster, R.P. (1973). Gills and kidneys in ureosmotic regulation in euryhaline skates. Am. J. Physiol. 224: 367-372.

Payan, P. and Maetz, J. (1971). Balance hydrique chez les Elasmobranches: arguments en faveur d'un contrôle endocrinien. Gen. Comp. Endocrin. 16: 535-554.

Payan, P. and Maetz, J. (1973). Branchial sodium transport mechanisms in Scyliorhinus canicula: evidence for Na^+/NH_4^+ and Na^+/H^+ exchanges and for a role of carbonic anhydrase. J. Exp. Biol. 58: 487-502.

Payan, P., Matty, A.J. and Maetz, J. (1975). A study of the sodium pump in the perfused head preparation of the trout Salmo gairdneri in freshwater. J. Comp. Physiol. 104: 33-48.

Pickering, A.D. and Morris, R. (1970). Osmoregulation of Lampetra fluviatilis and Petromyzon marinus (Cyclostomata) in hypertonic solutions. J. Exp. Biol. 53: 231-243.

Potts, W.T.W. (1977). Fish gills. In: Transport of Ions and Water in Animals. Eds. B.L. Gupta, R.B. Moreton, J.L. Oschmand and B.W. Wall. London and New York, Academic Press, p. 453-480.

Potts, W.T.W. and Evans, D.H. (1967). Sodium and chloride balance in the killifish Fundulus heteroclitus. Biol. Bull. 133: 411-425.

Potts, W.T.W., Fletcher, C.R. and Eddy, B. (1973). An analysis of the sodium and chloride fluxes in the flounder Platichthys flesus. J. Comp. Physiol. 82: 21-28.

Potts, W.T.W. and Rudy, P.P. (1972). Aspects of osmotic and ionic regulation in the sturgeon. J. Exp. Biol. 56: 703-715.

Renfro, J.L. (1975). Water and ion transport by the urinary bladder of the teleost, Pseudopleuronectes americanus. Am. J. Physiol. 228: 52-61.

Richards, B.D. and Fromm, P.O. (1970). Sodium uptake by the isolated-perfused gills of rainbow trout (Salmo gairdneri). Comp. Biochem. Physiol. 33: 303-310.

Riegel, J.A. (1978). Factors affecting glomerular functions in the Pacific hagfish Eptatretus stouti (Lockington). J. Exp. Biol. 73: 261-277.

Rudy,P.P. and Wagner, R.C. (1970). Water permeability in the Pacific Hagfish Polistotrema stouti and the Staghorn Sculpin Leptocottus armatus. Comp. Biochem. Physiol. 34: 399-403.

Sawyer, W.H. (1970). Vasopressor, diuretic, and natriuretic responses by the lungfish to arginine vasotocin. Am. J. Physiol. 218: 1789-1794.

Schmidt-Nielsen, B. (1977). Volume regulation of muscle fibers in the killifish, Fundulus heteroclitus. J. Exp. Zool. 199: 411-418.

Schmidt-Nielsen, B., Truniger, B. and Rabinowitz, L. (1972). Sodium-linked urea transport by the renal tubule of the spiny dogfish Squalus acanthias. Comp. Biochem. Physiol. 42A: 13-25.

Shehadeh, Z.H. and Gordon, M.S. (1969). The role of the intestine in salinity adaptation of the rainbow trout, Salmo gairdneri. Comp. Biochem. Physiol. 30: 397-418.

Shuttleworth, T.J. (1972). A new isolated perfused gill preparation from the study of the mechanisms of ionic regulation in teleosts. Comp. Biochem. Physiol. 43A: 59-64.

Shuttleworth, T.J. (1978). The effect of adrenaline on potentials in the isolated gills of the flounder (Platichthys flesus L.). J. Comp. Physiol. 124: 129-136.

Shuttleworth, T.J. and Freeman, R.F. (1974). Factors affecting the net fluxes of ions in the isolated perfused gills of freshwater Anguilla dieffenbachii. J. Comp. Physiol. 94: 297-307.

Silva, P., Solomon, R., Spokes, K. and Epstein, F.H. (1977a). Ouabain inhibition of gill Na-K-ATPase: relationship to active chloride transport. J. Exp. Zool. 199: 419-427.

Silva, P., Stoff, J., Field, M., Fine, L., Forrest, J.N. and Epstein, F.H. (1977b). Mechanism of active chloride secretion by shark rectal gland: role of Na-K-ATPase in chloride transport. Am. J. Physiol. 233: F298-F306.

Skadhauge, E. (1974). Coupling of transmural flows of NaCl and water in the intestine of the eel (Anguilla anguilla). J. Exp. Biol. 60: 535-546.

Smith, H.W. (1930). The absorption and excretion of water and salts by marine teleosts. Am. J. Physiol. 93: 480-505.

Smith, H.W. (1931a). The absorption and excretion of water and salts by the elasmobranch fishes. I. Freshwater elasmobranchs. Am. J. Physiol. 98: 279-295.

Smith, H.W. (1931b). The absorption and excretion of water and salts by the elasmobranch fishes. II. Marine elasmobranchs. Am. J. Physiol. 98: 296-310.

Stensio, E. (1967). The cyclostomes with special reference to the diphyletic origin of the Petromyzontida and Myxinoidea. In: Current problems of lower vertebrate phylogeny. Ed. Tor Orvig. Nobel Symposium 4. New York, Interscience, p. 13071.

Strahan, R. (1962). Survival of the hag, Paramyxine atami (Dean) in diluted sea water. Copeia 1962: 471-473.

Sulya, L.L., Box, B.E. and Gunther, G. (1960). Distribution of some blood constituents in fish from the Gulf of Mexico. Am. J. Physiol. 199: 1117-1180.

Thorson, T.B. (1970). Freshwater stingrays, Potamotrygon spp.: failure to concentrate urea when exposed to saline medium. Life Sci. 9: 893-900.

Thorson, T.B., Cowan, C.M. and Watson, D.E. (1967). Potamotrygon spp.: elasmobranchs with low urea content. Science 158: 375-377.

Thorson, T.B., Wotton, R.M. and Georgi, T.A. (1978). Rectal gland of freshwater stingrays, Potamotrygon spp. (Chondrichthyes: Potamotrygonidae). Biol. Bull. 154: 508-516.

Urist,M.R. (1966). Calcium and electrolyte control mechanisms in lower vertebrates. In: Phylogenetic Approach to Immunity. Eds. R.T. Smith, R.A. Good, P.A. Miescher. Gainesville, Univ. of Florida Press, p. 18-28.

Vooys, C.B.N. de (1968). Formation and excretion of ammonia in Teleostei. I. Excretion of ammonia through the gills. Arch. Int. Physiol. Biochem. 76: 268-272.

Wong, T.M. and Chan, D.K.O. (1977). Physiological adjustments to dilution of the external medium in the lip-shark Hemiscyllium plagiosum (Bennett). J. Exp. Zool. 200: 85-96.

Zawodny, J.F. (1975). Osmoregulation in the Florida Spotted gar, Lepisosteus platyrhincus. MS. Thesis, University of Miami, Coral Gables, Florida USA.

REGULATION OF THE ACID-BASE STATUS IN FISHES

Norbert Heisler
Abteilung Physiologie
Max-Planck-Institut für experimentelle Medizin
D-34 Göttingen, FRG

Dedicated to the memory of my friend, George Holeton

INTRODUCTION

In spite of the fact that the classical articles on acid-base theory, physico-chemistry and measurement techniques were published more than 50 years ago (e.g. Arrhenius, 1887; Henderson, 1909; Sörensen, 1912; Hasselbalch, 1917, Van Slyke, 1917; Brønstedt, 1923; Van Slyke, 1922; see also Peters and Van Slyke, 1932), our knowledge about the acid-base physiology of lower vertebrates, especially fish, is still quite fragmentary. After the tools for scientific work in this field had been developed, interest was mainly focused on mammalian acid-base status, and scientists only rarely dedicated their work to the field of fish physiology (e.g. Smith, 1929; Dill, Edwards and Florkin, 1932). Systematic studies of acid-base regulation in fish were not undertaken before about 15 years ago, and most investigations were actually performed within the last eight years. The aim of this paper is to summarize the data about normal acid-base status in fish, the regulatory mechanisms which are probably involved in the maintenance of steady state and the effects on acid-base status from environmentally induced disturbances. The conclusions drawn from this review must remain tentative due to the small number of species and parameters studied, but may hopefully stimulate scientific activity in this field of fish physiology.

I. NORMAL ACID-BASE STATUS

Several investigations, dating back to 1927, have shown that poikilothermic animals do not maintain a constant extracellular pH (pHe) as homoiothermic species do, but that pHe can vary between 7.4 and values higher than 7.8 (Austin, Sunderman and Camack, 1927; Dill and Edwards, 1935; Edwards and Dill, 1935; Dill et al., 1935). Accordingly, it was thought for a long time that pH regulation in poikilothermic animals was rather poor or non-existent, until Robin (1962) could reproduce specific extracellular pH values (pHe) in turtles by proper control of the environmental temperature. His experiments showed that pHe falls with rising temperature. Subsequent experiments by Rahn (1966b, 1967) along with Robin's data led to the suggestion that extracellular pH in poikilothermic animals is regulated in such a way that, with changes in body temperature, a constant relative alkalinity (i.e. a constant difference between pHe and pH of neutral water, or a constant OH^-/H^+ ratio) is maintained. Since this concept was introduced, several additional studies have in general confirmed the rule of a constant relative alkalinity in air breathing species (Tucker, 1966; Baumgardner and Rahn, 1967; Howell et al., 1970; Wood and Moberly, 1970; Howell et al., 1972; Reeves, 1972; Crawford and Gatz, 1974; Jackson et al., 1974; Jackson and Kagen, 1976; Malan, Wilson and Reeves, 1976) and have shown that the observed pH behavior is achieved by adjustment of P_{CO_2} alone, while the bicarbonate concentration remains essentially constant (in those of the above cited studies where bicarbonate was determined).

Plotting blood pH measurements from 11 different fish species against their respective environmental temperatures, Rahn and Baumgardner (1972) showed that these fish species maintained their normal pHe within a range of 0.25 pH units around a constant relative alkalinity and thus confirmed the overall picture of decreasing pH with rising temperature also for fish. Since then, some systematic studies have been performed in fish on the correlations between environmental temperature and plasma pH, P_{CO_2}, bicarbonate concentration and, in three species, the intracellular pH of muscle tissues. The extracellular parameters were determined in the water breathing species *Salmo gairdneri* (rainbow trout, Randall and Cameron, 1973), *Scyliorhinus stellaris* (larger spotted dogfish, Heisler et al., 1976a, 1980a), *Cynoscion arenarius* (marine seatrout, Cameron, 1978) *Cyprinus carpio* (carp, Heisler and Neumann, unpublished) and in the air-breathing *Synbranchus marmoratus* (Heisler, 1980a). In all five species, extracellular pH falls with rising acclimation temperature (Fig. 1 and 2). A constant relative alkalinity, however, is maintained only in the air breathing *Synbranchus* ($\Delta pHe/\Delta t$ = -0.017 U/°C), whereas the correlation between pHe and temperature is smaller in the water breathers ($\Delta pHe/\Delta t$ = - 0.012 U/°C) and significantly different from that expected on the basis of a constant relative alkalinity ($\Delta pN/\Delta t$ = - 0.019 U/°C, range 5-20°C; - 0.016 U/°C, range 20 to 37°C; Handbook of Chemistry and Physics, 56th ed.). At any given temperature, the mean pHe values of the five species never differ by more than 0.15 pH units. This similarity in plasma pH values is the result of variable combinations of P_{CO_2} and bicarbonate concentration. The levels of P_{CO_2} in these fish ranged from 1 mm Hg (dogfish, 10°C) to 30 mm Hg (*Synbranchus*, 37°C); the bicarbonate concentrations ranged from 3mM in juvenile dogfish to 35 mM in *Synbranchus*, both at 20°C.

In all species studied, P_{CO_2} increases with rising temperature, but the changes in P_{CO_2} are too small (in adult dogfish too large) to account entirely for the observed changes in pHe. Hence, bicarbonate concentration is decreased with rising temperature (increased in adult dogfish) in order to

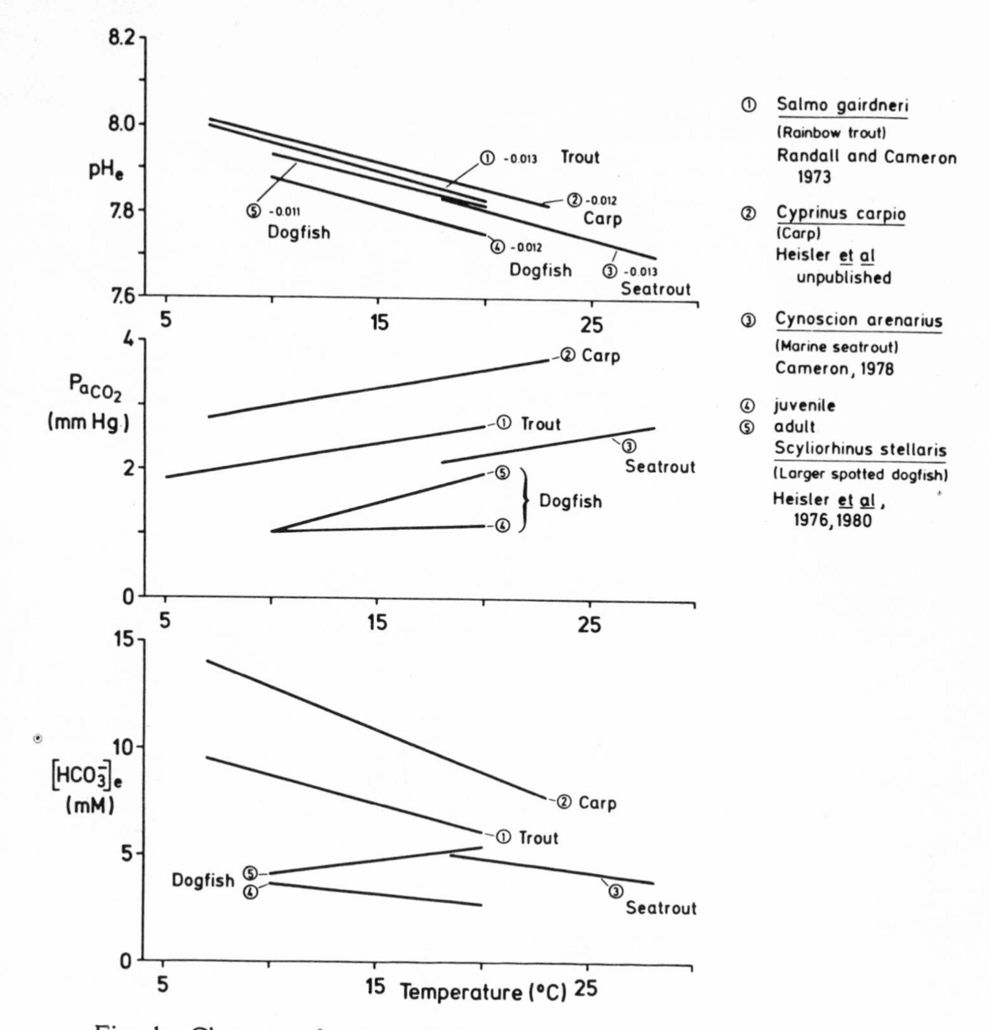

Fig. 1: Changes of extracellular pH, P_{CO_2} and bicarbonate concentration with environmental temperature in four water breathing fish species.

achieve the observed changes in pHe. This type of regulation is different from that in amphibians and reptiles (references above), where pHe is adjusted only by changes in P_{CO_2} induced by ventilation changes (Jackson, 1970). Fish are handicapped in such P_{CO_2} adjustments because of physical limitations (Rahn, 1966a) or the energetic problems involved with long-term hyperventilation of the viscous breathing medium water. Thus, changes in bicarbonate appear to be the more important, and probably the ultimate regulatory measure for acid-base regulation in fish. This hypothesis is confirmed by the fact that reversal of the normally observed pattern of rising P_{CO_2} with increasing temperature ($\Delta P_{CO_2}/\Delta t$ positive) by moderately

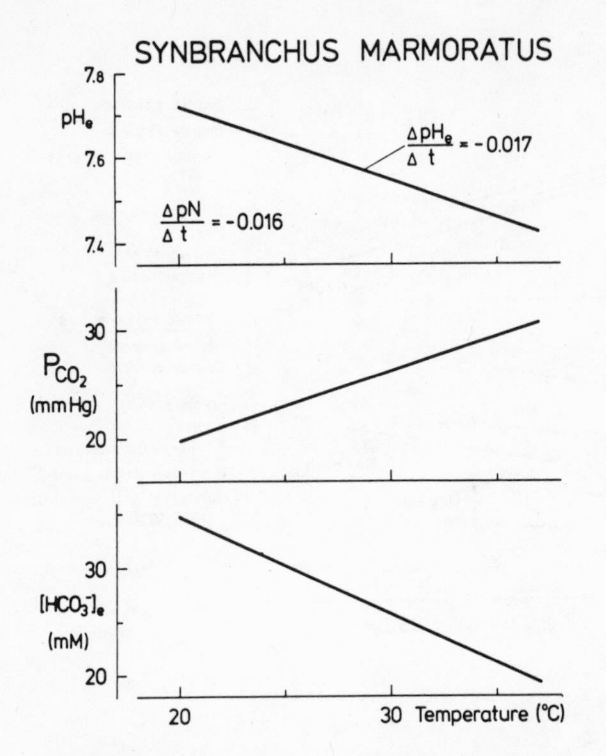

Fig. 2: Changes of extracellular pH, P_{CO_2} and bicarbonate concentration with changes of the environmental water temperature in the air breathing teleost fish Synbranchus marmoratus (Heisler, 1980a).

increased inspiratory P_{CO_2} at low acclimation temperature ($\Delta P_{CO_2}/\Delta t$ negative) does not change the pH/temperature regulation in dogfish (fig. 3, Heisler, Neumann and Holeton, 1980).

Intracellular pH (pHi) as a function of acclimation temperature has been determined in muscle tissues of dogfish, carp and the air-breathing Synbranchus (fig. 4, 5 and 6). In all eight muscle tissues investigated, only in the white muscle of dogfish a constant relative alkalinity was maintained (fig. 4), whereas pHi in white muscle of the other two species changes only by two-thirds of the extent required to match a constant relative alkalinity (figs. 5 + 6). The changes in pH with acclimation temperature in the heart muscle of dogfish, carp and Synbranchus are also much smaller and in red muscle of dogfish and carp are much larger than required for a constant relative alkalinity.

Imidazole Alphastat Regulation in Fish

Reeves (1972) attempted to explain the observed acid-base regulation in poikilothermic vertebrates. Based on the fact that the pK value of histidine imidazole changes with temperature more or less in parallel with a constant relative alkalinity, he proposed that ventilation and, thus, P_{CO_2} are regulated such that the dissociation of imidazole (Alpha) was kept constant. As histidine imidazole is the predominant nonbicarbonate buffer in the blood and extracellular space and probably also in some intracellular

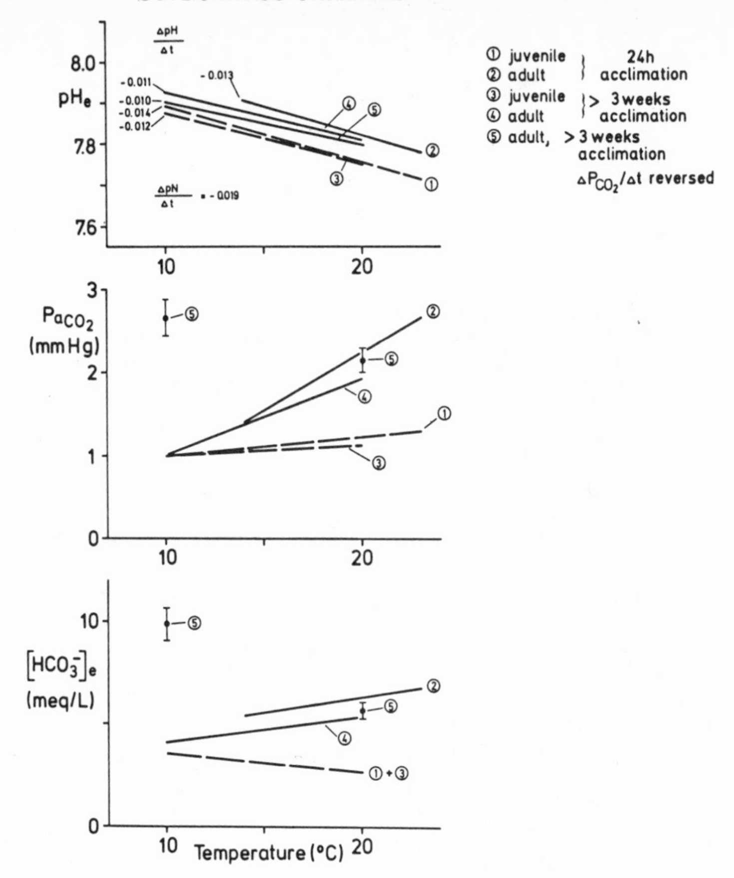

Fig. 3: Correlations between environmental temperature and extracellular pH, P_{CO_2} and bicarbonate concentration after 24h (1 and 2) or more than 3 weeks of acclimation (3, 4 and 5) in the larger spotted dogfish. The pH regulation remains unaffected also when the normally observed pattern of increasing P_{CO_2} with rising temperature ($\Delta P_{CO_2}/\Delta t$ positive) is reversed by increased P_{CO_2} at low acclimation temperature (5). 1 and 2 from Heisler, H. Weitz, A.M. Weitz (1976a); 3 , 4 and 5 from Heisler, Neumann and Holeton, 1980.

compartments (Reeves and Malan, 1976), regulation of P_{CO_2} towards constant imidazole dissociation would result in only small changes in bicarbonate concentration. Apparently, the conditions of Reeves Alphastat hypothesis, constant relative alkalinity and more or less constant bicarbonate concentration, are met in the extracellular body compartments of several air breathing ectotherms (references above) and also in the intracellular space of frog and turtle muscle (Malan, Wilson and Reeves, 1976).

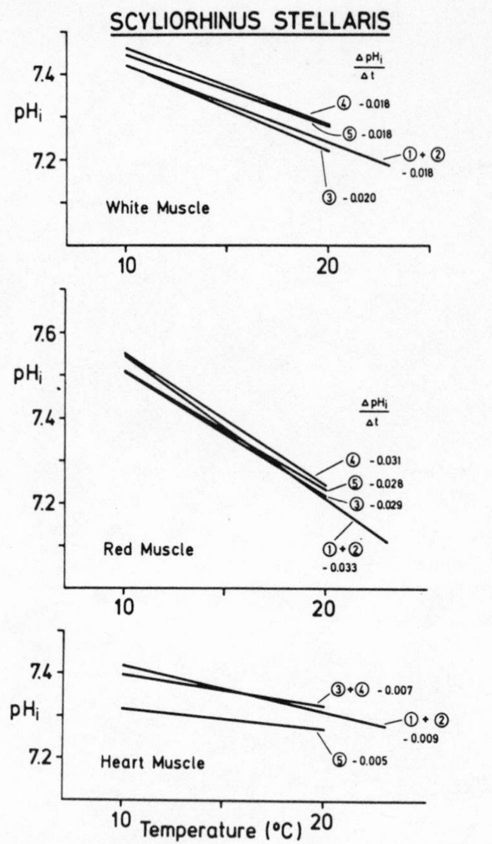

① juvenile } 24h
② adult } acclimation
③ juvenile } >3 weeks
④ adult } acclimation
⑤ adult, > 3 weeks acclimation $\Delta P_{CO_2}/\Delta t$ reversed

Fig. 4: Intracellular pH as a function of temperature in muscle tissues of the larger spotted dogfish (Heisler, Weitz, and Weitz, 1976a; Heisler, Neumann and Holeton, 1980).

In fish, however, the regulation of P_{CO_2} is limited by physical factors (see Mechanisms). This is probably the reason why the recorded acid-base status in relation to temperature is always achieved by changes of both P_{CO_2} and bicarbonate concentration in fish. As recently shown, these bicarbonate concentration changes are, at least in Scyliorhinus, due to considerable transfer of bicarbonate (or equivalent H^+) across cell membranes and gill epithelium (Heisler, 1978). The amounts transferred after temperature changes can be several times as large as the actual content of intracellular body compartments (Heisler and Neumann, 1980). According to these results, the bicarbonate transfer appears to be a very potent mechanism which should assure the adjustment of a constant relative alkalinity if the dissociation of imidazole was the trigger for this mechanism. But in only two of thirteen body compartments in five species studied, the extracellular space of Synbranchus and the intracellular compartment of white muscle in Scyliorhinus (Figs. 2, 4), pH is regulated towards a constant relative alkalinity. Thus, imidazole Alphastat regulation is of less importance in fish. In contrast to the conditions in air breathing animals the imidazole mechanism can only be considered as one of several contributors to a rather complex system of mechanisms for acid-base regulation in fish.

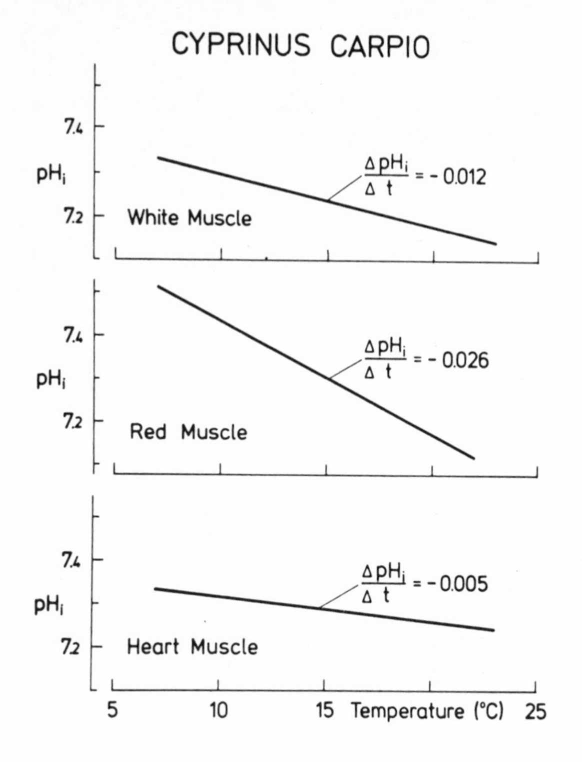

Fig. 5: Intracellular pH as a function of environmental temperature in carp (Heisler and Neumann, unpublished).

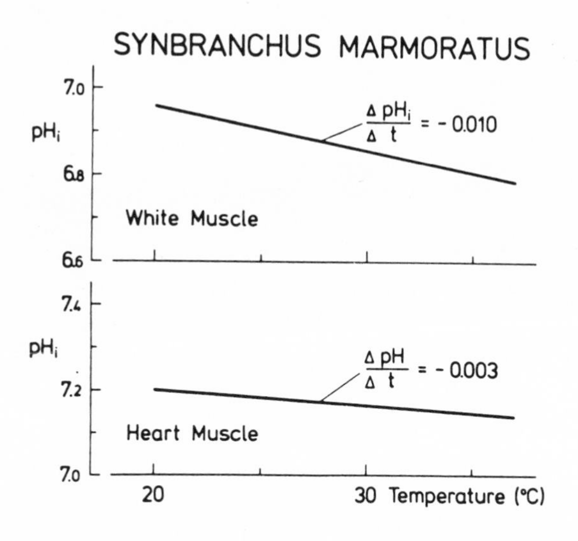

Fig. 6: Intracellular pH as a function of environmental water temperature in the air breather Synbranchus marmoratus (Heisler, 1980).

II. REGULATORY MECHANISMS

The acid-base status in fish, which has been described in some of its main features in the preceding section, is a steady-state in which continuous endogenous production of surplus H^+ or OH^- ions resulting from metabolism

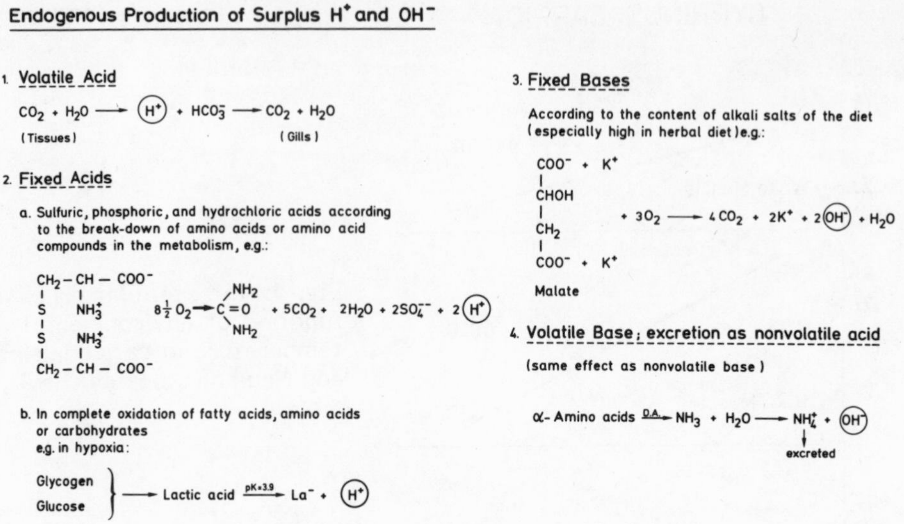

Endogenous Production of Surplus H^+ and OH^-

1. Volatile Acid

$CO_2 + H_2O \longrightarrow (H^+) + HCO_3^- \longrightarrow CO_2 + H_2O$

(Tissues) (Gills)

2. Fixed Acids

a. Sulfuric, phosphoric, and hydrochloric acids according to the break-down of amino acids or amino acid compounds in the metabolism, e.g.:

$CH_2-CH-COO^-$ / S / NH_3^+ / S / NH_3^+ / $CH_2-CH-COO^-$ $\quad 8\frac{1}{2}O_2 \longrightarrow C(=O)(NH_2)_2 + 5CO_2 + 2H_2O + 2SO_4^{--} + 2(H^+)$

b. In complete oxidation of fatty acids, amino acids or carbohydrates e.g. in hypoxia:

Glycogen, Glucose $\longrightarrow$ Lactic acid $\xrightarrow{pK=3.9}$ $La^- + (H^+)$

3. Fixed Bases

According to the content of alkali salts of the diet (especially high in herbal diet) e.g.:

$COO^- + K^+$ / CHOH / CH_2 / $COO^- + K^+$ $\quad + 3O_2 \longrightarrow 4CO_2 + 2K^+ + 2(OH^-) + H_2O$

Malate

4. Volatile Base; excretion as nonvolatile acid

(same effect as nonvolatile base)

α- Amino acids $\xrightarrow{D.A.}$ $NH_3 + H_2O \longrightarrow NH_4^+ + (OH^-)$

excreted

Fig. 7: Production of surplus H^+ or OH^- ions in the metabolism.

(fig. 7) is counterbalanced by continuous removal of these ions from the "milieu interieur" of the fish. The balance between H^+ or OH^- ions produced and those eliminated is maintained even with changing metabolic output due to changing diet, periodic feeding and activity periods. Probably not all of the regulatory mechanisms underlying these fine adjustments are known, but at least the following are considered to be involved.

1. Changes of P_{CO_2}

Under resting conditions at a constant temperature, P_{CO_2} is a fairly constant parameter and is a function of CO_2 production, P_{CO_2} in the inspired water, gill ventilation and the CO_2 diffusion resistances in the gill epithelium including rate limitation of HCO_3 dehydration (see also: Cameron, 1979). Due to the low oxygen content in water as compared to air, fish have to ventilate their gills much more than airbreathing animals ventilate their lungs in order to take up sufficient amounts of oxygen. This larger ventilation is mainly responsible for the much smaller inspired/arterial P_{CO_2} differences (Rahn, 1966) which are at least a factor of 10, and in some cases, a factor of 60, smaller (see chapter 4) than in air breathing species. Even with this high gill ventilation, the O_2 extraction from the water is generally larger than 50% whereas about 20% O_2 extraction from the ventilating air is typical for mammals. This comparison clarifies the point that a reduction of ventilation and increases in P_{CO_2} for acid-base regulatory purposes must be very limited in fish (see also Rahn, 1966a). Additionally, hyperventilation in water creates energetic problems for the fish. Therefore, it is not surprising that gill ventilation is either affected very little by even severe hypercapnia (Babak, 1907; Van Dam, 1938; Saunders, 1962;

Peyraud and Serfaty, 1964; Dejours, 1973), or only transiently increases until the plasma pH is partially restored by accumulation of bicarbonate due to other mechanisms (Janssen and Randall, 1975; Randall, Heisler and Drees, 1976). Only when the oxygen supply of the animals is compromised by environmental hypoxia fish do hyperventilate considerably, whereas hyperoxia results in hypoventilation (Babak, 1907; Peyraud and Serfaty, 1964; Dejours, 1973; Dejours, Toulmond and Truchot, 1977). Concomitantly, P_{CO_2} is lowered in hypoxia and increased in hyperoxia (Dejours, 1973). Based on this evidence, the steady-state P_{CO_2} in fish cannot be considered as the result of a regulatory process responsible for the acid-base balance, but as a parameter which is predetermined according to the oxygen content of the environment, the gas exchange conditions and probably other parameters.

2. Physico-Chemical Buffering

Physico-chemical buffering can be characterized as the transduction of added H^+ or OH^- ions into a non-dissociated state. Biological buffer systems consist of CO_2/bicarbonate buffer and nonbicarbonate buffers (B^-, HB), which are interlocked via the pool of free H^+ ions (fig. 8). Regardless of the different reactions of such combined buffer systems with CO_2 and nonvolatile H^+ or OH^- ions (see fig. 8) with respect to the changes in bicarbonate concentration, buffering is, according to the Henderson-Hasselbalch equation, always coupled with a shift in pH. The extent of this shift is dependent for buffering of CO_2, on the nonbicarbonate buffer value (β_{NB}, fig. 8), or for buffering of nonvolatile H^+ or OH ions, on the total buffer value (β_{tot}, fig. 8).

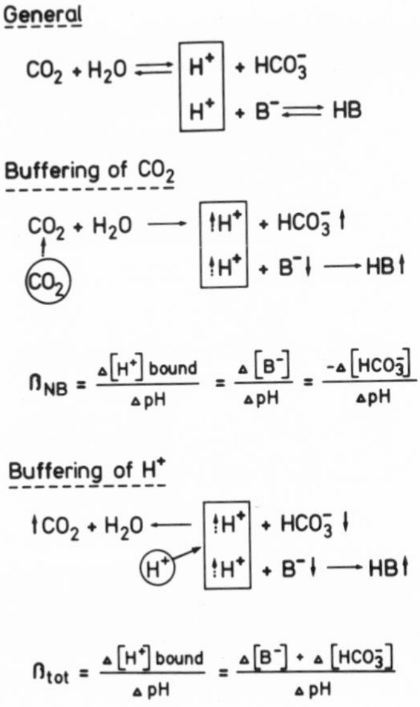

General

$$CO_2 + H_2O \rightleftharpoons \boxed{H^+} + HCO_3^-$$
$$\boxed{H^+} + B^- \rightleftharpoons HB$$

Buffering of CO_2

$$\overset{\uparrow}{CO_2} + H_2O \longrightarrow \boxed{\uparrow H^+} + HCO_3^- \uparrow$$
$$\boxed{\downarrow H^+} + B^- \downarrow \longrightarrow HB \uparrow$$

$$\beta_{NB} = \frac{\Delta [H^+]\,\text{bound}}{\Delta pH} = \frac{\Delta [B^-]}{\Delta pH} = \frac{-\Delta [HCO_3^-]}{\Delta pH}$$

Buffering of H^+

$$\uparrow CO_2 + H_2O \longleftarrow \boxed{\uparrow H^+} + HCO_3^- \downarrow$$
$$\boxed{\downarrow H^+} + B^- \downarrow \longrightarrow HB \uparrow$$

$$\beta_{tot} = \frac{\Delta [H^+]\,\text{bound}}{\Delta pH} = \frac{\Delta [B^-] + \Delta [HCO_3^-]}{\Delta pH}$$

Fig. 8: Buffering in biological buffer systems. Horizontal arrows indicate the direction of pathways utilized, vertical arrows increase (↑) or decrease (↓) in the concentration of the respective substance. B^- and HB designate the base form or acid form of nonbicarbonate buffers respectively.

The progressive shift in pH and the consumption of buffer bases (B^- and HCO_3^-) excludes the utilization of buffering as a mechanism for the final elimination of surplus H^+ or OH^-. Buffering is, however, extremely valuable for CO_2 transport in the blood from tissues to gills, for the transient removal of surplus H^+ or OH^- from the body fluids during sporadic periods of large endogenous production, or when the acid-base status is stressed by H^+ or OH^- ions originating in the environment (see chapter III, 4). While buffer values for blood have been determined in a number of fish species (5 - 20 meq/pH·L blood, see Albers, 1970), the buffering ability of intracellular tissue compartments has been studied only in the larger spotted dogfish (β_{NB} = 36 - 52 meq/pH·L cell-water for white, red and heart muscle, Heisler and Neumann, 1980). The data obtained together with data on bicarbonate concentrations, on the volume of the extracellular space and on the buffer value of whole blood (Heisler, Weitz, H., Weitz, A.M., 1976a; Heisler, Neumann and Holeton, 1980a; Heisler, 1978; Albers and Pleschka, 1967; Heisler and Holeton, 1980), allow an estimation of the mean total buffering capacity of dogfish: 38 meq/pH·L body water. About 85-90% of this capacity is attributable to non-bicarbonate buffers, 98% of which are situated in the intracellular compartments. Similar patterns can be expected for other fish species as well.

3. Excretion of H^+ Ions

Hydrogen ions resulting from metabolism are eliminated from the internal fluids of the fish via the gill epithelium, after tubular secretion with the urine and probably with coelomic fluid released through the abdominal pores in elasmobranchs. In the gills, the H^+ extrusion appears to be linked to the sodium uptake mechanism in form of an active carrier mediated H^+/Na^+ exchange mechanism (fig. 9) (e.g. Kerstetter, Kirschner and Rafuse, 1970; Payan and Maetz, 1973; Maetz, 1973, see also Maetz, 1974; Evans, 1979). The extent of linkage between the transfer of H^+ and Na^+ is rather variable between different species and, moreover, appears to be dependent on a variety of factors such as ammonia production and excretion (Maetz, 1973), see also Evans, 1979).

In the kidney, excretion of H^+ ions into the tubular lumen may serve two tasks: first, the final elimination of H^+ ions from body fluids, which are then buffered in the tubule as titratable acid mainly by secondary phosphate (fig. 9) or as ammonium ions (NH_4^+, see: Ammonia excretion). Second, it probably mediates the conservation of bicarbonate which is filtered from the plasma into the primary urine (fig. 9). The amount of bicarbonate filtered can roughly be estimated to be about 1.5 mmole/kg.day for rainbow trout (calculated from data of Holmes and Steiner, 1966 and Heisler and Holeton, unpublished) and 0.4 mmole/kg.day for spiny dogfish (calculated from data of Shannon, 1940; Burger, 1967 and Robin et al., 1966). These amounts, representing about 60 or 25% respectively of the total bicarbonate content of the animals, are large in comparison with the net excretion of H^+ ions in the urine as metabolic endproducts (e.g. rainbow trout: 0.02 mmole/kg.day, Kobayashi and Wood, 1979). Only a small fraction of the filtered

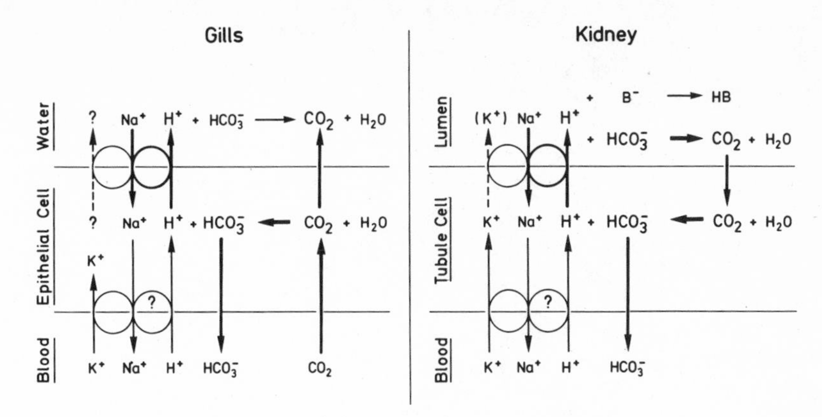

Fig. 9: Mechanisms of H^+ ion excretion in gills and kidney (see text).

bicarbonate occurs in the released urine, the larger part being reabsorbed during the passage of the tubular system. How much of the bicarbonate reabsorption is attributable to the non-ionic back diffusion of CO_2 produced by secreted H^+ (fig. 9) or to direct absorption of bicarbonate, which is postulated to be the predominant mechanisms in the elasmobranch Raja erinacea (Deetjen and Maren, 1974), is still unknown.

The coelomic fluid of the spiny dogfish has been reported to be acid (pH 5.0) and to contain no bicarbonate (Smith, 1929 b; Murdaugh and Robin, 1967). Since elasmobranchs have direct connections between the coelomic cavity and environmental water via the abdominal pores (Gans and Parsons, 1965), release of coelomic fluid could be an additional mechanism for the elimination of H^+. In Scyliorhinus stellaris, however, collection of the fluid coming from the abdominal pores and the rectal gland in a finger of a rubber glove with was glued around the cloaca, showed that no significant amounts of H^+ are excreted via abdominal pores and the rectal gland in this species under a variety of different conditions (Heisler, Weitz, Neumann, unpublished data).

4. Ammonia Excretion

Ammonia represents the main nitrogenous waste product of ammoniotelic fish (60-90%, Smith, 1929a; Wood, 1958; Fromm, 1963) and is produced by deaminiation of α - amino acids in liver, kidneys and gills (Goldstein, Forster and Fanelli, 1964; Pequin, 1962; Pequin and Serfaty, 1963). Immediately after its formation more than 97% of the highly toxic ammonia is ionized by combination with H^+ (forming NH_4^+) according to the high pK value of the NH_3/NH_4^+ buffer system, (see Emerson et al., 1975; Bower and Bidwell, 1978). In this form, it is transported to the site of

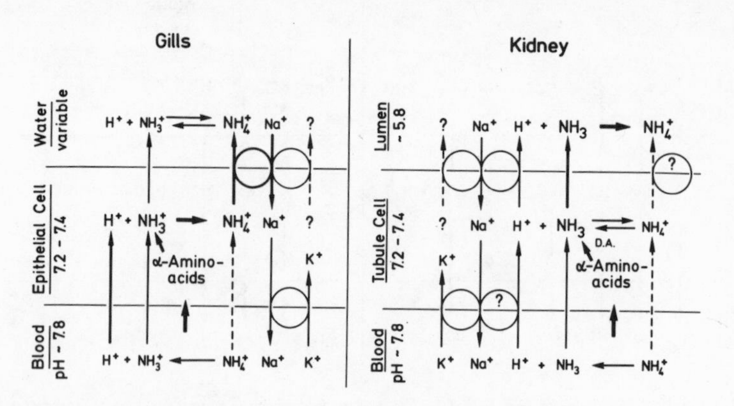

Fig. 10: Mechanisms of H^+ elimination via the ammonia mechanism in gills and kidney (see text).

excretion. In terrestrial air breathing animals, ammonia is produced in only relatively small amounts, probably because of the problem of elimination and is almost exclusively excreted in the kidneys. In contrast, it is predominantly eliminated by the gills in fish (90%, Smith, 1929a).

In the gills (see fig. 10), ammonia appears to be transferred through the epithelium mainly in form of NH_4^+ by a carrier mediated active Na^+/NH_4^+ ion exchange mechanism similar to that for H^+ ions (e.g. Maetz, 1973; Kerstetter, Kirschner and Rafuse, 1970; Evans, 1977; Payan and Maetz, 1973; see also Maetz, 1974 and Evans, 1979), but also according to the high lipid solubility, and therefore high cell-membrane diffusivity, by non-ionic diffusion through the epithelium (Maetz, 1973). The latter takes place especially when the ammonia production exceeds the active resorption of Na^+ ions, or at lowered ambient pH because of the decreased concentration of the free base NH_3 (Maetz, 1973).

Ionic and non-ionic elimination of ammonia in the gills have different effects on the acid-base status of the fish. Excretion of NH_4^+ is effectively the same as removal of H^+ from the body fluids, whereas non-ionic outdiffusion of NH_3 is neutral with respect to the acid-base balance of the organism. Ammonia in the acidic urine is always released to the environment in the form of NH_4^+, regardless of the way in which ammonia is transferred across the tubular wall. Most probably, ammonia enters the tubule lumen as in the mammalian kidney (by non-ionic diffusion) and is trapped there as NH_4^+ by buffering of H^+ ions (fig. 10), which are secreted by the tubular cells (see: H^+ excretion). It is, however, possible that NH_4^+ is actively secreted as in the gill epithelium as well. The effect on the acid-base status of the fish is the same for both processes: alkalinization by removal of H^+ ions.

5. Bicarbonate Excretion

In ammoniotelic fish species, relatively large amounts of bicarbonate are produced when ammonia is ionized by uptake of H^+ ions (see Ammonia Mechanism). Krogh (1939) was the first to suggest that in order to compensate for the passive leakage of chloride to the environment along the electrochemical gradient in freshwater fishes, chloride is taken up in exchange for endogenous negatively charged ions, preferably bicarbonate. Such a HCO_3^-/Cl^- exchange mechanism is an advantageous combination of leakage compensation and metabolic endproduct elimination. It appears to be present in goldfish gills (Maetz and Garcia-Romeu, 1964, De Renzis and Maetz, 1973; De Renzis, 1975) in trout gills (Kerstetter and Kirschner, 1972), and was recently shown in the gills of the sea-water teleost Opsanus beta as well (Kormanik and Evans, 1979) in the opposite direction. Good evidence exists that in the goldfish (Maetz and Garcia-Romeu, 1964, injection of acetazolamide inhibits Cl^-/HCO_3^- exchange) as well as in the gulf toadfish (Kormanik and Evans, 1979, no effect of high pH in absence of HCO_3^-) bicarbonate and not OH^- is the main counter ion for the Cl^- exchange mechanisms.

Interrelationships Between H^+, NH_4^+ and HCO_3^- Exchange Mechanisms

Hydrogen ions and NH_4^+ ions appear to be at least partially excreted from the internal body fluids by exchange with Na^+ ions. The linkages between either H^+ or NH_4^+ extrusion alone and sodium uptake are reported to be dependent on the fish species and its internal and external conditions (references above). However, the combined amounts of H^+ and NH_4^+ excreted were fairly well correlated to the amount of Na+ taken up in a variety of different conditions in goldfish (Maetz, 1973). This tight correlation suggests that H^+ and NH_4^+ are competitors in the exchange with Na^+. Also, the transfer of bicarbonate from fish to water seems closely correlated to chloride uptake in this species (De Renzis and Maetz, 1973). For both exchange mechanisms, the same ion exchange ratio of 4 Na^+ or Cl^- against 3 endogenous ions ($H^+ + NH_4^+$ or HCO_3^-, respectively) was found in goldfish gills (De Renzis and Maetz, 1973). This coherence suggests that the two mechanisms are interdependent, a hypothesis which is supported by experimental evidence: hindrance of one of the two ion exchange mechanisms by manipulations of ion concentrations in the environment, or by specific ion exchange inhibitors caused a considerable impact on the other ion exchanges in most species (for references, see: Maetz, 1974; Evans, 1979).

Linkages and interrelationships of ion exchange processes in fish gills have been very much emphasized during recent years. There is now clear experimental evidence for the interdependence of ion transfer processes. It should, however, be considered that this interdependence must be limited. A rigid and absolute linkage between the ion exchange mechanisms would clearly limit the fish in its ability to regulate both electrolyte and acid-base status. Since both are closely regulated at certain set values under a wide range of environmental conditions, there must be some flexibility of the whole regulatory system. This has to be so since, under steady state conditions, not only HCO_3^-, Cl^-, H^+, NH_4^+ and Na^+ have to be exchanged between fish and environment, but anions like SO_4^{--}, HPO_4^{--} or cations like K^+, Mg^{++}, Ca^{++} are also taken up with the food or are produced in the metabolism and, consequently, have to be excreted into the environment. These aspects of the acid-base regulation in fish have rarely been considered in ion exchange studies and require additional investigation.

III. ENVIRONMENTALLY INDUCED DISTURBANCES OF THE ACID-BASE STATUS AND RESPONSES OF THE REGULATORY SYSTEM

The acid-base status of fish is, in addition to the endogenously produced steady state load of H^+ or OH^- ions, stressed by various environmental factors. Some of these are changes in water temperature, hypercapnia, low water pH, and lactacidosis due to hypoxia or severe muscular activity induced by emergency situations. This list of stress factors is far from complete, but these are the only ones which have been imposed upon fish in order to study acid-base regulation. Generalizations are difficult as only a limited number of species was subjected to specific environmental stress factors and as some of the studies are incomplete with respect to certain parameters. Additional investigations are required.

1. Temperature Changes of the Environmental Water

Most fish species encounter relatively large temperature changes in their environments throughout the year. This is also true for so-called stenothermal species like trout and other salmonids, which are subjected to temperatures up to 22°C in the summer and to temperatures near the freezing point in the winter. In addition to these slow seasonal changes, fish are increasingly subjected to sudden temperature changes, especially due to the release of cooling water from industrial plants.

The only fish species which has been studied with respect to the influences of sudden temperature changes on acid-base status is the small shark species Scyliorhinus stellaris (larger spotted dogfish, Heisler, 1978). When the environmental temperature is step changed from 10 to 20°, plasma pH drops immediately by about 0.4 pH units (fig. 11) exclusively due to a considerable increase in plasma P_{CO_2} (3 fold). P_{CO_2} returns rapidly into the range of later observed steady-state values for 20°C (see also section I). pH recovers as well, reaching 20°C steady-state values after 5h and, after a slight overshoot, remains steady from the 15th hour on. Most of the pH recovery is attributable to a considerable rise in plasma bicarbonate during the first 15h after the temperature change. At the same time, bicarbonate is also released to the sea-water. Both the amount of bicarbonate added to the extracellular space as well as the amount released from the fish to the environment originates in the intracellular compartments (fig. 13). The changes of plasma and water bicarbonate concentrations after the opposite

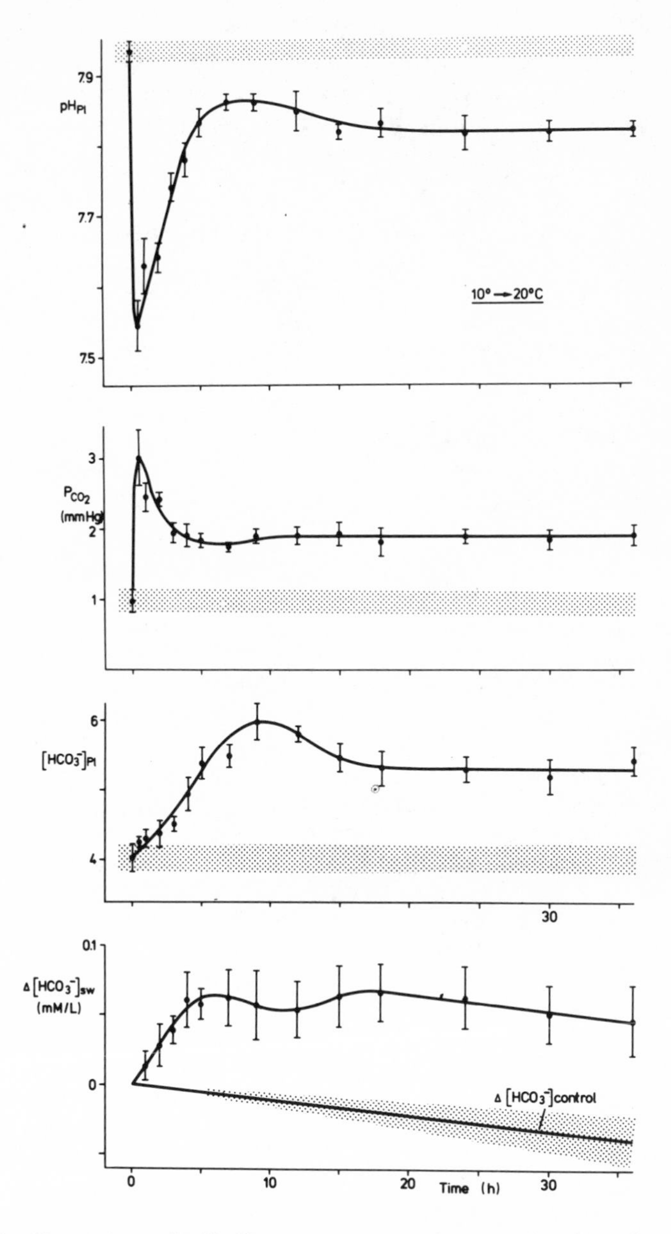

Fig. 11: Reaction of pH, P_{CO_2} and bicarbonate in dorsal aortic blood and of the bicarbonate concentration in the environmental sea-water of Scyliorhinus stellaris after a temperature step change from 10 to 20 °C (from Heisler, 1978).

temperature step from 20-10°C are almost mirror images of the patterns described above; the overshoots and undershoots of pH and P_{CO_2}, however, are smaller (fig. 12 and fig. 13). The recorded bicarbonate (or equivalent H^+ or OH^-) transfer across cell membranes and gill epithelium appears to be an important factor for the acid-base regulation after temperature changes in dogfish. As recently shown, large proportions of the bicarbonate formed or decomposed intracellularly by titration of the nonbicarbonate buffers induced by temperature changes has to be transferred in one or the other direction across the cell membranes in order to achieve the intracellular pH values determined in vivo (Heisler and Neumann, 1980). Calculations modelling the intracellular compartments as closed buffer systems (i.e. exclude bicarbonate transfer across the cell membranes) indicate that under such conditions, large deviations in pH from the observed steady state values would be expected (Heisler and Neumann, 1980). This regulatory pattern is different from the conditions in the air breathing species frog and turtle where no bicarbonate transfer, but only adjustment of P_{CO_2} appears to account for the adjustment of intracellular pH to the steady state value at the new temperature (Reeves and Malan, 1976). In dogfish, temperature induced changes in P_{CO_2} are of only minor importance. Even reversal of the normally observed $\Delta P_{CO_2}/\Delta t$ value (positive) by artificially increased plasma P_{CO_2} in the animals acclimated to low temperatures does not change the pH regulation of intracellular and extracellular compartments; the bicarbonate concentration is adjusted such that the same pH values are achieved as under normal conditions (Heisler, Neumann and Holeton, 1980; figs. 3 and 4).

The mechanism of bicarbonate (or equivalent H^+ or OH^-) transfer across cell membranes between intracellular and extracellular body compartments after temperature changes is considered to be active (Heisler, Neumann and Holeton, 1980), but the form of the carrier as well as the predominant counter ion is still unknown. The site of exchange with the environment is in Scyliorhinus almost exclusively the gill epithelium, and only 3% of the bicarbonate exchange can be attributed to kidney action (Heisler, unpublished). The small ammonia output of the ureotelic dogfish does not change significantly with temperature, so that the acid-base regulation in the gills must be due to either transfer of H^+, probably in exchange with Na^+ or to transfer of HCO_3^- in exchange with Cl^-. Some evidence pointing to the HCO_3^-/Cl^- exchange mechanism is the observation that, in the arctic grayling, temperature increases are accompanied by an increase in branchial Cl^- influx (Cameron, 1976). In these experiments, however, bicarbonate efflux was not determined. For clarification of the mechanisms involved, further experiments with measurements of both electrolyte and acid-base ion fluxes are necessary.

2. Environmental Hypercapnia

Hypercapnia frequently occurs in natural waters due to inhibition of CO_2 exchange with the air by thermo-stratification or by surface plant

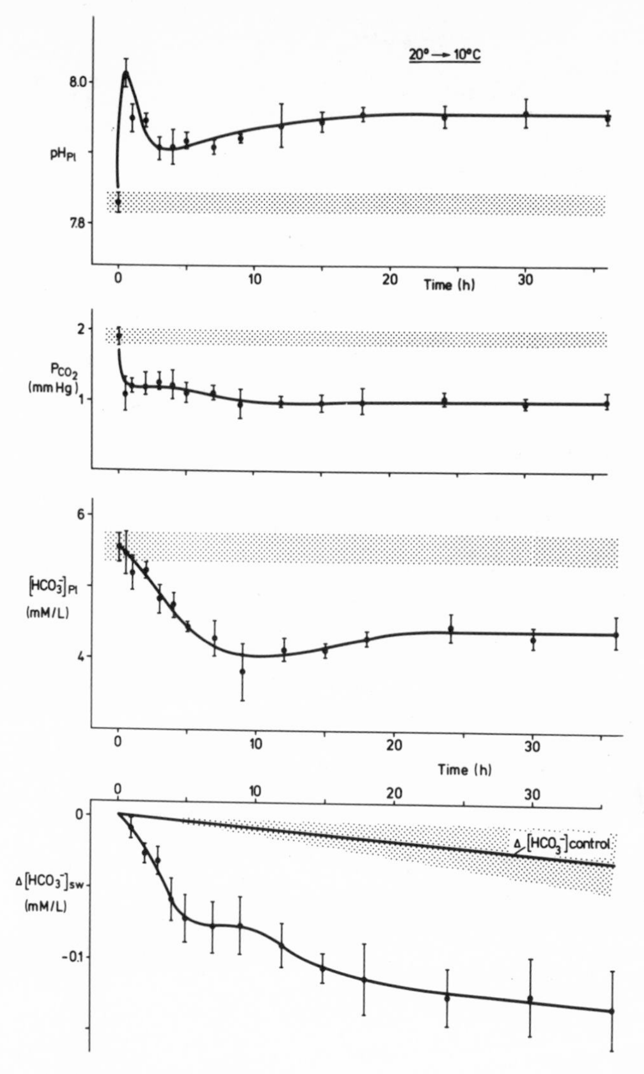

Fig. 12: As for fig. 11 for a step change in temperature from 20 to 10°C (from Heisler, 1978).

layers like water hyacinth mats (P_{CO_2} values up to 50-60 mmHg, Ultsch and Antony, 1973; Heisler and Ultsch, unpublished). In sea-water, P_{CO_2} can also rise considerably in comparison with the surface value of 0.20 to 0.25 mmHg, reaching values of more than 5-10 mmHg in water depths of 100-500 m (Harvey, 1974). This rise in P_{CO_2} is due to the absence of photosynthesis combined with respiration and anaerobic glycolysis of animals, and bacteria attached to falling particles of organic debris.

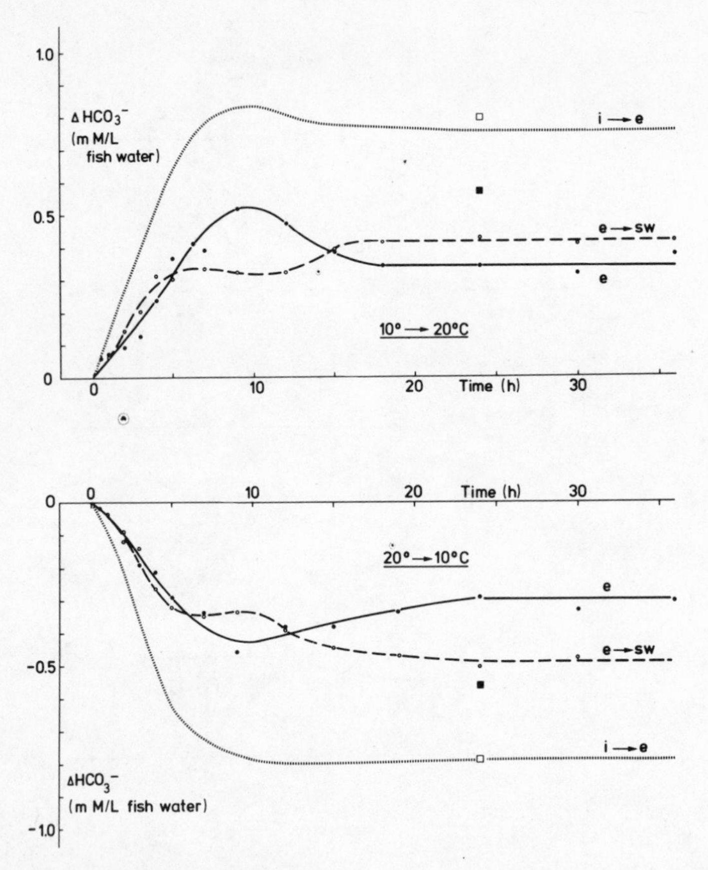

Fig. 13: Changes in bicarbonate amount in the extracellular space (e) and bicarbonate (or equivalent H^+ or OH^-) transfer from intracellular to extracellular compartment (i→e), and from extracellular compartment to sea-water (e→sw) after temperature step changes from 10 to 20°C and from 20 to 10°C in Scyliorhinus stellaris (from Heisler, 1978).

The effect of environmental hypercapnia on the blood acid-base status has been studied in a number of water breathing fish species (fig. 14). The response in acid-base status is similar in all species and is characterized by an initial drop in plasma pH, which starts to recover rapidly and which is finally compensated, reaching almost control values, in spite of continued hypercapnia, by considerable increases in plasma bicarbonate. Also, intracellular pH of three muscle types in dogfish (fig.15) and the mean intracellular pH of channel catfish is largely compensated to less than - 0.05 pH units difference from control values (Heisler, Weitz, Weitz and Neumann, 1978; Cameron, pers. comm.).

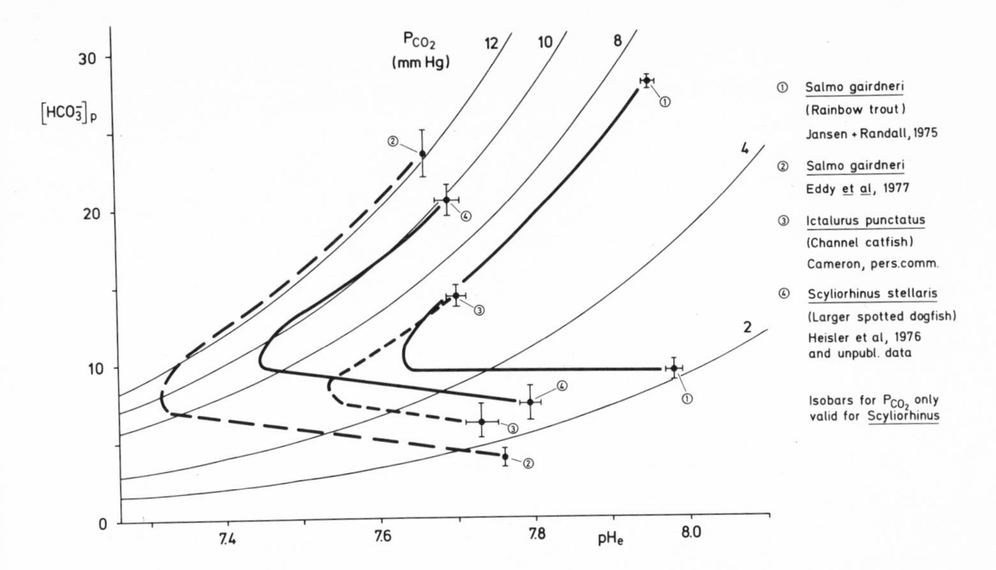

Fig. 14: Compensatory changes of plasma bicarbonate concentration after exposure to environmental hypercapnia in three water breathing fish species.

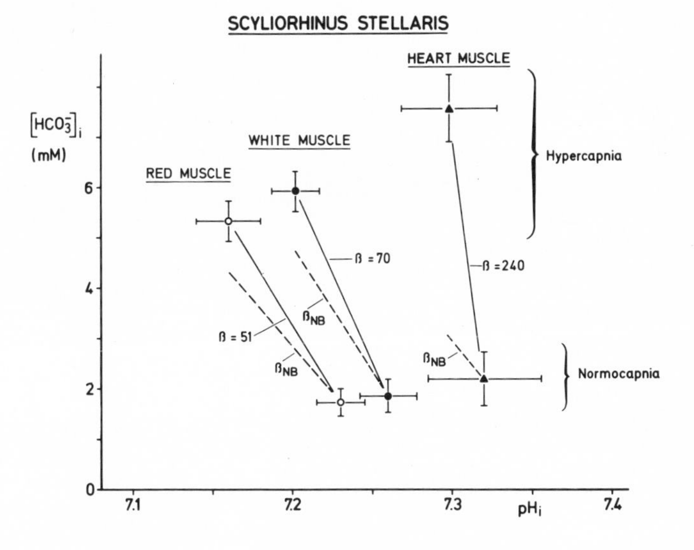

Fig. 15: Changes in intracellular pH and bicarbonate concentration after exposure to environmental hypercapnia for 5 days in the larger spotted dogfish. Dashed lines labelled 'β_{NB}' indicate the rise in bicarbonate attributable to intracellular nonbicarbonate buffering (Heisler and Neumann, 1980).

Non-bicarbonate buffering of CO_2 suggested by Cross et al. (1969) to be the main source of additional bicarbonate cannot be responsible for the observed bicarbonate accumulation since the extracellular nonbicarbonate buffer value (Albers and Pleschka, 1967; Heisler and Holeton, 1980) is negligible in comparison with the observed bicarbonate increases. Also, in the intracellular compartments, the bicarbonate increases are much larger than those attributable to nonbicarbonate buffering (fig. 15, Heisler and Neumann, 1980). Thus, the bicarbonate could only have been produced in the metabolism or, more likely, have been taken up from the environment. This latter possibility was studied in Scyliorhinus kept in a closed seawater recirculation system (Heisler, Weitz and Weitz, 1976b). Accumulating changes in the water bicarbonate concentration were determined by very precise pH measurements in samples of the recirculating water which was equilibrated to a constant P_{CO_2} similar to the method described by Heisler (1978). Under these conditions, changes in pH are exclusively due to bicarbonate (or equivalent H^+ or OH^-) transfer between environmental water and fish. After an initial phase where the extracellular bicarbonate increase was due to intracellular/extracellular transfer, the compensatory bicarbonate accumulation was due to bicarbonate uptake from the water into both extracellular and intracellular compartments. Separation of the urine and of the fluid excreted from the abdominal pores and rectal glands from the recirculating water (via catheter or a finger of a rubber glove glued around the cloaca, respectively) showed that the bicarbonate transfer was performed at the gills, whereas the compensatory contribution of kidneys, abdominal pores and rectal gland (increased excretion of H^+ or NH_4^+) was of only minor importance (Heisler, Weitz and Weitz, 1976b). In channel catfish (Ictalurus punctatus), the hypercapnia-induced increase in the total body bicarbonate pool was due only to a minor extent to increased H^+ excretion in the kidneys, and was presumably also performed via branchial transfer (Cameron, pers. comm.).

The time course of compensatory bicarbonate accumulation is rather diverse in different fish species. While the marine elasmobranch Scyliorhinus stellaris compensates the extracellular pH towards the control value (to less than - 0.1 units difference) within 8 hours (Heisler, Weitz and Weitz, 1976b), the fresh water teleost Salmo gairdneri needs more than 50 hours for the same degree of compensation (Janssen and Randall, 1975). This may be due to the larger bicarbonate concentration gradients between plasma and environment for the fresh water fish in comparison with the marine dogfish. For Scyliorhinus, it has been shown that, at the same level of hypercapnia, an increased water pH (i.e. increased bicarbonate concentration) speeds up the compensation considerably; on the other hand lowering the environmental pH reduces the speed of compensation or stops the compensation completely (fig. 16) according to the fact that the bicarbonate uptake rate in hypercapnia is, in a certain range, dependent upon the pH gradient (or, at constant P_{CO_2}, upon the bicarbonate gradient) between plasma and seawater ($pH_{Pl} - pH_{SW} > 0.6$ pH units; Heisler and Neumann, 1977).

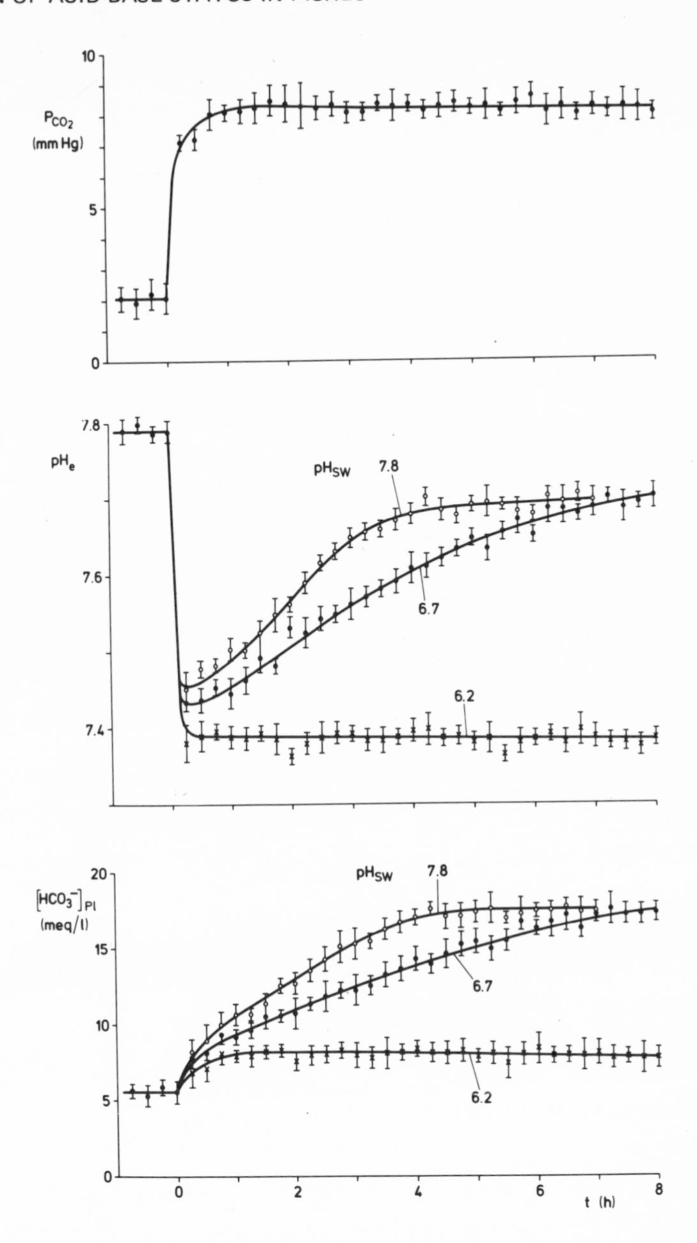

Fig. 16: Dependence of the time course of compensation of the extracellular pH by increases in bicarbonate concentration upon the environmental sea-water pH (pH_{SW}).

The mechanisms utilized for the net branchial bicarbonate uptake in hypercapnia are not quite clear. In some of his experiments on the arctic grayling (Thymallus arcticus Pallas) subjected to hypercapnia, Cameron

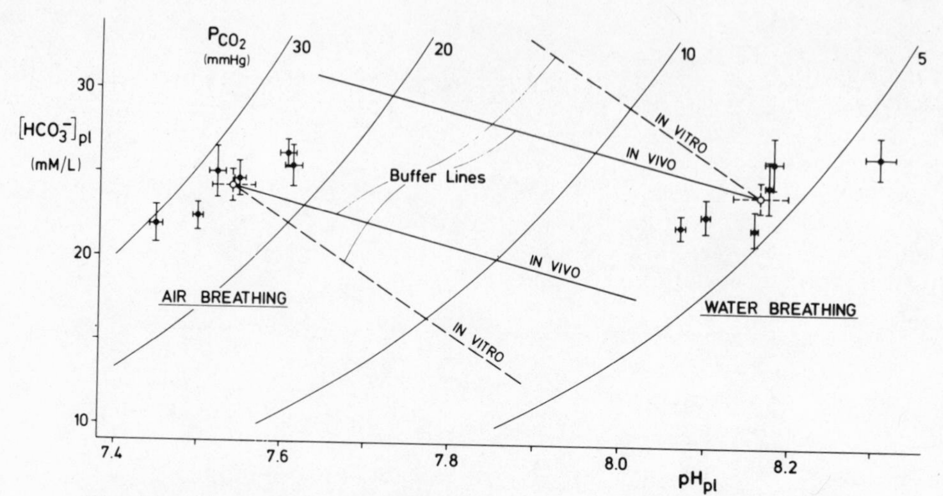

Fig. 17: Changes in plasma pH, P_{CO_2} and bicarbonate concentration during water breathing and air breathing in <u>Synbranchus marmoratus</u> (Heisler et al., 1978; Heisler, 1980b).

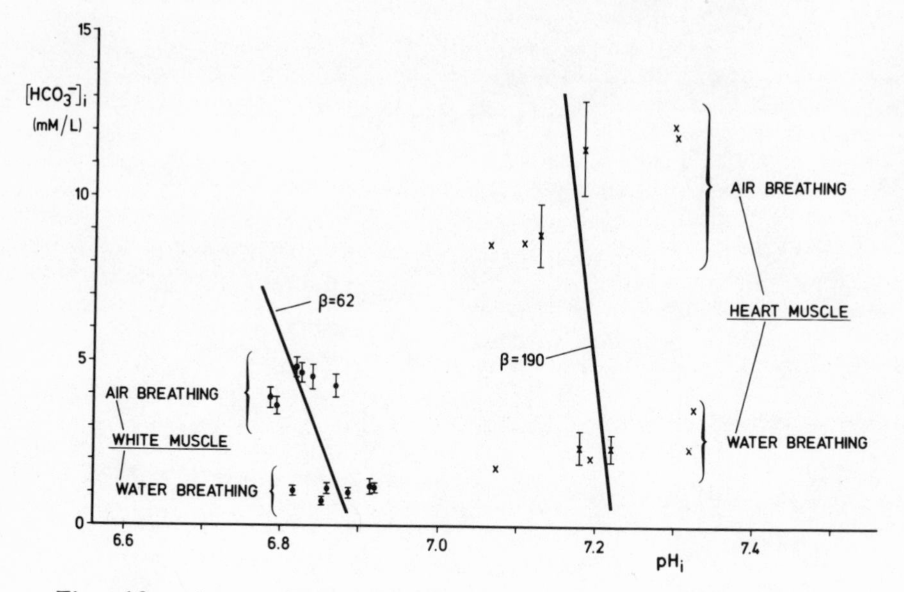

Fig. 18: Intracellular pH and bicarbonate concentration in <u>Synbranchus marmoratus</u> during water breathing and air breathing (Heisler et al., 1978; Heisler, 1980b).

(1976) could show a reduction in branchial chloride influx and an increase in sodium influx. These data suggest that the Cl^-/HCO_3^- exchange and the $Na^+/(H^+ + NH_4^+)$ exchange mechanisms are involved in the net bicarbonate uptake in the gills during hypercapnia.

3. Environmentally Induced Hypercapnia

Cessation of aquatic oxygen uptake through gills and skin and transition to exclusive air breathing causes considerable increases in blood P_{CO_2} in air breathing fishes (DeLaney, Lahiri and Fishman, 1974; DeLaney, Lahiri, Hamilton and Fishman, 1977; Lenfant, Johansen and Grigg, 1966/67; Heisler, Weitz, H., Weitz, A.M. and Neumann, 1978; Heisler, 1980b). This is due to a reduction of ventilation according to the much higher oxygen content of the breathing medium air and/or to low activity of carbonic anhydrase in the aerial gas exchange structure.

In the South American facultative air breather *Synbranchus marmoratus*, transition from water to air breathing causes a rise in plasma P_{CO_2} from 5.6 to 26 mmHg within 2 to 3 days (fig. 17). This considerable increase in P_{CO_2} is, in contrast to the observed regulation in water breathing species, not followed by a compensatory increase in plasma bicarbonate so that pH drops by more than 0.6 units showing no tendency to recover for at least 5 days (Heisler et al., 1978; Heisler, 1980). Also in the aestivating African lungfish *Protopterus*, the twofold increase of P_{CO_2} is followed by only a small rise in bicarbonate, which is actually not the manifestation of a regulatory process, but due to loss of water as indicated by concentration changes of the other electrolytes (DeLaney, Lahiri, Hamilton and Fishman, 1977). In contrast to the extracellular pH, the intracellular pH of *Synbranchus* is tightly regulated. The increased P_{CO_2} is almost completely balanced by increases in bicarbonate (fig. 18), partly due to nonbicarbonate buffering of CO_2, but also to transmembrane bicarbonate (or equivalent H^+ or OH^-) transfer. Of these relatively small amounts of bicarbonate (~ 1meq/l cell water for white muscle) transferred, about half is produced by buffering of CO_2 by the hemoglobin of the blood and half is taken up from the environment (Heisler, 1980b). This completely different strategy of acid base regulation in hypercapnia in comparison with water breathing fishes is probably closely related to the very special environment of *Synbranchus*. The Amazonian waters (environment for *Synbranchus*) are extremely electrolyte-poor (table 1) so that the uptake of bicarbonate after transition to air breathing has to be performed against a large ionic gradient. But even if enough bicarbonate could be taken up for a compensation of the extracellular pH, most of the chloride of the animal would have to be excreted in order to maintain electroneutrality. Return to water breathing, then, would imply release of the accumulated bicarbonate and rapid uptake of large quantities of chloride, a process which appears to be extremely difficult in the face of chloride concentration ratios between plasma and water of 3,000 to 30,000 (table 1). These problems in ion exchanges between fish and environment may have led to the observed strategy of *Synbranchus* in which they handle acid-base disturbances mainly by intracorporal buffering, and maintaining the more important intracellular pH by redistribution of the total bicarbonate pool.

Table 1. Ion concentrations in plasma and environmental waters of Synbranchus marmoratus (mM)

	$[Na^+]$	$[K^+]$	$[Cl^-]$	$[HCO_3^-]$
Plasma	120	3	90	25
Solimoes water basin	0.030	0.010	0.030	0.2 - 2
Rio Negro water basin	0.003	0.002	0.003	0.02 - 0.06

4. Low Environmental Water pH

Low environmental water pH is naturally observed in some electrolyte poor water-basins, such as the Amazon basin, where the typical pH in most of the studied waters was determined to be between 4 and 5, and, in some areas, up to 6 (Sioli, 1954, 1955, 1957). Such low water pH values were rarely found in Europe and North America until, during the last decades, coal mine drainages and acid precipitation lowered the pH of originally neutral or slightly alkaline (pH 7-9) water basins to values which were incompatible with sustained fish life (e.g. Parsons, 1968, 1976; Kinney, 1964; Leivestad et al., 1976; Leivestad and Muniz, 1976; Beamish, 1976; Schofield, 1976).

The effects of such rapidly occurring environmental acidosis on the acid-base status in fish has been studied by a number of investigators. Contradictory results have often been obtained. In brook trout (Salvelinus fontinalis) exposed to an environmental pH of 4.2 for 5 days (Dively et al., 1977), and in brown trout (Salmo trutta) exposed to pH 4.0 for 8 days (Leivestad et al., 1976), arterial pH remained unaffected; in contrast Janssen and Randall (1975) reported a fall in plasma pH in rainbow trout (Salmo gairdneri) exposed to water pH of 5.0. Neville (1979) also found that exposure of rainbow trout to a water pH of 4.0 resulted in a progressive reduction of plasma pH during the five days of the experiment, but when the fish were exposed to the same pH at elevated water P_{CO_2} (~ 4.5mmHg), i.e. at increased bicarbonate concentration, then no acidemia occurred (Neville, 1979).

Recently the effects of different levels of environmental acidosis have been studied in carp (Cyprinus carpio, Heisler, Ultsch and Gillespie, 1980).

status, shifting plasma pH and bicarbonate to new, slightly lower steady state values. Na^+, Cl^- and K^+ concentrations remained unaffected. The impact of a pHw of 4.0 was larger: plasma bicarbonate and pH, as well as plasma $[Na^+]$ and $[Cl^-]$ were considerably reduced. Lowering water pH to 3.5 led to rapid decompensation of the acid-base status followed by the death of the animals within several hours (Heisler et al., 1980). While the integrated net base loss (bicarbonate loss minus ammonia excretion) during exposure to pHw 5.1 accounted roughly for the bicarbonate and pH changes in the fish, the integrated net base loss during the first 80 h at pHw 4.0 was at least 3 times larger than the amount attributable to bicarbonate changes and physico-chemical buffering in the animal. This large discrepancy suggests that loss of bicarbonate under these conditions stimulated the release of carbonate from the bony structures of the fish and/or the compensatory production of a nonidentified base in the metabolism.

The apparent bicarbonate losses, determined as bicarbonate concentration increases in the environmental water at low pH, are not necessarily an expression of real bicarbonate loss (or H^+ influx). While at a pHw of 7.4, 1% of the ammonia buffer system is still present as non-ionized NH_3, at a pHw of 5.1, the relative concentration is already so low that the environment can be considered as infinite sink for NH_3. This makes it likely that most of the ammonia produced is eliminated by non-ionic diffusion at low pHw according to the high diffusivity of NH_3 (similar to the conditions in the mammalian kidney, Pitts, 1964). The immediate ionization of NH_3 to NH_4^+ by uptake of H^+ results then in the production of bicarbonate in the water. Thus, the increased production of ammonia at lower pHw probably does not contribute to the acid-base regulation under these conditions.

The postulated linkage between H^+ and NH_4^+ extrusion and Na^+ uptake (see: Mechanisms) suggests a reasonable explanation for the fall in plasma $[Na^+]$. The concentration ratios between external and internal fluid of the fish are much larger at low pH for the H^+ ions, and very likely for NH_4^+ as well, because of the fast outward diffusion of NH_3 and subsequent ionization in the water. At equilibrium the external/internal concentration ratio at pHw of 4.0 is about 2,500 times that at pHw of 7.4. Clearly equilibrium conditions for the extrusion mechanisms are never reached; but evidently, the back pressure against which H^+ and NH_4^+ must be extruded in exchange with Na^+ is considerably enhanced and is probably too high for the pump mechanism to balance passive Na^+ efflux. This hypothesis is supported by the observation of several investigators that low ambient pH interferes with Na^+ uptake in fish (Bentley et al., 1976, Maetz et al., 1976, Packer and Dunson, 1970). However, the reduction in plasma Cl^- cannot be adequately explained. Since bicarbonate is considered to be the counterion for the Cl^- pump mechanism, the observed decrease in plasma bicarbonate concentration may lower the active Cl^- influx (Ehrenfeld and Garcia-Romeu, 1978). Together with the observation that the transepithelial potential at low pHw is reversed (from negative to positive, McWilliams and Potts, 1978), these observations seem to indicate that the fall in plasma Cl^- is due to a reduced observations seem to indicate that the fall in plasma Cl^- is due to a reduced influx rather than an enhanced efflux of Cl^-. These explanations, however, cannot easily be brought into agreement with the observation that in trout exposed to water of pH 4.3 with a low calcium content, the observed acid-base changes are small in comparison with the electrolyte disturbances. On the other hand, high levels of calcium have a conserving effect on the electrolyte status, but result in considerable disturbances of the acid-base status (McDonald, Wood and Hobe, pers. comm.). These results show that our knowledge about the interrelationships of electrolyte and acid-base regulation is still insufficient and that additional work should be done in this field.

5. Lactacidosis

Lactic acid, as the most important metabolic end-product of anaerobic glycolysis, is produced under conditions of tissue anoxia mainly induced by environmental hypoxia, disturbances of oxygen uptake and transport in the blood, or to relative hypoperfusion of tissues. The principle effect on acid-base status, the production of nonvolatile H^+ ions, is the same in all cases, but is especially pronounced according to strenuous exercise in emergency situations. Up to 30-84 mmol/kg tissue (Wardle, 1978) can be accumulated in the poorly perfused white muscle of fish.

In comparison with mammalian muscle, lactate efflux from the muscle cells of fish is delayed. While in man maximal blood lactate concentrations are already attained a few minutes after the end of exercise (Margaria et al., 1933, 1934, 1963, 1964), peak values are not observed in salmonid fishes before 2-4 h (Black et al., 1957a, b, c; 1959, 1962; Holeton, Neumann and Heisler, 1980), and in tench and dogfish not before 6 h after cessation of exercise (Secondat and Diaz, 1942, Piiper et al., 1972, Heisler and Holeton, 1980). The maximal concentrations observed were higher than 20 mmol/l blood, whereas in the starry flounder (Platichthys stellatus), blood lactate rose with similar kinetics only to values of 1.8 mmol/l (Wood, McMahon, McDonald, 1977). The delayed approach to maximal blood lactate concentrations in fish cannot be attributed to additional lactate formation in the recovery period. At least in rainbow trout, the muscle lactate concentration decreases continuously from the beginning of the recovery period (Black et al., 1962) and rises only in response to additional exercise periods (Stevens and Black, 1966).

In contrast to the slow attainment of steady-state for blood lactate concentrations, the efflux of H^+ ions from the muscle cells is apparently much faster. Total blood CO_2 or plasma bicarbonate, as indicators of the extracellular H^+ load, are maximally depressed less than 20 min after exercise in trout (Black et al., 1959, Holeton, Neumann and Heisler, 1980) and dogfish (Heisler and Holeton, 1979, 1980) or 1 h after exercise in the starry flounder (Platychthys stellatus; Wood, McMahon and McDonald, 1977). But even though a steady-state for the extracellular H^+ load is attained much earlier than for lactate, the absolute amounts of surplus H^+ and lactate in the extracellular space 15 min after exercise are similar and later

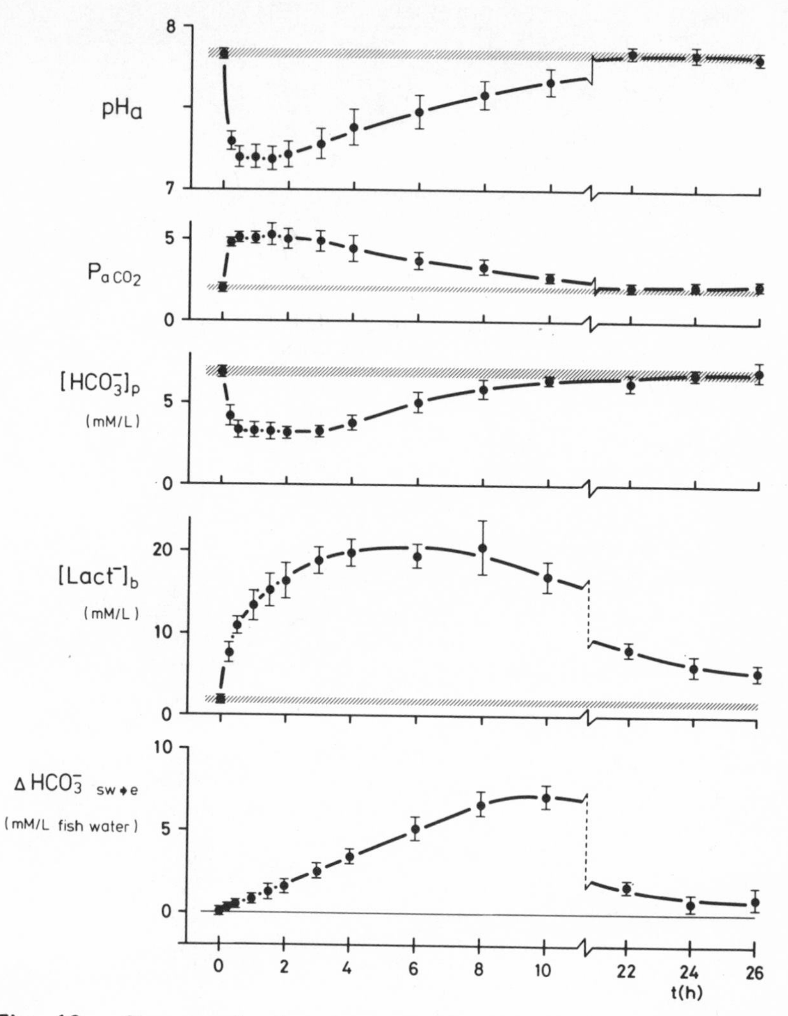

Fig. 19: Changes in dorsal aortic pH, P_{CO_2}, bicarbonate and lactate concentrations, and the amount of bicarbonate (or equivalent H^+ or OH^-) transferred from sea-water to extracellular space (sw→ e) as a function of time after a 10 min period of severe exercise in Scyliorhinus stellaris (Holeton and Heisler, 1978; Heisler and Holeton, 1980).

the lactate load even exceeds the H^+ load by a considerable amount (Piiper et al., 1972, Heisler and Holeton, 1980). Evidently, stoichiometrically equivalent amounts of H^+ and lactate are formed from the dissociation of lactic acid. When only a fraction of lactate is balanced by H^+ ions in the extracellular space, then the remaining H^+ ions must be buffered in intracellular body compartments or excreted into the environmental water (Piiper et al., 1972).

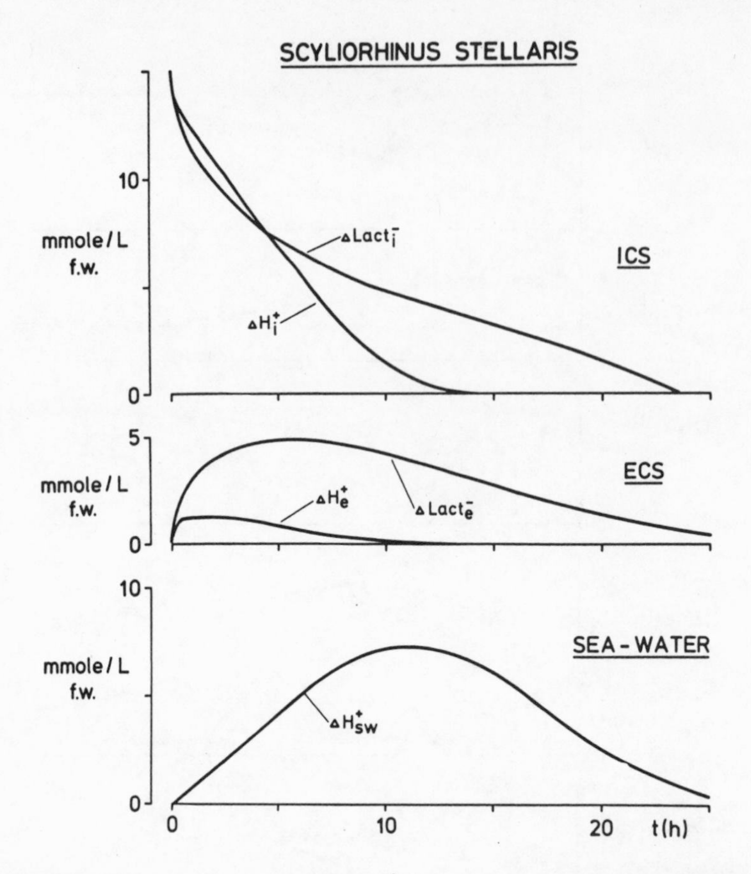

Fig. 20: Amounts of surplus H^+ and lactate ions in the intracellular (ICS) and extracellular (ECS) body compartments and in the environmental sea-water of the Larger Spotted Dogfish as a function of time after a 10 min period of strenuous activity (Heisler and Holeton, 1979, 1980).

The handling of surplus H^+ ions after exercise has been the main subject of recent studies in dogfish and trout (Holeton and Heisler, 1978; Heisler and Holeton, 1980; Holeton, Neumann and Heisler, 1980). As reported by other investigators (Piiper et al., 1972), arterial pH and bicarbonate in dogfish dropped considerably, immediately after the end of a 10 min activity period and attained steady values after 15 to 30 min (Fig. 19). The blood lactate concentration rose until the fourth hour after exercise. During the recovery time of the plasma bicarbonate concentration (10 h) the fish took up a considerable amount of bicarbonate (fig. 19, lower panel) from the water, which was repaid to the environment from the 12th hour on. At no time in the experiment did the extracellular space contain more than 10% of the total amount of surplus H^+ ions in the fish. During the first hours after exercise the largest proportion of surplus H^+ was buffered in intracellular body compartments (fig. 20, upper panel, ΔH^+_i), whereas from the tenth hour on, almost all remaining surplus H^+ ions were stored in

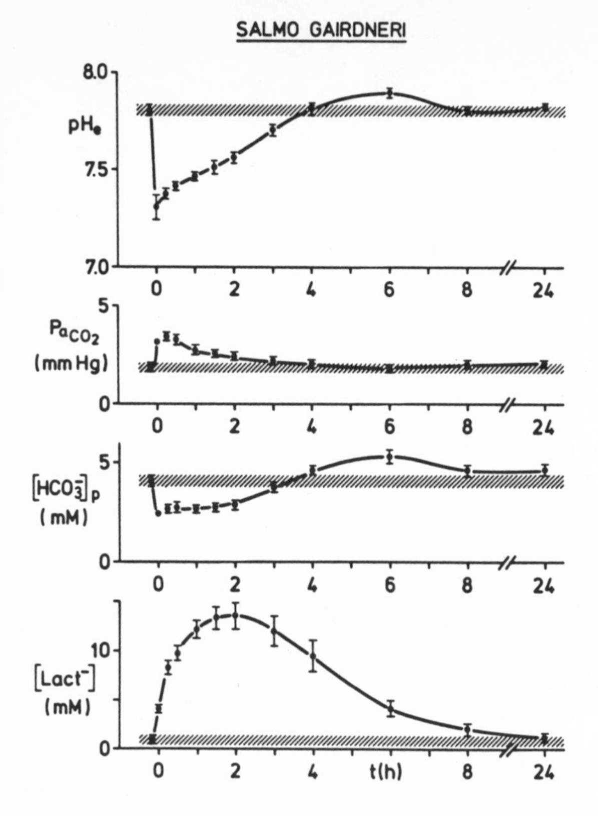

Fig. 21: Dorsal aortic pH, P_{CO_2}, bicarbonate and lactate concentration as a function of time after a 10 min period of severe muscular activity in rainbow trout (Holeton, Neumann and Heisler, 1980).

the environmental water (fig. 20, lower panel) until they were taken up once again for the further metabolic processing as lactic acid. Lactate was not transferred to the environment in detectable amounts at any time in the experiment.

In trout, the response in blood acid-base status was similar to that observed in dogfish, but plasma pH and bicarbonate were maximally depressed 5 min and lactate 2 h after the end of exercise (fig. 21).

Also in trout, large proportions of the total amount of surplus H^+ ions were transferred to the environment, the mechanisms involved, however, are different. The reduction in bicarbonate excretion after exercise is relatively small and accounts for only one fourth of the H^+ transferred (fig. 22, upper panel). The major portion is excreted utilizing the ammonia mechanisms (see above). During the first hours after exercise, the excretion of ammonia is largely increased (fig. 22, lower panel). When blood pH and bicarbonate concentration are restored, from the 6th hour on the ammonia excretion is lowered below the control rate in order to gain the H^+ ions needed for further metabolic utilization of lactic acid in the organisms (fig. 21, Holeton, Neumann and Heisler, 1980). These results show that as in the mammalian kidney, the ammonia production and excretion in fish gills can be modulated according to the requirements of the acid-base regulation

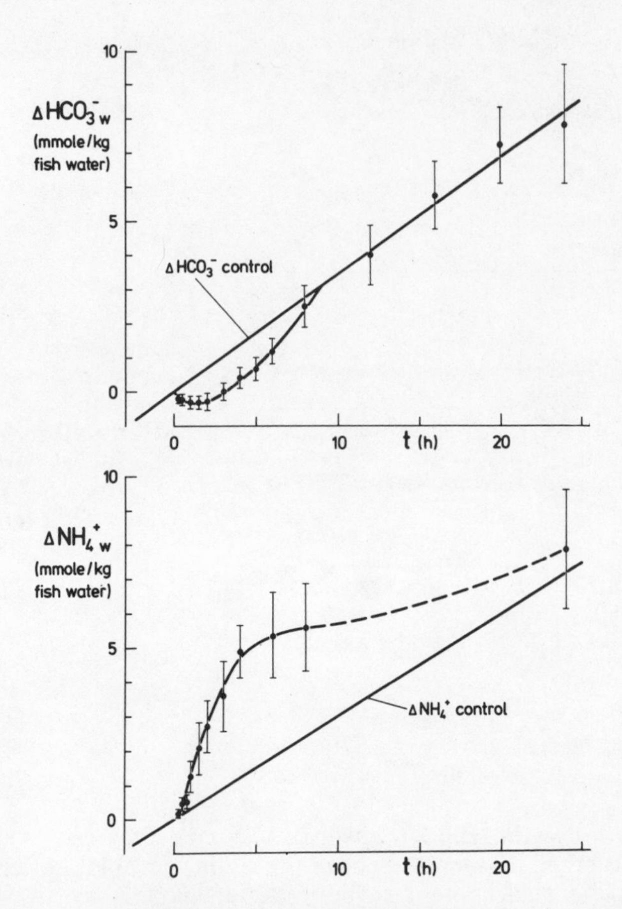

Fig. 22: Changes in the amounts of bicarbonate ($HCO_3^-{}_w$) and ammonium (NH_4^+) in the environmental water of rainbow trout as a function of time after a 10 min exercise period. Deviations from the control lines are equivalent to net bicarbonate transfer between fish and environment or to net H^+ excretion via the ammonia mechanism (Holeton, Neumann and Heisler, 1980).

and is not entirely determined by the demands of the protein metabolism. The question whether the changes in ammonia excretion are due to shifts in the relative production of ammonia and of other nitrogenous waste products, or to quantitative changes in protein metabolization remains for future studies.

Physiological Significance of Mechanisms for the Acid-base Regulation in Fish

The phenotypes of environmentally induced disturbances of the acid-base status have been described for a number of species in the preceding chapter, but the mechanisms probably involved have not been thoroughly investigated. Therefore, it is difficult to perform a generalized evaluation of any single mechanism involved with acid-base regulation in fish, and the following attempt should be taken with reserve.

Changes of body temperature in fish are accompanied by negatively correlated changes of pH (Δ pH/ Δ t negative) in all body compartments studied. Changes in P_{CO_2} contribute little to the observed pH changes, whereas bicarbonate concentration changes appear to be the much more important regulatory measure. They are, at least in the single species studied so far, not attributable to the titration of body buffers, but rather to transmembrane transfer of HCO_3^-(or equivalent H^+ of OH^-) between body compartments and the environmental water mainly via the gills.

Environmentally imposed hypercapnia regularly results in initial large decreases of the extracellular and intracellular pH values, which are then compensated by increases in the bicarbonate concentration of the respective compartment. A portion of the compensatory mechanism is attributable to bicarbonate formation by nonbicarbonate buffering, but the larger proportion is again due to bicarbonate transfer between body compartments and environment.

In air breathing fishes, cessation of aquatic gas exchange and the concomitant increase in plasma P_{CO_2} is not followed by a compensatory increase in extracellular bicarbonate. Thus, extracellular pH remains lowered. However, the intracellular pH is, at least in the one species studied, completely compensated by increased bicarbonate which is primarily produced by intracellular buffers and by the hemoglobin of the blood, and then transferred into the intracellular space. Exchange of ions with the environmental water takes place to only a small extent.

In fish exposed to moderately low water pH, arterial pH and bicarbonate concentration are slightly lowered, but attain new steady state values. At more pronounced water acidosis, the fish are unable to establish a steady state and arterial pH and bicarbonate progressively fall due to a significantly increased net base loss. In this stage, the plasma concentrations of Cl^- and Na^+ are also significantly decreased. These disturbances are probably attributable to the failure of the $Na^+/H^+ + NH_4^+$ and the Cl^-/HCO_3^- exchange mechanisms in the gills due to the large concentration gradients for H^+, NH_4^+ and HCO_3^- between plasma and environmental water.

After severe activity, with substantial production of lactic acid, the extracellular pH and bicarbonate concentration are lowered to a new steady-state within about 15 min. The blood lactate concentration approaches steady values only after 2-6 h following the exercise period. During the first phase of the H^+ load, most of the surplus H^+ ions are buffered in the intracellular space, and are transferred to the extracellular space at about the same rate as H^+ ions and NH_4^+ ions are excreted into (or HCO_3^- taken

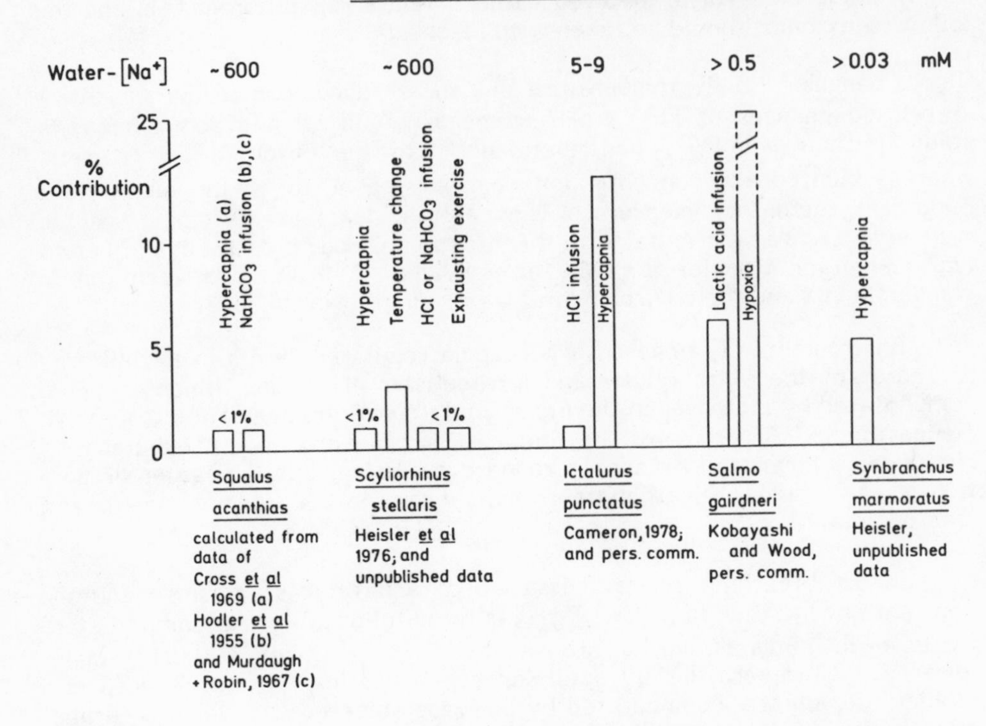

Fig. 23: Relative contribution of the kidneys in various fish species to the net regulatory transfer of HCO_3^- (or equivalent H^+ or OH^-) between fish and environmental water.

up from) the environmental water. Later, when H^+ ions are used up during the elimination of lactate from the body fluids by oxidation or resynthesis to glycogen, H^+ ions are taken up again and excretion of ammonium is reduced. In this way the environment is utilized as a store for H^+ ions, and the extracellular pH is already normalized at times when the lactate levels in the blood are still at peak values.

The preceding short summary, with special emphasis on the mechanisms clearly shows that acid-base regulation in environmental stress situations is mainly performed by ion transfer processes between the intracellular and the extracellular compartments and the environment. Buffering is undoubtedly an important mechanism as well, especially under the extraordinary conditions in air breathing fishes living in extremely electrolyte poor waters, but also after extreme work loads and in hypercapnia, before the ion exchange processes contribute significantly.

The majority of ion exchanges between organism and environment which are significantly involved in the observed regulation appear to be performed in the gills, since the contribution of the kidneys (fig. 23) seems to be small according to the available data. Which ions (H^+, OH^- or HCO_3^-) are actually transferred across cell membranes between the intracellular and the extracellular space and the environment under acid-base stress situations is still unclear, in spite of a large number of ion exchange studies in the field of osmoregulation. Simultaneous measurements of acid-base parameters and ion fluxes under such conditions appear to be necessary to increase our knowledge in this important area of fish physiology.

ACKNOWLEDGEMENTS

The author gratefully acknowledges the editorial work of Prof. M.A. Ali, as well as his co-workers' perfect organization of the NATO-ASI at Lennoxville/Quebec. The staff members of Bishop's University at Lennoxville succeeded in doing the best for accommodation of the participants of the conference and layed the basis for the important experiments by Dr. G.F. Holeton about the maximal H^+ load of the acid-base status that can be produced by break-down of muscle proteins from six steaks.

REFERENCES

Albers, C. and Pleschka, K. (1967). Effects of temperature on CO_2 transport in elasmobranch blood. Respir. Physiol. 2: 261-273.

Albers, C. (1970). Acid-base balance. In: Fish Physiology, Vol. IV: 173-208, ed. W.S. Hoar and D.J. Randall, New York, Academic Press.

Arrhenius, S. (1887). Über die Dissociation der in Wasser gelösten Stoffe. Z. Physikal. Chemie 1: 631-648.

Austin, J.H., Sunderman, F.W. and Camack, J.G. (1927). Studies in serum electrolytes. II. The electrolyte composition and the pH of a poikilothermous animal at different temperatures. J. Biol. Chem. 72: 677-685.

Babák, E. (1907). Uber die funktionelle Anpassung der äußeren Kiemen beim Sauerstoffmangel. Zbl. Physiol. 21: 97-99.

Baumgardner, F.W. and Rahn, H. (1967). Normal blood pH of the toad at different temperatures as a function of the ionization constant of water. Physiologist 10: 121.

Beamish, R. (1976). Acidification of lakes in Canada by acid precipitation and the resulting effects on fishes. In: Proceedings of the first international symposium on acid precipitation and the forest ecosystem. USDA Forest Service Technical Report NE 23. eds. L. Dochinger and T. Seliga, p. 479-498.

Bentley, P., Maetz, J. and Payan, P. (1976). A study of the unidirectional fluxes of Na and Cl across the gills of the dogfish _Scyliorhinus canicula_ (Chondricthyes). J. Exp. Biol. 64: 629-637.

Black, E.C. (1957a). Alterations in the blood level of lactic acid in certain salmonid fishes following muscular activity. I. Kamloop trout, _Salmo gairdneri_. J. Fish. Res. Board Can. 14: 117-134.

Black, E.C. (1957b). Alterations in the blood level of lactic acid in certain salmonid fishes following muscular activity. II. Lake trout, _Salvelinus_

namacush. J. Fish. Res. Board Can. 14: 645-649.
Black, E.C. (1957c). Alterations in the blood level of lactic acid in certain salmonid fishes following muscular activity. III. Sockeye salmon, Oncorhynchus nerka. J. Fish. Res. Board Can. 14: 807-814.
Black, E.C., Chiu, W.G., Forbes, F.D. and Hanslip, A. (1959). Changes in pH, carbonate and lactate of the blood of yearling Kamloops trout, Salmo gairdneri, during and following severe muscular activity. J. Fish. Res. Board Can. 16: 391-402.
Black, E.C., Connor, A.R., Lam, K.C. and Chiu, W.G. (1962). Changes in glycogen, pyruvate and lactate in rainbow trout (Salmo gairdneri) during and following muscular activity. J. Fish. Res. Board Can. 19: 409-436.
Bower, C.E. and Bidwell, J.P. (1978). Ionization of ammonia in seawater: Effects of temperature, pH and salinity. J. Fish. Res. Board Can. 35: 1012-1016.
Braekke, F. (ed.) (1976). Impact of acid precipitation on forest and fresh water ecosystems in Norway. Fag rapport FR 6176, SNSF Project, NISK, 1432, Aas-NLN, Norway, 111 p.
Brønsted, J.N. (1923). Einige Bemerkungen über den Begriff der Säuren und Basen. Rec. Trav. Chim. Pays-Bas 42: 718-728.
Burger, J.W. (1967). Problems in the electrolyte economy of the spiny dogfish Squalus acanthias. In: Sharks, Skates and Rays, eds. P.W. Gilbert, R.F. Mathewson and D.P. Rall, p. 177-185, Baltimore, John Hopkins Press.
Cameron, J.N. (1976). Branchial ion uptake in arctic grayling: Resting values and effects of acid-base disturbance. J. Exp. Biol. 64: 711-725.
Cameron, J.N. (1978). Regulation of blood pH in teleost fish. Respir. Physiol. 33: 129-144.
Cameron, J.N. (1979). Excretion of CO_2 in water-breathing animals. Marine Biology Letters 1: 3-13.
Crawford, E.C. and Gatz, R.N. (1974). Carbon dioxide tension and pH of the blood of the lizard Sauromalus obesus at different temperatures. Comp. Biochem. Physiol. 47A: 529-534.
Cross, C.E., Packer, B.S., Linta, J.M., Murdaugh, H.V. Jr., and Robin, E.D. (1969). H^+ buffering and excretion in response to acute hypercapnia in the dogfish Squalus acanthias. Am. J. Physiol. 216: 440-452.
Deetjen, P. and Maren, T. (1974). The dissociation between renal HCO_3^- reabsorption and H^+ secretion in the skate Raja erinacea. Pflügers Arch. 346: 25-30.
Dejours, P. (1973). Problems of control of breathing in fishes. In: Comparative Physiology, Locomotion, Respiration, Transport and Blood, eds. L. Bolis, K. Schmidt-Nielsen and S.H.P. Maddrell, p. 117-133. North Holland/American Elsevier: Amsterdam, New York.
Dejours, P., Toulmond, A. and Truchot, J.P. (1977). The effects of hyperoxia on the breathing of marine fishes. Comp. Biochem. Physiol. 58A: 409-411.
DeLaney, R.G., Lahiri, S. and Fishman, A.P. (1974). Aestivation of the African Lungfish Protopterus aethiopicus: cardiovascular and respiratory functions. J. Exp. Biol. 61: 111-128.

DeLaney, R.G., Lahiri, S., Hamilton, R. and Fishman, A.P. (1977). Acid-base balance and plasma composition in the aestivating lungfish (Protopterus). Am. J. Physiol. 232: R10-R17.

De Renzis, G. (1975). The branchial chloride pump in the goldfish Carassius auratus: Relationship between Cl^-/HCO_3^- and Cl^-/Cl^- exchanges and the effect of thiocyanate. J. Exp. Biol. 63: 587-602.

De Renzis, G. and Maetz, J. (1973). Studies on the mechanism of the chloride absorption by the goldfish gill: Relation with acid-base regulation. J. Exp. Biol. 59: 339-358.

Dill, D.B., Edwards, H.T. and Florkin, M. (1932). Properties of the blood of the skate (Raia oscillata). Biol. Bull. 62: 23-36.

Dill, D.B. and Edwards, H.T. (1935). Properties of reptilian blood. IV. The alligator (Alligator mississippiensis Daudin). J. Cell. Comp. Physiol. 6: 243-254.

Dill, D.B., Edwards, H.T., Bock, A.V. and Talbott, J.H. (1935). Properties of reptilian blood. III. The chuckwalla (Sauromalus obesus Baird). J. Cell. Comp. Physiol. 6: 37-42.

Dively, J.L., Mudge, J.E., Neff, W.H. and Antony, A. (1977). Blood P_{O_2}, P_{CO_2} and pH changes in brook trout (Salvelinus fontinalis) exposed to sublethal levels of acidity. Comp. Biochem. Physiol. 57A: 347-351.

Eddy, F.B., Lomholt, J.P., Weber, R.E. and Johansen, K. (1977). Blood respiratory properties of rainbow trout (Salmo gairdneri) kept in water of high CO_2 tension. J. Exp. Biol. 67: 37-47.

Edwards, H.T. and Dill, D.B. (1935). Properties of reptilian blood. II. Gila monster (Heloderma suspectum Cope). J. Cell. Comp. Physiol. 6: 21-35.

Ehrenfeld, J. and Garcia-Romeu, F. (1978). Coupling between chloride absorption and base excretion in isolated skin of Rana esculenta. Am. J. Physiol. 235: F33-F39.

Emerson, K., Russo, R.C., Lund, R.E. and Thurston, R.V. (1975). Aqueous ammonia equilibrium calculations: effect of pH and temperature. J. Fish. Res. Board Can. 32: 2379-2383.

Evans, D.H. (1977). Further evidence for Na^+/NH_4^+ exchange in marine teleost fish. J. Exp. Biol. 70: 213-220.

Evans, D.H. (1979). Kinetic studies of ion transport by fish gill epithelium. Am. J. Physiol. (in press).

Fromm, P.O. (1963). Studies on renal and extrarenal excretion in a freshwater teleost, Salmo gairdneri. Comp. Biochem. Physiol. 10: 121-128.

Gans, C. and Parsons, T.S. (1965). A photographic atlas of shark anatomy. The gross morphology of Squalus acanthias. New York, Academic Press.

Goldstein, L., Forster, G. and Fanelli, G.M. Jr. (1964). Gill blood flow and ammonia excretion in the marine teleost, Myoxocephalus scorpius. Comp. Biochem. Physiol. 12: 489-499.

Harvey, H.W. (1974). The chemistry and fertility of sea waters. Cambridge University Press. 240 p.

Hasselbalch, K.A. (1917). Die Berechnung der Wasserstoffzahl des Blutes aus der freien und gebundenen Kohlensäure desselben und die Sauersttoffbindung des Blutes als Funktion der Wasserstoffzahl. Biochem. Z. 78: 112-144.

Heisler, N. (1978). Bicarbonate exchanges between body compartments after changes of temperature in the larger spotted dogfish (Scyliorhinus stellaris). Respir. Physiol. 33: 145-160.

Heisler, N. (1980a). Acid-base regulation with temperature changes in the air breathing fish Synbranchus marmoratus. In preparation.

Heisler, N. (1980b). Regulation of the acid-base status in the facultative air breathing fish Synbranchus marmoratus after transition from water breathing to air breathing. In preparation.

Heisler, N., Holeton, G.F. (1979). Hydrogen and lactate ion elimination from muscle tissues after exhausting exercise in the larger spotted dogfish (Scyliorhinus stellaris). Pflügers Arch. 379: Suppl. R22.

Heisler, N. and Holeton, G.F. (1980). Acid-base regulation after exhausting exercise in the larger spotted dogfish (Scyliorhinus stellaris). In preparation.

Heisler, N. and Neumann, P. (1977). Influence of sea-water pH upon bicarbonate uptake induced by hypercapnia in an elasmobranch fish (Scyliorhinus stellaris). Pflügers Arch. 368: Suppl. R19.

Heisler, N. and Neumann, P. (1980). The role of physico-chemical buffering and of bicarbonate transfer processes in intracellular pH regulation in response to changes of temperature in the larger spotted dogfish (Scyliorhinus stellaris). J. Exp. Biol. 85: 89-98.

Heisler, N., Neumann, P. and Holeton, G.F. (1980). Mechanisms of acid-base adjustment in dogfish (Scyliorhinus stellaris) subjected to long-term temperature acclimation. J. Exp. Biol. 85: 99-110.

Heisler, N., Ultsch, G.R. and Ott, M.E. (1980). Acid-base and electrolyte status in Carp (Cyprinus carpio) exposed to low environmental water pH. In preparation.

Heisler, N., Weitz, H. and Weitz, A.M. (1976a). Extracellular and intracellular pH with changes of temperature in the dogfish Scyliorhinus stellaris. Respir. Physiol. 26: 249-263.

Heisler, N., Weitz, H. and Weitz, A.M. (1976b). Hypercapnia and resultant bicarbonate transfer processes in an elasmobranch fish (Scyliorhinus stellaris). Bull. Europ. Physiopath. Resp. 12: 77-85.

Heisler, N., Weitz, H., Weitz, A.M. and Neumann, P. (1978). Comparison of the acid-base regulation between a marine elasmobranch fish (Scyliorhinus stellaris) and a facultative air breathing fish (Synbranchus marmoratus). Physiologist 21: 52.

Henderson, L.J. (1909). Das Gleichgewicht zwischen Basen und Säuren im tierischen Organismus. Erg. Physiol. 8: 254-325.

Holeton, G.F. and Heisler, N. (1978). Acid-base regulation by bicarbonate exchange in the gills after exhausting exercise in the larger spotted dogfish Scyliorhinus stellaris. Physiologist 21: 56.

Holeton, G.F., Neumann, P. and Heisler, N. (1980). The role of changes in bicarbonate and ammonia excretion for the acid-base regulation after severe activity in rainbow trout (Salmo gairdneri). In preparation.

Holmes, W.N. and Stainer, I.M. (1966). Studies on the renal excretion of electrolytes by the trout (Salmo gairdneri) J. Exp. Biol. 44: 33-46.

Howell, B.J., Baumgardner, F.W., Bondi, K. and Rahn, H. (1970). Acid-base balance in cold blooded vertebrates as a function of body temperature. Am. J. Physiol. 218: 600-606.

Howell, B.J., Goodfellow, D., Rahn, H. and Herreid, C. (1972). Acid-base balance in selected vertebrates as a function of body temperature. Physiologist 15: 175.

Janssen, R.G. and Randall, D.J. (1975). The effect of changes in pH and P_{CO_2} in blood and water on breathing in rainbow trout, Salmo gairdneri. Respir. Physiol. 25: 235-245.

Jackson, D.C. (1970). The effect of temperature on pulmonary ventilation in the turtle. Physiologist 13: 230.

Jackson, D.C. and Kagen, R.D. (1976). Effects of temperature transients on gas exchange and acid-base status of turtles. Am. J. Physiol. 230: 1389-1393.

Jackson, D.C., Palmer, S.E. and Meadow, W.L. (1974). The effect of temperature and carbon dioxide breathing on ventilation and acid-base status of turtles. Respir. Physiol. 20: 131-146.

Kerstetter, T.H., Kirschner, L.B. and Rafuse, D.D. (1970). On the mechanism of sodium ion transport by the irrigated gills of rainbow trout (Salmo gairdneri). J. Gen. Physiol. 56: 342-359.

Kerstetter, F.H. and Kirschner, L.B. (1972). Active chloride transport by the gills of rainbow trout (Salmo gairdneri). J. Exp. Biol. 56: 263-272.

Kinney, E. (1964). Extent of acid mine pollution in United States affecting fish and wildlife. US Dept. Int. Bureau Sport Fisheries and Wildlife, Circular No. 191: 27 pp.

Kobayashi, K.A. and Wood, C.M. (1980). The response of the kidney of the fresh water rainbow trout to true metabolic acidosis. J. Exp. Biol. (in press).

Kormanik, G.A. and Evans, D.H. (1979). HCO_3^--stimulated Cl efflux in the gulf toadfish acclimated to sea-water. J. Exp. Zool. 208: 13-16.

Krogh, A. (1939). Osmotic regulation in aquatic animals. Cambridge, Univ. Press.

Leivestad, H., Hendry, G., Muniz, I. and Snevik, E. (1976). Effects of acid precipitation on fresh water organisms. In: Impact of acid precipitation on forest and fresh water ecosystems in Norway. Ed. F. Braekke, p. 87-111, Research Report 6/76, SNSF Project NISK, 1432 Aas-NLN, Norway.

Leivestad, H.and Muniz, I. (1976). Fish kill at low pH in a Norwegian river. Nature 259: 391-392.

Lenfant, C., Johansen, K. and Grigg, G.G. (1966/67). Respiratory properties of blood and pattern of gas exchange in the lungfish Neoceratodus forsteri (Krefft). Respir. Physiol. 2: 1-21.

Maetz, J. (1973). Na^+/NH_4^+, Na^+/H^+ exchanges and NH_3 movement across the gill of Carassius auratus. J. Exp. Biol. 58: 255-275.

Maetz, J. (1974). Aspects of adaptation to hypo-osmotic and hyperosmotic environments. In: Biochemical and biophysical perspectives in marine biology. Ed. D.C. Malins and J.R. Sargent. New York, Academic Press, p. 1-167.

Maetz, J. and Garcia-Romeu, F. (1964). The mechanism of sodium and chloride uptake by the gills of a fresh water fish, Carassius auratus. II. Evidence for NH_4^+/Na^+ and HCO_3^-/Cl^- exchanges. J. Gen. Physiol. 47: 1209-1227.

Maetz, J., Payan, P. and DeRenzis, G. (1976). Controversial aspects of ionic uptake in freshwater animals. In: Perspectives in experimental biology, Vol. I: Zoology. Ed. by P. Spencer Davies, p. 77-91, New York, Pergamon Press.

Malan, A., Wilson, T.L. and Reeves, R.B. (1976). Intracellular pH in cold-blooded vertebrates as a function of body temperature. Respir. Physiol. 28: 29-47.

Margaria, R., Cerretelli, P., di Prampero, P.E., Massari, C. and Torelli, G. (1963). Kinetics and mechanisms of oxygen debt after muscle contraction in man. J. Appl. Physiol. 19: 623-628.

Margaria, R., Cerretelli, P. and Mangili, F. (1964). Energy balance and kinetics of anaerobic energy release during strenuous exercise in man. J. Appl. Physiol. 19: 623-628.

Margaria, R., Edwards, H.T. and Dill, D.B. (1933). The possible mechanisms of contracting and paying the oxygen debt and the role of lactic acid in muscular exercise. Am. J. Physiol. 106: 689-715.

Margaria, R. and Edwards, H.T. (1934). The removal of lactic acid from the body during recovery from muscular exercise. Am. J. Physiol. 107: 681-686.

McWilliams, P. and Potts, W. (1978). The effects of pH and calcium concentrations on gill potentials in the brown trout (Salmo trutta). J. Comp. Physiol. 126: 277-286.

Murdaugh, H.V., Jr. and Robin, E.D. (1967). Acid-base metabolism in the dogfish shark. In: Sharks, Skates and Rays. Ed. W. Gilbert, R.F. Mathewson and D.P. Rall. Baltimore, John Hopkins Press, p. 249-264.

Neville, C.M. (1979). Sublethal effects of environmental acidification on rainbow trout (Salmo gairdneri). J. Fish. Res. Board Can. 36: 84-87.

Neville, C.M. (1979). Influence of mild hypercapnia on the effects of environmental acidification on rainbow trout (Salmo gairdneri). J. Exp. Biol. 83: 345-349.

Packer, R. and Dunson, W. (1970). Effects of low environmental pH on blood pH and sodium balance of brook trout. J. Exp. Zool. 174: 65-72.

Parsons, J.D. (1968). The effects of acid- strip-mine effluents on the ecology of stream. Arch. Hydrobiol. 65: 25-50.

Parsons, J.D. (1976). Effects of acid-mine wastes on aquatic ecosystems. In: Proc. First International Symposium on Acid Precipitation and the Forest Ecosystem. Ed. L.S. Dochinger and T.A. Seliga, 1074 p. US Dept. Agriculature Forest Service General Technical Report NE-23.

Payan, P. and Maetz, J. (1973). Branchial sodium transport mechanism in Scyliorhinus canicula: evidence for Na^+/NH_4^+ and Na^+/H^+ exchanges - and for a role of carbonic anhydrase. J. Exp. Biol. 58: 487-502.

Pequin, L. (1962). Les teneurs en azote ammoniacal du sang chez la carpe (Cyprinus carpio L.). Compt. Rend. 255: 1795-1797.

Pequin, L. and Serfaty, A. (1963). L'excretion ammoniacale chez un Téléostéen dulcicole: Cyprinus carpio L. Comp. Biochem. Physiol.

10: 315-324.

Peters, J.P. and Van Slyke, D.D. (1932). Quantitative clinical chemistry. Vol. II. Methods. Baltimore, Williams and Wilkins.

Peyraud, C. and Serfaty, A. (1964). Le rythme respiratoire de la carpe (Cyprinus carpio L.) et ses relation avec le taux de l'oxygène dissous dans le biotope. Hydrobiologia 23: 165-178.

Piiper, J., Meyer, M. and Drees, F. (1972). Hydrogen ion balance in the elasmobranch Scyliorhinus stellaris after exhausting activity. Respir. Physiol. 16: 290-303.

Pitts, R.F. (1964). Renal production and excretion of ammonia. Am. J. Med. 36: 720-742.

Rahn, H. (1966a). Aquatic gas exchange: Theory. Respir. Physiol. 1: 1-12.

Rahn, H. (1966b). Evolution of gas transport systems in vertebrates. Proc. Roy. Soc. Med. 59: 493-494.

Rahn, H. (1967). Gas transport from the environment to the cell. In: Development of the Lung. Ed. A.V.S. de Reuck and R. Porter, London, Churchill, p. 3-23.

Rahn, H. and Baumgardner, F.W. (1972). Temperature and acid-base regulation in fish. Respir. Physiol. 14: 171-181.

Randall, D.J. and Cameron, J.N. (1973). Respiratory control of arterial pH as temperature changes in rainbow trout. Am. J. Physiol. 225: 997-1002.

Randall, D.J., Heisler, N. and Drees, F. (1976). Ventilatory response to hypercapnia in the larger spotted dogfish Scyliorhinus stellaris. Am. J. Physiol. 230: 590-594.

Reeves, R.B. (1972). An imidazole alphastat hypothesis for vertebrate acid-base regulation: tisue carbon dioxide content and body temperature in bullfrogs. Respir. Physiol. 14: 219-236.

Reeves, R.B. and Malan, A. (1976). Model studies of intracellular acid-base temperature responses in ectotherms. Respir. Physiol. 28: 49-63.

Robin, E.D. (1962). Relationship between temperature and pH and carbon dioxide tension in the turtle. Nature (Lond). 195: 249-251.

Robin, E.D., Murdaugh, H.V., Jr. and Millen, J.E. (1966). Acid-base, fluid and electrolyte metabolism in the elasmobranch. III. Oxygen, CO_2, bicarbonate and lactate exchange across the gill. J. Cell. Physiol. 67: 93-106.

Saunders, R.L. (1962). The irrigation of the gill in fishes. II. Efficiency of oxygen uptake in relation to respiratory flow, activity, and concentration of oxygen and carbon dioxide. Can. J. Zool. 40: 817-862.

Schofield, G. (1976). Acid precipitation: effects on fish Ambio 5: 228-230.

Secondat, M. and Diaz, D. (1942). Recherches sur la lactacidémie chez le poisson d'eau douce. C.R. Acad. Sc. 215: 71-73.

Shannon, J.A. (1940). On the mechanisms of the renal tubular excretion of creatinine in the dogfish Squalus acanthias. J. Cell Comp. Physiol. 16: 285-291.

Sioli, H. (1954). Beiträge zur regionalen Limnologie des Amazonasgebietes. II. Der Rio Arapiuns. Arch. f. Hydrobiol. 49: 448-518.

Sioli, H. (1955). Beiträge zur regionalen Limnologie des Amazonasgebietes. III. Uber einige Gewässer des oberen Rio Negro-Gebietes. Arch. f.

Hydrobiol. 50: 1-32.

Sioli, H. (1957). Beiträge zur regionalen Limnologie des Amazonasgebietes. IV. Limnologische Untersuchungen in der Region der Eisenbahnlinie Belem - Branganca (Zona Brangantina) im Staate Para, Brasilien. Arch. f. Hydrobiol. 53: 161-222.

Smith, H.W. (1929a). The excretion of ammonia and urea by the gills of fish. J. Biol. Chem. 81: 727-742.

Smith, H.W. (1929b). The composition of the body fluids of elasmobranchs. J. Biol. Chem. 81: 407-419.

Sørensen, S.P.L. (1912). Uber die Messung und Bedeutung der Wasserstoffionenkonzentration bein biologischen Prozessen. Z. physikal. Chemie 12: 393-532.

Stevens, D.E. and Black, E.C. (1966). The effect of intermittent exercise on carbohydrate metabolism in rainbow trout (Salmo gairdneri). J. Fish. Res. Board Can. 23: 471-485.

Tucker, V.A. (1966). Oxygen transport by the circulatory system in green iguana (Iguana iguana) at a different body temperature. J. Exp. Biol. 44: 77-92.

Ultsch, G.R. and Antony, D.S. (1973). The role of the aquatic exchange of carbon dioxide in the ecology of the water hyacinth (Eichhornia crassipes). Florida Scient. 36: 16-22.

Van Dam, L. (1938). On the utilization of oxygen and regulation of breathing in some aquatic animals. Groningen, Volharding. 143 p.

Van Slyke, D.D. (1917). Studies of acidosis. II. A method for the determination of carbon dioxide and carbonates in solution. J. Biol. Chem. 30: 347-368.

Van Slyke, D.D. (1922). On the measurement of buffer values and on the relationship of buffer value to the dissociation constant of the buffer and the concentration and reaction of the buffer system. J. Biol. Chem. 52: 525-570.

Wardle, C.S. (1978). Non-release of lactic acid from anaerobic swimming muscle of plaice Pleuronectes platessa L.: A stress reaction. J. Exp. Biol. 77: 141-155.

Weast, R.C. (Ed.) (1975). Handbook of Chemistry and Physics. 56th edition, CRC Press, Cleveland, Ohio.

Wood, C.M. (1977). An analysis of changes in blood pH following exhausting activity in the starry flounder Platichthys stellatus. J. Exp. Biol. 69: 173-185.

Wood, J.D. (1958). Nitrogen excretion in some marine teleosts. Can. J. Biochem. Physiol. 36: 1237-1242.

Wood, S.C. and Moberly, W.K. (1970). The influence of temperature on the respiratory properties of iguana blood. Respir. Physiol. 10: 20-29.

EFFECTS OF HIGH PRESSURE ON ION TRANSPORT
AND OSMOREGULATION

A. Péqueux

Laboratoire de Physiologie Animale

Université de Liège
22 Quai Van Beneden
B-4020 Liège, Belgique

I. INTRODUCTION - DELIMITING THE MATTER

Any organism in an aquatic environment is subjected to varying degrees of hydrostatic pressure. This is particularly true in marine environments where hydrostatic pressure ranges from less than one atmosphere at the surface to around one thousand atmospheres at the bottom of the deep sea trenches. In fresh waters too, pressures up to 40 atm may be encountered and Lake Baikal in Siberia, which is 1620 meters deep, even exhibits hydrostatic pressures of the order of 162 atmospheres (Gordon, 1970). The knowledge of how hydrostatic pressure affects biological systems therefore becomes extremely important to a better and thorough understanding of the physiology of aquatic organisms.

Despite the pioneering contributions of Regnard (1884, 1885, 1891), Ebbecke (1935, 1944), Fontaine (1928), the subject has been rather neglected by biologists and progress in this field of study has remained very slow. In recent years, the role of hydrostatic pressure in the physiology and ecology of marine organisms has been renewed with increasing interest. Beyond descriptive reports of behavioural observations done on whole animals (Naroska, 1968; Brauer, 1972; MacDonald & Teal, 1975), physiological and biochemical experiments have been initiated throwing some light on the way in which pressure effects live processes (for review, see: Zimmerman, 1970; Hochachka, 1971, 1975; Hochachka, Moon & Mustafa, 1972; MacDonald, 1972; MacDonald & Miller, 1976).

Reviewing the effects of high pressure on ionic transport and osmoregulation in fishes is a very exciting but also a very difficult task. From the standpoint of a thorough understanding of the biology of deep-sea species, such a review should deal first with the ecological significance of the effects of hydrostatic pressure on osmo and iono-regulation and pressure sensitivity of mechanisms involved in ions translocation across cell membranes. However, concomitantly with increase of hydrostatic pressure, conditions for life also change considerably with increase in depth in the ocean, and the review should also deal with possible synergistic effects or interactions of the absence of sunlight and cold temperature. Moreover, when dealing with "ecological significance" of the effects of pressure, a distinction has to be made between the effects of more or less rapid transient compressions and decompressions encountered by free-swimming animals during cyclic vertical migrations, and the effects of the constant pressure supported by benthic organisms in the deep sea. Constant pressures, in addition to darkness and low temperature, might be responsible for the relative scarcity of fauna in depths exceeding 400 m (Zenkevitch, 1963). In turn they must have evolved a wide variety of adjustments and adaptation mechanisms which are difficult to study in the laboratory due to the difficulty of obtaining good healthy abyssal animals. Most of the pressure experiments thus deal with rather rapid changes in hydrostatic pressure applied to animals which normally live in the upper sea layers.

As a consequence, and in consideration of the scarcity, even the lack, of experimental data on deep-sea fishes, it appeared reasonable to concentrate in this review on the physiological effects of high hydrostatic pressures on membrane permeability and osmoregulation of surface animals normally submitted to pressure range of a few, even a few tens, atmospheres. This choice was also dictated by the fact that the general incidence of hydrostatic pressure is reviewed by Blaxter in another chapter of this book.

It will be shown that, beyond an ecological parameter whose importance is evident, pressure also appears as a very useful tool to study the normal functioning of biological membranes. By extension, hydrostatic pressure will appear as a powerful tool for the exploration of many other fundamental problems in biology among which the interplay between structure and function.

According to MacDonald & Miller (1976), from the physiological standpoint, three pressure ranges may be recognised. A first pressure range of a few atmospheres characterises sea shore areas and surface waters. It has been well established that organisms inhabiting this type ot biotope are sensitive to the ambient pressure and more particularly to small pressure changes. In fishes, such a sensitivity has long been related to the occurrence of a swimbladder, although it appears now clear that small fishes without swimbladders are pressure sensitive too (for review, see Knight-Jones & Morgan, 1966). It is very likely that such sensitivity involves an action of pressure on specific nervous receptors or very localised membrane phenomena of electrical nature, and must determine a wide variety of behavioural reactions. Several indirect evidences however point out that osmoregulatory processes must remain quite unaffected. The second pressure range, up to 100 atmospheres, corresponds to the upper 1000 m of the ocean which provide biotopes to the majority of marine organisms. This pressure range is therefore particularly suitable for studies of pressure effects on upper layer or surface animals which may undertake rapid or slow vertical migrations in this deeper range. Such studies still present a concrete physiological interest from the standpoint of pressure considered as an important ecological parameter.

The bulk deep-sea environment extends from 1000 till 10,000 m, providing the third pressure range 100-1,000 atm. According to a table of Zenkevitch (1963), fishes are still represented at more than 7,000 m where they have to cope with pressures of more than 700 atm. Moreover, some of them still migrate vertically in this depth range. These deep sea species must have developed adaptive features to tolerate this ambient high pressure. Physiological investigation on such animals is however very complicated due to the difficulty of collecting healthy specimens. As will be shown in the following, pressures up to 1,000 atm are known to disturb reversibly a number of biological systems among which permeability processes. However, from the standpoint of pressure considered as a tool for exploration of fundamental biological problems, this pressure range can still be considered as moderate. We have nevertheless decided to limit our review to the effects of applied pressures up to 1,000 atm in order to remain in the range which can be experienced in nature.

II. EFFECTS OF HYDROSTATIC PRESSURE ON FISHES. LABORATORY STUDIES.

A. Whole animals

1. Behavioural effects

In early experiments done by Regnard (1884), the effects of pressure steps up to 1,000 atm have been investigated on invertebrates and fishes under laboratory conditions. These studies essentially deal with descriptions of various behavioural events, some of them like convulsions and paralysis thresholds being selected as criteria for measuring pressure tolerance.

Later on, Ebbecke (1935-1944) completed early works of Regnard and explained the observed changes in locomotor activity paralysis and tetany states in species of *Gobius*, *Pleuronectes* and *Spinachia* as the result of a pressure induced nervous reaction.

More recently, Naroska (1968) reported the effects of pressures ranging from 100 to 800 atm in 1 h experiments on several fishes by determining, after a 24-hour recovery period, the magnitude of pressure which killed 50% of the animals (LD50). The LD50 was about 370 atm for *Zoarces viviparus*, 150 atm for *Pleuronectes platessa* and of 130 atm for *Platichthys flesus*.

In 1972, Menzies, George & Avent found *Hippocampus* to experience tetany at 3,160 psi (about 215 atmospheres), whereas the American eel Anguilla rostrata exhibits tremor and convulsions respectively upon slow compression at 55 and 105 atm (compression rate: 24 atm/h) (Brauer, 1975). The review of behavioural changes and tolerance for high pressures could be continued in that way for a variety of both freshwater and marine shallowwater fishes. Although details may vary somewhat between species, the basic pattern of response to application of high hydrostatic pressure steps to surface fishes could be summarised by considering pressure to have a stimulatory effect on activity up to 100-150 atm, an inhibitory effect from 150-300 atm and a lethal effect beyond 400 atm.

2. Oxydative metabolism

Several attempts have been made to correlate the above events with a possible action of pressure on fish metabolism. As an example, Fontaine (1928) has established that the oxygen consumption of the plaice Pleuronectes platessa increases by 28% at 25 atm., by 39% at 50 atm and by 58% at 100 atm, whereas it falls down at pressures exceeding 100 atm, which is in rather good agreement with the observations on locomotor activity. This is however in contrast with the results of Meek & Childress (1973) on the bathypelagic species Anopologaster cornuta which does not exhibit any significant changes of oxygen consumption upon application of a pressure step of 1,000 psi (68 atm). While oxygen consumption data of A. cornuta fit with the range described for other mesopelagic organisms, they were comparatively low (3.14 mg O_2/KgDW/min). These results however corroborates the fact that respiratory rates of deep-living animals are significantly slower than the rates of similar shallow-living forms (Childress, 1971). This is in accordance with more recent in situ measurements done at 1230 m on the benthopelagic fishes Coryphaenoides acrolepis and Eptatretus deani (Smith & Hessler, 1974) (Table 1). By comparison with respiration rates of the phylogenetically related gadid and particularly the shallowwater cod Gadus morhua, the macrourid C. acrolepis indeed exhibits respiration rates of two orders of magnitude lower.

These findings, which are consistent with other in situ measurements (Smith & Teal, 1973), indicate that metabolic oxydative activity is reduced in deep-sea fish. This has been related to a synergistic effect of food availability, pressure and temperature (Smith & Hessler, 1974). In constrast with the situation encountered in shallowwater species, pressure changes have moreover been shown to have a negligible effect on oxydative metabolism of bathypelagic species. These observations obviously raise the problem of physiological adaptation of deep-living forms to their hyperbaric environment.

3. Osmoregulation

Despite the increasing interest of biochemists and physiologists in biological effects of high hydrostatic pressures, informations on osmoregulatory processes at work in deep-sea fishes, as well as on their pressure

TABLE 1.

Comparative respiration measurements of shallow-water, meso- and bentho-pelagic fishes.

Animal species	Depth (m)	Respiratory rate	Temperature
		mgO_2/KG DW/min (1)	
-	0-400	12.60 ± 3.70 (n=19)	-
-	400-900	4.50 ± 2.50 (n=53)	-
Anopologaster cornuta	600-700	3.14 ± 0.59 (n=5)	-
-	900-1300	1.20 ± 0.30 (n=8)	-
		mlO_2/Kg WW/h (2)	
Coryphaenoides acrolepis	1230 (3)	2.4	3.5
Gadus morhua	-	55.6	3.0
Eptatretus deani	1230 (3)	2.2	3.5
Eptatretus stoutii	-	9.4	4.0
Petromyzon marinus	-	75.5	5.0

(1) Compiled data from Childress (1971) and Meek and Childress (1973) mean ± SD (n experiments).

(2) Compiled data from Smith and Hessler (1974).

(3) in situ measurements.

to the difficulty of collecting healthy specimens able to survive at atmospheric pressure or even to be kept in a high pressure aquarium. Most deepwater organisms are indeed extremely fragile so that fishes caught by deep hauls of long duration are often in poor condition. Provided fishes are sampled immediately after capture, certain analyses can however be done and give rather significant results. As an example, Denton & Marshall (1958) have investigated the water content and biochemical composition of the tissues from three bathypelagic species, Gonostoma elongatum, Xenodermichthys copei and Chauliodus sloanei by comparison with other species among which Ctenolabrus rupestris, Diaphus rafinesquii and Gadus minutus. The more striking point of that work is the particularly high water content (87-90%) of Gonostoma and Xenodermichthys bodies, compared with that of inshore species (70-75%). The protein content was also low, arising 4-7% against 16% in inshore fishes. These findings have been later on sensitivity are still extremely scanty or even practically lacking, likely due corroborated by Blaxter, Wardle & Roberts (1971) in an investigation dealing with a wider range of species. They indeed confirmed that fishes with no swimbladder,with soft bodies and poorly developed skeletons, had mostly more than 90% water. According to these authors, the soft, watery fish without swimbladder might use skeletal reduction combined with high water content was a buoyancy mechanism. The understanding of the different mechanisms which have allowed this type of adaptive reaction implicate the knowledge of all the processes at work in control of the thermodynamic activity of water in biological fluids. By the way, it also includes the knowledge of any kind of metabolic and transport process involved in the regulation of the amount of inorganic and organic compounds acting as osmotic effectors both in intracellular and body fluids.

Blaxter, Wardle & Roberts (1971) have measured freezing-point depressions (FPD) of the body fluids in a wide range of species divided into four groups (A: soft body - no swimbladder; B: harder body - no swimbladder; C: harder body - with swimbladder; D: very hard body - with swimbladder - epipelagic). They found FPD values to lie between - 0.7 and - 1.4°C with no consistent differences between watery fishes (group A) and other ones. By comparison with data obtained with surface marine teleosts (-0.6 to -0.8°C; Black, 1957) and with seawater (-1.8 to -2.2°C), FPD values of Blaxter et al (1971) are high and not far from isosmoticity with environmental seawater. Moreover, there is a slight tendency for the muscle to have a higher FPD than the plasma. In that study, no attempt was unfortunately made to identify and locate inorganic and organic compounds which act as intracellular and extracellular osmotic effectors, and there is a lack of information on biological processes involved in their regulation. Nevertheless, it can be argued that such a situation is propitious to survive in environments where temperature is low, oxygen considerably depleted and food availability limited. A high blood osmolarity indeed implicates a reduced concentration gradient needing less energy to be maintained. That idea is furthermore in agreement with the data of reduced metabolic activities found to characterise deepsea fishes (see the beginning of this part), but need to be corroborated by comparatively studying more experimental results collected on bathypelagic, abyssal, mesopelagic and epipelagic species. Such a systematic study however will still fail to provide information on the effects of absolute pressure or pressure changes on osmoregulation and related processes, but in that field, almost everything remains to be done. This is also true as far as surface and shallow water species able or not of vertical migrations are concerned.

From the results reported above, it clearly appears that interpretation of the effects of pressure on whole animals, either bathypelagic or not, remains very complicated. Most of the time, technical difficulties render the experimental approach difficult, ineffective or even impossible. Moreover, nervous responses can hardly be separated from other cell responses. All these difficulties therefore initiated investigations on isolated organs and tissues.

B. Isolated tissues

In isolated muscle preparations, moderate pressure applications result in an increase of the twitch tension. If the pressure is increased, twitches become smaller and slower, whereas a reduction in efficiency occurs. Pressure steps up to 800-1,000 atm abolish or considerably reduce the action potentials, and long term compression lead to complete loss of muscle contractility (Flügel & Schlieper, 1970; Gordon, 1970; Brown, 1934-1935; Johnson, Eyring & Polissar, 1954).

Early investigations support the view that increased pressure accelerates transiently heart beating in vitro and in vivo. That acceleration phase is however generally followed by a slowing down phase (Draper & Edwards, 1932) and pressures higher than 400-500 atm may even completely abolish heart beating (Ebbecke, 1935). Pressure furthermore induces cardiac arrhythmia, local block and isolated fibrillary activity. As these effects also occur in still non-innervated hearts of young embryos and in dennervated hearts of older ones, they have been ascribed to a direct influence on the muscle cell itself (Gordon, 1970). In a recent work performed on the eel Anguilla anguilla, Belaud et al. (1976) have analysed the combined effects of pressure and temperature on heart beating. They have identified the specific effects of pressure per se as a bradycardia above 24.5°C and as a tachycardia below this temperature. These authors have moreover established that, besides a direct action on autonomic cardiac cells, pressure also acts through extrinsic mechanisms of heart control. Down to the cell level, Marsland (1944) studied the effects of pressure on protoplasmic movements within single melanophores of isolated scales of the killifish, Fundulus heteroclitus. Upon slow compression up to 500 atm, the contracted melanophores displayed an increased expansion of the pigment. This effect was shown to be enhanced at low temperature, and completely independent of the cell innervation. It has been discussed in terms of a pressure induced shift of the gel-sol equilibrium toward sol formation.

In addition to an action at the cell level, it has been reported, in the preceeding, that all kinds of changes in locomotor activity had to be partly related to pressure induced nervous reaction. To the best of my knowledge, nothing has even been done on the effects of pressure on fish nerves but, on amphibian motor nerve bundles, Ebbecke & Schaefer (1935) (see also Grundfest, 1936) have established that pressure steps lower and higher than 500 atm respectively, enhanced and diminished excitability. Later on, experiments were carried out on squid giant axons by Spyropoulos (1957, a, b). Most axons generated spontaneous spikes in the pressure range 200-460 atm and a decrease in threshold membrane current at lower pressures. The duration of electric response, the conduction velocity and the height of the action potential showed little changes too. These results nevertheless remain extremely difficult to relate exactly to the high pressure nervous syndromes, likely due to the complexity of the possible interactions at work in whole living organisms and the eventual presence of other structures much more sensitive to pressure.

Among all the pressure effects reported above, a lot of indirect evidence suggests osmoregulatory processes and ion transport properties of cell membranes to be pressure sensitive. Most of the effects on isolated tissues, as well as on whole animals, might indeed be discussed in terms of changes in thermodynamic activity of water inside the cells, in terms of changes in permeability and transport or, in terms of changes in molecular structure.

The question of the pressure sensitivity of membrane phenomena has been more particularly investigated in our laboratory by means of a gill preparation isolated from seawater acclimated eels *Anguilla anguilla*. Experiments carried out under pressure were based on the findings of Bellamy (1961) (see also Kamiya, 1967) according to which "when isolated gills from seawater eels are incubated in seawater, a constant level of sodium chloride is maintained in the gills during the incubation period against the diffusion gradient" (figure 1). As the maintenance of constant ionic level appeared to be very sensitive to the presence of oxygen in the medium and to the action of inhibitors like ouabain and 2-4 (α) - dinitrophenol, it has been explained in terms of an active NaCl extrusion coupled with K^+ entry at work to face the passive NaCl influx along the concentration gradient. These passive and active processes can very satisfactorily be integrated to the general physiological features which are known to be responsible of the osmo-and ionoregulation in marine fishes. It is not my aim to include in this chapter a status report on that question which will be reviewed in another chapter of this book by Evans, but, by using selected examples, I will try to bring more insight on the pressure sensitivity of ion transport processes known to be bound to mineral regulation in marine fishes.

The effects of pressure steps up to 750 atm on the tissue Na^+, K^+ and Cl^- contents of isolated gills are summarised in figure 2. The dotted line represents the ionic level reached when gills are incubated at atmospheric pressure in the pressure chamber where no oxygenation can be provided. As shown by the figure, a slight increase in Na^+ and Cl^- tissue content has occurred at atmospheric pressure but it has been explained as corresponding to the new steady-state level determined by the method of supplying oxygen to the tissue *in vitro* (Péqueux & Gilles, 1977).

Application of pressure steps higher than 250 atm induces an increase in Na^+ content which is the higher the higher the pressure. On the contrary, at lower pressures of 100 and 250 atm., the Na^+ tissue concentration appears to be lower than control values at atmospheric pressure but nevertheless higher than the Na^+ content measured on isolated gills incubated in the presence of oxygen. Similarly, pressure steps of small amplitudes (100 and 250 atm) induce a decrease in Cl^- content, while pressures higher than 350 atm lead to an increased Cl^- concentration. In such conditions of high hydrostatic pressure, there is also a decrease in K^+ concentration (Péqueux, 1972; Péqueux & Gilles, 1977). The possibility that ionic changes observed under pressure in the gill tissue could be indirect and only related to pressure induced modifications in physical and chemical properties of the

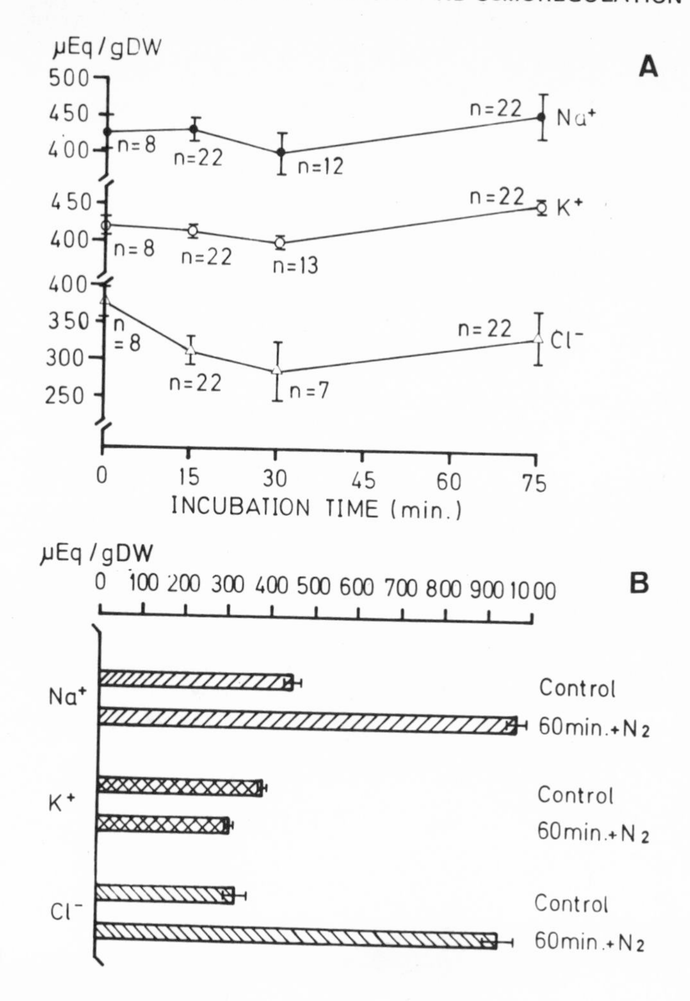

Fig. 1: Na^+, K^+ and Cl^- content (μ Eq/gDW) in gills tissue from sea-water (SW) acclimated eels Anguilla anguilla. Isolated non perfused gills are (A) incubated in oxygenated SW and (B) incubated for 60 min in anoxic SW (60 min + N_2). (Péqueux, unpublished).

incubation medium, has been questioned. Pressure is indeed known to affect the dissociation of weak electrolytes and the pH of natural seawater (Distèche, 1959, 1964, 1972; Distèche & Distèche, 1967; Distèche & Dubuisson, 1960). Incubations have therefore been carried out in artificial seawater buffered with phosphate and bicarbonate known to be pressure sensitive, and with trishydroxymethylaminomethane. Tris buffer is particularly suitable to be used in such experiments as it does not combine Ca^{++} and Mg^{++} ions; furthermore, its pH remains almost unaffected by pressure. Comparison of data obtained with the various incubation media allowed to assess that ionic changes are indeed essentially due to hydrostatic pressure.

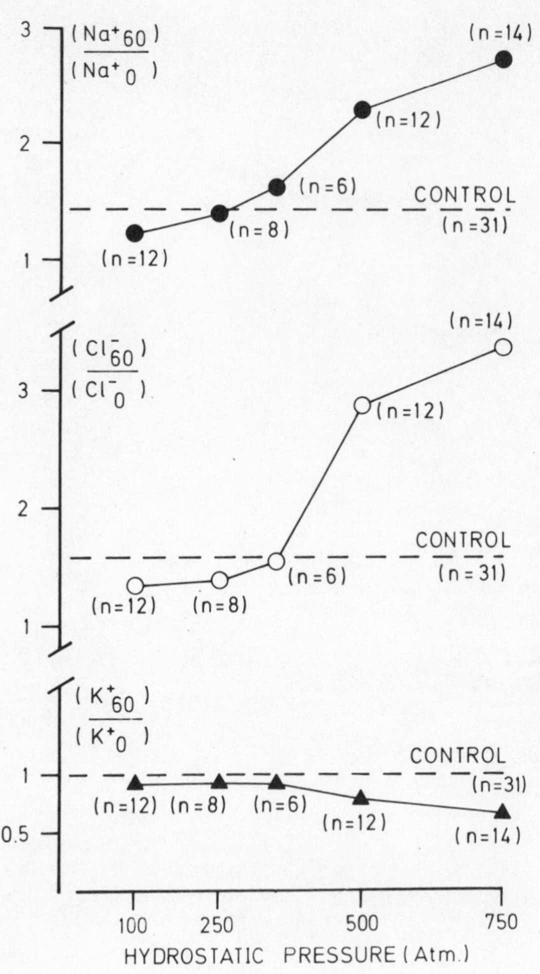

Fig. 2: Effects of hydrostatic pressure (abscissa) on Na^+, K^+ and Cl^- content (ordinate) of gills tissue from SW acclimated eels Anguilla anguilla. Results are expressed as the ratio between ions content measured in the gill tissue after (X_{60}) and before (X_o) 60 min of pressure application. Dotted line = control ionic level at atmospheric pressure. n = number of experiments. SEM always lower than ± 0.23 (from Péqueux and Gilles, 1977, modified, by permission of Birkhauser Verlag).

Pressure effects on ion content in tissues thus appears very complex and dependent on the applied magnitude. As an example, all ions species, more particularly Na^+ and Cl^-, do not seem to be affected in the same way by the pressure application. Higher pressures indeed are needed to increase Cl^- content ($P > 350$ atm for Cl^- instead of 250 atm for Na^+), but the magnitude of the Cl^- increase measured after 60 min pressure application ($P > 350$ atm) is much larger than for Na^+ ions. Moreover, by studying the magnitude of ions changes as a function of the duration of pressure application (500 atm), Na^+ content appeared to be much more rapidly affected than Cl^- and K^+ ones. These findings raise the idea that pressure could selectively act on the various permeability processes and moreover indicate that Na^+ and Cl^- movements are governed by independent mechanisms.

In the following section, an attempt will be made to explain these pressure induced changes in net ion fluxes by considering the possible action of pressure on their various components either active or passive, and on the molecular processes involved in membrane phenomena.

III. EFFECTS OF HYDROSTATIC PRESSURE ON MEMBRANE TRANSPORT AND RELATED ENZYMES

A. Passive and active transport

1. Isolated eel gills

It has been reported in the preceeding section that pressure steps up to 250 atm diminish the increase in NaCl content observed at atmospheric pressure in similar conditions as in the pressure chamber. Such reducing in the net influx of both Na^+ and Cl^- ions may result either from a pressure effect on the passive Na^+ and Cl^- entrance or from a stimulation of their active extrusion. In the last hypothesis, however, an increase in K^+ content concomitant to the decrease in Na^+ concentrations might be expected since part of the K^+ in the eel gills (Kamiya, 1967; Maetz, 1969, 1971). At this stage of the study it could thus be tentatively concluded that pressure acts in this experimental situation by decreasing the membrane permeability to Na^+ and Cl^-, but this remains of course to be confirmed by further experiments.

On the other hand, pressure steps of larger amplitude (500 atm and more) affect in the opposite way the tissue concentrations of Na^+ and K^+. That situation strangely corresponds to what happens when nitrogen is supplied instead of oxygen. By analogy, it could therefore be tentatively argued that pressure induced Na^+ increase reflects an inhibition of Na^+ active permeability processes involved in Na^+/K^+ coupled transport. This is moreover supported by our data of (Na^++ K^+) ATPase activity measured in gill extracts subjected to high hydrostatic pressure. That question will be developed in detail in a following section, but it is now worth noticing that pressure exerts an inhibitors effect which increases with the magnitude of the applied pressure step (Table 2) in agreement with the inhibition of Na^+ active transport suggested by our results. That hypothesis however does not exclude an effect of pressure at another level as for example the passive permeability. That idea would indeed agree with the observations made by me that, further important ions movements still occur under pressure even when Na^+ active transport processes had previously been abolished by inhibitors like ouabain or dinitrophenol. In order to easier discriminate between active and passive mechanisms, experiments varying the compositi-on of the artificial seawater bathing the gills were carried out. Results of figure 3 show that application of a pressure step of 500 atm on gills incubated in Na^+ - free SW induces a large decrease in tissue Na^+ content, which can reasonably be considered as the consequence of a pressure action on the Na^+ passive permeability. In such incubation conditions when Na^+ has been substituted in seawater by choline or lithium,the Na^+ concentration gradient is reversed, hence the passive Na^+ entrance is abolished and the

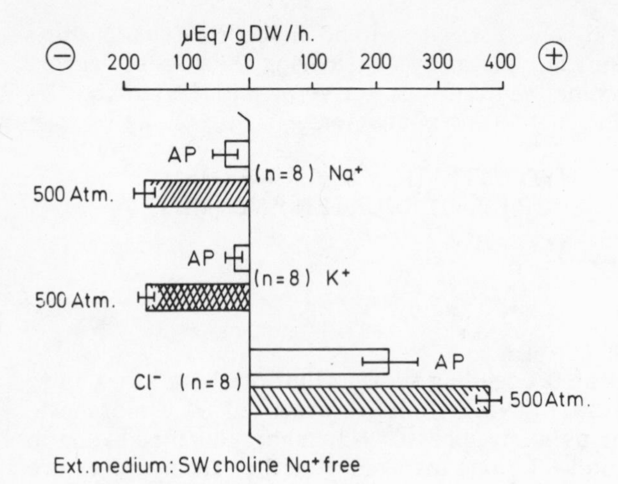

Fig. 3: Effect of a pressure step of 500 atm on ions net fluxes (µ Eq/g DW/h) at the level of isolated eel gills (Anguilla anguilla) incubated for 60 min. in Na^+-free SW (Na^+ substituted by Choline). Mean ± S.D. (n experiments). AP: atmospheric pressure. (Péqueux, unpublished).

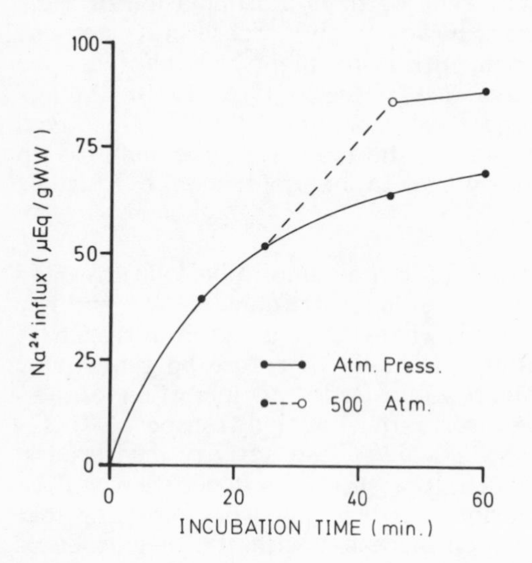

Fig. 4: Effect of a pressure step of 500 atm on the radio sodium influx (µ Eq Na^{24} /gWW) in isolated eel gills (Anguilla anguilla) incubated in artificial SW. (From Péqueux, 1979b, in press, by permission of Pergamon Press).

Na^+ active transport mechanism does no more need working. To support this last statement, preliminary experiments done at atmospheric pressure have shown that substitution of O_2 supply by N_2 do no more result in a larger inhibition of the Na^+ efflux mainly considered as active. Moreover such a drastic drop in Na^+ efflux had already been demonstrated to occur when living fishes are suddenly transfered in a Na^+ - poor medium (freshwater) (Motais, Garcia-Romeu & Maetz, 1966; Maetz, 1971). According to the results of figure 3, furthermore corroborated by the curves of Figure 4, a pressure step of 500 atm indeed induces enhancement of the passive Na^+ permeability in addition to inhibition of the Na^+ pump. Concomitantly, there is also a drop in K^+ and a very large increase in Cl^- contents.

Another argument supporting the findings that Na^+ passive permeability is pressure sensitive, also arises from experiments carried out in a

TABLE 2

Effect of high hydrostatic pressure on the activity of the $(Na^+ + K^+)$ ATPase extracted from eel gill epithelium.†

ATPase activity of gill extracts from SW acclimated eels, at atmospheric pressure *		Pressure (atm)	% variation due to pressure
$(Mg^{++}, Na^+ + K^+)$ ATPase	0.65±0.02 (n=6)	100	no significant effect
Mg^{++} ATPase	0.34±0.01 (n=6)	250	-25.5
$(Na^+ + K^+)$ ATPase	0.31±0.01 (n=6)	500	-52.8
		750	-71.9
		1,000	-83.7

*μMP_i/mg prot/h ± SEM

† After Péqueux and Gilles (1977)

physiological saline isotonic to the blood. Experiments done at atmospheric pressure in such conditions where there is no concentration gradient across the gill epithelium have established that active transport of ions is extremely reduced, or even absent. Application of ouabain, dinitrophenol and anoxic conditions indeed result in no significant changes in tissue ions content (Table 3). Pressure application however induces an increase in Na^+ content of about 40% which can therefore be reasonably ascribed to an effect on Na^+ passive permeability only. Concomitantly and despite the absence of active components, there is a decrease in tissue K^+ content of about 25%.

The large decrease in tissue K^+ content which still occurs at 500 atm when the supposed Na^+/K^+ coupled transport seems to be ineffective, is in agreement with the results obtained by incubating the gills in Na^+ free sea water (fig. 3) but raises the problem of the real nature of the pressure induced K^+ movements.

Up to now, that question is far from being completely solved even at atmospheric pressure. According to Bellamy (1961), the gill epithelium is little permeable to K^+ ions. On the other hand, we have seen at the beginning of this chapter, that a K^+ lack in the surrounding severely disturbs the maintenance of the blood Na^+ balance (Maetz, 1969; Kamiya & Utida, 1968). That K^+ ions may be involved in active exchange processes against Na^+ appear therefore as very likely but the importance and the exact modalities of the procedure remain to be established. At this step of our investigation on pressure effects, it is thus reasonable to consider that high

TABLE 3.

Changes in tissue ions content of isolated gills from SW acclimated eels Anguilla anguilla upon 60 minutes of incubation in isotonic saline.

Incubation conditions	Na^+	K^+	Cl^-
A. Atmospheric pressure			
Control (O_2 supply)	0.97*	0.98	1.05
Anoxy (N_2 supply)	1.09	0.95	1.14
Ouabain 10^{-3}	1.04	0.89	0.95
DNP 5.10^{-4}	1.03	0.98	1.03
B. Hydrostatic pressure			
500 atm.	1.39	0.73	1.34

* Results are expressed as ratio of ionic content means (μEq/gDW) measured before and after incubation (X60/Xo). After Péqueux, unpublished.

pressures (500 atm and more) mainly act by enhancing the passive K^+ permeability when producing the above mentioned effects.

2. Other isolated tissues and cells

Due to the complexity of the various transport processes at work in fish gills (Kirschner, 1979), experiments with isolated eel gills do not always allow good discrimination between the various transport components. This is particularly true when several kinds of disturbance contribute to the same final effect (for example, the pressure induced increase of tissue Na^+ content). In consideration to the relative universality of transport processes, reference will therefore be made in the following to works done on other biological materials more accessible than fish epithelia to the peculiar technics of analysis in use.

a) Isolated frog skin: Very popular in transport studies, amphibian skin is particularly convenient to investigate the effects of high hydrostatic pressures on transepithelial potential differences.

In early work, Okada (1954) reported that the voltage of isolated frog skin was considerably reduced under pressure. On the contrary, Brouha et al. (1970), later on corroborated by Woodhouse (1973), established that a pressure step of 100 atm induces a small and transient depolarisation phase followed by a much larger and sustained hyperpolarisation, while decompression is accompanied by a short hyperpolarisation before the skin potential slowly resumes its initial level.

Recently, that question has been more systematically studied over a wide range of high pressures (Péqueux, 1976a). It has been demonstrated that application of a pressure step to the skin of the frog Rana temporaria induces potential difference (PD) modifications of two types (fig. 5). First, there are short and apparently transient variations which accompany application and release of pressure. Being changeable in magnitude, not reproducible and variably affected by changes in ionic composition of the salines, these variations have not been related to a specific effect of pressure on ionic permeabilities (Péqueux, 1976a). More attention has been paid to the second and more significant change of potential which persists as long as pressure is applied. Figures 5 and 6 indeed show that pressure steps up to 500 atm induce a sustained hyperpolarisation phase which is maximum at 200-300 atm. At about 400 atm, a potential drop follows the hyperpolarisation. At higher pressures, the hyperpolarisation is no longer seen and the depolarisation immediately prevails. Several attempts have been made to explain the origin of the sustained potential modifications, particularly by considering a modification of the components which generate the skin potential according to the Koefoed-Johnsen & Ussing (1958) model. Accordingly, any pressure effect on the skin potential would have to be related to an action on concentration gradients or on relative permeability coefficients. As far as the sustained hyperpolarisation is concerned, it does not seem likely that the bulk concentration gradients are affected quickly enough to account for the large potential variations observed. The second possibility of an effect on the passive permeability coefficients has therefore been more particularly investigated and some relevant conclusions have been drawn from experiments where the authors varied the ion composition of the salines bathing the skin. For example, upon substitution of Cl^- ion by the poor diffusible SO_4^{2-} ion, pressure fails to enhance the skin potential. Such an observation raised the idea that pressure might hyperpolarise the skin by reducing the Cl^- shunt, which was moreover in agreement with the fact that there is a relation between the sustained hyperpolarisation obtained in NaCl Ringer's solution and the increase of potential observed when the skin is transferred at atmospheric pressure in Na_2SO_4 saline. Since this increase represents the shunt of the Cl^- through the skin, it was indeed considered as measuring the actual permeability of the skin to Cl^-. In addition to an inhibitory effect of pressure on the Cl^- permeability, the hyperpolarisation phase might also result from a change in Na^+ permeability as predicted by the Goldman equation. This idea has been proved to be consistent with results obtained by removing the Na^+ from the outer medium, by changing the pH of the solutions bathing the skin (Péqueux 1976b) and by submitting the skin to the action of oxytocin, an agent assumed to increase the Na^+ permeability of the outer side of the skin. Several evidences moreover suggest that, beside a decrease in skin Cl^- permeability and an increase in passive Na^+ permeability of the outer-facing cell membranes, rather low pressure (< 500 atm) also enhances the passive K^+ permeability of the inner cell membranes. Concomitantly to the sustained hyperpolarisation, the skin short-circuit current has been demonstrated by Péqueux (1976a) to increase significantly (Table 4). According to Ussing's theory, this would correspond to an increase in the net active Na^+ flux, therefore substantiating the idea that the skin secondarily adjusts the activity of its Na^+ active transport to the pressure induced changes of passive permeability in order to maintain the intracellular Na^+ concentration at a normal level. At variance with these last results, Woodhouse (1973) remained unable to measure, at 100 atm, any current increase concomitant to hyperpolarisation of skins isolated from the frog Rana pipiens. The transient depolarisation seen immediately following compression was however matched by a brief decrease in current. Woodhouse (1973) therefore argued that the most likely explanation for the sustained hyperpolarisation of R. pipiens skin's hyperpolarisation seemed to be a reduced permeability in the outer layer of the skin to Cl^- ions. Although the origin of this apparent contradiction needs further investigation to be elucidated it is worth noticing that both amphibian species were different and submitted to pressure steps of different durations. Moreover, the fact that the sustained hyperpolarisation observed at 100 atm decreases and generally disappears when the active transport of Na^+ has been inhibited by agents like ouabain and 2-4 (α) dinitrophenol and by acidification of the solutions bathing the skin (Péqueux 1976b) is another argument supporting the idea that the integrity of the active transport mechanism needs to be maintained (Brouha et al. 1970; Péqueux, 1976b). It could be suggested that the potential variations result from the action of hydrostatic pressure on a Na^+ pump which would operate electrogenically. This view remains however much contested and there is a lack of evidence to extend such a model to the present experimental conditions.

The idea that the Na^+ pump is stimulated in response to an increase in passive permeability, but remains unaffected by pressure itself is further-

TABLE 4. Effect of a pressure step of 100 atm on potential difference PD and short-circuit current SCC across the skin of the frog Rana temporaria. (1)

PD (mV)		SCC (μA/cm^2)	
initial PD	max hyperpolarization under pressure	initial SCC	max variation under pressure
39.2	14.0	36.7	8.3
23.0	17.0	12.5	10.8
42.2	14.0	24.7	7.0
26.9	34.0	52.5	9.2
47.9	17.8	23.7	6.3

(1) After Péqueux (1976)

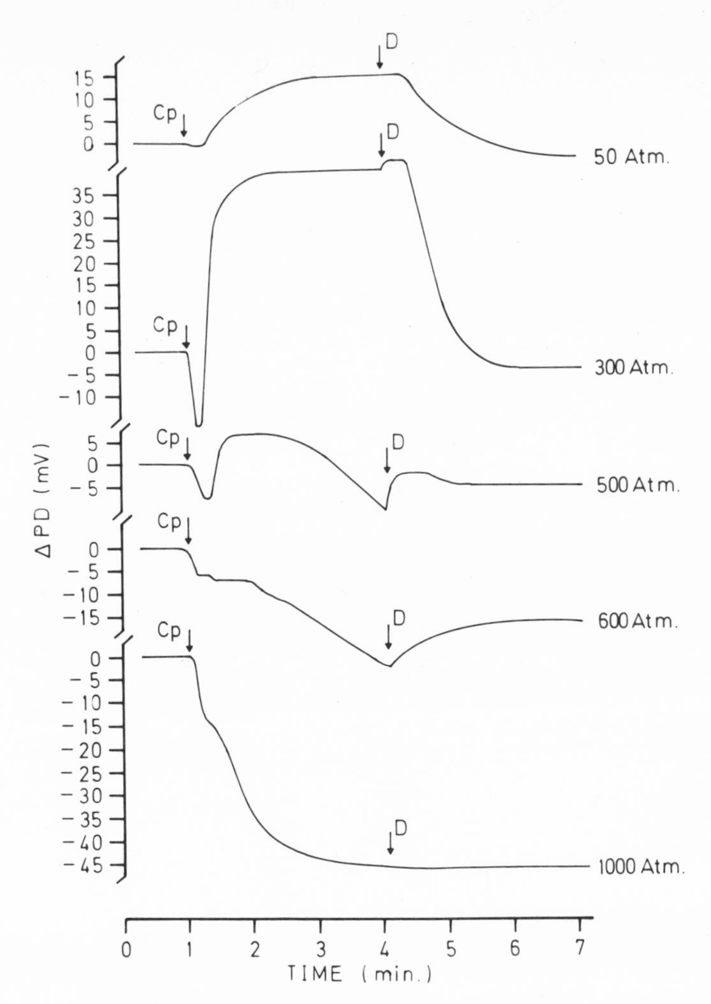

Fig. 5: Typical records of the effects of hydrostatic pressure (atm) on the electrical potential difference (PD) of isolated skins from the frog Rana temporaria. Skins are bathed on both sides with ordinary Ringer's saline.
Cp = pressure application.
D = pressure release.
(From Péqueux 1976a, modified, by permission of Cambridge University Press).

more substantiated by the fact that the activity of the $(Na^{+}+K^{+})$ ATPase is not modified by pressure steps of 100 and 250 atm (fig. 7).

At higher pressures, however, another kind of perturbation has to be considered.

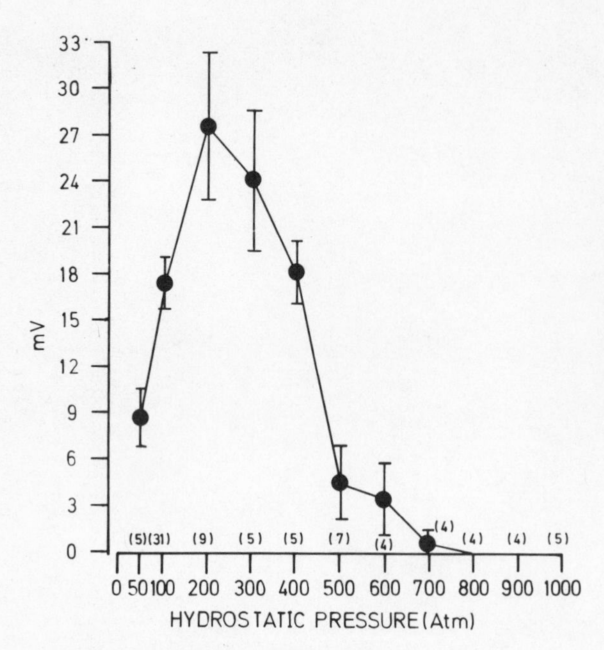

Fig. 6: Pressure induced long lasting hyperpolarization at the level of isolated skins from the frog Rana temporaria. Skins are bathed with ordinary Ringer's saline. Mean of (n) experiments ± SEM. (From Péqueux, 1976a, by permission of Cambridge University Press).

At 500 ± 100 atm., a reversible depolarisation indeed follows a transient hyperpolarisation. This might be explained by considering that pressure immediately acts on passive permeabilities so as to induce the hyperpolarisation phase; the depolarisation would then reflect an action at another level which is closely bound to the maintenance of the skin potential. The mechanism most likely affected is the Na^+ pump. The pressure induced potential fall is indeed of the order of magnitude of the change resulting from an increase in the intracellular Na^+ content when the active output is inhibited at atmospheric pressure by chemical inhibitors like ouabain and 2-4 (α)-dinitrophenol. According to that hypothesis, a transient hyperpolarisation first occurs at 500 atm but subsequently the Na^+ pump activity is directly inhibited by pressure, causing the skin potential to decrease. At higher pressures, the drop in potential difference is immediate and irreversible, likely due to the fact that the pump is more drastically inhibited. This explanation is consistent with the observed effects of pressure on the $(Na^+ + K^+)$ ATPase activity (fig. 7), i.e. a significant and severe decrease of enzyme activity from 500 atm on. In this high range, pressure would thus directly affect the Na^+ active transport by acting on the pump-bound enzyme.

The above explanations obviously do not exclude the possibility that metabolic factors, and hormonal factors too, might play some part in the skin responses to pressure application (Woodhouse, 1973; Macdonald & Miller, 1976). In agreement with this idea are the findings of Fenn & Boschen (1969) that oxygen consumption of the isolated frog skin is reduced at 300 atm although it is not significantly changed at 100 atm.

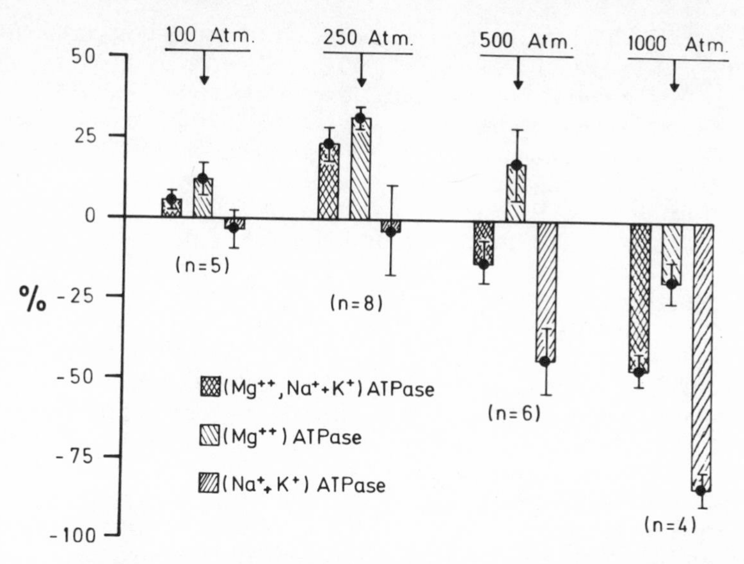

Fig. 7: Effects of hydrostatic pressure on the frog skin ATPase system. Incubation is done in ATP 4 mM, $MgCl_2$ 5 mM, NaCl 100 mM, KCl 25 mM (+ EGTA and Tris buffer pH 7.4). Ouabain 0.2 mM is added in order to estimate the ouabain-sensitive activity of the $(Na^+ + K^+)$ ATPase.
Ordinate: percentage of variation under pressure.
Mean of (n) experiments ± SEM
(From Péqueux, 1976a, reproduced by permission of Cambridge University Press.)

b) Erythrocytes: In early studies, Fontaine (1927a, b) investigated the effects of pressure steps up to 700 atm on red blood cells (RBCs) volume. Failing to demonstrate any significant effect, he concluded that erythrocyte permeability is not changed under pressure. This is at variance with the result of Yamato (1952a, b) who reported that hemolysis of mammalian red blood cells (RBCs) in isotonic saline increased at 300 atm.

Later on, Miyatake (1957) reported that the amount of ^{32}P incorporation into erythrocytes increases at pressures between 197 and 973 atm, and that electrical conductivity increases concomitantly at 973 atm and following decompression.

In 1956, Podolsky demonstrated that 80 atm reduced the Na^+ efflux of cat erythrocytes. That pressure induced flux drop being of the same order of magnitude as the flux reduction achieved by substituting NO_3 ions for Cl^- ions in the incubation medium at 1 atmosphere, lead this author to suggest that 80 atm altered the hydration of Cl^- ions so that they appeared as NO_3^- ions in the RBCs membrane.

More recently, Murphy & Libby (1976) investigated the effects of pressures up to 3,000 atm on the incorporation of radiophosphate into rabbit erythrocytes. Pressures of 200-1,000 atm were found to decrease significantly the amount of phosphate transported into RBCs although no irreversible morphological changes or membrane disruption could be observed. That drop in incorporation reached its minimum at 500-600 atm. At higher pressures, the exchange rose more or less linearly until 2,000 atm at which lysis occurred. Pressures in excess of 2,000 atm gave very high exchange values likely due to irreversible damage of the cell membrane.

The effects of pressure steps up to 1,000 atm on the Na^+ and K^+ permeability of mammalian erythrocytes have been more extensively studies by us (Péqueux, 1979; Péqueux, Gilles, Pilwat & Zimmermann, MS; Zimmermann, Pilwat, Péqueux & Gilles, MS; Péqueux & Parisis, unpublished results). By incubating pig intact RBCs in a Na^+-free medium containing 25 mM KCl at atmospheric pressure and upon application of high hydrostatic pressures, it has been demonstrated that the Na^+ total efflux diminishes when pressure is raised (Péqueux, 1979). This effect, which is higher the higher the pressure, reached about 80% at 1,000 atm. A very similar picture was obtained by considering the "ouabain-sensitive" component of the Na^+ flux, which may reasonably be related to the Na^+ pump. A K^+ "ouabain-sensitive" influx has also been observed to be concomitantly inhibited. The flux decrease reached about 20% at 250 atm but appeared to be dependent on the K^+ content of the incubation medium. These findings are consistent with the idea that both the "ouabain-sensitive" Na^+ and K^+ fluxes might be linked and implicated in Na^+/K^+ exchange pumping process. This is furthermore supported by measurements of (Na^++K^+)ATPase activity under pressure (Péqueux, 1979). As far as the "ouabain-insensitive" Na^+ efflux of pig RBCs is concerned, the situation does not look so clear and needs further investigation. It is worth noticing that experiments carried out on

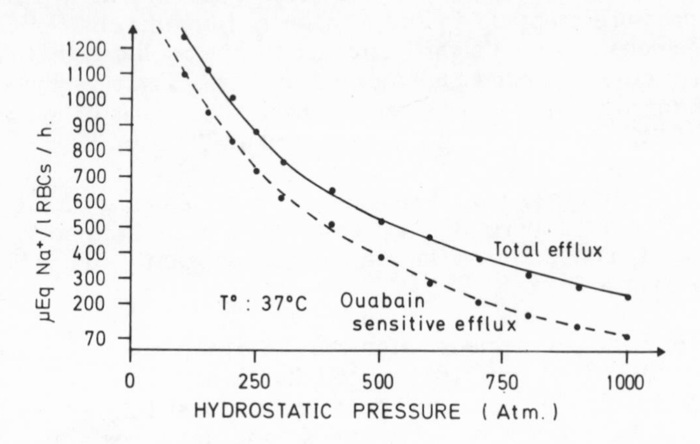

Fig. 8: Effect of hydrostatic pressure (atm) on total and ouabain sensitive Na^+ efflux from human erythrocytes. Results are expressed in $\mu Eq\ Na^+/l\ RBC_s/h$. (Péqueux and Parisis, unpublished results).

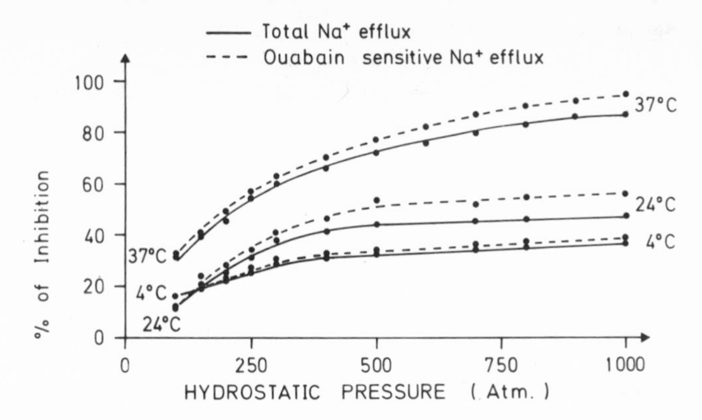

Fig. 9: Pressure induced inhibition of total and ouabain-sensitive Na^+ efflux in human erythrocytes, as a function of temperature. (Péqueux and Parisis, unpublished results.)

depigmented pig erythrocytes (ghosts) at atmospheric and at high pressure give identical results as for intact RBCs. This implies that the inhibition of the ionic active transport results from a direct effect of pressure on membrane-bound mechanisms. As ATP is artificially supplied when preparing the ghosts, an effect of pressure on the metabolism producing energy can indeed be discarded.

Despite some quantitative differences, experiments done on ox and human erythrocytes confirm and complete these findings. Figure 8 shows the effects of pressure step up to 1,000 atm on the total and "ouabain-sensitive" Na^+ efflux from human erythrocytes. The particularly drastic inhibition of the ouabain-sensitive component, which reaches almost 95% at 37°C and at 1,000 atm, still supports the hypothesis that, in human erythrocytes too, pressure mainly acts on the Na^+ active transport processes. This is moreover consistent with the fact that decreasing the ambient temperature results in a severe drop in flux magnitude and pressure effects (fig. 9). At lower temperatures, cell metabolism, active transport mechanisms and related enzyme processes must indeed be considerably slowed.

Data shown in fig. 10 moreover show that pressure in the low range do not affect the percentage of ouabain-sensitive Na^+ flux while it severely slows down the flux rate.

500 atm and more are indeed needed to significantly decrease this percentage and the effect then becomes the more drastic the higher the pressure. This picture is strikingly close to the effects of pressure on the activity of the $(Na^+ + K^+)$ATPase extracted from several organs and tissues implied in ion transport (see fig. 7 and part III B.1 of the present chapter). An attempt will be made in the following to explain such a two-step response and to correlate it with other processes, but, by now on, it can

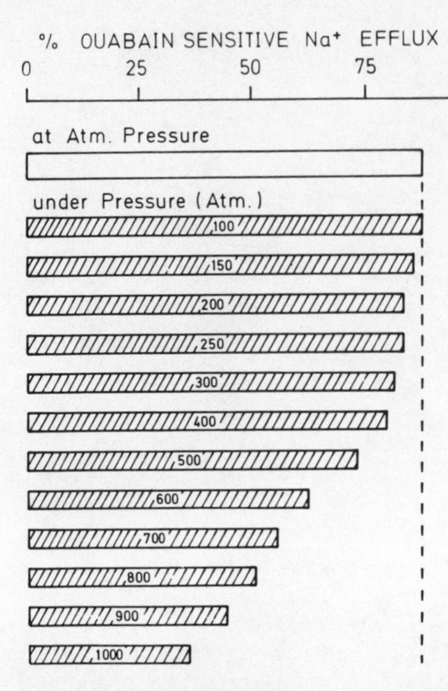

Fig. 10: Effects of hydrostatic pressure on the percentage of "ouabain-sensitive" Na^+ efflux in human erythrocytes. Total Na^+ efflux is always considered as being 100% for each pressure condition. (Péqueux and Parisis, unpublished results.)

reasonably be considered as reflecting the existence of two different sensibility levels.

Data of Table 5 show that ouabain-insensitive Na^+ efflux does not seem to be significantly affected by pressure when incubation is carried out

TABLE 5.

Effects of high hydrostatic pressures on ouabain-insensitive Na^+ efflux from human erythrocytes. Influence of temperature.

Applied pressure (atm)	Incubation temperature : 37°C. Ouab.insensitive Na^+ efflux $\mu Eq/lRBC_s/h$ (under pressure)	% of the total efflux at atm.press.	% of the total efflux under press.	Incubation temperature : 4°C. Ouab.insensit. Na^+efflux $\mu Eq/lRBC_s/hr$ (under pressure)	% of the total efflux at atm.press.	% of the total efflux under pressure.
100	157.5	8.7	12.6	87.5	38.9	46.1
150	160.5	8.8	14.4	86.5	37.9	46.8
200	168.9	9.2	16.7	84.0	36.5	46.8
250	145.5	7.7	16.6	84.0	35.9	48.0
300	141.0	7.4	18.5	78.8	34.2	47.8
400	131.0	6.8	20.3	80.0	33.6	48.8
500	142.4	7.4	26.8	77.0	33.6	49.7
600	174.9	9.1	37.3	-	-	-
700	174.0	9.1	44.4	77.0	33.5	50.3
800	160.5	8.2	48.9	75.0	32.9	50.3
900	153.0	8.0	55.1	-	-	-
1000	156.3	8.2	63.8	73.1	31.9	49.7

After Péqueux and Parisis, unpublished.

at 37°C. At lower temperature, the magnitude of ouabain-insensitive efflux is much reduced and pressure in the high range appears to be slightly inhibitory. At variance with these results, a significant and severe increase in ouabain-insensitive Na^+ permeability at high pressure (800-1,000 atm) can been argued by calculating the relative Na^+ release of the RBCs due to pressure from the following equation which takes into consideration the RBCs intracellular Na^+ content (lytic sample) (Schaeffer & Péqueux, unpublished):

$$C_{Na^+} = \frac{C_s - C_c}{C_l} \cdot 100$$

C = Na^+ concentration in :
S : compressed sample
C = control (atm press)
l = lytic sample.

The effect of pressure on ouabain-insensitive Na^+ efflux from human erythrocytes thus appears to be much more complex to explain than on the ouabain-sensitive one. Evidence in the literature is consistent with the idea that it results in fact from combination of at least two components i.e. a passive one and another which might be carrier mediated or even related to the cell metabolism while insensitive to ouabain (Beauge, 1975; Passow, 1964).

If this is true, further investigation would be necessary to discriminate between each component and clearly establish the pressure sensitivity of the Na^+ passive permeability. More accessible to experimentation is the passive permeability to K^+ ions. In collaboration with the German group of Zimmermann, I have investigated the effects of pressure steps up to 1,000 atm on the K^+ permeability of human erythrocytes incubated at 22°C in a saline containing 145mM NaCl, 5 mM $MgCl_2$, 10 mM glucose and 5 mM Tris buffer. Addition of ouabain to this incubation medium does not result in significant changes in K^+ release from the cell thus excluding involvement of any active component in the observed K^+ net efflux. Figure 11 demonstrates that the net K^+ efflux increases almost linearly with the pressure until 600-700 atm. At higher pressure, a significant rise in K^+ efflux occurs. The observed permeability changes, which have been established to be fully reversible, are almost linearly dependent on time over the whole pressure range up to 1,000 atm (Péqueux, Gilles, Pilwat & Zimmermann; Zimmermann, Pilwat, Péqueux & Gilles, MS). These results show that a rather "narrow" pressure range exists, at which a pronounced increase in membrane K^+ permeability occurs. This pressure range furthermore has been demonstrated to be temperature dependent.

c) Other cells: Several more scanty investigations have also been carried out on other biological materials, and establish, most of the time indirectly, that pressure acts on membrane permeability. For example, in onion epidermal cells (*Allium cepa*), Murakami (1963) demonstrated that the time for plasmolysis in hypertonic solutions was increased under high pressure (476 atm), while, on the other hand, the time for deplasmolysis was decreased. The permeability of *Arbacia* eggs has been investigated under

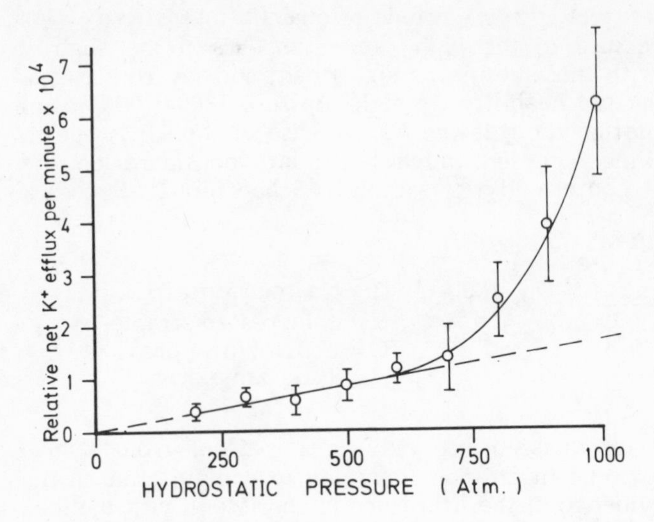

Fig. 11: Effect of hydrostatic pressure (atm) on the relative net K^+ efflux from human erythrocytes. Efflux values are referred to the amount of K^+ released by complete osmotic lysis of the same number of cells in distilled water. Mean of 5 experiments ± SD. (From Zimmermann, Pilwat, Péqueux and Gilles, unpublished, submitted for publication.)

pressure by Murakami & Zimmerman (Zimmerman, 1970) and pressure effects were shown to be very dependent of the cationic environment. Paul & Morita (1971) have studied the ability of the shallow water vibrio MP 38 to accumulate labelled amino acids at high pressure. In the absence of preincubation in the ^{14}C-labelled amino-acid solution, incorporation into the cell was shown to be considerably reduced at pressures in the range 100-500 atm. Hydrostatic pressure has also been shown to induce striking behavioural and physiological effects in protozoa, in relation with electrical events and membrane phenomena (for details, see Zimmerman, 1969, 1970; Murakami & Zimmerman, 1970; Kitching, 1969, 1970).

B. Enzymes

1. (Mg^{2+}, Na^++K^+) ATPase

Most of the tissues pumping Na^+ ions are known to have high activities of the enzyme ouabain-sensitive Na^+-K^+ activated ATPase (ATP phosphohydrolase, E.N. 3.6.1.3). Several lines of evidence indicate that the activity of this enzyme is closely related to the active transport of Na^+ across membrane systems (Skou, 1965; Kamiya & Utida, 1968; Jampol & Epstein, 1970). In Teleostean gills, ultrastructural studies have localised the (Na^++K^+) ATPase activity in the chloride cells (Skirai, 1972) and much evidence has been presented to implicate the enzyme (Na^++K^+) ATPase in hydromineral regulation of marine teleosts. The interest to analyse the effects of hydrostatic pressure on the (Na^++K^+)ATPase therefore becomes evident in a review of pressure effects on membrane permeability and osmoregulation. Some scanty reports on the pressure sensitivity of the membrane bound enzyme extracted from eel gills, frog skin and red cells have already been mentioned in this chapter. A more complete and systematic study has been

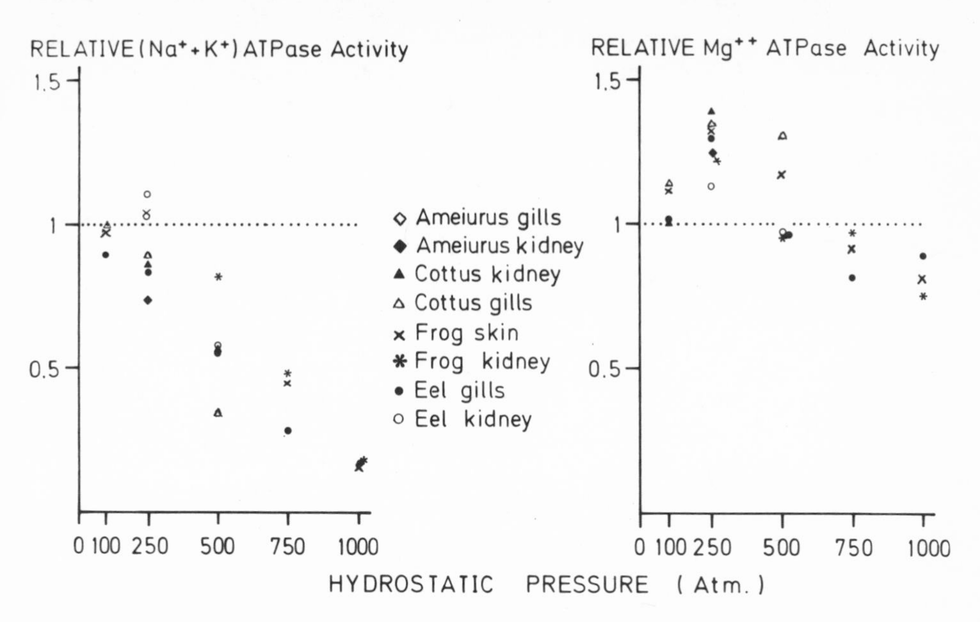

Fig. 12: Effects of hydrostatic pressure on the activity of ouabain-sensitive and insensitive (respectively $(Na^{+}+K^{+})$ and Mg^{2+}activated) ATPases.
Ordinate: relative enzyme activity expressed as the ratio between the activity measured when under pressure and the control activity at atmospheric pressure.
Abscissa: hydrostatic pressure (atm)
(From Péqueux and Gilles, 1978, by permission of Pergamon Press.)

initiated to characterise and compare the effects of pressure on the membrane ATPase isolated from tissues implicated in iono and osmoregulation in aquatics animals which may or may not be subjected to important changes in hydrostatic pressure (Péqueux & Gilles, 1978).

As a function of the enzyme considered ($Mg^{2+}+Na^{+}+K^{+}$, $Na^{+}+K^{+}$ or Mg^{2+} activated) and of the pressure, the activity was enhanced, decreased or not affected. If most of the $(Na^{+}+K^{+})$ ATPases appeared to remain relatively insensitive to pressure steps of 100 and 250 atm, higher hydrostatic pressures drastically inhibited their activity (figure 12; see also figure 7 and Table 2). Most of the extracts were more than 80% inhibited at 1,000 atm. Pressure qualitatively affected all considered tissues in the same way; only quantitative differences occurred, either in the percentage of inhibition or in the magnitude of pressure needed to induce inhibition. Even at 1,000 atm, the ouabain-insensitive Mg^{2+} ATPase was only slightly inhibited and an increase in activity even occurred in the low pressure range (fig. 12). This last stimulation could not result from an action of pressure on the ouabain-enzyme interaction since a similar effect was obtained by measuring the activity in the absence of Na^+, K^+, instead of being measured in the presence of Na^+, K^+ and ouabain. This result was in agreement with the findings of Moon (1975) who reported no pressure effect on the inhibitory power of ouabain, at least on gill extracts from the coho salmon (Oncorhynchus kisutch) upon compression at 562 atm.

Information on specific activity of $(Na^+ + K^+)$ ATPase, even on its presence, in deep-sea fishes is almost lacking. To the best of my knowledge, that question has, up to now, only been touched by Moon (1975) in a comparative study of gill $(Na^+ + K^+)$ ATPases isolated from the surface SW adapted coho salmon, Oncorhynchus kisutch, and the benthic marine teleost, Antimora rostrata. Specific activities in gills in both species were shown to be similar and within the same range (5-25 μ MP_i/mg proteins/h) as most of $(Na^+ + K^+)$ ATPase activities in gill of marine and sea water acclimated teleosts. At variance with the immediate and drastic pressure induced inhibition of the coho salmon enzyme (80% of inhibition at 612 atm), the Antimora enzyme is pressure activated within the physiological range of pressures (1-272 atm). At pressures exceeding 272 atm, a severe inhibition however occurs with ΔV values exceeding those found for the coho enzyme (61.4 cm^3/mole in Antimora against 46.8 cm^3/mole in coho). In addition to these observations, differences in ouabain sensitivity and enzyme stability support the idea that both enzymes have distinctive catalytic properties which might result in the particular case of the benthic Antimora in a significant advantage to survive in a moderately high pressure habitat. More comparative data are however necessary to firmly establish the existence of marked alterations in the enzyme response of deep-sea fishes. At this step of the investigations, it must be kept in mind that the functioning of the gill $(Na^+ + K^+)$ ATPase is essentially impaired at high pressure. Such an effect may be of great physiological significance as it might be directly related to the effects that pressure exerts on the active ionic movements across epithelia and cell membranes reviewed above. A rather close parallelism can indeed be established between the pressure sensitivity of both processes in most of the tissues studied, either when considering the lack of direct effect at low pressure, or the progressive inhibition at higher pressure. According to the present view, high pressure would thus induce inhibition of the active transport by acting directly on the enzyme linked to the pumps.

2. Other enzymes

The action of pressure upon metabolic processes is among those which have been the best documented, and a large piece of information is to be found particularly on the effects of pressure on enzymes.

In a review dealing with membrane permeability and osmo-regulation under pressure, the knowledge of how enzymes are affected may be of importance. It is evident that any effect on enzymes involved in metabolic energy production will in turn implicate more or less rapidly an effect on transport processes linked to energy supply. This effect however remains essentially indirect and, when it occurs, always results in an inhibition by deficiency in energy provision. I therefore consider that reviewing pressure action on enzymes by presenting selected examples is not exactly the subject of the present chapter, and I shall suggest the readers who might be interested in the question to refer to the extensive work and reviews of Hochachka & Somero (1973), and Hochachka, Moon & Mustafa (1972). When considering kinetics and thermodynamics of pressure effects on enzyme reaction the key parameters which establish the direction and magnitude of pressure effects involve volume changes of the reaction (ΔV). If the volume of the system containing the reactants is larger than the volume of the system containing the products, pressure will shift the equilibrium toward product formation, and conversely, according to the following relationship.

$$\Delta V = 2.3\, R\, T_2 \, \frac{\log_{10} K_{P_1} - \log_{10} K_{P_2}}{P_1 - P_2}$$

(ΔV : volume change of the reaction; R : gas constant; T : absolute temperature; K : equilibrium constant at pressures P_1 and P_2 atmospheres). On the other hand, the key factor which determines the pressure effect on reaction rate is the volume change accompanying formation of the transitory activated complex (volume difference between system containing non activated reactants and system containing activated reactants). Accordingly, the activation volume change can be described as follows:

$$\Delta V^{\ddagger} = 2.3\, R\, T_2 \, \frac{\log_{10} k_{P_1} - \log_{10} k_{P_2}}{P_1 - P_2}$$

(k = rate constants at pressures P_1 and P_2). Once formed, the activated complex may dissociate into reactants or products and the reaction rate is regarded as dependent on an equilibrium between reactants and transition state complex. When the volume of the activated complex exceeds the average volume of its constituents outside the complex, pressure is inhibitory. If the volumes are equal, pressure does not induce any change, but, when the volume of the activated complex is less than volume of reactants, pressure then enhances reaction rate. Accordingly, three main categories of pressure sensitive biochemical reactions have been identified: pressure activated, pressure inhibited and pressure independent ones.

In enzyme catalysed reactions, the structures of reactants and of transitory activated complex are moreover temperature sensitive. This implies that temperature will also interfere with, or at least influence pressure effects with the following combined consequences of differential pressure effects among different enzymes, and differential pressure responses by a single enzyme at different temperatures. This explains the high complexity and variety of pressure effects on enzymes and emphasizes the particular vulnerability of metabolic processes to pressure.

From the preceeding, it appears that a rather close parallelism can be drawn between the way to treat temperature and pressure effects. Accordingly, beyond an effect on metabolic reaction rates, pressure may also act on weak-bond dependent structures and processes involved in determining the final conformation and aggregation state of proteins.

Pressure induced changes in macromolecular structure and phase transitions may indeed have a significant influence upon macromolecular functions among which enzyme catalysis. It is evident that both categories of effects may interact with each other. Considering the effect of pressure on the structure and functional form of the enzymes, Penniston (1971) has argued that pressure determines reaction rates in vivo according to their oligomeric and multimeric natures. According to this view, multi-subunit enzymes tend to be pressure inactivated by dissociation into monomeres, while monomeric enzymes are either pressure insensitive or pressure activated.

Reactions which are catalysed by enzymes located on/in cell membranes may also be affected by pressure if the architecture of the membrane is pressure sensitive. Conformational changes in the enzyme protein implicating activity modifications, might indeed result, among other things, from changes in protein-protein or protein-membrane lipid interactions.

C. Conclusions

Laboratory investigations have clearly established that hydrostatic pressure affects the functioning of biological membranes by modifying selectively their properties of passive and active ion transport in a way depending of the magnitude of the applied pressure.

From the physiological point of view, that conclusion is of importance and emphasises that hydrostatic pressure may moreover be considered as a valuable tool in the study of mechanisms involved in ion transfer across biological membranes. The selectivity of its effects on Na^+ and Cl^- for example, indicates that movements of both ion species are governed by independent mechanisms and therefore suggests different and spatially separated sites for the passage of anions and cations in the cell membrane. Further investigation by means of these methods, on Na^+ and Cl^- transports across teleostean gill epithelium might likely help to bring more insight on the very disputed problem of the exact relationships binding or not both mechanisms at atmospheric pressure. Little can still be said at present as to the molecular aspect of the pressure induced changes in membrane permeability. However, as pressure is known to favour phenomena involving a volume decrease such as the ionisation of weak electrolytes including proteins, it seems logical to a priori consider that pressure will act on passive permeability by modifying the structure of "channels" at the level of which ion transport occurs. Several evidences at atmospheric pressure support the idea that fixed charges in the membrane play an important part in regulating the ionic permeability under normal conditions. Among other things, pH changes in salines bathing isolated organs induce selective modifications in Na^+ and Cl^- cell permeability, probably in relation with the state of ionisation of acidic or basic groups of proteins of the membrane. In acid medium, proteins are positively charged and their NH_3^+ groups must attract any negatively charged ion but repel the cations; conversley, at alkaline pH, the negative COO^- groups favour the passage of cations. At the level of the isolated frog skin, the magnitude of the potential variations induced by a pressure step of 100 atm is affected also by pH (Péqueux, 1976b) and this effect is consistent with the idea of an additive or a competitive action between pH and pressure on passive permeability to Na^+ ions. The fact that pressure is known to increase the passive permeability to Na^+ ions and to favour phenomena accompanied by a volume decrease suggests therefore that Na^+ passages across cell membranes most likely have to be related to mechanisms involving a volume decrease. As a consequence, it seems logical to believe that pressure enhances the Na^+ passive permeability by increasing ionisation of fixed charged groups in the membrane. Pressure induced ionisation changes however do not seem to be solely responsible for the passive permeability modifications. We believe that pressure action on transport processes is more likely the net result of serial effects at several sensitive levels. It has been reported that pressure acts on proteins by affecting the amount of ionisation but it further may disturb their native conformation. Protein structure is stabilised by weak secondary bonds between hydrophobic groups, polar groups and peptide hydrogen bonds or side-chain hydrogen bonds. Ionic transfers may be thought to imply changes in the molecular architecture of the passage sites involving the formation or disruption of weak secondary bonds. The disruption of hydrophobic bonds or electrostatic bonds in aqueous solution is accompanied by water molecules rearrangements around the exposed groups with the consequent decrease in volume (about - 15 $cm^3 . mole^{-1}$ according to Hochachka, Moon & Mustafa, 1972). Pressure could thus favour such a mechanism. On the other hand, the formation of hydrophobic or electrostatic bonds, as the disruption of hydrogen bonds involving a volume increase will be retarded by pressure. Besides an effect on proteins, pressure also could act on the lipidic compounds of the cell membrane and on the mucopolysaccharides which are known to bear strongly negatively charged groups (Seaman & Heard, 1960). Pressure induced changes of the water structure, of the hydration shells around ions (electrostriction) and of the pH of salines likely may be neglected at the physiological pressure used when separately considered although it is not impossible that the summation of several small effects may become sufficient to account for a measurable change.

Alternatively, pressure induced action on K^+ permeability has been recently explained in terms of a reversible mechanical breakdown of the membrane (Zimmermann, Pilwat, Péqueux & Gilles, MS). This interpretation is based on an electro-mechanical model formulated for plant and animal cell membranes. It postulates for certain finite membrane areas that the actual membrane thickness depends on the voltage across the membrane and the applied pressure (Zimmermann, 1977; Zimmermann, Pilwat & Riemann,

1974). The magnitude of the expected change in membrane thickness in response to the electrical and mechanical compressive forces depends on the compressibility and the relative dielectric constant of the membrane area where electrical breakdown occurs. The electrical breakdown is associated with marked increase in membrane permeability. The model explains the mechanism by which small pressure signals of less than 1 atm can be transformed into changes of intrinsic electrical field and, in turn, in membrane processes. According to that model, pressure would compress certain areas of the membrane to the critical thickness at which the instrinsic membrane potential is sufficiently high to induce reversible breakdown, hence pronounced permeability increase. Although the experimental results are consistent with the predictions of the electro-mechanical model, other possible mechanisms and explanations cannot be excluded. For example, several evidences are consistent with the idea that permeability changes could be related to pressure induced phase transition in the lipid bilayer of the membrane (Murphy & Libby, 1976; Wattiaux-de Coninck, Dubois & Wattiaux, 1977; Heremans, 1979). According to that view, the disturbance in the lipid part of the membrane might be transmitted to some functionally vital proteins in such a way as to modify the passive ions permeability properties. Moreover, a phase transition in lipids might also affect the conformation of the enzyme proteins associated with active permeability. This last hypothesis is supported by the results obtained with the (Na^++K^+) ATPase system. Particularly, the enzyme is known to require phospholipids and the temperature of the break in the Arrhenius plot is pressure dependent (Tanaka & Teruya, 1973; Dahl & Hokin, 1974; Ceuterick, Peeters, Heremans, De Smedt & Albrechts, 1976; Péqueux, 1979). According to that hypothesis, pressure induced phase transition too might be responsible for the direct inhibition of Na^+ active transport observed at high hydrostatic pressures. More insight on that question might arise from experiments on artificial lipid bilayers but, up to now, the effects of pressure on lipid bilayers has not yet been extensively investigated. The few results available however emphasise the importance of such experiments. Trudel, Hubbell & Cohen (1973) indeed established by means of a spin-labelled probe, that the motion of the fatty acid chains of phospholipids decreases with increase in pressure, which results in a drop in membrane fluidity and likely consecutive effects on its permeability properties. That observation is of importance as it allows to shed some light on the mechanisms involved in pressure effects and overall in the well known mutual antagonism between pressure and anaesthetics. Anaesthetics have indeed been demonstrated to "protect" from the effects of pressure and this might well be related to the fluidifying effect they exert on membrane's phospholipid bilayer (Johnson & Miller, 1970; Lever, Miller, Paton & Smith, 1971). Experiments in this field are still nevertheless in their early phase and more results are needed for a better understanding of this phenomenon which might lead to perfect new means of conquest of the deep sea.

IV. GENERAL CONCLUSIONS AND PERSPECTIVE

The results reviewed in this chapter emphasise that membrane function at elevated pressure is a fundamental problem in marine biology.

From the ecological point of view, hydrostatic pressure must indeed act as a limiting factor in vertical migrations of surface and diving species. Accordingly, Flügel & Schlieper (1970) have regarded pressure sensitivity as being directly related to penetration into the deep-sea. Several evidences however indicate that animal groups which show the least resistance to hydrostatic pressure are represented in the deep-sea, and conversely that genera showing the greatest pressure resistance may essentially remain intertidal in depth distribution (Menzies, George & Avent, 1972).

Without denying the possibility that deep-sea species may have greater pressure tolerance than shallow-water species, survival and colonisation of the deep-sea must likely be only related to long term ecologic adaptation both at the physiological and biochemical levels. Experiments done on deep-living fishes indeed suggest that they are relatively insensitive to hydrostatic pressure changes (Meek & Childress, 1973). Pressure resistance in relation with adaptation to life in the deep-sea is particularly evident, at least qualitatively, at the biochemical level (Hochachka, Schneider & Moon, 1971; Moon, Mustafa & Hochachka, 1971). For example, an important factor in pressure adaptation of enzymes might be the minimising of volume changes of activation. By minimizing $\Delta V^{\ddagger}$, pressure inhibition of enzyme substrate complex formation should decrease, causing a general insensitivity to variations over a wide range of pressures. That idea is in good agreement with experimental results obtained by Moon, Mustafa & Hochachka (1971) on abyssal rattail fishes (Coryphaenoides sp) and on the vertically migrating midwater sea bass (Ectreposebastes imus). In many respects, mechanisms of enzyme adaptation to pressure strikingly resemble to the patterns of temperature adaptation (Hochachka & Somero, 1973).

An alternative explanation to pressure adaptation has been proposed by Penniston (1971). He found that the activities of multimeric enzymes were inhibited and the activities of monomeric enzymes stimulated by pressure. From these observations, he argued that enzyme systems of deep-sea organisms had to be either monomeric, or to bear stronger noncovalent interactions between multimers in order to remain operational under high pressure.

From the preceeding, it clearly appears that the field of pressure studies dealing with ions transport and osmoregulation is full of lacunae. In consideration of adaptive processes developed at the enzyme level and of the fact that deep-sea organisms have likely evolved special membrane properties to cope with the ambient pressure and with other ecological parameters specific to the deep-sea environment, we consider that more insight in pressure biology could only arise from comparative studies on abyssal animals. Experiments on surface animals, even terrestrial diving ones, most of the time indeed lead to dramatise the types of problems which hydrostatic pressure impose on marine organisms and fail to throw some light on adaptive processes they have developed.

Investigation in that direction of course involves the perfection of new technics of capture and maintenance in the laboratory.

ACKNOWLEDGEMENTS

I am greatly indebted to Professor R. Gilles for helpful discussions and criticism throughout this work.

Part of it has been aided by grants "Crédit aux chercheurs" from the FNRS to the author and to Prof. R. Gilles and by a grant n° 1.5.733.79 from the FRFC to Prof. R. Gilles.

REFERENCES

Beauge, L. (1975). Non-pumped sodium fluxes in human red blood cells. Evidence for facilitated diffusion. Biochem. Biophys. Acta 401: 95-108.

Belaud, A., Barthelemy, L., Le Saint, J. and Peyraud, C. (1976). Trying to explain an effect of "per se" hydrostatic pressure on heart rate in fish. Aviat. Space Environ. Med. 47: 252-257.

Bellamy, D. (1961). Movements of potassium, sodium and chloride in incubated gills from the silver eel. Comp. Biochem. Physiol. 3: 125-135.

Black, V.S. (1957). Excretion and osmoregulation. In: The Physiology of Fishes. Ed. M.E. Brown. New York, Academic Press, p. 163-205.

Blaxter, J.H.S., Wardle, C.S. and Roberts, B.L. (1971). Aspects of the circulatory physiology and muscle systems of deep-sea fish. J. Mar. Biol. Ass. U.K. 51: 991-1006.

Brauer, R.W. (1972). Parameters controlling experimental studies of deep sea biology. In: Barobiology and the Experimental Biology of the Deep Sea. Ed. R.W. Brauer. North Carolina Sea Grant Program, Univ. of North Carolina, Chapel Hill, p. 1-13.

Brauer, R.W. (1975). The high pressure nerves syndrome: animals. In: Physiology and Medicine of Diving and Compressed Air Work. Eds. P.B. Bennett and D.H. Elliot - London, Ballières and Tindall, p. 231-247.

Brown, D.E.S. (1934-1935). Cellular reactions to high hydrostatic pressure. Ann. Rept. Tortugas Lab. Washington, D.C., Carnegie Inst. p. 76-77.

Brouha, A., Péqueux, A., Schoffeniels, E. and Distèche, A. (1970). The effects of high hydrostatic pressure on the permeability characteristics of the isolated frog skin. Biochem. Biophys. Acta 219: 455-462.

Ceuterick, F., Peeters, J., Heremans, K., De Smedt, H. and Olbrechts, H. (1976). Involvement of lipids in the break of the Arrhénius plot of Azotobacter nitrogenase. Arch. Int. Physiol. Biochem. 84: 587.

Childress, J.J. (1971). Respiratory rate and depth of occurence of midwater animals. Limnol. Oceanogr. 16: 104-106.

Dahl, J.L. and Hokin, L.E. (1974). The sodium and potassium adenosine triphosphatase. Ann. Rev. Biochem. 43: 327-328.

Denton, E.J. and Marshall, N.B. (1958). The buoyancy of bathypelagic fishes

without a gas-filled swimbladder. J. Mar. Biol. Ass. U.K. 36: 753-767.
Distèche, A. (1959). pH measurements with a glass electrode withstanding 1,500 Kg/cm 2 hydrostatic pressure. Rev. Scient. Instrum. 30: 474-478.
Distèche, A. (1964). Nouvelle cellule à électrode de verre pour la mesure directe de pH aux grandes profondeurs sous-marines. Bull. Instr. Océanogr. Monaco 64: 1-10.
Distèche, A. (1972). Effects of pressure on the dissociation of weak acids. In: The effects of pressure on organisms. Symp. Soc. Exp. Biol. 26: 27-60. Cambridge University Press.
Distèche, A. and Distèche, S. (1967). The effect of pressure on the dissociation of carbonic acid from measurements with buffered glass electrode cells. The effect of NaCl, KCl, Mg^{2+}, Ca^{2+}, SO_4^{2-} and boric acid with special reference to sea-water. J. Electrochem. Soc. 114: 330-340.
Distèche, A. and Dubuisson, M. (1960). Mesures directes de pH aux grandes profondeurs sous-marines. Bull. Inst. Océanogr. Monaco 57: 1-8.
Draper, J.W. and Edwards, D.J. (1932). Some effects of high pressure on developing marine forms. Biol. Bull. 63: 99-107.
Ebbecke, U. (1935). Uber die Wirkung hoher Drucke auf Herzschlag und Elektrokardiogram. Pflügers Arch. ges. Physiol. 236: 416-426.
Ebbecke, U. (1935). Uber die Wirkungen hoher Drucke auf marine Lebewesen. Pflügers Arch. Ges. Physiol. 236: 648-657.
Ebbecke, U. (1944). Lebensvorgänge unter der Einwirkung hoher Drucke. Ergeb. Physiol. Biol. Chem. Exptl. Pharmakol. 45: 34-183.
Ebbecke, U. and Schaefer, H. (1935). Uber den Einfluss hoher Drucke auf den Aktionsstrom von Muskeln und Nerven. Plügers Arch. Ges. Physiol. 236: 678-692.
Fenn, W.O. and Boschen, V. (1969). Oxygen consumption of frog tissues under high hydrostatic pressure. Respir. Physiol. 7: 335-340.
Flügel, H. and Schlieper, C. (1970). The effects of pressure on marine invertebrates and fishes. In: High Pressure Effects on Cellular Process. Ed. A.M. Zimmerman. New York and London, Academic Press, p. 211-234.
Fontaine, M. (1927a). Influence des fortes pressions sur le volume globulaire. C.R. Soc. Biol. 97: 1656-1657.
Fontaine, M. (1927b). Sur la compressibilité comparée du sérum et des globules du sang de cheval. C.R. Acad. Sci. 184: 627-628.
Fontaine, M. (1928). Les fortes pressions et la consommation d'oxygène de quelques animaux marins. Influences de la taille de l'animal. C.R. Soc. Biol. 99: 1789-1790.
Gordon, M.S. (1970). Hydrostatic pressure. In: Fish Physiology, vol. 4. Ed. W.S. Hoar and D.J. Randall, New York and London, Academic Press, p. 445-464.
Grundfest, H. (1936). Effects of hydrostatic pressures upon the excitability, the recovery, and the potential sequence of frog nerve. Cold Spring Harbor Symp. Quant. Biol. 4: 179-187.
Heremans, K. (1979). High-pressure biochemistry: a survey. In: High-Pressure Science and Technology, Six AIRAPT Conferences, vol. 1.

Ed. K.O. Timmerhaus and M.S. Barber, New York Plenum Press, p. 699-705.

Hochachka, P.W. (1971). Enzyme mechanisms in temperature and pressure adaptations of off-shore benthic organisms: the basic problem. Ann. Zool. 11: 425-435.

Hochachka, P.W. (1975). Biochemistry at depth. Pressure effects on biochemical systems of abyssal and midwater organisms. The 1973 Kona Expedition of the Alpha Helix. Ed. P.W. Hochachka. Pergamon Press. p. 202.

Hochachka, P.W., Moon, T.W. and Mustafa, T. (1972). The adaptation of enzymes to pressure in abyssal and mid-water fishes. Symp. Soc. Exp. Biol. 26: 175-195.

Hochachka, P.W., Schneider, D.E. and Moon, T.W. (1971). The adaptation of enzymes to pressure. I. A comparison of trout liver fructose diphosphatase with the homologous enzyme from an off-shore benthic fish. Am. Zool. 11: 479-490.

Hochachka, P.W. and Somero, G.N. (1973). Strategies of biochemical adaptation. Toronto, Ed. Saunders Company. p. 358.

Jampol, L.M. and Epstein, F.H. (1970). Sodium-potassium-activated adenosine triphosphatase and osmotic regulation by fishes. Am. J. Physiol. 218: 607-611.

Johnson, F.H., Eyring, H. and Polissar, M.J. (1954). Activity of narcotized amphibian larvae under hydrostatic pressure. J. Cell. and Comp. Physiol. 37: 15-25.

Johnson, S.M. and Miller, K.W. (1970). Antagonism of pressure and anaesthesia. Nature (Lond.) 228: 75-76.

Kamiya, M. (1967). Changes in ion and water transport in isolated gills of the cultured eel during the course of salt adaptation. Annotnes Zool. Jap. 40: 123-129.

Kamiya, M. and Utida, S. (1968). Changes in activity of sodium-potassium-activated adenosinetriphosphatase in gills during adaptation of the japanese eel to sea water. Comp. Biochem. Physiol. 26: 675-685.

Kirschner, L.B. (1979). Control mechanisms in crustaceans and fishes. In: Mechanisms of osmoregulation in animals. Ed. R. Gilles. New York, A. Wiley-Interscience publication Chichester. pp. 157-222.

Kitching, J.A. (1969). Effects of high hydrostatic pressure on the activity and behaviour of the ciliate Spirostomum. J. of Expertl. Biol. 51: 319-324.

Kitching, J.A. (1970). Some effects of high pressure on protozoa. In: High Pressure Effects on Cellular Processes. Ed. A.M. Zimmerman. New York, Academic Press. pp. 155-177.

Knight-Jones, E.W. and Morgan, E. (1966). Responses of marine animals to changes in hydrostatic pressure. Oceanog. Marine Biol. Ann. Rev. 4: 267-299.

Koefoed-Johnsen, V. and Ussing, H.H. (1958). The nature of the frog skin potential. Acta Physiol. Scand. 42: 298.

Lever, M.J., Miller, K.W., Paton, W.D.M. and Smith, E.B. (1971). Pressure reversal of anaesthesia. Nature 231: 368-371.

Macdonald, A.G. (1972). The role of high hydrostatic pressure in the

physiology of marine animals. Symp. Soc. Exp. Biol. 26: 209-232.
Macdonald, A.G. and Miller, K.W. (1976). Biological membranes at high hydrostatic pressure. In: Biochemical and Biophysical Perspectives in Marine Biology, Vol. 3. Eds. D.C. Malins and J.R. Sargent. London, New York, Academic Press. pp. 117-147.
Macdonald, A.G. and Teal, J.M. (1975). Tolerance of oceanic and shallow water crustacea to high hydrostatic pressure. Deep-Sea Res. 22: 131-144.
Maetz, J. (1969). Seawater teleosts: evidence for a sodium-potassium exchange in the branchial sodium-excreting pump. Science, 166: 613-615.
Maetz, J. (1971). Fish gills: mechanisms of salt transfer in fresh water and sea water. Phil. Trans. Roy. Soc. Lond. B 262: 209-249.
Marsland, D.A. (1944). Mechanism of pigment displacement in unicellular chromatophores. Biol. Bull. 87: 252-261.
Meek, R.P. and Childress, J.J. (1973). Respiration and the effect of pressure in the mesopelagic fish Anopologaster cornuta (Beryciformes). Deep-Sea Res. 20: 1111-1118
Menzies, R.J., George, R.Y. and Avent, R. (1972). Responses of selected aquatic organisms to increased hydrostatic pressure: preliminary results. In: Barobiology and the experimental biology of the deep sea. Ed. R.W. Brauer. Univ. of North Carolina. pp. 37-57.
Miyatake, T. (1957). Studies on effects of high hydrostatic pressure on blood cell. II. On the exchange of ions in the erythrocyte. Okayama Igakkai Zasshi 69: 461-471.
Moon, T.W. (1975). Effects of hydrostatic pressure on gill Na-K ATPase in an abyssal and a surfacing-dwelling teleost. Comp. Biochem. Physiol. 52B: 59-65.
Moon, T.W., Mustafa, T. and Hochachka, P.W. (1971). The adaptation of enzymes to pressure. II. A comparison of muscle pyruvate kinases from surface and midwater fishes with the homologous enzyme from an off-shore benthic species. Am. Zool. 11: 491-502.
Motais, R., Garcia-Romeu, F. and Maetz, J. (1966). Exchange diffusion effect and euryhalinity in teleosts. J. Gen. Physiol. 50: 391-422.
Murakami, T.H. (1963). Effect of hydrostatic pressure on the permeability of membrane under the various temperatüre. Symp. Cellular Chem. 13: 147-156.
Murakami, T.H. and Zimmerman, A.M. (1970). A pressure study of galvanotaxis in Tetrahymena. In: High Pressure Effects on Cellular Processes. Ed. A.M. Zimmerman. New York and London, Academic Press. pp. 139-153.
Murphy, R.B. and Libby, W.F. (1976). Inhibition of erythrocyte phosphate transport by high pressures. Proc. Natl. Acad. Sci. 73: 2767-2769.
Naroska, V. (1968). Vergleichende Untersuchungen über den Einfluss des hydrostatischen Druckes auf Uberlebensfähighkeit und Stoffwechselin - tensität mariner Evertebraten und Teleosteer. Kiel Meeresforsch. 24: 95-123.
Okada, K. (1954). Effects of hydrostatic high pressure on the permeability of plasma membrane. V. On plasmolysis. Okayama Igakkai Zasshi 66: 2095-2099.

Passow, H. (1964). Ion and water permeability of the red blood cell. In: The Red Blood Cell. New York, Academic Press. pp. 30-32.

Paul, K.L. and Morita, R.Y. (1971). Effects of hydrostatic pressure and temperature on the uptake and respiration of amino acids by a facultatively psychrophilic marine bacterium. J. Bacteriol. 108: 835-843.

Penniston, J.T. (1971). High hydrostatic pressure and enzymic activity inhibition of multimeric enzymes by dissociation. Arch. Biochem. Biophys. 142: 322-332.

Péqueux, A. (1972). Hydrostatic pressure and membrane permeability. Symp. Soc. Exp. Biol. 26: 483-484.

Péqueux, A. (1976a). Polarization variations induced by high hydrostatic pressures in the isolated frog skin as related to the effects on passive ionic permeability and active Na^+ transport. J. Exp. Biol. 64: 587-602.

Péqueux, A. (1976b). Effects of pH changes on the frog skin electrical potential difference and on the potential variations induced by high hydrostatic pressures. Comp. Biochem. Physiol. 55A: 103-108.

Péqueux, A. (1979a). Ionic transport changes induced by high hydrostatic pressures in mammalian red blood cells. In: High Pressure Science and Technology, Vol. I. Ed. K.D. Timerhaus and M.S. Barber. New York, Plenum Press, pp. 720-726.

Péqueux, A. (1979b). Effects of high hydrostatic pressures on ions permeability of isolated gills from sea water acclimated eels Anguilla anguilla. Proc. 1st congress of the ESCPB, Liège 1979, Pergamon Press, In Press.

Péqueux, A. and Gilles, R. (1977). Effects of high hydrostatic pressures on the movements of Na^+, K^+ and Cl^- in isolated eel gills. Experientia 33: 46-48.

Péqueux, A. and Gilles, R. (1978). Effects of high hydrostatic pressures on the activity on the membrane ATPases of some organs implicated in hydromineral regulation. Comp. Biochem. Physiol. 59B: 207-212.

Péqueux, A., Gilles, R., Pilwat, G. and Zimmermann, U. Pressure induced variations of K^+ permeability as related to a possible reversible mechanical breakdown in human erythrocytes. Submitted for publication.

Péqueux, A. and Parisis, M. In preparation.

Podolsky, R.J. (1956). A mechanism for the effect of hydrostatic pressure on biological systems. J. Physiol. 132: 38P-39P.

Regnard, P. (1884). Recherches expérimentales sur l'influence des très hautes pressions sur les organismes vivants. Compt. Rend. 98: 745-747.

Regnard, P. (1885). Phénomènes objectifs que l'on peut observer sur les animaux soumis aux hautes pressions. Compt. Rend. Soc. Biol. 37: 510-515.

Regnard, P. (1891). Recherches expérimentales sur les conditions physiques de la vie dans les eaux. Paris, Masson, 500 p.

Schlieper, C., Flügel, H. and Theede, H. (1967). Experimental investigations of the cellular resistance ranges of marine temperate and tropical bivalves: results of the Indian Ocean Expedition of the German

Research Association. Physiol. Zool. 40: 345-360.

Seaman, G.V. and Heard, D.H. (1960). The surface of the washed human erythrocyte as a polyanion. J. Gen. Physiol. 44: 251-268.

Shirai, N. (1972). Electronmicroscope localization of sodium ions and adenosinetriphosphatase in chloride cells of the Japanese eel, Anguilla japonica. J. Fac. Sci. Tokyo Univ. Sec. IV, 22: 385-403.

Skou, J.C. (1965). Enzymatic basis for active transport of Na^+ and K^+ across cell membranes. Physiol. Rev. 45: 596-617.

Smith, K.L. and Hessler, R.R. (1974). Respiration of benthopelagic fishes: in situ measurements at 1230 meters. Science 184: 72-73.

Smith, K.L. and Teal, J.M. (1973). Deep-sea benthic community respiration: an in situ study at 1850 meters. Science 179: 282-283.

Spyropoulos, C.S. (1957a). Response of single nerve fibers at different hydrostatic pressures. Am. J. Physiol. 189: 214-218.

Spyropoulos, C.S. (1957b). The effects of hydrostatic pressure upon the normal and narcotized nerve fiber. J. Gen. Physiol. 40: 849-857.

Tanaka, R. and Teruya, A. (1973). Lipid dependence of activity-temperature relationship of (Na^++K^+)-activated ATPase. Biochem. Biophys. Acta 323: 584-591.

Trudell, J.R., Hubbell, W.L. and Cohen, E.N. (1973). Pressure reversal of inhalation anaesthetic induced disorder in spin-labelled phospholipid vesicles. Biochem. Biophys. Acta 291: 328-334.

Wattiaux-de Coninck, W., Dubois, F. and Wattiaux, R. (1977). Lateral phase separations and structural integrity of the inner membrane of rat-liver mitochondria. Biochim. Biophys. Acta 471: 421-435.

Woodhouse, B.J. (1973). Thesis. University of Aberdeen cited by Macdonald, A.G. and Miller, K.W. (1976).

Yamato, H. (1952) Effects of high hydrostatic pressure on the action of erythrocyte. III. On the kalium contents. Okayama Igakkai Zasshi 64: 874-881.

Yamato, H. (1952) Effects of high hydrostatic pressure on the action of erythrocyte. V. On the lysis. Okayama Igakkai Zasshi 64: 888-900.

Zenkevitch, L.A. (1963) Biology of the Seas of the U.S.S.R. London, Allen & Unwin.

Zimmerman, A.M. (1969) Effects of high pressure on macromolecular synthesis in synchronized Tetrahymena. In: The Cell Cycle. Eds. G.M. Padilla, G.L. Whitson, and I.L. Cameron. New York, Academic Press, p. 203-225.

Zimmerman, A.M. (1970) High Pressure Effects on Cellular Processes. Ed. A.M. Zimmerman. New York and London, Academic Press, Inc., pp. 324.

Zimmermann, U. (1977) Cell turgor regulation and pressure mediated transport processes. In: Integration of Activity in the Higher Plant. Ed. D.H. Jennings. Proc. 31st Symp. Soc. Exp. Biol. Cambridge, Univ. Press, p. 117-154.

Zimmermann, U., Pilwat, G., Péqueux, A. and Gilles, R. Electro-mechanical properties of human erythrocyte membranes: the pressure-dependence of potassium permeability. Submitted for publication.

Zimmermann, U., Pilwat, G. and Riemann, F. (1974) Dielectric breakdown of cell membranes. Biophys. J. 14: 881-899.

LES HORMONES DANS L'OSMOREGULATION DES POISSONS

B. Lahlou

Laboratoire de Physiologie Comparée

Faculté des Sciences et des Techniques

Parc Valrose, 06034 NICE Cedex, France

INTRODUCTION

La fonction osmorégulatrice des poissons présente une extrême diversité liée tant à la variété des biotopes qu'à celle de la composition du milieu intérieur, dissemblable d'un groupe à l'autre.

Le maintien des concentrations internes dans un environnement donné ou leur réajustement à la suite d'un changement de milieu résulte essentiellement des échanges d'eau, de sodium et de chlore. A ces derniers ions il convient d'ajouter cependant le calcium qui, malgré sa faible teneur dans le plasma, joue un rôle considérable dans le maintien de la cohésion cellulaire et de l'intégrité fonctionnelle des membranes et a fait l'objet d'importantes études récentes. Ces échanges. s'effectuent principalement au niveau des branchies, de l'intestin et des reins et, accessoirement, au niveau d'organes osmorégulateurs particuliers à certains groupes ou à certaines espèces: vessie urinaire, tégument, glande rectale (voir article de D.H. EVANS dans le présent volume).

Les échanges osmorégulateurs sont eux-mêmes modulés par de nombreuses hormones dont l'étude chez les poissons a fait l'objet d'un nombre considérable de travaux. Les poissons possèdent, en effet, un système endocrinien complexe qui inclut non seulement toutes les glandes des Vertébrés supérieurs, à l'exception des parathyroides, mais également des organes fonctionnels qui ne sont guère développés dans les autres classes: système neurosécrétoire caudal, corpuscules de Stannius, corps ultimobranchiaux (Figure 1).

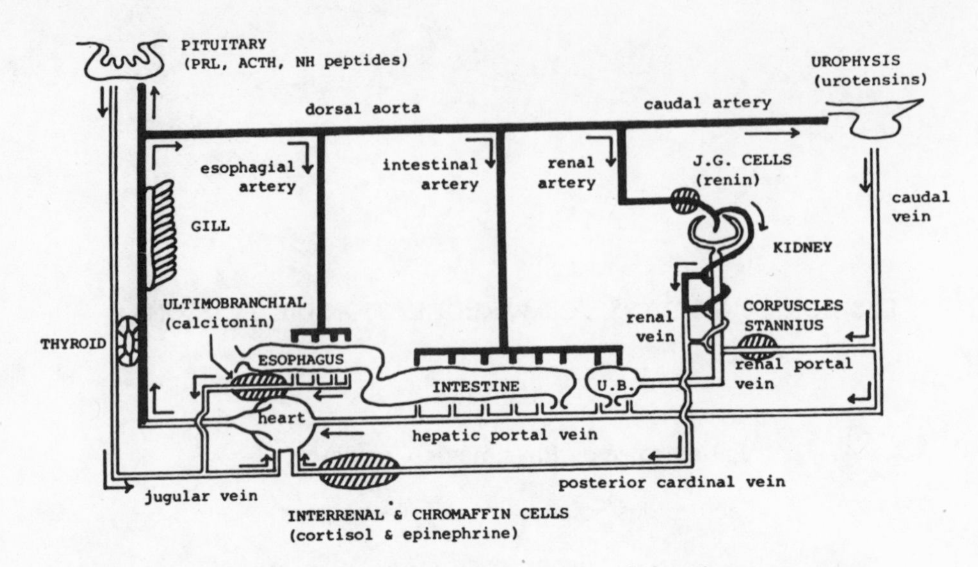

Fig. 1. Le système endocrinien des Téléostéens et ses relations avec l'appareil circulatoire et les organes osmorégulateurs.
NH: neurohypophyse - J.G. cells: cellules juxta-glomérulaires - U.B.: vessie urinaire (urinary bladder) (d'après T. Hirano et N. Mayer-Gostan, 1978)

De ces glandes, la THYROIDE est celle qui a d'abord suscité le plus d'intérêt chez les physiologistes. De son intervention bien connue dans la métamorphose ou la croissance des Tétrapodes, on pouvait inférer qu'elle devait présider aux transformations morphologiques et fonctionnelles qui accompagnent la migration des poissons, en particulier la smoltification des Salmonidés (Hoar, 1939; travaux de M. Fontaine et ses collaborateurs). De fait, on observe par exemple un pic important de thyroxine plasmatique au printemps chez le Coho (Dickhoff et al, 1978). Cependant de telles corrélations n'établissent que de façon indirecte un contrôle par les hormones thyroidiennes des échanges osmorégulateurs d'eau et d'électrolytes. Il est manifeste qu'il existe dans ce domaine de l'endocrinologie des poissons une lacune injustifiée qui devrait être rapidement comblée.

Dans l'exposé qui suit, nous considérerons à tour de rôle les autres glandes, groupes d'hormones ou hormones qui ont été étudiées du point de vue de leur intervention dans l'osmorégulation. Un accent particulier sera mis d'une part sur les changements endocriniens associés à un changement de la salinité de l'environnement, d'autre part sur les tendances actuelles de la méthodologie dans ce domaine. Nous avons cru nécessaire, par ailleurs, d'inclure dans cette étude le cas particulier des Cyclostomes qui présentent avec les poissons de nombreuses similitudes physiologiques.

PROLACTINE

Alors que chez les Mammifères la prolactine (PRL) est associée aux fonctions de reproduction, il est établi que chez les poissons elle est d'une importance primordiale dans le contrôle hormonal de l'osmorégulation.

Les effets physiologiques attribuables à la prolactine chez les Téléostéens ont été revus à plusieurs reprises (Ensor et Ball, 1972 - Bern, 1975 - Hirano, 1977). Leur mise en évidence repose sur l'utilisation de la prolactine ovine purifiée injectée aux animaux intacts ou hypophysectomisés.

Récemment des tentatives sérieuses ont été conduites en vue de caractériser la structure chimique de l'hormone chez les poissons. Une séparation préliminaire a été réalisée par électrophorèse sur gel de polyacrylamide chez le flet (Chadwick, 1970), l'anguille (Knight, *et al.*, 1978), les Cichlides *Tilapia* et *Cichlasoma* (Clarke, 1973). Une fraction hautement purifiée a été isolée chez *Tilapia* (Farmer *et al.*, 1977). La substance présente une certaine similarité de structure avec l'hormone de croissance. Ces 2 hormones de poisson pourraient provenir d'une même molécule ancestrale. La spécificité zoologique apparaît dans le fait que la PRL de poisson ne stimule pas les sécrétions du jabot de pigeon et de la glande mammaire ou ne provoque que des effets atypiques.

PRL et adaptation aux milieux aquatiques:

L'intervention de l'hypophyse dans l'adaptation des Téléostéens à l'eau douce (ED) a été démontrée pour la première fois par Burden (1956), montrant que *F. heteroclitus* privé d'hypophyse ne survit pas dans ce milieu. Pickford et Phillips (1959) ont démontré que seules des injections de PRL (ovine) pouvaient assurer la survie dans ces conditions. Les observations entreprises sur de nombreuses espèces montrent que les Téléostéens se répartissent en 2 groupes (Tableau I), l'hypophyse étant indispensable pour l'un mais non pour l'autre. Dans tous les cas cependant, la suppression de la glande provoque une chute plus ou moins marquée de la pression osmotique et des teneurs du plasma en électrolytes (P_{Na} et P_{Cl}). Ainsi chez F. heteroclitus, ces paramètres tombent de 40 à 60% en ED (Burden, 1956; Pickford et al., 1970). Cette diminution des concentrations internes s'observe à un degré important même chez les espèces telles que Carassius auratus qui résistent à l'opération. Il semble que la survie éventuelle soit liée à la capacité des cellules de supporter une variation importante des teneurs en électrolytes sanguins (Lahlou et Sawyer, 1969). Dans tous les cas, l'effet de PRL injectée de façon répétée aux animaux opérés est de maintenir à un niveau suffisant les concentrations plasmatiques.

Tableau I

Effets de l'hypophysectomie
sur la survie en eau douce

Survivent:	
Anguilla anguilla	(Fontaine et al., 1949)
Carassius auratus	(Chavin, 1956)
Fundulus kansae	(Stanley et Fleming, 1966)
Platichthys flesus	(MacFarlane, 1974)
Salmo trutta	(Oduleye, 1975)
Ne survivent pas:	
Fundulus heteroclitus	(Burden, 1956; Pickford et Phillips, 1959)
Gambusia sp	(Chambolle, 1967)
Ictalurus melas	(Chidambaram et al., 1972)
Oryzias latipes	(Utida et al., 1971)
Peocilia latipinna	(Ball et Ensor, 1965)
Tilapia mossambica	(Dharmamba et al., 1967)
Xiphophorus maculatus, X. helleri, X. milleri	(Schreibman et Kalman, 1966)

Chez les larves de Téléostéens, la capacité de survivre en ED paraît nécessiter la présence de PRL, mais peu d'espèces ont été jusqu'à présent étudiées: truite (Nozaki et al., 1974), Lebistes reticulatus (Ichikawa et al., 1973).

Chez les animaux marins il ne semble pas que la PRL soit indispensable bien que l'hypophysectomie ou l'injection de l'hormone entraînent des perturbations dans les teneurs et les échanges d'électrolytes chez les Téléostéens. En outre, aucune corrélation n'a été établie entre la présence de PRL et l'adaptation osmotique au milieu ambiant chez les Cyclostomes et les Elasmobranches.

Contrôle de la sécrétion de PRL par la salinité externe:

De nombreuses études histologiques établissent que l'état physiologique de la glande pituitaire est affecté par le niveau des concentrations d'électrolytes internes ou externes. Cependant il n'existe pas une concordance parfaite entre les observations disponibles jusqu'à présent.

En règle générale, la dimension et l'activité sécrétoire des cellules rostrales et de l'hypophyse antérieure augmente lorsqu'un poisson euryhalin acclimaté en EM est transféré en ED. De telles variations ont été trouvées chez S. salar en cours de migration (Fontaine et Olivereau, 1954), Anguilla anguilla (Olivereau, 1968), Gasterostus aculeatus (Leatherland, 1970; Schreibmann et al., 1973; Wendelaar Bonga, 1978), Gillichthys mirabilis passant d'EM en ED et Tilapia transféré d'ED en EM (Nagahama et al., 1975; Dharmamba et Nishioka, 1968). In vitro, la sécrétion de PRL est affectée directement et dans le même sens par la salinité du milieu d'incubation chez Xiphophorus (Sage, 1968) et chez d'autres espèces comme Tilapia (Nagahama et al., 1975).

Certains auteurs, cependant, considèrent que cette relation n'est ni permanente, ni exclusive. Ainsi chez Tilapia (Clarke, 1973) et Anguilla (Hall et Chadwick, 1978), la PRL diminue d'abord après un transfert en EM puis retourne progressivement à son niveau d'eau douce. En outre Wendelaar Bonga suggère que la teneur en Ca^{++} (et en Mg^{++}) constitue le facteur principal de la régulation. Cet auteur utilise des animaux (Gasterosteus) privés de leurs corpuscules de Stannius (voir plus loin) en vue de dissocier l'une de l'autre les teneurs plasmatiques en Na^+ et Ca^{++}. Chez les animaux ainsi opérés, le Ca interne étant élevé, les cellules à PRL ne sont plus affectées par le transfert d'ED et EM ou inversement. Il observe que les cellules à PRL ont une activité reliée de façon inverse à P_{Ca} (et non à P_{Na} ou à l'osmolarité externe). Par ailleurs, la réduction du Ca externe entraîne l'élargissement des cellules à PRL, l'enrichissement en Ca produisant l'effet inverse. Selon l'auteur, les teneurs externe et interne en Ca^{++} interviennent simultanément pour contrôler l'activité PRL, mais par des mécanismes distincts.

Compte tenu des autres informations actuellement disponibles, il apparaît que le contrôle de la sécrétion de la prolactine revêt une grande complexité chez les Téléostéens (Figure 2).

PRL et échanges branchiaux:

Les méthodes radioisotopiques appliquées aux ions Na^+ et Cl^- ont montré que le déséquilibre ionique résultant de l'hypophysectomie en eau douce provenait d'une augmentation du flux sortant, le flux entrant étant pratiquement inchangé. L'effet correcteur de la PRL porte donc sur la composante passive des échanges (Maetz et al., 1967; Dharmamba et Maetz, 1972).

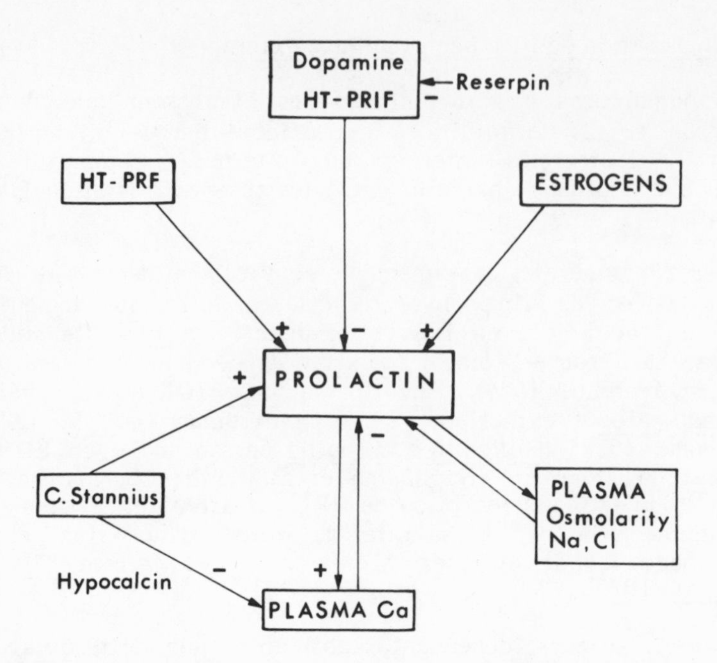

Fig. 2. Actions régulatrices principales concernant la prolactine (HT: facteurs hypothalamiques de stimulation (PRF) ou d'inhibition (PRIF) de la libération de prolactine - Le signe (+) indique une stimulation, le signe (-) une inhibition).

En eau de mer, la prolactine réduit de façon importante le taux de renouvellement de sodium, mesuré en tant que flux d'excrétion branchiale, et augmente les électrolytes plasmatiques chez plusieurs espèces, en particulier Tilapia (Dharmamba et Maetz, 1972). La signification physiologique de ces effets apparaît pour le moment obscure.

Chez Carassius, l'ablation de l'hypophyse réduit de façon importante la perméabilité diffusionnelle et osmotique des surfaces externes à l'eau, la PRL ayant un effet inverse (Lahlou et Giordan, 1970). Ces modifications, opposées à celles des échanges ioniques branchiaux montrent que les mouvements de l'eau et des électrolytes sont indépendants au niveau branchial. Des résultats analogues ont été obtenus chez F. kansae (Potts et Fleming, 1970). Des observations opposées aux précédentes ont été recueillies sur des branchies isolées, sur lesquelles la PRL inhibe l'entrée osmotique de l'eau chez Carassius

(Ogawa et al., 1973), et A. japonica (Ogawa, 1977). Il est cependant difficile de comparer ces conditions expérimentales à la situation in vivo, du fait que la circulation sanguine est supprimée sur les organes isolés.

Un facteur important qui conditionne l'estimation de flux isotopiques d'eau est constitué en effet par l'épaisseur des barrières de diffusion ou "unstirred layers" qui bordent les membranes perméables. Au niveau des branchies et du tégument, l'hypophysectomie entraîne une régression des cellules à mucus, tandis que la PRL provoque leur prolifération et stimule leur sécrétion (Burden, 1956; Bern et Nicoll, 1968; Olivereau et Olivereau, 1971; Leatherland et Lam, 1969; Marshall, 1976). Par ailleurs le mucus est considéré habituellement comme jouant un rôle protecteur du milieu interne en réduisant les perméabilités passive aux ions et à l'eau. Les résultats ci-dessus de Lahlou et Giordan (1970) sur Carassius et de Potts et Fleming (1970) sur F. kansae ne peuvent s'expliquer que si la PRL modifie la perméabilite per se des parois cellulaires, indépendamment d'un effet sur le mucus.

PRL et fonction rénale:

1- Nephrons: la plupart des observations ont été obtenues chez les poissons adaptés à ED, l'excrétion rénale étant, dans ce milieu, caractérisée par un GRF et un débit urinaire élevés et une faible teneur de l'urine en électrolytes du fait que ceux-ci sont largement réabsorbés par le tubule. L'ablation de l'hypophyse entraîne en général une réduction prolongée du débit et une augmentation des concentrations ioniques. Chez Carassius l'injection de PRL aux animaux opérés provoque par des effets inverses le retour à la normale (Lahlou et Giordan, 1970). D'intéressantes études histologiques prouvent que la PRL induit des modifications morphologiques profondes au niveau des cellules du nephron de Téléostéens.

Chez l'Anguille, Olivereau et Olivereau (1977) ont mis en évidence des transformations cellulaires lentes après hypophysectomie, concernant principalement une réduction de la hauteur cellulaire des tubules distal et collecteur principal. Il est remarquable que la PRL injectée à dose suffisante induise la formation de nouveaux tubules.

Chez Gasterosteus aculeatus transféré d'EM en ED on observe une augmentation du diamètre glomérulaire (non affecté par des injections préalables de PRL), un développement de l'épithélium tubulaire et uretéral (qui est accéléré par cette hormone) et un ralentissement (non influencé par PRL) de l'activité des cellules juxta-glomérulaires (Wendelaar Bonga, 1976).

2- Vessie urinaire: la vessie urinaire des Téléostéens peut être considérée comme une extension de l'appareil rénal du fait qu'elle résulte d'un élargissement distal des uretères. Elle effectue une réabsorption d'eau et de NaCl sous forme d'une solution généralement fortement hyperosmotique par comparaison au plasma. Chez divers Téléostéens, notamment chez le flet et la truite, les vessies des animaux d'ED transportent moins d'eau et plus de sodium que les poissons d'EM (Hirano et al., 1971; Lahlou et Fossat, 1971; Hirano et al., 1973). Hirano et al. (1971) ont mis en évidence un contrôle remarquable de ces absorptions par la prolactine. Cette hormone, injectée à des poissons d'EM, imperméabilise leur vessie à l'eau et la transforme fonctionnellement en une vessie de poisson d'ED. En même temps, ces vessies réabsorbent *in vitro* davantage de Na^+ et Cl^- à partir de solutions hypo ou isotoniques que les vessies d'animaux non traités (Hirano, 1975).

PRL et fonction intestinale:

C'est en EM que l'intestin des poissons Téléostéens exerce une fonction osmorégulatrice fondamentale en absorbant l'eau nécessaire au maintien de l'équilibre hydrique. Les effets de la PRL ont été encore peu étudiés sur cet organe. L'intestin d'anguille répond à la PRL par une diminution de sa perméabilité à l'eau (Utida et al., 1972). En revanche l'intestin de flet ne semble pas sensible à l'hormone (Hirano, 1975).

En utilisant une technique rapide de mesure de flux ioniques sur l'intestin isolé de *Carassius* et de *Platichthys flesus*, Ellory et al. (1972) et Lahlou (1975) ont montré que la perméabilité apicale des entérocytes aux ions Na^+ et Cl^- n'était pas modifiée par l'hypophysectomie quelle que soit la salinité. Ces résultats montrent qu'un contrôle éventuel par la prolactine ne peut concerner que l'activité de la pompe ionique basolatérale qui transfère les ions vers le milieu intérieur.

La prolactine chez les autres Poissons:

Des données expérimentales précises n'existent que chez les Elasmobranches. Ces derniers présentent un taux de renouvellement de l'eau (mesuré à l'aide d'eau tritiée) extrêmement élevé (environ 100 à 160% par heure chez *Scyliorhinus canicula*, *Raia montagu* et *Torpedo marmorata* (Payan et Maetz, 1971). L'ablation de l'hypophyse provoque une réduction de 50% de la perméabilité branchiale. Comme chez les Téléostéens, l'injection de PRL corrige cette réduction.

Mode d'action cellulaires de la prolactine:

Très peu de recherches ont été effectuées jusqu'à présent dans ce domaine. Le contrôle de la perméabilité à l'eau dans les vessies urinaires du Gobie *Gillichthys mirabilis* en culture d'organe constitue le seul exemple d'action obtenue exclusivement *in vitro* (Doneen et Bern, 1974). Les effets morphogénétiques et à long terme observés au niveau des divers organes indiqués ci-dessus montrent que la PRL intervient par l'intermédiaire de synthèses protéiques dans les cellules cibles.

L'administration de prolactine chez le flet *Kareius bicoloratus* adapté à EM entraîne une augmentation de l'incorporation de ^{3}H - thymidine dans la fraction DNA de la vessie urinaire et de l'intestin (Hirano *et al.*, 1973).

Il est également admis que l'activité enzymatique Na-K-ATPase qui est présente dans tous les organes osmorégulateurs des poissons conditionne le niveau du transport de sodium. La régulation de cette activité paraît complexe, mais la PRL l'augmente dans le rein et la diminue dans les branchies (Pickford *et al.*, 1970; Kamiya, 1972). Une telle stimulation s'observe au niveau de la vessie urinaire du flet euryhalin *Platichthys stellatus* transféré d'EM en ED, mais non chez *Kareius* qui ne survit pas à ce transfert (Utida *et al.*, 1974).

Une approche cellulaire différente a été effectuée récemment par Horseman et Meier (1978). Ces auteurs observent chez *Fundulus grandis* que l'élévation des concentrations plasmatiques de chlore induite par la PRL est empêchée par des inhibiteurs de la synthèse de prostaglandines, telles que l'aspirine ou l'indomethacine. Ils en concluent que certaines actions physiologiques au moins de la PRL s'exercent par l'intermédiaire d'une stimulation préalable de la synthèse des prostaglandines. De façon indirecte, par conséquent, la prolactine ferait intervenir le système adénylcyclase.

Comme chez les autres groupes de Vertébrés, le mécanisme d'action cellulaire de la PRL reste à élucider chez les poissons qui constituent, pour cette hormone, un excellent modèle d'étude.

NEUROHYPOPHYSE

La neurohypophyse des poissons contient un assortiment de peptides différents de ceux des Mammifères. Les travaux de Du Vigneaud, Acher, Heller et Sawyer ont permis de connaître la distribution phylétique, les propriétés pharmacologiques et physiologiques et la structure chimique. La revue récente de Maetz et Lahlou (1975) décrit en détail les actions physiologiques de ces hormones chez les poissons. Seules les données récentes en rapport avec l'osmorégulation sont indiquées ci-après.

La neurohypophyse des Vertébrés contient typiquement au moins deux peptides différents. Seuls les Cyclostomes possèdent une seule hormone, l'arginine-vasotocine (AVT). Celle-ci est considérée comme une hormone ancestrale car on la rencontre chez les adultes de toutes les classes de Vertébrés, sauf les Mammifères. Chez ces derniers, elle existe de façon transitoire pendant la vie foetale. En outre la glande pinéale contient un principe ressemblant à l'AVT mais qui ne semble pas être identique à cette substance (Pevet *et al.*, 1979).

Le contenu hormonal de la neurohypophyse est complexe et encore incomplètement connu chez les Elasmobranches, bien que des peptides particuliers aient été identifiés: valitocine et aspartocine notamment (Acher *et al.*, 1972). Les Téléostéens renferment de l'AVT et de l'isotocine (ou ichthyotocine). Les poissons pulmonés offrent une plus grande diversité, avec possibilité de la présence d'AVT, d'isotocine, de mésotocine et peut-être d'ocytocine (Sawyer, 1969; Sawyer, 1977).

Les effets de ces peptides sur les échanges ioniques peuvent être *a priori* masqués ou modifiés par leurs actions vasomotrices, en particulier au niveau de la branchie où la circulation sanguine est intense. Cette interférence n'a cependant pu être évaluée jusqu'à présent pour ces hormones. Les réponses hémodynamiques ont été surtout étudiées sur l'anguille, les Elasmobranches et le Protoptère (Chan et Chester-Jones, 1969; Maetz et Rankin, 1969; Sawyer, 1970; Chan, 1977). Un effet vasopresseur important et de longue durée est observé au niveau de l'aorte ventrale. L'isotocine et la mésotocine sont les hormones les plus puissantes sur ce test, étant actives à 10 exp (-11) mole/kg chez le requin *Hemiscyllium* et l'anguille *A. japonica* (Chan, 1977). Sur la branchie d'*A. anguilla* isolée et perfusée à pression constante le seuil d'action de l'isotocine serait de 10 exp (-11) M (Maetz et Rankin, 1969). Au niveau de l'aorte dorsale on observe un effet vasodépresseur chez l'anguille (Chan et Chester Jones, 1969), mais c'est un effet vasopresseur caractéristique qui est obtenu avec d'autres espèces, telles que *Protopterus* et *Opsanus*, l'AVT étant cette fois la plus active et la dose injectée de l'ordre du ng/kg (Sawyer, 1970; Lahlou *et al.*, 1969; Babiker et Rankin, 1979).

La sensibilité du système cardíovasculaire des poissons aux hormones neurohypophysaires est impressionnante. *Protopterus* et plus encore *Neoceratodus*, répondent à des quantités d'AVT inférieures au ng/kg qui ne déclenchent pas d'effet vasopresseur chez le rat privé des centres nerveux supérieurs qui est la préparation mammalienne la plus sensible à ces peptides. Aussi Sawyer et al. (1976) considèrent-ils que l'AVT, hormone primitive, a pu avoir comme fonction originelle de réguler le système hémodynamique.

Chez le poisson d'eau douce *Carassius auratus*, Maetz et al., (1964) ont observé un effet positif de l'isotocine sur la balance sodique, mais pour des doses d'hormones élevées (de l'ordre de ng/kg). Chez le flet préalablement adapté à ED, Motais et Maetz (1964) ont montré que l'AVT hâtait l'apparition de l'augmentation du flux sortant de Na à la suite du transfert des animaux en EM. Chez *Carassius*, enfin, l'AVT a été décrite comme réduisant les échanges branchiaux d'eau mesurés à l'aide d'eau tritiée, sans que cet effet puisse être séparé des effets hémodynamiques (Lahlou et Giordan, 1970). Récemment cependant, Sawyer et Pang (1980) n'ont pu retrouver aucun effet "natriférique" sur les branchies de Protopterus et Lepidosiren pour des doses hormonales actives au niveau du rein. Comme le suggèrent ces derniers auteurs, il est possible que les modifications décrites précédemment concernant les échanges ioniques branchiaux ne soient nullement physiologiques, eu égard aux doses hormonales très élevées utilisées.

En revanche, des effets nets et caractéristiques ont été recueillis au niveau de la fonction rénale des poissons. Les peptides neurohypophysaires de poisson sont diurétiques et matriurétiques chez plusieurs espèces: Carassius intact (Maetz et al., 1964) ou hypophysectomisé (Lahlou et Giordan, 1970), Anguilla anguilla (Chester-Jones et al., 1969). Les résultats les plus démonstratifs sont ceux obtenus par Sawyer et al., (1976) chez Protopterus aethiopicus et Neoceratodus forsteri. La diurèse est induite chez ces animaux à des doses voisines de lng/kg soit 10 exp (-12) mole/kg. Chez Neoceratodus (Sawyer et al., 1976) la réponse diurétique à l'AVT est exactement contemporaine de l'effet hypertensif. De même, chez Anguilla anguilla les effets diurétique et presseur s'obtiennent simultanément pour des doses supérieures à lng/kg (Henderson et Wales, 1974). Au contraire chez le poisson aglomérulaire Opsanus tau (Lahlou et al., 1969) les peptides neurohypophysaires n'agissent pas sur la diurèse, alors qu'ils élèvent (l'AVT particulièrement) la pression sanguine dorsale. Ces résultats ont conduit les divers auteurs à conclure que les hormones concernées agissent d'abord par un effet glomérulaire pur, c'est-à-dire en modifiant, par l'intermédiaire de l'élévation de la pression sanguine dans les vaisseaux rénaux, le nombre des glomérules (et donc des nephrons) fonctionnels. Ce phénomène de "recrutement" n'est cependant probablement pas exclusif, les fonctions tubulaires pouvant être également altérées.

La présence d'un effet diurétique des peptides neurohypophysaires chez les poissons revêt probablement une grande importance pour l'osmorégulation en eau douce. Dans ce milieu, en effet, l'animal doit excréter une quantité appréciable d'eau libre pour compenser l'entrée osmotique branchiale. L'absence d'une telle action rénale chez les poissons marins est probablement liée au fait que ces animaux éliminent un volume réduit d'urine, l'eau excrétée représentant la quantité minimale nécessitée par l'évacuation des ions polyvalents et des substances organiques. Sawyer (1972) suggère que le passage de la vie aquatique à la vie terrestre, qui s'est accompagnée chez les Vertébrés de la transition de la réponse diurétique à la réponse antidiurétique des Tétrapodes, a pu également correspondre au déplacement du site d'action hormonale d'une localisation pré- à une localisation post-glomérulaire. Ainsi les poissons répondent à une expansion du volume de fluide extracellulaire ou à sa dilution, les Tétrapodes à une rétraction de cet espace (ou à son hyperconcentration osmotique).

Deux types de données récentes paraissent cependant devoir compliquer ces interprétations. D'abord, il a été observé que de très faibles quantités d'hormone (de l'ordre du pg d'AVT/kg) provoquent un effet antidiurétique chez l'anguille (Babiker et Rankin, 1972; Henderson et Wales, 1974). Ensuite, Sawyer et Pang (1980) doutent même que les peptides neurohypophysaires circulent dans l'espace vasculaire des poissons car le dosage radioimmunologique de l'AVT n'a pas permis de caractériser cette substance dans le sang. Aussi, dans l'état actuel de nos connaissances, considèrent-ils que les effets presseurs, diurétiques et natrifériques observés peuvent être seulement de nature pharmacologique.

Aucune autre action positive des peptides neurohypophysaires n'a été mise en évidence sur les autres organes de poisson.

SYSTEME NEUROSECRETOIRE CAUDAL

Cette formation, située à l'extrémité postérieure du névraxe des poissons rappelle de façon frappante le système hypothalamo-neurohypophysaire. Elle comporte des cellules neurosécrétoires volumineuses (cellules de Dahlgren) éparpillées dans la moelle épinière au niveau des dernières vertèbres et du filum terminale. Les axones se rendent pour la plupart vers une expansion ventrale de la moelle désignée sous le terme d'urophyse (Fridberg, 1962) qui constitue une importante zone neurohémale où les produits de sécrétion peuvent entrer dans le sang. Le premier travail étendu concernant à la fois la morphologie de ce système et sa signification physiologique est dû à Enami (1956). Plusieurs revues récentes ont fait le point des progrès réalisés depuis concernant cet organe complexe, exclusif des poissons (Arvy, 1966; Chan, 1969; Fridberg et Bern, 1968; Lederis, 1977; Bern et Lederis, 1978).

Parmi les fonctions physiologiques proposées par Enami pour ce système endocrinien, celle qui a été le plus souvent explorée concerne précisément son intervention dans l'osmorégulation.

Enami, Miyashita et Imai (1956) observent que l'ablation de l'urophyse ou la section du nevraxe en avant du renflement entraîne une élévation du sodium plasmatique chez le poisson euryhalin *Oryzias latipes*. Enami signale que des modifications morphologiques affectent le système caudal de *Misgurnus* soumis à des stimuli osmotiques. Takasugi et Bern (1962) observent que *Tilapia* privé d'urophyse résiste mal aux solutions hypertoniques, le taux de chlore sérique s'élevant et provoquant une augmentation de la mortalité. Ireland (1969) indique que l'urophysectomie compromet la survie de *Gasterosteus aculeatus* à la suite du transfert d'ED en EM et inversement. Bennett et Fox (1962) chez les Elasmobranches et Yagi et Bern (1963) chez les Téléostéens constatent que les neurones du système caudal conduisent des potentiels d'action comme ceux du système hypothalamohypophysaire. Des changements de salinité ou des injections d'eau distillée, utilisés comme stimuli osmotiques externes et internes, provoquent des modifications de l'activité électrique des neurones enregistrée in situ. Deux types cellulaires différents paraissent répondre soit au stimuli hypotonique soit aux stimuli hypertoniques respectivement.

Meatz et al., (1964) analysent les échanges périphériques d'eau et de sodium chez le poisson d'eau douce Carassius auratus. L'injection d'un extrait hypophysaire stimule le flux branchial entrant de sodium. En même temps on observe, au niveau rénal, une augmentation du GRF et de l'excrétion d'eau libre. Ces deux effets, qui déterminent un gain de sodium pour l'organisme, rappellent ceux produits chez le même animal par les peptides neurohypophysaires. Cependant l'extrait d'urophyse est dépourvu des activités caractéristiques de ces derniers: natriférique (transport de sodium par la peau de Rana esculenta) et ocytocique (contractions de l'utérus de rate) notamment. Les substances actives paraissent donc différentes des peptides neurohypophysaires.

Chez l'anguille, des extraits d'urophyse provoquent aussi une augmentation du GRF, du débit urinaire et, paradoxalement, de l'excrétion sodique (Chester-Jones et al., 1969).

Des observations histologiques démontrent également une corrélation entre l'aspect et l'activité apparente de la glande d'une part et la salinité externe d'autre part. Chez Albula vulpes, Fridberg et al., (1966) ont étudié minutieusement les modifications ultrastructurales du système caudal. Les granules électron-denses de neurosécrétion sont plus abondants en milieu dessalé que chez les poissons pêchés en plein océan. On observe également en plus grand nombre des "inclusions nucléaires" d'origine cytoplasmique indiquant une grande activité de synthèse de produit sécrétoire.

Chez le Salmonidé Salvelinus fontinalis, Chevalier (1976) observe au contraire un maximum d'activité neurosécrétoire en eau désionisée tandis que le transfert en EM 100% entraîne une réduction de l'activité cellulaire.

Chez la même espèce, l'injection in vivo d'acides aminés tritiés montre que la leucine et la tyrosine s'incorporent beaucoup plus dans les cellules du système caudal chez les animaux maintenus en eau désionisée que chez ceux d'eau douce (Chevalier, 1978).

Ces observations morphologiques, parmi d'autres du même type, indiquent une grande variabilité des réponses cellulaires qui peut s'expliquer en supposant que les neurones concernés peuvent être hérérogènes et différemment spécialisés, répondant positivement ou négativement aux fluctuations de la salinité.

Les données actuellement disponibles ne permettent pas d'évaluer l'importance exacte du système caudal dans l'osmorégulation. L'ablation de l'urophyse n'est pas mortelle. Ni cette opération, ni l'injection d'extraits urophysaires n'induisent de changement significatifs dans les teneurs en électrolytes plasmatiques, sauf dans quelques rares cas comme Gillichthys ou Catostomus (Fryer et al., 1978).

D'importantes recherches biochimiques et pharmacologiques ont, en revanche, permis de caractériser de façon très encourageante les principes actifs contenus dans le système neurosécrétoire caudal. Ces travaux ont été récemment revus par Lederis (1977) qui aura été l'un des principaux investigateurs dans ce domaine de recherche. Les deux tests biologiques les plus significatifs utilisés in vitro ont été: la stimulation du transport osmotique d'eau par la vessie urinaire du crapaud hawaien Bufo marinus (Lacanilao, 1969; 1972) et la contraction musculaire de la vessie urinaire de la truite Salmo gairdneri (Lederis, 1970). Quatre activités (ou peptides?) dénommées urotensines (I à IV) ont été mises en évidence.

L'urotensine I possède la propriété spécifique d'abaisser la pression artérielle chez le rat (Kobayashi et al., 1968). Elle paraît correspondre à 3 peptides différents qui ne peuvent être distingués par leurs propriétés pharmacologiques. Chez le poisson, U I est diurétique et augmente l'excrétion de Ca^{++} et de Mg^{++} (Chan, 1975).

L'urotensine III serait le principe responsable de la régulation des échanges de sodium. Son existence en tant que substance chimique distincte n'est cependant pas établie.

L'urotensine IV est responsable des effets hydrosmotiques provoqués notamment sur la vessie isolée d'Amphibien. Ses propriétés pharmacologiques et chimiques n'ont pas permis de la dissocier de l'AVT (Lacanilao et Bern, 1972). Des essais de localisation in situ à l'aide de techniques immunohistochimiques chez Catostomus montrent cependant que l'AVT est absente du système caudal et donc que l'urotensine IV en est distincte.

Parmi les glandes endocrines "mineures" des poissons, l'urophyse est celle que l'on a étudiée avec le plus de précision, mais son rôle dans l'osmorégulation demeure encore incertain.

CORTICOSTEROIDES

La glande interrénale des poissons, équivalente de la corticosurrénale des Mammifères et secrétrice des hormones corticostéroides, se présente rarement sous l'aspect d'un organe compact chez les poissons et n'est pas associée au tissu chromaffine comme chez les Vertébrés Tétrapodes. Chez les Téléostéens, elle est logée essentiellement dans la partie antérieure du rein (head kidney), mais des cellules éparses se disséminent aussi le long des veines cardinales jusqu'au voisinage du coeur. L'ablation chirurgicale (adrénalectomie) complète est pratiquement impossible à réaliser. Elle peut être remplacée, mais de façon moins satisfaisante, par la suppression de l'hypophyse (source de l'hormone corticotrope, ACTH) ou par l'utilisation de substances pharmacologiques antagonistes de la formation, de la libération ou de l'utilisation périphérique des corticostéroides.

La distribution et le rôle des hormones corticostéroides des poissons ont fait l'objet de multiples mises au point récentes: Gottfried (1964), Idler (1972), Chester-Jones et Henderson (1976), Chester-Jones (1976). Des nombreux travaux de recherche pharmacologiques et biochimiques consacrés à ces substances, il ressort que les poissons participent au même schéma stéroidogénique que tous les autres Vertébrés avec, cependant, d'intéressantes variantes concernant les produits terminaux. Ainsi l'aldostérone qui est présente et qui joue un rôle physiologique considérable chez les Tétrapodes n'a été mise en évidence chez les poissons que dans un nombre très limité d'espèces. On peut donc s'interroger sur sa signification et son importance effective chez les poissons. Tous les Téléostéens possèdent le cortisol comme principale hormone circulante (Sandor et al., 1976). Il en est de même pour les autres poissons osseux et probablement aussi pour les Cyclostomes. Les poissons cartilagineux fabriquent principalement la corticostérone accompagnée de quantités variables d'une hormone spécifique à ce groupe, la 1α-hydroxycorticostérone (Truscott et Idler, 1968).

La présence d'un axe hypothalamus-hypophyse-interrénale a été démontré chez un nombre consistant d'espèces de Téléostéens (voir références in Maetz, 1969) et paraît générale chez les poissons et les Cyclostomes. Ainsi l'hypophysectomie entraîne une régression morphologique du tissu interrénal et abaisse le cortisol plasmatique à un niveau de base minimal. L'injection d'ACTH (de mammifère) augmente considérablement le cortisol circulant, notamment chez les animaux privés d'hypophyse, sauf chez l'Holostéen Amia calva où l'effet est inhibiteur (Hanson et Fleming, 1979). L'existence d'une rétroaction négative est montrée par l'hypertrophie des cellules hypophysaires à ACTH à la suite d'une adrénalectomie, même incomplète, ou par l'emploi de substances pharmacologiques qui bloquent la stéroidogenèse (metopirone) ou interfèrent avec l'action périphérique des stéroides (aldactone).

Dans l'impossibilité de pratiquer de façon certaine une interrénactomie totale chez les Téléostéens, les effets des injections d'ACTH sont attribués à la stimulation préalable de la libération des corticostéroides par cette hormone. Chez Gambusia, et seulement chez cette espèce, l'ACTH est capable autant que la prolactine de maintenir en survie les animaux d'eau douce hypophysectomisés (Chambolle,

1967). Chez le poisson rouge, Lahlou et Giordan (1970) constatent que l'ACTH augmente de façon importante le renouvellement de l'eau interne par action sur la perméabilité branchiale. Chez l'Elasmobranche S. canicula, Payan et Maetz (1971) ont observé des résultats analogues. En outre, chez cette dernière espèce, le débit urinaire est augmenté ce qui indique que chez ces animaux, qui ne boivent pas, la perméabilité osmotique périphérique doit être élevée par cette hormone.

C'est cependant l'utilisation des corticostéroides qui a montré l'intervention directe probable de ces substances dans les processus osmorégulateurs. L'un des résultats les plus démonstratifs a été obtenu sur l'anguille européenne (Mayer et al., 1967) adaptée en EM (Figure 3). L'interrénalectomie aussi complète que possible se traduit chez ces animaux par une réduction importante de l'efflux de Na, l'élévation de la concentration plasmatique de sodium et la mort. L'injection répétée de cortisol produit les effets opposés chez les animaux opérés et permet de les maintenir en vie. Le cortisol apparaît ainsi comme un facteur d'excrétion sodique chez les Téléostéens en eau de mer. Un effet similaire ou opposé n'a pas été démontré de façon péremptoire chez les poissons d'eau douce. Des précisions complémentaires ont été apportées par des études in vitro.

Utida, Hirano et leurs collègues ont analysé de façon approfondie le transport de l'eau et des ions effectué par l'intestin isolé de l'anguille japonaise, Anguilla japonica, les animaux ayant été soumis au préalable à divers traitements (hypophysectomie, injection d'hormones). Chez l'anguille comme chez les autre Téléostéens euryhalins l'adaptation à l'eau de mer est marquée par une augmentation importante de la réabsorption intestinale d'eau et d'ions. La présence de cortisol est nécessaire au maintien de ce taux élevé. Chez les poissons d'eau douce également, le cortisol stimule l'absorption (Hirano et Utida, 1968; Ando, 1974; Hirano et Utida, 1971). Cependant, chez une espèce stérohaline d'eau douce comme Carassius (Ellory et al., 1972), le passage dans un milieu salé hypertonique (à la limite de concentration que le poisson peut tolérer) n'entraîne pas une augmentation, mais au contraire une réduction du transport intestinal, et le cortisol ne peut corriger cette déficience.

Une étude comparative de la spécificité des stéroides montre que le cortisol est l'hormone naturelle la plus active. Son rôle minéralocorticoide est ainsi plus accusé que pour l'aldostérone. Toutefois des stéroides de synthèse tels que la dexamethasone peuvent présenter une plus grande potentialité (Porthé-Nibelle et Lahlou, 1974).

Le mode d'action du cortisol a été analysé au regard de l'évolution de l'activité Na-K-ATPase, l'enzyme majeure impliquée dans le transport actif de sodium et, par voie de conséquence, de celui de

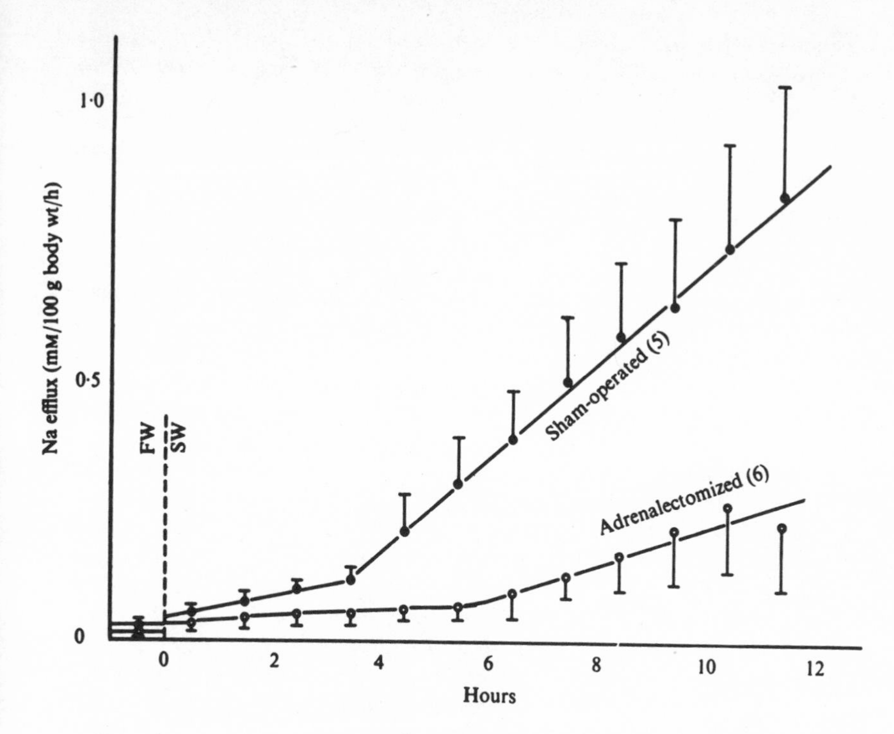

Fig. 3. Effet de l'interrénalectomie sur le flux sortant total, ou efflux de sodium chez l'anguille transférée de l'eau douce (FW: freshwater) à l'eau de mer (SW: sea water). Adrenalectomized: animaux opérés - Sham-operated: témoins - Nombre d'animaux entre parenthèses. (d'après I. Chester Jones, 1976)

chlore qui lui est couplé. Cette activité augmente dans les cellules épithéliales de l'intestin et d'A. rostrata (Jampol et Epstein, 1970) après adaptation en EM.

L'ouabaine qui inhibe spécifiquement la Na-K-ATPase supprime le transport intestinal d'ions et d'eau. Epstein, Cynamon et Mackay (1971) ont montré que le cortisol augmentait cette enzyme dans la branchie et l'intestin d'A. rostrata adaptée à l'eau douce. Cependant les variations de l'efflux de sodium et de la Na-K-ATPase ne se superposent pas (Forrest et al., 1973). Chez le poisson rouge, Ellory, Lahlou et Smith (1972) ont tenté de dissocier au niveau de l'intestin in vitro les rôles respectifs des faces apicale et basolatérale des cellules dans le transport ionique. Ils ont suggéré que le cortisol ne modifiait pas l'entrée apicale de sodium, mais agissait sur l'étape du transport sérosal dépendant de l'activité Na-K-ATPasique.

La situation pourrait être plus simple que chez les Tétrapodes puisque, au niveau de la vessie urinaire d'Amphibien par exemple, des effets de l'aldostérone apicaux et basolatéraux ont été tour à tour proposés (voir Crabbé, 1977).

Chez les Vertébrés Tétrapodes, le taux circulant d'aldostérone dépend de la teneur plasmatique en sodium, c'est-à-dire des besoins immédiats de l'osmorégulation. Chez les Amphibiens, en outre, la réactivité des tissus épithéliaux aux hormones minéralocorticoides varie avec la salinité d'adaptation. Une telle relation n'a pas été mise en évidence chez les poissons. Les taux circulants de cortisol sont essentiellement les mêmes en eau douce et en eau de mer chez les anguilles (Hirano, 1969; Ball et al., 1971; Forrest et al., 1973) et en eau douce et en milieu salé chez le poisson rouge (Porthé-Nibelle et Lahlou, 1975). En revanche, des variations transitoires sont observées immédiatement après le changement de salinité chez toutes ces espèces. Bien qu'il soit difficile de séparer dans ce cas l'effet d'un stress non spécifique de celui de la salinité sur la cortisolémie, il apparaît que ces variations du taux hormonal doivent participer aux réajustements des mécanismes d'échanges ioniques. Ceci est démontré par le fait que les animaux interrénalectomisés modifient beaucoup plus lentement que les témoins leur efflux de Na après passage en EM (Mayer et al., 1967).

Si les taux plasmatiques d'hormone sont identiques chez les animaux pleinement adaptés, d'importantes différences apparaissent entre les deux milieux lorsque l'on considère les paramètres dynamiques. Chez l'anguille européenne, Leloup-Hatey (1974) et Henderson et al., (1974) ont démontré que le taux de clearance métabolique du cortisol était plus élevé chez les animaux EM que chez ceux d'ED, ce qui, compte tenu des concentrations plasmatiques, indique un taux de production plus élevé en EM qu'en ED. Il apparaît donc que des différences existent au niveau de l'utilisation périphérique de l'hormone.

Il paraît donc indispensable de s'orienter désormais vers une approche cellulaire du comportement des corticostéroides chez les poissons. Peu de travaux ont été jusqu'à présent consacrés à cet aspect. Chez Anguilla rostrata, Goodman et Butler (1972) ont montré que le cortisol marqué injecté par l'aorte ventrale était retenu davantage dans les cellules branchiales que la progrestérone donnée dans les mêmes conditions. La liaison du cortisol à l'épithélium branchial ne diffère pas entre ED et EM. En outre une chromatographie du plasma montre que la radioactivité se situe à la place de la cortisone, indiquant que le cortisol injecté a été converti en cette dernière hormone. L'importance de ce processus dans l'osmorégulation n'est cependant pas évidente, la cortisone ne stimulant pas le transport ionique intestinal (Porthé-Nibelle et Lahlou, 1974).

Récemment, Porthé-Nibelle et Lahlou (1978) ont tenté de démontrer la présence de récepteurs spécifiques du cortisol (ou des autres corticostéroides) dans divers types cellulaires (branchie, intestin, rein, foie) de la truite et du poisson rouge. Il est frappant de constater qu'un binding spécifique de haute affinité est virtuellement inexistant dans le cytosol obtenu à partir de ces organes, à la différence de ce que l'on observe généralement chez les Vertébrés supérieurs. L'hormone (cortisol ou dexamethasone) pénètre également dans le noyau mais la spécificité du binding n'a pu être analysée. Par ailleurs, ces auteurs suggèrent que l'entrée de l'hormone dans les cellules s'effectue à la fois par diffusion simple et diffusion facilitée. Des différences entre ED et EM pourraient ainsi être cherchées au niveau de la cinétique d'entrée dans les cellules, de la liaison et de l'utilisation nucléaire, de la conversion du cortisol en cortisone.

Il conviendra également de préciser la nature des interrelations physiologiques évidentes observées entre la prolactine et le cortisol chez les poissons. En culture d'organe, la vessie urinaire de *Gillichthys* est sensible aux deux hormones. Le cortisol y maintient une perméabilité à l'eau élevée qui est inhibée par la prolactine (Doneen, 1976). S'agissant de deux hormones stimulatrices de la synthèse protéique, on peut supposer que leur interaction intervient au niveau du noyau ou des ribosomes. En revanche, il ne semble pas y avoir d'association précoce entre le cortisol et le système adénylcyclasique bien que des relations puissent exister au-delà de l'étape de la libération de l'AMP cyclique.

Au total, la spécificité du cortisol dans le contrôle de l'osmorégulation des poissons paraît établie, mais de nombreux points obscurs subsistent quant aux mécanismes cellulaires impliqués dans cette régulation.

SYSTEME RENINE-ANGIOTENSINE

L'existence du système rénine-angiotensine est bien établie chez les Mammifères et son importance physiologique est considérable puisqu'il participe au contrôle de la sécrétion de l'aldostérone, l'hormone de réabsorption ou d'absorption sodique la plus active de l'organisme chez les Tétrapodes. Dans le contrôle de l'osmorégulation, ce système se situe donc à proximité immédiate des corticostéroides. Cependant un tel dispositif n'existe pas chez les poissons, la présence constante de l'aldostérone n'étant pas établie de façon définitive chez ces derniers. Quant au système rénine-angiotensine, il ne paraît pas exister chez les Cyclostomes et les Elasmobranches (Nishimura *et al.*, 1970; Crockett *et* al., 1973). La morphologie comparée de l'appareil justaglomérulaire producteur de rénine, ainsi que les caractéristiques des angiotensines produites sous l'action de la rénine, ont été revues en détail (Sokabe et Ogawa, 1974).

Au niveau des reins, la rénine est associée à des granules observables en microscopie électronique dans la paroi des vaisseaux afférents glomérulaires. Chez les Téléostéens, elle existe même chez les espèces dépourvues de glomérules (poissons aglomérulaires) et ne possédant pas de ce fait, une telle vascularisation (Mizogami et al., 1968). La rénine et les granules ont été mis en évidence même chez les poissons osseux les plus primitifs, y compris la coelacanthe Latimeria (Nishimura et al., 1973). La production de rénine chez les poissons ne nécessite donc ni la présence de cellules juxtaglomérulaires caractérisées, ni celle de la macula densa différenciée par le tubule distal des Mammifères.

L'intervention du système rénine-angiotensine dans l'osmorégulation des poissons a été suggérée pour la première fois par Sokabe et al., (1966). Chez Anguilla japonica, la rénine rénale est environ deux fois plus abondante en eau douce qu'en eau de mer. Dans cette même espèce, la rénine plasmatique augmente, de façon transitoire, chez les animaux transférés d'eau douce en eau de mer (Sokabe et al., 1973). D'autres travaux, réalisés sur d'autres espèces, ne confirment cependant ni l'une ni l'autre de ces observations. Chez Tilapia mossambica et plusieurs autres Téléostéens euryhalins l'activité rénine du plasma ne change pas avec la salinité ambiante (Malvin et Vander, 1967; Nolly et Fasciolo, 1972). Elle reste également inchangée chez l'aglomérulaire Opsanus tau et diminue en eau douce chez A. rostrata (Nishimura et al., 1976). La rénine rénale, qui est beaucoup plus abondante que celle du sang, ne varie pas ou diminue chez ces espèces en milieu hypo-osmotique. En outre, le maximum d'angiotensine formée dans le plasma en présence de rénine homologue, et représentant le niveau d'angiotensinogène, est le même d'un milieu à l'autre. Chez Anguilla anguilla, c'est encore en EM que la rénine plasmatique est plus abondante: plus du double des teneurs en ED (Henderson et al., 1976) comme le montrent les transferts dans les deux sens.

Nishimura et ses collaborateurs considèrent donc que la sécrétion de rénine doit être déterminée par d'autres facteurs que la salinité externe d'adaptation.

Les figures histologiques montrent par ailleurs que les cellules à granules de rénine augmentent généralement d'activité en eau de mer, contrairement à ce que laisserait supposer l'hypothèse de Sokabe (Krishnamurthy et Bern, 1973; Wendelaar Bonga, 1973).

Les variations de rénine peuvent refléter d'autres changements tels que ceux de la pression hémodynamique systémique. Chez l'anguille, en effet, la pression artérielle diminue à la suite du transfert d'eau douce en eau de mer (Chester-Jones et al., 1969). Il est donc concevable que le fonctionnement des cellules à granules soit lié à ce paramètre. Des extraits de rein d'anguille ainsi que l'angiotensine II de synthèse élèvent la pression sanguine chez A. anguilla normal ou hypophysectomisé (Henderson et al., 1976). Il est néanmoins difficile de préciser si ce mécanisme, même si on ignore la variabilité des résultats obtenus, intervient dans l'osmorégulation.

Le système rénine-angiotensine pourrait participer indirectement aux mécanismes osmorégulateurs en agissant sur d'autres fonctions que les échanges branchiaux. Henderson et ses collègues ont montré chez la truite Salmo gairdneri que l'angiotensine II infusée dans le système circulatoire provoquait une réduction de la filtration rénale, mais avec une sensibilité différente des animaux suivant la salinité externe. En ED, c'est le GRF total qui est abaissé; en EM la réduction affecte individuellement les glomérules (diminution du SNGFR), le nombre de nephrons fonctionnels étant très réduit dans ce milieu (Jackson et al., 1977). La signification de ces résultats reste cependant à établir.

La présence de rénine dans les corpuscules de Stannius (Sokabe et al., 1970) ajoute enfin un élément de plus à l'énigme que pose le système rénine-angiotensine des poissons. L'hypothèse de son rôle osmorégulateur doit être, pour le moment, accueillie avec prudence.

CORPUSCULES DE STANNIUS

Décrits au siècle dernier (Stannius, 1839) les corpuscules de Stannius (CS) sont associés au tissu rénal des poissons, mais ne se rencontrent pas dans tous les groupes. Ils ont été le plus souvent assimilés au tissu glandulaire adrénocortical, mais leur production de substances stéroides paraît très limitée. En revanche, plusieurs rôles fonctionnels semblent pouvoir leur être attribués (Chester-Jones et al., 1969).

Leur intervention dans l'osmorégulation des poissons Téléostéens a d'abord été suggérée par Rasquin (1956). L'immersion prolongée d'Astyanax mexicanus dans une solution de 1% NaCl provoque en effet des changements histologiques au niveau des cellules des corpuscules. Chez l'anguille adaptée à l'eau de mer, ces cellules présentent l'apparence d'une grande activité (Olivereau, 1964). Leur ablation chez les poissons d'ED provoque une chute de la natrémie et une élévation de la kaliémie et de la calcémie (Fontaine, 1964; Leloup-Hatley, 1964). L'effet le plus consistant de la stanniectomie demeure l'hypercalcémie, bien que celle-ci soit transitoire (Fontaine, 1964; Chan, 1972). Aussi est-il généralement considéré que ces glandes contiennent un facteur hypocalcémiant que Pang et Sawyer (1974) ont proposé d'appeler "hypocalcine". L'effet hypocalcémiant a été montré par ces auteurs chez Fundulus heteroclitus dans les conditions où la sensibilité de la réponse a été augmentée en adaptant au préalable les poissons à un milieu d'eau de mer et à une nourriture appauvris en calcium. Les figures histologiques montrent que l'activité glandulaire des corpuscules est pratiquement inexistante en eau de mer sans calcium. Dans ces conditions, la diminution du Ca sérique peut être utilisée comme test biologique de l'activité hypocalcique. Cette dernière a été mise en évidence également chez Heteropneustes fossilis (Dubewar et Suryawanshi, 1978).

Les sites d'action ont été précisés avec des méthodes plus discriminantes que la seule mesure de la calcémie. Fontaine et al., (1972) ont démontré, en utilisant le radioisotope ^{45}Ca pour marquer le milieu ambiant, que l'ablation des corpuscules chez l'anguille d'eau douce provoquait un accroissement de l'absorption de calcium, par suite d'une augmentation du flux entrant. Ces changements affectaient probablement les échanges branchiaux.

Plus récemment, en utilisant une préparation de branchies isolées d'A. rostrata, Fenwick et So (1974) ont montré que la stanniectomie entraînait une augmentation considérable de l'entrée nette de calcium à travers l'épithélium. Le phénomène paraît résulter d'un processus actif car le calcium est accumulé dans le liquide de perfusion. De surcroît l'augmentation de l'influx branchial de Ca^{++} consécutif à la stanniectomie coincide avec une augmentation de l'activité ATPasique branchiale calcium-dépendante (Fenwick, 1976; So et Fenwick, 1979).

Le rein constitue un autre candidat à la qualité d'organe osmorégulateur dépendant des corpuscules de Stannius (Chester-Jones et al., 1969). Ces auteurs ont établi que la suppression des corpuscules provoquait une rétention rénale de calcium. Il n'existe cependant aucune analyse précise permettant d'indiquer sur quelle fonction du nephron s'exerce cette action.

Le rôle physiologique des CS paraît d'autant plus énigmatique que "l'hypocalcine" ne semble pas être le seul facteur présent dans ces glandes. Chez plusieurs espèces, l'étude histologique des CS a révélé la présence d'au moins 2 types cellulaires distincts. Chez Salmo gairdneri, Meats et al., (1978) ont découvert la présence de 2 populations cellulaires apparemment actives et répondant à des stimuli différents; les cellules C_1 seraient sensibles à des teneurs élevées en Ca du milieu externe et produiraient une hormone hypocalcémiante, les cellules C_2 aux milieux à faible concentration ionique et osmotique. Par l'intermédiaire de ces dernières, les CS pourraient être associés directement ou par l'intermédiaire d'une régulation de la sécrétion de prolactine au contrôle de la natrémie. Cette dernière hypothèse, suggérée particulièrement par les observations de Wendelaar Bonga sur Gasterosteus aculeatus stanniectomisé (Wendelaar Bonga, 1978), permet d'associer dans un même schéma de régulation hormonale les électrolytes majeurs extracellulaires: Na^+, Cl^- et Ca^{++} (voir Figure 2).

CORPS ULTIMOBRANCHIAUX

Les glandes parathyroides, qui constituent chez les Tétrapodes les organes principaux de la régulation du métabolisme calcique, n'ont pas été mises en évidence chez les poissons. Ces derniers, en revanche, possèdent comme tous les autres Vertébrés (sauf les Cyclostomes) un autre tissu glandulaire, individualisé, associé à cette régulation, les corps ultimobranchiaux (CUB). Ceux-ci secrètent la calcitonine (CT), hormone agissant puissamment sur l'équilibre phosphocalcique, mise en évidence et nommée par Copp et ses collaborateurs (1962).

Le problème de la régulation endocrine du calcium a été revu notamment par Fleming (1967), Copp (1969), Chan (1972), Pang (1973).

Il n'existe guère de documents concernant la régulation du calcium chez les Cyclostomes. Ces animaux ne possèdent pas de tissu minéralisé jouant le rôle d'organe de stockage du calcium et seraient pourtant un matériel particulièrement favorable à l'analyse cinétique des échanges de cet ion. Urist (1963) a étudié le métabolisme calcique chez la myxine (hagfish, Polistotrema stoutii) et la lamproie (lamprey, Petromyzon marinus). Chez la première, le calcium sérique est de 5.4 mM, contre 10 dans l'eau de mer. Chez la lamproie potamodrome, les concentrations plasmatique et externe sont de 2.6 et 0.9 mM respectivement. Ces gradients de concentration suggèrent la présence d'une pompe à Ca^{++} qui puisse les maintenir, mais aucune information n'existe concernant le contrôle hormonal éventuel.

Les corps ultimobranchiaux d'Elasmobranches contiennent de la calcitonine (Copp et al., 1967). Le plasma de ces poissons renferme des teneurs en calcium relativement élevées: 4 - 5 mM (Urist, 1961).

Les Dipneustes et les Chondrostéens présentent les plus faibles teneurs sériques en calcium (moins de 2mM). Chez Lepidosiren paradoxa, les substances connues pour modifier la calcémie telles que la vitamine D, la PTH et la calcitonine ne modifient en aucune façon le calcium plasmatique (Urist et al., 1972). Ces derniers auteurs en concluent que les mécanismes de transport membranaires doivent suffire à assurer l'homéostasie calcique chez ces animaux. Il semble cependant difficile d'admettre qu'il n'existe aucune modulation hormonale de ces mécanismes, d'autant qu'une activité hypocalcémiante (chez le rat) a été mise en évidence dans les CUB de L. paradoxa (Pang et al., 1971).

Les résultats obtenus chez les Téléostéens sont apparemment contradictoires. La calcitonine, qui est extrêmement active sur la calcémie du rat, n'a pas d'action évidente chez le poisson. Paradoxalement, la CT de porc est hypocalcémiante chez Ictalurus melas (Louw et al., 1967) et A. anguilla (Chan et al., 1968).

La CT de saumon injectée à des anguilles européennes (en ED) réduit l'efflux de ^{45}Ca sans affecter celui de ^{22}Na (Dacke, 1974). Sur la branchie isolée de saumon, la norépinephrine et la CT présentent des effets opposés. La CT synthétique de saumon diminue le débit de perfusion et l'influx de Ca^{++}, les branchies des animaux ED étant plus sensibles à l'hormone que celles d'EM (Milhaud *et al.*, 1977). A nouveau chez l'anguille européenne, l'ablation chirurgicale du CUB provoque une hypercalcémie transitoire, moins marquée que celle induite par le stanniectomie (Lopez *et al.*, 1978). Bien que discordants et sujets à critique au point de vue méthodologique, les résultats précédents établissent dans leur ensemble que la branchie des Téléostéens constitue un organe-cible pour la CT des corps ultimobranchiaux.

Le tubule rénal représente un important site d'action à la fois pour la parathormone et pour la calcitonine chez le Mammifère. Pang et Sawyer (1975) ont montré que la PTH de synthèse (fraction 1-34) et la CT de saumon produisaient également des effets rénaux opposés chez *Lepidosiren paradoxa*, la première étant antidiurétique, la seconde diurétique.

Au total, si l'on tient compte du fait que les corpuscules de Stannius et d'autres substances agonistes ou antagonistes interviennent simultanément, il est concevable que les effets spécifiques de la CT ne soient pas nécessairement reflétés chez les poissons par une variation de la calcémie. Il est curieux de constater que les poissons peuvent être pourvus de deux organes hypocalcémiants alors qu'une glande hypercalcémiante, analogue aux parathyroides des Tétrapodes, fait défaut. L'équilibre est probablement rétabli par la mise en jeu du transport branchial (actif ou passif) de calcium, de la stimulation de l'absorption intestinale (par les dérivés de la vitamine D), de la prolactine (voir Figure 2).

HORMONES INTERVENANT PAR L'INTERMEDIAIRE DE L'AMP CYCLIQUE

L'intervention directe de l'AMP cyclique (AMPc) dans la physiologie des poissons n'a été que très peu explorée, si l'on excepte la fonction de reproduction et quelques effets métaboliques (Menon et Smith, 1971; Fontaine *et al.*, 1972).

L'équipement enzymatique associé habituellement à la présence de l'AMP cyclique existe chez les poissons comme chez les autres Vertébrés. Une protéine-kinase stimulable par ce second messager existe dans le tissu adipeux de la truite (Fontaine-Bertrand *et al*, 1974). La phosphodiestérase qui hydrolyse l'AMPc est présente dans le cerveau de truite (Yamamoto et Massey, 1969). Les taux d'AMP cyclique ont été mesurés dans le plasma de truite (Terrier et Perrier, 1975; Porthé-Nibelle et Lahlou, 1978), du poisson rouge et du flet

(Porthé-Nibelle et Lahlou, 1978), dans le muscle de truite (Nakano et Tomlinson, 1967), la branchie de mullet (Cuthbert et Pic, 1973), l'intestin de la truite, du poisson rouge et du flet (Porthé-Nibelle et Lahlou, 1978).

Les niveaux de base de l'AMP cyclique, obtenus en l'absence de toute stimulation externe, sont comparables à ceux que l'on trouve dans les tissus de Mammifères. La théophylline, qui inhibe la phosphodiestérase, provoque une forte accumulation de l'AMPc (qui atteint 600% du taux témoin) dans la muqueuse intestinale de poisson (Porthé-Nibelle et Lahlou, 1978). Les taux plasmiques sont augmentés à la suite d'un simple stress de manipulation et multipliés plusieurs fois après une injection d'adrénaline. De même, la concentration branchiale est fortement augmentée chez le mullet par l'anesthésie au MS222 ou par l'injection intrapéritonéale d'adrénaline (Cuthbert et Pic, 1973).

Au regard de l'osmorégulation, l'intervention de l'AMP cyclique n'apparaît pas évidents si l'on considère seulement les taux tissulaires chez les animaux adaptés. Les concentrations sont, en effet, comparables entre l'ED et l'EM (Porthé-Nibelle et Lahlou, 1978). Il est donc probable que les hormones stimulatrices de l'adénylcyclase n'exercent leurs effets que lors de situations brutales, telles qu'un changement appréciable de la salinité externe, agissant comme un stress et nécessitant un réajustement maximal immédiat du niveau des échanges ioniques.

Il est bien connu que les états de choc stimulent chez les Vertébrés la libération immédiate de l'ensemble ACTH-corticostéroides d'une part et des catécholamines d'autre part.

Dans le cas de l'ACTH, un effet direct sur les organes osmorégulateurs n'est pas exclu, mais il est admis que, pour l'essentiel, cette hormone agit par l'intermédiaire des corticostéroides dont elle stimule la libération.

Le rôle spécifique des catécholamines est bien documenté grâce à des travaux récents concernant les échanges branchiaux. Keys et Bateman (1932) qui ont mis au point la première préparation coeur-branchies avaient déjà montré que l'adrénaline provoquait une élévation de la concentration de chlore dans le liquide de perfusion. Plus récemment, Pickford _et al._, (1971) chez _Fundulus heteroclitus_ et Pic (1972) chez _Mugil capito_ ont observé que l'injection d'adrénaline provoquait une augmentation de l'osmolarité plasmatique. D'un point de vue méthodologique, la mesure des flux branchiaux sous l'effet des catécholamines peut être altérée du fait que ces hormones produisent une vasodilatation branchiale en même temps qu'elles modifient la perméabilité ou le transport cellulaires. Il est donc indispensable de supprimer ou de limiter suffisamment les erreurs provenant des actions purement hémodynamiques.

Chez Mugil capito, maintenu en EM et injecté d'adrénaline, Pic et al., (1973) ont observé une forte réduction des efflux de Na et Cl, une diminution du sodium ingéré avec l'eau de boisson et une perte de poids suggérant une augmentation de la perméabilité omotique. L'adrénaline inhibe la pompe branchiale d'excrétion de Na, cet effet impliquant l'intervention de récepteurs α. L'action facilitatrice sur la perméabilité à l'eau met en jeu les récepteurs β de la branchie. Les échanges ioniques passifs (leak pathway) demeurent insensibles à l'adrénaline. Ces effets paraissent enfin indépendants des modifications hémodynamiques.

L'utilisation de préparations in vitro pour d'autres tissus que la branchie a permis de montrer des effets de l'AMP cyclique isolés de toute action vasculaire. La glande rectale de l'Elasmobranche Squalus acanthias, isolée et perfusée, répond à la théophylline et à l'addition d'AMP cyclique (ou mieux, de son dérivé dibutyrylé davantage liposoluble) par une augmentation de sa sécrétion ionique (Stoff et al., 1977). Chez Scyliorhinus canicula, Shuttleworth et Thomson (1978) ont montré que le db - AMP cyclique et la théophylline mis en présence de la glande rectale in vitro augmentaient la liaison d'ouabaine tritiée dans le tissu. Ceci suggère que l'AMP cyclique doit normalement augmenter l'activité Na-K-ATPase, c'est-à-dire le nombre de sites disponibles pour la pompe à sodium.

Des résultats différents ont été obtenus sur l'intestin de poisson. Dans les conditions in vitro, la muqueuse intestinale des Téléostéens marins présente une ddp électrique négative du côté séreux. Celle-ci a été attribuée à la présence d'un transport actif prépondérant de chlore (et non de sodium comme dans l'intestin des autres Vertébrés en général). Chez le flet américain en EM (Frizzell et al., 1979) et chez le flet européen (Mackay et al., 1978), l'AMP cyclique réduit fortement le courant de court-circuit et le transport de chlore, augmente largement la perméabilité passive au chlore, mais n'agit pas sur les flux de sodium. Contrairement au cas des Mammifères, il n'apparaît pas de sécrétion d'ions et d'eau. On peut supposer que les effets de l'AMP cyclique sont ici indépendants de la Na-K-ATPase. En effet, ils ne concernent pas le transport de sodium et, de plus, la perméabilité apicale des cellules pour le chlore seul est modifiée (Mackay et al., 1978). Du fait, en outre, que l'activité Na-K-ATPasique est localisée du côté basolatéral, on peut conclure que la face muqueuse des entérocytes constitue le (ou un) site d'action de l'AMP cyclique.

L'adrénaline et l'isoprotérénol (celui-ci agissant spécifiquement sur les récepteurs β) n'augmentent pas l'AMP cyclique tissulaire dans l'intestin des Téléostéens (Porthé-Nibelle et Lahlou, 1978). Celui-ci semble donc pourvu uniquement de récepteurs α. Il n'a pas été établi si une stimulation de ces derniers, en conduisant à une réduction de l'AMP cyclique intracellulaire, pouvait entraîner une stimulation de l'absorption ionique.

Ces nouvelles approches expérimentales des mécanismes cellulaires délimitent probablement une tendance qui marquera désormais de façon profonde les recherches sur l'osmorégulation des poissons. En dehors des hormones déjà connues pour agir sur l'adénylcyclase chez les autres Vertébrés (catécholamines, prostaglandines) il sera intéressant de tester aussi les substances spécifiques des poissons (hypocalcine, urotensines, peptides neurohypophysaires, etc...).

Dans le cadre du présent ouvrage, il convient de constater que si de nombreuses actions hormonales sur les échanges osmorégulateurs ont été mises en évidence chez les poissons, il n'a pas été établi, dans la plupart des cas, de relation simple et univoque entre la salinité de l'environnement et l'activité des glandes endocrines.

REMERCIEMENTS

L'auteur remercie les Editions Elsevier et les Drs T. Hirano et N. Mayer Gostan pour l'autorisation de reproduire le figure 1 et la Journal of Endocrinology et le Professeur I. Chester Jones pour l'autorisation de reproduire la figure 3 du présent article.

REFERENCES

Acher, R., Chauvet, J. and Chauvet, M.T. (1972) Identification de deux nouvelles hormones neurohypophysaires, la valitocine (Val^{8}-ocytocine) et l'aspartocine (Asn^{4}-ocytocine) chez un poisson sélacien, l'Aiguillat (*Squalus acanthias*). C.R. Acad. Sci. 274: 313-316.

Ando, M. (1974) Effects of cortisol on water transport across the eel intestine. Endocrinol. Jap. 21: 539-546.

Arvy, L. (1966) Le système neurosécréteur spinal des poissons ou "système neurosécrétoire caudal" d'Enami (1955). Bull. Soc. Zool. 91: 217-249.

Babiker, M.M. and Rankin, J.C. (1979) Renal and vascular effects of neurohypophysial hormones in the African lungfish, *Protopterus annectens*. Gen. Comp. Endocrinol. 37: 26-35.

Ball, J.N., Chester-Jones, I., Forster, M.E., Hargreaves, G., Hawkins, E.F. and Milne, K.P. (1971) Measurement of plasma cortisol levels in the eel *Anguilla anguilla* in relation to osmotic adjustments. J. Endocrinol. 50: 75-96.

Ball, J.N. and Ensor, D.M. (1965) Effect of prolactin on plasma sodium in the teleost *Poecilia latipinna*. J. Endocrinol. 32: 269-270.

Bennett, M.V.L. and Fox, S. (1962) Electrophysiology of caudal neurosecretory systems of the fluke and skates. Gen. Comp. Endocrinol. 2: 77-95.

Bern, H.A. (1975) Prolactin and osmoregulation. Am. Zool. 15: 937-949.

Bern, H.A. and Lederis, K. (1978) The caudal neurosecretory system of fishes. In: Neurosecretion and neuroendocrinology, Ed. W. Bargmann. Berlin, Springer-Verlag.

Bern, H.A. and Nicoll, C.S. (1968) The comparative endocrinology of prolactin. Recent Progr. Hormone Res. 24: 681-720.

Burden, C.E. (1956) The failure of hypophysectomized Fundulus heteroclitus to survive in fresh water. Biol. Bull. 110: 8-28.

Chadwick, A. (1970) Pigeon crop sac-stimulating activity in the pituitary of the flounder (Pleuronectes flesus). J. Endocrinol. 47: 463-469.

Chambolle, P. (1967) Influence de l'injection d'ACTH sur la survie de Gambusia sp (Poisson Téléostéen) privé d'hypophyse. C.R. Acad. Sci. 264: 1464-1466.

Chan, D.K.O. (1972) Hormonal regulation of calcium balance in teleost fish. Gen. Comp. Endocrinol. 3: 411-420.

Chan, D.K.O. (1975) Cardiovascular and renal effects of urotensins I and II in the eel, Anguilla rostrata. Gen. Comp. Endocrinol. 27: 52-61.

Chan, D.K.O. (1977) Comparative physiology of the vasomotor effects of neurohypophysial peptides in the Vertebrates. Am. Zool. 17: 751-761.

Chan, D.K.O. and Chester-Jones, I. (1968) The effect of mammalian calcitonin on the plasma levels of calcium and inorganic phosphate in the european eel (Anguilla anguilla L.). Gen. Comp. Endocrinol. 11: 243-245.

Chan, D.K.O. and Chester-Jones, I. (1969) Pressor effects of neurohypophysial peptides in the eel Anguilla anguilla L., with some reference to their interaction with adrenergic and cholinergic receptors. J. Endocrinol. 45: 161-174.

Chavin, W. (1956) Pituitary adrenal control of melanization in Xanthic goldfish, Carassius auratus L. J. Exp. Zool. 133: 1-36.

Chester-Jones, I. (1976) Evolutionary aspects of the adrenal cortex and its homologues. J. Endocrinol. 71: 1P-31P.

Chester-Jones, I., Chan, D.K.O. and Rankin, J.C. (1969) Renal function in the European eel (Anguilla anguilla L.): changes in blood pressure and renal function of the freshwater eel transferred to sea-water. J. Endocrinol. 43: 9-19.

Chester-Jones, I., Chan, D.K.O. and Rankin, J.C. (1969) Renal function in the European eel (Anguilla anguilla L.): effects of the caudal neurosecretory system, corpuscules of Stannius, neurohypophysial peptides and vasoactive substances. J. Endocrinol. 43: 21-31.

Chester-Jones, I. and Henderson, I.W. (1976) General, Comparative and Clinical Endocrinology of the Adrenal Cortex, Vol. 1. Ed. Chester-Jones, I. and Henderson, I.W. New York, Academic Press.

Chevalier, G. (1976) Ultrastructural changes in the caudal neurosecretory cells of the trout Salvelinus fontinalis in relation to external salinity. Gen. Comp. Endocrinol. 29: 441-454.

Chevalier, G. (1978) In vivo incorporation of (^{3}H) leucine and (^{3}H) tyrosine by caudal neurosecretory cells of the trout Salvelinus fontinalis in relation to osmotic manipulations. A radioautographic study. Gen. Comp. Endocrinol. 36: 223-229.

Chidambaram, S., Meyer, R.K. and Hasler, A.D. (1972) Effects of hypophysectomy, pituitary autograft, prolactin, temperature and salinity of the medium on survival and natremia in the bullhead, Ictalurus melas. Comp. Biochem. Physiol. 43A: 443-457.

Clarke, W. (1973) Sodium-retaining bioassay of prolactin in the intact teleost Tilapia mossambica acclimated to sea-water. Gen. Comp. Endocrinol. 21: 498-512.

Clarke, W. (1973) Disc-electrophoretic identification of prolactin in the cichlid teleosts Tilapia and Cichlasoma and densitometric measurements of its concentration in Tilapia pituitaries during salinity transfer experiments. Can. J. Zool. 51: 687-695.

Copp, D.H. (1969) Endocrine control of calcium homeostasis. J. Endocrinol. 43: 137-161.

Copp, D.H., Cameron, E.C., Cheney, B.A., Davidson, A.G.F. and Henze, I.G. (1962) Evidence for calcitonin - a new hormone from the parathyroid that lowers blood calcium. Endocrinol. 70: 638-649.

Copp, D.H., Cockcroft, D.W. and Kueh, Y. (1967) Calcitonin from ultimobranchial glands of dogfish and chickens. Science. 158: 924-925.

Crabbe, J. (1977) The mechanism of action of aldosterone. In: Receptors and mechanism of action of steroid hormones. Ed. R. Pasqualini. New York, M. Dekker, p. 513-568.

Crockett, D.R., Jeffery, W.G. and Blankenship, S. (1973) Absence of juxtaglomerular cells in the kidneys of elasmobranch fishes. Comp. Biochem. Physiol. 44A: 673-675.

Cuthbert, A.W. and Pic, P. (1973) Adrenoceptors and adenyl cyclase in gills. J. Pharmacol. 49: 134-137.

Dacke, C.G. (1974) Effects of salmon calcitonin on calcium and sodium efflux in the European eel (Anguilla anguilla L.). J. Physiol. 246: 75-76P.

Dharmamba, M., Handin, R.J., Nandi, J. and Bern, H.A. (1967) Effect of prolactin on freshwater survival and on plasma osmotic pressure of hypophysectomized Tilapia mossambica. Gen. Comp. Endocrinol. 9: 295-302.

Dharmamba, M. and Maetz, J. (1972) Effects of hypophysectomy and prolactin on the sodium balance of Tilapia mossambica in fresh water. Gen. Comp. Endocrinol. 19: 175-183.

Dharmamba, M. and Nishioka, R.S. (1968) Response of prolactin-secreting cells of Tilapia mossambica to environmental salinity. Gen. Comp. Endocrinol. 10: 409-420.

Dickhoff, W.W., Folmar, L.C. and Gorbman, A. (1978) Changes in plasma thyroxine during smoltification of coho salmon Oncorhynchus kisutch. Gen. Comp. Endocrinol. 36: 229-233.

Doneen, B.A. (1976) Water and ion movements in the urinary bladder of the Gobiid teleost Gillichthys mirabilis in response to prolactin and to cortisol. Gen. Comp. Endocrinol. 28: 33-41.

Doneen, B.A. and Bern, H.A. (1974) In vitro effects of prolactin and cortisol water permeability of the urinary bladder of the teleost Gillichthys mirabilis. J. Exp. Zool. 187: 173-179.

Dubewar, D.M. and Suryanwanshi, S.A. (1978) Evidence of hypocalcemic factor from corpuscles of Stannius of the teleost Heteropneustes fossilis (Bloch). Endokrinologie 71: 210-215.

Ellory, J.C., Lahlou, B. and Smith, M.W. (1972) Changes in the intestinal transport of sodium induced by exposure of Goldfish to a saline environment. J. Physiol. 222: 497-509.

Enami, M. (1956) Changes in the caudal neurosecretory system in the loach (Misgurnus anguillii caudatus) in response to osmotic stimuli. Proc. Jap. Acad. 32: 759-764.

Enami, M., Miyashita, S. and Imai, K. (1956) Possibility of occurence of a sodium regulating hormone in the caudal neurosecretory system of teleosts. Endocrinol. Jap. 3: 280-290.

Ensor, D.M. and Ball, J.N. (1972) Prolactin and osmoregulation in fishes. Fed. Proc. 31: 1615-1623.

Epstein, F.E., Cynamon, M. and Mackay, W. (1971) Endocrine control of Na-K-ATPase and seawater adaptation in Anguilla rostrata. Gen. Comp. Endocrinol. 16: 323-328.

Farmer, S.W., Papkoff, H., Bewley, T.A., Hayashida, T., Nishioka, R.S., Bern, H.A. and Hao Li C. (1977) Isolation and properties of teleosts prolactin. Gen. Comp. Endocrinol. 31: 60-71.

Fenwick, J.C. (1976) Effect of Stanniectomy on calcium activated adenosinetriphosphatase activity in the gills of freshwater adapted North American eels, Anguilla rostrata Lesueur. Gen. Comp. Endocrinol. 29: 383-387.

Fenwick, J.C. and So, Y.P. (1974) A perfusion study of the effect of Stanniectomy on the net influx of calcium 45 across an isolated eel gill. J. Exp. Zool. 188: 125-131.

Fleming, W.R. (1967) Calcium metabolism in teleosts. Am. Zool. 7: 835-842.

Fontaine, M. (1964) Corpuscules de Stannius et régulation ionique (Ca, K et Na) du milieu intérieur de l'anguille (Anguilla anguilla L.). C.R. Acad. Sci. 259: 875-878.

Fontaine, M., Callamand, O. and Olivereau, M. (1949) Hypophyse et euryhalinité chez l'anguille. C.R. Acad. Sci. 228: 513-514.

Fontaine, M., Delerue, N., Martelly, E., Marchelidon, J. and Milet, C. (1972) Rôle des corpuscules de Stannius dans les échanges de calcium d'un poisson téléostéen, l'anguille (Anguilla anguilla L.) avec le milieu ambiant. C.R. Acad. Sci. 275: 1523-1528.

Fontaine, M. and Olivereau, M. (1947) La glande thyroide du saumon atlantique (Salmo salar L.) ♀ au cours de sa vie en eau douce. C.R. Acad. Sci. 224: 1660-1662.

Fontaine, Y.A., Salmon, C., Fontaine-Bertrand, E., Burzawa-Gérard, E. and Donaldson, E.M. (1972) Comparison of the activities of two purified fish gonadotropins on adenyl activity in the goldfish ovary. Can. J. Zool. 50: 1673-1676.

Fontaine-Bertrand, E., Delerue-Le Belle, N. and Fontaine, Y.A. (1974) Existence d'une activité protéine kinase stimulable par le 3', 5' -adénosine monophosphate cyclique dans le tissu adipeux d'un poisson téléostéen, la truite (Salmo gairdnerii R.). Biochimie 56: 1223-1228.

Forrest, J.N., Mackay, W.C., Gallagher, B. and Epstein, F.H. (1973) Plasma cortisol response to saltwater adaptation in the American eel Anguilla rostrata. Am. J. Physiol. 224: 714-717.

Fridberg, G. (1962) Studies on the caudal neurosecretory system in teleosts. Acta Zool. 43: 1-77.

Fridberg, G. and Bern, H.A. (1968) The urophysis and the caudal neurosecretory system of fishes. Biol. Rev. 43: 175-199.

Fridberg, G., Bern, H.A. and Nishika, R.S. (1966) The caudal neurosecretory system of isospondylous teleost, Albula vulpes, from different habitats. Gen. Comp. Endocrinol. 6: 195-212.

Frizzell, R.A., Smith, P.L., Vosburgh, E. and Field, M. (1979) Coupled sodium-chloride influx across brush border of flounder intestine. J. Membr. Biol. 46: 27-39.

Fryer, J.N., Woo, N.Y.S., Gunther, R.L. and Bern, H.A. (1978) Effect of urophysial homogenates on plasma ion levels Gillichthys mirabilis (Teleostei: Gobiidae). Gen. Comp. Endocrinol. 35: 238-244.

Goodman, J.H. and Butler, D.G. (1972) Localization of 4 - 14C cortisol in the gills of the North American eel Anguilla rostrata Lesueur). Comp. Biochem. Physiol. 42A: 277-296.

Gottfried, H. (1964) The occurence and biological significance of steroids in lower vertebrates. Steroids 3: 219-242.

Hall, T.R. and Chadwick, A. (1978) Control of prolatin and growth hormone secretion in the eel Anguilla anguilla L. Gen. Comp. Endocrinol. 36: 388-396.

Hanson, R.C. and Fleming, W.R. (1979) Serum cortisol levels of juvenile bowfin, Amia calva: effects of hypophysectomy, hormone replacement and environmental salinity. Comp. Biochem. Physiol. 63A: 467-471.

Henderson, I.W., Jotisankasa, V., Mosley, W. and Oguri, M. (1976) Endocrine and environmental influences upon plasma cortisol concentrations and plasma renin activity of the eel, Anguilla anguilla L. J. Endocr. 70: 81-95.

Henderson, I.W., Sa'Di, M.N. and Hargreaves, G. (1974) Studies on the production and metabolic clearance rates of cortisol in the European eel (Anguilla anguilla L.). J. Steroid Biochem, 5: 701-708.

Henderson, I.W. and Wales, N.A. (1974) Renal diuresis and antidiuresis after injections of arginine vasotocin in the freshwater eel (Anguilla anguilla). J. Endocrinol. 61: 487-500.

Hirano, T. (1969) Effects of hypophysectomy and salinity changes on plasma cortisol concentration in the Japanese eel, Anguilla japonica. Endocrinol. Jap. 16: 557-561.

Hirano, T. (1975) Effects of prolactin on osmotic and diffusion permeability of the urinary bladder of the flounder, Platichthys flesus. Gen. Comp. Endocrinol. 27: 88-94.

Hirano, T. (1977) Prolactin and hydromineral metabolism in the vertebrates. Gunma Symp. Endocrinol. 14: 45-59.

Hirano, T., Hayashi, S. and Utida, S. (1973) Stimulatory effect of prolactin on incorporation of (3H) thymidine into the urinary bladder of the flounder (Kareius bicoloratus). J. Endocrinol. 56: 591-597.

Hirano, T., Johnson, D.W. and Bern, H.A. (1971) Control of water movement in flounder urinary bladder by prolactin. Nature 230: 469-471.

Hirano, T., Johnson, D.W., Bern, H.A. and Utida, S. (1973) Studies on water and ion movements in the isolated urinary bladder of selected fresh water, marine and euryhaline teleosts. Comp. Biochem. Physiol. 45A: 529-540.

Hirano, T. and Utida, S. (1968) Effects of ACTH and cortisol on water movement in isolated intestine of the eel (Anguilla anguilla). Gen. Comp. Endocrinol. 11: 373-380.

Hirano, T. and Utida, S. (1971) Plasma concentration and the rate of intestinal water absorption in the eel, Anguilla japonica, Endocrinol. 18: 47-51.

Hoar, W.S. (1939) The thyroid gland of the Atlantic salmon. J. Morph. 65: 257-295.

Horseman, N.D. and Meier, A.H. (1978) Prostaglandin and the osmoregulatory role of prolactin in a teleost. Life Sci. 22: 1485-1490.

Ichikawa, T., Kobayash, H., Zimmermann, P. and Muller, U. (1973) Pituitary response to environmental osmotic changes in the larval guppy, Lebistes reticulatus (Peters). Z. Zellforsch. 141: 161-180.

Idler, D.R. (1972) Steroids in non-mammalian vertebrates. New York, Academic Press.

Ireland, M.P. (1969) Effect of urophysectomy in Gasterosteus aculeatus on survival in freshwater and sea-water. J. Endocrinol. 43: 133-134.

Jackson, B.A., Brown, J.A., Oliver, J.A. and Henderson, I.W. (1977) Actions of angiotensin on single nephron filtration rates of trout, Salmo gairdneri, adapted to fresh and sea-water environment. J. Endocrinol. 75: 32P-33P.

Jampol L.M. and Epstein, F.H. (1970) Sodium-potassium-activated adenosine triphosphatase and osmotic regulation by fishes. Am. J. Physiol. 218: 607-611.

Kamiya, M. (1972) Hormonal effect on Na-K-ATPase activity in the gill of the Japanese eel, Anguilla japonica, with special reference to seawater adaptation. Endocrinol. Jap. 19: 489-493.

Keys, A. and Bateman, J.B. (1932) Branchial responses to adrenaline and to pitressin in the eel. Biol. Bull. 63: 327-336.

Knight, P., Chadwick, A. and Ball, J.N. (1978) Biological tests of eel "prolactin" separated by gel electrophoresis. Gen. Comp. Endocrinol. 36: 30-33.

Kobayashi, H., Matsui, T., Hirano, T., Iwata, T. and Ishii, S. (1968) Vasodepressor substance in fish urophyses. Annot. Jap. 41: 154-158.

Krishnamurthy, V.G. and Bern, H.A. (1973) Juxtaglomerular changes in the euryhaline fresh-water fish, Tilapia mossambica, during adaptation to sea water. Acta Zool. 54: 9-14.

Lacanilao, F. (1969) Teleostean urophysis: stimulation of water movement across the bladder of the toad Bufo marinus. Sci. 163: 1326-1327.

Lacanilao, F. (1972) The urophysial hydrosmotic factor of fishes. II. Chromatographic and pharmacologie indications of similarity to arginine vasotocin. Gen. Comp. Endocrinol. 19: 413-420.

Lacanilao, F. (1972) The urophysial hydrosmotic factor of fishes. III. Survey of fish caudal spinal cord regions for hydrosmotic activity. Proc. Exp. Biol. Med. 140: 1252-1253.

Lahlou, B. (1975) Ionic permeability of fish intestinal mucosa in regulation to hypophysectomy and salt adaptation. In: Intestinal Ion Transport. Ed. J.W.L. Robinson, London, MTP, p. 318-328.

Lahlou, B. and Fossat, B. (1971) Mécanisme du transport de l'eau et du sel à travers la vessie urinaire d'un poisson téléostéen en eau douce, la truite arc-en-ciel. C.R. Acad. Sci. 273: 2108-2110.

Lahlou, B. and Giordan, A. (1970) Le contrôle hormonal des échanges et de la balance de l'eau chez le téléostéen d'eau douce Carassius auratus intact et hypophysectomisé. Gen. Comp. Endocrinol. 14: 491-509.

Lahlou, B., Henderson, I.W. and Sawyer, W.H. (1969) Renal adaptation by Opsanus tau, a euryhaline aglomerular teleost to dilute media. Am. J. Physiol. 216: 1266-1272.

Lahlou, B. and Sawyer, W.H. (1969) Electrolyte balance in hypophysectomized goldfish Carassius auratus L. Gen. Comp. Endocrinol. 12: 370-377.

Leatherland, J.F. and Lam, T.J. (1969) Effect of prolactin on the density of mucous cells on the gill filaments of the marine form (trachurus) of the three-spine stickleback Gasterosteus aculeatus L. Can. J. Zool. 47: 787-792.

Leatherland, J.F. (1970) Histological investigation of pituitary homotransplants in the marine form (Trachurus) of the treespine stickleback, Gasterosteus aculeatus L. Z. Zellforsch. 104: 337-344.

Lederis, K. (1970) Teleost urophysis. II Biological characterization of the bladder-contracting activity. Gen. Comp. Endocrinol. 14: 427-437.

Lederis, K. (1977) Chemical properties and the physiological and pharmacological actions of urophysial peptides. Am. J. Zool. 17: 823-832.

Leloup-Hatey, J. (1964) Modification de l'équilibre minéral de l'anguille (Anguilla anguilla L.) consécutive à l'ablation des corpuscules de Stannius. C.R. Soc. Biol. 158: 711-715.

Leloup-Hatey, J. (1974) Influence de l'adaptation à l'eau de mer sur la fonction interrénalienne de l'anguille (Anguilla anguilla L.). Gen. Comp. Endocrinol. 24: 28-37.

Lopez, E., Peignoux-Deville, J., Lallier, F., Martelly, E. and Milet, C. (1976) Effects of calcitonin and ultimobranchialectomy (UBX) on calcium and bone metabolisms in the eel, Anguilla anguilla L. Calcif. Tiss. Res. 20: 173-186.

Louw, G.N., Sutton, W.S. and Kenny, A.D. (1967) Action of calcitonin in the teleost fish Ictalurus melas. Nature 215: 888-889.

Macfarlane, N.A.A. (1974) Effects of hypophysectomy on osmoregulation in the euryhaline flounder, Platichthys flesus (L.), in sea water and in fresh water. Comp. Biochem. Physiol. 47: 201-217.

Mackay, W.C., Lahlou, B. and Porthé-Nibelle, J. (1978) AMP cyclique et contrôle des échanges ioniques au niveau de l'intestin de poisson. C.R. Acad. Sci. 287: 1239-1242.

Maetz, J. (1969) Observations on the role of the pituitary interrenal axis in the ion regulation of the eel and other teleosts. Gen. Comp. Endocrinol. Suppo. 2: 299-316.

Maetz, J., Bourguet, J., Lahlou, B. and Hourdry, J, (1964) Peptides neurohypophysaires et osmorégulation chez Carassius auratus. Gen. Comp. Endocrinol. 4: 508-522.

Maetz, J. and Lahlou, B. (1975) Actions of neurohypophysical hormones in fishes. In: Handbook of physiology, Endocrinology section, IV. Ed. R.O. Greep and E.B. Astwood, Baltimore, Williams and Wilkins, p. 521-544.

Maetz, J. and Rankin, J.C. (1969) Quelques aspects du rôle biologique des hormones neurohypophysaires chez les poissons. Coll. Intern. CNRS 177: 45-54.

Maetz, J., Sawyer, W.H., Pickford, G.E. and Mayer, N. (1967) Evolution de la balance minérale du sodium chez Fundulus hereroclitus au cours du transfert d'eau de mer en eau douce: effets de l'hypophysectomie et de la prolactine. Gen. Comp. Endocrinol. 8: 163-176.

Malvin, R.L. and Vander, A.J. (1967) Plasma renin activity in marine teleosts and Cetacea. Am. J. Physiol. 213: 1582-1584.

Marshall, W.S. (1976) Effects of hypophysectomy and ovine prolactin on the epithelial mucus-secreting cells of the pacific staghorn sculpin, Leptocottus armatus (Teleostei: Cottidae). Can. J. Zool. 54: 1604-1609.

Mayer, N., Maetz, J., Chan, D.K.O., Forster, M. and Chester-Jones, I. (1967) Cortisol, a sodium excreting factor in the eel (Anguilla anguilla L.) adapted to sea water. Nature 214: 1118-1120.

Meats. M., Ingleton, P.M., Chester-Jones, I., Garland, H.O. and Henyon, C.J. (1978) Fine structure of the corpuscule of Stannius of the trout, Salmo gairdneri: structural changes in response to increased environmental salinity and calcium ions. Gen. Comp. Endocrinol. 36: 451-461.

Menon, K.M.J. and Smith, M. (1971) Characterization of adenyl cyclase from the testis of chinook salmon. Biochem. 10: 1186-1190.

Milhaud, G., Rankin, J.C., Bolis, L. and Benson, A.A. (1977) Calcitonin: its hormonal action on the gill. Proc. Nat. Acad. Sci. 74: 4693-4696.

Mizogami, S., Oguri, M., Sokabe, H. and Nishimura, H. (1968) Presence of renin in the glomerular and aglomerular kidney of marine teleosts. Am. J. Physiol. 215: 991-994.

Motais, R. and Maetz, J. (1964) Action des hormones neurohypophysaires sur les échanges de sodium (mesurés à l'aide du radio sodium 24Na chez un Téléostéen euryhalin, Platichthys flesus L. Gen. Comp. Endocrinol. 4: 210-224.

Nagahama, Y., Nishioka, R.S., Bern, H.A. and Gunther, R.L. (1975) Control of prolactin secretion in teleosts, with special reference to Gillichthys mirabilis and Tilapia mossambica. Gen. Comp. Endocrinol. 25: 166-188.

Nakano, T. and Tomlinson, N. (1967) Catecholamine and carbohydrate concentrations in rainbow trout (Salmo gairdnerii) in relation to physical disturbance. J. Fish. Res. Board Can. 24: 1701-1715.

Nishimura, H., Ogawa, M. and Sawyer, W.H. (1973) Renin-angiotensin system in primitive bony fishes and a holocephalian. Am. J. Physiol. 224: 950-956.

Nishimura, H., Ogurs, M. Ogawa, M., Sokabe, H. and Imar, M. (1970) Absence of renin in kidneys of elasmobranches and cyclostomes. Am. J. Physiol. 218: 911-915.

Nishimura, H., Sawyer, W.H. and Nigrelli, R.F. (1976) Renin, cortisol and plasma volume in marine teleost fishes adapted to dilute media. J. Endocrinol. 70: 47-59.

Nolly, H.L. and Fasciolo, J.C. (1972) The renin-angiotensin system through the phylogenetic scale. Comp. Biochem. Physiol. 41A: 249-254.

Nozaki, M., Tatsumi, Y. and Ichikawa, T. (1974) Histological changes in the prolactin cells of the rainbow trout, Salmo gairdneri irideus, at the time of hatching. Annot. Zool. Jap. 47: 15-21.

Oduleye, S.O., (1975) The effect of hypophysectomy and prolactin therapy on water balance of the brown trout Salmo trutta. J. Exp. Biol. 63: 357-366.

Ogawa, M. (1977) The effect of hypophysectomy and prolactin treatment on the osmotic water influx into the isolated gills of the Japanese eel (Anguilla japonica). Can. J. Zool. 55: 872-876.

Ogawa, M., Yagasaki, M. and Yamazaki, F. (1973) The effect of prolactin on water influx in isolated gills of the goldfish, Carassius auratus L. Comp. Biochem. Physiol. 44: 1177-1183.

Olivereau, M. (1964) Corpuscules of Stannius in seawater eels. Am. J. Zool. 4: 415.

Olivereau, M. (1968) Functional cytology of prolactin secreting cells. Gen. Comp. Endocrinol. 2: 32-41.

Olivereau, M. and Olivereau, J. (1971) Influence de l'hypophysectomie et d'un traitement prolactinique sur les cellules à mucus de la branchie chez l'Anguille. C.R. Soc. Biol. 165: 2267-2271.

Olivereau, M. and Olivereau, J. (1977) Long term effect of hypophysectomy and prolactin treatment on kidney structure in the eel in fresh water. Cell. Tissue Res. 176: 349-359.

Pang, P.K.T. (1973) Endocrine control of calcium metabolism in teleost. Am. J. Zool. 13: 775-792.

Pang, P.K.T., Clark, N.B. and Thomson, K.S. (1971) Hypocalcemic activities in the ultimobranchial bodies of lung fishes, Neoceratodus forsteri and Lepidosiren paradoxa and teleosts, Fundulus heteroclitus and Gadus morhua. Gen. Comp. Endocrinol. 17: 583-585.

Pang, P.K.T., Pang, R.K. and Sawyer, W.H. (1974) Environmental calcium and the sensitivity of killifish (Fundulus heteroclitus) in bioassays for the hypocalcemic response to Stannius corpuscles from killifish and cod (Gadus morhua). J. Endocrinol. 94: 548-555.

Pang, P.K.T. and Sawyer, W.H. (1975) Parathyroid hormone preparations, Salmon calcitonin, and urine flow in the South American lungfish, Lepidosiren paradoxa. J. Exp. Zool. 193: 407-412.

Payan, P. and Maetz, J. (1971) Balance hydrique chez les Elasmobranches: arguments en faveur d'un contrôle endocrinien. Gen. Comp. Endocrinol. 16: 535-554.

Peignoux-Deville, J., Milet, C. and Martelly, E. (1978) Effets de l'ablation du corps ultimobranchial et de la perfusion de calcitonine sur les flux de calcium au niveau des branchies de l'anguille (Anguilla anguilla L.). Ann. Biol. Anim. Biochem. Biophs 18: 119-126.

Pevet, P., Dogterom, J., Buijs, R.M. and Reinharz, A. (1979) Is it the vasotocin or a vasotocin-like peptide which is present in the mammalian pineal and subcommissural organ? J. Endocrinol. 80: 49.

Pic, P. (1972) Hypernatrémie consécutive à l'administration d'adrénaline chez Mugil capito (Téléostéen Mugilidés) adapté à l'eau de mer. C.R. Soc. Biol. 166: 131-136.

Pic, P., Mayer-Gostan, N. and Maetz, J. (1973) Sea-water teleosts: presence of α and β adrenergic receptors in the gill regulating salt extrusion and water permeability. In: Comparative physiology. Ed. L. Bolis, K. Schmidt-Nielsen and S.H.P. Maddrell. North Holland Publ., p. 292-320.

Pickford, G.E., Griffith, R.W., Torretti, J., Hendler, E. and Epstein, F.H. (1970) Branchial reduction and renal stimulation of (Na^+, K^+)-ATPase by prolactin in hypophysectomized killifish in freshwater. Nature 228: 378-379.

Pickford, G.E. and Phillips, J.G. (1959) Prolactin, a factor in promoting the survival of hypophysectomized killifish in fresh water. Sci. 130: 454-455.

Pickford, G.E., Srivastava, A.K., Slicher, A.M. and Pang, P.K.T. (1971) The stress response in the abundance of circulating leucocytes in the killifish, Fundulus heteroclitus, II the role of catecholamines. J. Exp. Zool. 177: 97-108.

Porthé-Nibelle, J. and Lahlou, B. (1974) Plasma concentrations of cortisol in hypophysectomized and sodium chloride-adapted goldfish (Carassius auratus). J. Endocrinol. 63: 377-387.

Porthé-Nibelle, J. and Lahlou, B. (1975) Effects of corticosteroids and inhibitors of steroids on sodium and water transport by goldfish intestine. Comp. Biochem. Physiol. 50: 801-806.

Porthé-Nibelle, J. and Lahlou, B. (1978) Uptake and binding of glucocorticoids in fish tissues. J. Endocrinol. 78: 407-417.

Porthé-Nibelle, J. and Lahlou, B. (1978) Cyclic AMP levels in plasma and in intestinal mucosa of fishes, with special reference to the effects of theophylline and catecholamines. Gen. Comp. Endocrinol. 36: 609-618.

Potts, W.T.W. and Fleming, W.R. (1970) The effects of prolactin and divalent cations on the permeability to water of Fundulus kansae. J. Exp. Biol. 53: 317-327.

Rasquin, P. (1956) Cytological evidence for a role of the corpuscules of Stannius in osmoregulation of teleosts. Woods Hole 111: 399-409.

Sage, M. (1968) Responses to osmotic stimulii of Xiphophorus prolactin cells in organ culture. Gen. Comp. Endocrinol. 10: 70-74.

Sandor, T., Fazekas, A.G. and Robinson, B.H. (1976) The biosynthesis of cortocosteroids throughout the vertebrates. In: General, comparative and clinical endocrinology of the adrenal cortex, Ed. I. Chester-Jones, and I.W. Henderson, New York, Academic Press, p. 25-142.

Sawyer, W.H. (1969) The active neurohypophysial principles of two primitive bony fishes, the bichir (Polypterus senegalis) and the African lungfish (Protopterus aethiopicus). J. Endocrinol. 44: 421-435.

Sawyer, W.H. (1970) Vasopressor, diurectic and natriuretic, responses by lungfish to arginine vasotocin. Am. J. Physiol. 218: 1789-1794.

Sawyer, W.H. (1972) Lungfish and amphibians: endocrine adaptation and the transition from aquatic to terrestrial life. Fed. Proc. 31: 1609-1614.

Sawyer, W.H. (1977) Evolution of active neurohypophysial principles among the Vertebrates. Am. J. Physiol. 17: 727-737.

Sawyer, W.H., Blair-West, J.R., Simpson, P.A. and Sawyer, M.K. (1976) Renal responses of Australian lungfish to vasotocin, angiotensin II, and NaCl infusion. Am. J. Physiol. 231: 593-602.

Sawyer, W.H. and Pang, P.K.T. (1980) Neurohypophysial peptides and epithelial sodium transport. In: Epithelial transport in the lower vertebrates. Ed. B. Lahlou. Cambridge University Press (in press).

Schreibman, M.P. and Kallman, K.D. (1966) Endocrine control of freshwater tolerance in teleost. Gen. Comp. Endocrinol. 6: 144-155.

Schreibman, M.P., Leatherland, J.F. and Mc Keown, B.A. (1973) Functional morphology of the teleost pituitary gland. Am. J. Zool. 13: 719-742.

Shuttleworth, T.J. and Thompson, J.L. (1978) Cyclic AMP and ouabain-binding sites in the rectal gland of the dogfish Scyliorhinus canicula. J. Exp. Zool. 206: 297-302.

So, Y.P. and Fenwick, J.C. (1979) In vivo and in vitro effects of stannius corpuscle extract on the branchial uptake of ^{45}Ca in stanniectomized north American eels (Anguilla rostrata). Gen. Comp. Endocrinol. 37: 143-149.

Sokabe, H., Mizogami, S., Murase, T. and Sakai, F. (1966) Renin and euryhalinity in the japanese eel, Anguilla anguilla. Nature 212: 952-953.

Sokabe, H., Nishimura, H., Ogawa, M. and Oguri M. (1970) Determination of renin in the corpuscules of Stannius of the teleost. Gen. Comp. Endocrinol. 14: 510-516.

Sokabe, H. and Ogawa, M. (1974) Comparative studies in the juxtaglomerular apparatus. Int. Rev. Cytol. 37: 271-327.

Sokabe, H., Oide, H., Ogawa, M. and Utida, S. (1973) Plasma renin activity in Japanese eels (Anguilla japonica) adapted to seawater in dehydration. Gen. Comp. Endocrinol. 21: 160-167.

Stanley, J.G. and Fleming, W.R. (1966) The effect of hypophysectomy on sodium metabolism of the gill and kidney of Fundulus kansae. Biol. Bull. 131: 155-165.

Stannius, H. (1839) Die nebennieren bei knochenfischen. Arch. Anat. Physiol. 8: 233-271.

Stoff, J.S., Sulva, P., Field, M., Forrest, J., Stevens, A. and Epstein, F.H. (1977) Cyclic AMP regulation of active chloride transport in the rectal gland of marine elasmobranchs. J. Exp. Zool. 199: 443-448.

Takasugi, N. and Bern, H.A. (1962) Experimental studies on the caudal neurosecretory system of Tilapia mossambica. Comp. Biochem. Physiol. 6: 289-303.

Terrier, M. and Perrier, H. (1975) Cyclic 3', 5'-adenosine monophosphatase level in the plasma of the rainbow trout (Salmo gairdnerii Richardson) following adrenaline administration and constrained exercise. Experientia 31: 196.

Truscott, B. and Idler, D.R. (1968) The widespread occurrence of a corticosteroid 1 α -hydroxylase in the interrenals of Elasmobranchii. J. Endocrinol. 40: 515-526.

Urist, M.R. (1961) Calcium and phosphorus in the blood and skeleton of the elasmobranchii. Endocrinol. 69: 778-801.

Urist, M.R. (1963) The regulation of calcium and other ions on the serums of hagfish and lampreys. Ann. Acad. Sci. 109: 294-311.

Urist, M.R., Uyeno, S., King, E., Okada, M. and Applegate, S. (1972) Calcium and phosphorus in the skeleton and blood of the lungfish Lepidosiren paradoxa, with comment on humoral factors in calcium homeostasis in the osteichthyes. Comp. Biochem. Physiol. 42A: 393-408.

Utida, S., Hatai, S., Hirano, T. and Kamemoto, F.I. (1971) Effect of prolactin on survival and plasma sodium levels in hypophysectomized medaka Oryzias latipes. Gen. Comp. Endocrinol. 16: 566-573.

Utida, S., Hirano, T., Oide, H., Ando, M., Johnson, D.W. and Bern, H.A. (1972) Hormonal control of the intestine and urinary bladder in teleost osmoregulation. Gen. Comp. Endocrinol Suppl. 3: 317-327.

Utida, S., Kamiya, M., Johnson, D.W. and Bern, H.A. (1974) Effects of freshwater adaptation and of prolactin on sodium-potassium-activated adenosine triphosphatase activity in the urinary bladder of two flounder species. J. Endocrinol. 62: 11-14.

Wendelaar Bonga, S.E. (1973) Morphometrical analysis with the light and electron microscope of the kidney of the anadromous 3-spined stickleback, Gasterosteus aculeatus form trachurus from fresh water and sea water. In: Comparative Endocrinology. Ed. P.J. Gaillard and H.H. Boer. Amsterdam, Elsevier 1978, p. 259-262.

Wendelaar Bonga, S.E. (1976) The effect of prolactin on kidney structure of the euryhaline teleost Gasterosteus aculeatus during adaptation to fresh water. Cell Tiss. Res. 166: 319-338.

Wendelaar Bonga, S.E. (1978) The role of environmental calcium and magnesium ions in the control of prolactin secretion in the teleost Gasterosteus aculeatus. Zeit. Zell. Mikr. Anat. 137: 563-588.

Yagi, K. and Bern, H.A. (1963) Electrophysiologic indications of the osmoregulatory role of the teleost urophysis. Sci. 142: 491-493.

Yamamoto, M. and Massey, K.L. (1969) Cyclic 3', 5'-nucleotide phosphodiesterase of fish (Salmo gairdneri) brain. Comp. Biochem. Physiol. 30: 941-954.

ABSTRACT

HORMONES IN OSMOREGULATION OF FISHES

A large number of endocrine glands participate in the osmoregulatory patterns of fishes (and Cyclostomes) in their various biotopes. The present review summarizes some well established facts and attempts to delineate the present trends of research in this field.

Prolactine remains the most studied hormone. It controls passive permeability to water and ions separately but its effects are best defined in freshwater adapted fishes. It induces morphological changes, including newly formed structures in gill and kidneys, and modifies Na-K-ATPase activity in epithelia. Thus it appears as versatile in its actions in fishes as it is in higher Vertebrates.

Neurohypophyseal peptides produce striking effects such as a large diuresis, but their occurrence as circulating hormones is questioned. The caudal neurosecretory system, which includes urophysis, is clearly sensitive to changes in environmental salinity. It produces specific peptides (named urotensins I to IV) which are currently under chemical, pharmacological and physiological study.

The interrenal gland produces typical corticosteroids, the most abundant of which is cortisol. As the general presence of aldosterone is still debated, hormones of predominant mineralocorticoid potency are missing. Much evidence suggests that this role is played by cortisol wich increases sodium exchanges and Na-K-ATPase activity. While the effects of prolactin are more obvious in freshwater, those of cortisol are better shown in sea water-adapted animals. Both hormones act antagonistically in various circumstances. A renin - angiotensin system is associated to the interrenal gland as in higher Vertebrates but the data obtained on it are too conflicting to consider it conclusively as an osmoregulatory organ.

Calcium metabolism has attracted much attention in the recent years. New evidence suggests that it is closely linked to sodium homeostasis both directly and through prolactin secretion. Hypocalcemic glands are represented by corpuscles of Stannius (secreting a "hypocalcin") and ultimobranchial bodies (producing calcitonin). Strikingly, there are no parathyroids nor other specific hypercalcemic organs.

Recent works demonstrate that adenylcyclase activity exists in the osmoregulatory epithelia of fishes and may be stimulated by short-term acting hormones such as catecholamines.

The following aspects of fish endocrinology are expected to receive much attention in the near future:

- chemical isolation of fish hormones in view of subsequent development of relevant radioimmunoassays.

- cellular behavior such as binding to specific receptors, which all provide a more significant picture than the mere circulating levels of hormones.

- establishment of precise relationships between external salinity and changes in hormonal secretion, the latter including biorythms.

COMPONENTS OF THE HEMATOLOGICAL RESPONSE OF FISHES TO ENVIRONMENTAL TEMPERATURE CHANGE: A REVIEW

Arthur H. Houston

Biological Sciences, Brock University

St. Catherines, Canada L2S 3A1

INTRODUCTION

Few north temperate zone freshwater fishes enjoy the benefits of life under constant thermal conditions. Virtually all must accomodate to relatively large seasonal variations in environmental temperature conditions. Some, notably those occupying relatively shallow, slow-flowing habitats, must also adjust to substantial diurnal variations under circumstances which often preclude behavioural avoidance of thermal extremes. Furthermore, in much of the Great Lakes region, as in many other parts of Canada, what might be termed 'natural' temperature circumstances are in the process of alteration as a consequence of heat incrementation through a variety of municipal and industrial activities. Seasonal baseline temperatures are often significantly increased, as are the magnitudes of diurnal temperature fluctuations and-episodically-indigenous fauna may be confronted by shock heat discharges of lethal, or near-lethal proportions. Not surprisingly, then, the thermal biology of aquatic organisms, and particularly fishes, has been the subject of renewed interest and emphasis.

Increases in environmental temperature impose a variety of stresses upon the teleost, not least among them those associated with respiratory requirements. Routine or standard oxygen demand in these animals can ordinarily be approximated in relation to temperature by means of a simple parabolic function, the well-known Belehradek relationship:

$$\dot{V}_{0_2} = K_o T^{Ki}$$

or its linear equivalent, the Krogh metabolism-temperature equation:

$$\dot{V}_{0_2} = K_o + K_1 T$$

where,

$$\dot{V}_{O_2} = O_2 \text{ consumption,}$$

T = temperature, K_o is a limiting value and K_1 is essentially the van't Hoff Q_{10}. Depending upon the species under consideration, its prior thermal history and the temperature conditions to which it is exposed values for K_1 may range from less than 2 to as high as 10 or more. The effect of temperature upon oxygen solubility in water is, however, such that the teleost challenged by an elevation in environmental temperature must satisfy enhanced oxygen requirements under circumstances of reduced oxygen availability.

SYSTEMIC RESPONSE PATTERNS

Broadly speaking, fishes exposed to circumstances in which oxygen demand has been increased and/or oxygen availability reduced may resort to either or both of two general response modalities. These involve, on the one hand, the branchial exchanger complex, and on the other, the blood gas transport system. Although the primary concern here is with the latter, it is worth examining some of the principal systemic responses invoked under these circumstances and, particularly, their inherent limitations. Analyses of respiratory system function in fishes commonly take as their starting point the analogy between the gill as a gas transfer system and the operation of compact, counter-current heat exchangers. Thus, emphasis is placed upon variables such as flow over and through the exchanger system, exchange surface area, the oxygen capacities of the respiratory medium and blood, oxygen tension and content in inspired and expired water, venous and arterial blood oxygen content and tension and translamellar driving gradients. The relationships between these have been extensively considered under static and dynamic circumstances, both theoretically and under a variety of experimental circumstances and need not be formally developed in the present context (e.g. Hughes, 1964; Rahn, 1966; Holeton & Randall, 1967 a, b; Randall _et al._, 1967; Stevens & Randall, 1967; Taylor _et al._, 1968; Hughes & Saunders, 1970; Randall, 1970; Heath, 1973). Consequently, comment will be restricted to the effects of altering only three basic variables: ventilatory flow, cardiac output and effective exchange surface.

Ventilatory Flow. Responsive increases in ventilatory flow offer an obvious means of amplifying oxygen uptake. This constitutes a principal mechanism of adjustment to increases in oxygen demand by terrestrial vertebrates, and one which imposes relatively modest metabolic requirements. The teleostean respiratory medium, however, is dense, viscous and relatively oxygen-poor, and the movement of it in adequate volumes imposes a not insignificant metabolic load. Estimates of this vary widely, ranging from as high as 70% of overall metabolism to the more reasonable figure of 5 to 15% suggested by Cameron and Cech (1970). Increases in flow are, however, associated with some reduction in the proportion of oxygen extracted which is subsequently made available to tissues (Jones, 1971). The cost of operating the branchial pump may, in fact, impose significant limitations at higher temperatures, and particularly in larger animals. Increases in flow are, of course, also associated with reductions in lamellar exposure time, and the shunting of increasing proportions of total flow through non-exchange areas. Because of factors of this kind a good deal of the potential benefit of increased ventilation is dissipated as a consequence of the increased cost of operating the system, and reductions in its overall efficiency.

Branchial Perfusion. A second means of amplifying oxygen uptake lies in increasing gill perfusion; in essence, by increasing cardiac output since virtually all blood passed through the heart is delivered to the gills. Output can be modified as a function of cardiac rate and stroke volume, and although the costs of doing so have again not been well defined, the overall metabolic load imposed upon the animal appears to be less than that associated with increased ventilation (Jones, 1971). Jones (1971) suggests that cardiac pump operation does not impose major limitations upon oxygen uptake at higher temperatures, although this may well be the case under cold conditions. However, to be effective, changes in cardiac output must be closely coupled to those in ventilatory flow. Maximally effective blood oxygenation is achieved - in theory at least - as the capacity-rate ratio,

$$\left[\dot{Q} \cdot \alpha B_{0_2} (P_{a0_2} - P_{V0_2})\right] / \left[\dot{V}_I \cdot \alpha W_{0_2} (P_{I0_2} - P_{E0_2})\right]^1$$

approaches zero, for it is under these circumstances that P_{a0_2} will approximate P_{I0_2}.

Accordingly, the extent to which cardiac output can be usefully increased is limited by the degree to which it is practical and efficient to increase ventilatory flow.

1

$\dot{Q}$	= cardiac output	$\dot{V}_1$	= ventilatory flow
αB_{0_2}	= blood oxygen capacity	αW_{0_2}	= water oxygen capacity
P_a0_2	= arterial oxygen tension	P_{I0_2}	= inspiratory oxygen tension
P_V0_2	= venous oxygen tension	P_{E0_2}	= expiratory oxygen tension
A	= exchange area	d	= coefficient of oxygen diffusion across the lamellar barrier

Exchange Surface Area. The final factor to be considered is gill exchange area. Effectiveness of blood oxygenation can be related to this by means of several derived parameters including the branchial transfer unit, defined as:

$$d\ A \cdot \dot{Q} \cdot B_{O_2} \cdot$$

Consequently, other factors being constant, any increase in the effective exchange area of the gill will improve oxygen uptake. Although there is little question that teleost fishes have the capacity to do this, some controversy remains as to the means by which area increases are accomplished. Steen and Kruyyse (1964) and Richards and Fromm (1969), for example, suggest that area amplification is achieved by the proportioning of branchial blood flow through respiratory, as compared to non-respiratory lamellar pathways. Alternatively, increases in area can be accomplished through control of the number of lamellae perfused (Davis, 1972; Cameron, 1974), and recent studies favour the latter mechanism. In any event, simulation analyses indicate that gill area amplification constitutes a critical feature of response to exercise-induced increase in oxygen demand as well as to hypoxic circumstances (Taylor et al., 1968). Presumably such is the case with temperature-related increases in oxygen requirements as well.

Again, however, this is a form of response embodying a number of disadvantageous features. Transfer relationships similar to those which obtain for oxygen also hold for water and diffusible solutes. Because of this, increases in exchange area (and also increases in ventilatory flow and cardiac output as well) will prompt increases in endosmosis and branchial ion efflux (Randall, et al., 1972). Two obvious avenues of compensation for these secondary stresses are open to freshwater fishes: (1) reductions in lamellar permeabilities, and (2) enhancement of branchial absorption and/or renal recovery of electrolytes. The former response may well operate in the case of some ions, but does not appear to function as far as water is concerned. The effects of temperature upon oxygen consumption, ventilatory flow and diffusional water influx are comparable (Evans, 1969; Isaia, 1972; Motais & Isaia, 1972), and it is difficult to reconcile such correspondence with the occurrence of any major reduction in lamellar water permeability at higher temperatures. Compensation for water-loading under these circumstances is achieved primarily through increases in urine output, and although urinary electrolyte concentrations tend to be reduced, the depletion rates of almost all ion species rise (Mackay & Beatty, 1968; Lloyd & Orr, 1969; Motais & Isaia, 1972; Houston, 1973; Mackay, 1974). The effects of temperature upon branchial electrolyte fluxes, on the other hand, are substantially less than those seen with respect to endosmosis; suggesting that adaptive reductions in lamellar permeability may occur (Maetz, 1972; Cameron, 1976). Despite this, overall loss rates rise with temperature. Consequently, iono-regulatory compensation must include some means of amplifying electrolyte recruitment and, consistent with this requirement, increases in gill, kidney and blood (Na^+/K^+) - and (HCO_3^-) - stimulated

ATPase and carbonic anhydrase activities are associated with the acclimatory process (McCarty & Houston, 1977; Houston & McCarty, 1978; Houston & Mearow, 1979 a, b; Smeda & Houston, 1979).

In short, virtually all responses at the branchial exchanger level involve increased metabolic costs, are inherently self-limiting and/or induce secondary stresses which, in turn, require some form of compensatory response. This is not to suggest, of course, that the teleost does not rely heavily upon systemic response systems in meeting temperature-induced increases in oxygen requirements. The occurrence of these is well-documented, and their magnitude can be reasonably well illustrated by reference to recent studies upon rainbow trout acclimated to, and tested at 2°, 10° and 18°C (Fig. 1, Henry & Houston, 1978-1979, unpublished observations). By comparison with cold-acclimated specimens trout maintained at 18°C exhibited an approximately five-fold increase in resting oxygen consumption. Ventilation rate, ventilatory flow and ventilatory stroke volume rose by 125%, 280% and 85% respectively. Cardiac rate increased by a factor of almost 3, while electrocardiogram P-Q, Q-S and S-T intervals dropped to 35%, 55% and 40% of the values characteristic of 2°C animals; observations which point to increases in cardiac output, and thus branchial perfusion.

The question then, is not whether systemic responses are involved, but rather whether adaptive alterations at the hematological level can be employed to diminish reliance upon more metabolically costly systemic adjustments, and thus also the need to respond to secondary osmo- and ionoregulatory (and presumably acid-base) perturbances. Some evidence pertinent to this question has been provided by the work of Cameron and

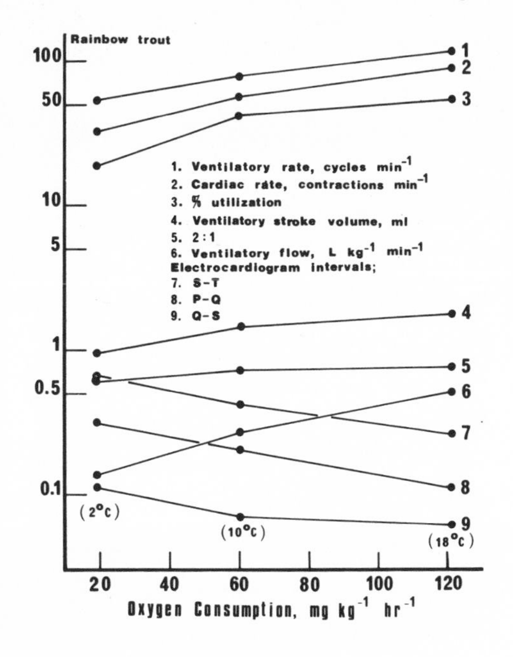

Fig. 1. Ventilatory and cardiac activities in rainbow trout acclimated to 2°, 10° and 18°C as a function of oxygen consumption.

Davis (1970) who rendered rainbow trout anemic by phenylhydrazine treatment, and subsequently carried out an extensive assessment of cardiovascular-respiratory function. Despite reduction of hemoglobin levels to ~ 22.5% of normal, ventilation rate, flow and stroke volume were little altered. Oxygen consumption and utilization fell slightly; ~12% and ~15% respectively. Cardiac output, however, was elevated by a factor of almost 3. The ventilation-perfusion ratio dropped to 35% of normal, and capacity-rate ratio rose by almost 50%. In short, reduction in blood oxygen-carrying capacity had a profound effect upon cardiac function, and presumably upon the costs of operating this element of the branchial exchanger complex.

RESPONSES AT THE HEMATOLOGICAL LEVEL

Three potential modes of response at the hematological level can be envisaged on the part of teleosts confronted by the necessity of satisfying temperature-induced increases in oxygen demand.

Blood Oxygen-Carrying Capacity. Oxygen consumption can be viewed, for example, as a product function of cardiac output, blood oxygen-carrying capacity and the arteriovenous oxygen differential (Hughes, 1964), i.e.,

$$\dot{V}_{O_2} = \dot{Q}\alpha B_{O_2}(P_{aO_2} - P_{VO_2})$$

Thus, even relatively modest increases in hemoglobin concentration might, in conjunction with increases in the other variables prompt substantial increases in oxygen uptake.

Hemoglobin Variants. With rather few exceptions teleostean hemoglobins occur as families of electrophoretically-distinguishable components. If these differ significantly in their transport characteristics and/or their sensitivities to hemoglobin-oxygen affinity modulators, a form of response involving selective modifications favouring hemoglobin variants having transport properties appropriate to particular conditions can be envisaged. For example, in animals exposed to higher temperatures under normoxic conditions hemoglobins of reduced oxygen affinity might well offer considerable advantage. These would saturate, or approach saturation at the gill, and more readily release oxygen at the microcirculatory level. On the other hand, coupling of temperature increases with hypoxic circumstances might well present conditions in which advantage accrued from the utilization of hemoglobins of relatively high affinity, for this would at least facilitate oxygen loading.

Intraerythrocytic Conditions. The transport characteristics of many, though by no means all teleostean hemoglobins can be altered by temperature, *per se*, pH, a variety of organophosphates and several inorganic electrolytes. These influence hemoglobin-oxygen affinity, and thereby the tensions at which oxygen uptake and release take place. Consequently, thermoacclimatory alterations in the intraerythrocytic milieu in which hemoglobin functions, particularly if they are linked to changes in the abundancies of modulator-sensitive hemoglobins, offer a potent means for reducing the need for potentially system-destabilizing alterations in branchial exchanger activities.

These response modalities, which clearly correspond to the 'quantitative', 'qualitative', and 'modulatory' adaptive strategies defined by Hochachka and Somero (1973) are considered in more detail in succeeding sections.

I. Responses Involving Overall Oxygen Transport Capacity

Several expressions of adaptive response to temperature-imposed increases in oxygen requirements can be visualized at the hematological level. Amplification of overall hemoglobin content, for example, would be of obvious utility. Reductions in mean erythrocytic volume also offer some advantage, since rate of oxygen combination with hemoglobin varies inversely with red cell volume (Holland, 1970). Within limits imposed by concommitant changes in blood viscosity, increases in erythrocyte numbers might well be of value. Elevation of mean erythrocytic hemoglobin content has also been suggested as a possible compensatory response. However, as Riggs (1976) has pointed out, hemoglobin concentrations are frequently close to their solubility maxima in the red cells of normal animals.

The results of several studies on this aspect of the acclimatory process - selected as representative of relatively stenothermal, moderately eurythermal and eurythermal species - are summarized in Table I. An immediately obvious feature of these is the lack of consistency apparent in many studies. The goldfish, *Carassius auratus*, is noteworthy in this respect, for animals held under ostensibly similar conditions have been reported to exhibit: (1) obviously adaptive responses (e.g., increases in hemoglobin content, hematocrit, red cell numbers and mean erythrocytic hemoglobin and/or reductions in mean erythrocytic volume), (2) no response, or (3) responses which are apparently anti-adaptive in nature (e.g., reduction in hemoglobin content, hematocrit and/or erythrocyte abundance). In part, at least, variability stems from pre-sampling stress. Handling, anesthetization and related procedures prompt rapid alterations in hematological status, with the magnitude of response being related to the extent of the imposed stress (Houston *et al.*, 1969, 1971; Hattingh & van Pletzen, 1974; Fletcher, 1975; Oikari & Soivio, 1975; Soivio & Oikari, 1976). Nutritional status may also significantly alter the hematological picture (Kamra, 1966; Smith, 1968; Weinberg *et al.*, 1973; Johansson-Sjöbeck *et al.*, 1975). Seasonal variations have been encountered in many species (Denton & Yousef, 1975; Bridges *et al.*, 1976; van Vuren & Hattingh, 1978), and these and other studies (e.g., Weinberg *et al.*, 1976) underscore our present lack of understanding of the factors which regulate erythropoeisis in fishes, and the extent to which these are influenced by environmental variations.

The foregoing comments notwithstanding, a general pattern of hematological response to increased temperature is frequently seen, and this is reasonably well depicted in Figs. 2 and 3. These summarize results obtained from studies on the rainbow trout (DeWilde & Houston, 1967;

TABLE I. Thermoacclimatory modifications in some primary hematological indices of representative eurythermal, moderately eurythermal and relatively stenothermal freshwater, or freshwater-adapted teleost fishes. Hb, hemoglobin g 100 ml^{-1}; PCV, packed cell volume, %; RBC, red blood cells, millions mm^{3}; MEV, mean erythrocytic volume, μ^{3}; MEHbC, mean erythrocytic hemoglobin content, $\mu\mu g$ $cell^{-1}$.

Species	upper incipient lethal temp.	Acclimation temperature	Hb	PCV	RBC	MEV	MEHbC	Reference
Carassius auratus	38.6°C							Fry, et al., 1942
	36.6°C							Fry, et al,, 1946
		13.9°C			1.6			Spoor, 1951
		30 °C			2.2			
		5 °C		34.2	2.0	172		Anthony, 1961
		6 °C		36.0	2.1	181		
		26 °C		37.5	1.7	219		
		30 °C		36.9	1.8	176		
		10 °C		37.7,39.9				Linn, 1965
		15.6 °C		39.6,39.6				
		2 °C	4.6 ± 0.17	30.7 ± 1.04				Houston & Cyr, ,1974
		20 °C	7.6 ± 0.21	35.3 ± 0.49				
		35 °C	8.4 ± 0.04	44.7 ± 0.43				
		5 °C	7.9 ± 0.34	23.6 ± 1.37				Houston, et al.,1976
		30 °C	6.7 ± 0.26	28.3 ± 0.97				
		3 °C	7.7 ± 1.3	31.1 ± 5.5				Houston & Rupert,1976
		23 °C	5.6 ± 0.8	21.1 ± 3.6				

Species	upper incipient lethal temp.	acclimation temperature	Hb	PCV	RBC	MEV	MEHbC	Reference
Cyprinus carpio	35.7							Black, 1953
		4°C (fall)	6.2 ± 1.5	25.7 ± 3.7	1.43 ± 0.26	183 ± 17.2	23.6 ± 3.0	Houston & DeWilde, 1968
		17°C	6.4 ± 2.2	27.1 ± 7.5	1.43 ± 0.43	186 ± 22.6	23.4 ± 3.2	
		27°C	8.2 ± 1.3	31.6 ± 4.3	1.74 ± 0.23	182 ± 9.6	26.1 ± 2.0	
		33°C	7.3 ± 0.7	30.9 ± 3.6	1.77 ± 0.16	175 ± 16.7	23.8 ± 1.0	
		7°C winter	7.9 ± 1.2	30.6 ± 3.8	1.71 ± 0.26	175 ± 16.3	26.4 ± 1.6	
		27°C	8.8 ± 1.1	31.7 ± 3.0	1.78 ± 0.13	183 ± 12.1	27.6 ± 3.2	
		5°C	9.7 ± 0.6	35.2 ± 2.5				Houston, *et al.*, 1976
		30°C	10.6 ± 0.6	31.3 ± 1.4				
		2°C	7.5 ± 0.2	32.9 ± 1.0	1.48 ± 0.05	266 ± 9.9		Houston & Smeda, 1979
		16°C	7.5 ± 0.3	32.3 ± 1.2	1.48 ± 0.07	222 ± 8.4		
		30°C	8.2 ± 0.4	33.4 ± 1.5	1.67 ± 0.08	202 ± 5.5		
C. auratus x *C. carpio*	?	5°C	10.0 ± 0.4	33.4 ± 0.8				Houston, *et al.*, 1976
		30°C	8.5 ± 0.4	25.6 ± 0.8				
Ictalurus nebulosa	34.5°C							Brett, 1944
	36 °C							Cairns, 1956.
		9°C		32.5 ± 5.7				Grigg, 1969
		24°C		26.0 ± 5.4				
Lepomis gibbosus	34.5°C							Brett, 1944
	30.2°C							Black, 1953
		3°C	10.3 ± 0.4	37.0 ± 1.3				Houston, *et al*, 1976
		10°C	8.8 ± 0.4	27.8 ± 1.6				
		20°C	9.5 ± 0.7	28.6 ± 1.8				

TABLE I. (continued)

Species	upper incipient lethal temp.	acclimation temperature	Hb	PCV	RBC	MEV	MEHbC	Reference
Catostomus	31.2°C							Brett, 1944
commersoni	29.3°C							Hart, 1947
		3°C	9.3 ± 0.5	33.5 ± 0.7				Houston, et al., 1976
		10°C	9.7 ± 0.4	35.3 ± 1.3				
		20°C	10.4 ± 0.4	37.4 ± 0.9				
Fundulus	28.0°C							Garside & Jordan, 1968
heteroclitus	31.3°C							Garside & Chin Yuen Kee, 1972
		10°C			1.13 ± 0.72			Slicher & Pickford, 1968
		20°C			3.36 ± 0.67			
Lagodon	?	10°C	10.2 ± 1.5; 10.4 ± 2.3	35.4 ± 5.2; 35.6 ± 7.6;				Cameron, 1970
rhomboides		25°C	11.7 ± 0.9	37.5 ± 2.6				
Mugil	?	7°C	6.7	25.9	2.84	91	23.7	Cameron, 1970
cephalus		25°C	7.6	30.1	3.13	96	24.2	
Salvelinus	26.6°C							Brett, 1944
fontinalis	25.3°C							Fry, et al., 1946
	25.5°C							McCauley, 1958
		2°C	7.2 ± 0.2	34.1 ± 0.9				Houston & DeWilde, 1968
		5°C	8.3 ± 0.2	37.6 ± 0.5				
		8°C	8.5 ± 0.2	36.8 ± 0.6				
		10°C	7.1 ± 0.2	35.2 ± 0.8				
		20°C	7.6 ± 0.2	37.2 ± 1.1				

Species	upper incipient lethal temp.	acclimation temperature	Hb	PCV	RBC	MEV	MEHbC	Reference
Salmo gairdneri	24.0°C							Black, 1953
	26.2°C							Kaya, 1978
	25.5°C							Threader & Houston (unpub. obs.)
		3 °C(summer)	6.4 ± 0.8	29.0 ± 2.1	1.09 ± 0.08	259 ± 22.1	22.2 ± 1.7	DeWilde & Houston, 1967
		7 °C	6.4 ± 0.6	30.3 ± 2.9	1.14 ± 0.13	268 ± 28.2	20.8 ± 1.3	
		11 °C	6.8 ± 1.2	32.3 ± 2.9	1.36 ± 0.13	253 ± 48.6	21.5 ± 3.2	
		14 °C	7.3 ± 0.7	30.6 ± 2.0	--	273 ± 23.8	23.7 ± 2.5	
		17 °C	6.8 ± 0.8	31.7 ± 0.2	1.26 ± 0.16	253 ± 37.7	21.7 ± 2.7	
		21 °C	8.3 ± 1.1	33.6 ± 2.5	1.44 ± 0.12	237 ± 19.0	24.6 ± 2.2	
		4 °C(winter)	6.5 ± 0.6	28.4 ± 2.9	1.19 ± 0.14	241 ± 18.6	22.1 ± 1.4	
		7 °C	7.3 ± 0.6	29.7 ± 1.5	1.25 ± 0.12	239 ± 18.8	24.3 ± 1.6	
		11 °C	8.2 ± 0.5	34.5 ± 1.7	1.44 ± 0.10	240 ± 11.5	23.8 ± 1.4	
		14 °C	8.2 ± 1.4	33.4 ± 3.8	1.30 ± 0.14	252 ± 15.8	24.9 ± 1.7	
		17 °C	8.0 ± 0.5	32.4 ± 1.4	1.40 ± 0.09	232 ± 12.8	24.7 ± 1.6	
		18 °C	8.3 ± 0.9	34.5 ± 2.0	--	--	23.4 ± 1.7	
		21 °C	8.5 ± 0.8	34.7 ± 2.6	1.51 ± 0.11	232 ± 23.1	24.2 ± 0.9	
		2 °C	6.3 ± 0.2	32.7 ± 0.7				Houston & Cyr, 1974
		10 °C	7.3 ± 0.1	42.9 ± 0.5				
		18 °C	8.3 ± 0.02	46.6 ± 0.7				
		2 °C(summer)	8.1 ± 0.2	32.8 ± 1.1	1.19 ± 0.05	277 ± 9.8		Houston & Smeda, 1979
		10 °C	8.2 ± 0.2	33.7 ± 0.7	1.36 ± 0.04	252 ± 6.3		
		18 °C	8.1 ± 0.3	33.2 ± 1.0	1.33 ± 0.06	255 ± 12.1		
		2 °C(winter)	8.1 ± 0.3	31.5 ± 0.9	1.32 ± 0.05	241 ± 6.7		
		10 °C	8.2 ± 0.3	31.6 ± 1.1	1.23 ± 0.06	263 ± 8.6		
		18 °C	7.6 ± 0.4	30.8 ± 1.6	1.28 ± 0.05	242 ± 6.5		
		2 °C(summer)		34.0				Murphy & Houston (unpub.)
		10 °C		35.8				
		18 °C		44.1				
		2 °C(winter)		32.1				
		10 °C		34.6				
		18 °C		41.7				

Houston & Cyr, 1974; Houston & Smeda, 1979) and carp (Houston & DeWilde, 1968; Houston et al., 1976; Houston & Smeda, 1979); species generally regarded as representative of the relatively stenothermal and eurythermal conditions respectively. Two studies on the former species (DeWilde & Houston, 1967; Houston & Cyr, 1974) indicated that acclimation to higher temperatures was accompanied by some elevation in both total hemoglobin and red cell numbers. Moderate reductions in cell volume and minor changes in mean erythrocytic hemoglobin content were also observed. In the third study, however, cold-acclimated (2°C) specimens possessed hematological characteristics not unlike those of animals adapted to warmer conditions, and these were not markedly altered following acclimation (Houston &

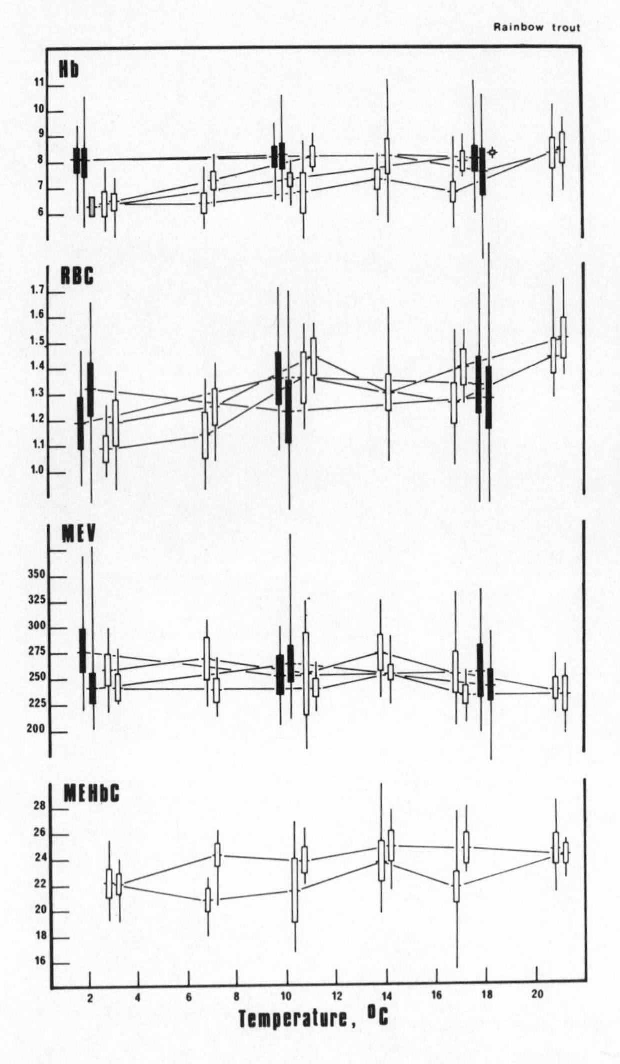

Fig. 2. Hematological responses of the relatively stenothermal rainbow trout following thermal acclimation. Hb- hemoglobin content, g 100 ml^{-1}; RBC - erythrocyte numbers, 10^6 mm^{-3}; MEV - mean erythrocytic volume, μ^3; MEHbC - mean erythrocytic hemoglobin content, $\mu\mu$g cell^{-1}. Horizontal line - mean; vertical bar - 95% confidence interval of the mean; vertical line - range.

Smeda, 1979). Somewhat more consistent responses were seen in carp, with hemoglobin levels, red cell numbers and mean erythrocytic volume rising at higher acclimation temperatures while mean erythrocytic volume declined.

In the carp, and one suspects, in other species as well, specimen size has considerable bearing on both the occurrence and extent of hematological response. Smaller carp, for example, exhibit substantially greater increases in hemoglobin levels and red cell numbers than do larger animals exposed to the same temperature conditions, coupling this with proportionally larger reductions in erythrocytic volume (Smeda & Houston, 1979). All of these

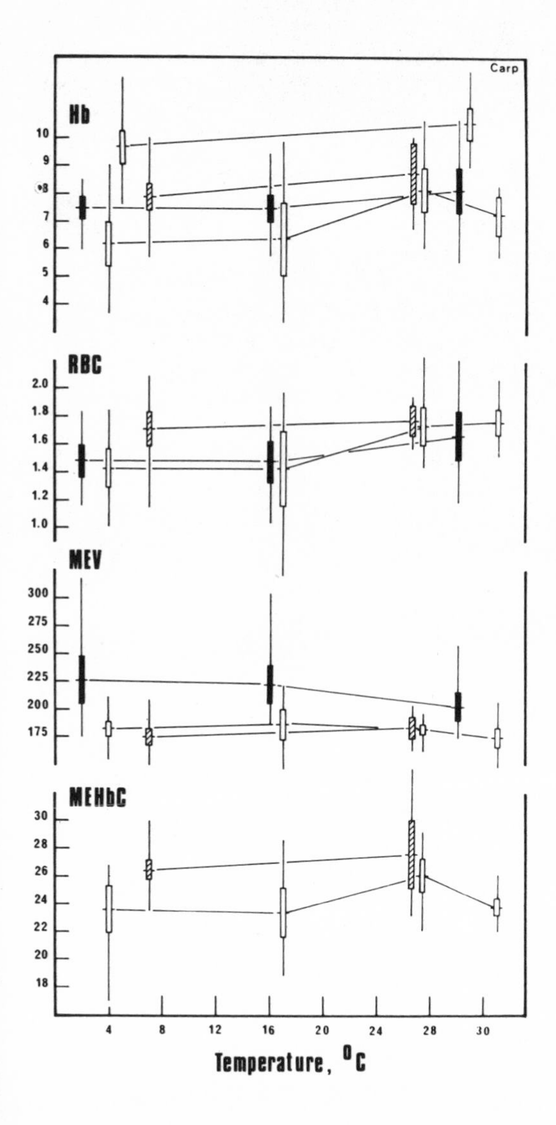

Fig. 3. Hematological responses of the relatively eurythermal carp following thermal acclimation. Hb - hemoglobin content, g 100 ml^{-1}; RBC - erythrocyte numbers, 10^6 mm^{-3}; MEV - mean erythrocytic volume, μ^3; MEHbC - mean erythrocytic hemoglobin content, μμg cell^{-1}. Horizontal line - mean; vertical bar - 95% confidence interval of the mean; vertical line - range.

variables proved to be power or logarithmic functions of weight. In contrast to the situation described by Smith (1977) for the plaice, Hippoglossoides platessoides, all tended to vary inversely with weight, and increasingly so at higher temperatures. On the other hand, they correspond well with observed relationships between weight, oxygen consumption and temperature in this species (Beamish, 1964). Although branchial exchange surface area is also a negative power function of weight in most fishes (Hughes & Morgan, 1973) the amplification of oxygen-carrying capacity suggested by these observations would seem to offer positive advantages to smaller fishes.

Interestingly, weight-specific differential hematological responses of the type described may also account, in part at least, for the puzzling discrepancies in response alluded to earlier. For example, Houston and Cyr (1974) reported substantial increases in the hemoglobin levels of relatively small goldfish (13.4 $\pm$ 1.9 to 15.6 $\pm$ 0.6 g) following acclimation to temperatures ranging from 2°C to 35°C. Larger specimens (22.2 to 38.0 g) exhibited only moderate changes (Houston & Rupert, 1976). In still larger animals (36.9 $\pm$ 8.8 to 64.3 $\pm$ 11.4 g) no significant variations were discernable, and mean hemoglobin concentrations were sometimes reduced at higher acclimation temperatures (Houston et al., 1976).

The principal conclusion arising from these and similar studies is, however, that hematological response to increased temperature -when it takes place - ordinarily involves only modest increases in overall hemoglobin content. This is usually associated with the occurrence of somewhat larger numbers of red cells of reduced volume, and modest elevations in cellular hemoglobin content. Clearly, changes of this character are adaptively appropriate. Equally clearly, however, their somewhat erratic occurrence in response to temperature increases and their magnitude are such as to preclude any major contribution toward the resolution of temperature-oxygen demand problems.

II. Acclimatory Reorganization of the Hemoglobin System

The existence of multiple-component hemoglobin systems in fishes provides another potential means of adaptive response. The existence of this phenomenon was first inferred by Manwell (1957) on the basis of observed differences in the alkali denaturation rates of the postlarval and adult hemoglobins of Scorpanenicthys marmoratus; differences which clearly pointed to distinctions in molecular identity. Ontogenetic changes in hemoglobin types and relative abundancies have since been confirmed in numerous species (e.g., Vanstone et al., 1964; Wilkins & Iles, 1966; Iuchi & Yamagami, 1969; Perez & Maclean, 1974, 1975 a, b, 1976). Adults are also normally characterized by multiple hemoglobin systems. Of the 23 species examined by Yamanaka et al. (1965) none possessed single-component systems. This was true as well of the 14 cypriniform and perciform fishes considered by De Smet (1978). Only 10% of the 31 species of marine scombroid and related species surveyed by Sharp (1973) possessed single hemoglobins. The most recent and extensive study of this phenomenon is, however, that of Fyhn et al. (1979) who examined some 94 species of Amazonian teleosts representing 28 families and 77 genera. Up to 12 mobilities were seen in some species, with the average being 4. Only about 8% of the species considered proved to have single-component hemoglobin systems. In general, however, these estimates should be regarded as conservative for they are based upon electrophoretic mobilities, and thus discriminate only on the basis of net charge in relation to molecular weight and configuration.

Whatever the real complexity of teleostean hemoglobin system may prove to be, the point at issue is whether the hemoglobins found in a particular species differ sufficiently in their transport characteristics to offer some adaptive potential. Should such differences exist, physiological advantage might well accrue from selective adjustment in fractional abundancies during response to circumstances which alter either, or both of oxygen demand or oxygen availability. There are, however, several examples in the human literature of hemoglobins possessing residue substitutions in 'non-critical' regions of the molecule. These lead to differences in electrophonetic mobility but do not greatly alter hemoglobin oxygen affinity (Natelson & Natelson, 1978). Consequently only relatively minor effects upon oxygen uptake and release are seen. Quite obviously, such hemoglobins would provide little scope for adaptive response. It is increasingly apparent, however, that both situations occur in teleosts. Individual species may possess hemoglobin variants of distinctly different oxygen affinities and modulator sensitivities. Others are characterized by multiple-component systems whose elements are functionally indistinguishable. In yet others both phenomena are seen, i.e., groups of functionally-distinct and functionally-equivalent fractions co'exist. A few selected examples clearly demonstrate some of the complexities which have been recently encountered.

Carp typically possess three principal hemoglobins with a fourth minor fraction being occasionally reported (Yamanaka et al., 1965; Tan et al., 1972; Gillen & Riggs, 1972; Houston et al., 1976; Weber & Lykkeboe, 1978). All exhibit sensitivities to pH and other modulating agents but, to a first approximation at least, they appear to constitute a functionally-homogeneous group (Gillen & Riggs, 1972). This appears to be true of the four hemoglobins of the killifish, Fundulus heteroclitus as well (Mied & Powers, 1978). Interestingly, however, Weber and Lykkeboe (1978) have recently reported that altering the proportions of component hemoglobins in carp (without change in total hemoglobin) significantly influences affinity; an observation pointing to the possibility of a subtle, and heretofor largely unanticipated form of affinity adjustment through component interactions.

The rainbow trout has a complex hemoglobin system variously estimated as consisting of 7 to 11 component fractions (Houston et al., 1976). These appear to fall into two braod categories (Brunori, 1975); those which have virtually no sensitivity to hydrogen ion, are little influenced by inorganic and organic phosphate and exhibit minimal thermosensitivity, and others whose affinity for oxygen is profoundly effected by these factors.

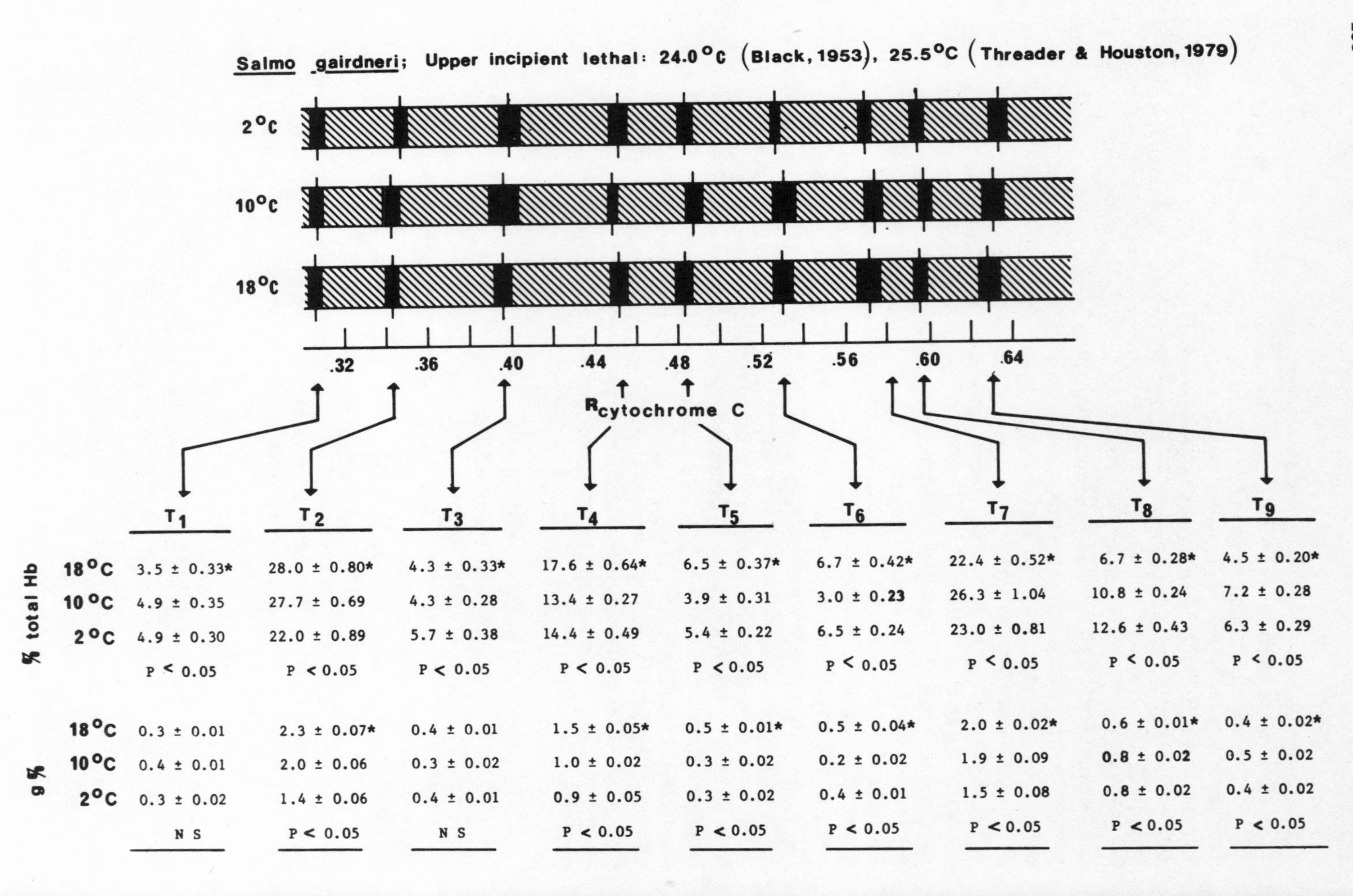

		T1	T2	T3	T4	T5	T6	T7	T8	T9
% total Hb	18 °C	3.5 ± 0.33*	28.0 ± 0.80*	4.3 ± 0.33*	17.6 ± 0.64*	6.5 ± 0.37*	6.7 ± 0.42*	22.4 ± 0.52*	6.7 ± 0.28*	4.5 ± 0.20*
	10 °C	4.9 ± 0.35	27.7 ± 0.69	4.3 ± 0.28	13.4 ± 0.27	3.9 ± 0.31	3.0 ± 0.23	26.3 ± 1.04	10.8 ± 0.24	7.2 ± 0.28
	2 °C	4.9 ± 0.30	22.0 ± 0.89	5.7 ± 0.38	14.4 ± 0.49	5.4 ± 0.22	6.5 ± 0.24	23.0 ± 0.81	12.6 ± 0.43	6.3 ± 0.29
		$P < 0.05$	$P < 0.05$	$P < 0.05$	$P < 0.05$	$P < 0.05$	$P < 0.05$	$P < 0.05$	$P < 0.05$	$P < 0.05$
g %	18 °C	0.3 ± 0.01	2.3 ± 0.07*	0.4 ± 0.01	1.5 ± 0.05*	0.5 ± 0.01*	0.5 ± 0.04*	2.0 ± 0.02*	0.6 ± 0.01*	0.4 ± 0.02*
	10 °C	0.4 ± 0.01	2.0 ± 0.06	0.3 ± 0.02	1.0 ± 0.02	0.3 ± 0.02	0.2 ± 0.02	1.9 ± 0.09	0.8 ± 0.02	0.5 ± 0.02
	2 °C	0.3 ± 0.02	1.4 ± 0.06	0.4 ± 0.01	0.9 ± 0.05	0.3 ± 0.02	0.4 ± 0.01	1.5 ± 0.08	0.8 ± 0.02	0.4 ± 0.02
		N S	$P < 0.05$	N S	$P < 0.05$	$P < 0.05$	$P < 0.05$	$P < 0.05$	$P < 0.05$	$P < 0.05$

Brunori (Brunori et al., 1979) has also examined Pterygoplichthys pardalis, an Amazonian catfish commonly challenged by hypoxic conditions. Four main hemoglobins are present. Three of these, which collectively account for approximately half of the total, are functionally similar. All exhibit relatively large Bohr effects, with hemoglobin-oxygen affinity decreasing almost 20-fold over the 8.5 to 6.2 pH range. Reductions in pH are accompanied by increases in O_2 dissociation rate, and a decrease in the CO combination rate constant. Modulatory interactions with ATP were negligible over the pH range considered. By contrast, the major fraction, which makes up ~50% of total hemoglobin, has a comparatively small Bohr effect, and exhibits ATP-dependent changes in affinity at reduced pH.

The facultatively air-breathing catfish, Hoplosternum littorale, possesses two functionally-heterogeneous hemoglobin variants (Garlick et al., 1979). One of these is characterized by a relatively high affinity for oxygen. No Root effect is apparent. The reverse Bohr effect seen in the striped fraction reverts to the normal alkaline Bohr pattern upon addition of ATP. The more abundant component (~70% of the total) has a relatively lower affinity for O_2 at reduced pH, and a relatively higher affinity under more alkaline conditions. A Root effect is present at reduced pH, and is enhanced by ATP. On the whole, however, ATP influence upon oxygen affinity is somewhat less pronounced in the case of this fraction.

There can be little doubt, then, as to the occurrence within at least some multiple hemoglobin systems of components of very different functional properties. Several questions then arise. Can the teleost affect quantitative (i.e., variations in absolute or relative component abundancies) or qualitative alterations (i.e., additions or deletions of specific components) in the hemoglobin system during the thermo-acclimatory process? Are such changes, if they occur, adaptive in character?

In addressing the first of these questions consideration was given to five species selected for differences in thermal tolerance, metabolic activity and its thermal dependence and inherent hemoglobin system complexity. These were the goldfish, Carassius auratus, carp, Cyprinus carpio, pumpkinseed, Lepomis gibbosus, white sucker, Catostomus commersoni, and rainbow trout, Salmo gairdneri. In addition, a population of C. auratus x C. carpio hybrids was examined. In each instance polyacryla-

Fig. 4. The hemoglobin system of the rainbow trout following acclimation to 2°, 10° and 18°C. Mobility relative to cytochrome C; mean mobility ± 95% confidence interval of the mean. Fractional contribution of each hemoglobin variant to total hemoglobin (% total Hb); mean ± 95% confidence interval of the mean. Fractional concentration of each hemoglobin variant (g%); mean ± 95% confidence interval of the mean.

mide gel electrophoresis was employed in conjunction with scanning densitometry, the techniques used being outlined in Houston et al. (1976).

Salmo gairdneri. The rainbow trout was the most thermosensitive of the species examined. Black (1953) has estimated the upper incipient lethal of this species as 24.0°C. More recent studies (Kaya, 1978; Threader & Houston, 1979, unpublished observations) have tended to produce higher values; 26.2°C and 25.5°C respectively. Fig. 4 summarizes the results of studies upon animals acclimated to 2°, 10° and 18°C. In this, as in succeeding figures, mean fractional mobilities are reported with 95% confidence intervals relative to a marker protein (in this instance, cytochrome C) with data upon the relative contribution of each fraction to total hemoglobin (%Hb), and its actual concentration (g%Hb). Recent estimates of system complexity in the trout suggest that from 7 to 11 hemoglobin variants occur (Yamanaka et al., 1965; Ronald & Tsuyuki, 1971; Iucki, 1973; Yoshiyasu, 1973; Houston & Cyr, 1974; Braman et al., 1977); values well below that reported by Tsuyuki & Gadd (1963). Acclimation did not qualitatively alter the hemoglobin system; each specimen was characterized by 9 well-defined mobilities at each temperature. All fractions, however, exhibited significant changes in proportional abundance. Seven, including the major components, T_2 and T_7, were also noteworthy for significant changes in actual concentration as well.

Catostomus commersoni. The white sucker is somewhat more eurythermal than the rainbow trout. Brett (1944) reported the upper incipient lethal temperature of this species as 31.2°C, while Hart (1947) gave a value of 29.3°C. Electrophoresis revealed 8 fractions (Fig. 5), suggesting a level of complexity comparable to that observed by Powers (1972) in this genus. Again, as evidenced by $R_{albumin}$ values and coelectrophoresis studies, there was little evidence of qualitative alteration in hemoglobin types as a consequence of acclimation.

Assessment of quantitative variations was complicated by the inability of the scanning system used (Gilford model 2400 spectrophotometer) to provide acceptable resolution of two pairs of mobilities (S_3 and S_4; S_6 and S_7) in all specimens, although these were normally distinguishable by visual inspection. Accordingly, these pairs have been treated as single elements. By comparison with rainbow trout, sucker proved to be relatively conservative. Only modest alterations in the relative abundancies of S_1 and the principal components S_2 and S_8 were encountered. Four fractions were characterized by significant changes in actual concentration. S_{3+4} and S_8 increased in amount at higher temperature. Hemoglobins S_1 and S_{6+7}, on the other hand, exhibited minima at 10°C by comparison with 2°C and 18°C.

Lepomis gibbosus. The pumpkinseed is a relatively eurythermal species, and like other members of this genus (L. macrochirus, L. megalotis) tolerates temperatures in the mid-to-upper 30°C range (Brett, 1944; Hart, 1952; Cairns, 1956; Neil et al., 1966). The hemoglobin system proved to be of intermediate complexity, consisting of 6 well-defined components (Fig.

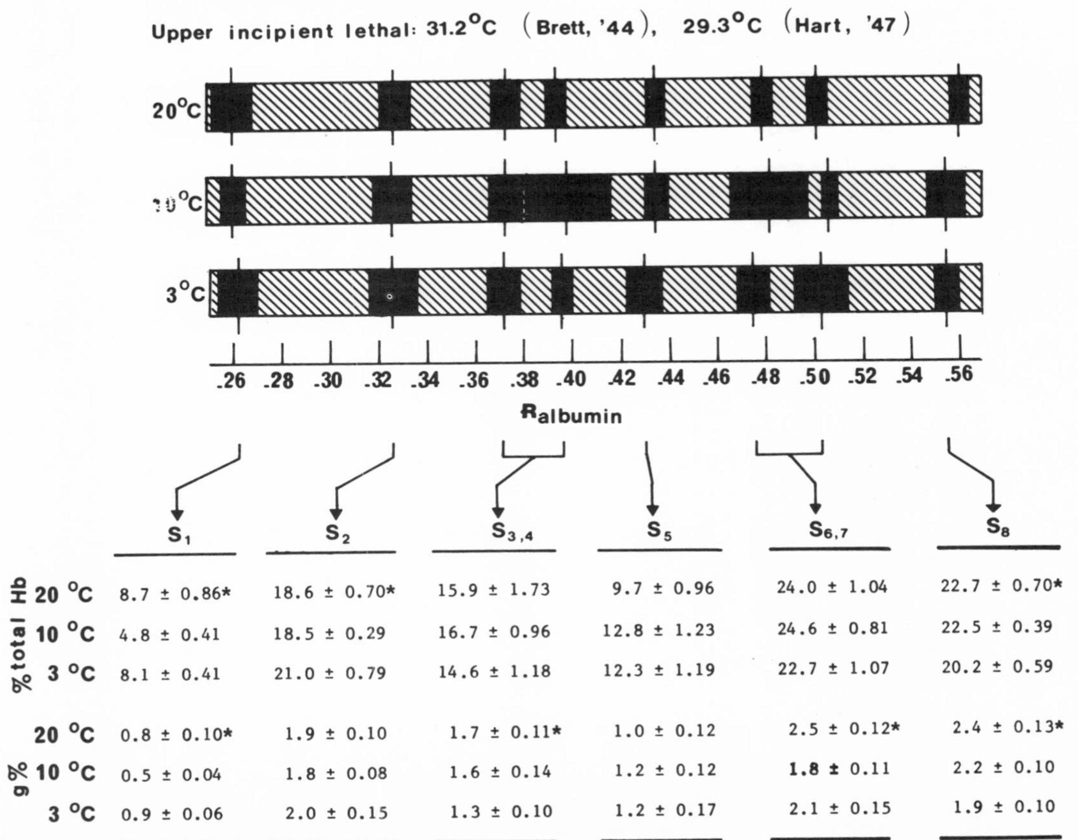

		S_1	S_2	$S_{3,4}$	S_5	$S_{6,7}$	S_8
% total Hb	20 °C	8.7 ± 0.86*	18.6 ± 0.70*	15.9 ± 1.73	9.7 ± 0.96	24.0 ± 1.04	22.7 ± 0.70*
	10 °C	4.8 ± 0.41	18.5 ± 0.29	16.7 ± 0.96	12.8 ± 1.23	24.6 ± 0.81	22.5 ± 0.39
	3 °C	8.1 ± 0.41	21.0 ± 0.79	14.6 ± 1.18	12.3 ± 1.19	22.7 ± 1.07	20.2 ± 0.59
g%	20 °C	0.8 ± 0.10*	1.9 ± 0.10	1.7 ± 0.11*	1.0 ± 0.12	2.5 ± 0.12*	2.4 ± 0.13*
	10 °C	0.5 ± 0.04	1.8 ± 0.08	1.6 ± 0.14	1.2 ± 0.12	1.8 ± 0.11	2.2 ± 0.10
	3 °C	0.9 ± 0.06	2.0 ± 0.15	1.3 ± 0.10	1.2 ± 0.17	2.1 ± 0.15	1.9 ± 0.10

Fig. 5. The hemoglobin system of the sucker following acclimation to 3°, 10° and 20°C. Mobility relative to albumen; mean mobility ± 95% confidence interval of the mean. Fractional contribution of each hemoglobin variant to total hemoglobin (% total Hb); mean ± 95% confidence interval of the mean. Fractional concentration of each hemoglobin variant (g%); mean ± 95% confidence interval of the mean. (NOTE, fractions S_3 and S_4, S_6 and S_7 have been treated as single fractions).

6), and comparable in this sense to the 5-fraction system of the closely-related bluegill, L. macrochirus (Manwell & Baker, 1970). Comparisons of relative mobilities, and coelectrophoresis of pooled hemolyzates indicated that no elements of the system were entirely lost as a consequence of acclimation. Two, however, were sharply reduced in abundance upon exposure to cold (P_3) and warm conditions (P_1) respectively. All hemoglobin variants exhibited significant changes in relative abundance with temperature. In addition, hemoglobins P_1 through P_5 were characterized by

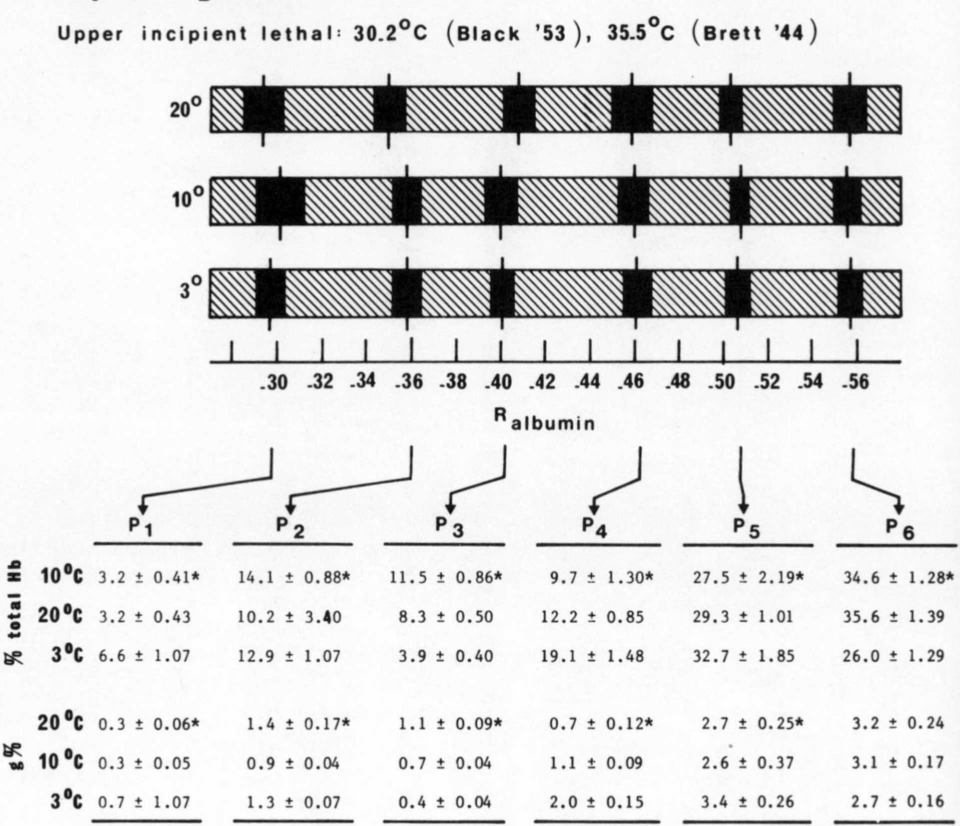

		P1	P2	P3	P4	P5	P6
% total Hb	10°C	3.2 ± 0.41*	14.1 ± 0.88*	11.5 ± 0.86*	9.7 ± 1.30*	27.5 ± 2.19*	34.6 ± 1.28*
	20°C	3.2 ± 0.43	10.2 ± 3.40	8.3 ± 0.50	12.2 ± 0.85	29.3 ± 1.01	35.6 ± 1.39
	3°C	6.6 ± 1.07	12.9 ± 1.07	3.9 ± 0.40	19.1 ± 1.48	32.7 ± 1.85	26.0 ± 1.29
g%	20°C	0.3 ± 0.06*	1.4 ± 0.17*	1.1 ± 0.09*	0.7 ± 0.12*	2.7 ± 0.25*	3.2 ± 0.24
	10°C	0.3 ± 0.05	0.9 ± 0.04	0.7 ± 0.04	1.1 ± 0.09	2.6 ± 0.37	3.1 ± 0.17
	3°C	0.7 ± 1.07	1.3 ± 0.07	0.4 ± 0.04	2.0 ± 0.15	3.4 ± 0.26	2.7 ± 0.16

Fig. 6. The hemoglobin system of the pumpkinseed following acclimation to 3°, 10° and 20°C. Mobility relative to albumen; mean mobility ± 95% confidence interval of the mean. Fractional contribution of each hemoglobin variant to total hemoglobin (% total Hb); mean ± 95% confidence interval of the mean. Fractional concentration of each hemoglobin variant (g%); mean ± 95% confidence interval of the mean.

significant variations in actual concentration as well. P_1, P_4 and P_5 decreased in abundance as temperature increased, the first notably so. P_2 was characterized by reduced abundance at 10°C as compared to 3° and 20°C. The concentration of P_3 rose, and a near-significant ($P < 0.10$, but not < 0.05) elevation in P_6 also took place at higher temperatures.

Cyprinus carpio. Carp are among the most temperature- resistant of the freshwater temperate zone teleosts, tolerating temperatures in the 36° - 38°C range (Black, 1953; Houston, unpublished observations). As is usually the case, three principal hemoglobin fractions were seen (Fig. 7). The fourth minor component alluded to earlier was not present although, as subsequently noted, a fraction of this type was observed in C. carpio x C. auratus hybrids. Acclimation lead to no qualitative alteration in hemoglobin system complexity. All fractions, however, were characterized by significant modifications in both relative abundance and actual concentration between 5° and 30°C. Two fractions, C_2 and C_3, were increasingly prevalent at the higher temperature, while the principal component, C_1, declined in abundance under these conditions.

Carassius auratus. Like the carp, the goldfish displays marked heat resistance. It has been suggested (Fry, 1947) that the ultimate upper incipient lethal temperature may exceed 40°C. Recent (unpublished) studies in this laboratory have failed to confirm this. Nevertheless, this species readily adjusts to temperatures exceeding 38°C. Thus far, the goldfish appears to be unique in its expression of both qualitative and quantitative changes in the hemoglobin system following acclimation. Specimens maintained under cold conditions (< ~10°C) normally have a 2-component system, whereas those acclimated to warmer temperatures typically have 3 distinguishable hemoglobins (Fig. 7). Apparently first reported by Falkner and Houston (1966) this phenomenon has subsequently been confirmed in several studies (Houston & Cyr, 1974; Houston et al., 1976; Houston & Rupert, 1976). Quantitative variations were comparable to those observed in carp, i.e., the principal fraction (G_2) declined significantly in both actual and proportional concentration at higher temperature. A second persistent fraction (G_3) increased under these conditions, and this was accompanied by the appearance of the thermolabile variant, G_1.

C. carpio x C. auratus Hybrid. Carp and goldfish hybridize freely, and several populations of both parental species and their hybrid offspring occur in the Niagara region of Southern Ontario. Thermal tolerance studies have not as yet been carried out but - not surprisingly - the hybrid appears to be highly eurythermal. Four hemoglobins, three of which comprise 80-90% of the total hemoglobin present, can be distinguished (Fig. 7). Of these two are common to carp and hybrid, one is seen in goldfish and hybrid and one is found in carp, goldfish and hybrid (Fig. 8). The minor thermolabile fraction of the goldfish (G_1) does not occur in the hybrid. One principal (H_3), and one less abundant hemoglobin (H_1) exhibited significant changes in relative abundance during acclimation, but only the latter varied significantly in actual concentration. No notable changes in the major fraction H_2 of in H_4 were apparent.

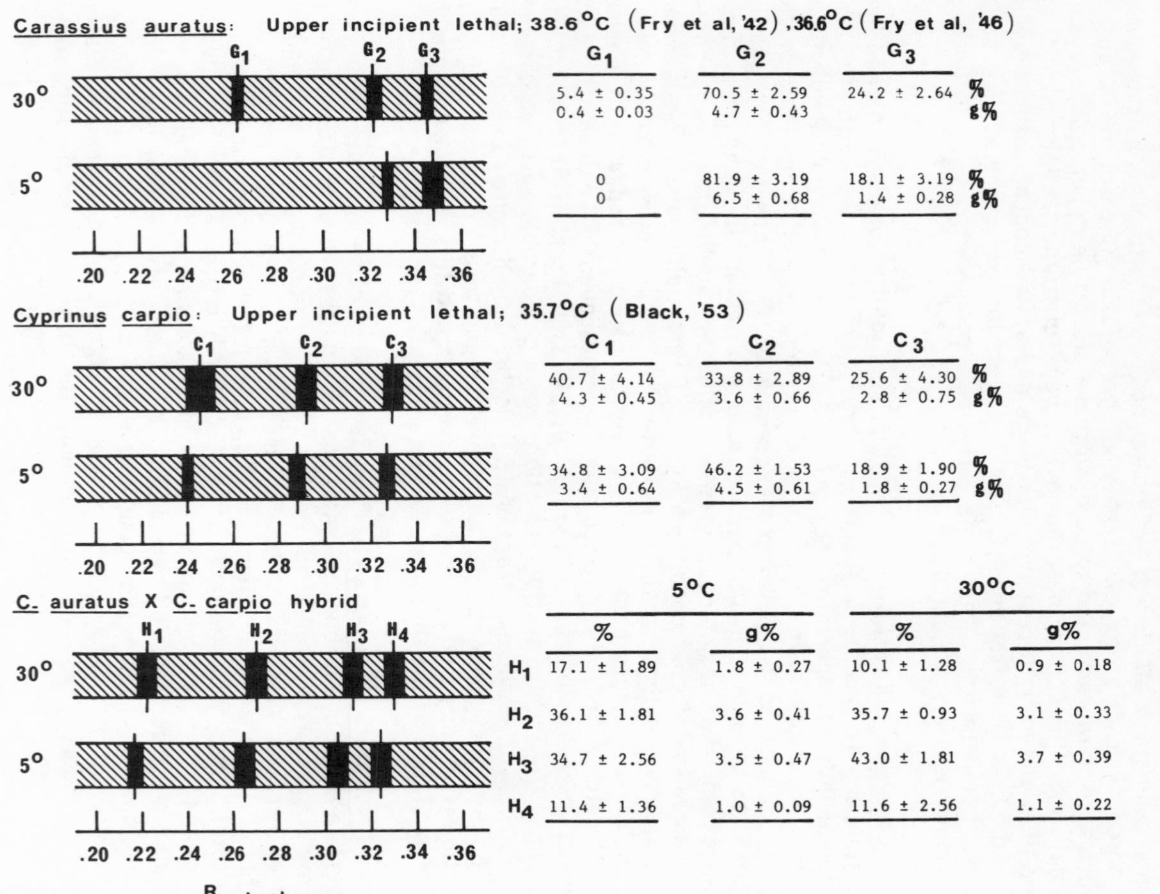

Carassius auratus

		G_1	G_2	G_3	
30°		5.4 ± 0.35	70.5 ± 2.59	24.2 ± 2.64	%
		0.4 ± 0.03	4.7 ± 0.43		g%
5°		0	81.9 ± 3.19	18.1 ± 3.19	%
		0	6.5 ± 0.68	1.4 ± 0.28	g%

Cyprinus carpio

	C_1	C_2	C_3	
30°	40.7 ± 4.14	33.8 ± 2.89	25.6 ± 4.30	%
	4.3 ± 0.45	3.6 ± 0.66	2.8 ± 0.75	g%
5°	34.8 ± 3.09	46.2 ± 1.53	18.9 ± 1.90	%
	3.4 ± 0.64	4.5 ± 0.61	1.8 ± 0.27	g%

C. auratus X *C. carpio* hybrid

	5°C		30°C	
	%	g%	%	g%
H_1	17.1 ± 1.89	1.8 ± 0.27	10.1 ± 1.28	0.9 ± 0.18
H_2	36.1 ± 1.81	3.6 ± 0.41	35.7 ± 0.93	3.1 ± 0.33
H_3	34.7 ± 2.56	3.5 ± 0.47	43.0 ± 1.81	3.7 ± 0.39
H_4	11.4 ± 1.36	1.0 ± 0.09	11.6 ± 2.56	1.1 ± 0.22

Fig. 7. The hemoglobin systems of the goldfish, carp and goldfish x carp hybrid following acclimation to 5° and 30°C. Mobility relative to cytochrome C; mean mobility ± 95% confidence interval of the mean. Fractional contribution of each hemoglobin variant to total hemoglobin (% total Hb); mean ± 95% confidence interval of the mean. Fractional concentration of each hemoglobin variant (g%); mean ± 95% confidence interval of the mean.

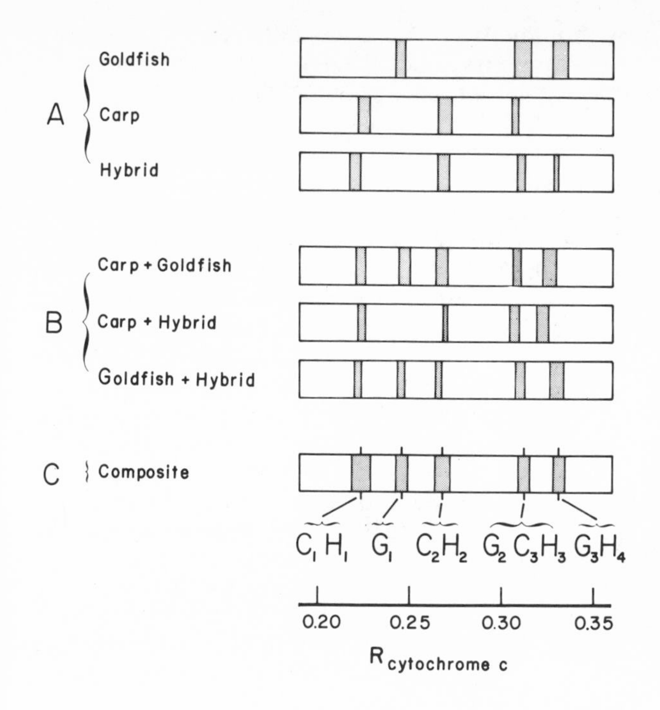

Fig. 8. Hemoglobin equivalencies in the goldfish, carp and goldfish x carp hybrid. A. $R_{cytochrome\ C}$ ranges obtained for pooled hemolyzates of all goldfish, all carp and all hybrids. B. $R_{cytochrome\ C}$ ranges obtained with pooled hemolyzates of all carp plus all goldfish, all carp plus all hybrids, all goldfish plus all hybrids. C. Mean $R_{cytochrome\ C}$ mobilities $\pm$ 95% confidence intervals for the hemoglobin components of the goldfish, carp and goldfish x carp hybrid.

In summary, it is apparent that the teleostean hemoglobin system possesses a considerable degree of thermolability. It is also apparent, however, that major species differences exist in this regard. The white sucker, for example, exhibits only limited modifications in the relative and absolute abundancies of a restricted range of hemoglobins, resembling in this regard the bullhead, Ictalurus nebulosus (Grigg, 1969). In other species, of which the rainbow trout is representative, substantial variation takes place, but no single hemoglobin component is markedly altered. The carp and goldfish are distinguished by substantial variations following acclimation. The latter species is also unique in exhibiting temperature-dependent qualitative alteration in hemoglobin system complexity. There is some possibility, however, that the pumpkinseed might also undergo hemoglobin additions and deletions if challenged by thermal extremes.

Prior to any consideration of thermoacclimatory alterations in the modulatory microenvironment of the hemoglobin molecule, and the adaptive character (if any) of observed modifications in the hemoglobin system, two related points warrant some comment. These concern the relationship between hemoglobin system complexity and thermal tolerance, and the means whereby thermoacclimatory variations in hemoglobin abundance take place.

Hemoglobin System Complexity and Thermal Tolerance. As noted previously teleost fishes, in common with most vertebrates, exhibit a remarkable degree of hemoglobin polymorphism. On the basis of a survey of some 300 species (Osteichthyes: 10, Amphibia: 23, Reptilia: 164, Aves: 48, Mammalia: 55) De Smet (1978) has calculated probabilities of occurrence of multiple hemoglobin systems in the various vertebrate classes as: fishes-100%, amphibians-61%, reptiles-44%, birds-89% and mammals-47%. The figure cited for fishes obviously disagrees with the findings of Sharp (1973) and Fyhn et al. (1979), but the discrepancy is not of any fundamental importance. It is often assumed that molecular, like morphological complexity, must necessarily have some functional significance. In the instance of the teleosts, the inference has been often made that hemoglobin multiplicity may provide capacity for response to environmental variations (Grigg, 1974; Riggs, 1970). In the specific instance of temperature, however, the relationship between overt hemoglobin system complexity and thermal influences is not as obvious as might, a priori, be anticipated. Hemoglobin polymorphism occurs, for example, in the Antarctic nototheniids, *Trematomus bernacchii* and *T. borchgrevinki*; species inhabiting notably thermostable habitats. The salmonid hemoglobin systems are among the most complex which have thus far been described. Yet these animals are not remarkable for their ability to adapt to alterations in oxygen demand and/or availability. Carp and goldfish, species which do display noteworthy adaptive capabilities, have relatively simple systems.

The relationship of hemoglobin system complexity and thermal tolerance - at least insofar as can be inferred from the limited data now available - actually appears to be inverse in nature; increased heat tolerance is apparently correlated with reduced hemoglobin system complexity (Fig. 9). Some relationship can also be seen between system complexity and the magnitudes of the changes in specific hemoglobin abundancies which accompany acclimation. These have also been depicted in Fig. 9 for the six species and carp-goldfish hybrid previously considered. For purposes of comparison the various hemoglobin types have been arbitrarily designated as major components (>30% of total hemoglobin), moderately abundant components (15-30% of total hemoglobin) and minor components (<15% of total hemoglobin). While the number of species considered is far too limited to warrant generalization, these data suggest that substantial changes in the concentrations of major hemoglobin variants take place only in species possessing the limited-variability systems which appear to be characteristic of eurythermal species. The obvious inference is that adaptive advantage accrues from a restricted form of polymorphism which permits significant alterations in the abundance of specific hemoglobins. From this viewpoint much of the complexity of, for example, the salmonid hemoglobin system

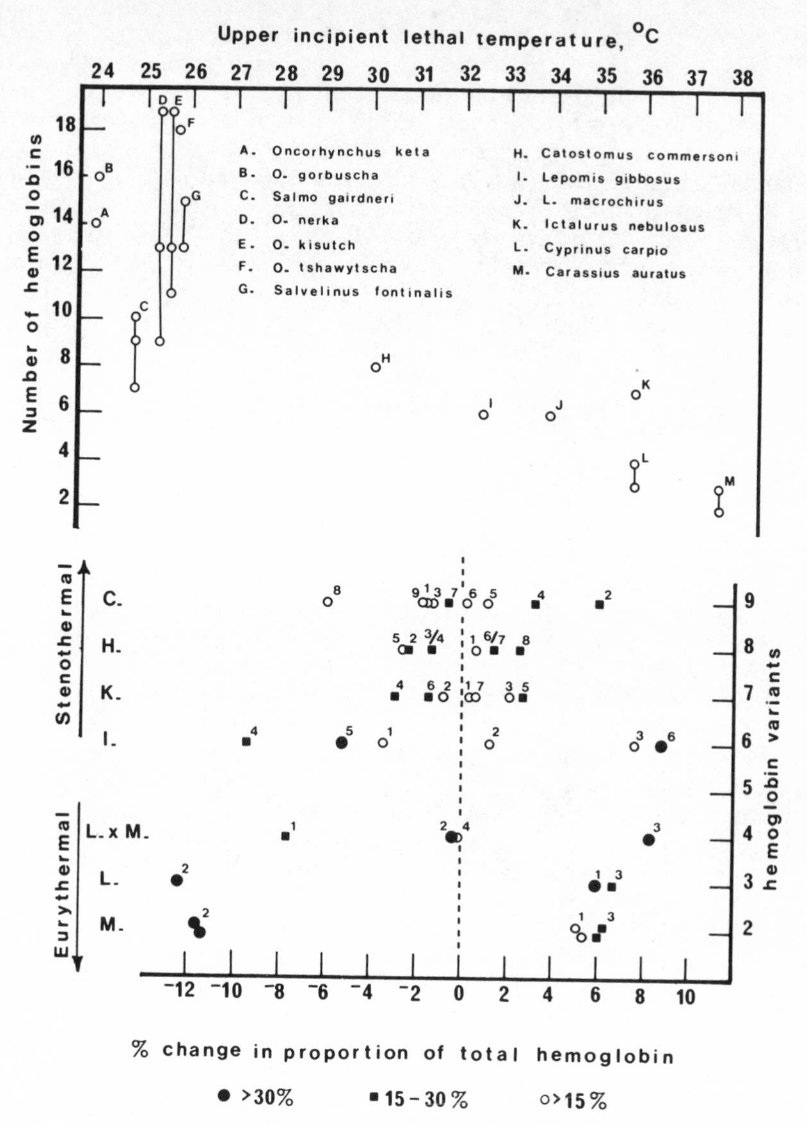

Fig. 9. Upper panel. Hemoglobin system complexity (number of electrophoretically-distinct hemoglobin variants) in 13 species of freshwater teleosts in relation to thermal tolerance (upper incipient lethal temperature). Lower panel. Changes in hemoglobin variant abundancies between the highest and lowest temperatures tested in Salmo gairdneri (c), Catostomus commersoni (H), Ictalurus nebulosus (K), Lepomis gibbosus (I) C. auratus x C. carpio hybrid (L x M), Cyprinus carpio (L) and Carassius auratus. Data from Figs. 4-7 and Grigg (1969). Fractions have been categorized as major (●, >30% of total hemoglobin), moderately abundant (■, 15 - 30% of total hemoglobin) and minor (○, > 15% of total hemoglobin). Numerals refer to component mobilities as designated in Figs. 4-7 and by Grigg (1969) for I. nebulosus.

would have to be regarded as essentially neutral variation around a more limited range of basic hemoglobin types. This contention is compatible, in many respects, with the definition of functionally-heterogeneous hemoglobin groups in the rainbow trout (Brunori, 1975). Therefore, however, it has been difficult to reconcile the functional homogeneity of, for example, the carp hemoglobins (Gillen & Riggs, 1972) with thermoacclimatory variations in fractional concentrations (Houston *et al.*, 1976; Houston & Smeda, 1979). The recent suggestion by Weber and Lykkeboe (1978) that alterations in component proportions lead to functionally-significant interactions between hemoglobin variants which, in turn, modify oxygen affinity is, however, highly pertinent and will be further considered in the context of potentially adaptive adjustments in red cell composition.

Immediate Response of the Hemoglobin System to Temperature Change. While there is little question that many species have the ability to alter hemoglobin system organization during the thermoacclimatory process it is by no means clear how this is achieved. There are, however, two obvious possibilities. Changes in component abundancies may, of course, result from differential effects of temperature upon synthesis and/or assembly of particular subunits. Alternatively, such changes could stem from the aggregation of existing subunits in combinations not possible, or not stable at other temperatures.

Initial studies upon the goldfish tend to favour the second of these hypotheses - at least to the extent that initial changes in the hemoglobin system are involved. Goldfish provide a useful system for investigations of this kind. The thermolabile fraction (G_1) serves as a convenient marker, whose appearance or disappearance following temperature changes under *in vivo* and *in vitro* conditions can provide insights concerning some features of hemoglobin system reorganization. For example, if the first hypothesis is valid relatively long intervals prior to changes in G_1 would be expected, for hemoglobin formation in fishes apparently proceeds slowly (Hevesy *et al.*, 1964). Presumably synthesis would require erythrocyte integrity. Hemoglobin formation admittedly occurs under cell-free circumstances in embryonic avian red cells' but does so at markedly reduced rates (Henderson & Lee, 1976). Whether this is the case in the teleostean erythrocyte has apparently not yet been determined. If the alternative hypothesis is valid, the reorganization process must be non-metabolic in nature. Lag phases following temperature change should be abbreviated, and the process should occur at appreciable rates under cell-free circumstances as well as in the intact cell.

To examine these possibilities several temperature treatments were applied to intact animals and to *in vitro* erythrocyte suspensions. In the instance of whole animals 3 sets of conditions were employed:

(1) acclimation to two constant temperatures (3° and 23°C) known to be associated with 2- and 3-component hemoglobin systems respectively,

(2) acclimation to diurnally-cycling temperatures with maxima and minima of ~ 3° and ~ 23^oC,

(3) acclimation to 3° and 23^oC followed by abrupt reciprocal transfers, i.e., $3^o \rightarrow 23^oC$, $23^o \rightarrow 3^oC$.

All specimens were acclimated to these conditions for not less than three weeks before use. Samples from the cycling series were taken at ~3° and ~ 23^oC, with equivalent numbers of animals from the corresponding constant-temperature groups. In the case of transfer experiments samples were taken at hourly intervals for 12 hours. The results obtained are summarized in Fig. 10.

As previously noted goldfish tend to be inconsistent in their response to temperature and in this instance, as in a number of earlier studies (e.g., Anthony, 1961; Houston *et al.*, 1976), overall hemoglobin content actually decreased at the higher constant temperature. In the cycling series of animals hemoglobin levels were of intermediate magnitude, and no significant differences were seen at 3° amd 23^oC. Animals on both constant and cycling regimes possessed hemoglobins of similar electrophoretic mobilities. Those acclimated to 23^oC, or sampled at 23^oC were characterized by three fractions corresponding to G_1, G_2 and G_3. Specimens maintained at 3^oC, or sampled at that temperature possessed G_2 and G_3, but lacked the G_1 variant seen at higher temperatures. Accordingly, these observations suggest that hemoglobin system reorganization, with the introduction or elimination of the thermolabile component can occur within a 12-hour period.

The abrupt transfer experiments confirmed this, and revealed that reorganization could take place in still shorter time intervals. Specimens shifted from 23° to 3^oC lost the G_1 fraction within 3 hours. Similarly, a component of this mobility was present 3 hours following transfer from 3° to 23^oC. Thus, the time course of reorganization, taken in context with rates of temperature change under natural conditions, suggests a response of sufficient rapidity to have physiological relevance.

Because the cycled fish had comparable hemoglobin levels, variations in the relative abundancies of the different polymorphs were of particular interest. The increase in G_3 (from 11.5 ± 2.1 to $16.3 \pm 5.3\%$; $+ 4.8\%$) was roughly equivalent to that in G_1 (0 to $5.1 \pm 0.7\%$), and the sum of the two were almost equal to the decline in G_2 (from 88.4 ± 2.1 to $78.6 \pm 5.7\%$; $- 9.7\%$). A similar situation was previously observed in this species, and in carp acclimated to 5° and 30^oC (Houston *et al.*, 1976). In the case of the goldfish, acclimation to 30^oC was accompanied by a decrease of some 11.4% in the abundance of G_2 by comparison with cold-acclimated specimens. G_1 and G_3 each rose by 5.4% under these circumstances. In the carp the C_2 variant decreased in abundance by 12.4%, while C_1 and C_3 rose by 5.9 and 6.7% respectively. These data suggest that the loss of two G_2 (or C_2) hemoglobin molecules is ultimately associated with the appearance of one of each of G_1 and G_3 (or C_1 and C_3). The simplest model which will account

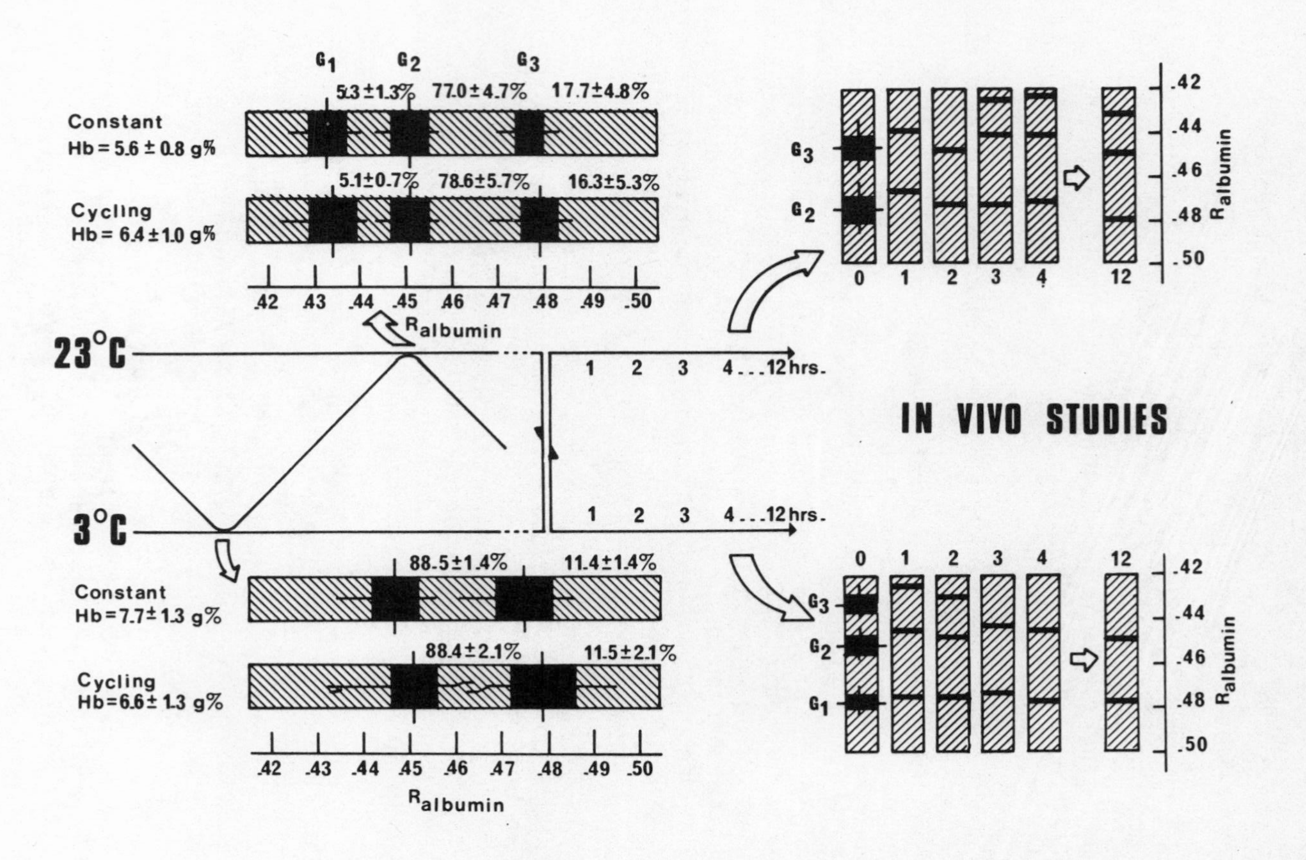
IN VIVO STUDIES
23°C
3°C
G1 G2 G3
5.3±1.3% 77.0±4.7% 17.7±4.8%
Constant
Hb = 5.6 ± 0.8 g%
5.1±0.7% 78.6±5.7% 16.3±5.3%
Cycling
Hb = 6.4 ± 1.0 g%
.42 .43 .44 .45 .46 .47 .48 .49 .50
R albumin
1 2 3 4 . . . 12 hrs.
88.5±1.4% 11.4±1.4%
Constant
Hb = 7.7 ± 1.3 g%
88.4±2.1% 11.5±2.1%
Cycling
Hb = 6.6 ± 1.3 g%
0 1 2 3 4 12
.42 .44 .46 .48 .50

for this requires that at least one of the subunit pairs of G_2 (or C_2) be heterogenous, and that the corresponding subunits of G_1 and G_3 (or C_1 and C_3) be homogeneous, i.e., $2(\alpha_1\alpha_2: - -) \rightarrow 1(\alpha_1\alpha_1: - -) + 1(\alpha_2\alpha_2: - -)$, or the corresponding situation with respect to β subunits. Again, in terms of the time scale of response involved these observations tend to favour the passive reorganization hypothesis.

Further support was obtained during the course of <u>in vitro</u> studies. In this instance 15 animals were acclimated to 23°C for a 3 week period, and a pooled blood sample collected. Half of this was kept intact and hemolyzate immediately prepared from the remainder. About two-thirds of the red cell suspension and hemolyzate were cooled to 3°C, and incubated at that temperature for 4 hours. At this time portions of the suspension were hemolyzed, and these as well as aliquots of the original hemolyzate were subjected to electrophoresis. The remainder of the suspension and hemolyzate were then warmed to, and incubated at 23°C for 4 hours prior to electrophoresis. Because of the readiness with which teleostean hemoglobins undergo reactions leading to the formation of 'spurious' fractions, one-third of the original erythrocyte suspension and hemolyzate were continuously incubated at 23°C, and sampled at 4 and 8 hours. It should be noted that stabilization with CO was carried out subsequent to incubation, but prior to electrophoresis.

As indicated in Fig. 11 the control (continuous incubation at 23°C) red cell suspension and hemolyzate exhibited 3 fractions similar in mobility to G_1, G_2 and G_3 at 0,4 and 8 hours, suggesting that auto-oxidative and other processes had not seriously compromised the system during the 8 hour incubation period. Red cells and hemolyzates cooled from 23°C to 3°C and held at 3°C for 4 hours lacked a mobility corresponding to G_1, but possessed G_2 and G_3. Upon rewarming to 23°C for 4 hours all three fractions were seen in both the erythrocyte suspension and hemolyzate. These findings corroborate those of the abrupt transfer and cycling studies with respect ot the rapidity with which hemoglobin system reorganization takes place following temperature change. More importantly, however, the observation that change occurred in the cell-free, as well as cell-intact system further supports the view that the modifications seen stem from some form of aggregative rather than metabolic activity.

Fig. 10. The hemoglobin system of the goldfish exposed to constant acclimation temperatures (3°, 23°C) and to diurnally-cycling conditions (~3° to ~23°C) and sampled at 3° and 23° and abruptly shifted from 3° to 23°C or 23°C to 3°C and sampled hourly for 12 hours. Mobilities relative to albumen given as means or means ± 95% confidence intervals of the mean. Total hemoglobin values in g %. Fractional contributions given as means ± 95% confidence intervals of the mean.

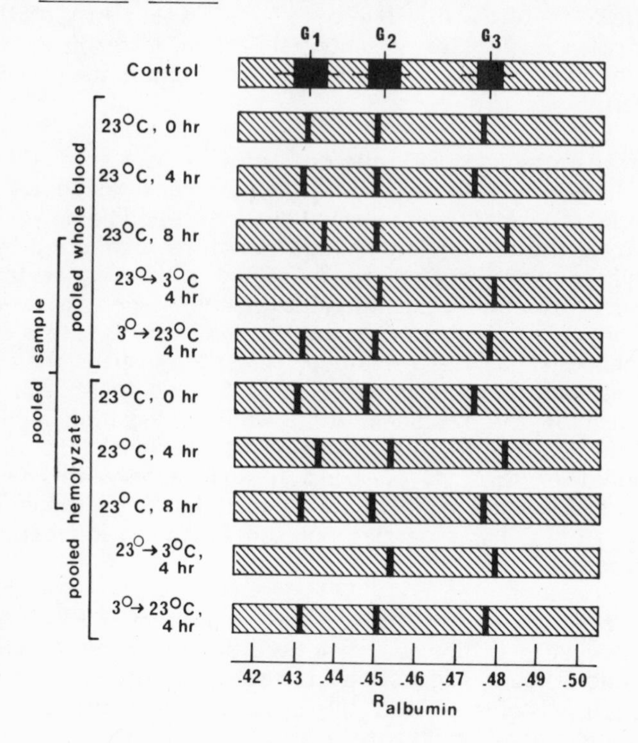

Fig. 11. In vitro studies on pooled blood samples from goldfish initially acclimated to 23°C. Mobilities relative to albumen given as mean ± 95% confidence interval of the mean, or as means. Incubation protocols as designated in Figure.

It would, of course, be unwise to generalize on the basis of these limited findings. Nevertheless, in the instance of the goldfish at least it appears that the hemoglobin system is amenable to reorganization by essentially passive processes at rates which are sufficiently rapid to have physiological significance. However, the existence of an immediate response system of this kind in no way precludes the possibility of metabolic involvements over the longer term, and this is currently under investigation.

III. Acclimatory Changes in the Modulatory Microenvironment of Hemoglobin.

The potential adaptive significance of acclimatory alternations in multiple hemoglobin systems can be appreciated only in relation to accompanying changes in the microenvironment in which these hemoglobins actually function. Four factors are particularly critical in the sense of modulating hemoglobin-oxygen affinity, and thereby whole blood oxygen equilibrium curves. These are temperature, pH, organophosphates and a number of inorganic electrolytes.

Temperature. Oxygen-hemoglobin interactions in vertebrates are typically exothermic in nature. Increases in temperature would therefore be expected to reduce affinity and facilitate oxygen release to tissues. Thermal effects upon the oxygen affinities of teleostean hemoglobins, however, present a variable picture (Johansen & Lenfant, 1972; Johansen & Weber, 1976). Weber et al. (1976), for example, have defined three general situations in fishes possessing multiple-hemoglobin systems: (1) all hemoglobins sensitive to both temperature and pH, (2) some hemoglobins of the foregoing type, others with little or no sensitivity to temperature and (3) hemoglobins sensitive to pH changes, but largely unaffected by temperature. Johansen and Lenfant (1972) have considered the distribution of these, and hypothesize that hemoglobin evolution in fishes has been such as to reduce thermal sensitivity in species normally encountering moderate-to-large fluctuations in environmental temperature. Powers et al, (1979), however, have noted a number of cases which are imcompatible with this generalization. For example, in the rainbow trout, a major element of the hemoglobin system (designated HbIV, but almost certainly consisting of several fractions) exhibits notable thermosensitivity. Under alkaline circumstances, which largely preclude Bohr and Root effects, the apparent enthalpy of this 'hemoglobin' is similar to those of many other species. This is true as well of Fundulus heteroclitus, a species whose thermal circumstances undergo major annual and even diurnal variations. These, and similar observations have lead Powers et al. (1979) to conclude that the evidence supporting "... the generalized evolutionary development of reduced thermal sensitivity of hemoglobin for fish in fluctuating thermal environments is not compelling ...". Thus, the probability exists that at least some hemoglobin types may exhibit thermally induced reductions in affinity as temperature rises; an effect which would facilitate oxygen release under circumstance elevating cellular oxygen requirements.

Acid-Base Status. The effects of temperature upon acid-base balance in ectothermic vertebrates has been extensively studied (Reeves, 1977), and the inverse nature of the relationship between body fluid pH and temperature is well-documented. Of particular interest in the context of the present review are studies upon Salmo gairdneri (Cameron & Randall, 1972; Randall & Cameron, 1973; Janssen & Randall, 1975) and Cyprinus carpio (Dejours, 1973; Dejours & Armand, 1973). Carp hemoglobins are acutely sensitive to reductions in pH; an approximately 5-fold increase in log P_{50} takes place between pH 7.0 and 7.5 (Riggs, 1970). The HbIV fraction of the rainbow trout referred to earlier is also characterized by a marked Bohr effect. In such instances, as is ordinarily the case with pH-sensitive hemoglobins, increases in red cell hydrogen ion concentration presumably exert their effects through strengthening salt bridges - favouring the tense configurational state of the deoxyhemoglobin molecule over the relaxed oxyhemoglobin state. In any event, reductions in erythrocytic pH accompanying exposure to higher temperatures would be expected to operate additively or synergistically with temperature in decreasing hemoglobin-oxygen affinity at higher temperatures.

Organophosphate Modulators. Since the demonstration by Benesch and Benesch (1967) and Chanutin and Curnish (1967) that the organophosphate

2.3-DPG affects hemoglobin-oxygen affinity in mammals considerable emphasis has been given to assessment of the functional role of organophosphates in fishes. The principal organic polyanions in these animals is not, however, diphosphoglycerate but the nucleoside triphosphates, ATP and GTP; the latter being, in a number of fish species both more prevalent and more effective than ATP (Geohegan & Poluhowich, 1974, Lykkeboe et al., 1975, Peterson & Poluhowich, 1976; Torracca et al., 1977; Bartlett, 1978 a, b, c). Binding in mammalian system occurs preferentially between the NH_2-terminal ends of the β-chains of unliganded hemoglobin molecules (Benesch et al., 1967; Arone, 1972). The consequent introduction of new salt bridges then stabilizes the deoxy T state of the molecule, prompts an increase in P_{50} and right-shift of the oxygen equilibrium curve. Thus, the final result is much like that induced by reduction in pH. Binding of phosphates at the NH_2-termini of proteins appears to be a general phenomenon (Hol et al., 1978), and it is generally believed that fish hemoglobins bind organophosphates at the same sites. The magnitude of the organophosphate effect can be inferred from the report by Tan and Noble (1973) that the addition of 0.7 mM inosital hexaphosphate to carp hemoglobin is equivalent to a pH decrease of 1.6 units. Similarly, addition of 1 mM ATP to a variety of teleostean hemoglobins was equivalent to a 0.5 unit reduction in pH (Gillen & Riggs, 1977).

Adaptive alternations in red cell organophosphate levels have been examined primarily in relation to hypoxic stress, and there is substantial evidence that under these circumstances significant decreases in concentration take place. (Wood & Johansen, 1972; Wood et al., 1975; Weber et al., 1976; Greaney & Powers, 1978). This contrasts sharply with the response normally expressed by mammals. Reductions in oxygen availability are associated with increases in erythrocytic 2.3-DPG levels. Hemoglobin-oxygen affinity therefore declines, and oxygen delivery to tissues is facilitated. In the instance of the teleost, of course, the opposite result is seen; reductions in organophosphate increased oxygen affinity, and thereby favours branchial oxygen loading. Less emphasis has been given to the question of organophosphate changes during thermoacclimatory response. Powers (1974) has, however, reported decreases in ATP: Hb ratio from ~1.0 at 20°C to ~0.6 at 30°C in catostomid fishes; a response comparable to that seen in hypoxically-stressed mammals. Weber et al. (1978), on the other hand, found no evidence of reductions in either the ATP or GTP content of red cells from trout acclimated to 5° and 22°C. Accordingly, the effect of acclimation is not yet clear in this case.

Ionic Composition and Hemoglobin-Oxygen Affinity Modulation. During the past decade attention has also been given to the modulating influences of a number of inorganic ions, and their significance has been well established in the case of mammalian hemoglobins. (Bunn, et al., 1971; de Bruin et al., 1974; Benesch & Benesch, 1974; Rollema et al., 1975; Laver et al., 1977). This now appears to be true of fish hemoglobins as well (Bonaventura et al., 1976; Weber & Lykkeboe, 1976).

Chloride apparently reduces affinity in much the same fashion as 2.3-DPG, ATP and GTP through the provision of additional salt bridges favouring the tense, deoxygenated state of the molecule. Magnesium (and calcium) have little direct influence upon striped hemoglobins (Bunn et al., 1971; Weber & Lykkeboe, 1978). Both, and particularly the former are, however, potent affinity modulators, for they compete with hemoglobin for nucleoside triphosphates. Bunn et al., (1971), for example, have estimated that approximately two-thirds of the magnesium present in the deoxygenated mammalian red cell is ATP-complexed, and thereby denied any substantial role in affinity modulation. Less attention has been given to other prominent red cell ions such as sodium and potassium. Bonaventura et al. (1976) have, however, shown the single hemoglobin of the Spot, Leiostomus xanthurus, responds to added NaCl; approximately doubling its P_{50} over that of striped hemoglobin at 20°C. Potassium, however, exerts a more profound influence. Both Rossi-Fanelli et al. (1961) and Bunn et al. (1971) have shown, for example, that at equimolar concentrations KCl reduces affinity more than does NaCl.

Relatively few studies on the ionic composition of teleostean erythrocytes have been reported (Monroe & Poluhowich, 1974; Fugelli & Zachariassen, 1976; Børjeson, 1977; Calla, 1977). Little attention has been given to divalent cations, and in only a few instances (Grigg, 1969; Catlett & Millich, 1976) have possible thermoacclimatory variations in composition been considered.

An investigation of red cell ionic composition was therefore carried out using the rainbow trout and carp (Houston & Smeda, 1979; Houston & Mearow, 1979). As before, these species were chosen on the basis of major differences in thermal tolerance, metabolic activity and hemoglobin system complexity. Hematological observations made in the study have been alluded to previously (Figs. 2 and 3). Reported cellular ion levels were based upon analyses of packed cell columns, corrected for plasma trapped in the column (2.8 ± 0.15% of packed cell volume, Houston & Smeda, 1979). In carrying out calculations of ionic concentration it has been assumed that all cell water is available to electrolyte. This assumption appears to be reasonable insofar as sodium, potassium and chloride are concerned, for Gary-Bobo and Salomon (1968) have provided convincing evidence that there is little if any exclusion of these ions from hemoglobin-associated water at physiologically-realistic hemoglobin concentrations. Whether this is also the case for calcium and magnesium is uncertain. Furthermore, it will be appreciated that the values reported are means for the entire aqueous phase of the red cell. They give no indication as to subcellular compartmentalization, specific binding or the consequences of non-specific interionic attraction upon ionic activities.

Acclimation of rainbow trout to higher temperature was accompanied by significant increases in cell potassium levels, and these were accompanied by approximately equivalent reductions in sodium (Fig. 12). Modest increases in chloride were apparent, as was some reduction in magnesium and calcium content. Marked concentration differences frequently distinguished summer and winter populations at equivalent temperatures.

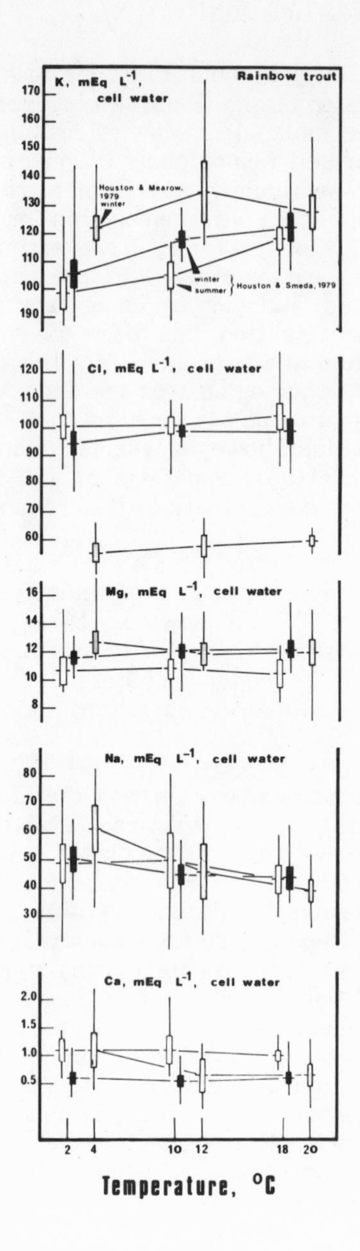

Fig. 12. Erythrocytic potassium, chloride, magnesium, sodium and calcium concentrations in rainbow trout acclimated to 2°, 4°, 10°, 12°, 18° and 20°C. All concentrations in mEq L^{-1}, cell water. Horizontal line - mean; vertical bar $\pm$ 95% confidence interval of the mean; vertical line - range.

In carp a somewhat different pattern of response was observed (Fig. 13). Potassium concentrations were similar to those of trout, but the modest increase seen at 30°C proved to be non-significant. Chloride levels, on the other hand, rose sharply. Concentrations at 30°C were comparable to those seen in trout, and some 40 mEq L^{-1} in excess of the values in cold-acclimated specimens. Sodium concentrations were far below the levels characteristic of trout. There was, in fact, little evidence of sodium in the

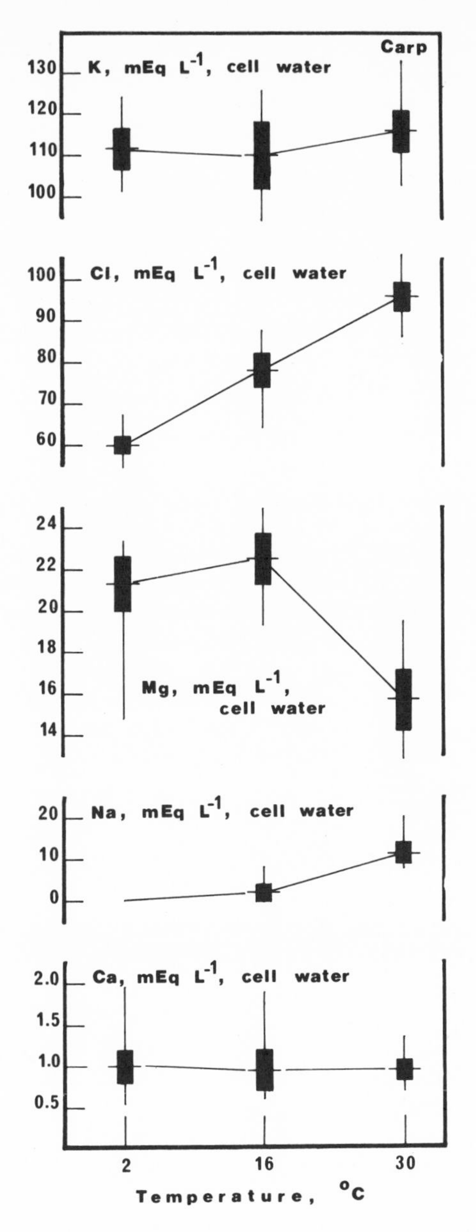

Fig. 13. Erythrocytic potassium, chloride, magnesium, sodium and calcium in carp acclimated to 2°, 16° and 30°C. All concentrations in mEq L^{-1}, cell water. Horizontal line -mean; vertical bar- ± 95% confidence interval of the mean; vertical line - range.

erythrocytes of animals held at 2°C. Detectable amounts were found at 16°C, and between that temperature and 30°C an approximately five-fold rise in concentration occurred. Even at the latter temperature, however, sodium content was only about one-quarter of that seen in trout. Acclimation was also accompanied by significant changes in magnesium at higher temperatures. Between 2° and 16°C erythrocytic magnesium levels were roughly twice those seen in trout. At 30°C, however, they fell sharply to values only slightly in excess of those seen in trout over the 2° to 18°C acclimation range to which that species was exposed. Finally, red cell calcium levels resembled those in trout, both in magnitude and in the absence of any obvious variation with acclimation temperature.

The changes in red cell ionic compositions seen in these species were not unlike those reported by Grigg (1969) in *Ictalurus nebulosus*, and in *Carassius auratus* by Catlett and Millich (1976). In the bullhead, for example, sodium rose by ~80% and potassium by ~50% between 9-10°C and 24-25°C. Calcium levels tended to be both low (~1mEq L^{-1}), and thermostable. In the goldfish erythrocyte potassium, sodium and chloride concentrations increased by ~20%, ~25% and ~15% respectively between 1° and 21.5°C.

In several test groups both erythrocytic ion and hemoglobin levels were altered following acclimation. Accordingly, ion: hemoglobin (mEq:mM) ratios were calculated, and are summarized in Fig. 14. Trout were noteworthy for the constancy with which proportionalities between hemoglobin and chloride, magnesium and calcium were maintained. Only in the case of K^+:Hb were substantial and significant changes encountered; the ratio rising sharply at higher temperatures. A modest, but significant decrease in Na^+:Hb was also observed in summer fish between 2° and 10°C.

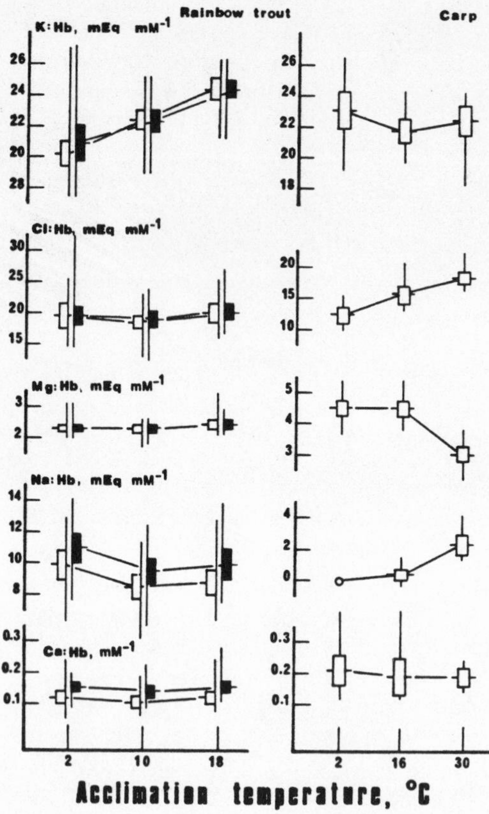

Fig. 14. Red cell K^+:Hb, Cl^-:Hb, Mg^{2+}:Hb, Na^+:Hb and Ca^{2+}:Hb ratios (mEq mM^{-1}) in rainbow trout acclimated to 2°, 10° and 18°C and carp acclimated to 2°, 16° and 30°C.

This was not, however, seen in the winter series of animals. Carp did not exhibit significant variations in K^+:Hb following acclimation. However, both Cl^-:Hb and Na^+:Hb rose markedly at higher temperatures, while Mg^{2+}:Hb fell sharply between 16° and 30^oC. By contrast, Ca^{2+}:Hb was thermostable at levels somewhat above those characterizing the trout.

Weber and Lykkeboe (1978) have recently examined several of these parameters in carp acclimated to hypoxic as compared to normoxic conditions; a circumstance which is associated with a striking increase in hemoglobin-oxygen affinity. Little change in Mg^{2+}:Hb and P_i:Hb was apparent. GTP:Hb and ATP:Hb ratios were, however, reduced under hypoxic conditions, as was $\Sigma GTP + ATP + P_i:/Mg^{2+}$, and it is apparent that the challenge of hypoxia is met in this species by increasing hemoglobin-oxygen affinity through reduction in nucleoside triphosphate content; both in absolute amount, and in relation to magnesium.

The observations reported herein suggest that response to temperature-induced increases in oxygen requirements under essentially normoxic conditions is quite different in character from that in response to hypoxia and, moreover, follows different patterns of expression in the relatively stenothermal trout and eurythermal carp. In the trout magnesium concentration, Mg^{2+}:Hb ratio and nucleoside triphosphate:Hb relationship are not significantly influenced by temperature (Weber *et al*., 1976; Houston & Smeda, 1979). Accordingly, it is probable that Mg^{2+}:NTP relationships are also thermostable. It is unlikely, therefore, that affinity relationships are adjusted through alterations in *effective* organophosphate concentrations. Chloride concentrations, too, were essentially thermostable in relation to hemoglobin, and there is little likelihood that hemoglobin-oxygen affinity was altered by this ion. On the other hand, increases in red cell potassium content was linked to reductions in sodium and, in view of the studies of Rossi-Fanelli *et al*., (1961) and Bunn *et al* (1971) some reduction in affinity might be anticipated.

On the other hand, by comparison with the human red cell (Natelson & Natelson, 1978), and the other tissues of rainbow trout (Murphy and Houston, 1977) erythrocytic chloride and potassium levels were relatively high, while those of magnesium were not unusually so (Weber & Lykkeboe, 1978; Smeda & Houston, 1979). Consequently, from the viewpoint of ionic contributions to affinity modulation, the rainbow trout appears to be adapted for effective reduction in hemoglobin-oxygen affinity over the whole of its thermal tolerance zone; a form of adaptation not inconsistent with the relatively high oxygen demands of these animals. Such a situation, however, provides relatively little scope for response to temperature-induced increases in oxygen requirements. Indeed, beyond some modest increases in overall hemoglobin and red cell numbers, and reduction in red cell volume coupled with hemoglobin system alterations which can be linked to reductions in oxygen affinity (DeWilde & Houston, 1967; Houston and Cyr, 1974; Weber *et al*., 1976) the trout appears to rely principally upon cardiovascular-respiratory responses in resolving the temperature-oxygen demand problem (Heath & Hughes, 1976; Henry & Houston, 1979, unpublished information).

By contrast, carp exhibit major increases in chloride and Cl^-:Hb which would be expected to reduce hemoglobin-oxygen affinity and facilitate oxygen release. Coupled with these are substantial reductions in magnesium and Mg^{2+}:Hb ratio; a situation which contrasts with the absence of any major change in the red cell magnesium content of hypoxic carp (Weber and Lykkeboe, 1978). Information on temperature-related variations in carp red cell organophosphate levels does not appear to be available. However, if this species, like the trout, maintains ATP and GTP concentrations constant during acclimation reductions in magnesium should lead to increases in _effective_ organophosphate content. On the other hand, Powers (1974) has reported a decline in the ATP:Hb ratio of catostomid fishes from ~1.0 at $20^{o}C$ to ~0.6 at $30^{o}C$. However, if these animals also reduce magnesium to the extent seen in carp the _effective_ ATP:Hb ratio might well be little altered by acclimation. If this were the case, increases in temperature, when coupled with concomittant changes in chloride and hydrogen ion would tend to prompt reductions in oxygen affinity. It is noteworthy in this regard that the cathodal components of the catostomid hemoglobin system are highly sensitive to pH (Powers, 1974). It should be appreciated however, that reductions in magnesium content need not necessarily lead to enhancement of ATP availability. The stability of the Mg-ATP complex, unlike that of Mg-2, 3-DPG, rises with temperature (Bunn _et al._, 1971), and notably so at higher temperatures. Consequently, decline in magnesium might serve only to limit reductions in organophosphate availability. However, given the influences of other affective agencies such changes might well contribute to the facilitation of oxygen delivery to the tissues.

Reciprocal Variations in Red Cell Ion Concentrations. In terms of possible influence upon hemoglobin-oxygen affinity the most obviously utilitarian response of the teleost lies in coupling increases in the concentrations of red cell ions which exert positive modulatory effects with reductions in those having the opposite influence. For example, reduced affinity might best be accomplished by linking increases in chloride and/or potassium with decreases in magnesium, calcium and sodium. To examine this, correlation analyses were carried out on data obtained from two winter and one summer populaion of rainbow trout and a more limited sample of summer carp (Houston & Smeda, 1979; Houston & Mearow, 1979). The results of this are summarized in Table II, in which the ratios, $[Mg^{2+}]/[Cl^-]$ and $[Ca^{2+}]/[Cl^-]$ have been related to changes in $[Cl^-]$, and $[Na^+]/[K^+]$ to variations in $[K^+]$. For purposes of this analysis acclimation groups were lumped. In general, best fits were obtained with a simple logarithmic function ($Y = a + b\ln X$), and this was used throughout.

With a single exception, $[Na^+]/[K^+]$ vs $[K^+]$ in carp, the relationships obtained were negative in nature; the relative concentrations of affinity-enhancing to affinity-reducing ions declined as the red cell concentrations of the latter increased. In the instance of Cl^- and Mg^{2+} highly significant correlations were encountered in all populations of both species. This was

TABLE II Variation in red cell electrolytes having antagonistic effects upon hemoglobin-oxygen affinity.

Ion pair	Species	Acclimation temperature C°	Relationship	N	r	Significance
Cl^- m Mg^{2+}	Salmo gairdneri					
	(winter)	4, 12, 20	$[Mg^{2+}]/[Cl^-] = 1.265-0.259 \ln [Cl^-]$	41	-0.715	$P<0.01$
	(winter)	2, 10, 18	$[Mg^{2+}]/[Cl^-] = 0.899-0.169 \ln [Cl^-]$	63	-0.869	$P<0.01$
	(summer)	2, 10, 18	$[Mg^{2+}]/[Cl^-] = 0.061-0.109 \ln [Cl^-]$	43	-0.548	$P<0.01$
	Cyprinus carpio					
	(summer)	2, 16, 30	$[Mg^{2+}]/[Cl^-] = 2.071-0.415 \ln [Cl^-]$	40	-0.893	$P<0.01$
Cl^-,Ca^{2+}	Salmo gairdneri					
	(winter)	4, 12, 20	$[Ca^{2+}]/[Cl^-] = 0.143-0.032 \ln [Cl^-]$	38	-0.373	$P<0.05$
	(winter)	2, 10, 18	$[Ca^{2+}]/[Cl^-] = 0.032-0.006 \ln [Cl^-]$	59	-0.256	NS
	(summer)	2, 10, 18	$[Ca^{2+}]/[Cl^-] = 0.069-0.013 \ln [Cl^-]$	43	-0.196	NS
	Cyprinus carpio					
	(summer)	2, 16, 30	$[Ca^{2+}]/[Cl^-] = 0.061-0.011 \ln [Cl^-]$	38	-0.417	$P<0.05$
K^+, Na^+	Salmo gairdneri					
	(winter)	4, 12, 20	$[Na^+]/[K^+] = 2.994-0.537 \ln [K^+]$	43	-0.381	$P<0.05$
	(winter)	2, 10, 18	$[Na^+]/[K^+] = 4.198-0.797 \ln [K^+]$	54	-0.792	$P<0.01$
	(summer)	2, 10, 18	$[Na^+]/[K^+] = 4.510-0.865 \ln [K^+]$	43	-0 681	$P<0.01$
	Cyprinus carpio					
	(summer)	2, 16, 30	$[Na^+]/[K^+] = 0.089 \ln [K^+] - 0.355$	22	0.172	NS

generally true of K^+ and Na^+ in rainbow trout as well. Both carp and trout also exhibited negative correlations between $[Ca^{2+}]/[Cl^-]$, and $[Cl^-]$, but in only two groups were these significant, and in one instance (trout) only barely so. Nevertheless, these data generally support the view that temperature-related increases in the concentrations of ions tending to reduce affinity are associated with reductions in those which would oppose this effects.

Acclimatory Variations in Red Cell Ion Transport Enzyme Activities. The occurrence of changes in red cell ionic composition with acclimation immediately raises a number of questions regarding the mechanism which bring this about. To examine this feature of the adaptive response assays of (Na^+/K^+) - and (HCO_3^-) - stimulated, (Mg^{2+}) - dependent adenosine triphosphatase activities were carried out with membrane preparations of trout and goldfish erythrocytes, as were assessments of 'membrane' and 'cytosol' carbonic anhydrase activities (Smeda & Houston, 1979; Houston & Mearow, 1979 a, b). In all cases assays were conducted at the actual acclimation temperatures of the specimens, and are regarded as providing a more physiologically-realistic indication of the actual activities of these systems than is given by determinations carried out at standardized (and usually high) temperatures. The very modest levels of (Na^+/K^+) - ATPase activity observed in the trout red cell are consistent with the low activities ordinarily encountered in mammalian erythrocytes (Cavieres, 1977), and also apparent in the nucleated avian red cell (Kregenow, 1977). Nevertheless, some increase in activity was seen at higher acclimation temperatures, and is consistent with the rise in red cell potassium content and reduction in sodium concentration which takes place in this species. The increased activity of carbonic anhydrase at elevated temperatures may presumably be related to the hydration of erythrocytic CO_2, and its export from the red cell as bicarbonate in exchange for chloride; a process which leads to enhancement of red cell chloride content (Fortes, 1977). The resulting decline in cellular bicarbonate would be expected to contribute to reduction in erythrocytic pH as well. The role of (HCO_3) - stimulated ATPase is less certain, but may well supplement that of the carbonic anhydrases. At present there is little evidence of active magnesium transport across the red cell membrane (Christensen, 1975). There is, however, at least some possibility that changes in erythrocytic magnesium content represent passive redistributions in response to temperature-related variations in membrane potential. The latter can be approximated by the chloride equilibrium potential (Dalmark, 1976). In the instance of carp highly significant reductions in potential accompanied increases in acclimation temperature; values at 2°, 16° and 18^oC being -14.7 ± 0.72, -10.2 ± 1.00 and -5.7 ± 0.31 mV respectively, when calculated as $E = (2.303\ RT/_3F) \log (\text{rbc } Cl^-/\text{plasma } Cl^-)$. Corresponding variations in summer rainbow trout were: -7.3 ± 0.70 (2^oC), -6.6 ± 0.47 (10^oC) and -7.0 ± 0.63 mV (18^oC), with equivalent values in winter fish being -9.8 ± 0.48, -9.4 ± 0.48 and -9.1 ± 0.44 mV. The relationships between red cell magnesium concentrations (mEq L^{-1}, cell water) and chloride equilibrium potential (MV) were:

carp: $[Mg^{2+}] = 5.68 + 6.30 \ln E_{Cl}$ $(r = 0.725, P < 0.01)$

trout: $[Mg^{2+}] = 9.23 + 0.25 E_{Cl}$ $(r = 0.25, P < 0.05)$

Thus, these relationships provide some support, particularly in the case of the carp, for the suggestion that magnesium content varies in relation to red cell membrane potential.

Intraerythrocytic changes in the erythrocytic milieu accompanying the acclimatory process, and possessing at least some potential for adaptive adjustment of hemoglobin-oxygen affinity can thus be summarized as follows:

1, Temperature. The exothermic character of hemoglobin-oxygen interaction is such that increases in temperature should lead to reduced affinity (in temperature-sensitive hemoglobin variants), and hence facilitate oxygen delivery to tissues.

2. pH. Reduction in erythrocytic pH, a probable accompaniment of acclimation of all aquatic organisms to higher temperatures, should also prompt reductions in affinity favouring oxygen release.

3. Organophosphates. Increases in organophosphate levels reduce affinity, while reductions in concentration have the opposite effect. Acclimation of catostomids to higher temperature is accompanied by decreases in red cell ATP content and ATP:Hb ratios; adjustments which would increase hemoglobin-oxygen affinity and favour branchial oxygen loading rather than tissue oxygen release (Powers, 1974). By contrast, in the rainbow trout acclimation leads to little or no change in ATP and GTP levels, or in ATP:Hb and GTP:Hb (Weber et al., 1976). Accordingly, no change in organophosphate modulation of affinity would be expected as a correlate of the acclimatory process.

4. Chloride. Since chloride mimics the actions of ATP, GTP and 2,3-DPG in favouring the tense configuration of deoxyhemoglobin elevation of chloride leads to reduced affinity, with the converse taking place upon reduction in chloride content. Modest rises in red cell chloride levels took place during adjustment of trout to warmer conditions (Houston & Smeda, 1979; Houston & Mearow, 1979). More pronounced increases were seen in the eurythermal carp and goldfish (Catlett & Millich, 1976; Houston & Smeda, 1979).

5. Magnesium and Calcium. By complexing nucleoside triphosphates, magnesium, and to a lesser degree calcium, reduce the effective concentrations of these modulators (Bunn et al., 1971). Accordingly, enhancement of the cellular magnesium level should increase hemoglobin-oxygen affinity and prompt oxygen uptake at the gill. Conversely,

reductions in magnesium - with organophosphate levels constant - should produce the converse effect. In the rainbow trout magnesium (and calcium) levels were relatively low, and essentially thermostable; resembling in this respect the situation in the bullhead (Gregg, 1969). In the carp, however, magnesium and Mg^{2+}:Hb ratio fell sharply as specimens were exposed to high temperatures.

6. Sodium and Potassium. At equivalent concentrations potassium reduces hemoglobin-oxygen affinity more than does sodium, although the effects are not notably large (Rossi-Fanelli et al., 1961; Bunn et al., 1971). In the goldfish and trout acclimation to increased temperature is accompanied by significant increases in red cell potassium content (Catlett and Millich, 1976; Houston & Smeda, 1979). Such, however, was not the case in either the bullhead or carp (Gregg, 1969; Houston & Smeda, 1979).

A striking feature of the variations in red cell composition accompanying acclimation was the inverse relation between the concentrations of electrolytes having opposed effects upon affinity. This was particularly noteworthy in the instance of chloride and magnesium/calcium in both trout and carp. A similar situation occurred in the former species insofar as potassium and sodium were concerned, but was absent in the carp.

In general, the variations which have been identified suggest that acclimation to increases in environmental temperature would be accompanied by reduction in affinity, increase in P_{50} and facilitation of oxygen delivery to tissues. Such a response would be adaptively consistent with the heightened oxygen demands seen under such circumstances. However, Eaton (1974) contends that increases in affinity favouring branchial oxygen loading are the more appropriate response, and this view finds support in the observation of Powers (1974) that organophosphate levels decline as temperature increases. The obvious discrepancy may prove more apparent than real. The species examined in this, and several of the cited studies were exposed to temperature increases under normoxic conditions. Whole blood oxygen equilibrium curves frequently plateau at tensions well below those ordinarily regarded as hypoxic (Cameron, 1971, 1973; Eddy, 1973; Hayden et al., 1975). In such cases there is no obvious advantage in increasing hemoglobin-oxygen affinity. Indeed, increase in affinity might well prove anti-adaptive, since oxygen release at the tissue level would presumably be impeded. On the other hand, increases in temperature frequently involve true hypoxia in the sense that oxygen tensions are reduced. Under these circumstances responses which maximize branchial oxygen loading might constitute the more appropriate strategy. As yet, however, little emphasis has been given to response to circumstances involving simultaneous modifications in affective environmental agencies. At the present time studies involving exposure to normoxic and hypoxic conditions as relatively high temperatures and under cold conditions are being initiated with a view to better definition of adaptative flexibilities.

Any assessment of the adaptive nature of the changes which have been observed in the hemoglobin system-intraerythrocytic microenvironment complex must necessarily include a substantial element of speculation at the present time. Nonetheless, information for two species, the relatively eurythermal carp and the relatively stenothermal rainbow trout, appears reasonably adequate and, moreover, points to the possibility that these species may resort to somewhat different strategies in their responses to the respiratory requirements of life at higher temperatures.

As noted previously the rainbow trout hemoglobin system includes at least two functionally-distinct hemoglobin groups (Brunori, 1974). Of these, one consists of fractions remarkably insensitive to temperature, pH and organophosphate modulators (Gillen & Riggs, 1972). The surprising absence of any significant thermal influence upon what is commonly an exothermic relationship is believed to be the result of an unusual balancing of endo- and exothermic processes, the net result of which is variation with temperature in equilibrium ratio of relaxed (oxyhemoglobin) and tense (deoxyhemoglobin) molecular configurations (Wyman et al., 1977). The most abundant fraction (HbIV) exhibits marked changes in oxygen affinity with increases in temperature, hydrogen ion concentration and organic and inorganic phosphate levels. While it is not possible to directly compare the four Brunori trout fractions, which were separated by starch gel electrophoresis and chromatographically purified, with the nine fractions visualized by polyacrylamide disc electrophoresis (Houston & Cyr, 1974; Braman et al., 1977) it is worth noting that all of the major fractions separated by the latter method increased in abundance between 2° and 18°C (Houston & Cyr, 1974). It is probable, particularly in view of the recent observations of Weber et al. (1976) on 5° and 15°C-acclimated trout, that these included Brunori's modulator -sensitive fraction IV complex. Thus, it seems clear that one feature of the response of the rainbow trout to increases in temperature lies in enhancement of modulator-responsive hemoglobin types.

The trout does not, however, exhibit very marked changes in the hemoglobin microenvironment. Obviously, temperature increases and pH decreases would be expected to reduce oxygen affinity. On the other hand, ATP and GTP levels and NTP:Hb ratios are not significantly altered during acclimation (Weber et al., 1976). Chloride, magnesium and calcium levels within the erythrocyte also appear to be relatively thermostable. Potassium increases to some extent, and this is linked to a decline in sodium content; variations which might be expected to prompt some reduction in affinity. However, the fact that few adaptive changes with temperature occur must be balanced against the observation that trout maintain a high chloride/potassium-low magnesium/calcium/sodium erythrocytic milieu over most of their thermal tolerance zone. Furthermore, NTP:Hb ratios exceed unity (1.21 $\pm$ 0.21 to 1.57 $\pm$ 0.30 mM/mM) over a 5° to 22°C range. Since the trout is among the more metabolically-active of freshwater teleosts, and commonly inhabits areas of high oxygen content it would appear that their primary strategy lies in the provision of circumstances which facilitate oxygen delivery to tissues at all temperatures. Adaptation to higher temperatures is accompanied by some increase in overall hemoglobin content, alterations in hemoglobin abundancies and, inevitably, increases in temperature and hydrogen ion concentration. Undoubtedly, it is these factors which collectively account for the adaptive reductions in hemoglobin - oxygen affinity identified by Weber et al. (1976) and Hughes et al. (1976). They also, however, underscore the reliance which must be placed by this species upon systemic responses in the satisfaction of temperature-induced increases in oxygen demand.

The carp presents a somewhat different pattern of response. This species possesses three major hemoglobin variants, with a fourth minor component occasionally reported. All major fractions are notably sensitive to reduced pH, increased temperature, ATP, chloride, calcium and magnesium (Gillen & Riggs, 1972; Weber & Lykkeboe, 1978). These findings appear to be compatible with the view(Pennelly et al, 1975) that in this species shifts between the R and T configurations, with uptake and release of oxygen, stem from the combined actions of a variety of modulating agencies rather than from any single predominant factor. Therefore, however, the hemoglobins of the carp, unlike those of the trout, have been regarded as a functionally-homogeneous group (Gillen & Riggs, 1972). The substantial changes in component abundancies observed following acclimation have therefore presented a puzzling anomaly. The recent suggestion (Weber & Lykkeboe, 1978) that alterations in the proportional representation of the different hemoglobin variants can effect significant changes in overall oxygen affinity consequently assumes considerable significance. The alterations in fractional abundancies encountered upon acclimation may well have substantial, though more subtle effects upon affinity than do those exhibited by the trout. In addition, the carp red cell undergoes several compositional changes with considerable potential effect upon affinity. It is to be presumed that temperature rises, while pH falls during acclimation, and that these changes probably lead to reductions in hemoglobin-oxygen affinity. In addition, chloride levels increase sharply, while those of magnesium decline. It is uncertain as yet whether ATP and/or GTP concentrations are altered during acclimation. Nevertheless, the picture in this metabolically-sluggish, but relatively eurythermal species is not unlike that of the trout. While different mechanisms apparently operate, the net result - establishment of circumstances under which oxygen delivery to tissues whose oxygen requirements have been increased is facilitated - appears to be much the same.

Erythrocytic Volume Change During Acclimation. Finally, some consideration should be given to the changes in mean erythrocytic volume which accompany acclimation. In both the carp and trout, response to increased temperature is associated with the accumulation of univalent ions within the red cell; chloride, and to a lesser degree sodium in the case of the carp, potassium in the rainbow trout. These are the predominant ions of the erythrocyte, and presumably account for much of the inorganic component of overall cell osmolarity. Increases in erythrocyte ion levels, in fact, exceed the changes encountered in plasma. Thus, in summer rainbow trout the ratio, Σ plasma electrolytes, mM l^{-1}: Σ red cell electrolytes, mM l^{-}, declined from 1.157 at 2°C to 1.091 and 1.052 at 10° and 18°C. Corresponding values in the winter series of trout were 1.077, 1.073 and 1.060. More pronounced decreases were exhibited by carp, with a reduction of ~20% occurring between 2° and 30°C, i.e., 2°C - 1.345; 16°C - 1.220; 30°C - 1.075. These observations suggest that the osmolarity of the red cell rises relative to that of plasma at higher temperatures and, assuming that erythrocyte and plasma are in osmotic equilibrium, increases in mean red cell volume would be anticipated. These would, however, be anti-adaptive for, as noted earlier, rate of oxygen combination with hemoglobin varies inversely with red cell volume (Holland, 1970) and increases in volume would presumably lead to reduced rates of oxygen loading at the gill surface. In fact, reductions in red cell volume represent a reasonably consistent feature of hematological response to increases in temperature (DeWilde & Houston, 1967; Houston & DeWilde, 1968), and are generally regarded as one of several means of compensation for reductions in branchial transit times.

Inorganic electrolytes have been widely implicated in cell volume regulation (MacKnight & Leaf, 1977), and it was therefore of interest to examine relationships between cell volume and plasma and cellular ion concentrations. Several such correlations proved to be significant

$$\text{(e.g., MEV vs rbc } [Cl^-],$$

$$\text{rbc } [\Sigma Na^+ + K^+ + Mg^{2+} + Ca^{2+}]$$

$$\text{rbc } [\text{cations} + Cl^-]$$

Those between mean erythrocytic volume and the sum of cations in trout and carp may be taken as representative, i.e.,

$$\text{Rainbow trout: MEV} = 351.7 - 0.60\ [\Sigma \text{ rbc Cations}]\ (r = -0.270,\ P < 0.05)$$

$$\text{Carp: MEV} = 436.0\ e^{-0.005}\ [\Sigma \text{ rbc Cations}]\ (r = -0.469,\ P < 0.01)$$

In this instance, as in all others in which significant correlations were obtained the relationship was negative in nature; increases in cell volume were associated with reductions in cellular ion content.

This apparent paradox suggests that carp and trout regulate red cell volume by a mechanism(s) similar to that employed by flounder, Pleuronectes flesus, following exposure to salinity changes altering plasma osmolarity (Fugelli, 1967; Fugelli & Zachariassen, 1976), i.e., adjustment of cellular organic solute concentrations. Volume-controlling systems of this type are common among euryhaline invertebrates, and frequently involve shifts of organic acids, usually non-essential amino acids (Florkin & Schoffeniels, 1969). In the instance of the flounder, taurine and gamma amino butyric acid account for much of the osmotically-acid organic solute transferred. Interestingly, in the nucleated duck erythrocyte volume control depends largely upon active potassium-passive chloride transfers (Kregenow, 1977), and in this sense resembles the situation in the mammalian red cell (MacKnight & Leaf, 1977) where inorganic ion shifts are utilized to adjust osmolarity. Such a response would, however, be inappropriate in the teleost if any reliance is to be placed upon affinity modulation through enhancement of chloride and/or potassium; an argument which might be loosely construed as supportive of this role. Some idea of the magnitude of the changes in solute content involved can be gained by comparison of the sums of plasma and red electrolytes. Teleostean plasmas are notably low in protein content (Feeney & Brown, 1974), and sums of electrolytes approximate osmolarities in many species (Holmes & Donaldson, 1969). Thus, if red cell and plasma are in osmotic equilibrium the difference between plasma and red cell ionic totals provide some indication of osmotically active cellular solute. It will be appreciated that such difference values are, at best, approximations and furthermore provide no indication as to whether the solute(s) involved is inorganic (e.g., HCO_3^-, P_i) or organic in nature. These restrictions notwithstanding, if cell volume is reduced by solute export, difference values should decrease at higher acclimation temperatures, and mean erythrocytic volume should be positively related to difference.

Both predictions are apparently satisfied. Difference values at 2°, 10° and 18°C were 46.8 ± 3.93, 26.3 ± 3.97 and 19.0 ± 3.44 mM in the summer series of trout. Corresponding values in winter fish were 25.0 ± 2.70, 19.6 ± 3.28 and 18.9 ± 4.23 mM. Those of carp were 67.7 ± 3.85, 48.7 ± 4.62 and 18.7 ± 4.86 mM at 2°, 16° and 30°C respectively. In both species mean erythrocytic volume was positively and significantly correlated with the difference between the sum of plasma and red cell ion levels, i.e.,

$$\text{Rainbow trout: MEV} = 227.9 + 0.759\,[\text{difference}]\ (r = 0.300,\ P < 0.05)$$

$$\text{Carp: MEV} = 190.2 + 0.619\,[\text{difference}]\ (r = 0.443,\ P < 0.05)$$

Thus, it appears that both have resolved the problem of enhancing ionic concentrations in a manner appropriate for reduction of hemoglobin-oxygen affinity, while simultaneously reducing red cell volume in a fashion which facilitates oxygen loading.

SUMMARY

The hematological response of the freshwater teleost fish to increased environmental temperature conditions, and the consequent elevation of oxygen requirements is complex in nature. Notable species differences are apparent, but among the more obvious adaptively appropriate features of the thermoacclimatory process the following can be included.

1) Increases in overall blood oxygen-carrying capacity are common, but not inevitable. These are normally expressed as elevations in

circulating hemoglobin content, and are ordinarily accompanied by some increase in red cell numbers. Minor increases in mean erythrocytic hemoglobin content have been observed in some instances. Erythrocytic volume typically declines at higher temperatures, possibly as a consequence of reductions in cell osmolarity through export of as yet unidentified, but probably organic solute or solutes.

2) Alterations in the hemoglobin system are a common accompaniment of acclimation, and there is some evidence at least, that these may take place rapidly enough to have physiological significance even in animals encountering diurnal fluctuations in temperature. In some species possessing functionally-heterogeneous hemoglobin types, thermal reorganization of the hemoglobin system appears to favour components whose transport characteristics are appropriate to the new conditions of temperature. In other species, exemplified by the carp, there is some evidence that alterations in the proportions of functionally-similar hemoglobin variants may lead to alterations in oxygen affinity. It is to be presumed, however, that in some cases at least, observed variations in the relative and/or absolute concentrations of specific hemoglobins will prove to be adaptively neutral.

3) The red cell milieu, the actual operating microenvironment of hemoglobin, also undergoes significant changes. The red cell must be in thermal equilibrium with the environment. It is to be presumed, therefore, that increases in erythrocytic temperature will prompt reductions in hemoglobin-oxygen affinity; at least among temperature-sensitive hemoglobin variants. This would also be the case with hemoglobin variants characterized by a significant Bohr effect. Increases in erythrocytic chloride content of greater-or-lesser magnitude are also reasonably typical of the warm-acclimated animal, and would be expected to reduce affinity. Reductions in magnesium (and calcium), when they accompany the acclimatory processes would, by enhancing effective ATP and/or GTP concentrations, be expected to have much the same effect. Increases in potassium relative to sodium levels should enhance the effects previously noted; always given that the hemoglobins involved are sensitive to these agents. The nature of acclimatory variations in the principal organophosphate modulators of hemoglobin-oxygen affinity remains unclear. Catostomid ATP:Hb ratios decrease during acclimation, whereas both ATP:Hb and GTP:Hb ratios were thermostable in the trout.

In general, then, the picture appears to be one in which most of the factors which have been considered are altered so as to enhance overall oxygen capacity and facilitate its release at the microcirculatory level. Individually, these variations are small in relation to temperature-induced increases in oxygen consumption. Collectively, however, they may have substantial physiological impact.

ACKNOWLEDGEMENTS

The author gratefully acknowledges the financial support provided by the National Science Foundation and the Natural Sciences and Engineering Research Council for various of the studies referred to in this review. In addition, several of the author's students, notably Mary Anne DeWilde, Jane Madden, Darlene Cyr, Robert Rupert, Lynn McCarty, James Henry, and foremost of all, Karen Mearow and John Smeda have made invaluable contributions to the work at various times.

REFERENCES

Anthony, E.H. (1961). The oxygen capacity of goldfish blood (Carassius auratus) in relation to thermal environment. J. Exp. Biol. 38: 93-107.

Arone, A. (1972). X-ray diffraction study of binding of 2,3 - dephosphoglycerate to human deoxyhaemoglobin. Nature (Lond.) 237: 146-149.

Bartlett, G.R. (1978a). Water soluble phosphates of fish red cells. Can. J. Zool. 56: 870-877.

Bartlett, G.R. (1978b). Phosphates in red cells of two South American osteoglossids: Arapaima gigas and Osteoglossum bicirrhosum. Can.J. Zool. 56: 878-881.

Bartlett, G.R. (1978c). Phosphates in red cells of two lungfish, the South American, Lepidosiren paradoxa, and the African, Protopterus aethiopicus. Can. J. Zool. 56: 882-886.

Beamish, F.W.H. (1964). Respiration of fishes with special emphasis on standard oxygen consumption. II. Influence of weight and temperature on respiration of several species. Can. J. Zool. 42: 177-188.

Benesch, R. & Benesch, R.E. (1967). The effect of organic phosphates from the human erythrocyte on the allosteric properties of hemoglobin. Biochem. Biophys. Res. Commun. 26: 162-167.

Benesch, R., Benesch, R.E. & Yu, C.I. (1968). Reciprocal binding of oxygen and diphosphoglycerate by human hemoglobin. Proc. Natn. Acad. U.S.A. 59: 526-532.

Benesch, R.E. & Benesch, R. (1974). The mechanism of interaction of red cell organic phosphates with hemoglobin. Adv. Protein Chem. 28: 211-237.

Black, E.C. (1953). Upper lethal temperatures of some British Columbia fresh water fish. J. Fish. Res. Board Can. 10: 196-210.

Bonaventura, C., Sullivan, B. & Bonaventura, J. (1976). Spot hemoglobin. Studies on the Root effect hemoglobin of a marine teleost. J. Biol. Chem. 251: 1871-1876.

Börjeson, H. (1977). Effects of hypercapnia on the buffer capacity and haematological values in Salmo Salar L. J. Fish. Biol. 11: 133-142.

Braman, J.C., Stalnaker, C.B., Farley, T.M. & Klar, G.T. (1977). Starch gel electrophoresis of rainbow trout, Salmo gairdneri, and cutthroat trout hemoglobins. Comp. Biochem. Physiol. 68: 435-437.

Brett, J.R. (1944). Some lethal temperature relations of Algonquin Parkfishes. Univ. Toronto Biol. Series, No. 52: Pub. Ont. Fish. Res. Lab. No. 63: 1-49.

Bridges, D.W., Cech, J.J. & Pedro, D.N. (1976). Seasonal hemotological changes in winter flounder, Pseudopleuronectes americanus. Trans. Am. Fish. Soc. 105: 596-600.

Brunori, M. (1975). Molecular adaptation to physiological requirements: the hemoglobin system of trout. Curr. Top. Cell. Regul. 9: 1-39.

Brunori, M., Bonaventura, J., Focesi, A., Galdames-Portus, M.I. & Wilson, M.T. (1979). Separation and characterization of the hemoglobin components of Pterygoplichthys pardalis, the Acaribodo. Comp. Biochem. Physiol. 62A: 173-178.

Bunn, H.F., Ransil, B.J. & Chao, A. (1971). The interaction between erythrocyte organic phosphates, magnesium ion and hemoglobin. J. Biol. Chem. 246: 5273-5279.

Cairns, J. (1956). Effects of heat on fish. Indust. Wastes 1: 180-183.

Calla, P.M. (1977). Volume regulation by flounder red blood cells: the role of the membrane potential. J. Exp. Zool. 199: 339-344.

Cameron, J.N. (1970). The influence of environmental variables on the hematology of pinfish (Lagodon rhomboides) and stripped mullet (Mugil cephalus). Comp. Biochem. Physiol. 32: 175-192.

Cameron, J.N. (1971). Oxygen dissociation characteristics of the blood of the rainbow trout, Salmo gairdneri. Comp. Biochem. Physiol. 38A: 699-704.

Cameron, J.N. (1973). Oxygen dissociation and content of blood from Alaskan burlot (Lota lota), pike (Esox lucius) and grayling (Thymallus arcticus). Comp. Biochem. Physiol. 46A: 491-496.

Cameron, J.N. (1974). Evidence for the lack of by-pass shunting in teleost gills. J. Fish. Res. Board Can. 31: 211-213.

Cameron, J.N. (1976). Branchial ion uptake in Arctic grayling: resting values and effects of acid-base disturbance. J. Exp. Biol. 64: 711-725.

Cameron, J.N. & Cech, J.J. (1970). Notes on the energy cost of gill ventilation in teleosts. Comp. Biochem. Physiol. 34: 447-455.

Cameron, J.N. & David, J.C. (1970). Gas exchange in rainbow trout (Salmo gairdneri) with varying blood oxygen capacity. J. Fish. Res. Board Can. 27: 1069-1085.

Cameron, J.N. & Randall, D.J. (1972). The effect of increased ambient CO_2 on arterial CO_2 tension, CO_2 content and pH in rainbow trout. J. Exp. Biol. 57: 673-680.

Catlett, R.H. & Millich, D.R. (1976). Intracellular and extracellular osmoregulation of temperature acclimated goldfish, Carassius auratus. Comp. Biochem. Physiol. 55A: 261-269.

Cavieres, J.D. (1977). The sodium pump in human red cells. In: Membrane Transport in Red Cells. Eds. J.C. Ellory & V.L. Law, p. 1-37, London, Academic Press.

Chanutin, A. & Curnish, R. (1967). Effect of organic and inorganic phosphate on the oxygen equilibrium of human erythrocytes. Arch. Biochem. Biophys. 121: 96-102.

Christensen, H.N. (1975). Biological Transport 2nd ed. Reading, Mass., W.A. Benjamin.

Dalmark, M. (1976). Chloride in the human erythrocyte. Prog. Biophys. Molec. Biol. 31: 145-164.

Davis, J.C. (1972). An infrared photographic technique useful for studying vascularization of fish gills. J. Fish. Res. Board Can. 29: 109-111.

DeBruin, S.H., Rollema, H.S., Janssen, L.H.M. & vanOs, G.A.J. (1974). The interaction of chloride ions with human hemoglobin. Biochem. Biophys. Res. Comm. 58: 210-215.

Dejours, P. (1973). Problems of control of breathing in fishes. In: Comparative Physiology. Eds. L. Bolis, K. Schmidt-Nielsen & S.H.P. Maddrell, p. 117-133, New York, Elsevier.

Dejours, P. & Armand, J. (1973). L'équilibre acide-base du sang chez la Carpe en fonction de la température. J. Physiol. Paris 67: 264A.

Denton, J.C. & Yousef, M.K. (1975). Seasonal changes in the hematology of rainbow trout, Salmo gairdneri. Comp. Biochem. Physiol. 51A: 151-153.

DeSmet, W.H.O. (1978). A comparison of the electrophoretic haemoglobin patterns of the vertebrates. Acta Zool. Pathol. Antverpiensia 70: 119-131.

DeWilde, A.H. & Houston, A.H. (1967). Haematological aspects of the thermo-acclimatory process in the rainbow trout. J. Fish. Res. Board Can. 24: 2267-2281.

Eaton, J.W. (1974). Oxygen affinity and environmental adaptation. Ann. N.Y. Acad. Sci. 241: 491-497.

Eddy, F.B. (1973). Oxygen dissociation curves of the blood of the tench, Tinca tinca. J. Exp. Biol. 58: 281-293.

Evans, D.H. (1969). Studies on the permeability to water of selected marine, freshwater and euryhaline teleosts. J. Exp. Biol. 50: 689-703.

Falkner, N.W. & Houston, A.H. (1966). Some haematological responses to sublethal thermal shock in the goldfish, Carassius auratus. J. Fish. Res. Board Can. 23: 1109-1120.

Feeney, R.E. & Brown, W.D. (1974). Plasma proteins in fishes. In: Chemical Zoology, Vol. VIII. Eds. M. Florkin & B.T. Scheer, p. 307-329 New York, Academic Press.

Fletcher, G.L. (1975). The effects of capture, 'stress' and storage of whole blood on the red blood cells, plasma proteins, glucose, and electrolytes of the winter flounder (Pseudopleuronectes americanus). Can. J. Zool. 53: 197-206.

Florkin, M. & Schoffeniels, E. (1969). Molecular Approaches to Ecology. London, Academic Press.

Fortes, P.A.G. (1977). Anion movements in red blood cells. In: Membrane Transport in Red Cells. Eds. J.C. Ellory & V.L. Lew, p. 175-195. London, Academic Press.

Fry, F.E.J. (1947). Effects of the environment on animal activity. University Toronto Stud., Biol. Ser., No. Pub. Cent. Fish. Res. Lab. 55: 5-62.

Fry, F.E.J., Brett, J.R. & Clawson, G.H. (1942). Lethal limits of temperature for young goldfish. Rev. Can. Biol. 1: 50-56.

Fry, F.E.J., Hart, J.S. & Walker, K.F. (1946). Lethal temperature relations for a sample of young speckled trout, Salvelinus fontinalis. Univ. Toronto Stud., Biol. Serv. No. 54, Pub. Ont. Fish. Res. Lab. 66: 9-35.

Fugelli, K. (1967). Regulation of cell volume in flounder (Pleuronectes flesus) erythrocytes accompanying a decrease in plasma osmolarity. Comp. Biochem. Physiol. 22: 253-260.

Fugelli, K. & Zachariassen, K.E. (1976). The distribution of taurine, gamma-amino butyric acid and inorganic ions between plasma and erythrocytes in flounder (Platichthys flesus) at different plasma osmolarities. Comp. Biochem. Physiol. 55A: 173-177.

Fyhn, U.E.H., Fyhn, H.J., Davis, B.J., Powers, D.A., Fink, W.L. & Garlick, R.L. (1979). Hemoglobin heterogeneity in Amazonian fishes. Comp. Biochem. Physiol. 62A: 39-66.

Garlick, R.L., Bunn, H.F., Fyhn, H.J., Fyhn, E.U.H., Martin, J.P., Noble, R.W. & Power, D.A. (1979). Functional studies on the separated hemoglobin components of an air-breathing catfish, Hoplosternum littorale (Hancock). Comp. Biochem. Physiol. 62A: 219-226.

Garside, E.R. & Jordan, C.M. (1968). Upper lethal temperature at various levels of salinity in the euryhaline cyprinodonts, Fundulus heteroclitus and F. diaphanus after isosmotic acclimation. J. Fish. Res. Board Can. 33: 36-42.

Gary-Bobo, C.M. & Solomon, A.K. (1968). Properties of hemoglobin solutions in red cells. J. Gen. Physiol. 52: 825-853.

Geoghegan, W.D. & Poluhowich, J.J. (1974). The major erythrocytic organic phosphates of the American eel, Anguilla rostrata. Comp. Biochem. Physiol. 49B: 281-290.

Gillen, R.G. & Riggs, A. (1972). Structure and function of the hemoglobin of the carp, Cyprinus carpio. J. Biol. Chem. 247: 6039-6046.

Gillen, R.G. & Riggs, A. (1977). The enhancement of the alkaline Bohr effect of some fish hemoglobins with adenosine triphosphate. Arch. Biochem. Biophys. 183: 678-685.

Greaney, G.S. & Powers, D.A. (1977). Cellular regulation of an allosteric modifier of fish hemoglobin. Nature (Lond.) 270: 73-74.

Greaney, G.S. & Powers, D.A. (1978). Allosteric modifiers of fish hemoglobins: in vitro and in vivo studies of the effect of ambient oxygen and pH or erythrocyte ATP concentrations. J. Exp. Zool. 203: 339-350.

Grigg, G.C. (1969). Temperature-induced changes in the oxygen equilibrium curve of the blood of the brown bullhead, Ictalurus nebulosus. Comp. Biochem. Physiol. 28: 1203-1223.

Grigg, G.C. (1974). Respiratory function of blood in fishes. In: Chemical Zoology, Vol. 1. Ed. N. Florkin & B.T. Scheer, p. 331-368. New York, Academic Press.

Hart, J.S. (1947). Lethal temperature relations of certain fish of the Toronto region. Trans. Roy. Soc. Can. 41: Series 3, pp. 57-70.

Hart, J.S. (1952). Geographic variations in some physiological and morphological characters in certain freshwater fish. Univ. Toronto Stud., Biol. Series, No. 60, Pub. Ont. Fish. Res. Lab. 72: 1-79.

Hattingh, J. & Van Pletzen, A.J.J. (1974). The influence of capture and transportation on some blood parameters of fresh water fish. Comp. Biochem. Physiol. 49A: 607-609.

Hayden, J.B., Cech, J.J. & Bridges, D.W. (1975). Blood oxygen dissociation characteristics of the winter flounder, Pseudopleuronectes americanus J. Fish. Res. Board Can. 32: 1539-1544.

Heath, A.G. (1973). Ventilatory responses of teleost fish to exercise and thermal stress. Am. Zool. 13: 491-503.

Heath, A.G. & Hughes, G.M. (1974). Cardiovascular and respiratory changes during heat stress in the rainbow trout (Salmo gairdneri). J. Exp. Biol. 59: 323-338.

Henderson, A.B. & Lee, J.C. (1976). Hemoglobin transition in erythrocytes of developing chick. Studies with cell-free protein-synthetizing systems. Arch. Biochem. Biophys. 174: 637-646.

Hevesey, G., Lockner, D. & Sletten, K. (1964). Iron metabolism and erythrocyte formation in fish. Acta Physiol. Scand. 60: 256-266.

Hockachka, P.W. & Somero, G.N. (1973). Strategies of Biochemical Adaptation. Philadelphia, Saunders.

Hol, W.G.J., van Duijen, P.T. & Berendsen, H.J.C. (1978). The α-helix dipole and the properties of proteins. Nature (Lond.) 273: 443-446.

Holeton, G.F. & Randall, D.J. (1967a). Changes in blood pressure in the rainbow trout during hypoxia. J. Exp. Biol. 46: 297-305.

Holeton, G.F. & Randall, D.J. (1967b). The effect of hypoxia upon the partial pressure of gases in the blood and water afferent and efferent to the gills of rainbow trout. J. Exp. Biol. 46: 317-327.

Holland, R.A.B. (1970). Factors determining the velocity of gas uptake by intracellular hemoglobin. In: Blood Oxygenation. Ed. D. Hershey, p. 1-23. New York, Plenum Press.

Holmes, W.N. & Donaldson, E.M. (1969). The body compartments and the distribution of electrolytes. In: Fish Physiology. Vol. I. Eds. W.S. Hoar & D.J. Randall, p. 1-89. New York, Academic Press.

Houston, A.H. (1973). Environmental temperature and the body fluid system of the teleost. In: Responses of Fish to Environmental Changes. Eds. W. Chavin, p. 87-162. Springfield, C.C. Thomas.

Houston, A.H. & Cyr, D. (1974). Thermoacclimatory variation in the haemoglobin systems of goldfish (Carassius auratus) and rainbow trout (Salmo gairdneri). J. Exp. Biol. 61: 445-446.

Houston, A.H. & DeWilde, M.A. (1968). Thermoacclimatory variations in the haematology of the common carp, Cyprinus carpio. J. Exp. Biol. 49: 71-81.

Houston, A.H. & DeWilde, M.A. (1969). Environmental temperature and the body fluid system of the freshwater teleost - III. Hematology and blood volume of thermally acclimated brook trout, Salvelinus fontinalis. Comp. Biochem. Physiol. 28: 877-895.

Houston, A.H., DeWilde, M.A. & Madden, J.A. (1969). Some physiological consequences of aortic catheterization in the brook trout, Salvelinus fontinalis. J. Fish. Res. Board Can. 26: 1847-1856.

Houston, A.H., Madden, J.A. & DeWilde, M.A. (1970). Environmental temperature and the body fluid system of the freshwater teleost - IV. Ionic regulation in the carp, Cyprinus carpio following thermal acclimation. Comp. Biochem. Physiol. 34: 805-818.

Houston, A.H., Madden, J.A., Woods, R.J. & Miles, H.M. (1971). Some effects of handling and tricaine methane sulphonate anesthetization upon the brook trout, Salvelinus fontinalis. J. Fish. Res. Board Can. 28: 625-631.

Houston, A.H. & McCarty, L.S. (1978). Carbonic anhydrase (acetazolamide-sensitive esterase) activity in the blood, gill and kidney of the thermally acclimated rainbow trout, Salmo gairdneri. J. Exp. Biol. 73: 15-27.

Houston, A.H. & Mearow, K.A. (1979a). Temperature-related changes in the erythrocytic carbonic anhydrase (acetazolamide-sensitive esterase) activity of goldfish, Carassius auratus. J. Exp. Biol. 78: 255-264.

Houston, A.H. & Mearow, K.M. (1979b). Temperature-related changes in the ionic composition and (HCO_3^-) - and (Na^+/K^+) - ATPase activities of the rainbow trout erythrocyte. J. Exp. Zool. (under review).

Houston, A.H., Mearow, K.M. & Smeda, J.S. (1976). Further observations on the hemoglobin systems of thermally-acclimated freshwater teleosts: pumpkinseed (Lepomis gibbosus), white sucker (Catostomus commersoni), carp (Cyprinus carpio), goldfish (Carassius auratus) and carp-goldfish hybrids. Comp. Biochem. Physiol. 54A: 267-273.

Houston, A.H. & Rupert, R. (1976). Immediate response of the hemoglobin system of the goldfish, Carassius auratus, to temperature change. Can. J. Zool. 54: 1737-1741.

Houston, A.H. & Smeda, J.S. (1979). Thermoacclimatory changes in the ionic microenvironment of haemoglobin in the stenothermal rainbow trout (Salmo gairdneri) and eurythermal carp (Cyprinus carpio). J. Exp. Biol. 80: 317-340.

Hughes, G.M. (1964). Fish respiratory homeostasis. In: Homeostasis and Feedback Mechanisms. Ed. G.M. Hughes, p. 81-107. Cambridge, The University Press.

Hughes, G.M. & Morgan, M. (1973). The structure of fish gills in relation to their respiratory function. Biol. Revs. 48: 419-475.

Hughes, G.M., O'Neill, J.G. & van Aardt, W.J. (1976). An electrolytic method for determining oxygen dissociation curves using small blood samples: the effect of temperature on trout and human blood. J. Exp. Biol. 65: 21-38.

Hughes, G.M. & Saunders, R.L. (1970). Response of the respiratory pumps to hypoxia in the rainbow trout, Salmo gairdneri. J. Exp. Biol. 53: 529-545.

Isaia, J. (1972). Comparative effects of temperature on the sodium and water permeabilities of the gills of a stenohaline freshwater fish (Carassius auratus) and a stenohaline marine fish (Serranus scriba, Serranus cabrilla). J. Exp. Biol. 57: 359-366.

Iuchi, I. (1973). Chemical and physiological properties of the larvae and adult haemoglobins in the rainbow trout, Salmo gairdneri irideus. Comp. Biochem. Physiol. 44B: 1087-1101.

Iuchi, I. & Yamagami, K. (1969). Electrophoretic pattern of larval haemoglobins of the salmonid fish, Salmo gairdneri irideus. Comp. Biochem. Physiol. 28: 977-979.

Janssen, R.G. & Randall, D.J. (1975). The effects of changes in pH and P_{CO_2} in blood and water on breathing in rainbow trout, Salmo gairdneri. Resp. Physiol. 25: 235-245.

Johansen, K. & Lenfant, C. (1972). A comparative approach to the adaptability of O_2-Hb affinity. In: Oxygen Affinity of Hemoglobin and Red Cell Acid-Base Status. Eds. P. Astrup & M. Rorth. Copenhagen, Academic Press.

Johansen, K. & Weber, R.E. (1976). On the adaptability of haemoglobin function to environmental conditions. In Perspectives in Experimental Biology. Ed. P. Spencer Davies, p. 212-234. Oxford, Pergamon Press.

Johansson-Sjöbeck, M.-L., Dave, G., Larsson, Å., Lewander, K. & Lidman, U. (1975). Metabolic and hematological effects of starvation in the European eel, Anguilla anguilla L. II. Hematology. Comp. Biochem. Physiol. 52A: 431-434.

Jones, D.R. (1971). Theoretical analysis of factors which may limit the maximum oxygen uptake of fish: the oxygen cost of the cardiac and branchial pumps. J. Theoret. Biol. 32: 341-349.

Kamra, S.K. (1966). Effect of starvation and refeeding on some liver and blood constituents of Atlantic cod (Gadus morhua). J. Fish. Res. Board Can. 23: 975-982.

Kaya, C.M. (1978). Thermal resistance of rainbow trout from a permanently heated stream, and of two hatchery strains. Prog. Fish-Cult. 40: 138-142.

Kregenow, F.M. (1977). Transport in avian red cells. In: Membrane Transport in Red Cells. Eds. J.C. Ellory & V.L. Lew, p. 383-426.

Laver, M.B., Jackson, E., Scherperel, M., Tung, C., Tung, W. & Radford, E.P. (1977). Hemoglobin-O_2 affinity regulation: DPG, monovalent anions, and hemoglobin concentration. J. Appl. Physiol., Resp. Environ. Exercise Physiol. 43: 632-642.

Linn, D.W. (1965). Take a one gallon jar... Prog. Fish-Cult. 27: 147-152.

Lloyd, R. & Orr, L.D. (1969). The diuretic response of rainbow trout to sublethal concentrations of ammonia. Water Res. 3: 335-344.

Lykkeboe, G., Johanson, K. & Maloiy, G.M.O. (1975). Functional properties of hemoglobins in the teleost, Tilapia grahami. J. Comp. Physiol. 104: 1-11.

Mackay, W.C. (1974). Effect of temperature on osmotic and ionic regulation in goldfish, Carassius auratus. J. Comp. Physiol. 88: 1-9.

Mackay, W.C. & Beatty, D.D. (1968). The effect of temperature on renal function in the white sucker fish, Catostomus commersoni. Comp. Biochem. Physiol. 26: 235-245.

MacKnight, A.D. & Leaf, A. (1977). Regulation of cellular volume. Physiol. Revs. 57: 510-573.

Maetz, J. (1972). Branchial sodium exchange and ammonia excretion in the goldfish, Carassius auratus. Effect of ammonia-loading and temperature changes. J. Exp. Biol. 56: 601-620.

Manwell, C. (1957). Alkaline denaturation of hemoglobin of postlarval and adult Scorpaenichthys marmoratus. Science 126: 1175-1176.

Manwell, C. & Baker, C.M.A. (1970). Molecular Biology and the Origin of Species. Seattle, Univers., Washington Press.

McCarty, L.S. & Houston, A.H. (1977). Na^{+}:K^{+}- and HCO_3^- - stimulated ATPase activities in the gills and kidneys of thermally acclimated rainbow trout, Salmo gairdneri. Can. J. Zool. 55: 704-712.

Mied, P. & Powers, D.A. (1978). The hemoglobins of the killifish, Fundulus heteroclitus: separation, characterization and a model for the subunit composition. J. Biol. Chem. 253: 3521-3528.

Motais, R. & Isaia, J. (1972). Temperature dependence of permeability to water and to sodium of the gill epithelium of the eel, Anguilla anguilla. J. Exp. Biol. 56: 587-600.

Munroe, V. & Poluhowich, J.J. (1974). Ionic composition of the plasma and whole blood of marine and freshwater eels, Anguilla rostrata. Comp. Biochem. Physiol. 49A: 541-544.

Murphy, P.G.F. & Houston, A.H. (1977). Temperature, photoperiod and water-electrolyte balance in the rainbow trout, Salmo gairdneri. Can. J. Zool. 55: 1377-1388.

Natelson, S. & Natelson, E.A. (1978). Principles of Applied Clinical Chemistry. Vol. 2. The Erythrocyte: Chemical Composition and Metabolism. New York, Plenum Press.

Neil, W.H., Strawn, K. & Dunn, J.E. (1966). Heat resistance experiments with the longear sunfish, Lepomis megalotis. Arkansas Acad. Sci. Proc. 20: 39-49.

Oikari, A.& Soivio, A. (1975). Influence of sampling methods and anaesthetization on various haematological parameters of several teleosts. Aquaculture 6: 171-180.

Pennelly, R.R., Tan-Wilson, A.L. & Noble, R.W. (1975). Structural states and transitions of carp hemoglobin. J. Biol. Chem. 250: 7239-7244.

Perez, J.E. & Maclean, N. (1974). Ontogenetic changes in roach, Rutilus rutilus (L.) and rudd, Scardinius erythrophthalmus (L.). J. Fish Biol. 6: 479-482.

Perez, J.E. & Maclean, N. (1975a). Multiple globins and haemoglobins in four species of grey mullets (Mugilidae, Teleosta). Comp. Biochem. Physiol. 538: 465-468.

Perez, J.E. & Maclean, N. (1975b). Multiple globins and haemoglobins in the bass, Dicentrarchus labrax (L.) (Serranidae, Teleosta). J. Fish Biol. 8: 413-417.

Perez, J.E. & Maclean, N. (1976). The haemoglobins of the fish, Sarotherodon mossambicus (Peters): functional significance and ontogenetic changes. J. Fish Biol. 9: 447-455.

Peterson, A.J. & Poluhowich, J.J. (1976). The effects of organic phosphates on the oxygenation behaviour of eel multiple hemoglobins. Comp. Biochem. Physiol. 55A: 351-354.

Powers, D.A. (1972). Hemoglobin adaptation for fast and slow water habitats in sympatric catostomid fishes. Science 117: 360-362.

Powers, D.A. (1974). Structure, function and molecular ecology fo fish hemoglobins. Ann. N.Y. Acad. Sci. 241: 472-490.

Powers, D.A., Martin, J.P., Garlick, R.L., Fyhn, H.J. & Fyhn, U.E.H. (1979). The effect of temperature on the oxygen equilibria of fish hemoglobins in relation to environmental thermal variability. Comp. Biochem. Physiol. 62A: 87-94.

Rahn,H. (1966). Aquatic gas exchange: theory. Resp. Physiol. 1: 1-12.

Randall, D.J. (1970). Gas exchange in fish. In: Fish Physiology, Vol. IV. Eds. W.S. Hoar & D.J. Randall, p. 253-292. New York, Academic Press.

Randall, D.J., Baumgarten, D. & Malyusz, M. (1972). The relationship between gas and ion transfer across the gills of fishes. Comp. Biochem. Physiol. 41A: 629-637.

Randall, D.J. & Cameron, J.D. (1973). Respiratory control of arterial pH as temperature changes in rainbow trout, Salmo gairdneri. Am. J. Physiol. 225: 997-1002.

Randall, D.J., Holeton, G.F. & Stevens, E.D. (1967). The exchange of oxygen and carbon dioxide across the gills of rainbow trout. J. Exp. Biol. 46: 339-348.

Reeves, R.B. (1977). The interaction of body temperature and acid-base balance in ectothermic vertebrates. Ann. Revs. Physiol. 39: 550-586.

Richards, B.D. & Fromm, P.O. (1969). Patterns of blood flow through filaments and lamellae of isolated-perfused rainbow trout (Salmo gairdneri) gills. Comp. Biochem. Physiol. 29: 1063-1070.

Riggs, A. (1970). Properties of fish hemoglobins. In: Fish Physiology, Vol. 4. Eds. W.S. Hoar & D.J. Randall, p. 209-252. New York, Academic Press.

Riggs, A. (1976). Factors in the evolution of hemoglobin function. Fed. Proc. Fed. Am. Soc. Exp. Biol. 35: 2115-2118.

Rollema, H.S., DeBruin, S.H., Janssen, L.H.M. & van Os, G.A.J. (1975). The effect of potassium chloride on the Bohr effect of human hemoglobin. J. Biol. Chem. 250: 1333-1339.

Ronald, A.P. & Tsuyuki, H. (1971). The subunit structures and the molecular basis of the multiple hemoglobins of two species of trout, Salmo gairdneri and S. clarki clariki. Comp. Biochem. Physiol. 39B: 195-202.

Rossi-Fanelli, A., Antonini, E. & Caputo, A. (1961). Studies on the relations between molecular and functional properties of hemoglobin. II. The effect of salts on the oxygen equilibrium of human hemoglobin. J. Biol. Chem. 236: 397-401.

Sharp, G.D. (1973). An electrophoretic study of hemoglobins of some scombroid fishes and related forms. Comp. Biochem. Physiol. 44B: 381-388.

Slicher, A.M. & Pickford, G.E. (1968). Temperature-controlled stimulation of hemopoesis in a hypophysectomized cyprinodont fish, Fundulus heteroclitus. Physiol. Zool. 41: 293-297.

Smeda, J.S. & Houston, A.H. (1979). Carbonic anhydrase (acetazolamide-sensitive esterase) in the red blood cells of thermally-acclimated rainbow trout, Salmo gairdneri. Comp. Biochem. PHysiol. 62A: 719-723.

Smeda, J.S. & Houston, A.H. (1979). Evidence of weight-dependent differential hematological response to increased environmental temperature by carp, Cyprinus carpio. Env. Biol. Fish. 4: 89-92.

Smith, C.E. (1968). Hematological changes in coho salmon fed a folic acid deficient diet. J. Fish. Res. Board Can. 25: 151-156.

Smith, J.C. (1977). Body weight and the haematology of the American plaice, Hippoglossoides platessoides. J. Exp. Biol. 67: 17-28.

Soivio, A. & Okari, A. (1976). Haematological effects of stress on a teleost, Esox lucius (L.). J. Fish Biol. 8: 397-441.

Spoor, W.A. (1951). Temperature and the erythrocyte count of goldfish. Fed. Proc., Fed. Am. Soc. Exp. Biol. 10: 131.

Steen, J.B. & Kruysse (1964). The respiratory function of teleostean gills. Comp. Biochem. Physiol. 12: 127-142.

Stevens, E.D. & Randall, D.J. (1967). Changes of gas concentrations in blood and water during moderate swimming activity in rainbow trout. J. Exp. Biol. 46: 329-337.

Tan, A.L., De Young, A. & Noble, R.W. (1972). The pH dependence of the affinity, kinetics, and cooperativity of ligand binding to carp hemoglobin, Cyprinus carpio. J. Biol. Chem. 247: 2493-2498.

Tan, A.L. & Noble, R.W. (1973). The effect of inosital hexaphosphate on the allosteric properties of carp hemoglobin. J. Biol. Chem. 24B: 7412-7416.

Taylor, W., Houston, A.H. & Horgan, J.D. (1968). Development of a computer model simulating some aspects of the cardiovascular-respiratory dynamics of the salmonid fish. J. Exp. Biol. 49: 477-494.

Torracca, A.M.V., Raschetti, R., Salvioli, R., Ricciardi, G. & Winterhalter, K.H. (1977). Modulation of the Root effect in goldfish by ATP and GTP. Biochim. Biophys. Acta 496: 367-373.

Tsuyuki, H. & Gadd, R.E.A. (1963). The multiple hemoglobins of some members of the Salmonidae family. Biochim. Biophys. Acta 71: 219-221.

VanVuren, J.H.J. & Hattingh, J. (1978). Seasonal changes in the haemoglobins of freshwater fish in their natural environment. Comp. Biochem. Physiol. 60: 265-268.

Vanstone, W.E., Roberts, E. & Tsuyuki, H. (1964). Changes in the multiple hemoglobin patterns of some Pacific salmon, genus Oncorhynchus during the parr-smolt transformation. Can. J. Physiol. Pharmacol. 42: 697-703.

Weber, R.E. & Lykkeboe, G. (1978). Respiratory adaptations in carp blood: influences of hypoxia, red cell organic phosphates, divalent cations and CO_2 on hemoglobin-oxygen affinity. J. Comp. Physiol. 128: 127-137.

Weber, R.E., Lykkeboe, G. & Johansen, K. (1976). Physiological properties of eel haemoglobin: Hypoxic acclimation, phosphate effects and multiplicity. J. Exp. Biol. 64: 75-88.

Weber, R.E., Wood, S.C. & Lomholt, J.P. (1976). Temperature acclimation and oxygen-binding properties of blood and multiple haemoglobins of rainbow trout. J. Exp. Biol. 65: 333-345.

Weinberg, S.R., LoBue, J., Siegel, C.D. & Gordon, A.S. (1976). Hematopoesis of the kissing gourami (Helostoma temmincki). Effects of starvation, bleeding and plasma-stimulating factors on its erythropoiesis. Can. J. Zool. 54: 1115-1127.

Weinberg, S.R., Siegel, C.D. & Gordon, A.L. (1973). Studies on the peripheral blood cell parameters and morphology of the red paradise fish, Macropodus opercularis. Effect of food deprivation on erythropoesis. Anat. Rec. 175: 7-14.

Wilkins, N.P. & Iles, T.D. (1966). Haemoglobin polymorphism and its ontogeny in herring (Clupea harengus) and sprat (Sprattus sprattus) Comp. Biochem. Physiol. 17: 1141-1158.

Wood, S.C. & Johansen, K. (1972). Adaptation to hypoxia by increased HbO_2 affinity and decreased red cell ATP concentration. Nature, New Biol. 237: 278-279.

Wood, S.C., Johansen, K. & Weber, R.E. (1975). Effects of ambient P_{O_2} on hemoglobin-oxygen affinity and red cell ATP concentrations in a benthic fish, Pleuronectes platessa. Resp. Physiol. 25: 259-267.

Wyman, J., Gill, S.J., Noll, L., Giardina, B., Colosimo, A. & Brunnori, M. (1977). The balance sheet of a hemoglobin: thermodynamics of CO binding by hemoglobin trout. J. Molec. Biol. 109: 195-208.

Yamanaka, H., Yamaguchi, K. & Matsuura, F. (1965). Starch gel electrophoresis of fish hemoglobins - II. Electrophoretic patterns of various species. Bull. Jap. Soc. Sci. Fish. 31: 833-839.

Yoshiyasu, K. (1973). Starch-gel electrophoresis of hemoglobins of freshwater salmonid fishes in northeast Japan. Bull. Jap. Soc. Sci. Fish. 39: 449-459.

ACQUISITION OF ENERGY BY TELEOSTS:
ADAPTIVE MECHANISMS AND EVOLUTIONARY PATTERNS

Karel F. Liem

Museum of Comparative Zoology, Harvard University

Cambridge, Massachusetts 02138, USA

INTRODUCTION

Organisms are self-regulating highly integrated systems capable of autoduplication (Waterman, 1967). As such they must acquire energy from the environment permitting growth, locomotion, reproduction, and all other biological processes. Although acquisition, transfer and utilization of energy in fishes have been studied rather extensively, more emphasis has been placed on the latter two. Because environmental physiologists are mainly concerned with the physiological adaptations to various environmental parameters, it seems appropriate to focus on the proximate interactions between the fish and trophic resources at the level of acquisition. Transfer and utilization of energy are beyond the scope of this A.S.I. and will not be discussed.

It has been proposed that evolution has maximized the exploitation and control of trophic energy (Van Valen, 1967). Trophic energy is considered the determining factor among competing strategies of adaptation, because natural selection maximizes expansive energy, which is the energy used for growth and reproduction. The phylogenetic and evolutionary patterns of the actinopterygian fishes (Liem and Lauder, 1980) have been derived mostly from data of the feeding apparatus. Ever since the publications of Alexander (1967, 1969, 1970), Osse (1969), and Liem (1970), we are witnessing a rapid proliferation of papers dealing with the experimental analysis of the feeding apparatus of teleosts (Ballintijn et al., 1972; Elshoud-Oldenhave and Osse, 1976; Lauder 1979a; Liem, 1978; Nyberg, 1971). Aside from reducing the speculative element, which has dominated earlier efforts for so long, the vast body of experimental data represents a scientifically sound foundation upon which can be formulated: (a) more precise general theories on teleostean feeding; (b) kinematic models that are predictive and contribute to the concept adaptation; (c) hypotheses on evolutionary innovations and their effects on the operational width of food exploitation; (d) more meaningful statements on the limits of optimization and the nature of the limiting factors. By providing this review I hope not only to emphasize the recent advances, but also to identify those areas that need further inquiry, testing and expansion.

INERTIAL SUCTION: THE DOMINANT TELEOSTEAN MODE OF FEEDING

Functionally teleostean feeding can be grouped into three major categories: (1) Inertial suction, is employed by the vast majority of adult teleosts and by all teleosts sometime during their developmental history. Because this mode is virtually universal, it will be the major topic of this paper; (2) Ram feeding, in which the fish simply overtakes the food or prey by vigorous forward swimming with opened mouth and opercula. This mode is especially evident in those taxa (e.g. Scombridae) employing ram-jet ventilation as a respiratory mechanism; (3) Manipulation, which covers a broad range of feeding behaviors involving actual use of true or dermal teeth of the upper and lower jaws (e.g. clipping, gripping, rasping, scraping, biting). Of course it cannot be overemphasized that many teleosts can and do use a combination of the three modes, depending on the nature, position and behavior of the prey or food.

Empirical Data on Inertial Suction: Alexander (1969, 1970) has measured buccal cavity pressures in a wide variety of teleosts (Fig. 1: a-j). He found wide variations in the negative pressures ranging from 80-400 cm H_2O. However, the pressure profiles seem to be rather uniform. An initial sharp negative pressure is followed by a much smaller positive pulse ranging from 1-9 cm H_2O. Liem (1978) measured the buccal pressure in piscivorous cichlids by a chronically implanted cannula. With this method a variably occurring positive pressure has been recorded during the preparatory phase (Fig. 2: d), which is immediately followed by a sharp negative pressure peaking around 800 cm H_2O. The positive pressure following the pressure drop seemed to be either absent or minimal. Recently Lauder and Lanyon (1980) have measured the pressure in both the opercular and buccal cavity in *Lepomis macrochirus* (Fig. 2: a-c) using chronically implanted cannulae. For the opercular cavity Lauder has recorded peak negative pressures from 10-145 cm H_2O depending on the prey item. Positive pressure pulses ranged from 5-50 cm H_2O and were usually smaller than the preceeding negative pressure but, occasionally, were nearly equal in size. Lauder (personal communication and Fig. 2: a, b) has also found a wide range of variations in the buccal pressure. A preparatory positive pulse was rarely encountered, and the negative pressure down to a peak

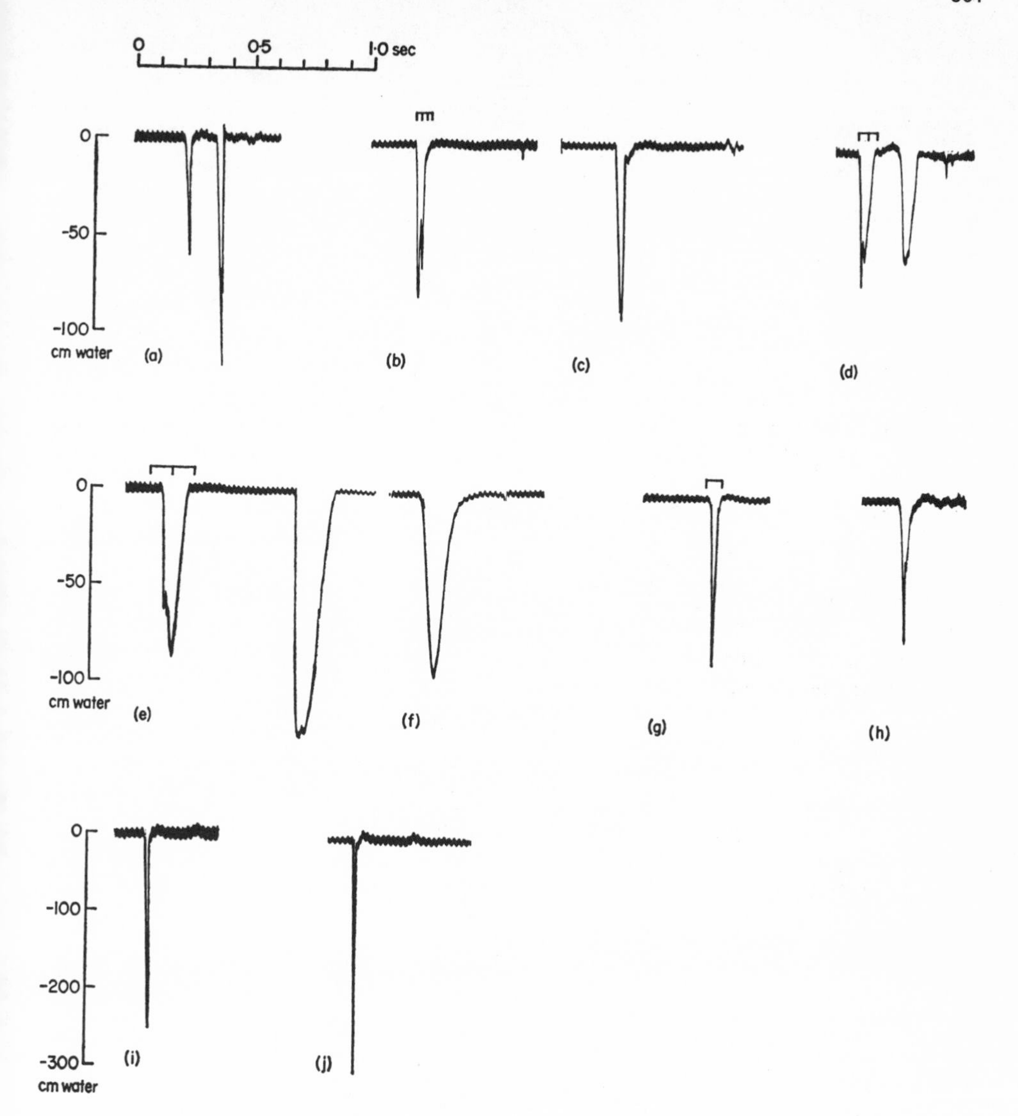

Fig. 1. Pressure records of suction action by several teleosts (From Alexander, 1970, courtesy of The Zoological Society of London). (a) Papyrocranus: (b), (c) Carassius; (d) Ictalurus; (e), (f) Anguilla; (g) Taurulus; (h) Blennius; (i) Macropodus; (j) Pterophyllum.

of 200 cm H_2O was mostly followed by a positive pressure pulse or even oscillations (Fig. 2: b), although, occasionally such a positive pressure does not occur (Fig. 2: a).

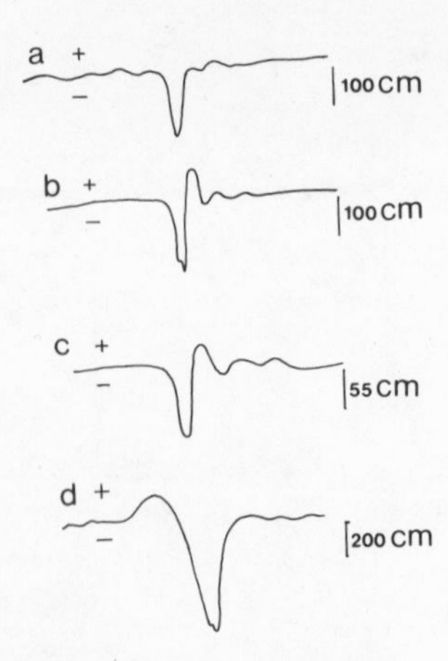

Fig. 2. Intra-oral pressure records obtained from chronically implanted polyethylene cannulae (a, b, c) and pressure record of the opercular cavity (c, courtesy of G.V. Lauder). a, b, intra-oral pressure record of Lepomis macrochirus during feeding showing variations within an individual (courtesy of G.V. Lauder); (c) pressure record of the opercular cavity of Lepomis macrochirus during feeding (courtesy of G.V. Lauder); d, pressure record of the buccal cavity in Serranochromis robustus during the capture of prey.

It is evident that the amount of theory and modelling on suction in teleosts far exceeds the empirical data. The picture emerging from the admittedly meagre body of empirical data is that the pressure profile is not an all-or-none physiological phenomenon determined by principles of optimal anatomical design. Suction is apparently finely regulated in respect to negative pressure, velocity, and perhaps, direction.

Biomechanical Basis of Inertial Suction: Prey capture by inertial suction in advanced teleosts may be divided into preparatory phase followed by expansive and compressive phases. The preparatory phase is not always present, but when it occurs, the adductor arcus palatini, adductor mandibulae and geniohyoideus muscles (Fig. 3, 4: AAP, AM_3, GHA, GHP) become active before mouth opening. In this way the buccal cavity is compressed to a small volume, to increase the volume in the next expansive phase.

Mouth opening (expansive phase) is started by activity of the levator operculi and sternohyoideus muscles (Fig. 3, 4: LO, SH). The levator operculi depresses the mandible by rotating the opercular series caudodorsally and transmitting a posterodorsal force to the

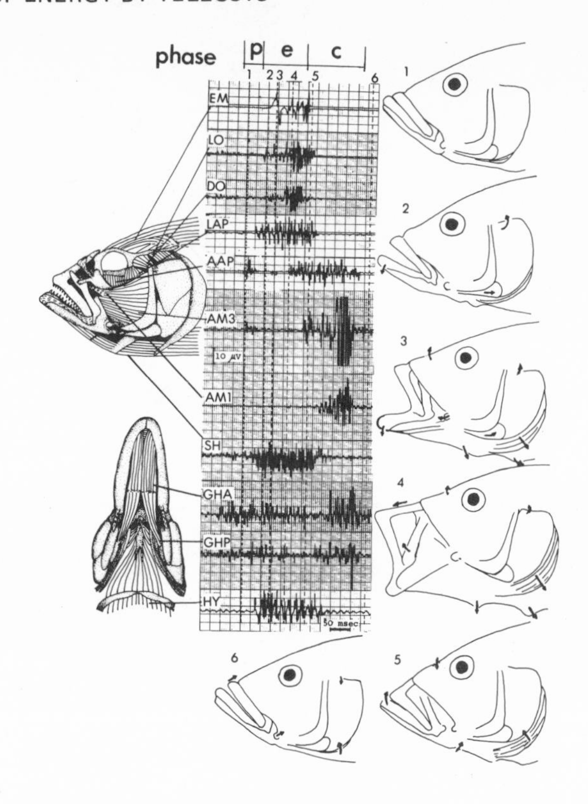

Fig. 3. On left, lateral and ventral aspects of the cephalic musculature of Serranochromis robustus is depicted. In the center, representative myograms taken during capture of prey. Surrounding the myograms are tracings of frames of a high speed motion picture. Frame numbers (1-6) accompanying the tracings correspond with the numbers indicated at the top of the myograms. The three phases are the preparatory (p), expansive (e), and compressive (c). Major movements of the cephalic components between successive frames are indicated by arrows. Abbreviations: AAP, adductor arcus palatini; AM_1, AM_3, adductor mandibulae parts A_1 and A_3; c, compressive phase; DO, dilatator operculi; e, expansive phase; EM, epaxial muscles; GHA, geniohyoideus anterior; GHP geniohyoideus posterior; HY, hypaxial muscles; LAP, levator arcus palatini; LO, levator operculi; p, preparatory phase; SH, sternohyoideus. (From Liem, 1978; courtesy of the Wistar Institute Press).

retroarticular process of the lower jaw by means of the interoperculomandibular ligament (Fig. 3, 4). A second mouth opening mechanism begins after the onset of activity: it is the sternohyoideus

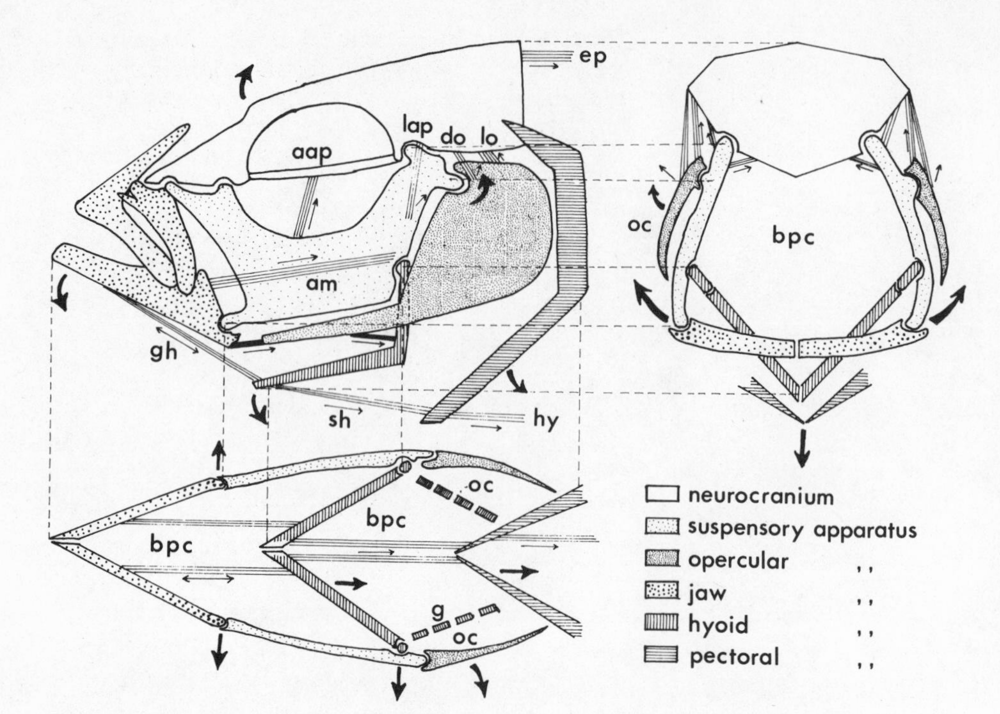

Fig. 4. Diagram of the major mechanical units, couplings and ligaments of the head of a piscivorous cichlid. Lateral, ventral and frontal views. Muscles and the principal direction of their forces are indicated respectively by light lines and light arrows. Heavy arrows depict major movements during the expansive phase of the feeding cycle. (From Liem, 1978; courtesy of the Wistar Institute Press).

muscle (Fig. 3, 4: SH) which depresses the lower jaw by pulling the hyoid posterodorsally. Posterodorsal hyoid movement is transmitted to the interoperculum by the interoperculohyoid ligament (Fig. 4, 11: ihl, iop). Both systems, which I have called, respectively, the levator operculi and sternohyoideus couplings (Liem, 1970), are summated since both depend on the interoperculum and interoperculomandibular ligament as a vehicle for transmission of force. Other muscles active during the expansive phase are the epaxial, levator arcus palatini, sternohyoideus, hypaxial and dilatator operculi muscles (Fig. 3: EM, LAP, SH, HY, DO). The expansive phase consists of a rapid mouth opening immediately followed by an explosive expansion of the buccal cavity by lifting of the head, depression of the floor and lateral expansion of the head (Fig. 3: 4). All these events contribute to a sudden reduction in pressure in the buccal cavity which creates a water current into the mouth drawing the prey in. The water then passes out over the gills in the compressive phase and out to the compressive phase

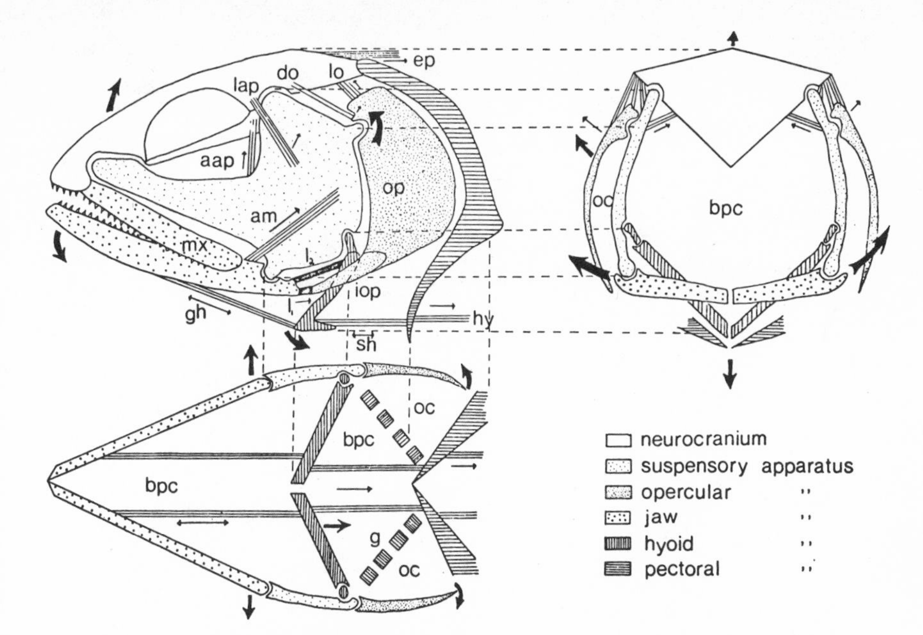

Fig. 5. Diagrammatic model of the feeding apparatus in primitive teleost fishes. The major mechanical units of the jaw are shown in different patterns (see key). Heavy arrows indicate the major movements of the bony elements during mouth opening, light arrows indicate the major action of the muscles. Abbreviations: aap, adductor arcus palatini; am, adductor mandibulae; bpc, buccopharyngeal cavity; do, dilatator operculi; ep, epaxial muscles; g, gills, gh; geniohyoideus; hy, hypaxial muscles; iop, interoperculum; l_1, interoperculomandibular ligament; l_2, mandibulohyoid ligament; lap, levator arcus palatini; lo, levator operculi; mx, maxilla; oc, opercular chamber; op, operculum; sh, sternohyoideus (modified from Lauder and Liem, 1980).

and out to the outside between the operculum and the pectoral girdle when the branchiostegal apparatus is opened. The compressive phase is accomplished by activity of the adductor mandibulae (Fig. 3: AM_1, AM_3), adductor arcus palatini and geniohyoideus muscles (AAP, GHA, GHP). In advanced teleosts the upper jaw is protrusible (Fig. 3: 3, 4). Its symphysis moves forward as the mandible is lowered, thereby creating a circular mouth opening, enlarging the volume of the buccal cavity and directing the suction.

The biomechanical pattern in primitive teleosts deviates from that of advanced teleosts in several salient features (Lauder and Liem, 1980). Anatomically, the sternohyoideus coupling is different since posterodorsal movement of the hyoid is transmitted directly to the mandible via the mandibulohyoid ligament (Fig. 5: l_2). Thus the

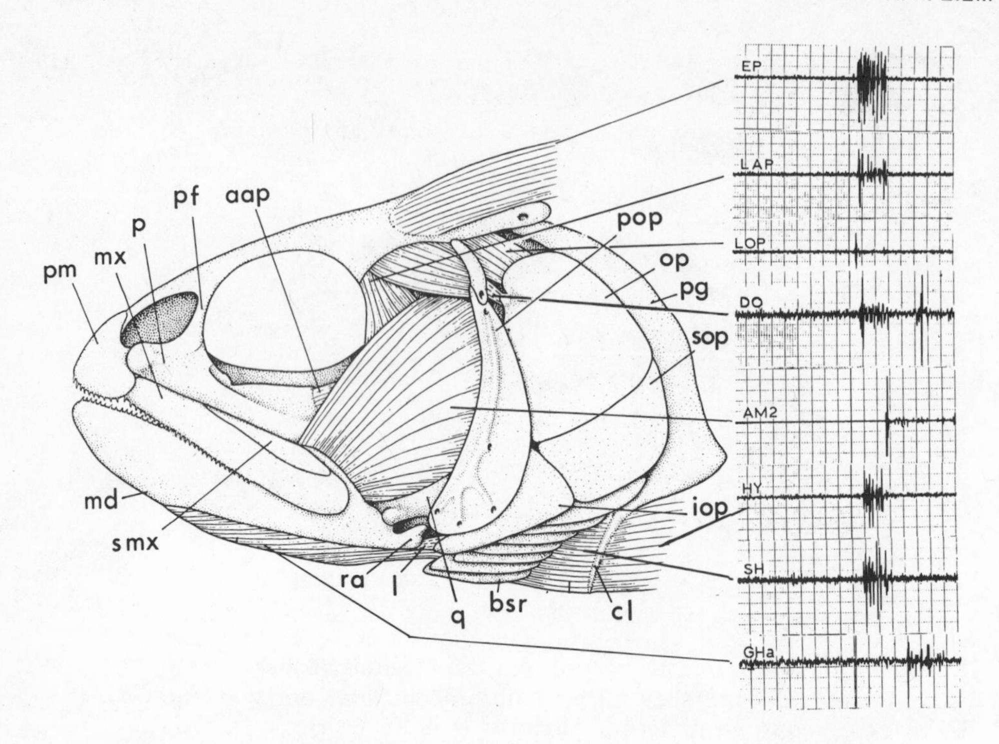

Fig. 6. On left, the left lateral view of the superficial cephalic muscles of Savelinus fontinalis. On right electromyographic recordings of muscle activity of 8 cranial muscles during the strike at the prey. Abbreviations: aap, adductor arcus palatini; AM_2, A_2 part of adductor mandibulae; bsr, branchiostegal rays; cl, cleithrum; DO, dilatator operculi; EP, epaxial muscles; GH_a, geniohyoideus anterior; iop, interoperculum; HY, hypaxial muscles; l, interoperculomandibular ligament; LAP, levator arcus palatini; LO, levator operculi; md, mandible; mx, maxilla; op, operculum; p, palatine; pf, prefrontal; pg, pectoral girdle; pm, premaxilla; pop, preoperculum; q, quadrate; ra, retroarticular process; SH, sternohyoideus; smx, supramaxilla; sop, suboperculum (modified from Lauder and Liem, 1980).

interoperculum is not used as an intermediate element in the transmission of force in Salvelinus. Furthermore, the upper jaw of primitive teleosts is not prostrusible, although the maxilla does swing forward using its attachment to the neurocranium as a pivot (Fig. 5) and using the combination of suspensoral and neurocranial movements as the driving force to cause rotation. Maxillary swing in primitive teleosts is important since it brings the buccal chamber forward towards the prey, and increases the velocity of flow (Lauder, 1979a). Primitive teleosts lack a typical preparatory phase (Fig. 6). Thus prey

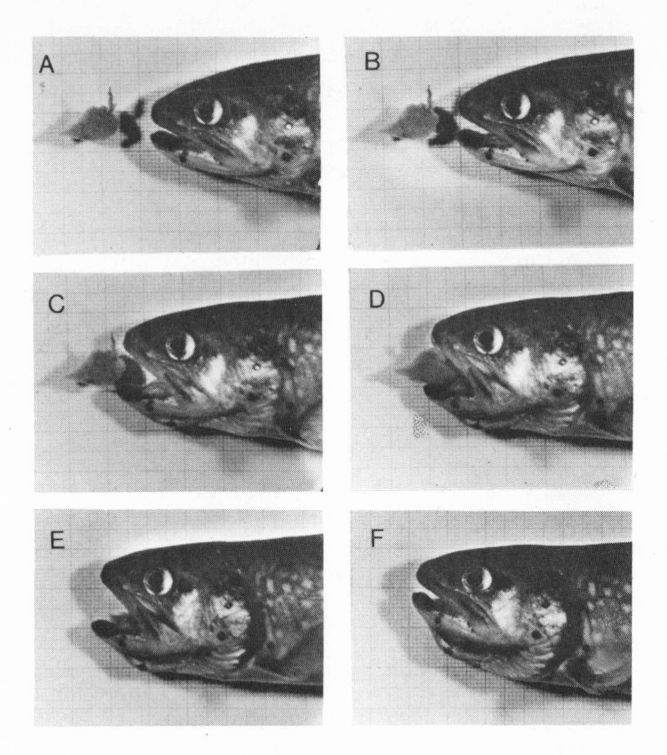

Fig. 7. Selective prints of a high speed (200 frame sec^{-1}) motion picture of feeding Salvelinus fontinalis (17.8 cm standard length). Time between successive frames is 5 m sec.

capture begins with the expansive phase during which the following events occur (Fig. 7: A-F): (1) The mouth begins to open, and hyoid depression starts. (2) As the mouth continues to open, the maxilla swings anteriorly and the hyoid is depressed further. At this time suspensory abduction and opercular dilation begin. (3) Mouth opening, maxillary swing, and hyoid depression reach a near simultaneous maximum while suspensorial abduction and opercular dilation reach their maximum value somewhat later. At this time the pressure in the buccal cavity drops to a minimum. The compressive phase starts when the mouth closes, however, in sharp contrast to the pattern in advanced teleosts, the neurocranium maintains an elevated position, the hyoid remains in a relatively depressed condition, and the suspensorium and operculum remain abducted. After a delay, the hyoid is retracted and the suspensorium and operculum slowly return to their initial positions. The kinematic and electromyographic profiles (Figs. 6, 7) clearly indicate the existence of a levator operculi coupling. Action of the levator operculi (Fig. 5, 6: LO) rotates the opercular apparatus posterodorsally. This movement is transmitted to the retroarticular process of the mandible via the interoperculum and interoperculomandibular ligament (Fig. 5: LIM). Although delayed, the sternohyoideus does play a role in lowering the mandible via the mandibulohyoid ligament (Fig. 5: l_2). Thus the linkage of the sternohyoideus coupling in primitive teleosts involves fewer elements resulting in a more direct transmission of force than in higher teleosts.

Although differences separate the biomechanical patterns of inertial suction of primitive teleosts from those of advanced ones, in both groups the levator operculi coupling plays a key role in depressing the mandible.

BIOMECHANICAL MODELS AND OPTIMIZATION

On the basis of the above mentioned empirical data, various biomechanical models have been formulated (Anker, 1974; Barel et al., 1977; and Lauder, 1979a) of which two have become pervasively influential in the interpretation of teleostean mechanisms (e.g. Dullemeijer and Barel, 1977).

Levator Operculi Coupling as a Four-Bar-Linkage-System: Anker (1974) and Barel et al. (1977) have transformed the levator operculi coupling (Liem, 1970) into a mathematical operational model. According to this mechanical model, the coupling is reduced into a four-bar-linkage-system (Fig. 8). The four bars being: The gill-cover bar (Fig. 8: AD), interopercular bar (CD), mandibular bar (BC) and the suspensorial bar (AB). The important connecting points are: A, the operculohyomandibular joint; D, the ligamentous connection of suboperculum to interoperculum; C, attachment of the interoperculomandibular ligament to the mandible; and B, the quadratomandibular joint. In this four-bar-linkage model, the angle of mandibular depression is a mathematical function only of the opercular levation with the bar lengths as constants (Fig. 8). The perpendicular distances between the operculohyomandibular (A) and quadratomandibular joint (B), to the interopercular bar (CD) are designated respectively q and r. As proved by Anker (1974), the relation between input-rotation of the gill cover and output-rotation of the mandible is the kinematic transmission coefficient k, which can be expressed by the ratio q/r (Fig. 8). A second mathematical function of the four-bar-linkage-system is the ratio between output-moment and input-moment. The ratio is the force efficiency coefficient f. The

---

Arrow indicates the shallow retroarticular area below the jaw joint B, to minimize the horizontal bar CD. ih and sh are respectively the interhyal and sternohyoideus. (B, modified from Dullemeijer and Barel, 1977).

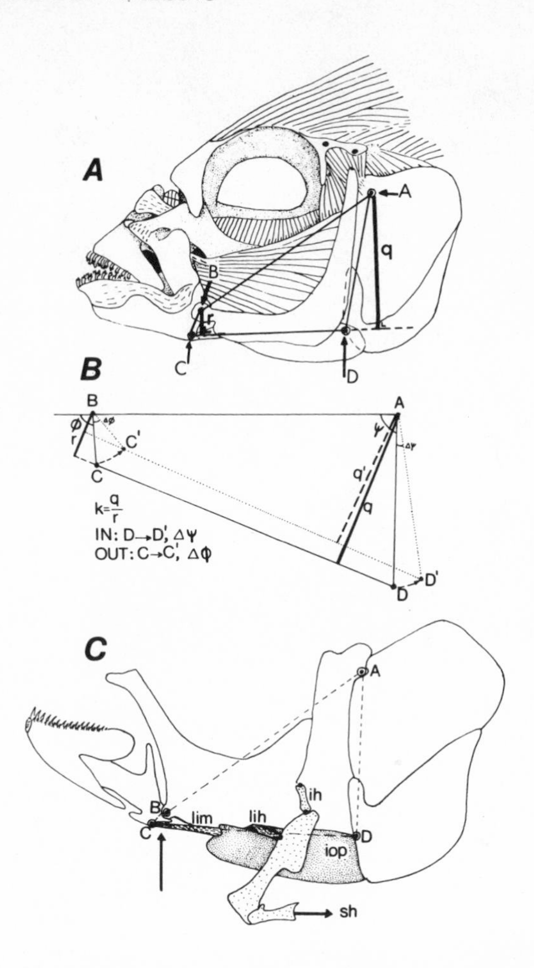

Fig. 8. A. Lateral view of dissected head of Petrotilapia tridentiger showing the key positions (A, B, C, D) of the four-bar-linkage system. A is the operculohyomandibular joint; B the quadratomandibular joint, C, the insertion site of the interoperculomandibular ligament on the mandible; D, the junction of interoperculum and suboperculum. q and r are the perpendicular distances between A and B to the horizontal bar CD. B. The four-bar-linkage system reduced into a mechanical model. Symbols as in A. k expresses the kinematic (movement) transmission efficiency coefficient. In and Out designate respectively input and output rotations. C. Medial view of the suspensory, opercular and mandibular apparatus of a piscivorous cichlid to show how the hyoid apparatus is linked to the interoperculum (iop) via the ineroperculohyoid ligament (lih) and indirectly to the mandible via the interoperculomandibular ligament (lim).

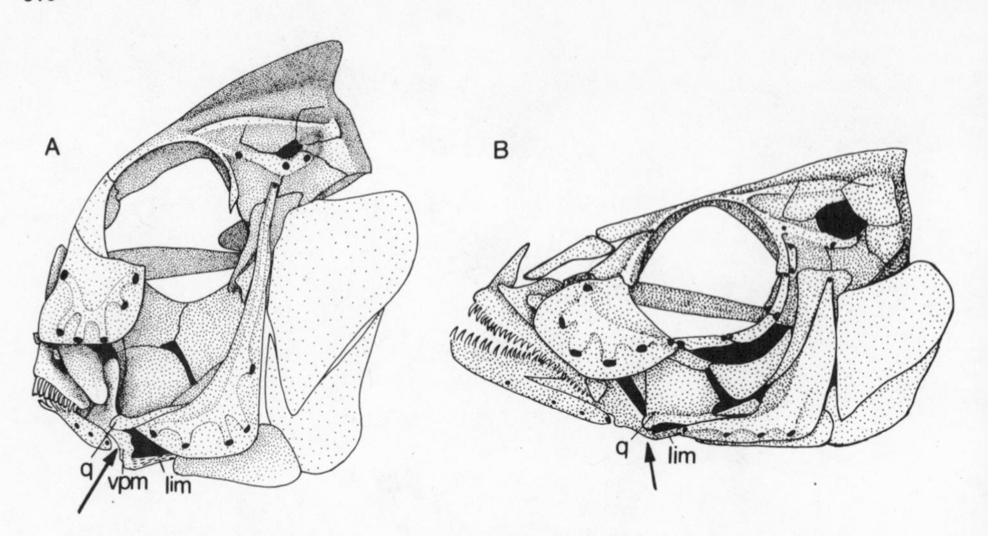

Fig. 9. Lateral view of the skulls of two endemic cichlids from Lake Tanganyika, Simochromis curvifrons (A), and Bathybates ferox (B). In Simochromis (A) the retroarticular projects ventrally to form a prominent ventral process of the mandible (vpm) to serve as a ventrally placed attachment for the interoperculomandibular ligament (lim). In Bathybates (B) the retroarticular is excavated into a notch (see arrow) bringing the attachment of the interoperculomandibular ligament as closely as possible to the quadratomandibular joint (q). Abbreviations: lim, interoperculomandibular ligament; q, quadrate; vpm, ventral process of mandible.

input-moment refers to the moment of lowering the mandible. Mathematically the force efficiency coefficient f equals the ratio r/q (Fig. 8). As discussed by Barel et al., (1977), a higher k (kinematic transmission coefficient) means that movement is transmitted more efficiently, while a higher f (force efficiency coefficient) indicates that force is transmitted more efficiently. Thus the functional characteristics of the levator operculi coupling are determined by just two parameters: q and r, which is the perpendicular distance between the quadratomandibular joint B and the interopercular bar (CD). It is clear, that according to this four-bar-linkage model, differences in the ratios of q and r in various teleosts, may have far reaching biological implications related to the velocity of jaw movement, energetics etc. For example, the model predicts that when k is optimized by a high q and a low r value, a relatively small and short levator operculi (e.g. Hemibates stenosoma, Liem, 1978; and Fig. 9) would be sufficient to lower the mandible over the physiologically determined angle. On the other hand, a fish with a low k value would need a longer and more massive levator operculi (e.g. Spathodus, Liem, 1979a and Fig. 9) to lower the mandible over a comparable angle.

The Four-Bar-Linkage Model, Optimal Design and Adaptation: Since the four-bar-linkage model is applicable to virtually all teleosts, it may offer a plausible argument about how the levator operculi coupling functions as a mechanical device. In general adaptation is regarded as the process of evolutionary change by which the organism provides a more and more efficient solution to a problem. The four-bar-linkage model can predict the efficiency of the lowering of the mandible in teleosts. Because the four-bar-linkage model emerged from an experimental analysis in which problems are posed and characters are understood as being functional design solutions it may predict in advance which taxa will be fitter to cope with known environmental problems.

In Lake Tanganyika and Lake Malawi cichlid fishes, piscivores (e.g. Boulengerochromis, Hemibates, Bathybates; Liem, 1978) have high values of k ranging from 5 to 17, while algae scrapers (e.g. Simochromis, Fig. 9) and invertebrate pickers (e.g. Eretmodus, Liem and Osse, 1975; Spathodus, Liem, 1979) have relatively low values of k ranging from 2.5 to 4. Among cichlids, the varying figures for k are closely correlated with the presence or absence of a ventral process of the mandible (Figs. 9, 11: vpm) to which the interoperculomandibular ligament (lim) attaches. With the appearance of the ventral process of the mandible, the value of r is enlarged, thereby reducing k and increasing f. Thus the system in algae scrapers is a force efficient mechanism rather than a movement-transmission efficient apparatus. The contrast between piscivores and algae scraping cichlids is illustrated in Figures 9 and 12. In piscivores r is minimized by having the interoperculomandibular ligament (lim) insert immediately below the quadratomandibular joint. In algae scrapers the insertion site is far removed from the joint.

From data obtained by cinematography it is evident that among cichlids, piscivores exhibit much higher velocities in mandibular depression than other trophic types (Liem, 1978, 1979). Thus, the high kinematic transmission coefficients (k) of piscivorous cichlids may be a reliable indicator that movements are transmitted very efficiently, enhancing the velocity of mandibular depression. Of course the velocity of mandibular depression has a pronounced effect on inertial suction.

Within the framework of the four-bar-linkage system, the functional features can be varied simply by varying the q/r ratio. As evident from the data on cichlids (Liem and Osse, 1975; Liem and Stewart, 1976; Dullemeijer and Barel, 1977; Liem, 1978), the r and q values can be changed by very simple morphogenetic mechanisms, i.e. an increase in depth of the operculum (increase in q), development of a ventral process of the mandible (increase in r, Fig. 9), differentiation of a retroarticular notch (decrease in r, Fig. 8, C, see arrow). Thus simple changes in the lengths and angles of the vertical gill-cover and mandibular bars (Fig. 8, 9: BC, AD) will appreciably alter the mechanical properties of the levator operculi coupling.

More importantly, the integration of the structures into a four-bar-linkage system is such that within certain limits the lengths of the suspensory and interopercular bars (Figs. 8, 14: AB, CD) can be varied greatly without affecting k (Dullemeijer and Barel, 1977). As a result, teleosts can vary the lengths of their heads over a wide-range, and still rely on the levator operculi coupling to depress the lower jaw. Thus a rather radical change in form of the horizontal bars may evolve without appreciably altering the function of the levator operculi coupling.

For non-cichlid teleosts, Barel et al., (1975) have found that the efficiency coefficients correlate with the size of the food. They have found that predators on large prey have a low k and thus possess a force-efficient rather than a movement transmission efficient levator operculi coupling. On the contrary fishes feeding on small food or prey tend to have a high k. A comparison of the charr (*Salvelinus*), a predator feeding on large prey, and two cichlids (the omnivorous *Pseudotropheus* and herbivorous *Labeotropheus*) support the generalization (Fig. 10 C) made by Barel et al., (1975). However, piscivorous cichlids feeding on large prey represent remarkable exceptions with high k values (*Hemibates* 17, *Boulengerochromis* 9.5, *Serranochromis* 7). Furthermore, the k values change during ontogeny (Fig. 10 A, B). Small and juvenile *Salvelinus* and *Labeotropheus* possess significantly higher k's than their large and adult counterparts. This pattern seems to support the notion that as the fishes grow, the size of their food become larger and their kinematic transmission coefficients (k's) decrease. These preliminary ontogenetic observations should caution us in making interspecific comparisons of k values without considering size, and possibly allometric factors.

The basic principle that has led to the formulation of the four-bar-linkage model is that of the optimal design or minimum principle (see e.g. Dullemeijer, 1974). Although the application of this principle is independent of evolutionary factors, it strongly implies "adaptedness," "adaptation," and "adaptability." It fits one view of adaptation by visualizing that the external world sets certain

Fig. 10. Graphs expressing the relations between the kinematic (transmission) efficiency coefficient k and body size (A and B), and the velocity with which the mandible is depressed maximally (C). A, *Salvelinus*; B, *Labeotropheus* (cichlid); C, *Salvelinus* and the cichlids *Pseudotropheus* and *Labeotropheus*.

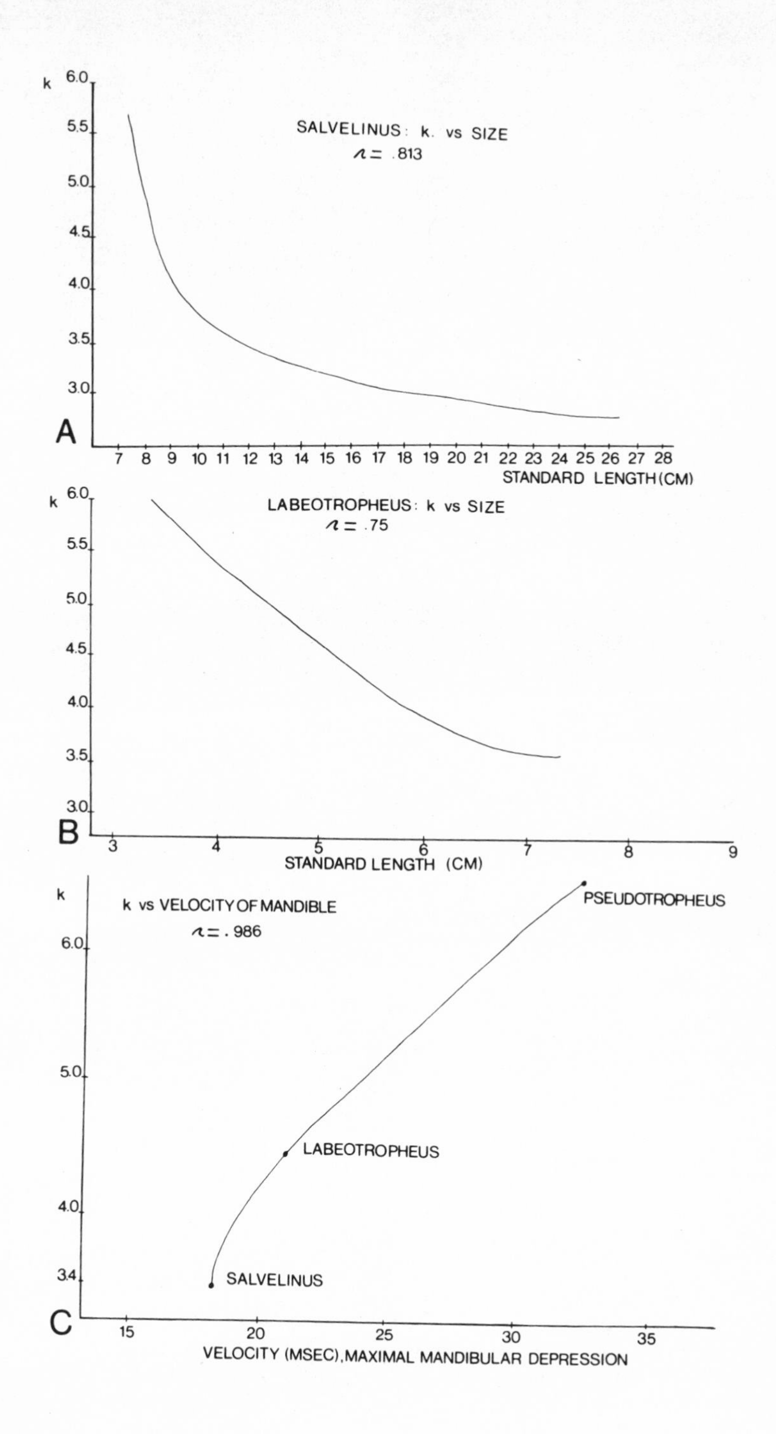
k
6.0
5.5
5.0
4.5
4.0
3.5
3.0
SALVELINUS: k. vs SIZE
r = .813
A
7 8 9 10 11 12 13 14 15 16 17 18 19 20 21 22 23 24 25 26 27 28
STANDARD LENGTH (CM)
k
6.0
5.5
5.0
4.5
4.0
3.5
3.0
LABEOTROPHEUS: k vs SIZE
r = .75
B
3 4 5 6 7 8 9
STANDARD LENGTH (CM)
k
k vs VELOCITY OF MANDIBLE
r = .986
6.0
5.0
4.0
3.4
PSEUDOTROPHEUS
LABEOTROPHEUS
SALVELINUS
C
15 20 25 30 35
VELOCITY (MSEC), MAXIMAL MANDIBULAR DEPRESSION

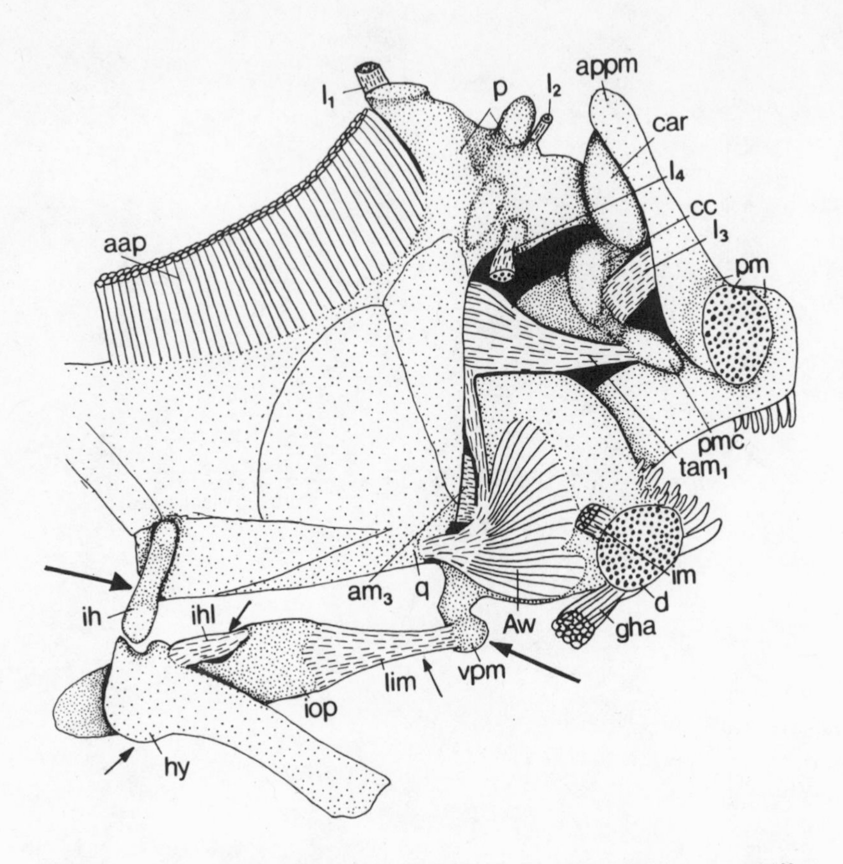

Fig. 11. Medial aspect of the left mandible, maxilla, premaxilla, mandible, part of the suspensory apparatus, hyoid apparatus (except urohyal) and interoperculum with associated muscles and ligaments of the cichlid Tropheus moorii from Lake Tanganyika. Arrows point to key features of the mechanical properties of the hyoid apparatus in depressing the mandible. Abbreviations: aap, adductor arcus palatani; am_3, A_3 part of adductor mandibulae; A_w, intramandibularis portion of adductor mandibulae; appm, ascending process of the premaxilla; car, rostral cartilage; cc, cranial condyle of maxilla; gha, geniohyoideus anterior; hy, hyoid; ih, interhyal; ihl, interoperculohyoid ligament; im, intermandibularis; iop, interoperculum; l_1, palatoethmoid ligament; l_2, ligament; l_3, ligament associated with premaxillary condyle of maxilla; l_4, palatovomerine ligament; lim, interoperculomandibular ligament; p, palatine; pm, premaxilla; pmc, premaxillary condyle of the maxilla; q, quadrate; tam_1, tendon of A_1 portion of adductor mandibulae; vpm, ventral process of the mandible.

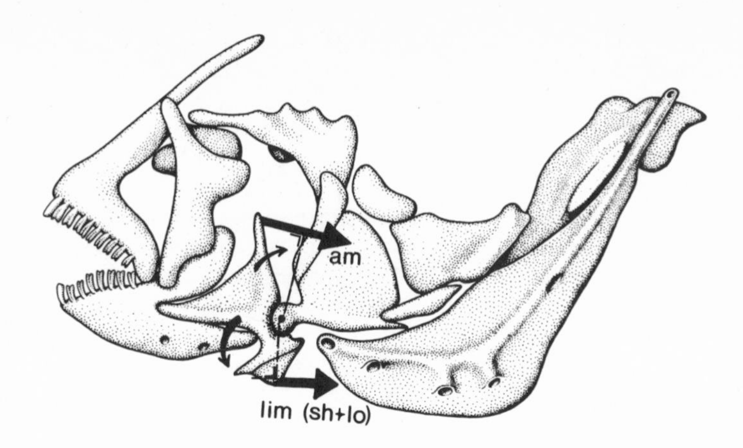

Fig. 12. Lateral view of the suspensory and jaw apparatus of Lobochilotes labiatus showing the torques around the quadratomandibular (jaw) joint of the adductor mandibulae (am) and the sternohyoideus (sh) and levator operculi (lo) muscles that act indirectly on the retroarticular process of the mandible via the interoperculomandibular ligament (lim).

"functional demands" that the organisms need to "meet," and that evolution by means of natural selection is the mechanism for creating these solutions in the direction of optimization. The four-bar-linkage paradigm predicts the efficiencies by which the mandible can be lowered by the levator operculi muscle. Although efficiency is not defined any further than as a mathematical coefficient, it implies that ultimately a particular function can be correlated with the expenditure of less energy.

The Sternohyoideus Coupling: Limits in Functional Design: As we have seen, the sternohyoideus coupling can depress the lower jaw by means of the interoperculohyoid ligament (Figs. 4, 8, 11: ihl), the interoperculum (iop) and the interoperculomandibular ligament. Thus, the action of the sternohyoideus is transmitted to the mandible by using the interopercular and mandibular bars (Fig. 8: CD, BC) of the four-bar-linkage system of the levator operculi coupling. The levator operculi and sternohyoideus couplings are mechanically linked. As a result, the mechanical features of the sternohyoideus coupling is partially determined by the same r value (Fig. 8, 12) of the four-bar-linkage system. Of course, other important parameters are the length of the interhyal and its position in respect to the interopercular and mandibular bars (Figs. 8, 11). It is evident from the topography and morphology of the hyoid apparatus that the sternohyoideus coupling has a low kinematic transmission coefficient (k), but a high force coefficient, f (an in-depth analysis by C. Elshoud of the Zoologisch Laboratorium of the University of Leiden, The Netherlands, is forthcoming). Thus an increase in r will augment the torque available for mandibular depression by means of the posteriorly directed force from the sternohyoideus via the interoperculum and interoperculomandibular ligament (Figs. 8, 11, 12). The differentiation of a large ventral process of the mandible in some cichlids (Figs. 9, 11: vpm) will increase the torque around the quadratomandibular joint enhancing the force efficiency of the sternohyoideus muscle in depressing the lower jaw. Yet, as we have seen, the ventral process of the mandible reduces the kinematic transmission efficiency of the levator operculi muscle in lowering the mandible. The torque transmission is mathematically the reverse of movement transmission (Anker, 1974). Consequently, the levator operculi and sternohyoideus couplings to depress the lower jaw cannot be optimized simultaneously by a specific r value (Fig. 12). Contrasting functional demands may put a selective premium on a mixed strategy or compromise of the mechanically linked integrations.

In addition to its role in depressing the lower jaw, the sternohyoideus also plays a role in abducting the suspensory apparatus (Fig. 13; Osse, 1969; Liem and Osse, 1975). As shown in Fig. 13 the force along the hyoid rami can be resolved in posteriorly and laterally directed components. The latter acts as a synergist of the levator arcus palatini muscle (Figs. 3, 6: lap) in abducting the suspensory apparatus. The lateral component of the force generated by the sternohyoideus can be optimized by an increased angle between the hyoid rami (compare A and B in Fig. 13). More acute angles between the hyoid rami will reduce the lateral component, and therefore decrease the efficiency of the sternohyoideus in abducting the side walls of the head. During the adaptive radiation of teleosts, a great array of interhyoid angles have evolved. In those species in which lateral expansion of the buccal cavity is emphasized, the angles between the hyoid rami are indeed greatly increased.

Because of the intimate linkage of the levator operculi and sternohyoideus couplings via the mandibular and interopercular bars, the two systems to depress the mandible, cannot be optimized simultaneously. Furthermore, the sternohyoideus coupling plays an indirect yet important role in the abduction of the side walls of the head. A search for optima becomes an exercise in futility in the majority of the cases. Instead, the integration of the levator operculi and sternohyoideus couplings may be viewed as a major adaptive breakthrough because it allows multiple kinematic pathways in the mandibular depression mechanism and thus a mixed strategy. Although the mixed strategy may limit optimization of a narrowly defined function, its greater built-in adaptive versatility can have a determining influence on: (a) the direction of evolutionary change of any of its properties when the environment changes; (b) functional

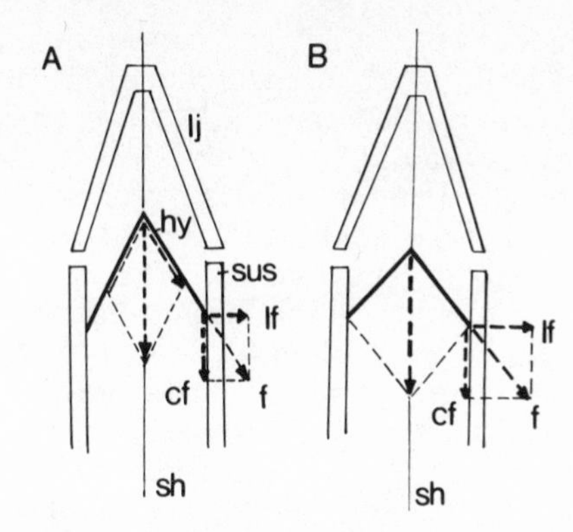

Fig. 13. Diagrams depicting some of the mechanical properties of the hyoid coupling in depressing the lower jaw and abducting the suspensory apparatus. In A the model possesses an acute angle between the hyoid (hy) rami; in B the angle is 90 degrees. The view is from ventral. Abbreviations: cf, caudal component of the force f generated by the sternohyoideus (sh); f, force along the hyoid (hy) ramus originating from the sternohyoideus (sh); hy, hyoid apparatus indicated by heavy lines; lf, lateral component of the force f, which is generated by the sternohyoideus along the hyoid ramus; lj, lower jaw; sh, sternohyoideus. See text for further explanation.

and ecological plasticity; (c) the evolutionary tempo during which seemingly major functional changes evolve (cf. Sage and Selander, 1975; Kornfield and Koehn, 1975).

LEVATOR OPERCULI COUPLING: A KEY EVOLUTIONARY INNOVATION

The most primitive actinopterygians (paleoniscoids) below the Halecostomi possessed a jaw structure adapted to a primarily "gripping" and "biting" habit rather than to inertial suction (Schaeffer and Rosen, 1961). They could open their mouth widely by lifting the neurocranium at a substantial angle to the vertebral axis by action of the epaxial musculature (Lauder, 1979a). However, expansion of the buccal cavity was very limited, and thus they must have primarily used their body velocity to overtake and capture prey (ram feeding). Besides lifting of the neurocranium, paleoniscoids must have opened their mouths by lowering the mandible by contractions of the sternohyoideus, and hypaxial muscles. Mandibular depressions was thus probably effected by actions of the ventral head and body musculature (Lauder, 1979b).

The appearance of the Halecostomi is accompanied by the emergence of three important biomechanical improvements in the feeding mechanism (Lauder, 1979a): (1) A free maxilla that can be propelled forward by neurocranial lifting creating a tunnel which directs water flow from the region immediately in front of the mouth into the buccal cavity (Fig. 7); (2) Increased lateral suspensorial mobility enabling the development of suction; (3) The differentiation of the interoperculum, a key element in the new mechanism to depress the lower jaw by levation of the operculum by action of the levator operculi muscle. Once this new integration becomes fully differentiated, the halecostomes (Liem and Lauder, 1981) were endowed with three modes of mouth opening, i.e. lifting of the neurocranium, lowering the mandible by action of the ventral head muscles and depression of the mandible by opercular levation.

The biological corollaries of the emergence of the interoperculum and subsequent differentiation of the levator operculi coupling as a mouth opening device are many and far reaching. It is probable that multiple functional components to depress the mandible play an important role in broadening the exploitation of a much wider array of trophic resources and that multiplicity may serve changing trophic needs during development and growth (Fig. 10). The addition of the levator operculi coupling as a mechanism of mandibular depression mechanism may have freed the sternohyoideus coupling to assume new functions, i.e. suspensory abduction (Fig. 13; and Lauder, 1979b) to enhance suction, and food preparation and mastication by the pharyngeal jaws during which the sternohyoideus plays a very active role (Liem, 1970, 1974, 1978). Based on the four-bar-linkage model, the levator operculi coupling is efficient in respect to movement-transmission (Fig. 8). Furthermore, the kinematic transmission coefficient can be optimized by minimizing r and maximizing q, both of which can probably be accomplished by simple morphogenetic mechanisms affecting relative growth rates of the vertical components (Fig. 8, mandibular and opercular bars, BC, AD) of the four-bar-linkage-model. In more advanced teleosts, the sternohyoideus coupling to depress the mandible becomes linked to the levator operculi coupling by the appearance of the interoperculohyoid ligament (Fig. 11: ihl), thereby providing a mechanism to summate the mechanical effects of a movement-transmission efficient with those of a force efficient system. Because of the mathematical characteristics of the four-bar-linkage system, the horizontal bars (Fig. 8, suspensory and interopercular bars, AB, CD) can vary in length over a wide range without affecting the jaw opening function. As a result, numerous and drastic changes in the shape of the head and other functional components (e.g. compare Fig. 3 with Figs. 9, 14) can occur within the framework and basic functional configuration of the four-bar-linkage system.

Since the appearance of the interoperculum together with the levator operculi coupling is central to the exploitation of new and diverse trophic resources, we may consider this halecostome specialization a key evolutionary innovation (Mayr, 1960; Liem, 1974). From the viewpoint of trophic biology, the Halecostomi represents a fundamentally distinct type of major taxon, if compared with prehalecostomes (Liem and Lauder, 1981). Although no major shift of habitat is discernible during the emergence of the Halecostomi, the levator operculi coupling with its built-in versatility (plasticity) and efficiency enabled the group to exploit vast trophic resources, which were not accessible to the prehalecostome actinopterygians. Once the levator operculi coupling was fully differentiated new functional demands may have produced an evolutionary avalanche: The Teleostei, the greatest radiation known among vertebrates. During the spectacular evolutionary radiation of the teleosts involving at least 20,000 species, the levator operculi coupling is maintained except in a very few taxa (e.g. *Mormyrus*, (Lauder, personal communication), and possibly the true eels). The integration of the levator operculi coupling is so basic and its adaptive versatility potentially so rich, that it is rigidly retained throughout the adaptive radiation even though numerous nondisruptive evolutionary changes have taken place (e.g. lengths of the horizontal bars (compare Figs. 3, 9, 14 B).

FEEDING REPERTOIRES, ADAPTATION AND EVOLUTION

Two contrasting strategies of the feeding apparatus of two advanced, unrelated, acanthopterygian teleosts, which have been studied experimentally, will be discussed here.

Specialization and Narrowing of the Feeding Repertoire: Specialization for a very narrow niche is perhaps a common evolutionary trend. To illustrate this pervasive evolutionary concept I will discusss the feeding mechanism of an amphibious synbranchiform fish, *Monopterus albus* (Liem, 1967, 1969). It is a highly specialized air-breathing, eel-shaped tropical fish, which practices parental care by building and guarding a foam nest. As most synbranchiforms, *Monopterus* is a burrower and engages in terrestrial excursions. In nature, *Monopterus albus* feeds on small fishes, prawns, crayfish, snails, insect larvae, insects, fish, and frogs eggs, frogs and tadpoles. Thus, its feeding habits seem very similar to those of *Synbranchus marmoratus* (Breder, 1927; Luling, 1959; Conner, 1966). Laboratory studies have revealed two modes of feeding; (1) When capturing small prey or food high speed inertial suction draws the food into the esophagus; (2) When the prey or food item is large, *Monopterus* strikes with an equally high speed, firmly grasping the prey/food and subsequently wedging it into the corner of the aquarium or into crevices between stones and spinning itself violently like a corkscrew. The large prey or food item is thus twisted and often broken into two or more parts. The smaller parts are then swallowed by high speed inertial suction. Such specialized feeding behavior has also been reported for Synbranchus marmoratus (Luling, 1958; Conner, 1966) and Monopterus (Amphipnous) cuchia.

Anatomically, several major specializations can be found. The opercular apparatus is drastically reduced, but the branchiostegal apparatus and its musculature are hypertrophied (Fig. 14 A, B, C). The k value (kinematic transmission coefficient) are very low: 1.5-1.7. The levator operculi muscle (Fig. 14: lo) is elongate. Without doubt, the adductor mandibulae muscle complex dominates the head (Fig. 14, A, B, C: am 1-3) to an extent not found in any other group of teleosts. The adductor mandibulae am_1 (Fig. 14 A: am_1) has its insertion on the lateral surface of the ascending process of the dentary, leaving the upper jaw (maxilla and premaxilla) devoid of any muscle, a unique specialization among acanthopterygians. Thus the non-protrusible upper jaw has become a passive element stabilized only by ligaments. The insertion of all three major parts of the adductor mandibulae complex (Fig. 14, A, B: am 1-3) on the ascending (coronoid) process of the dentary is also unusual. In most acanthopterygians, am_2 inserts on the ascending process of the articular (Fig. 3).

Monopterus opens its mouth at a high velocity ((Fig. 14 D). Peak gape is accomplished in 25 msec. More importantly, small prey or food is sucked in completely in less than 5 msec. (F. 14, D, frame 4, 5, between vertical arrows). Adduction of the mandible is not only slower but also more variable, being completed between 40 and 200 msec. depending on the feeding situation. Of course when feeding on very large prey, the kinematic profile becomes more complex and drawn out. However, opening of the jaws during the initial strike proceeds invariably at a very high velocity. Depression of the hyoid lags behind jaw opening (Fig. 14, D). The time lag between peak gape of the mouth and maximal depression of the hyoid varies between 25 and 100 msec.

The electromyographic profile (Fig. 14) is characterized by a wide overlap of the lower jaw depressor (levator operculi, lo) and its chief adductor (am_2). Synchronous firings of antagonists seem to accompany high velocity mechanisms in teleosts (e.g. Liem, 1978; Lauder, 1980). Other expansive muscles, the sternohyoideus (sh) and hypaxial (hy) fire synchronously with the levator operculi but then for longer durations (Fig. 14). Exceptionally short and high amplitude bursts of the adductor mandibulae am_1 and am_3, geniohyoideus anterior (gha) and geniohyoideus posterior ghp) conclude the strike.

From the kinematic and electromyographic profiles it appears that mouth opening is caused mainly by the levator operculi muscle, since movement of the hyoid lags behind that of the mandible by a considerable margin (Fig. 14, D), and the levator operculi is the first muscle to show activity (Fig. 14, D: lo). Because the velocity of mandibular depression is one of the highest on record among teleosts,

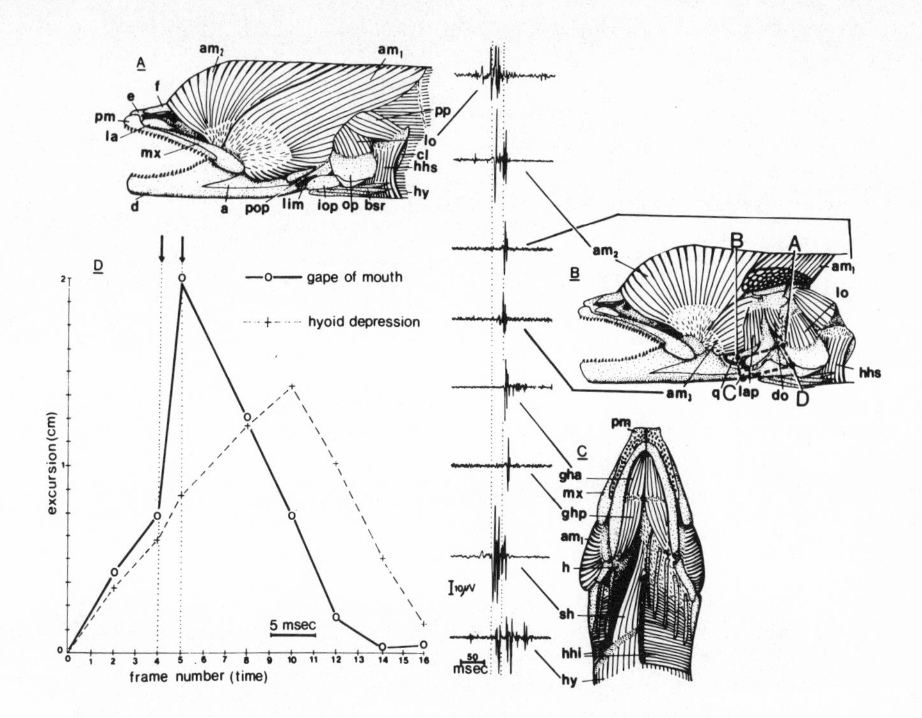

Fig. 14. Lateral view of superficial cephalic musculature (A) and deep cephalic (B) and ventral musculature (C) of Monopterus albus. D, kinematic profile of the capture of small food; in centre, the electromyographic profile of some key muscles during capture of small prey. In B, most of the adductor mandibulae part A_1 (am_1) has been dissected away; in C, the hyohyoideus inferior (hhi) and branchiostegal rays from the right side have been dissected away. In B, the letters A, B, C and D indicate the four bars of the four-bar-linkage model. In D, the two vertical arrows indicate the time span within which the small food is sucked in completely i.e. in frame 4 the food was still outside the mouth, in frame 5 the food has entered the mouth and is no longer visible. Abbreviations: a, articular; am_{1-3}, adductor mandibulae parts A_1, A_2, and A_3; bsr, branchiostegal rays; cl, cleithrum; d, dentary; do, dilatator operculi; e, ethmoid; f, frontal; gha, geniohyoideus anterior; ghp, geniohyoideus posterior; h, hyoid; hhi, hyohyoideus inferior; hhs, hyohyoideus superior; hy, hypaxial muscles; iop, interoperculum; la, lachrymal; lap, levator arcus palatini; lim, interoperculomandibular ligament; lo, levator operculi; mx, maxilla; pm, premaxilla; pop, preoperculum; pp, protractor pectoralis; q, quadrate; sh, sternohyoideus.

one would expect a very high movement-transmission efficiency (k), but Monopterus has one of the lowest k valves (1.5 in adults) recorded for teleosts! The discrepancy between prediction derived from the four-bar-linkage model and the empirical data presented here, may indicate that the model as it now stands may be too simple (for a discussion see Barel et al., 1977). Of course the functional significance of the synchronous high amplitude firings of the antagonists, the adductor mandibulae am_2 and levator operculi, so prevalent in high velocity teleostean feeding mechanisms remains an enigma. The electromyogram of the adductor mandibulae part am_2, indicates that it is the dominant jaw closing muscle. It is the first adductor to fire and because its fibers are perpendicular to the long axis of the mandible (Fig. 14, A, B) it is in a mechanically advantageous position to do so. After the mandible has begun its closing movement, the fibers of the other two adductors (am_1, am_3) are in a mechanically favorable position in respect to the coronoid process and the jaw joint. This may explain why am_1 and am_3 fire toward the end of the strike.

One of the most striking features is the constancy of the kinematic and electromyographic profiles in Monopterus. Such a constancy supports the notion that we are dealing with a preprogrammed oscillator, regardless whether the fish is capturing small, large, live or dead prey. The stereotyped motor activity may enchance velocity. Once prey is captured, the highly hypertrophied adductor mandibulae complex plays a key role in conjunction with the corkscrew-like twisting motion of the body in breaking up the prey or food into pieces. Swallowing is invariably accomplished by inertial suction. Large pieces of food are raked in by movements of the upper pharyngeal jaws caused by action of the retractores dorsales and levators interni as in nandids (Liem, 1970). Although Monopterus is a truly amphibious fish, attempts in the laboratory to feed the fish out of water failed. Monopterus seems to depend on inertial suction, which is only effective in the aquatic medium. According to the data gathered in this study, Monopterus is probably inefficient in exploiting terrestrial trophic resources. Although Monopterus possesses many remarkable adaptations for amphibious life (Liem, 1961, 1967, 1969), its feeding apparatus remains narrowly tied to the aquatic medium. The preprogrammed high velocity oscillator is so narrowly specialized and rigidly maintained that it is deployed even when capturing a stationary snail or an immobile piece of frozen smelt!

Judging from the many similarities in morphology of the feeding apparatus, all Synbranchiforms share the narrowly specialized feeding mode of Monopterus. The evolutionary origin of the Synbranchiforms remains unknown, although it has been hypothesized that the Channiformes (Ophicephaliformes, snake heads) are the closest related group (Liem and Lauder, 1981). The evolutionary emergence of the adaptations correlated with the amphibious and highly peculiar feeding modes may have been under strong and continuous selection pressure. Because the evolutionary trend was highly adaptive, it led to a group of fishes with a very narrow and specialized adaptive zone, with no intermediate stages persisting.

Although it has been claimed that some amphibious freshwater forms may occasionally feed out of water (Graham, 1976, p. 183), none seem to exploit terrestrial trophic resources and amphibious behavior is thought to play only a role in reaching new aquatic habitats. Graham (1976) invokes an argument favored by most ecologists because it is based on the principle of competitive exclusion. His hypothesis is: "The presence of amphibians and a host of other terrestrial organisms in and around freshwaters seem to have been important in preventing the evolution of a highly amphibious freshwater fish fauna like that of marine habitats." However amphibians (frogs are a part of the regular diet!) and many terrestrial invertebrates represent an enormous potential food resource, which remains unexploited by the highly amphibious synbranchiform fishes. Instead of viewing the organisms as black boxes it seems justifiable to focus on the internal constraints, or the internal environment of the organism. The built-in functional constraints in the narrowly specialized feeding apparatus of the amphibious synbranchiform fishes may have played a far greater role in the prevention of the exploitation of terrestrial trophic resources than the alleged competition with other terrestrial organisms, i.e. the lack of an empty ecospace. Its heavy reliance on a high speed inertial suction system is clearly incompatible with an environment devoid of water. As shown by my experiments, the synbranchiform feeding apparatus possesses an extremely narrow functional repertoire and lacks plasticity altogether.

Specialization and Broadening of the Feeding Repertoire: Previous studies on the invertebrate-picking cichlids (Liem, 1979) and a herbivorous cichlid (Liem, 1980) have revealed that phylogenetically and morphologically more specialized forms are not only remarkable specialists in a narrow sense, but also formidable jacks-of-all-trades. Thus, the functional and consequently ecological patterns accompanying specialization in cichlids seem to differ drastically from those in synbranchiforms. Here the intraspecific trophic repertoire of *Lobochilotes labiatus* is analyzed experimentally. *Lobochilotes* is distinctly omnivorous (Poll, 1956) feeding on a wide array of invertebrates (Insect larvae, crustaceans, clams, snails), diatoms (*Navicula* and *Cocconeis*), algae and other plants. Phylogenetically it is considered a derived and specialized member of the cichlids endemic to Lake Tanganyika (Fryer and Iles, 1972). Morphologically, *Lobochilotes* is specialized in having very large fleshy lobes on the lips, which are presumably used to detect prey by touch (Fryer and Iles, 1972) and crushing type pharyngeal jaw apparatus (cf. Hoogerhoud and Barel, 1978).

In the following experiments, the same individual is presented different foods and prey in different positions. Five distinct kinematic and electromyographic profiles have been found in Lobochilotes (Figs. 15, 16).

Inertial Suction (IS): Slow and Horizontal. As discussed in the beginning of this paper, inertial suction is the universal mode of feeding strategy in teleosts. The profiles of Lobochilotes (Fig. 15) closely resemble those of Serranochromis (Fig. 3). Moving prey or food in the water column is captured by this mechanism. Live fish, Daphnia, crickets, frozen brine shrimp, dried fish food (Tetramin) are all collected by slow horizontal suction. Even swiftly swimming goldfish can be engulfed with this method combined with a forward swimming movement.

Inertial Suction (IS): Slow and Upward. Any food or prey floating on the surface of the water is collected by a dorsally directed suction, causing a flow of surface water to enter the mouth. The characteristic features are the silence of the epaxial muscles (Fig. 15: ep) and the restriction of upper jaw protrusion. The adductor mandibulae am_1 is active during mouth opening and expansion, thereby minimizing upper jaw protrusion. As a result, the mouth opening is directed upwards as suction is created by strong abduction of the suspensorium and lowering of the buccal floor by actions of respectively the levator arcus palatini and sternohyoideus muscles. The compressive phase proceeds as in slow horizontal IS by unopposed actions of all adductors.

Inertial Suction (IS): Slow and Downward. When collecting prey and food from the bottom, e.g. Tubifex, midgefly larvae, earthworms, snails, bivalves, frozen brine shrimp, both the kinematic and electromyographic profiles deviate significantly from those during slow horizontal IS. In sharp contrast to the pattern during IS slow upward, full upper jaw protrusion occurs early and mandibular depression is very limited. Early activity in the epaxial muscles (Fig. 15: ep) and synchronous actions of the levator operculi, sternohyoideus, and their antagonist, the entire adductor mandibulae depression results in a ventrally directed small and round mouth opening. The degree of mouth opening is continuously modulated by cocontraction of the adductor mandibulae complex and levator operculi and sternohyoideus muscles (Fig. 15: am 1-3, lo, sh). Upper jaw protrusion is regulated extensively by synchronous firings of the epaxial and adductor A_1 muscles. Adduction of the jaws proceeds according to the preprogrammed profile as in slow horizontal IS by unopposed actions of occasionally all adductors.

Biting. Lobochilotes can bite pieces of lettuce and fins of other fish. Bites can be characterized by extreme opening of the mouth and extreme jaw protrusion. The epaxial muscles, and levator operculi

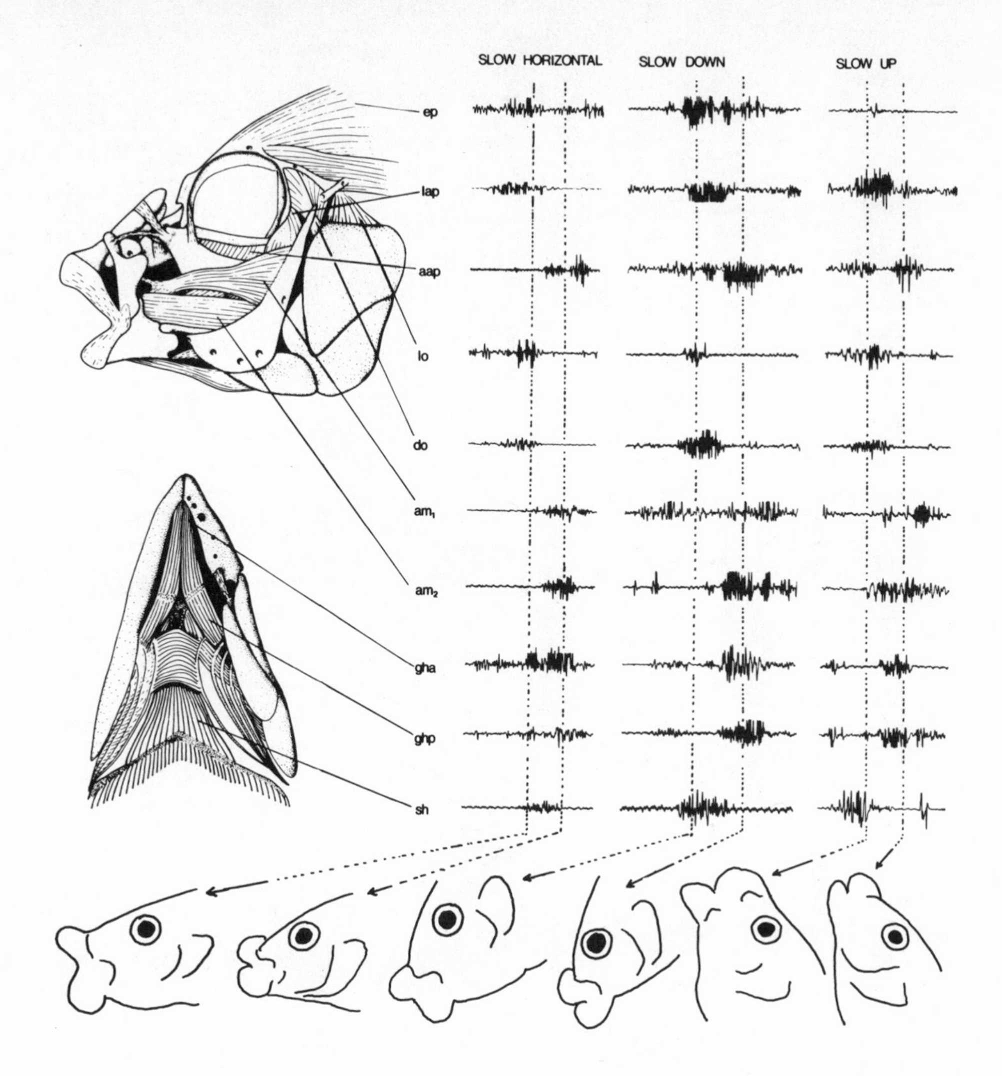

Fig. 15. On left, lateral and ventral aspects of the cephalic musculature of the cichlid fish Lobochilotes labiatus, with accompanying electromyograms to the right. Three modes of feeding with their respective electromyographic and kinematic profiles are presented. Kinematic events are tracings from a motion picture. Vertical, stippled lines correlate electromyographic event with kinematic event. Abbreviations: AM_{1-2}, parts A_1 and A_2 of adductor mandibulae; aap, adductor arcus palatini; do, dilator operculi; ep, epaxial muscles; gha, ghp, respectively geniohyoideus anterior and posterior; lap, levator arcus palatini; lo, levator operculi; sh, sternohyoideus.

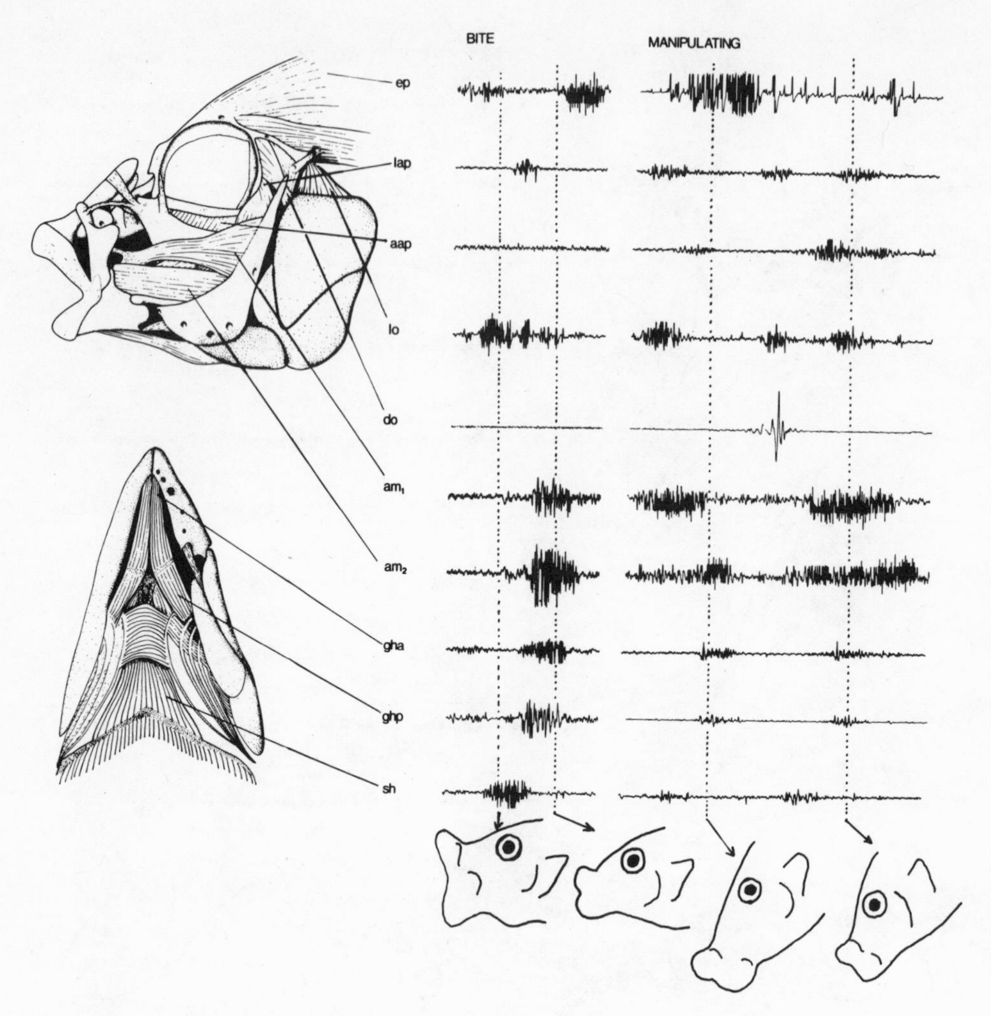

Fig. 16. See legend of Fig. 15 for explanation and abbreviations.

fire at a very high amplitude (Fig. 16). The ventral muscles, geniohyoideus and sternohyoideus are also active alternately. During the bite, there is little suction generated. The levator arcus palatini, adductor arcus palatini and dilatator operculi are virtually silent (Fig. 16). Closing of the jaws is accomplished by unopposed and strong (high amplitude bursts) actions of the entire adductor mandibulae complex and the geniohyoideus (Fig. 16). As the jaws close, the upper jaw is retracted much earlier than during inertial suction.

Manipulation. This continuously modulated, non cyclical pattern is perhaps the most complex one in teleosts. Jaw protrusion takes place in varying degrees and is regulated independently from the continuously varied gape of the mouth. To elicit manipulating jaw movements, Lobochilotes was offered arthropod limbs protruding from the openings of a plastic screen (Liem, 1979). By varying the activity of the epaxial (Fig. lb: ep) and adductor mandibulae complex (am 1-3), Lobochilotes changes the extent of upper jaw protrusion continuously during the extended feeding sequence. Protrusive and retrusive movements of the upper jaw can take place while the mouth is completely closed. The electromyographic profile is exceedingly complex and variable and resembles that of Petrotilapia (Liem, 1980) and Eretmodus (Liem, 1979) during the dislodging of sessile prey.

Meaning of intraspecific functional repertoires. Lobochilotes resembles other specialized cichlids in performing a rather broad spectrum of dissimilar functions (Fig. 17, Category IV). Are the five functions really that different? Or is the subdivision in Fig. 17 unnecessarily complex and without meaning? The five functions found in Lobochilotes are very different indeed. Each is correlated consistently with a well defined activity, and the differences between the categories are as great as or greater than those between different and unrelated taxa, e.g. Perca (Osse, 1969), Gymnocephalus (Elshoud-Oldenhave and Osse, 1976), Cyprinus (Ballintijn et al., 1972), Salvelinus (Lauder and Liem, 1980), and Lepomis (Lauder and Lanyon, 1980). Lobochilotes lacks three patterns known from the repertoire of category IV cichlids (Fig. 17). They lack the ability to engage in high speed inertial suction, and scraping from horizontal and vertical surfaces. At present there is no empirical evidence to support the notion that prey or food collected from the surface or bottom by respectively "IS-slow-up" and "IS-slow-down" cannot be collected by the same movements and structures of "IS-slow-horizontal." At the moment we have no quantitative data on the differential efficiency of the various feeding patterns. Intuitively it seems that "IS-slow-up" is more "efficient" in collecting prey/food from the surface of the water than "IS-slow-horizontal." But how can efficiency be measured? Traditional parameters such as number of bites per time unit or handling time appear much too simplified and are very poor methods to apply to cichlids. To date there is no method to compartmentalize and determine the energy balance of the various feeding patterns. Yet the concept of adaptation is closely linked with efficiency (e.g. Bock and von Wahlert, 1965) and optimal design. The intraspecific occurrence of multiple feeding patterns indicates that the feeding apparatus in cichlids can meet multiple problems. Consequently the functional design of the feeding apparatus of cichlids may represent a compromise, in which multiple and diverse functions are realized suboptimally.

Currently we must view the intraspecific multiplicity of feeding patterns simply as variations on a basic theme. All teleosts will depend on "IS-slow-horizontal" sometime during their life cycle. From this basic theme, "IS-slow-up" and "IS-slow-down" may have evolved by varying the activity of the epaxial, and adductor mandibulae am_1 muscles (Figs. 15, 16). Therefore the different patterns may simply reflect the ability of the central nervous system of the fish to modulate and thus to vary the degree and timing of the activity of some muscles. This ability to modulate may improve and broaden progressively as ontogeny advances. However, biting and manipulation deviate rather significantly from "IS-slow-horizontal" (Figs. 15, 16). Thus they may represent true specializations, which originated from the "IS-slow-horizontal" pattern by more elaborate modifications than just an ability to modulate the actions of two muscles. If this hypothesis is correct, it may clarify the seeming paradox (Liem, 1980) that morphologically and functionally specialized cichlids are not only remarkable specialists in a narrow sense, but also jacks-of-all-trades (Liem, 1980). As their ability to vary, modulate, and eventually alter the basic theme to a very specialized pattern, they do not lose the ability to perform the "intermediate" patterns. Support for this hypothesis can be derived from the fact that so many of the phylogenetically and morphologically specialized cichlids have developed an improved function for a narrowly defined trophic niche but without sacrificing less specialized operational fields (Liem, 1979, 1980, and Fig. 17). Recently, Lauder (1980) has discovered intraspecific functional repertoires in the feeding mechanism of the characoid fishes *Lebiasina* and *Chalceus*. It is interesting to note that these rather extensive intraspecific functional repertoires have been found only in two groups that have undergone dramatic adaptive radiations, i.e. Cichlidae and Characoidei.

CONCLUDING REMARKS

The basic premises of functional morphologists are so permeated by the recognition that many, perhaps most (Dullemeijer, 1974), features have functions that they often forget that organisms can be suboptimal. However, natural selection does not shape the best possible organisms, it merely replaces existing genotypes with whatever better (not best) ones arise (Futuyma, 1979). The four-bar-linkage model as originally conceived (Anker, 1974; Dullemeijer and Barel, 1977) is an example of the application of the theory of optimization or minimum principle. This form of theorizing does not take into consideration any evolutionary factors and asks what the "best" mechanical construction would be to perform a particular function (i.e. lowering the mandible by opercular rotation). On the basis of the optimization principle the model is constructed. When a reasonable correspondence between predicted and observed characteristics is

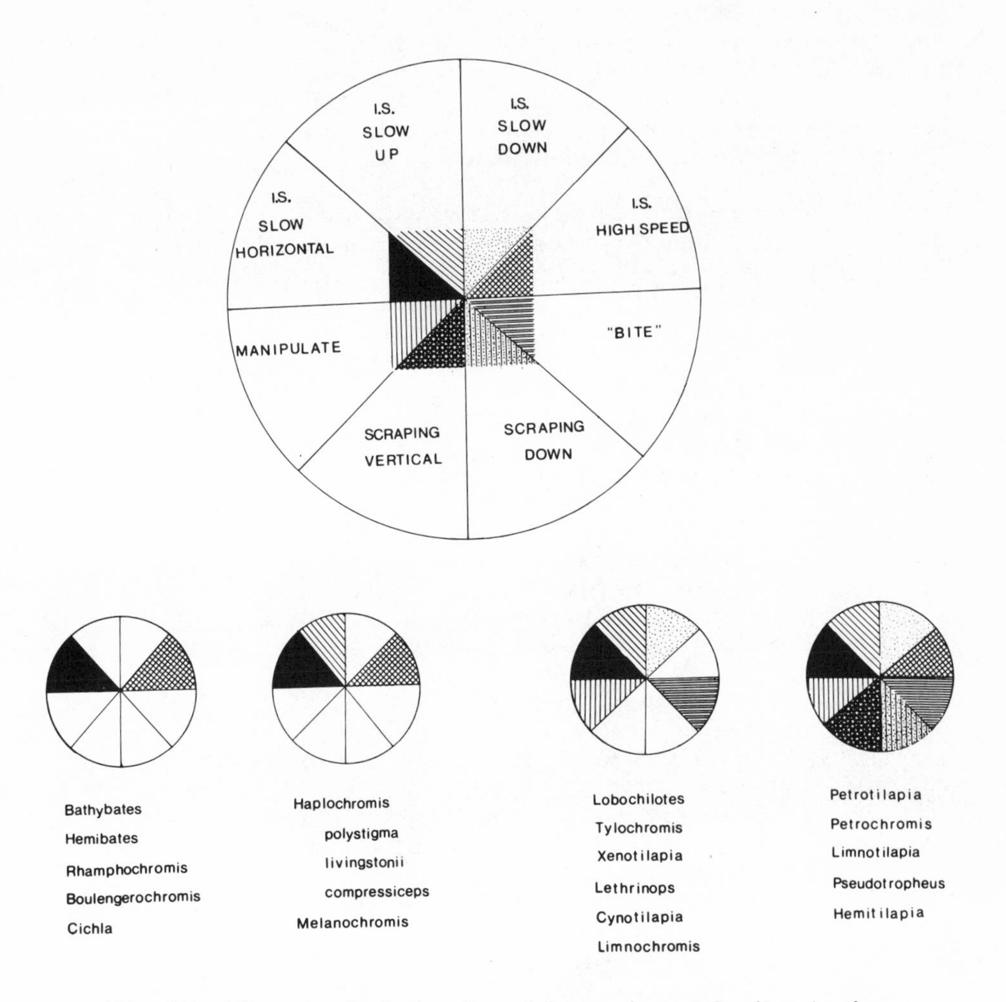

Fig. 17. Diagram depicting the eight modes of feeding in the functional trophic repertoire of cichlids. Pattern of shading in the large circle depict the conventions used in the smaller circles. Each of the eight modes is characterized by specific electromyographic and kinematic profiles. Each distinct profile is usually correlated with a particular kind and position of the food. Four different categories are recognized among the cichlids surveyed, depicted by the four smaller circles. Taxa representing the respective categories are listed under each appropriate circle (from Liem, 1980, courtesy of the American Zoologist).

found, it is assumed that the theory is "right" and, therefore, that evolution should go only in the direction of optimalization. It is not surprising that strict adherence to the deductive method in which evolutionary factors are disregarded a priori, has led to the overly pessimistic statement (Dullemeijer and Barel, 1977, p. 113): "Logically evolutionary theory would not be expected to play an essential role in the explanation of functional morphology."

However, organisms do not live in the best of all possible worlds, the world of Platonic ideas, but in a continuously changing world of materials and history (e.g. Dobzhansky et al., 1977; Lewontin, 1978; Futuyma, 1979). If we view the four-bar-linkage system (levator operculi coupling) as a functional integration in which some parameters (e.g. the horizontal bars AB, CD in Fig. 8) can vary independently, without changing its basic function (Dullemeijer and Barel, 1977) we are dealing with a more meaningful adaptive phenomenon. Within this inherently versatile pattern, it is possible to broaden the range of functions of other structures by allowing the use of more variables without significantly reducing the efficiency of the original function of the integration (i.e. the four-bar-linkage system). In this way biological versatility is achieved but not mechanical or energetic optimality. A posteriori the levator operculi coupling can be viewed as a versatile adaptation since it has been maintained throughout the long, complex and rich evolutionary history of the Teleostei.

Environmental physiologists, ecologists and functional morphologists are concerned with the adaptive sensitivity or adaptability of organisms to changes in the external environment. The versatility and adaptability of the organism depend on the "internal" environment. The study of the organism as a network of interacting constraints (e.g. Liem, 1979b) is becoming increasingly more important.

In this chapter I have tried to show that the preoccupation with optimization may divert attention from other equally important aspects of evolution. For example, an understanding of factors limiting optimization but enhancing versatility and therefore the matrix upon which natural selection can act, should be an important goal to environmental physiologists and functional morphologists. Such a goal can be achieved by a reciprocative illumination of the internal and external environments.

ACKNOWLEDGEMENTS

My understanding of the theoretical topics covered and the experimental approaches used in this paper have been improved greatly as a result of correspondence and discussions with many friends and colleagues. I wish to express my gratitude to: P.H. Greenwood, G.V.

Lauder, John Cosgrove, L. Radinsky, S. Emerson, W.L. Fink, S. Fink, D. Kramer, J.S. Levine, and William McFarland. I am grateful to the following persons whose generosity, active cooperation, and special skills have made this work possible: Karsten Hartel, L. Nobrega, David Kraus, William Winn, Al Coleman, Christine Fox and Robert Schoknecht. I am further indebted to Lou Garibaldi of the New England Aquarium for much help, and to the British Museum (Natural History) and the Field Museum of Natural History for the loan of specimens and for making available research facilities. This research was supported in part by a fellowship from the John Simon Guggenheim Memorial Foundation, New York, and grants from the National Science Foundation.

LITERATURE CITED

Alexander, R. McN. (1967). The functions and mechanisms of the protusible upper jaws of some acanthopterygian fish. J. Zool. (Lond.)., 151: 43-64.

Alexander, R. McN. (1969). Mechanics of the feeding action of a cyprinid fish. J. Zool. (Lond.)., 159: 1-15.

Alexander, R. McN. (1970). Mechanics of the feeding action of various teleost fishes. J. Zool. (Lond.)., 162: 145-156.

Anker, G. Ch. (1974). Morphology and kinetics of the head of the stickle-back, Gasterosteus aculeatus. Trans. Zool. Soc. Lond., 32: 311-416.

Ballintijn, C.M., van den Burg, A. and Egberink, B.P. (1972). An electromyographic study of adductor mandibulae complex of a free-swimming carp (Cyprinus carpio L.) during feeding. J. Exp. Biol., 57: 261-283.

Barel, C.D.N., Berkhoudt, H., and Anker, G. Ch. (1975). Functional aspects of four bar systems as models for mouth-opening mechanism in teleost fishes. Acta. Morph. Neerl. --Scand. 1975: 228-229.

Barel, C.D.N., van der Meulen, J.W., and Berkhoudt, H. (1977). Kinematischer transmissionskoeffizient und Vierstangensystem als Funktions parameter und Formmodell fur mandibulare Depressionsapparate bei Teleostiern. Ant. Anz., 142: 21-37.

Bock, W.J. and von Wahlert, G. (1965). Adaptation and the form-function complex. Evolution 19: 269-299.

Breder, C.M. Jr. (1927). The fishes of the Rio Chucunague drainage, eastern Panama Bull. Am. Mus. Nat. Hist., 57: 91-176.

Conner, J.V. (1966). Morphology of the neotropical swamp eel, Synbranchus marmoratus (Pisces: Teleostei), with emphasis on adaptive features. Master of Science Thesis, Texas A and M University, College Station, Texas. p. 1-71.

Dobzhansky, T., Ayala, F.J., Stebbins, G.L., and Valentine, J.W. (1977). Evolution. San Francisco, W.H., Freeman, p. 1-572.

Dullemeijer, P. (1974). Concepts and Approaches in Animal Morphology. Netherlands; Van Gorcum, Assen, p. 1-264.

Dullemeijer, P. and Barel, C.D.N. (1977). Functional morphology and evolution. In: Major Patterns in Vertebrate Evolution. Eds. M.K. Hecht, P.C. Goody and B.M. Hecht. New York Plenum Press., p. 83-117.

Elshoud-Oldenhave, M.J.W., and Osse, J.W.M. (1976). Functional morphology of the feeding system in the ruff (Gymnocephalus cernua L. 1758) (Teleostei, Percidae). J. Morph., 150: 399-422.

Frazzetta, T.H. (1975). Complex adaptations in evolving populations. Sunderland, Massachusetts, Sinauer Publishers, p. 1-267.

Fryer, G. and Iles T.D. (1972). The Cichlid Fishes of the Great Lakes of Africa: Their Biology and Evolution. Edinburgh, Oliver and Boyd, p. 1-641.

Futuyma, D.J. (1979). Evolutionary Biology. Sunderland, Massachusetts, Sinauer Assoc., p. 1-565.

Graham, J.B. (1976). Respiratory adaptations of marine air-breathing fishes. In: Respiration of Amphibious Vertebrates. Ed. G.M. Hughes, New York, Academic Press, p. 165-187.

Hoogerhoud, R.J.C. and Barel, C.D.N. (1978). Integrated morphological adaptations in Piscivorous and mollusc-crushing Haplochromis species. In: Proceedings of the Zodiac Symposium on Adaptation. Netherlands, Centre for Agricultural Publishing and Documentation, Wageningen, p. 52-56.

Kornfield, I.L. and Koehn, R.K. (1975). Genetic variation and evolution in some New World Cichlids. Evolution 29: 427-437.

Lauder, G.V. (1979a). Feeding mechanics in primitive teleosts and in the halecomorph fish Amia calva. J. Zool. (Lond.) 187: 543-578.

Lauder, G.V. (1979b). Evolution of the feeding mechanism in primitive actinopterygian fishes: a functional anatomical analysis of Polypterus Lepisosteus, and Amia. J. Morph. (in press).

Lauder, G.V. (1980). Intraspecific functional repertoire in the feeding mechanism of the characoid fishes Lebiasina, Hoplias, and Chalceus. (in press).

Lauder, G.V. and Lanyon, L.E. (1980). Functional anatomy of feeding in the bluegill sunfish, Lepomis macrochirus: In vivo measurement of bone strain. J. Exp. Biol. 80: 33-55.

Lauder, G.V. and Liem, K.F. (1980). The feeding mechanism and cephalic myology of Salvelinus fontinalis: form, function, and evolutionary significance. In: Charrs: Salmonid Fishes of the Genus Salvelinus. Ed. E.K. Balon, Netherlands, Junk Publishers. p. 365-390.

Lewontin, R.C. (1978). Adaptation. Sci. Am., 239: 212-230.

Liem, K.F. (1961). Tetrapod parallelism and other features in the functional morphology of the blood vascular system in Fluta alba Zuiew (Pisces; Teleostei). J. Morph., 108: 131-143.

Liem, K.F. (1967). Functional morphology of the head of the Anabantoid teleost fish Helostoma temmincki. J. Morph., 121: 135-158.

Liem, K.F. (1967). Functional morphology of the integumentary, respiratory and digestive systems of the synbranchoid fish, Monopterus albus. Copeia, 1967: 375-388.

Liem, K.F. (1969). Adaptive morphological features correlated with the invasion of terrestrial habitats by the amphibious fish order Synbranchiformes. Am. Zool., 9: 1147.

Liem, K.F. (1970). Comparative functional anatomy of the Nandidae (Pisces: Teleostei). Fieldiana, Zool., 56: 1-166.

Liem, K.F. (1974). Evolutionaary strategies and morphological innovations: cichlid pharyngeal jaws. Syst. Zool., 22: 425-441.

Liem, K.F. (1978). Modulatory multiplicity in the functional repertoire of the feeding mechanism in cichlid fishes. J. Morph., 158: 323-360.

Liem, K.F. (1979). Modulatory multiplicity in the feeding mechanism in cichlid fishes, as exemplified by the invertebrate pickers of Lake Tanganyika. J. Zool. (in press).

Liem, K.F. (1980). Adaptive signifiance of intra- and interspecific differences in the feeding repertoires of cichlid fishes. Am. Zool. (in press).

Liem, K.F. and Lauder, G.V. (1981). The evolution and interrelationships of the actinopterygian fishes. In: Neurobiology of Fishes. Ed. R. Davis and R.G. Northcutt. Ann Arbor, The University of Michigan Press (in press).

Liem, K.F. and Osse, J.W.M. (1975). Biological versatility, evolution and food resource exploitation in African cichlid fishes. Am. Zool., 15: 427-454.

Liem, K.F. and Stewart, D.J. (1976). Evolution of the scale-eating cichlid fishes of Lake Tanganyika. A generic revision with a description of a new species. Bull. Mus. Comp. Zool., 147: 319-350.

Luling, K.H. (1958). Uber die Atmung, amphibische Lebensweise und Futteraufnahme von Synbranchus marmoratus. Bonn. Zool. Beitr., 9: 68-94.

Mayr, E. (1960). The emergence of evolutionary novelties. In: The evolution of life. Ed. S. Tax. Chicago, Univ. Chicago Press, 349-380.

Muller, M. and Osse, J.W.M. (1978). Structural adaptations to suction feeding in fish. In: Proceedings of the Zodiac Symposium on Adaptation, Netherlands Centre for Agricultural Publishing and Documentation, Wageningen, p. 57-60.

Nyberg, D. (1971). Prey capture in the largemouth bass. Am. Midl. Nat., 86: 128-144.

Osse, J.W.M. (1969). Functional anatomy of the head of the Perch (Perca fluviatilis L.): An electromyographic study. Netherlands J. Zool., 19: 289-392.

Poll, M. (1956). Poissons Cichlidae. Résultats scientifiques exploration hydrobiologiques du Lac Tanganika (1946-1947). 3: 1-619.

Sage, R.D. and Selander, R.K. (1975). Trophic radiation through polymorphism in cichlid fishes. Proc. Natn. Acad. Sci. U.S.A. 72: 4669-4673.

Schaeffer, B. and Rosen, D.E. (1961). Major adaptive levels in the evolution of the actinopterygian feeding mechanism. Am. Zool. 1: 187-204.

Van Valen, L. (1967). Energy and evolution. Evolutionary Theory; 187-204.

Waterman, T.H. (1967). System Theory and Biology--View of a Biologist. Symposium on Systems Approach in Biology. Springer Verlag, p. 1-45.

A MODEL OF SUCTION FEEDING IN TELEOSTEAN FISHES

WITH SOME IMPLICATIONS FOR VENTILATION

Jan W.M. Osse and Mees Muller

Department of Experimental Morphology and Cell Biology

Agricultural University, Marijkeweg 40,
6709 PG Wageningen, The Netherlands

INTRODUCTION

Although no concise definition of environmental physiology can be given and any strict separation from that approach with the feedback regulatory processes inside the animal has to be rejected, the interaction between organism and environment consists of sets of interrelated direct and indirect effects. The present volume amply shows that the effects of temperature, light, oxygen, pH, etc. on fishes, their survival, growth, behaviour and reproduction are profound and complexly interwoven.

This paper deals with the mechanical effects of water on the way fishes capture their food. Since water with its extraordinary physical properties forms (not surprisingly) the main constituent of their environment there is a close link between environmental physiology and the functional morphological approach to be discussed. The understanding of fish head structures can be approached by comparing knowledge of the detailed ecological conditions of their habitat with information on structures and functional demands. An alternative is to consider the general aspects of water flow and pressure variations during aquatic feeding in fish, and formulate on the basis of such information hypotheses on necessary constructional details.

The logical connection, due to physical laws, between flow properties and structural details is similar as met in pumps designed by man for transporting fluids. We try to formulate some hypotheses based on a hydrodynamical model of suction feeding in fish, which can be tested in the actual specimen. Some physical properties of water and their ratio to the same parameters in air show profound differences (see Table I). Liquid and

TABLE I

Comparison of some physical differences between water and air (Modified after Dejours 1975).

Factor	Unit	Dimensions	Water	Air	Water/Air
viscosity (15°C)	η	$kg.m^{-1} sec^{-1}$	1.14×10^{-3}	1.8×10^{-5}	60
density (15°C)	ρ	$kg.m^{-3}$	999	0.001226 (1 atm.)	800
kinematic viscosity (15°C)	$\nu = \frac{\eta}{\rho}$	$m^2 s^{-1}$	1.14×10^{-6}	14.7×10^{-6}	1/13

gas flow although essentially similar differ substantially in their effect on objects in their path of motion. The compressibility of air in contrast to liquids starts to be of influence only at velocities of several hundreds m/sec.

This approach does not mean that we expect to derive all the morphological specialisations from this feeding function alone, being aware of the multiple demands imposed on fish throughout life. As will be shown good evidence exists for the severe demands of suction feeding on fish head architecture as compared to demands for respiration, protection of delicate structures and locomotion.

TYPES OF FEEDING IN AQUATIC HABITATS

The food of fishes can be in a fixed or semifixed position; suspended and slowly moving, fast swimming and turning, it can be large, small or of similar size as the predator.

The predator's movement in the water may do little or nothing to affect the position of fixed food items. Pieces can easily be bitten off as is seen in coral eating parrot fishes or sharks feeding on bulky preys. Suspended food or slow nekton can be filtered from the water by fish moving rather slowly with open mouth and gill slits and specialised branchiospinae. However, this causes a greatly increased drag through increase of the critical surface area of the predator. As the power needed for locomotion (Alexander & Goldspink, 1977) increases with velocity to the third power this method is unsuitable to obtain fast preys. It is used in combination with inertial suction. Planktivorous fishes, like alewifes pump water through their gillrakers and so filter water, remaining themselves rather stationary (Janssen, 1976).

Fast and slow preys, on bottom surface or in midwater are usually caught by suction. When a predator approaches a floating or swimming object in a straight line the moving profile as a whole causes a pressure difference between front and backside of the prey. Therefore the prey will move away from the predator due to its approach and the stagnation point on the predator's snout tip must be eliminated through opening of the mouth. This always means suction, which can be increased by movements of the headparts. These movements are fast and at the same time the forward motion of the predator can reach accelerations of 30-50 m/sec^2. due to tailbeats. A high drag coefficient of the prey will increase its chance of being overtaken.

Another type of feeding is seen in long snouted fishes like Lepisosteus (Lauder, 1979), Belone or Belonesox (Karrer, 1967). The predator grasps the prey with its long jaws after a sudden dart from quite a small distance. This grasping is sometimes done by lateral movements of the head. In the initial grasping phase suction appears to play only a very minor role in these specialized long snouted fishes.

Many fishes combine a fast approach with closed mouth, with a sudden gape during which the buccal and opercular cavities distend rapidly, preparing for a transient suction or engulfing of the fluid mass containing the prey.

The important influence of the predator's movement on the position and velocity of the prey in aquatic feeding as compared to feeding in air is due to the high density of water. This effect makes suction an essential process in aquatic feeding. In view of the fact that the acanthopterygians and many other fish have adopted a feeding system with suction as a central process, we examined the details of the hydrodynamic events accompanying this feeding movement. The forward dart of the predator to the prey resulting from suction was included in later stages of model constrution to provide an overall picture of the feeding event.

ARGUMENTS FOR SELECTING THE MODEL APPROACH OF FISH FEEDING

The reason for selecting a hydrodynamic model as a tool in understanding fish feeding mechanisms and consequently their morphological features, is the need for a generalised theory of construction principles.

a) Available approaches

Previous authors (Woskoboinikoff, 1932; Van Dobben, 1935; Tchernavin, 1948, 1953) describe respiratory and feeding events in certain fishes. They mostly used preserved material to characterise separate movements of neurocranium, pectoral girdle, upper and lower jaws, suspensorium and hyoid during feeding. Perciforms, cyprinids, Salmo and Chauliodus were studied in this way. Living animals were included in studies of head struture and function in the herring (Kirchhoff, 1958). Ammodytes (Kayser, 1962), Anarhichas and Blennius (Thiele, 1963) and Belonesox (Karrer, 1967). Observations, physiological

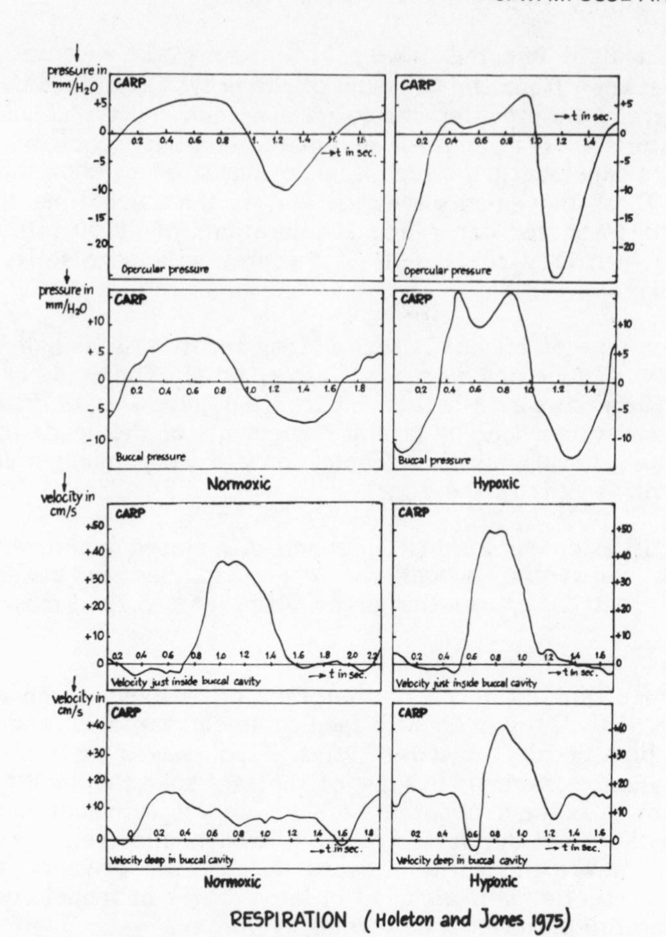

Fig. 1 Graphs of pressures and velocities measured during respiration in normoxic and hypoxic carp by Holeton and Jones, 1975. Compare curves and especially scales with Fig. 2.

apparatus and rather simple mathematical models led Alexander (1967, 1970) to document suction feeding in a number of fishes. Other authors were Liem: motion analysis of feeding in Helostoma (1967): Osse: combination of motion analysis with electromyography of head muscles (1969). Later papers deal with the Nandidae (Liem, 1970), Gasterosteus (Anker, 1974), Haplochromis (Hoogerhoud and Barel, 1978) and Gymnocephalus (Elshoud-Oldenhave & Osse, 1976). Most studies characterise a single species or a group of related types, but none lead to generalisations on the feeding mechanism.

b) Demands on head and body construction during feeding

Electromyograms of head and body muscles suggest that the muscular

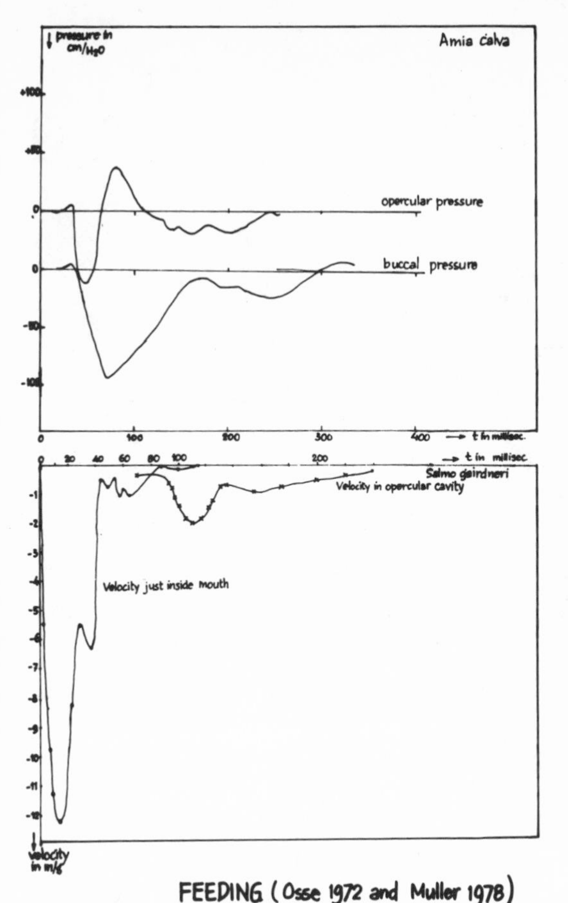

Fig. 2 Graphs of pressures in buccal and opercular cavities measured with Statham pressure transducers and 60 cm polyethylene tubes during feeding in Amia calva. Graphs of flow velocity measured with a hot film micro anemometer during feeding in Salmo gairdneri.

forces exerted during feeding are 6-10 times greater than during respiration (Osse, 1969, Elshoud-Oldenhave & Osse, 1976). This is documented in Fig. 1 and 2 showing pressures and velocities during quiet and hypoxial respiration (Holeton & Jones, 1975) and during feeding. The high velocities of the movement of headparts (jaws, suspensoria, neurocranium, hyoid, operculars, etc.) shown in Fig. 3, demonstrate that the mouth in this lionfish is protruded within 10 milliseconds. The prey is found within the predator's mouth within 20 milliseconds from the start of the feeding action. The body curves formed in larval fishes prior to darting at prey, are again extreme and comparable to the curves obtained from rapid starting slender fish (Weihs, 1973) showing accelerations up to 50 m/sec ^{2} in pike.

It is very likely that the demands of feeding impose the highest stresses on cephalic structures of fishes with predatorial feeding habits.

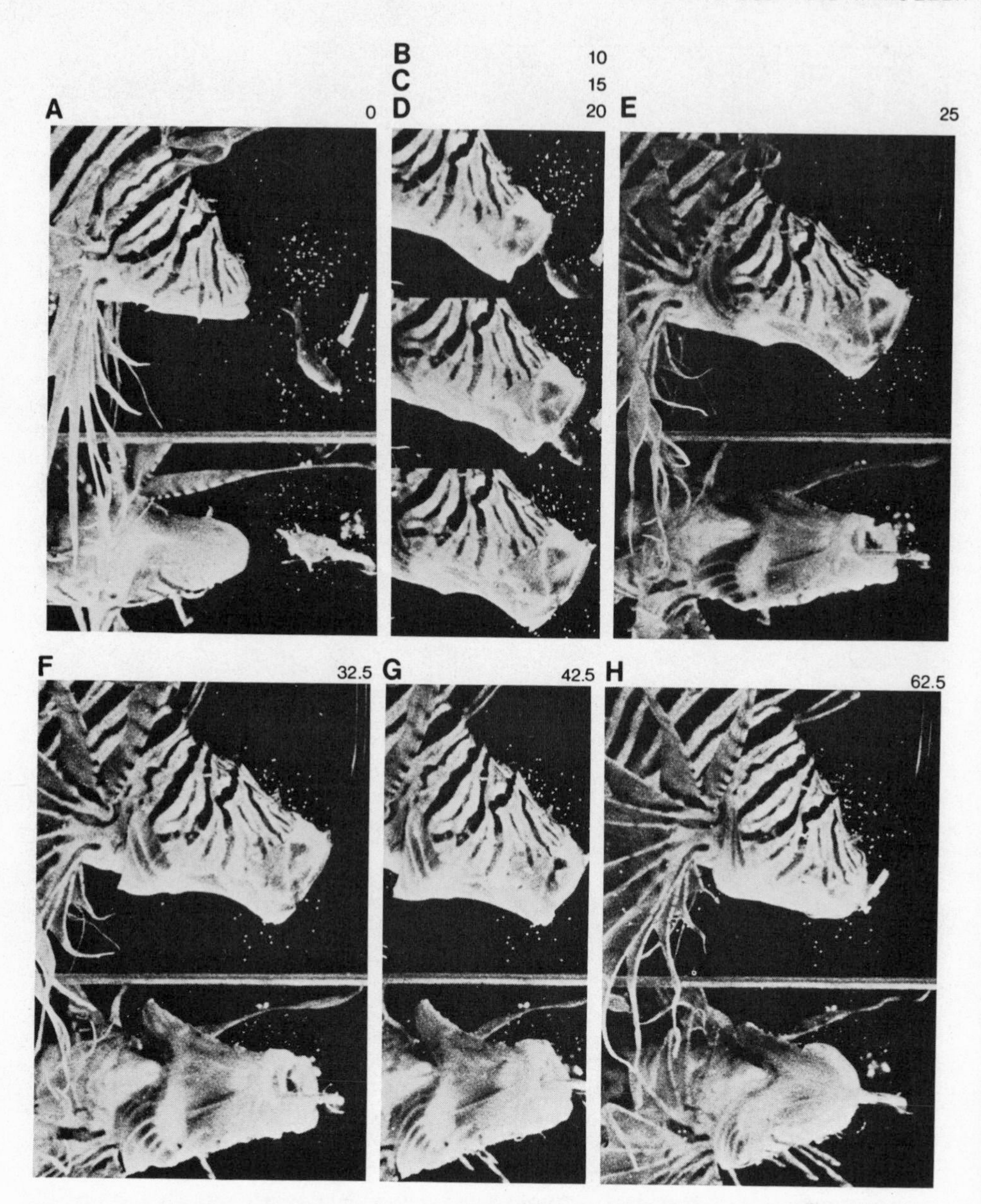

Fig. 3a Lateral and simultaneous ventral views of enlarged frames of a 16 mm 400 frames/sec movie of feeding Pterois russelli. The light dots are silverplated polystyrene spheres introduced to demonstrate the flow. Note the presence of these spheres in flow from the gill slits at later stages in the movement. Time in msec. In B, C and D the ventral views are omitted. Note the approximate rotational symmetry.

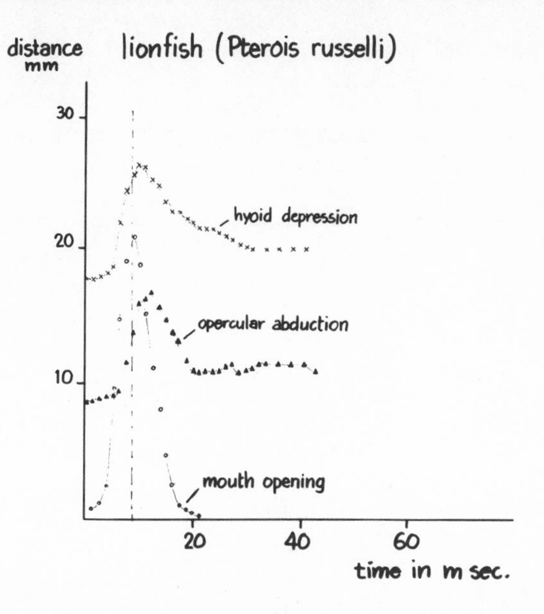

Fig. 3b Graphs of moving head parts from a feeding action of the lionfish. The vertical line indicates the time of opening of the opercular valves.

c) The peak utilisation of structures during feeding suggests that feeding will play a critical role in the ultimate construction of the involved systems. However, other factors, such as respiration, protection of internal structures and defense, as well as cranial streamlining for routine locomotion should not be neglected.

The question can be raised as to what extent the demands of feeding on fast preys will be reflected in the details of the mechanical construction. From the work of Connell (1961) and Schulz (1930) we have indications of the absence of important safety factors in mechanical strength of animal constructions in general. Very soon the disadvantages (such as being heavier, spending too much energy by constructing excessive solid bones, etc.) of being stronger than required to prevent accidental death will be apparent.

A detailed analysis of feeding events and related structural demands is therefore an adequate approach to track constructional demands in the head of fishes.

d) Calculations of the velocity of water entering the mouth and involved pressures have been made in the past by applying Bernouilli's equations for steady flow (Osse, 1969). Reservoir models (Lauder, 1979) showing the influence of the form of opening on the flow, are also based on steady flow conditions. However, during actual feeding the highest flow velocity, starting from zero, is achieved within 25 milliseconds. Errors, of almost an order of magnitude, are made when the continuous changing accelerations are not considered. Therefore, the model approach must include differential equations with continuously changing velocity depending on time and place on the long axis of the fish head (unsteady flow).

e) The model approach, based on hydrodynamic laws, allows the formulation of time independent hypotheses because the properties of water are considered to be constant. Therefore these hypotheses can be applied for measuring the specialisations of the feeding apparatus in recent as well as in extinct fishes.

METHODS

Information is needed on the exact time course of the movements during feeding, the velocity of the inflowing current, the hydrodynamic pressures involved and the movement of the prey resulting from the predators actions.

A Teledyne 54 DB 16 mm camera provided motion pictures of 400 frames per second and enabled us to measure the movements. Fishes were trained to feed at a specific place of the tank allowing simultaneous pictures of lateral and ventral views with the aid of a mirror. Exposure time was 0.25 milliseconds.

The velocity of the current was determined by suspending silver plated polystyrene spheres (0.5 mm diameter) with the exact density of the water (cf Fig. 3). A hot film micro anemometer (TSI Inc., St. Paul, Minn.) was used to measure flow inside the buccal cavity of a free swimming fish. The system is accurate to better than 5% up to 10 KHz. Pressures inside the buccal and opercular cavities were measured with Statham P 23 Db pressure transducers with a response time of 10 msec. Entran and Gaeltec pressure transducers were also used. Besides being fragile the dynamic properties of none of these are completely adequate for measuring rapidly changing pressures. Our curves (Fig. 2) together with those of Alexander (1970), Casinos (1973) and Lauder (in press) may however be considered as the best available indications of events.

Initially *Pterois russelli* (lionfish, Fig. 3) was selected as our experimental animal because of its voracity, its tolerance for high light intensities and the rather stationary way of feeding. Later feeding events were also studied and measured in trout, cod, flatfish and several other types.

CONSTRUCTION OF THE MODEL

Although a mechanical model of an expanding fish head is possible, we chose for a hydrodynamic model as with it the possibility of simulating several situations seemed more favourable. As the pressure and velocity at every point characterize the streaming motion (Milne-Thomson, 1976) we aimed at calculating these parameters. We used a form, cylinder or cone, performing a prefixed movement with respect to its amplitude and time course.

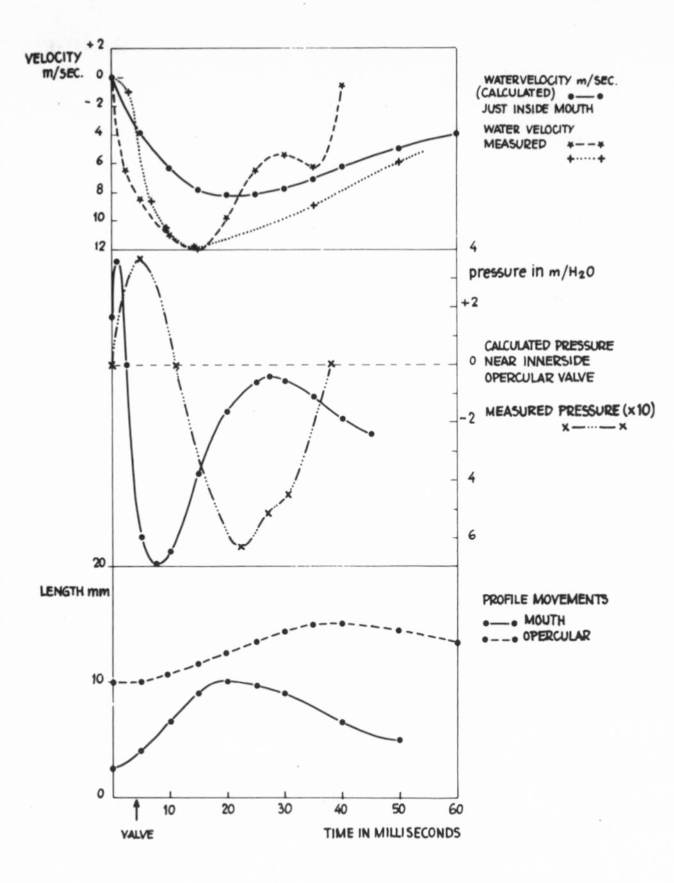

Fig. 4 Profile movements of the cone model with calculated pressure and velocity (graphs with black dots). Graphs of measured pressure (x) and velocities (+, ✱) are from separate experiments with Salmo gairdneri. The model movements differ from actual feeding in Salmo in the initially too small mouth opening and the slope of the mouth movement. The sensitivity of the model for parameter variation is therefore apparent. Further explanation in the text.

The concept of prefixed movements is acceptable because many of these suction processes are so fast that no between times neuromuscular adjustments appear to be possible.* The expanding structure has an anterior entrance, opening and closing in the course of a suction act. At its posterior end we envisaged a valve allowing an outflow of the current and opening at some pre-set conditions. The whole set of conditions determining the operating mode of the model are the boundary conditions. The velocities and pressures are calculated as variables of the place on the axis in the moving profile (x) and time (t) using the equation of continuity and the equation of motion.

* The resonance frequency of the inertial mass of the fish head was calculated from known movement curves assuming a simultaneous, step-like contraction of the involved muscles. This resonance frequency appears to be in the order of 10 Hz and therefore halfway adjustments in the fast movements would require 10 to 100 fold energy increase.

Measurements in a feeding fish permitted comparison of magnitude and time course of calculated and real values (Fig. 4). This served to improve the model in order to apply it for simulation of different situations and to generate hypotheses about expected parameter values. Calculations, especially of expected hydrodynamic pressures (checked in actual specimen), to which head structures will be subjected during suction feeding, have direct consequences for the necessary solidity of local head structures. This line of thought will be illustrated with some examples.

THE MODEL

Three simplifying assumptions are made:

1. Friction is neglected. A boundary layer gradually forms when a flow builts up. This boundary layer starts at the point of the object first meeting the flow and increases from that point on. The dimensions of the boundary layer can be estimated from the Navier-Stokes equation when the characteristic size and time scales for motion as encountered in fish sucking movements are substituted. The thickness of the boundary layer appears to be a fraction of a millimeter. Therefore the use of formulae without friction can be expected to give a close approximation of the actual events.

2. The prey moves as an element of the water. This is clear for a tiny prey having a density equal to water. For every prey the ratio between water and prey velocity is equal to the ratio of their densities. For most preys no appreciable difference exists with water. We have not observed differences in the suction movements between successful and unsuccessful feeding acts. This assumption is therefore justified.

3. In an expanding model due to the continuity equation flow will occur in three dimensions. The calculation of the velocity and pressure in directions perpendicular to the long axis of a fish becomes very

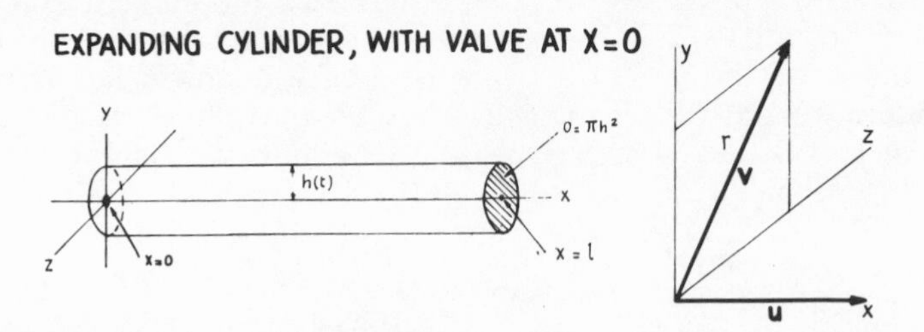

Fig. 5 Scheme of an expanding cylinder. Water velocity can be considered as uniform over the surface of the cross section when $l \geq 2.5h$, max. is water velocity in x-direction. Further explanation in the text.

complicated when y and z calculations are different. They are equal when the profile is rotational symmetric (Fig. 5), leading to a two dimensional system. The obtained films show the acquisition of rotational symmetry of the expanding profile at a very early stage of the suction movement, at least at its anterior end (Fig. 3).

The model calculations are therefore based on an expanding rotational symmetric profile. Details of the hydrodynamic model will be published elsewhere (Muller and Osse in preparation).

A short version of the results is given below.

THE CYLINDER MODEL AND THE MODEL OF THE EXPANDING CONE

Our first model consists of an expanding cylinder (Fig. 5) with a length l along the x-axis and a radius of h. At $x = l$ the mouth opening will be formed. When the radius is small compared to the length the flow in a direction perpendicular to the x-axis is negligible. The main boundary conditions are: $p(l,t) = 0$ and $u(0,t) = 0$, the valve at the caudal side opens when the pressure there is equal to P_0 (P_0, the ambient hydrostatic

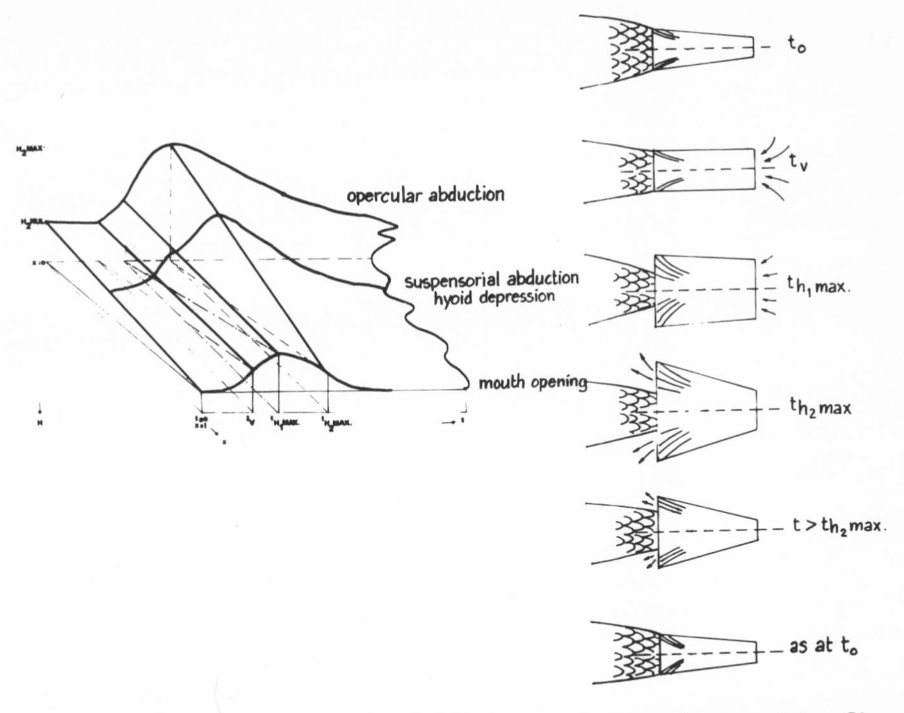

Fig. 6 Three dimensional view of an expanding conical profile. Six stages of the movements of mouth and opercular valve are given in ventral view. Arrows indicate water flow.

pressure, d = actual hydrodynamic pressure, u = velocity, t = time). The water velocity is proportional to the first derivative, the pressure curves are shaped by the second derivative of the profile motion. So the time course of these parameters can be judged from the movement curves. Although some of the results obtained resemble the curves from actual measurements, several improvements were necessary. The cylinder expanding uniformly along its length does not allow a retardation of the posterior expansion in the gill region with respect to the mouth. This is possible with an expanding conical profile (Fig. 6). This also is a better approximation of the real form of a fish mouth cavity. The boundary condition stating that outside the fish no pressure change occurs ($p = P_0$ at $x = l$, i.e. at the mouth opening) is inadequate because quite soon after the start of the movement a considerable flow enters the mouth. Therefore the flow outside the expanding profile must be considered. In fact the flow through the opened mouth is the result from water flowing into the enlarging buccal cavity, from a sudden dart at the prey (the velocity of the predator due to its swimming motion towards the prey) and from its forward motion as a result of suction. The importance of the latter movement depends on the ratio between the sum of the masses of sucked prey and water and the mass of the predator.

A good approximation of the total flow pattern is obtained when a circular vortex filament, laying at the lips, is visualised in combination with a parallel flow. In this way a funnel shaped flow profile, external to the fish, is created. This is of particular importance because it confirms the need of the predator to aim its suction at the prey. In the previous cylinder model every fluid particle in a sphere around the centre of the mouth cavity had an equal probability of being sucked which is in contrast with the observations. When the predator shows hardly any forward motion, as in the case of a flatfish, a particle lateral to its cheek was seen entering the mouth.

The dimensions of the profile, the amount and rates of rostral and caudal expansion, the forward velocity of the profile and the opening time of the caudal valve determine the velocity and pressure curves in time over the axis of the profile. These curves were calculated with a computer program because the complex functions prevent analytical solutions. Comparison of calculated and measured curves in a feeding fish show good agreement. Therefore the model has been used to simulate the effect of increasing length of an expanding profile on maximal pressure and velocity while the other parameters were constant. Fig. 7 shows the approximately parabolic increase in hydrodynamic pressure in such a case.

HEAD STRUCTURE AND GILL OPENING IN SYNGNATHIDAE

In a number of Syngnathids: Nerophis, Entelurus and Syngnathus, the length of the mouth tube increases (Table II, Fig. 8).

These fishes feed with a fast upward movement of the head (acceleration 43.10^3 rad/sec^2 in Entelurus) combined with a fast depression of

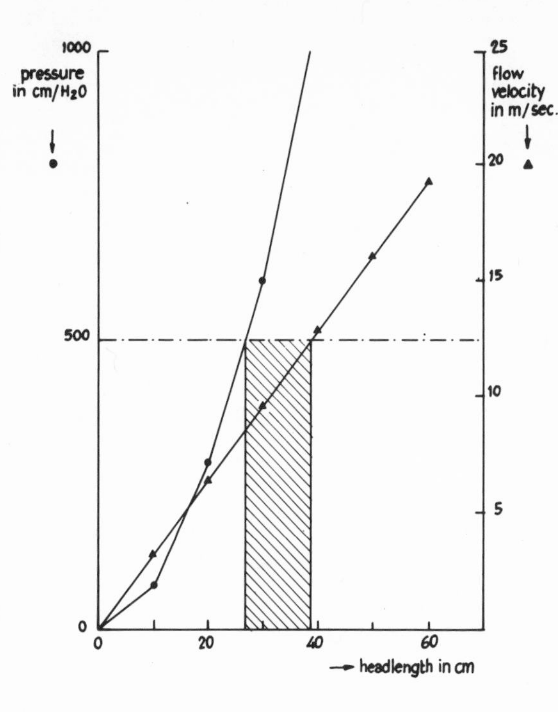

Fig. 7 Graphs showing the influence of head length variation on expected hydrodynamic pressure and velocity. Calculations from a model with l = 10 cm, h_1 = 0.5 cm, h_2 = 3.5 cm at t = 0, h_{1max} = 1.5 cm, h_{2max} = 4.5 cm, $t_{h_{1max}}$ = 27 msec, $t_{h_{2max}}$ = 37 msec (cf Fig. 6). The hatched area indicates lengths where pressure and velocity reach such high values that alteration of at least one of the involved parameters is expected.

TABLE II

Dimensions of head and opercular in three Syngnathidae. Curvature and sculpture of the opercular increase with absolute headlength. This indicates an increased ability to resist high pressure differences.

	Nerophis	Entelurus	Syngnathus
bodylength (mm)	103	230	415
headlength (mm)	9.4	20.1	56
snoutlength (eye-snouttip,mm)	3.3	9.3	32
ratio snoutlength-headlength	35%	46%	57%
opercular curvature *	0.22	0.29	0.33
opercular sculpture	absent	low ridges	high ridges

* Measured as the size independent ratio of perpendicular and chord in cross-section at half length

SYNGNATHIFORMES

Syngnathus acus

Entelurus aequoraeus

Nerophis lumbriciformis

Macrorhamphosus scolopax

Fig. 8 Lateral views of the head of three Syngnathidae and of Macrorhamphosus.

the slender hyoid followed by a rather slow abduction of the opercular region. Assuming other parameters constant, the model predicts an approximately parabolic increase of negative hydrodynamic pressure in the gill region. The increased pressure is reflected in an increased curvature and surface sculpture of the opercular. The gill slits are nearly closed by a firm sheet of connective tissue. Only at the extreme dorsoposterior tip a small funnel-like opening is left. The gills are lobed and lack the streamline profile in horizontal section found in many other fish. No quantitative data on oxygen consumption are available, but the ventilation mechanism does not differ essentially from other teleosts (Hughes, 1960). The normal vertical position of pipe-fishes among leaves of seaweed and the absence of active swimming may indicate that their oxygen consumption is rather low. The construction of the opercular region of Entelurus and Syngnathus can therefore be considered as a consequence of a pipette feeding type causing a low hydrodynamical pressure in the gill area. The firm sheet of connective tissue uniting the gill arches at their medial sides with small slits to the lateral opercular cavities, the lobed construction of the gills and nearly completely blocked gill slits with grown together branchiostegal rays in combination with the strongly ossified gill cover are considered as adaptations to the type of feeding. The efficiency of a counter-current flow of blood and water for respiration might well be reduced as a result of the type of feeding.

The pipe-fishes suck over a relative long period a great volume of water with a low velocity. This type of feeding can be described as volume-suction. Another solution to a long pipette-like snout is expected in Macrorhamphosus. This fish appears to feed in a vertical position with a downwards pointing head in muddy areas. A thin gill cover and streamlined gill arches and gills allow the passage of a fast flow for a very short time because the opercular valve probably opens very early after hyoid and suspensorial abduction. This type of velocity-suction might prevent that detritus enters the gill area.

DISCUSSION AND CONCLUSION

Capture of food by suction is a common method in aquatic feeding. The use of inertial forces to swallow the food in terrestial animals (Gans, 1969) is another example of applied physics used by animals during food intake.

Head structures of fishes applying suction are very efficient for this feeding type. The intrinsic head muscles are used during feeding for the positioning of the system of mandibular, hyoid and branchial arches. The epaxial and hypaxial muscles are at the same time used for supplying the power for the fast six to ten fold volume increase of the buccal cavity within 50 milliseconds (Osse, 1969). In this way the buccal and opercular pump mechanisms used for respiration are adapted to perform the feeding function. The gradual appearance of long slender and curved branchiostegal rays (Mc Allister, 1968) in the opercular valve in teleosts together with an increased surface area of the opercular allows enlargment of the lateral opercular cavities with closed gill slits (Osse, 1976). Thus more water can be sucked during one feeding act. The concommitant evolution of preys for rapid starting and turning probably necessitated this line of the development in predators specializing for this type of food.

Several other specializations like the formation of a round mouth opening (Osse, 1969, p. 377, Lauder, 1979) and protrusion (Alexander, 1967) are generally found in fishes. The remarks of O'Brien (1979, p. 577) about notched mouth in salmonids are inaccurate because he compares the rest situation in a salmonid with the round mouth during feeding in white fish or smelt.

The adaptive levels in the evolution of actinopterygian feeding mechanism described by Schaeffer and Rosen (1961) are mainly based on size increase of the mandibular adductor muscle, grasping and biting functions of the jaws and increasing movability of maxilla and premaxilla. Increased efficiency for suction feeding also asks for adaptive changes of the opercular and branchiostegal systems to increase the volume of sucked water. A more complete picture will result when these aspects are included in the designation of adaptive levels in fish evolution. Although the example of the syngathids points to the possible incompatibilities between structural demands for different functions (c.f. Hoogerhoud & Barel, 1978) more information is needed to increase the evidence.

SUMMARY

Feeding by suction in fishes is studied starting from the laws of hydrodynamics. A model has been constructed which describes the feeding act as the unfolding and refolding of a conical rotational symmetric profile. In this model different movements of mouth and opercular can be substituted as well as the delay between them. The predictions of the model have been controlled by actual measurements of velocity and pressure. Calculations suggest a high negative hydrodynamic pressure in the gill area of longsnouted pipefishes due to headlength and relative slow suspensorial and opercular movement. The tiny circular dorsal gill opening as well as other structural details of the opercular region can be regarded as adaptations to this type of feeding. These structural modifications limit the possibilities of countercurrent flow in gill irrigation.

A major improvement in the distinction of the adaptive levels in the evolution of actinopterygian fishes is possible when the evolution of the opercular and branchiostegal apparatus for improved suction feeding is added to the classical ideas of Schaeffer and Rosen (1961).

(The investigations were supported by the Foundation for Fundamental Biological Research (BION), which is subsidized by the Netherlands Organization for the Advancement of Pure Research (ZWO)).

ACKNOWLEDGEMENTS

The expert advice of two engineers, Ir. J.H.G. Verhagen and Ir. K.D. Maiwald from the Delft Hydraulics Laboratory appeared indispensable in the course of the development of the model. We are grateful for the contributions made by Johan van Leeuwen, Albert Ramakers and Arie Terlouw. Thanks are also due to Wim Valen for his drawings and to Nel van Cappelle for her typework. Comments on a previous version of this paper by Dr. C. Gans and Dr. K.F. Liem are gratefully acknowledged.

REFERENCES

Alexander, R. McN. (1967) The functions and mechanisms of the protrusible upper jaws of some acanthopterygian fish. J. Zool. (Lond.) 151: 43-64.

Alexander, R. McN. (1970) Mechanics of the feeding of various teleost fishes. J. Zool. (Lond.) 162: 145-156.

Alexander, R. McN. & Goldspink, G. (1977) Mechanics and Energetics of Animal Locomotion. Chapman and Hall, London, p. 346.

Anker, G. Ch. (1974) Morphology and kinetics of the head of the stickleback, Gasterosteus aculeatus. Trans. Zool. Soc. Lond. 32: 311-416.

Casinos, A. (1973) El mecanisme de deglucio de l'aliment a Gadus callarias, Linnaeus 1758 (Dades preliminary). Bull. Soc. Catalana Biol. I., p. 43-52.

Connell, J.H. (1961) Effects of competition, predation by Thais lapillus and other factors on natural populations of the barnacle, Balanus balanoides. Ecol. Monogr. 31: 61-104.

Dejours, P. (1975) Principles of Comparative Respiratory Physiology. Amsterdam, North-Holland Pub., 253 p.

Dobben, W.H. van (1935) Uber den Kiefermechanismus der Knochenfische, Arch. Neerl. Zool. 2: 1-72.

Elshoud-Oldenhave, M.J.W. & Osse, J.W.M. (1976) Functional Morphology of the Feeding System in the Ruff Gymnocephalus cernua (L. 1758) - (Teleostei, Percidae). J. Morph. 150: 399-422.

Gans, C. (1969) Comments on inertial feeding. Copeia 4: 855-857.

Holeton, G.F. & Jones, D.R. (1975) Water flow dynamics in the respiratory tract of the carp (Cyprinus carpio L.). J. Exp. Biol. 63: 537-549.

Hoogerhoud, R.J.C. & Barel, C.D.N. (1978) Integrated Morphological Adaptations in Piscivorous and Mollusc-Crushing Haplochromis Species. Proc. Zodiac Symposium (on adaptation). Pudoc, Wageningen, The Netherlands, p. 52-57.

Hughes, G.M. (1960) A comparative study of gill ventilation in marine teleosts. J. Exp. Biol. 37: 28-45.

Janssen, J. (1976) Feeding modes and prey size selection in the alewife (Alosa pseudoharengus). J. Fish. Res. Board Can. 33: 1972-1975.

Karrer, Chr. (1967) Funktionell-anatomische und vergleichende Untersuchung des Schädels vom Hechtkärpfling, Belonesox belizanus Kner (Teleostei, Cyprinodontiformes, Poeciliidae). Zool. Jb. Anat. 84: 191-248.

Kayser, H. (1962) Vergleichende Untersuchung uber Vorstreckmechanismen der Oberkiefer bei Fischen. Der Bau und die Function des Kiefer - und Kiemenapparates von Knochenfischen der Gattungen Ammodytes und Callionymus. Zool. Beitr. (n. ser.) 7: 321-445.

Kirchgoff, H. (1958) Funktionell-anatomische Untersuchung des Visceralapparates von Clupea harengus L. Zool. Jb. Anat. 76- 461-540.

Lauder, G.V. (1979) Feeding mechanics in primitive teleosts and in the halecomorph fish Amia calva. J. Zool. Lond. 187: 543-578.

Lauder, G.V., Jr. & Lanyon, L.E. (1980) Functional anatomy of feeding in the bluegill sunfish, Lepomis Macrochirus: in vivo measurement of bone strain. J. Exp. Biol. (in press).

Liem, K.F. (1967) Functional morphology of the head of the anabantoid Teleost fish Helostoma temmincki. J. Morph. 121: 135-158.

Liem, K.F. (1970) Comparative functional anatomy of the Nandidae (Pisces, Teleostei). Fieldiana. Zool. 56: 1-166.

McAllister, D.E. (1968) Evolution of branchiostegals and classification of Teleostome fishes. Bull. Nat. Mus. Can. 221, I-XIV: 1-239.

Milne-Thomson, L.M. (1976) Theoretical Hydrodynamics, 5th ed, London, Macmillan.

Muller, M. & Osse, J.W.M. (1978) Structural adaptation to suction feeding in fish. Proc. Zodiac Symp. (on adaptation), p. 57-60. Pudoc Wageningen, The Netherlands.

O'Brien, W.J. (1979) The predator-prey interaction of planktivorous fish and zooplankton. Am. Sci. 67: 572-581.

Osse, J.W.M. (1969) Functional morphology of the Perch (Perca fluviatilis L.). An electromyographic study. Neth. J. Zool. 19: 289-392.

Osse, J.W.M. (1976) Mécanismes de la respiration et la prise des proies chez Amia calva Linnaeus. Rev. Trav. Inst. Pêches Marit. 40: 701-702.

Schaeffer, B. & Rosen, D.E. (1961) Major adaptive levels in the evolution of the actinopterygian feeding mechanism. Am. Zool. 1: 187-204.

Schultz, A.H. (1930) The skeleton of the trunk and limbs of higher primates. Human Biol. II: 303-438.

Tchernavin, V.V. (1948) On the mechanical working of the head in bony fishes. Proc. Zool. Soc. Lond. 118: 129-143.

Tchernavin, V.V. (1953) The feeding mechanisms of a deep sea fish. British Museum (Nat. Hist.), Lond. p. 1-101, pls. 1-10.

Thiele, H. (1963) Vergleichend-morphologische Untersuchungen über die Funktion der Nahrungserwerbsapparate von Anarrhichas Lupus L. und einigen Blenniidae (Teleostei, Blennioidei). Zool. Beitr. (n. ser.) 9: 275-440.

Weihs, D. (1973) The mechanism of rapid starting of slender fish. Biorheology 10: 343-350.

Woskoboinikoff, M.M. (1932) Der Apparat der Kiemenatmung bei den Fischen. Zool. Jb. Anat. 55: 316-448.

CHEMICAL STIMULI: ROLE IN THE BEHAVIOR OF FISHES

Logi Jónsson
Department of Physiology
University of Iceland
Grensasvegur 12, Reykjavik, Iceland

INTRODUCTION

Chemoreception plays a major role in important areas of fish behavior such as feeding, migration, spawning, care of young, schooling and defence. Chemosensory adaptations to a variety of environmental conditions exist. These are mainly adaptations of receptor sites or of central processing systems rather than morphological changes of the chemosensory organs. It is not my intention to present a complete review of the literature on chemical sensation in fish, but rather to discuss briefly the properties of chemical stimuli in water, and to point out the importance of certain chemicals in fish behavior.

CHEMICAL STIMULI IN THE AQUATIC ENVIRONMENT

Compounds utilized as chemical signals in the aquatic environment are in solution rather than in the gas phase. Therefore the most obvious requirement a chemical must fulfill in order to be a chemical stimulus for fish is that it must be water soluble. Chemical stimuli for fishes range in size from inorganic ions such as Ca^{++} (Bodznik, 1978) to macromolecules, including proteins and polypeptides (Hara and MacDonald, 1976; Atema et al., 1973).

The specificity of a chemical stimulus is determined by its molecular structure, including its stereo-isometric configuration. Given the large variety of molecules that can be used as chemical signals, the possibility for different, specific signals are nearly endless. Most chemical signals turn out to be mixtures of compounds, each mixture carrying specific information. We may think of such mixtures as chemical "pictures" that fish can identify and react to.

One important property of chemical pictures is their persistence; when a chemical picture is produced it lasts beyond the moment of production. This is important in territorial and shelter marking in fish. (P. Rubec, pers. obs.; Müller-Schwartze and Atema, pers. obs., 1978; cited in Atema, 1979).

The distribution of a chemical stimulus in the aquatic environment depends upon diffusion which is very slow in water and is therefore of minor importance. The diffusion will, however, result in a concentration gradient which is steep close to the source, but which decreases with distance away from the source. Movement of the medium relative to the source or of the source relative to the medium, will result in a chemical trail. Chemical stimuli may, however, be intermittently released from an irregularly moving source in a turbulent environment. This will result in an uneven distribution of the chemical.

Chemical signals can be of various origin. They can for example be compounds entering the environment from animal metabolism in faeces or urine, specific pheromones released into the environment, compounds released from plants or microorganisms, or decomposition products.

CHEMICAL ORIENTATION

Locating a stimulus source (e.g. prey) is obviously of great survival value for fishes. Chemical stimuli are non-directional *per se*, but steep gradients and polarized trails contain sufficient information for localization.

Tropotaxis or comparison of chemical concentration between two points has been shown in some species. It depends upon very sensitive or widely spaced receptors (Little, 1979). The goldfish (*Carassius auratus*) is sensitive to steep gradients close to the source but is not able to orient within shallow gradients (Kleerekoper et al., 1972). The European eel (*Anguilla anguilla*) can follow a chemical trail by successive comparison of concentration along the trail (Teichmann, 1959). Bullheads (*Ictalurus nebulosus*) are able to find a distant source of chemical stimuli by taste, and they do this more easily in still water than in currents (Bardach et al., 1967).

Most often, however, the chemical stimulus triggers rheotaxis, or orientation to a water current. Rheotaxis is maintained as long as the chemical stimulation persists and adaptation does not occur. Food stimuli elicit rheotaxis in the lemon shark (*Negaprion brevirostris*) and the yellowfin tuna (*Thunnus albacares*) (Matthewson and Hodgson, 1972; Atema et al., 1979). Kleerekoper (1967) showed that a flow rate 10% greater than ambient water flow was needed for the *Diplodus sargus* to localize the source of food odors (see also Tesch, this volume).

Frequently, chemical signals in water are directionally indeterminate, and in such cases it is impossible for fish to pinpoint their source by chemical clues alone. The only way to locate the source in such cases is to search the water mass containing the stimulus and to rely upon other, more directional, clues (e.g. visual, auditory or electrical) in finding the exact source location.

Thus the general rule seems to be that fishes are alerted by chemical stimuli and initiate search, often in an upstream direction. Precise localization is accomplished through other stimuli. However, catfishes and eels can pinpoint sources by following chemical gradients or trails (see Atema, 1979, for further discussion).

OLFACTION AND TASTE

The distinction between olfaction and taste in fish has been a controversial topic. Chemical receptors are generally called olfactory if they are stimulated by volatile substances, are highly sensitive and specific, and are distance receptors. On the other hand, chemical receptors which are stimulated by solutions of moderate concentration and which are contact receptors are called taste receptors. These descriptions do not successfully distinguish between olfaction and taste in fish, where both sensory systems are stimulated by chemicals dissolved in water, and where the same chemical can stimulate both olfactory and taste receptors (Caprio, 1977).

The peripheral receptors and the neuroanatomy of the olfactory and taste systems in fish seem to be very similar to those of all other vertebrates. Olfactory receptors are ciliated primary receptors situated on the olfactory rosette in the nasal chamber. Their axons form the olfactory nerve (I) which enters the olfactory bulb. Fibers from the olfactory bulb form the olfactory tract and make connections to different parts of the forebrain (for review see Kleerekoper, 1969; Bardach and Villars, 1974; Hara, 1975). Taste receptors are microvillar epithelial cells grouped together to form taste buds. They are located in the oral cavity and in some species spread over the body and are innervated by the facial (VII), glossopharyngeal (IX) and vagal (X) nerves. These three nerves enter the medulla area of the brain (for review see Bardach and Atema, 1971; Bardach and Villars, 1974).

Strieck (1924) demonstrated that there is a functional difference between olfaction and taste in fish. Trained minnows (*Phoxinus phoxinus*) can discriminate between classical olfactory stimuli -- coumarin, muscone and skatole -- and taste substances -- sucrose, sodium chloride, acetic acid and quinine. When the forebrain was removed, the fishes were no longer able to discriminate between the olfactory stimuli while they still could discriminate between the taste substances.

Atema (1969) demonstrated that in yellow bullhead (*Ictalurus natalis*) dead bait is found by taste only. The facial part of the taste system is thought to function as a distance receptor and also controls the picking up of food. The vagal and glossopharyngeal nerve portion of the taste system is essential for swallowing of food. The olfactory system, on the other hand, is involved in intraspecific communication. This sharing of labor between the two sensory systems is confirmed by work on the Hawaiian goatfish (Parupeneus porphyreus) (Holland, 1978). In another study Atema (1977) showed that in the brown bullhead and the yellowfin tuna the olfactory sense is responsible for the mediation of learned chemical cues and serves to initiate feeding behavior. In general, responses to olfactory stimuli seem to vary with the environmental conditions, while responses to taste stimuli appear to be less plastic.

The existence of chemical search images in the olfactory system, analogus to visual pattern recognition, has been postulated (Atema, 1977, 1979; Atema et al., 1979). These search images are thought to be neutral filters at the level of the olfactory bulb and serve to enhance the reception of important chemical cues such as prey odors and homestream odors in the case of anadromous fishes. Thus olfaction is programmed to recognize specific chemical stimuli in changing chemical environments. It is involved in various appetitive behaviours, including prey detection, social recognition and homestream recognition. Taste search images may exist but they seem to be limited to food stimuli and possible poisons. Thus taste is involved in consummatory phases of feeding behavior (Atema, 1979).

Caprio (1977) has shown electrophysiologically that olfactory and taste receptors in catfish have overlapping response spectra for amino acids. Little (1977) using a heart rate conditionning technique and Atema (1977) and Herbert and Atema (1977) using behavioral training have shown that behavioral responses evoked by olfaction differ from those mediated by taste even when both are stimulated by the same amino acid. Therefore the behavioral meaning of a stimulus seems to depend on whether it is received by olfaction or taste.

To summarize: The distinction between olfaction and taste in fish is anatomically and functionally as clear in fish as it is in terrestrial vertebrates.

SENSITIVITY OF CHEMICAL SENSES

Olfactory sensitivity shows variation among species. Teichmann (1959) found the threshold concentration for β -phenylethanol in the eel to be $3x10^{-18}$M which is equal to $2x10^{3}$ molecules/ml or only a few molecules at a time in the 1-2 mm^{3} nasal cavity of the fish. This threshold concentration is comparable to that of the sex pheromone, bombykol, in the silk worm (Bombyx mori) which is $3.1x10^{4}$ molecules/ml (Kaissling, 1971). In minnows, Teichmann found the threshold concentration for β -phenylethanol to be ca 10^{-7}M, and in rainbow trout (Salmo irideus) $8x10^{-10}$M.

Glaser (1966) used a behavioral technique to determine the threshold concentration for various taste substances (different sugars, saccharin, quinine, sodium chloride and acetic acid) in minnows. The threshold concentration for quinine was the lowest at $4x10^{-8}$M; for saccharin it was $6.5x10^{-7}$M. These threshold concentrations are 24-2560 times lower than threshold concentrations for the same substances in humans. Elimination of the olfaction did not affect the sensitivity for quinine. Since the threshold concentration for quinine was lower than the threshold value for β - phenylethanol mentioned earlier, Glaser concluded that taste receptors in fish are as sensitive as their olfactory receptors. This result is in contrast with findings in terrestrial vertebrates, where olfactory receptors are more sensitive than taste receptors.

While the above mentioned experiments are of some interest, it should be noted that many of the stimulants utilized rarely -- or never-- occur in the natural environment of fish.

As a class, amino acids are by far the most powerful stimulants among the compounds tested. They stimulate both olfactory and taste receptors (Caprio, 1977). Those with the amino group in α-position are the most effective and the L-isomer is more effective that the D-form. Electrophysiological recordings from the olfactory epithelium of the young Atlantic salmon (Salmo salar) showed these threshold concentrations: for L-alanine 3.2×10^{-9}M, L-threonine 2.5×10^{-7}M and L-proline 3.2×10^{-5}M (Sutterlin and Sutterlin, 1971).

From recordings from the olfactory nerve in catfish (Ictalurus catus), threshold concentrations were estimated to be 10^{-8} to 10^{-7}M for the five most potent amino acids (Suzuki and Tucker, 1971). Olfactory bulb EEG recordings in Pacific salmon (Oncorhyncus kisutch) showed concentration thresholds 10^{-7} to 10^{-6}M for the most potent amino acids (Hara, 1972). Similar values for artic char (Salvelinus alpinus) rainbow trout and brook trout (Salvelinus fontinalis) were 10^{-8} to 10^{-7}M (Belghaug and Döving, 1977; Hara, 1973; Hara et al., 1973). Behavioral olfactory thresholds for amino acids using a heart rate conditioning technique were 10^{-8} to 10^{-7} M in the carp (Carassius carassius), 10^{-9} to 10^{-8}M in the channel catfish (Ictalurus punctatus) and 10^{-9} to 10^{-7}M in the artic char (Suzuki, 1973; Little, 1979; Jonsson, unpublished data) Atema et al. (1979) found, by looking at feeding behavior, that yellowfin tuna can detect L-trypthophane at concentrations as low as 10^{-11}M.

The taste system of catfish is extremely sensitive to amino acids. Electrophysiological recordings from the facial nerve showed threshold concentrations to be 10^{-12} to 10^{-9}M, which are the lowest taste thresholds reported in a vertebrate and are lower than the olfactory threshold mentioned above. Because of this high sensitivity Caprio (1975a, 1975b) concluded that the taste receptors in catfish must be considered distance receptors. Using a conditioned - escape response, thresholds for amino acids in the same species were found to range from 10^{-10} to 10^{-8}M (Little, 1975; cited in Little 1979). The development of two very sensitive chemosensory systems is very interesting, but the catfish is probably special in this respect due to the extremely numerous taste buds spread over the entire body and especially dense on the barbels.

The taste systems in other species show low sensitivity to amino acids. In puffer fish (Fungu pardalis) and Atlantic salmon parr, only few of several amino acids tested were effective at concentrations as high as 10^{-2} M (Hidaka and Kiyohara, 1975; Sutterlin and Sutterlin, 1970).

The threshold concentrations mentioned above show great variation. This is due to differences among species and to differences among the techniques used in the various studies. Seasonal fluctuation in sensitivity poses further complications. The olfactory threshold for β-phenylethanol in the European eel is a million times higher in late fall and early winter than the rest of the year (Teichmann, 1959). The seasonal fluctuation may be due to endocrine changes, since elevated levels of sex hormones associated with the pre-spawing period indicate increased olfactory sensitivity in the goldfish (Hara, 1967; Partridge et al., 1976). It is further suggested that thyroid hormones increase olfactory sensitivity in the lungfish (Protopterus annectens) and increased level of thyroid hormones associated with the onset of the wet season causes an "awakening reaction of olfactory origin" (Godet and Dupe, 1965; Dupe, 1973; cited in Atema, 1979). Whether the effect of these hormones is a general increase in sensitivity or a selective increase in sensitivity for seasonally important odors (such as sex pheromones, or homestream odors in anadromous fishes) is an open question.

CHEMICAL STIMULI AND BEHAVIOR

Amino acids

Some effort is currently being devoted to the study and identification of the active chemical components in natural food stimuli, a practical consequence of which could be the production of artificial baits for commercial fishing.

Johannes and Webb (1970) showed that several aquatic invertebrates release free amino acids into the environment in specific ratios. The specific ratio of amino acids, as well as other compounds released from each animal, may constitute a chemical picture specific for that particular animal. Different fish species appear to respond differently to identical chemical pictures. This is thought to be because each species has its own characteristic chemical search image and it is through matching of the chemical search image with a chemical picture that behavioral responses are elicited (Atema, 1979). Yellowfin tuna for example respond to prey school odor, the odor of fresh dead prey and prey tissue extract. The single most prevalent amino acid in the prey odor, tryptophan, elicits a slight response (Atema et al., 1979; Atema, 1977).

Most studies however have been done on tissue extracts rather than naturally released chemicals. Extracts from the clam (Tapes japonica) contain seven amino acids and stimulate feeding responses in eels (Anguilla japonica). A synthetic mixture of these amino acids also elicits feeding behavior but not as strongly as the original clam extract. Eels are most sensitive to arginine, but a combination of amino acids is more effective than single amino acid stimuli (Konosu et al., 1967).

Fractions of mullet, crab, shrimp, oyster and sea urchin extracts which stimulate feeding responses in the pinfish (Lagodon rhomboides) and the pigfish (Orthopristis chrysopterus) show that betaine is the most effective stimulus in these species. However other amino acids in specific concentrations, and in some cases other compounds, have to be added to obtain stimulatory effect equal to the original extract. Species differences are evident since in pigfish a synthetic mixture of 20 amino acids and betaine is as effective a stimulus as shrimp extract while in the pinfish it has only about 25% of the potency of shrimp extract (Carr, 1976; Carr and Chaney, 1976; Carr et al., 1976; Carr et al., 1977).

Pawson (1977) showed that the fraction of the lugworm (Arenicola marina) which stimulated feeding in the whiting (Merlangius merlangius) and the cod (Gadus morhua) contained seven amino acids. An artificial mixture of these amino acids in concentrations similar to those found in lugworms elicited a strong response but not as strong as the original extract. Whiting failed to respond to this mixture without glycine and alanine. In this study, the response to the artificial mixture could be attributed solely to the concentration of glycine in the mixture. However other non-amino compounds were required for stimulating potency equal to the original extract. In the lugworm extract glycine and alanine were found in concentrations 100 times higher than the other five amino acids. This suggests that whiting and cod focus narrowly on the prey's most common amino acids.

Extensive field studies are needed to support laboratory results. To my knowledge only one field study has been carried out so far. It showed that Atlantic silversides (Menidia menidia) were most attracted to alanine, winter flounder (Pseudopleuronectes americanus) to glycine and mummichogs (Fundulus heteroclitus) to GABA. Winter flounders and mummichogs even picked up and rejected oyster shells and other substrate material close to the spot where the amino acids were released (Sutterlin, 1975).

The amino acids are particularly important in stimulating feeding behavior in several fish species. As mentioned earlier, L-α- amino acids as a class of compounds are the most potent chemical stimuli tested. This is not surprising considering that these amino acids are the major constituents of animal proteins and most fish tested are predatory species. Mixtures of compounds are always more efficient stimuli than a single compound. Each species seems to have its own search image of a particular food (for review see Atema, 1979).

Amino acids are also involved in chemically mediated avoidance behavior. Skin rinses of mammalian predators (bears, seals, humans, etc.) caused cessation of upstream migration in all five species of Pacific salmon (Brett and MacKinnon, 1954). The most effective fraction of the mammalian skin rinse contained amino acids, and serine was the most potent constituent. The observed reaction presumably protects fish against predation (Idler et al., 1956,1961).

It is worth noting species differences, since in contrast to Pacific salmon, whiting, cod and yellow bullheads are attracted to serine (Pawson, 1977; Atema, 1979).

Substances in the skin mucus enables fishes to discriminate between individual conspecifics (Todd et al., 1967; Atema et al., 1969). Amino acids may be involved since the amino acid composition in the skin mucus of brown bullheads changes after they have experienced the stress of a territorial fight (Bryant et al., 1978). Tucker and Suzuki (1972) analyzed the chemical composition of the alarm substance, Schreckstoff (see below), in catfish. It was shown to be a mixture of several compounds some of which are amino acids. This indicates that amino acids may play a role in social behavior in fishes.

Pheromones

A pheromone is, by definition, a substance which is "secreted to the outside of an individual and received by a second individual of the same species, in which it releases a specific reaction, for example a definite behavior or a developmental process" (Karlson and Lüscher, 1959). Communication by pheromones has been shown to be involved in many stages in the life cycles of fishes.

Nordeng (1971) suggested that pheromones are involved in homestream selection in salmonids. The standing fish population in the stream is thought to release the pheromone into the water. Individuals on their way back from the sea therefore recognize the stream in which their relatives are living. In a more recent paper this pheromone hypothesis is extended to include also the open sea navigation in salmonids. Smolts migrating from the river to the open sea are thought to leave pheromone trail behind them, which adult migrating individuals can follow back to the homestream (Nordeng, 1977). Other theories on fish migration are discussed by Tesch in a separate chapter in this volume.

Sex pheromones have been reported in some fish species. Small quantities of seawater holding a gravid female gobie (*Bathygobius soporator*) are sufficient to elicit courtship display by the male (Tavolga, 1956). Ovarian fluid of a gravid female was the only effective stimulus of the several body fluids tested (these included skin mucus). The male gobie also responds to freshly extruded eggs, but they lose their effectiveness within 10 minutes after extrution. Neither males nor females respond to male skin mucus or testicular fluids. Similar results were obtained from the guppy (*Poecilia reticulata*), and synthetic estrogen was found to have an effect similar to that of ovarian fluid (Almouriq, 1965; cited in Little, 1979). Secretions produced in an anal gland present only during the mating season in female Astyanax mexicanus induce the male to spawn (Wilkens, 1972; cited in Little, 1979). Similarly an anal gland secretion from the male characin (Corynopoma tenius) stimulates females to spawn (Nelson, 1964; cited in Little, 1979). Chemicals released from both sexes of the zebrafish (Brachydanio rerio) showed inter- and intrasexual attraction, when the chemical was present in a certain concentration (Bloom and Perlmutter, 1977). Insufficient or excessive concentrations showed no effect. It is hypothesized that this could regulate the proportion between males and females within a spawning area, or determine the size of a school.

The chemical senses appear to play a role in the care of the young in some species of fish. Young jewel fish (Hemichromis bimaculatus) emit chemical stimuli which influence parental behavior (Kühme, 1963). The parents prefer water containing their young fry. They are able to distinguish the water containing their brood from others of the same species or other species. Myrberg (1966) found in a cichlid (Cichlasoma nigrofasciatum) that they distinguish their own young from conspecific young and selectively eat the latter. The young Cichlasoma nigrofasciatum is also attracted by chemical stimuli from the adults (Myrberg, 1975). Similar results have been obtained from another cichlid (Cichlasoma citrinellum) (Noakes and Barlow, 1973).

Fish aggregation or schooling is, at least to some extent, regulated by pheromenes, although vision is the most significant sensory system in this respect. Kinosita (1972) demonstrated that a pheromone is involved in schooling in the marine catfish eel (Plotosus anguillaris). Although they will indiscriminately follow any moving elongate object, two schools do not join together when brought into the same aquarium. Small groups from a school remain selectively attracted to their own school even after a long separation. The pheromone is released from the body surface and is specific for each school. Schools of herring (Clupea harengus) are also maintained in the dark by chemical stimuli (Jones, 1962).

The existence of alarm pheromones have been demonstrated in many species belonging to the Ostariophysian and Gonorynchiform families (Pfeiffer, 1977). Von Frish (1938) discovered that when an injured minnow was introduced into a school of minnows they exhibited a fright reaction and fled away from the injured one. This fright reaction is caused by an alarm substance (Schreckstoff) which is released from an injured minnow skin. If an individual is attacked by a predator, the alarm substance released will warn other conspecifics of the danger. The alarm substance is produced and stored in club-shaped cells in the epidermis. These cells will only release their content to the body surface upon injury (Pfeiffer, 1960). Several Cyprinid species, all of which dig gravel nests, temporally lose their club cells at the onset of the spawning season (Smith, 1976). This allows them to dig the nests without releasing the alarm substance. The club cell loss is controlled by testosterone and can be induced with testosterone injections (Smith, 1973). The fright reaction is not species specific (Schutz, 1956). However it is always strongest in conspecifics and weakens in proportion to the taxonomic distance between the species.

Pollutants

Several fish species are able to detect and respond to sublethal concentrations of a variety of chemical pollutants (Little, in prep.). The rainbow trout for example avoids 10^{-4} mg/l copper sulfate solution, although the toxicity level is 1400 times higher or 0.14 mg/l (Folmar, 1976). Pulpmill effluents in a stream stop upstream migration in Pacific salmon (Brett and McKinnon, 1954). This avoidance is obviously of great survival value for the fishes, since the toxic effects are thereby minimized or averted.

Whitefish (*Coregonus clupeaformis*), on the other hand, is attracted to solutions of copper in high concentrations which necessarily increases the toxic effects (Hara and Scherer, 1977; cited in Little, in prep.).

Pollutants may be perceived through irritations of membranes in contact with the environment, such as sensory membranes, gill epithelium or the body surface. "However, perception of contaminants is largely mediated through olfaction and taste where the contaminants interact with sensory receptor sites as natural stimuli do. The perception of a chemical depends upon a confirmational match between the contaminant molecule and the receptor site. The receptor site has evolved for receiving chemicals of a particular molecular structure, thus perception of a contaminant may be accidental" (Little, in prep.).

Pollutants that do not act as olfactory or taste stimuli may still influence chemosensory function. They may block receptor sites, destroy sensory tissue, mask or mimic natural stimuli (Sutterlin, 1974). Heavy metals such as zinc have been shown to have detrimental effects on olfactory epithelium in several organisms (Smith, 1938; Moulton, 1974) and electrophysiological measurements showed that the function of the olfactory epithelium of Pacific salmon was impaired after it had been exposed to heavy metals (Hara, 1972). Similar effects were obtained in whitefish and Atlantic salmon using detergents (Hara and Thompson, 1978; Sutterlin and Sutterlin, 1971). Detergents in sublethal concentrations caused structural degeneration of the taste buds in yellow bullhead and feeding behavior was thereby altered (Bardach et al., 1965). Bloom et al. (1978) showed that female zebrafish do not respond to female pheromone in concentrations which normally attract them, after exposure to a sublethal concentration of zinc.

Behavioral responses of fishes to pollutants may be a very useful tool in monitoring aquatic pollution in the future. However, a lot of work remains to be done within this vast but interesting field.

ACKNOWLEDGEMENTS

I wish to thank Drs. Atema and E.E. Little for assistance in collecting the references for this work, Dr. M. Marlies for critically reading through the manuscript and to Miss. E. Sigurdardóttir for typing the manuscript. I also want to thank the University of Iceland for financial support.

REFERENCES

Atema, J. (1969) The chemical senses in feeding and social bahavior of the catfish (Ictalurus natalis). Ph.D. Thesis, Univ. Michigan. 134p.

Atema, J. (1977) Functional separaion of smell and taste in fish and crustacea. In: Olfaction and Taste VI. Eds. J. Le Magnen and P. MacLeod. London, Info. Retrieval Ltd., p. 165-174.

Atema, J. (1979) Chemical senses, chemical signals and feeding behavior in fishes. In: Fish Behavior and its Use in the Capture and Culture of Fishes. Eds. J.E. Bardach, J.J. Magnuson, R.C. May and J.M. Reinhart. Manila, Int. Center for Living Aquatic Resources Management.

Atema, J., Bylan, D. Jacobsen, S. and Todd, J. (1973) The importance of chemical signals in stimulating behavior of marine organisms: effects of altered environmental chemistry on animal communication. In: Bioassay Techniques andEnvironmental Chemistry, Ed. G.E. Glass, Ann Arbor Sci. Publ., p. 177-197.

Atema, J., Holland, K. and Ikehara, W. (1979) Chemical search image: olfactory responses of yellowfin tuna (Thunnus albacares) to prey odors. J. Chem Ecol. (in press).

Bardach, J.E. and Atema, J. (1971) The sense of taste in fishes. In: Handbook of Sensory Physiology, IV, 2. Ed. L.M. Beidler. Berlin, Springer-Verlag, p. 293-336.

Bardach, J.E., Fujiya, M. and Moll, A. (1965) Detergents: effects on the chemical senses of the fish Ictalurus nebulosus. Sci. 148: 1605-1607.

Bardach, J.E., Todd, J.H. and Crickmer, R. (1967) Orientation by taste in fish of the genus Ictalurus. Sci. 155: 1267-1278.

Bardach, J.E. and Villars, T. (1974) The chemical senses of fishes. In: Chemoreception in Marine Organisms. Eds. P.T. Grant and A.M. Mackie. New York, Academic Press, p. 49-104.

Belghaug, R. and Döving, K.B. (1977) Odour threshold determined by studies of the induced waves in the olfactory bulb of the char (Salmo alpinus L.). Comp. Biochem. Physiol. 57A: 327-330.

Bloom, H.D. and Perlmutter, A. (1977) A sexual aggregation pheromone system in the zebrafish, Brachydanio rerio. J. Exp. Zool. 199: 215-226.

Bloom, H.D., Perlmutter, A. and Seeley, R.J. (1978) Effect of a sublethal concentration of zinc on aggregating pheromone system in the zebrafish, Brachydanio rerio. Environ. Pollut. 17: 127-131.

Bodznik, D. (1978) Calcium iron: an odorant for water discriminations and the migratory behavior of sockeye salmon. J. Comp. Physiol. 127A: 157-166.

Brett, J.R. and MacKinnon, D. (1954) Some aspects of olfactory perception in migrating adult coho and spring salmon. J. Fish. Res. Board Can. 11: 310-318.

Bryant, B., Elgin, R. and Atema, J. (1978) Chemical communication in catfish: Stress-induced changes in body odor. Biol. Bull. 155: 429.

Caprio, J. (1975a) High sensitivity of catfish taste receptors to amino acids. Comp. Biochem. Physiol. 52A: 247-251.

Caprio, J. (1975b) Extreme sensitivity and specificity of catfish gustatory receptors to amino acids and derivatives. In: Olfaction and Taste V. Eds. D.A. Denton and J.P. Coghlan, New York, Academic Press, p. 157-161.

Caprio, J. (1977) Electrophysiological distinctions between the taste and smell of amino acids in catfish. Nature 266: 850-851.

Carr, W.E.S. (1976) Chemoreception and feeding behavior in the pigfish, Orthopristis chrysopterus: Characterization and identification of stimulatory substances in a shrimp extract. Comp. Biochem. Physiol. 55A: 153-157.

Carr, W.E.S., Blumenthal, K.M. and Netherton, J.C., III (1977) Chemoreception in the pigfish, Orthopristis chrysopterus: The contribution of amino acids and betaine to stimulation of feeding behavior by various extracts. Comp. Biochem. Physiol. 58A: 69-73.

Carr, W.E.S. and Chaney, T.B. (1976) Chemical stimulation of feeding behavior in the pinfish, Lagodon rhomboides: characterization and identification of stimulatory substances extracted from shrimp. Comp. Biochem. Physiol. 54A: 437-441.

Carr, W.E.S., Gondeck, A.R. and Delanoy, R.L. (1976) Chemical stimulation of feeding behavior in the pinfish, Lagodon rhomboides: a new approach to an old problem. Comp. Biochem. Physiol. 54A: 161-166.

Folmar, L.C. (1976) Overt avoidance reaction of rainbow trout fry to nine herbicides. Bull. Environ. Contam. Toxicol. 15: 509-514.

Glaser, D. (1966) Untersuchungen über die absoluten Gesmaksschwellen von Fischen. Z. Vergl. Physiol. 49: 492-500.

Hara, T.J. (1967) Electrophysiological studies of the olfactory system of the goldfish, Carassius auratus L. II. Effects of sex hormones on the electrical activity of the olfactory bulb. Comp. Biochem. Physiol. 22: 209-226.

Hara, T.J. (1972) Electrical responses of the olfactory bulb of Pacific salmon, Oncorhynchus nerka and Oncorhynchus kisutch. J. Fish. Res. Board Can. 29: 1351-1355.

Hara, T.J. (1973) Olfactory responses to amino acids in rainbow trout, Salmo gairdneri. Comp. Biochem. Physiol. 44A: 407-416.

Hara, T.J. (1975) Olfaction in fish. In: Progress in Neurobiology 5 (4). Eds. G.A. Kerkut and J.W. Phillis. Oxford, Pergamon Press, p. 271-335.

Hara, T.J., Law, Y.M.C. and Hobden, B.R. (1973) Comparison of the olfactory response to amino acids in rainbow trout, brook trout and whitefish. Comp. Biochem. Physiol. 45A: 969-977.

Hara, T.J. and MacDonald, S. (1976) Olfactory responses to skin mucus substances in rainbow trout, Salmo gairdneri. Comp. Biochem. Physiol. 54A: 41-44.

Hara, T.J. and Thompson, B.E. (1978) The reaction of whitefish Coregonus clupeaformius to the anionic detergent sodium lauryl sulphate and its effects on their olfactory responses. Water Res. 12: 893-897.

Herbert, P. and Atema, J. (1977) Olfactory discrimination of male and female conspecifics in the bullhead, catfish, Ictalurus nebulosus. Biol. Bull. 153: 429-430.

Hidaka, I. and Kiyohara, S. (1975) Taste responses to ribonucleotides and amino acids in fish. In: Olfaction and Taste V. Eds. D.A. Denton and J.P. Coghlan. New York, Academic Press, p. 147-151.

Holland, K. (1978) Chemosensory orientation to food by an Hawaiian goatfish (Parupeneus porphyreus, Mullidae). J. Chem. Ecol. 4: 173-186.

Idler, D.R., Fagerlund, J.H. and Mayoh, H. (1956) Olfactory perception in migrating salmon I. L-Serine a salmon repellent in mammalian skin. J. Gen. Physiol. 39: 889-892.

Idler, D.R., McBride, J.R., Jonas, R.E.E. and Thomlinson, N. (1961) Olfactory perception in migrating salmon II. Studies on a laboratory bioassay for homestream water and mammalian repellent. Can. J. Biochem. Physiol. 39: 1575-1584.

Johannes, R.E. and Webb, K.L. (1970) Release of dissolved organic compounds by marine and fresh water invertebrates. In: Symposium on Organic Matter in Natural Waters. Ed. D.W. Hood. Inst. of Mar. Sci., Univ. of Alaska, College Alaska, p.257-273.

Jones, F.R.H. (1962) Further observation on the movements of herring (Clupea harengus L.) schools in relation to the tidal current. J. Const. Int. Explor. Mer. 27: 52-76.

Kaissling, K.E. (1971) Insect olfaction, In: Handbook of Sensory Physiology 4 (1). Ed. L.M. Beidler. Heidelberg, Springer Verlag, p. 351-431.

Karlson, P. and Lüscher, M. (1959) "Pheromones": a new term for a class of biologically active substances. Nature. 183: 55-56.

Kinosita, H. (1972) Schooling behaviour in marine catfish eel, Plotosus anguillaris. Zool. Mag. 81: 241.

Kleerekoper, H. (1967) Some effects of olfactory stimulation on locomotor patterns in fish. In: Olfaction and Taste II. Ed. T. Hayashi. Oxford, Pergamon Press, p. 625-645.

Kleerekoper, H. (1969) Olfaction in Fishes. Bloomington, Indiana Univ. Press, 222p..

Kleerekoper, H., Westlake, G.F., Matis, J.M. and Gensler, P. (1972) Orientation of goldfish (Carassius auratus) in response to a shallow gradient of a sublethal concentration of copper in an open field. J. Fish. Res. Board Can. 29: 45-54.

Konsu, S.N., Fusetani. N., Nose, T. and Hashimoto, Y. (1968) Attractants for eels in the extracts of short-necked clam II. Survey of constituents eliciting feeding behavior by fractionation of the extracts. Bull. Jap. Soc. Sci. Fish. 34: 84-87.

Kühme, W.V. (1963) Chemish Ausgelöste Brutpflege - und Schwarmverhalten bei Hemichromis bimaculatus (Pisces). Z. Tierpsychol. 20: 688-704.

Little, E.E. (1977) Conditioned heart rate response to amino acids in channel catfish. Proc. Internatl. Union Physiol. Sci. XIII: 450.

Little, E.E. (1979) Behavioral function of olfaction and taste in fish. In: Fish Neurobiology and Behavior I. Eds. R.E. Davis and G. Northcutt.

Univ. of Michigan Press (in press).

Matthewsen, R.F. and Hodgson, E.S. (1972) Klinotaxis and rheotaxis in orientation of sharks toward chemical stimuli. Comp. Biochem. Physiol. 42A: 79-84.

Moulton, D.G. (1974) Dynamics of cell populations in the olfactory epithelium. Ann. N.Y. Acad. Sci. 273: 56-61.

Myrberg, A. (1966) Parental recognition of young in cichlid fishes. Animal Behav. 14: 565-571.

Myrberg, A. (1975) The role of chemical and visual stimuli in the preferential discrimination of young by cichlid fish, Cichlastoma nigrofasciatum (Günter). Z. Tierpsych. 37: 274-297.

Noakes, D.L.G. and Barlow, G.W. (1973) Cross-fostering and parent- off - spring responses in Cichlostoma citrinellum (Pisces, cichlidae). Z. Tierpsychol. 33: 147-152.

Nordeng, H. (1971) Is the local orientation of anadromous fishes determined by pheromones? Nature 233: 411-413.

Nordeng, H. (1977) A pheromone hypothesis for homeword migration in anadromous salmonids. Oikos 28: 155-159.

Partridge, B.L., Liley, N.R. and Stacey, N.E. (1976) The role of pheromones in the sexual behavior of the goldfish. Anim. Behav. 24: 291-299.

Pawson, M.G. (1977) Analysis of a natural chemical attractant for whiting, Merlangius merlangius L., and cod, Gadus morhua L., using a behavioural bioassay. Comp. Biochem. Physiol. 56A: 129-135.

Pfeiffer, W. (1960) Über die Schreckreaktion bei fischen und die Herkunft des Schrecstoffes. Z. vergl. Physiol. 43: 587-614.

Pfeiffer, W. (1977) The distribution of fright reaction and substance cells in fishes. Copeia 4: 653-665.

Schutz, F. (1956) Über die Schreckreaktion bei Fischen und die Herkunft des Schreckstoffes. Z. vergl. Physiol.. 38: 84-135.

Smith, C.G. (1938) Changes in the olfactory mucosa and the olfactory nerves following internasal treatment with 1% zinc sulfate. Can. Med. J. 39: 135-140.

Smith, R.J.F. (1973) Testosterone eliminates alarm substance in male fathead minnows. Can. J. Zool. 51: 875-876.

Smith, R.J.F. (1976) Seasonal loss of alarm substance cells in North American Cyprinoid fishes and its relation to abrasive spawning behavior. Can. J. Zool. 54: 1172-1182.

Strieck, F. (1924) Untersuchungen uber den Geruchs - und Geschmackssinn der Elritzen. Z. vergl. Physiol. 2: 122-154.

Sutterlin, A.M. (1974) Pollutants and the chemical senses of aquatic animals, perpectives and review. Chem. Senses Flavor 1: 167-178.

Sutterlin, A.M. (1975) Chemical attraction of some marine fish in their natural habitat. J. Fish. Res. Board Can. 32: 729-738.

Sutterlin, A.M. and Sutterlin, N. (1970) Taste responses in Atlantic salmon (Salmo salar) parr. J. Fish. Res. Board Can. 27: 1927-1942.

Sutterlin, A.M. and Sutterlin, N. (1971) Electrical responses of the olfactory epithelium of Atlantic salmon (Salmo salar). J. Fish. Res. Board Can. 28: 565-572.

Suzuki, N. (1973) Olfactory conditioning of fish by amino acids. Zool.

Mag. 82: 285-290.

Suzuki, N. and Tucker, D. (1971) Amino acids as olfactory stimuli in freshwater catfish, Ictalurus catus (Linn.). Comp. Biochem. Physiol. 40A: 399-404.

Tavolga, W.N. (1956) Visual, chemical and sound stimuli in sex-discriminatory behavior of the gobiid fish Bathygobius soporator. Zool. 41: 49-64.

Teichmann, H. (1959) Über die Leistung des Geruchssinnes beim Aal (Anguilla anguilla L.) Z. vergl. Physiol. 42: 206-254.

Todd, J.H., Atema, J. and Bardach, J.E. (1967) Chemical communication in social behavior of a fish, the yellow bullhead (Ictalurus natalis). Sci. 158: 672-673.

Tucker, D. and Suzuki, N. (1972) Olfactory responses to Schreckstoff of catfish. In: Olfaction and Taste IV. Ed. D. Schneider. Wissenschaftliche Verlagsgesellschaft MBH, Stuttgart, p. 121-127.

Von Frisch, K. (1938) Zur psychologie des fisch-schwarmes. Naturwissenschaften 26: 601-606.

THE EFFECT OF HYDROSTATIC PRESSURE ON FISHES

J.H.S. Blaxter

Dunstaffnage Marine Research Laboratory

Oban, Argyll, Scotland

INTRODUCTION

The effect of hydrostatic pressure on fish may be examined in two ways, by subjecting fish to different pressures and observing changes in their biochemistry, physiology and behaviour or by observing the characteristics of fish caught at different depths and inferring whether these characteristics are pressure-related or not.

It is not always easy to isolate primary pressure effects because of changes in other characteristics of the environment with depth, especially temperature and light penetration. For every 10 m depth increase the pressure increases by about 1 atm. The average depth of the ocean is about 4000 m and the maximum depth about 11000 m. An enormous volume of water at high pressure is thus available to organisms.

Pressure has the following direct effect on water (Kinne, 1972; Macdonald, 1975):

1. increases the density (e.g., by about 4% at 1000 atm) so that salt concentrations are higher;

2. depresses the freezing point, e.g., at surface -1.8°C and at 1000 atm -9°C.;

3. decreases the solubility of gases so raising their partial pressure. According to Henry's Law, M = PS where M is the mass of gas dissolved, P is the partial pressure and S the solubility;

4. increases conductivity;

5. increases viscosity.

Other highly significant changes with depth are, however:

1. a reduction in temperature to $<4^{o}$C in deep water;

2. an exponential fall in light intensity so that even in clear oceanic water no downwelling light will be present at 500-1000 m.;

3. often a reduction in oxygen in the "oxygen minimum layer", sometimes to levels as low as 1% saturation.

The changes may interact with pressure. For example the partial pressure of dissolved gases is raised by high pressures which reduce solubility and lowered by low temperatures which increase solubility.

Many of the adaptations of deep sea fish such as bioluminescence, sensitive eyes, black colouration and large mouths are clearly related to non-pressure aspects of depth and reflect the need to maintain intra-and inter-specific contact. Deep sea fish may have to obtain oxygen in water of low oxygen content and they may have to deal with the more pressure-related problem of maintaining buoyancy by means of a compliant gas-filled swimbladder. The maintenance of buoyancy will be made more difficult by many fish making substantial diel vertical migrations, towards the surface at night and away from the surface by day. The main theme of this paper will be the role of hydrostatic pressure in limiting the distribution of fish, working either through potential lethal or sub-lethal effects or through the limitations of the fishes physiology at different pressures.

PRESSURE TOLERANCE

1. Increases of pressure

Practically all the work on testing the ability of fish to withstand pressure increases (see Flügel, 1972) has been done on inshore species since deep sea fish usually arrive at the surface in moribund condition and are unfit for experimentation. Some 24 h LD_{50} values are given in Table 1 (Naroska, 1968). In addition Brauer (1972) reported the following convulsion pressures for shallow-water fish *Achirus fasciatus* 130 atm, *Paralichthys dentatus* 105 atm, *Anguilla rostrata* 102 atm and *Symphurus palguisa* 86 atm. Although ecologically somewhat meaningless these data suggest that deep water may act as a barrier to the migration of inshore species off the continental shelf.

In general high pressures inhibit ciliary activity and block cell division (Flügel, 1972). *Fundulus* embryos cease to divide at 330-400 atm. Heart rate in *Fundulus* embryos is at first increased by pressure but becomes reduced at 82 atm. The respiratory rates of flatfish such as plaice and flounders increase markedly after a change of pressure from 0 to 100 atm. (Fig. 1).

TABLE 1
Tolerance of fish to pressure (From Naroska, 1968)

Species	24 h LD_{50} atm
Platichthys flesus (flounder)	130
Pleuronectes platessa (plaice)	150
Zoarces viviparus (blenny)	370

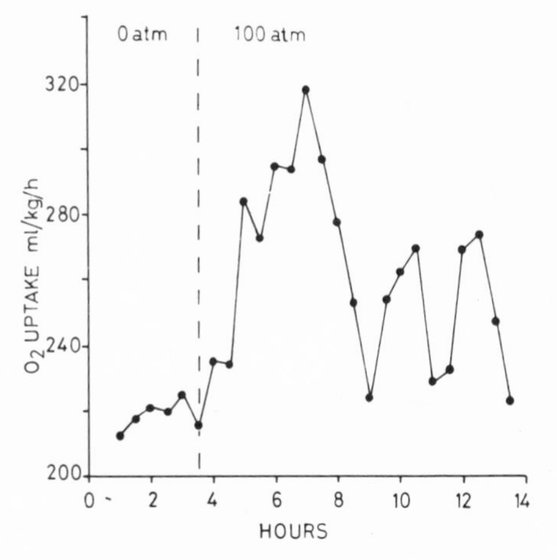

Fig. 1. Oxygen uptake of the flounder Platichthys flesus before and after a 100 atm pressure change (redrawn from Naroska, 1968).

Such physiological changes have a biochemical basis but much of the biochemical work has been done on small organisms such as bacteria and protozoa or excised portions of tissue (see Pequeux, this Volume), usually from invertebrates found in shallow depths. According to Le Chatelier's Principle (see Macdonald, 1975) pressure increases will favour conditions which occupy the least volume. Thus chemical or enzymic reactions will be stimulated by pressure if the reaction results in a volume decrease. Since enzymes "unfold" to accept the substrate an increase in volume is usually involved. The formation of enzyme - substrate complexes is thus inhibited by pressure (Kinne, 1972). In general pressures of 300-3000 atm inhibit biochemical reactions such as RNA and protein synthesis.

Some recent work done on a cruise of "Alpha Helix" is described in various papers in Comp. Biochem. Physiol. Vol. 52B, especially on the deep sea morid Antimora rostrata caught in about 2000 m. In this species enzymes such as acetylcholinesterase, lactate dehydrogenase, citrate synthase and liver isocitrate dehydrogenase are adapted to the low temperature: high pressure regime of the surroundings. It seems likely, therefore, that deep sea fish have evolved biochemical mechanisms which make them less vulnerable to high pressures.

2. Decreases of pressure

Pressure decreases do not lend themselves to the same type of study on inshore organisms since pressures below atmospheric are not normally experienced by fish except in the cooling systems of power stations. Fish will, however, undergo pressure decreases during vertical migration or during capture. Almost no physiological or biochemical work has been done on deep sea fish after capture. Prosser et al., (1975) showed changes in ion permeability with high Na and low K in the muscles, extensive haemolysis and inactivation of heart contractions when various species were brought to the surface from depths of 1400-2280 m. Such fish were, however, also experiencing rises of temperature of 20° C or more which might have been harmful to stenothermal deep sea species and it is difficult to isolate possible primary pressure effects.

A much more obvious direct pressure effect is the bursting of swimbladders and eversion of body cavity contents of fish brought up from quite modest depths. Tytler and Blaxter (1973) constructed a decompression schedule based on the internal pressure required to burst the swimbladders of various species and their ability to resorb gas (see later). Very little evidence is available of decompression sickness in fish. Since fish are not breathing gas at high pressures like human divers, this is hardly surprising (Casillas - personal communication).

BUOYANCY AND PRESSURE

Compliant gas-filled swimbladders are the most common buoyancy adaptation in fish. Maintaining their integrity is a major problem for fish living at depth or vertically migrating. Fish may be divided into four main groups:

a. Physostomes with swimbladders equipped with a duct to the exterior. Usually (apart from the eel Anguilla) there is little or no gas secretion and gas is obtained by swallowing and lost by voiding. This is the "primitive" condition present in clupeoids and salmonids and also in the Ostariophysi.

b. Physoclists with closed swimbladders, a gas secretion and a gas resorption mechanism. These include the gadoids and many species of deep water fish.

c. Fish with oil-filled swimbladders, especially a few deep sea species.

d. Fish without swimbladders, including many bottom-living flatfish and many deep sea species.

For physoclists the problem of maintaining the buoyancy function of a swimbladder at a constant depth, or when changing depth during vertical migration, depends on a number of factors:

a. The extent and depth of vertical movement. Near the surface a given excursion constitutes a greater % pressure change. For example moving from 0 to 10 m involves a doubling of pressure, but from 10 to 20 m only an increase of 1.5 times and from 20 - 30 m an increase of 1.33 times. The swimbladder is a compliant structure and will therefore, for a given vertical excursion, change more in volume near the surface than away from it. Loss of buoyancy will occur with pressure increase and danger of bursting with pressure decrease.

b. the ability to secrete gas against a gradient.

c. The ability to control gas loss from the swimbladder to the surrounding tissues.

d. The extent to which the swimbladder can "safely" expand.

1. Gas secretion

Under most conditions (except where there is an oxygen-minimum layer) the partial pressure of N_2 and 0_2 in sea water regardless of depth will be 0.8 and 0.2 respectively. The swimbladder is compliant and if it is to be inflated at depth high partial pressures will be required within the swimbladder.

The secretion mechanism has been fairly fully investigated (see review by Blaxter and Tytler, 1978) and takes place via rete mirabile (a network of blood capillaries) and an epithelial surface within the swimbladder called the gas gland. A counter current multiplication system operates as follows: the arterial and venous capillaries in the rete are closely opposed. Oxygen is unloaded from the arterial capillaries by the action of lactic acid and possibly carbonic anhydrase. The epithelial cap of the gas gland produces lactic acid from blood glucose which is released into the efferent capillaries. The resulting drop in pH releases 0_2 from oxyhaemoglobin (Root Effect and Bohr Shift) and the lactic acid causes "salting-out", a reduction in solubility of all dissolved gases in the blood through increased solute concentration of the plasma. A trans-retial gradient of partial pressures is created and gas will diffuse into the afferent capillaries. There will be a gradual build-up of gases in successive capillary loops as more lactate is added in the gas gland. The gases will diffuse into the swimbladder across the epithelial surface of the gas gland. The combination of salting-out, the Bohr Shift and Root

Effect set a limit to the PO_2 of 100-200 atm. Additional recent information also shows that lactate may effect the rates of association and dissociation of oxygen and haemoglobin so that oxygen dissociated by lactic acid in the rete has less time to recombine with haemoglobin during the short time spent in the efferent capillaries.

The above discussion refers especially to oxygen secretion against high gradients. Scholander and van Dam (1953) also found very high partial pressures of N_2 of 5-15 atm in 260 specimens brought up from 200-950 m. Hüfner (1892) showed that Coregonus (not a deep sea form) could secrete pure N_2 against a 6-8 atm gradient. The levels of N_2 are far above what would be expected by simple diffusion from the sea water as shown in Figs. 2 and 3 (Scholander and van Dam, 1953) and it may be assumed that N_2 secretion is achieved by the salting-out process.

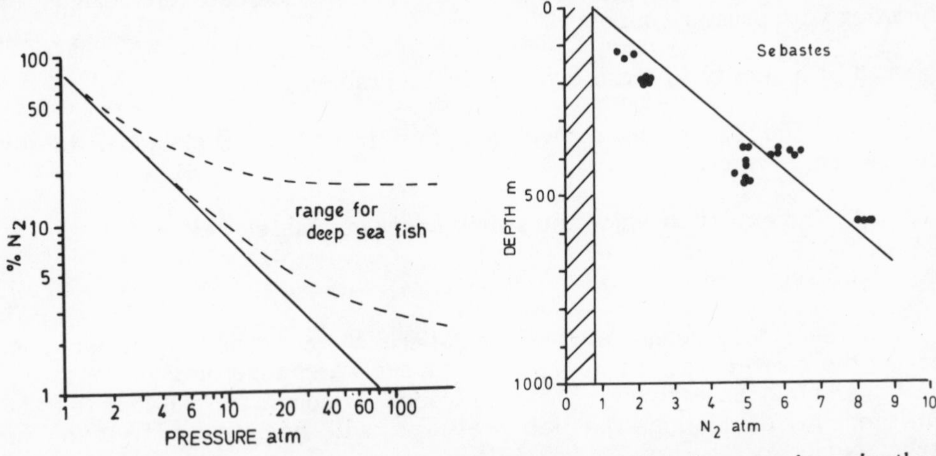

Fig. 2. % Nitrogen found in the swimbladder gas of fish from various depths. The dashed lines give the range of values for various species of deep sea fish. The diagonal line represents the percentage nitrogen which would give a partial pressure in the swimbladder of 0.8 atm. Nitrogen secretion appears to have taken place (redrawn from Scholander and van Dam, 1953).

Fig. 3. Partial pressure of nitrogen in the swimbladder of deep water fish Sebastes marinus. The shaded area represents nitrogen at a partial pressure of 0.8 atm. Nitrogen in excess of 0.8 atm increases linearly with depth.

Since secretion against high partial pressures of gas in the swimbladder is so important, what evidence is there that special mechanisms have evolved especially in deep sea fish? Scholander and van Dam (1953) showed that the blood of 4 species of marine physoclist caught between 400 and 700 m had a substantial reduction in oxygen affinity of haemoglobin (Bohr Shift) and oxygen-carrying capacity (Root Effect) following the addition of lactic acid. Marshall (1972) found very significant modifications in retial length in fish caught at different depths as shown in Table 2.

TABLE 2

Length of retial capillaries in deep sea teleosts
(from Marshall, 1972)

Group	Depth range m	Length of retial capillaries mm
Upper mesopelagic	150 - 600*	0.8 - 2.0
Lower mesopelagic	600 - 1200*	3.0 - 7.0
Bathypelagic	1000 - 4000	Usually no swimbladder
Benthopelagic	< 800	4.0 - 6.0
	800 - 2000	8.0 - 12.0
	> 1500	15.0 - 25.0

* usually migrate vertically by day and night

2. Gas loss

This is controlled in physoclists by an oval, an absorptive surface which can be exposed to the lumen of the swimbladder by a sphincter. Loss is also continuous by diffusion through the swimbladder wall, CO_2 being lost most rapidly, then O_2 and then N_2 (in the ratio 80:2:1).

Purines, especially guanine, in the swimbladder wall reduce diffusion very considerably (see Table 3). The density of purines seems to be related to the normal depth of the fish (Fig. 4). Despite this gas loss is serious at

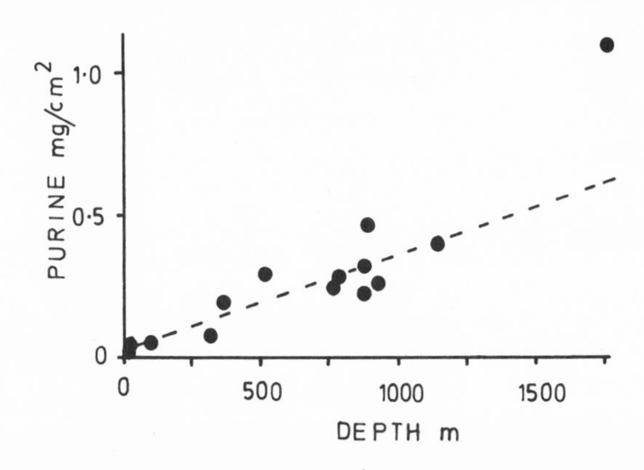

Fig. 4. Total purine (guanine and hypoxanthine) of the swimbladder wall of 14 species of deep sea fish from different depths (redrawn from Ross and Gordon, 1977).

TABLE 3

Guanine content and diffusion constants of oxygen in swimbladder walls

Species	Guanine $\mu g/cm^2$	KO_2 $ml/cm^2/\mu m$ min/atm	Author
Salmo trutta (trout)	24	-	Ross and Gordon (1978)
Pollachius virens (saithe)	47	0.053	Ross (1979)
Ceratoscopelus maderensis (lantern fish)	73	0.067	Ross (1976)
Anguilla anguilla (eel)	186	-	Denton *et al.* (1972)
	-	0.011	Kutchai & Steen (1971)
Conger conger (Conger eel)	239	0.001	Denton *et al.* (1972)
Clupea harengus (herring)	380-3680	-	Blaxter *et al.* (1979)
Synaphobranchus	3300	-	Denton *et al.* (1972)
Halosaurus	2370	-	Denton *et al.* (1972)

high pressures (Table 4) and secretion rates must be high to compensate. Estimates of the critical depth where diffusion and secretion are in equilibrium have been made by a number of authors. In *Conger conger* it is 1060 m and in *Anguilla anguilla* it is 250 m as shown by Lappenas and Schmidt-Nielsen (1977); in the myctophid *Ceratoscopelus maderensis* it is $<$ 100m (Ross, 1976) and in the saithe *Pollachius virens* 78 m (Ross, 1979).

TABLE 4

Estimated loss of O_2 by diffusion in *Anguilla* and *Conger* (from Blaxter and Tytler, 1978)

Depth (m)	Swimbladder PO_2 (atm)	PO_2 gradient across swimbladder wall (atm)	Gas loss ml/Kg/h *Anguilla*	*Conger*
10	0.9	0.7	0.16	0.02
100	9.0	8.8	1.96	0.19
1000	90.0	89.8	19.90	1.89

TABLE 5

Gas secretion and resorption rates (listed by Blaxter and Tytler, 1978) 1 ATA is atmospheric pressure

Species	Pressure stimulus	Secretion rate ml STP/Kg/h	Resorption rate ml STP/Kg/h
PHYSOCLISTS			
Lepomis macrochirus (sunfish)	swimbladder 50% deflated	1.36 - 1.65*	-
Pollachius virens (saithe)	1→2 ATA	1.67	-
	1→4 ATA	2.50	-
	2.2→1 ATA		7.80
Gadus morhua (cod)	1→4 ATA	1.08 - 6.42*	
	4→1.5 ATA		12.00 - 36.00
PHYSOSTOMES			
Anguilla anguilla (eel)		0.28	
Carassius auratus (goldfish)	swimbladder partly deflated	0.18 - 0.48	

* depending on temperature.

3. Vertical migration

Fish which make diel vertical migrations will tend to secrete gas when they sink at dawn and resorb gas when they rise to the surface at dusk. Secretion and resorption rates have been measured in a few species after subjecting fish to increased or decreased pressure or by deflating the swimbladder (Table 5). The rates measured are well below the amount required to maintain buoyancy during descent or ascent. Fish rising to the surface at dusk will be especially vulnerable to loss of depth control and need to monitor the degree of swimbladder inflation to prevent a catastrophe. The amplitude of vertical migration possible is almost certainly increased by the lag in secretion so that fish making regular dusk rises towards the surface will not have adapted to the day depth by the time of dusk and their swimbladder will not therefore contain so much gas as a fish remaining at that depth for a long period of time.

"Acceptable" rates of ascent will depend on the degree of adaptation of the swimbladder to the day time depth, the ability of the fish to compensate for increased flotation by swimming movements (bearing in mind that extra flotation will enhance vertical movement and be energy-sparing), the rate of gas loss and the decrease in pressure which would burst the swimbladder wall. Some data on "safe" excursions and bursting pressures are given in Table 6.

TABLE 6

Percentage pressure reductions (from atmospheric pressure) which are "safe" or which cause bursting of the swimbladder (all physoclists)

Species	"Safe" % pressure decrease	% increase in swimbladder volume	% pressure decrease to burst swimbladder
Perca fluviatilis* (perch)	28 - 33	34 - 43	58
Perca fluviatilis (perch)	32	40	71
Pungitius pungitius (stickleback)	36	45	54
Crenilabrus ocellatus (green wrasse)	63	120	71
Mugil cephalus (mullet)	72	150	74
Lebistes reticulatus (guppy)	-	-	89

* from Harden-Jones (1951, 1952) rest from Tsvetkov (1974).

A knowledge of gas resorption rates and bursting pressures (Fig. 5) allows a decompression schedule to be constructed (Fig. 6) based on the characteristics of the saithe (Pollachius virens). In effect stepwise halving of the pressure every 5 h is permissible.

4. Physostomes

This group of fish is presumed to swallow air at the surface before a descent (or merely to maintain their buoyancy) and to void it via a pneumatic or anal duct if gas is in excess. In fact a physostome performing a substantial vertical migration is never likely to be able to swallow enough gas to make it neutrally buoyant some tens of metres below the surface because the required inflation of the swimbladder within the body cavity at the surface would be excessive and because the extra buoyancy at the surface would prevent the fish from swimming downwards. It is therefore surprising that such habitual vertical migrants as the clupeoids are physostomes. Nevertheless in the herring one filling of the swimbladder at the surface should last the fish 53 days at 90 m, 128 days at 30 m and 238 days at 10 m (Blaxter et al., 1979). Why fish like herring habitually release gas on ascent (Sundnes and Bratland, 1972) is not clear. The herring seems to have little or no secretory activity in the swimbladder wall and so should never have excess gas.

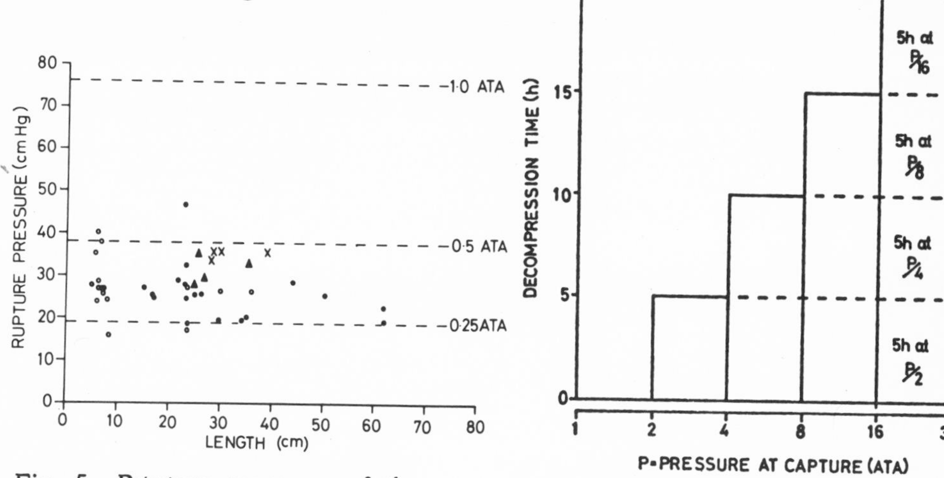

Fig. 5. Rupture pressures of the swimbladders of different species adapted to near surface pressures (1.0 ATA is atmospheric pressure). Filled circles cod, open circles saithe, triangles whiting, crosses haddock (from Tytler and Blaxter, 1973).

Fig. 6. Decompression schedule for saithe. For example fish caught at depths corresponding to pressure (P) of 16 - 32 atm should spend 5 h at P/2, the next 5 h at P/4, the next 5 h at P/8 and the last 5 h at P/16 before being brought to the surface. At lower pressures of capture, fewer stages are required as shown. (from Tytler and Blaxter, 1973).

Other physostomes can secrete gas (Table 5), Anguilla in particular having a rete. Gas secretion is, however, lower than in physoclists.

5. Other buoyancy adaptations

A low density liquid seems to provide an ideal source of flotation almost unaffected by pressure. The deep sea bathypelagic fish Gonostoma and Chauliodus both have oil-filled swimbladders while chimaeroids and squaloids have a light oil stored in the liver.

Many deep sea fish have watery bodies and regressed skeletons (Denton and Marshall, 1958; Blaxter et al., 1971) as shown in Table 7. These are almost certainly adaptations to reduce the metabolic cost of maintaining station. The low haematocrits and muscle systems of these fish suggest they are generally metabolically inactive.

6. Metabolic cost of buoyancy

Alexander (1972) estimated the metabolic cost of maintaining buoyancy by different means.

1. by hydrodynamic compensation 25 ml 0_2/kg/h,

2. by lipid storage 10-17 ml 0_2/kg/h,

3. by swimbladder 2 ml 0_2/kg/h plus a variable amount depending on the depth.

It certainly seems metabolically cheapest to maintain buoyancy near the surface with a swimbladder.

TABLE 7

Characteristics of oceanic fish
(from Blaxter et al. 1971)

Type	% water	Haematocrit % RBC	% Heart weight
Surface	64 - 74	48 - 57	0.06
Mesopelagic with swimbladder	70 - 83	14 - 35	0.02 - 0.04
Meso- or bathypelagic without swimbladder	88 - 95	5 - 9	0.01

PRESSURE SENSITIVITY

Pressure sensors are important for fish in monitoring their vertical movements. Since the swimbladder is an adapting organ it seems unlikely that an absolute pressure sense can exist and fish are more likely to respond to pressure change.

Pressure sensitivity has been determined by observing spontaneous locomotor or other behaviour after imposing pressure changes and by various types of conditioning experiment. Fish with swimbladders are more sensitive than those without (Table 8). The swimbladder wall is probably a site for pressure-operated stretch receptors (Qutob, 1962) and it is known that swimbladder deflation in saithe leads to loss of pressure sensitivity (Tytler and Blaxter, 1977). This is to be expected if the walls become relaxed after loss of gas - it might also be expected that pressure sensitivity would become less as a fish dived and would be restored as gas secretion reinflated the swimbladder. In fact the restoration of pressure sensitivity is much faster in saithe than can be explained by gas secretion. Little is known about the way in which swimbladders contract under pressure and whether certain parts of the swimbladder can be kept full of gas at the expense of other parts.

It is surprising that some species respond to pressure changes of the order of 1 cm H_2O, since it would seem that very small scale movements would create an undesirably high "noise" level. Most experiments have involved applying sudden changes of pressure which is quite unlike the gradual pressure change a fish imposes on itself by a vertical movement. Gibson (personal communication) observed the effects of slow tidal pressure changes on the plaice (which has no swimbladder) and found a high threshold, eg. 50 cm H_2O over 4 h rate of change of pressure being important.

Pressure sensitivity is not lacking in fish without swimbladders (Table 8). The thresholds are higher and the pressure sensor has not been identified.

THE PRO-OTIC BULLA

The clupeoids and mormyrids possess a gas-filled bony structure close to the labyrinth (see Allen et al. (1976), for a recent description). In the mormyrids it is a self-contained structure with its own rete which presumably keeps it full of gas. In larval mormyrids and in larval and adult clupeoids the lower part of the bulla is connected to the swimbladder by very fine gas ducts (radius 4-10 μm). Within the bulla a stiff elastic membrane divides the gas-filled lower part from the perilymph-filled upper part. Movements of this membrane, which can be caused by either hydrostatic or sound pressure, cause the perilymph to be forced in and out of a small fenestra in the upper wall of the bulla. This fenestra is just below the utricular macula. Thus the bulla membrane appears to be acting as a sort of amplification device for sound. It has a further function in that the

TABLE 8

Thresholds of pressure sensitivity

Species	Threshold cm H_2O	Method	Author
Phoxinus laevis (minnow)	0.5 - 1.0	Operant conditioning	Dijkgraaf (1941)
Phoxinus laevis (minnow)	5.0	Spontaneous behaviour. Classical conditioning.	Qutob (1962)
Lagodon rhomboides (pinfish) *Centopristus striatus* (bass)	1.0 - 2.0	Spontaneous behaviour	McCutcheon (1966)
Six freshwater species	0.4 - 2.0	Spontaneous behaviour and conditioning	Tsvetkov (1969)
Gadus morhua (cod) *Pollachius virens* (saithe)	5.0	Cardiac conditioning	Blaxter & Tytler (1972)
*Limanda limanda** (dab)	10 - 20	"	"
*Clupea harengus*** (herring)	<13	Spontaneous behaviour	Blaxter & Denton (1976)

* No swimbladder

** Bulla, no swimbladder

clupeoids have a head lateral line system radiating from a lateral recess on the side of the head. At the inner surface of the lateral recess there is a membrane in the wall of the skull which is just beside the bulla. Movements of the bulla membrane (and perilymph) also cause the lateral recess membrane to move, so stimulating the lateral line organs. This coupling of the bulla with the lateral line seems to be unique to clupeoids (Fig. 7).

The frequency response of this system is now known. What is of especial interest is its response to hydrostatic pressure changes. The swimbladder acts as a reservoir or sink of gas for the bulla (Denton and Blaxter, 1976). As the fish moves up and down in the water, gas passes to or from the swimbladder maintaining the membrane in an optimum flat position where it is most sensitive to sound. This is achieved by having a rather stiff bulla membrane and a compliant swimbladder. Much of the pressure difference caused by a vertical movement is "taken up" by the bulla membrane creating a pressure difference along the gas duct. The time constant for adaptation is of the order of 5-30 sec, fast enough to prevent the fish bursting the membrane even with a quick vertical movement. There is no reason why the bulla should not be a sensor for hydrostatic pressure changes even if it is adapting. In larval clupeoids (especially herring) where the bulla is functional before the swimbladder the bulla cannot adapt and could act as an absolute pressure sensor.

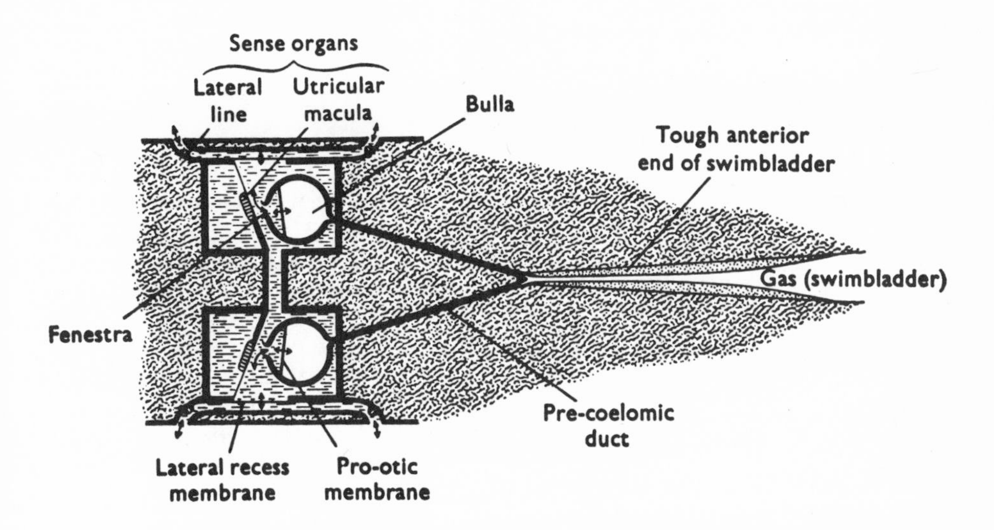

Fig. 7. Diagram of the clupeoid swimbladder, bulla, lateral line system (from Denton and Blaxter, 1976).

EPILOGUE

Pressure has a part to play in limiting both horizontal and vertical distribution of fish and in determining the rate of depth change possible within the normal depth range. It works at a biochemical and at a physiological level and most fish are probably equipped to monitor their changes in depth by pressure sensors and by other means such as the appreciation of ambient light levels. Some fish have very precise depth preferences, for example on inshore nursery grounds and on the continental slope.

The sensitivity of fish to pressure change is so high there is no reason why they should not respond to waves passing overhead (Fig. 8). The amplitude of meteorological pressure changes (20-30 mb) is well within their sensitivity although it is not certain whether slow changes of such magnitude could be appreciated. Nevertheless it seems likely that fish can perceive changes in barometric pressure. Indeed the weather loach Misgurnus fossilis is said to indicate the weather by changes in its behaviour.

The role of sound pressure has been dealt with by Popper in this Volume. Swimbladders are important in picking up sound pressures and re-radiating them as particle displacements to the ear. Hydrostatic pressure may have a subtle influence on hearing in that a change in volume of a swimbladder will change its resonance frequency. The importance of the swimbladder in buoyancy may have been overemphasised in the past. It may be rather inadequate as a buoyancy organ for a vertically migrating fish and it certainly has other important functions in hearing, sound production and perhaps as a source of oxygen.

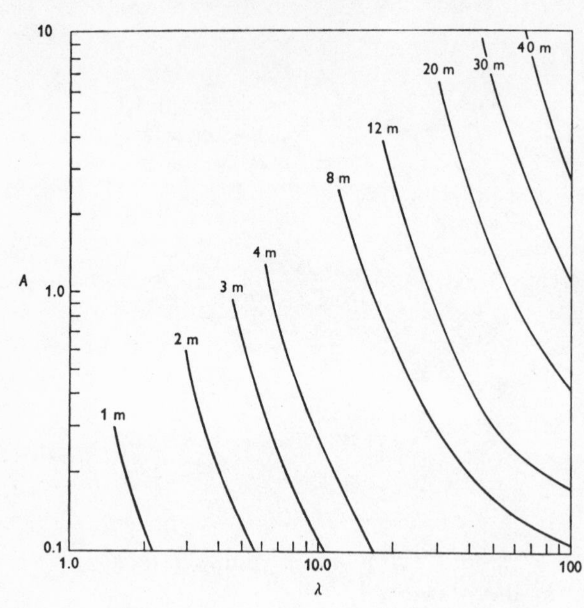

Fig. 8. The depth at which a 1% pressure change would be experienced for different wave heights in metres (A) and wave lengths in metres (λ). Fish resting on the sea bed should be able to perceive such a change of pressure (from Blaxter and Tytler, 1972).

REFERENCES

Alexander, R. McN. (1972). The energetics of vertical migration by fishes. Symp. Soc. Exp. Biol. 26: 273-294.

Allen, J.M., Blaxter, J.H.S. & Denton, E.J. (1976). The functional anatomy and development of the swimbladder inner ear - lateral line system in herring and sprat. J. Mar. Biol. Assoc. U.K. 56: 471-486.

Blaxter, J.H.S. & Denton, E.J. (1976). Function of the swimbladder - inner ear - lateral line system of herring in the young stages. J. Mar. Biol. Assoc. U.K. 56: 487-502.

Blaxter, J.H.S. & Tytler, P. (1972). Pressure discrimination in teleost fish. Symp. Soc. Exp. Biol. 26: 417-443.

Blaxter, J.H.S. & Tytler, P. (1978). Physiology and function of the swimbladder. Adv. Comp. Physiol. Biochem. 7: 311-367.

Blaxter, J.H.S., Denton, E.J. & Gray, J.A.B. (1979). The herring swimbladder as a gas reservoir for the acoustico-lateralis system. J. Mar. Biol. Assoc. U.K. 59: 1-10.

Blaxter, J.H.S., Wardle, C.S. & Roberts, B.L. (1971). Aspects of the circulatory physiology and muscle systems of deep-sea fish. J. Mar. Biol. Assoc. U.K. 51: 991-1006.

Brauer, R.W. (1972). Parameters controlling experimental studies of deep sea biology. In: Barobiology. Ed. R.W. Brauer. Chapel Hill N.C. University of North Carolina, p. 1-13.

Denton, E.J. & Blaxter, J.H.S. (1976). The mechanical relationships between the clupeid swimbladder, inner ear and lateral line. J. Mar. Biol. Assoc. U.K. 56: 787-807.

Denton, E.J. & Marshall, N.B. (1958). The buoyancy of bathypelagic fishes without a gas-filled swimbladder. J. Mar. Biol. Assoc. U.K. 37: 753-767.

Denton, E.J., Liddicoat, J.D. & Taylor, D.W. (1972). The permeability to gases of the swimbladder of the conger eel (Conger conger). J. Mar. Biol. Assoc. U.K. 52: 727-746.

Dijkgraaf, S. (1941). Weberian ossicles and hydrostatic pressure. Z. Vgl. Physiol. 28: 389.

Flügel, H. (1972). 8. Pressure 8.3 Animals. In: Marine Ecology, Vol.1 Pt. 3. Ed. O. Kinne. London, Wiley Interscience, p. 1407-1437.

Hüfner, G. (1892). Zur physikalischen Chemie der Schwimmblasengase. Arch. Anat. Physiol. Abt. 54-80.

Jones, F.R. Harden (1951). The swimbladder and the vertical movements of teleostean fishes. I. Physical factors. J. Exp. Biol. 28: 553-566.

Jones, F.R. Harden (1952). The swimbladder and the vertical movements of teleostean fishes. II. The restriction to rapid and slow movements. J. Exp. Biol. 29: 94-109.

Kinne, O. (1972). 8. Pressure 8.0. General Introduction. In: Marine Ecology, Vol. 1, Pt. 3. Ed. O. Kinne. London, Wiley Interscience, p. 1323-1360.

Kutchai, H. & Steen, J.B. (1971). The permeability of the swimbladder. Comp. Biochem. Physiol. A Comp. Physiol. 39A: 119-123.

Lappenas, G.N. & Schmidt-Nielsen, K. (1977). Swimbladder permeability to oxygen. J. Exp. Biol. 67: 175-196.

McCutcheon, F.H. (1966). Pressure sensitivity, reflexes and buoyancy responses in teleosts. Anim. Behav. 14: 204-217.

Macdonald, A.G. (1975). Physiological Aspects of Deep Sea Biology. Cambridge, University Press, 450 pp.

Marshall, N.B. (1972). Swimbladder organisation and depth ranges of deep sea teleosts. Symp. Soc. Exp. Biol. 26: 261-272.

Naroska, V. (1968). Vergleichende Untersuchungen über den Einfluss des hydrostatischen Drucks auf Uberlebensfahigkeit und Stoffwechselintensitat mariner Evertebraten und Teleosteer. Kiel. Meeresforsch. 24: 95-123.

Qutob, Z. (1962). The swimbladder of fishes as a pressure receptor. Arch. Neerl. Zool. 15: 1-67.

Prosser, C.L., Weems, W. & Meiss, R. (1975). Physiological state, contractile properties of heart and lateral muscles of fishes from different depths. Comp. Biochem. Physiol. B Comp. Biochem. 52B: 127-131.

Ross, L.G. (1976). The permeability to oxygen of the swimbladder of the mesopelagic fish Ceratoscopelus maderensis. Mar. Biol. (Berl.) 37: 83-87.

Ross, L.G. (1979). The permeability to oxygen and the guanine content of the swimbladder of a physoclist fish, Pollachius virens. J. Mar. Biol. Assoc. U.K. 59: 437-441.

Ross, L.G. & Gordon, J.D.M. (1978). Guanine and permeability in swimbladders of slope-dwelling fish. In: Physiology and Behaviour of Marine Organisms. Eds. D.S. McLusky and A.J. Berry. Oxford, Pergamon Press, p. 113-121.

Scholander, P.F. & van Dam, L. (1953). Composition of the swimbladder gas in deep sea fishes. Biol. Bull. (Woods Hole) 104: 75-86.

Sundnes, G. & Bratland, P. (1972). Notes on the gas content and neutral buoyancy in physostome fish. Fiskeridir Skr. Ser. Havunders 16: 89-97.

Tsvetkov, V.I. (1969). Sensitivity of certain freshwater fish to quick pressure changes (in Russian). Vopr. Ikhtiol. 9: 928-935.

Tsvetkov, V.I. (1974). Some patterns of hydrostatics in physoclistous fish (in Russian). Zool. Zh. 59(9): 1330-1340.

Tytler, P. & Blaxter, J.H.S. (1973). Adaptation by cod and saithe to pressure changes. Neth. J. Sea Res. 7: 31-45.

Tytler, P. & Blaxter, J.H.S. (1977). The effect of swimbladder deflation on pressure sensitivity in the saithe Pollachius virens. J. Mar. Biol. Assoc. U.K. 57: 1057-1064.

I am grateful to the Netherlands Journal of Sea Research for allowing the reproduction of Figs. 5 and 6, to Cambridge University Press for permission to reproduce Fig. 7 and to the Society of Experimental Biology for permission to reproduce Fig. 8.

SURFACE MORPHOLOGY OF THE ACOUSTICO-LATERALIS SENSORY ORGANS IN TELEOSTS: FUNCTIONAL AND EVOLUTIONARY ASPECTS

Tor Dale

Zoological Laboratory, University of Bergen

N-5000 Bergen, Norway

INTRODUCTION

Experimental studies show that the acoustico-lateralis system of teleosts responds to mechanical stimuli within a wide range of amplitudes and frequencies, from angular accelerations detected by the ampullary organs of the semicircular canals, to acoustic stimuli detected by the otolithic organs (see Lowenstein 1971). Thus there is a degree of specialization between the sensory organs of the acoustico-lateralis system, although the otolithic organs respond both to linear accelerations and to acoustic stimuli (Lowenstein 1971) and the lateral-line canal organs also respond to sound sources in the near field, in addition to hydrodynamic stimuli (Harris & van Bergeijk 1962).

Studies of the surface morphology of the acoustico-lateralis sensory organs in teleosts show that although the sensory epithelia have the same basic structure, there are distinct mutual and regional differences, especially in the length and polarization of the ciliary bundles (Dale 1976, Dale 1977, Enger 1976, Jenkins 1979, Platt 1977, Popper 1976, Popper 1977, Popper 1978a, Popper 1978b).

In the present paper, an attempt is made to correlate morphology with experimental data, to see if modifications in the surface morphology of the acoustico-lateralis sensory organs can be related to function. Among the questions that arise are the following. Has the variation in the length of the ciliary bundles any functional significance? How are the ciliary bundles coupled to the overlying tectorial structures? What is the role of the kinocilium and the stereocilia? What are the evolutionary aspects of modifications in the surface morphology of the acoustico-lateralis sensory organs?

MATERIALS AND METHODS

The species examined were: cod (Gadus morhua), saithe (Pollachius virens), herring (Clupea harengus) and pike (Esox lucius). The fishes were decapitated, and chilled (+4°C) primary fixative, consisting of 2.5% glutaraldehyde and 2% formaldehyde in Ringer's solution, buffered with 0.1M sodiumcacodylate/HCl at pH 7.4, was immediately injected into the perimeningeal space through the foramen magnum. Then both labyrinths were exposed by cutting the head sagitally through the parasphenoid bone, cutting the cranial nerves and removing the brain halves. The tissue blocks containing the labyrinths were trimmed down to minimum size and immersed in fresh fixative for minimum 2 hours at +4°C, or they were stored in this solution for later use.

Lateral-line canal organs from the hyomandibular branch of the pike were fixed by gently injecting primary fixative into the openings of the canal and immersing the whole hyomandibular arc into fresh fixative for minimum 2 hours at +4°C. The location of the sensory organs could easily be determined by tracing their nerves after the skin was removed from the bony canals.

Superficial neuromasts of the pike were initially fixed by dripping primary fixative on some of the "stitches" along the trunk of the fish. The superficial neuromasts are located in these "stitches" which could easily be seen under a binocular microscope. A piece of skin with some "stitches" on was isolated and immersed in fresh fixative for minimum 2 hours at +4°C.

After primary fixation, the sensory epithelia were carefully isolated by fine forceps and scissors and rinsed in 2 changes of buffered Ringer's solution before they were immersed for 1 hour at +4°C in secondary fixative which consisted of 1% OsO_4 in the same buffer solution as used for the primary fixative. After secondary fixation, the specimens were quickly rinsed in 2 changes of distilled water and dehydrated at room temperature in 30, 50, 70, 90 and 2 changes of 100% acetone.

After dehydration, the specimens were critical point-dried with CO_2, mounted on Al-stubs with double stick tape and colloidal silver, sputter-coated with Au/Pd and examined with a JEOL JSM-35 scanning electron microscope at 25 kV.

Cryo-fractured specimens were prepared according to the method described by Humphreys et al. (1978). Some ampullary organs of the semicircular canals were examined in the frozen-hydrated state at 20 kV, after the specimens had been fixed in primary fixative, frozen in liquid nitrogen, fractured, defrosted and sputter-coated with Au/Pd in a JEOL cryo-unit mounted on a JEOL JSM-35 scanning electron microscope.

RESULTS

General Morphology

Common to all sensory organs of the acoustico-lateralis system is a sensory epithelium which consists of ciliated sensory cells, called hair cells, and supporting cells which carry numerous short microvilli (Figs. 2-7). The cilia of the sensory cells are assembled in a bundle which consists of a single, true cilium, called the kinocilium, and about 30-50 shorter so-called stereocilia which are specialized microvilli, i.e. evaginations of the cell membrane (Figs. 4-6). Unlike the kinocilium and ordinary microvilli, the stereocilia taper towards their bases and are attached in a hexagonal pattern to a cuticular plate at the apical surface of the sensory cell (Figs. 5, 6). The kinocilium is not in contact with the cuticular plate, but is anchored to the apical region of the sensory cell by its own ciliary rootlet. The stereocilia increase stepwise in length towards the kinocilium which is eccentrically located in relation to the stereocilia (Figs. 2-7). This assymetry in the length of the stereocilia and in the position of the kinocilium polarizes the sensory cells both morphologically and functonally and is the basis for the directional sensitivity of the sensory cells (see discussion). The absolute and relative length of the kinocilium and stereocilia show great variation, both regionally within the same sensory organ and mutually between the different sensory organs of the acoustico-lateralis system.

Ampullary organs of the semicircular canals

The sensory epithelium of the ampullary organs of the semicircular canals is located on a crista and is covered by a cupula (Figs. 1, 2, 9). In critical point dried preparations, the cupula appears as a collapsed, fibrous mass on top of the crista. In the frozen-hydrated state, however, the cupula fills the ampullary cavity and consists of groups of densely packed parallel canals oriented in different directions (Fig. 9). Oblique cross walls divide the cupular canals into smaller compartments.

The ampullary organs have by far the longest ciliary bundles of the acoustico-lateralis sensory organs, with very long kinocilia (50-100 μm Fig. 2) and also relative long stereociliary bundles, which are shorter peripherally than centrally on the crista (Fig. 2).

Otolithic organs

The otolithic organs are located in the sacculus, utriculus and lagena of the membranous labyrinth (Fig. 1). Each otolithic organ consists of a disk or tongue of sensory epithelium, called a macula, covered by a thick, gelatinous otolithic membrane which is loaded with a single, massive otolith (Figs. 1, 8). The otolithic organ of the sacculus is the largest of the otholithic organs. Its otolith, called the sagitta, extends far beyond the margins of the sensory epithelium and fills the large cavity of the sacculus (Fig. 1). The otoliths of the utriculus and lagena, called the lapillus and astericus, respectively, are much smaller than the sagitta and do not cover the otolithic membrane completely (Fig. 1).

The sensory epithelia in the sacculus and lagena are oriented in the vertical plane in the medial wall of the saccular and lagenar cavities (Fig. 1). The sensory epithelium in the utriculus is oriented near the horizontal plane, but has a concave, spoon-shaped surface with steeply curved anterior and lateral walls. The spatial orientation of the sensory epithelia, together with the direction of polarization of the ciliary bundles of the sensory cells, determine the directional sensitivity of the sensory organs. The polarization maps of the sensory epithelia in the sacculus and lagena are shown in Fig. 1. The small arrows indicate the direction in which the kinocilium is located relative to the stereocilia in each ciliary bundle.

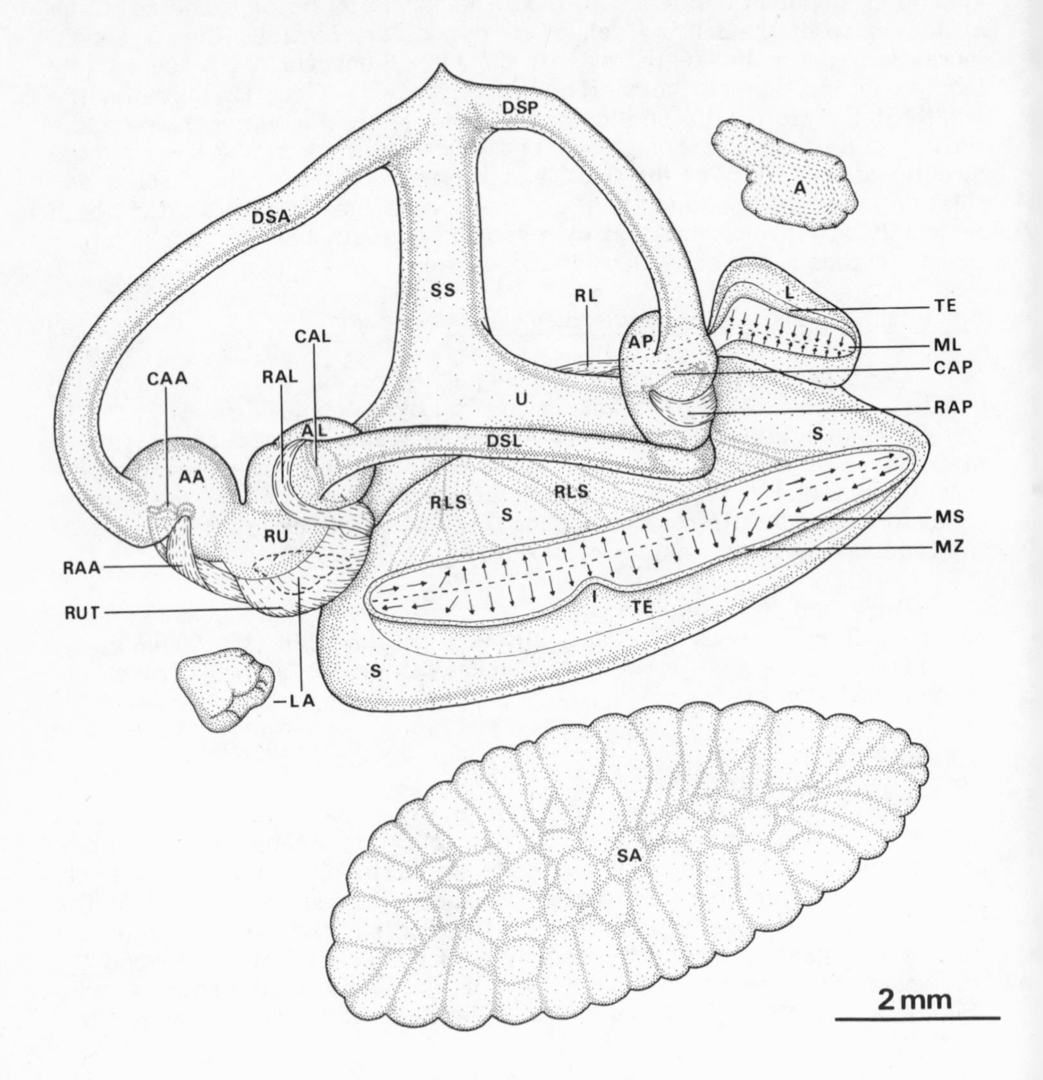

Fig. 1: Left labyrinth of the cod, lateral view. The otoliths are placed outside their respective maculae. The arrows indicate the direction of polarization of the sensory cells (from Dale 1976).

A = asteriscus
AA = ampulla anterior
AL = ampulla lateralis
AP = ampulla posterior
CAA = crista ampullaris anterior
CAL = crista ampullaris lateralis
CAP = crista ampullaris posterior
DSA = ductus semicircularis anterior
DSL = ductus semicircularis lateralis
DSP = ductus semicircularis posterior
I = indentation of macula sacculi
L = lagena
LA = lapillus
ML = macula lagena
MS = macula sacculi
MZ = marginal zone of macula sacculi
RAA = ramus ampullaris anterior
RAL = ramus ampullaris lateralis
RAP = ramus ampullaris posterior
RL = ramus lagenaris
RLS = ramulus saccularis
RU = recessus utriculi
RUT = ramus utricularis
S = sacculus
SA = sagitta
SS = sinus superior
TE = transitional epithelium
U = utriculus

The otolithic organs have much shorter ciliary bundles than the ampullary organs of the semicircular canals, with bundle length ranging from about 3-15 µm, depending on the length of the kinocilium. However, the length of the stereociliary bundles also show great variation. Based on the length of the kinocilium and the longest and shortest stereocilia in each bundle, the ciliary bundles can be grouped in 3 major groups:

1. Ciliary bundles with a long kinocilium (about 10-15 µm and a very short (about 2 µm), nearly ungraded stereociliary bundle (Fig. 3).
2. Ciliary bundles with a long kinocilium (about 10-15 µm) and a relatively long, steeply graded stereociliary bundle (about 8 µm, Fig. 7).
3. Ciliary bundles with a short kinocilium (about 3 µm) and a slightly shorter, graded stereociliary bundle (Figs. 4-6).

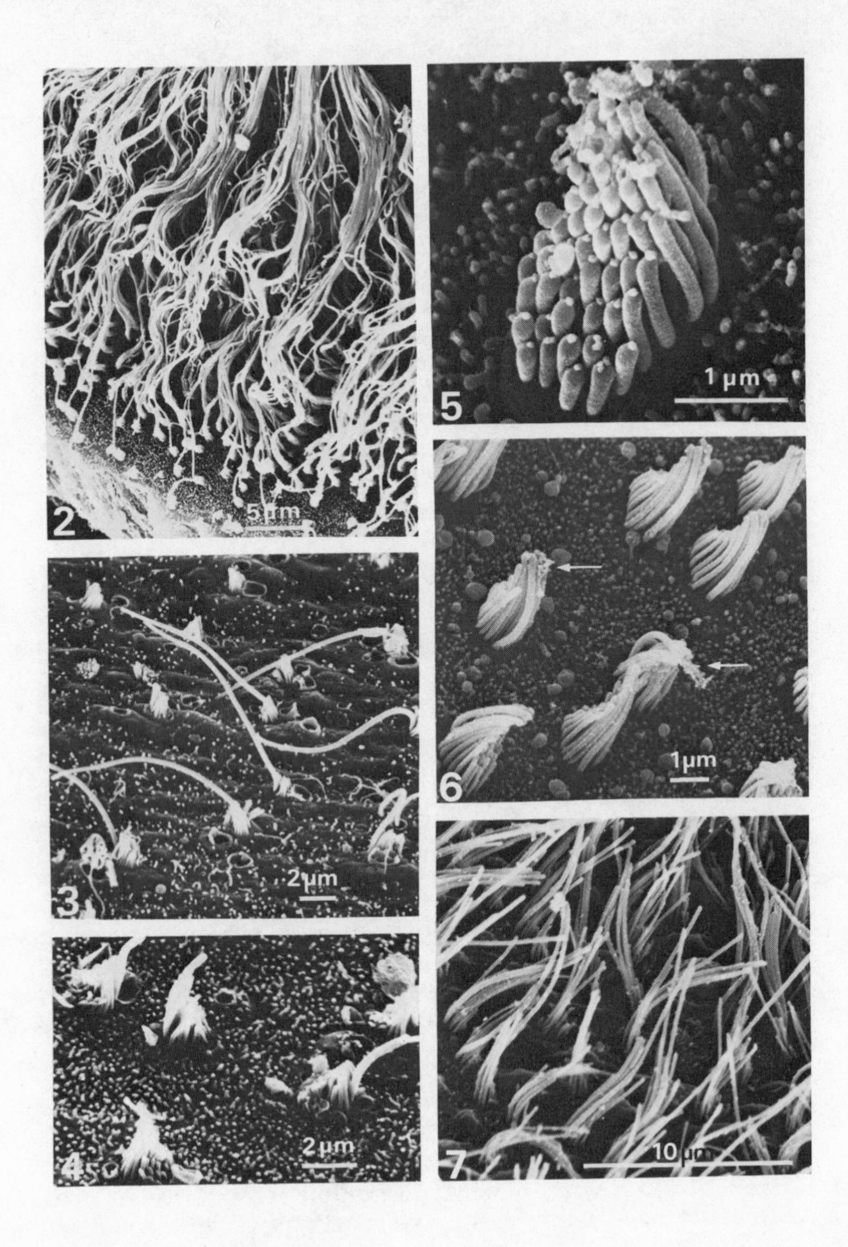
2
5 µm
3
2 µm
4
2 µm
5
1 µm
6
1µm
7
10 µm

Fig. 2 Ciliary bundles from a crista ampullaris of the semicircular canals of the saithe. Note the very long ciliary bundles.

Fig. 3 Ciliary bundles from the periphery of a lateral-line canal organ of the cod. Note the short stereociliary bundles and the relative long kinocilia.

Fig. 4 Ciliary bundles from the central region of the lateral-line canal organ of the cod. The kinocilia are much shorter than those at the periphery of the sensory epithelium (cf. Fig. 3). Note the oppositely polarized ciliary bundles in the upper left part of the picture.

Fig. 5 Ciliary bundle from the central region of the macula sacculi of the pike. Note the small "blebs" on the tips of the stereocilia.

Fig. 6 Ciliary bundles from the macula anterior in the utriculus of the herring. Note the shift in polarization and the remnants of tectorial material on the tips of some of the kinocilia (arrows).

Fig. 7 Ciliary bundles from the macula sacculi of the herring. Note the relative long kinocilia and stereocilia.

There are also transitions between the groups of ciliary bundles mentioned above. In the sacculus, short ciliary bundles (group 3) usually constitute the majority of the ciliary bundles, apart from at the periphery of the sensory epithelium, where the ciliary bundles are of group 1, grading into group 2 and further into group 3 across a narrow zone along the perimeter of the sensory epithelium. In the utriculus, short ciliary bundles (group 3) are usually present centrally and medially on the sensory epithelium, with transition from group 1 to group 2 at the anterior and lateral margins of the sensory epithelium. The lagena in general has long ciliary bundles all over the sensory epithelium, with transition from group 1 to group 2 around the perimeter.

The ciliary bundles project into cavities of the overlying otolithic membrane which rests upon the microvilli of the supporting cells (Figs. 8, 12). Ciliary bundles with a long kinocilium relative to the stereocilia seem to be coupled to the tectorial layer by the distal part of the kinocilium (Fig. 11), whereas ciliary bundles with a short kinocilium seem to be in contact with tectorial material also through the tips of the stereocilia (Fig. 12). Remnants of tectorial material is often present distally on the kinocilia (Fig. 6), and small "blebs" are occasionally present on the tips of the stereocilia of short ciliary bundles (Fig. 5).

Lateral-line organs

There are two types of lateral-line organs: superficial neuromasts and lateral-line canal organs, but not all teleosts have lateral-line canal organs.

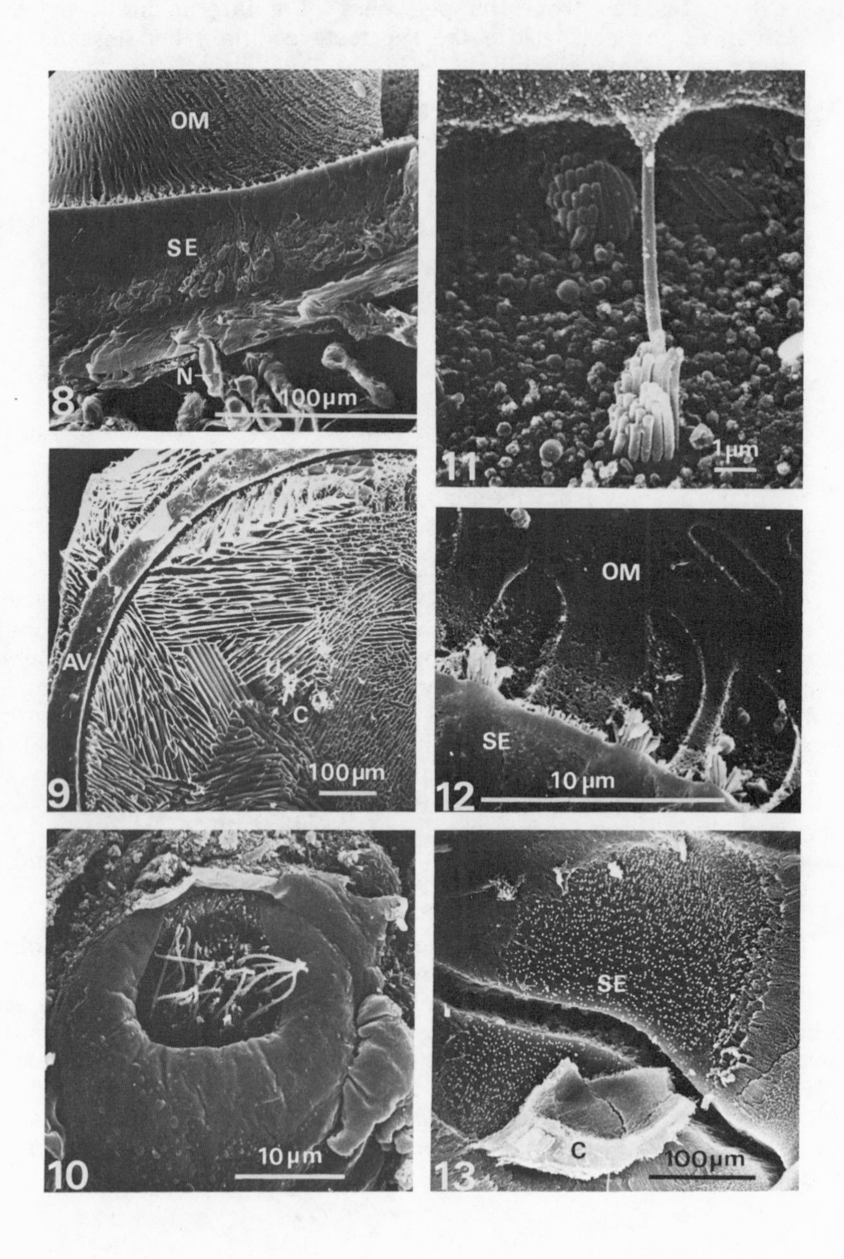
OM
SE
N
100μm
8
1μm
11
AV
C
100μm
9
OM
SE
10 μm
12
10μm
10
SE
C
100μm
13

Fig. 8 Transverse fracture of the lagena of the saithe. The otolithic membrane (OM) rests upon the sensory epithelium (SE). Ciliary bundles are located at the interface between the sensory epithelium and the otolithic membrane. N = nerve fiber. The specimen was fixed and dehydrated before it was cryofractured and critical point dried.

Fig. 9 Fractograph of an ampulla of the semicircular canals of the saithe, fractured transversely to the crista. The cupula (C) consists of groups of parallel canals oriented in different directions and fills the cavity of the ampulla completely. Oblique cross-walls divide the cupular canals into smaller compartments. The specimen was fixed and observed in the frozen-hydrated state. AV = ampullary wall.

Fig. 10 Superficial neuromast of the pike. Note the relative long kinocilia and short stereociliary bundles.

Fig. 11 Ciliary bundle from the periphery of a lateral-line canal organ of the pike. Note the coupling of the distal part of the kinocilium to the tectorial layer.

Fig. 12 Fracture through the central region of the utriculus of the saithe. The ciliary bundles protrude into cavities of the otolithic membrane (OM) which rests upon the microvilli of the supporting cells of the sensory epithelium (SE). The tips of the stereocilia seem to be in contact with reticulate material. The specimen was fixed and dehydrated before it was cryofractured and critical point dried.

Fig. 13 Survey picture of a lateral-line canal organ of the pike. C = remnant of cupula. SE = sensory epithelium.

The superficial neuromasts are located in epidermal pits and each sensory organ consists of a small disk of sensory epithelium covered by a cupula which is usually lost during preparation (Fig. 10). The ciliary bundles are uniform and quite few in number and consist of a kinocilium of moderate length (about 10 μm) and a very short, nearly ungraded stereociliary bundle (about 2 μm), like the ciliary bundles at the very periphery of the sensory epithelia of the otolithic organs (group 1) and of the lateral-line canal organs (Fig. 3).

The lateral-line organs are located at intervals in bony canals on the head and along the lateral line of the fish and communicate with the exterior environment through enterstitial pores. Each sensory organ consists of an oval disk of sensory epithelium, covered by a cupula (Fig. 13). The area of the sensory epithelium, and accordingly the number of sensory cells, is much larger than that of the superficial neuromasts (Figs. 10, 13). The ciliary bundles constitute two major groups, long peripheral ones (about 10 μm, Fig. 3) and short central ones (about 5 μm, Fig. 4), like the ciliary bundles of the groups 1 and 3 in the otolithic organs.

DISCUSSION

The sensory cells of the acoustico-lateralis system are mechano-electric transducers, transforming deflections of the ciliary bundles caused by shearing forces between the cilia and the overlying tectorial structures, to changes in the activity of the afferent nerves that innervate the sensory cells. Deflections of the cilia towards the kinocilium depolarize the sensory cells and increase the discharge frequency in the afferent nerves, whereas deflections of the cilia away from the kinocilium hyperpolarize the sensory cells and diminish the activity in the afferent nerves. The sensory cells of the acoustico-lateralis system are thus morphologically and functionally polarized, and this is the basis for their directional sensitivity (see Flock, 1971).

Although the sensory cells of the acoustico-lateralis system have the same basic structure and the adequate stimulus is the same, there are large variations in the frequencies and amplitudes of the stimuli, as well as in the lenght of the ciliary bundles. Thus the ampullary organs of the semicircular canals have very long ciliary bundles (cf Fig. 2) and respond to displacements caused by angular accelerations, whereas the otolithic organs that are concerned with hearing, have mainly short ciliary bundles (cf. Figs. 5, 6). This suggests that there is a relation between the lenght of the ciliary bundles and their frequency response, in accordance with basic mechanical principles. Thus, long ciliary bundles can functionally be classified as "slow" ciliary bundles, responding to slow or static displacements of relative large amplitudes, such as caused by angular and linear accelerations, whereas short ciliary bundles can be classified as "rapid" ciliary bundles, responding to minute, rapid displacements caused by acoustic stimuli.

The otolithic organ of the sacculus is usually considered as the main hearing organ, and that of the utriculus as the main equilibrium organ in fishes (Lowenstein, 1971). According to the hypothesis suggested above, the sacculus should then have mainly short ciliary bundles and the utriculus mainly long ones, and this is usually the case. In clupeids, however, the conditions are partly reversed, in that the sacculus has mainly long ciliary bundles (cf. Fig. 7) and the macula anterior of the utriculus has short ciliary bundles (cf. Fig. 6). This is in accordance with the fact that in clupeids, the utriculus is considered as the main hearing organ because extensions from the swimbladder are coupled to the utriculus rather than to the sacculus (Wohlfahrt, 1936, Gray & Denton, 1979).

The otolithic organ of the lagena has ciliary bundles with relative long kinocilia all over the sensory epithelium and should therefore, according to my hypothesis, be mainly a gravi-static sensory organ. Although the lagena has an important equilibrium function (see Lowenstein, 1971), experimental studies show that the lagena also responds to vibrations (Furukawa & Ishii, 1967, Sand, 1974), but within a lower frequency range than that of the sacculus, which has short ciliary bundles. This is in accordance with my hypothesis that long ciliary bundles should have a lower resonating frequency than that of short ones.

The fact that the lateral-line canal organs also respond to sound sources in the near-field, in addition to hydrodynamic stimuli (Harris & van Bergeijk, 1962) can, according to my hypothesis, be related to the presence of a majority of relative short ciliary bundles in these sensory organs (cf. Figs. 4, 13). This is also supported by the fact that the superficial neuromasts, which have relative long ciliary bundles (cf Fig. 10), are not sentitive to acoustic stimuli (see Schwartz, 1974). However, my hypothesis that short ciliary bundles should have a higher frequency response than that of long ciliary bundles does not preclude that short ciliary bundles also may respond to gravi-static stimuli. On the other hand, the presence of relative short ciliary bundles that are in close contact with the overlying tectorial layer, seems to be a prerequisite for the reception of acoustic stimuli. However, the frequency response and sensitivity not only depends on the lenght of the ciliary bundles, but also on the degree of coupling of the sensory organ to a gas-filled cavity, which acts as a pressure to displacement transducer (van Bergeijk, 1967a). Thus, the highest sensitivity and frequency response in teleosts is found in the Ostariophysi, where the labyrinth is directly coupled to the swimbladder by the Weberian ossicles (see Lowenstein, 1971).

As to the role of the kinocilium and the stereocilia in the stimulation and transduction process, Flock et al. (1977) have shown that when a ciliary bundle in the ampullary organ of the frog is deflected by a micropobe in vivo, the cilia pivot is stiffly around their base, with the shorter stereocilia following the longer ones. Further, Hudspeth & Jacobs (1979) have shown that it is the stereocilia and not the kinocilium that mediate transduction in vertebrate hair cells. This suggests that the kinocilium can be regarded as an auxiliary structure that couples the stereociliary bundle to the overlying tectorial layer. If the shearing forces between the tectorial layer and the ciliary bundles act at the tips of the kinocilia, as suggested in Fig. 11, the displacements at the tips of the cilia will be proportional to the length of the cilia, decreasing to zero at their base (Fig. 4). This means that short ciliary bundles with a close coupling to the tectorial layer, should have a higher sensitivity and frequency response than long ciliary bundles, because they have to be displaced for a shorter lateral distance for the same angle of deflection (Fig. 14). This supports my hypothesis that long ciliary bundles are adapted to respond to relative slow or static and ample displacements, caused by angular and linear accelerations, whereas short ciliary bundles are adapted to respond to minute, rapid displacements caused by acoustic stimuli.

However, the above hypothesis does not explain the ability to discriminate between slightly different frequencies, or so-called pitch discrimination, which is well developed in many fishes (see Lowenstein, 1971). Van Bergeijk (1967b) suggests that the otolithic organs may vibrate like bongo drums, with a higher pitch peripherally than centrally. Sand & Michelsen (1978) have shown experimentally that the otoliths vibrate in

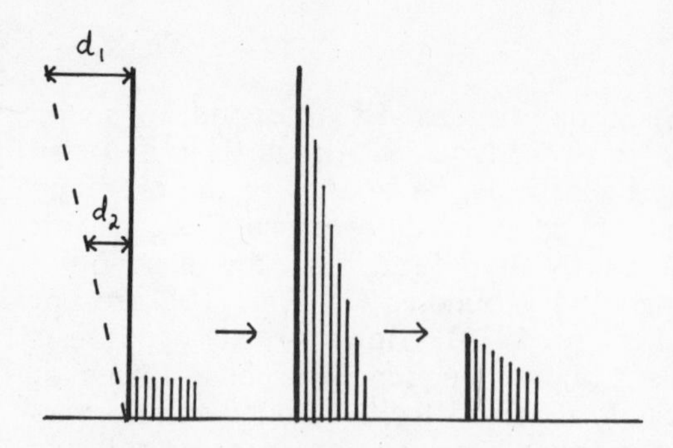

Fig. 14. Diagram illustrating the probable evolutionary and morphogenetic sequence between the different types of ciliary bundles. The deflection of the kinocilium in the left bundle show that for a given angle of deflection, the displacement at the tip of a cillium is proportional to the lenght of the cilium, as indicated by the distances d_1 and d_2.

different directions at different frequencies and suggest that this may be the mechanism of frequency analysis. Because the direction of polarization of the ciliary bundles changes over the sensory epithelium (cf. Fig. 1), different parts of the sensory epithelium will be stimulated at different frequencies. This suggests that there is after all a place mechanism of frequency analysis in teleosts, although not identical to that in the mammalian cochlea.

The otoliths of teleosts are unique among vertebrates because they are single, massive structures, in contrast to the numerous small otoconia of lower and higher vertebrates. Sand & Michelsen (1978) have shown that teleost otoliths respond to vibrations and thus in fact are "hearing stones" as their name implies, and not only have a passive inertial function. Morris & Kittleman (1967) have shown that teleost otoliths have piezo-electric properties and thus may function as frequency and pressure analysers. Sound waves obey the laws of diffraction and are used in acoustic microcospy at very high frequencies (Johnston et al., 1979). Perhaps also teleost otoliths function as acoustic lenses, diffracting and concentrating acoustic energy towards the underlying sensory epithelium?

Evolutionary aspects

Teleosts are of special interest in the evolution of hearing because they are the lowest vertebrates that can detect sound to some extent, although vibratory responses have been recorded from the labyrinth of *Lampetra* (see Lowenstein, 1971) and sharks are attracted to strong low-frequency sound sources (Nelson & Gruber, 1963). Ultrastructurally, the hearing ability of teleosts may be related to the presence of relative short ciliary bundles which are in close contact with the overlying tectorial layer, which is loaded with a single, massive otolith. These conditions are not present in cyclostomes which have ciliary bundles with relative long kinocilia (Hoshino, 1975) and no massive otoliths. The macula neglecta of sharks, however, which is considered as the most vibration-sensitive organ in elasmobranchs, has relative short ciliary bundles (Corwin, 1977), but lacks otoliths.

The hypothesis suggested above is supported by the fact that the ampullary organs of the semicircular canals, which are specialized equilibrium organs, have ciliary bundles with very long kinocilia (cf. Fig. 2), whereas the mammalian organ of Corti, which is a specialized hearing organ, has short ciliary bundles lacking a kinocilium. This shows that whereas kinocilia are indispensable in the equilibrium organs, they have lost their function in the hearing organs. This suggests that there has been an evolution from displacement-sensitive equilibrium organs to pressure-sensitive hearing organs, reflected by a shortening and final disappearance of the kinocilium, accompanied by a closer coupling of the stereocilia to the tectorial layer. The shortening of the ciliary bundles may have been induced by the appearance of a swimbladder which reinforced the acoustic stimulation of the otolithic organs by acting as a pressure-to-displacement transducer (van Bergeijk, 1967a).

According to the acoustico-lateralis theory, the labyrinthine sensory organs originate from the lateral-line sensory organs (see van Bergeijk, 1967a). The superficial neuromasts are considered as the most primitive of the acoustico-lateralis sensory organs and should accordingly have the most primitive ciliary bundles from which the other types of ciliary bundles have evolved. This type of ciliary bundle, consisting of a relative long kinocilium and a very short, almost ungraded stereociliary bundle, is also present at the very periphery of the otolithic and lateral-line canal organs (cf. Figs. 3, 10). From this indifferent, primitive type of ciliary bundle the other more specialized types of ciliary bundles may differientiate in sequence as the sensory epithelium and the tectorial layer grow laterally (Lewis & Li, 1973). Thus one may say that the morphogenesis of the ciliary bundles reflects their phylogeny (Fig. 14), and this supports the acoustico-lateralis theory.

ACKNOWLEDGEMENTS

This work was supported by the Norwegian Council for Science and Humanities (NAVF) grant D.61.46-16.

REFERENCES

Corwin, J.T. (1977) Morphology of the macula neglecta in sharks of the genus Carcharhinus. J. Morph. 152: 341-362.

Dale, T. (1976) The labyrinthine mechanorepter organs of the cod Gadus morhua L. (Teleostei: Gadidae). Norw. J. Zool. 24: 85-128.

Dale, T. (1977) Functional-morphological correlations in acoustico-lateralis sensory organs. In: Scanning Electron Microscopy 1977, Vol. II, Ed. O. Johari and R.P. Becker. IITRI, Chicago, Il., p. 445-452.

Enger, P.S. (1976) On the orientation of hair cells in the labyrinth of perch (Perca fluviatilis). In: Sound Reception in Fish, Ed. A. Schuijf and A.D. Hawkins. Amsterdam, Elsevier, p. 49-62.

Flock, A. (1971) Sensory transduction in hair cells. In: Handbook of Sensory Physiology, Vol. 1, Ed. W.R. Lowenstein. Berlin, Springer-Verlag, p. 396-441.

Flock, A., Flock, B. Murray, E. (1977) Studies on the hairs of receptor cells in the inner ear. Acta oto-lar. 83: 85-91.

Furukawa, T. and Ishii, Y. (1967) Neurophysiological studies on hearing in goldfish. J. Neurophysiol. 30: 1377-1403.

Gray, J.A.B. & Denton, E.J. (1979) The mechanics of the clupeid acoustico-lateralis system: Low frequency measurements. J. Mar. Biol. U.K. 59: 11-26.

Harris, G.G. & van Bergeijk, W.A. (1962). Evidence that the lateral-line organ responds to near-field displacements of sound sources in water. J. Acoust. Soc. Am. 34: 1831-1841.

Hoshino, T. (1975) An electron microcospic study of the otolithic maculae of the lamprey (Entosphenus japonicus). Acta oto-lar. 80: 43-53.

Hudspeth, A.J. & Jacobs, R. (1979) Stereocilia mediate transduction in vertebrate hair cells. Proc. Natl. Acad. Sci. U.S.A. 76: 1506-1509.

Humphreys, W.J., Spurlock, B.O. & Johnson, J.S. (1978) Critical-point drying of cryofractured specimens. In: Principles ans Techniques of Scanning Electron Microcospy, Vol. 6, Ed. M.A. Hayat. New York, Van Nostrand Reinhold, p. 136-158.

Jenkins, D.B. (1979) A transmission and scanning electron microscopic study of the saccule in five species of catfishes. Am. J. Anat. 154: 81-102.

Johnston, R.N., Atalar, A., Heiserman, J., Jipson, V. & Quate, C.F. (1979) Acoustic microscopy: Resolution of subcellular detail. Proc. Natl. Acad. Sci. U.S.A. 76: 3325-3329.

Lewis, E.R. & Li, C.W. (1973) Evidence concerning the morphogenesis of saccular receptors in the bullfrog (Rana catesbeiana). J. Morph. 139: 351-362.

Lowenstein, O. (1971) The Labyrinth. In: Fish Physiology, Vol. V, Ed. H.S. Hoar & D.J. Randall. New York, Academic Press, p. 207-240.

Morris, R.W. & Kittleman, L.R. (1967) Piezo-electric property of otoliths. Sci. 158: 368-370.

Nelson, D.R. & Gruber, S.H. (1963) Sharks: attraction by low-frequency sounds. Sci. 142: 975-977.

Platt,C. (1977) Hair cell distribution and orientation in goldfish otolithic organs. J. Comp. Neurol. 172: 283-298.

Popper, A.N. (1976) Ultrastructure of the auditory regions in the ear of the lake whitefish. Sci. 192: 1020-1022.

Popper, A.N. (1977) A scanning electron microscopic study of the sacculus and lagena in the ears of fifteen species of teleost fish. J. Morph. 153: 397-418.

Popper, A.N. (1978a) Scanning electron microscopic study of the otolithic organs in the bichir (Polypterus bichir) and shovel-nose sturgeon (Scaphirhynchus platorynchus). J. Comp. Neurol. 181: 117-128.

Popper, A.N. (1978b) A comparative study of the otolithic organs in fishes. In: Scanning Electron Microscopy, Vol. II, Ed. O. Johari, SEM Inc., AMF O'Hare, Il 60666, USA p. 405-416.

Sand, O. (1974) Directional sensitivity of microphonic potentials from the perch ear. J. Exp. Biol. 60: 881-899.

Sand, O. & Michelsen, A. (1978) Vibration measurements of the perch saccular otolith. J. Comp. Physiol. 123: 85-89.

Schwartz, E. (1974) Lateral-line mechano-receptors in fishes and amphibians. In: Handbook of Sensory Physiology, Vol. III/3. Ed. A. Fessard. Berlin, Springer-Verlag, p. 257-278.

van Bergeijk, W.A. (1967a) The evolution of vertebrate hearing. In: Contributions to Sensory Physiology, Vol. 2. Ed. W.D. Neff. New York, Academic Press, p. 1-49.

van Bergeijk, W.A. (1967b) Discussion of critical bands in hearing of fishes. In: Marine Bio-acoustics, Vol. 2. Ed. W.N. Tavolga. Oxford, Pergamon Press, p. 244-245.

Wohlfahrt, T.A. (1936) Das Ohrlabyrinth der Sardine (Clupea pilchardus Walb.) und seine Beziehungen zur Schwimmblase und Seitenline. Z. Morp. Okol. Tiere 31: 371-410.

ACOUSTIC DETECTION BY FISHES

Arthur N. Popper and Sheryl Coombs

Department of Anatomy, Georgetown University Schools of Medicine and Dentistry, 3900 Reservoir Road, N.W., Washington, D.C. 20007

INTRODUCTION

It is now widely known that many species of fish produce sounds as part of their behavioral repertoire involved with reproduction, territoriality, and general social interactions (see reviews by Demski, Gerald & Popper, 1973; Fine, Winn & Olla, 1977; Myrberg, Spanier & Ha, 1978; Tavolga, 1971, 1977a; Winn, 1964, 1972). Sound is of particular value under environmental conditions unsuitable for visual or chemical communication. For example, the effectiveness of visual stimuli in the aquatic environment is limited by low light levels and the rapid attenuation of signals over distance; while chemical signals propagate slowly, are non-directional, and are easily diffused by water currents. Sound in water, on the other hand, has a high speed of propagation, low rate of attenuation, and directional properties. Consequently, sound is useful for rapid, high speed communication over considerable distances. Further, depending upon the acoustic properties of the signals (e.g. frequency and transient characteristics), sounds can be used to direct an animal to the emitter, or to communicate in physically complex environments, such as coral reefs, where other signals would be blocked by barriers.

In addition to the ability to produce sound, we are becoming increasingly aware of the sound detection capabilities of fishes. Most species studied can detect sounds up to at least 600 Hz (with a low limit of 50 to 100 Hz), and a number of specialized hearing species can detect sounds to over 3000 Hz (see reviews by Fay, 1978b; Popper, in press a; Popper & Fay, 1973; Tavolga, 1976). Many fishes appear to have complex sound detecting mechanisms, and the morphology of fish auditory systems varies substantially among species (e.g Popper, 1977, 1978a, in press).

Our increased knowledge of inter-specific variability in sound detection, production mechanisms, and capabilities of fishes is perhaps the most exciting aspect of the recent growth in studies of fish acoustics. It is reasonable that a number of different selective pressures have been involved in the evolution of teleost acoustic mechanisms, although we know little about the nature of these pressures. In this chapter, we shall examine the teleost acoustic systems, including structure and function of auditory mechanisms; concurrently, we shall consider the variability in the systems; and finally, we shall consider the variability in the context of potential environmental selective pressures.

UNDERWATER SOUND

Aerial and underwater sound obey the same physical principles, but they differ quantitatively because of differences in density and compressibility of the two media (see Albers, 1967 for a general discussion of underwater acoustics). Quantitative differences considered most significant to aquatic bioacoustics include (a) speed of sound, which is 4.8 times faster in water than in air; (b) the associated difference in wavelength (λ) for any given frequency (4.8 times longer in water); and (c) the difference in the extent of the near-field (e.g. Harris & van Bergeijk, 1962; Siler, 1969; van Bergeijk, 1964, 1967).

The near-and far-field concepts, while useful in analysis of auditory capabilities, have proven to be significant problems in a general understanding of underwater acoustic communication. The reason for these problems is that the terms refer to changes in the relationship between certain properties of the sound field as a function of distance from the sound source, and not, as has often been thought, to actual qualitative differences in the sound field (see Siler, 1969 for a detailed discussion). Such misconceptions have resulted in a less than accurate understanding of the significance of the near- and far-fields to fish audition. Therefore, before any quantitative differences between sound in the near- and far-fields are considered, it is necessary to provide a brief qualitative description of basic properties associated with a propagated sound wave.

A propagated sound wave results from compression of particles which rebound or rarify after being compressed and impart their motional energy to neighboring particles. Each particle first moves in the direction of the wave propagation and then in the opposite direction during these pressure fluctuations. Thus, the sound pressure wave generated by the sound source causes individual particles to oscillate along the axis of wave propagation. However, it is important to note that the particles themselves have not moved from their original position once the sound wave has passed.

In considering the sound wave, we refer to it as having pressure (P) as well as three inter-related vector properties, velocity (u), displacement (d), and acceleration (a). Pressure refers to the change in the signal amplitude, or the amount of excess pressure above or below ambient as a result of compression and rarefaction of the medium. Pressure is a scalar quantity and provides no information about the direction of propagation. The vector components of the sound field refer to the actual distance (displacement), speed (velocity), and rate of change (acceleration) for the motion of individual particles which provide information about the amplitude of the sound and the direction of movement. As will be seen in a later section, information about the direction of movement of the propagated sound wave is inherent to our understanding of the mechanism of sound localization by fishes.

Since velocity is a function of displacement and frequency, the relationships between pressure and velocity are adequate for describing a propagated pressure wave. For a plane wave, the ratio between P and u (this ratio is the characteristic impedance of the medium) remains constant and both P and u are in phase, meaning that they reach their maximum values at the same time. Biological sound sources in water, however, may best be approximated by a dipole source such as a vibrating sphere (Harris, 1964) which generates spherical rather than planar waves. For spherical waves, the ratio between P and u (P/u) changes as a function of distance from the sound source. For any given frequency, P/u decreases as the sound source is approached, meaning that velocity and pressure are out of phase with one another and that velocity is increasing at a much faster rate than pressure. Another way of looking at this is that small pressure changes result in relatively high particle velocities close to the sound source. At greater distances from the sound source, the spherical wave front approaches that of a plane wave, and at some distance, P/u will be nearly constant, and pressure and velocity will be virtually in phase. This theoretical distance has been named the near-field/far-field boundary and its position relative to the sound source depends on the frequency of the sound: the lower the frequency, the further away the boundary and the larger the extent of the near-field (van Bergeijk, 1964; Siler, 1969). The near-field, then, describes an area relatively close to the sound source for which pressure and velocity are out of phase and for which velocity changes at a greater rate than pressure. As the far-field is approached, pressure and velocity come closer together in phase and their rates of change become more equivalent.

Because water is less compressible than air, the extent of the near-field in water is much greater than it would be in air for any given sound frequency. For a 100 Hz sound produced by a dipole source, the near-field extends approximately 5 m in water but only 1 m in air; whereas for a 1000 Hz signal those distances would be 0.5 m and 0.1 m respectively (estimates based on near-field/far-field boundary equal to $\lambda n/2\pi$ from Siler, 1969; where $n = 2$ for a dipole sound source). Thus, most terrestrial vertebrates are always in the acoustic far-field, since the extent of the near-field is very limited. However, aquatic animals often encounter near-field as well as far-field sounds, especially for low frequencies. Since most sounds encountered by fish are lower frequencies, the properties of the near-field must be considered important to auditory processing in fish.

There are a number of points that should be made with regard to teleost acoustic detection systems and to the various properties of an acoustic wave. It is of considerable importance that the nature of the acoustic signal, with respect to these properties, changes with distance from the sound source and, potentially, with the nature of the sound source. Further, the ratio between P and u will also change when there is a transition between media in which the sound travels, such as at the air-water or air-bottom interfaces. Thus, the ratio between P and u would be rather complex in an environment such as a coral reef where the water-bottom (and perhaps coral) interface is structurally complex. It is also worth pointing out that one of the most complex environments with regard to drastic changes in the ratio between P and u is in a tank of water bounded by air (e.g. a fish tank) where the acoustics are, according to Parvulescu (1964), mathematically undefinable.

An additional point is that the mechanical characteristics of the fish ear and lateral-line determine what aspect of the propagated wave will be detected. For example, the unaided inner ear in fishes, as in terrestrial vertebrates, is a detector of d or u (von Bekesy, 1960). However, with the aid of a P to u converter (such as the tympanic membrane in terrestrial vertebrates and the swimbladder in fishes) the auditory system becomes, in effect, a P detector. The lateral-line organs, on the other hand, rarely have access to P to u converters and are typically limited to being d or u detectors (Harris & Milne, 1966; Strelioff & Honrubia, 1978).

It follows from this discussion that, depending upon the position of the fish relative to the sound source, the animal will be receiving different types of stimuli. In the acoustic far-field, the pressure component of the signal may be greatest and, in fact, it has been suggested that fish can only detect pressure in the far-field (as will be shown below, this is now considered incorrect). In the near-field, however, the fish would receive a substantial d and/or u signal in addition to a signal with a P component. Thus, the nature of the signal impinging upon a fish would change in its acoustic components depending upon the distance from the sound source and the acoustic environment. Furthermore, it is clear that the fish, by being able to detect P, d, a, and u can obtain a good deal more information about the nature of the sound source than would be available if the animal only detected P, thus only getting information about stimulus frequency and amplitude.

SOUND RECEPTORS

Sound detection in fishes is often thought to involve both the inner ear and the lateral-line, although the functional distinction between the two is still difficult to assess, partly because of the paucity of data on lateral-line function. One distinction that may be made, however, is that the lateral-line probably responds to a lower range of frequencies (1 - 500 Hz) (Kuiper, 1956, 1967; Kroese, Van der Zalm & Van den Bercken, 1978; Strelioff & Honrubia, 1978) than the inner ear system, which is capable of responding to much higher frequencies (Fay & Popper, 1974, 1975).

Inner Ear Anatomy

The bulk of the data on fish audition relates to the inner ear which may function in conjunction with one or more peripheral structures such as the swimbladder and the Weberian ossicles (Fig. 1). The inner ear lies in the cranial cavity at about the level of the medulla, although the precise position of the ears varies in different species.

The ear consists of three semi-circular canals (Fig. 2) and three otolithic organs, the utriculus, sacculus and lagena. The utriculus is most frequently considered to be a vestibular organ, along with the semi-circular canals, while the sacculus and lagena are thought to be involved with audition. However, while this distinction between function of the three otolithic organs is commonly made, the differentiation is based upon data for only a few species and is still in need of verification.

Each of the otolithic organs consists of a membranous walled sac containing a single dense otolith which lies in close contact with a region of sensory epithelium, often referred to as the macula, which contains sensory hair cells and supporting cells. The surface of the macula contains ciliary bundles (fig. 3) projecting from the apical ends of the sensory cells. These ciliary bundles contain 30 to 70 stereocilia, which are often graded in size, and a single kinocilium located at one end of the stereociliary bundle. The stereocilia terminate in a base-like cuticle just under the apical cell membrane, while the kinocilium projects through the cell membrane, terminating at the basal bodies. Morphologically, the kinocilium closely resembles a typical flagella-like structure in having a 9+2 tubule pattern (see Flock, 1965; Hama, 1969; Popper, 1979a for discussion of sensory cell ultrastructure).

The morphological polarization of the ciliary bundle (Fig. 3) with the kinocilium displaced to one side of the cell is involved with physiological polarization of the cell. The cell undergoes maximum depolarization when the ciliary bundle is bent (sheared) towards the kinocilium, while there is maximum hyperpolarization when the bundle is bent in the opposite direction (e.g. Flock, 1965, 1971; Hudspeth & Corey, 1977; Hudspeth & Jacobs, 1979). Intermediate directions of deflection produces receptor potentials that vary as a cosine function of the stimulus direction (Flock, 1971).

Ultrastructure of the Ear

One of the most striking of the recent findings in studies of the anatomy and function of the fish ear has been the observations made of the sensory maculae with the scanning electron microscope (e.g. Dale, 1976; Enger, 1976; Jenkins, 1979; Jorgensen, 1976; Platt, 1977; Platt & Popper, in press; Popper, 1976, 1977, 1978b, 1979a,b; Popper & Platt, 1979). These investigations have demonstrated several features of the fish ear that may have substantial significance for our understanding of acoustic processing.

Briefly, recent observations have shown that the sensory cells on the otolithic maculae are divided into groups, with all of the cells in a particular group oriented in the same direction (Figs. 3 & 4). The cells on the saccular macula in the Ostariophysi are oriented bi-directionally (two orientation groups), with the cells on the dorsal side of the macula oriented dorsally (kinocilium on the dorsal side of the cell) and those on the ventral side of the macula oriented ventrally. A bi-directional pattern is also found on the ostariophysan lagena, although due to macula curvature, the cells wind up being oriented in a wide variety of different directions (see Hama, 1969; Jenkins, 1979; Platt, 1977).

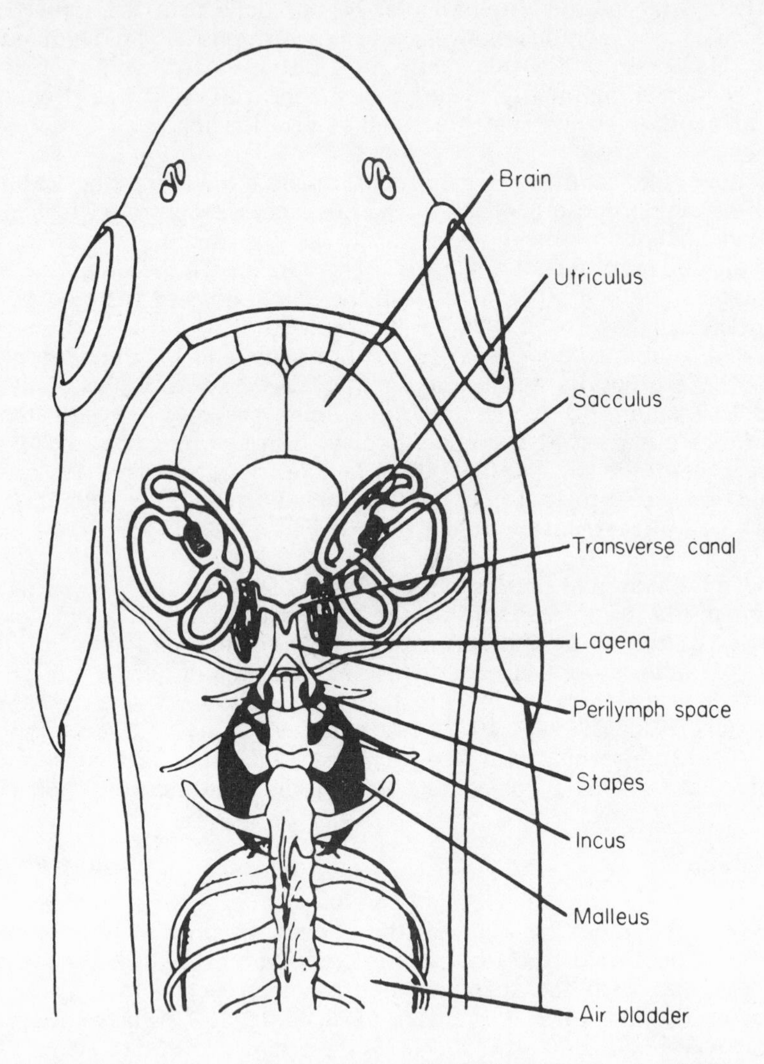

Figure 1. Dorsal view of the peripheral auditory system of the minnow Phoxinus laevis showing the swimbladder and inner ear connected by the Weberian ossicles in this ostariophysan. From von Frisch, 1936.

Hair cell orientation patterns on the saccular macula in non-ostariophysans is somewhat more complex than in the Ostariophysi, and there appears to be substantially greater inter-specific variation in the non-ostariophysans. With but one exception (a mormyrid, Popper, 1979b), all of the approximately 50 non-ostariophysans that have been studied have hair cells on the saccular macula organized into four orientation groups. The cells on the caudal end of the macula are oriented dorsally and ventrally, as in the Ostariophysi. However, the cells on the rostral end of the macula are oriented along the animal's horizontal axis. The most common pattern for the horizontally oriented cell groups is to have posteriorly oriented cells dorsal to anteriorly oriented cells, although a substantial number of variants on this pattern are now known (see Popper, 1978a, in press a, for reviews).

The second finding associated with the ultrastructure of the sensory macula is the variation in the size and shape of the ciliary bundles on the apical surfaces of the sensory cells. At least three different types of bundles are now known. One type, designated as F2 by Popper (1977), has a long kinocilium and short stereocilia. This bundle type is most often found at the very periphery of the different maculae (e.g. Platt, 1977; Popper, 1977, 1979a). Another bundle, F1 (Fig. 3), has graded stereocilia and a kinocilium slightly longer than the longest stereocilia. The F1 bundle is rather ubiquitous and is perhaps the most widely found bundle among different fishes and on different otolithic maculae. The third bundle type, F3, looks something like the F1 bundle but is substantially longer than that bundle. The F3 bundle is either found in small macula regions uncorrelated

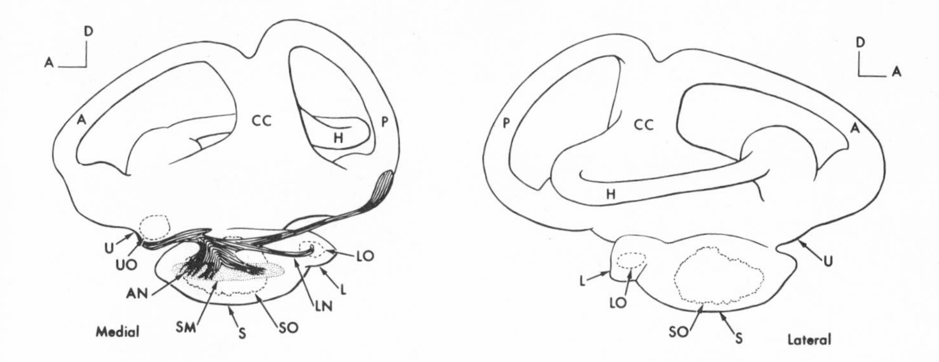

Figure 2. The ear from a deep-sea Halosaurid fish (species unidentified) to show the general pattern of the ear in non-ostariophysans. In the Ostariophysi, the same basic structures are present, although the sacculus is more elongate and somewhat smaller than the the lagena. A-anterior semi-circular canal; AN -auditory (eighth) nerve branch to the sacculus; CC-crus commune; H-horizontal semi-circular canal; L-lagena; LN-lagenar branch of the posteior ramus of the eighth nerve; LO-lagenar otolith; P-posterior semi-circular canal; S-sacculus; SM-saccular macula; SO-saccular otolith; U-utriculus; UO-utricular otolith.

with any other ear structure, or throughout the saccular maculae in a number of different deep-sea species (Popper, 1977; in press b) and clupeids (Platt & Popper, in press; Popper & Platt, 1979).

The functional differences associated with the various ciliary bundle types are not clear, although there is some evidence that the F3 bundles are found in regions of the ostariophysan saccular and lagenar maculae (Platt, 1977; Popper, unpublished) associated with detection of lower frequency vibratory signals (Furukawa & Ishii, 1967; Fay & Olsho, 1979). It should be pointed out that these data contradict observations in terrestrial vertebrates where long ciliary bundles are associated with detection of higher frequency signals (Weiss _et al._, 1976). However, suggestions on ciliary bundle function are purely speculative at this point and not based

Figure 3. Scanning electron micrograph of Type I sensory hair cell ciliary bundles (see text) from _Trichogaster trichopterus_, the blue gourami. Note that the cells on the top of the figure have their kinocilia oriented dorsally while the kinocilia in the bottom of the figure are oriented ventrally. Thus, the cells shown here represent two opposite orientation groups. K-kinocilia, S-stereocilia.

upon direct experimental studies of their function.

Inner Ear Stimulation

It is probable that the stimulation of the sensory hair cells in the fish otolith maculae result from relative movements between the otolith and the sensory maculae, resulting in a shearing action between the cilia and the otoliths. Further, it is likely that there are at least two different ways in which the relative motion between the sensory cell and otolith could be produced.

One path for sound to the ear involves direct stimulation of the otolith maculae as a result of vibration of the fish in response to the incident sound wave (e.g. Dijkgraaf, 1960). Since the fish is approximately the same density as that of the surrounding water (Cushing, 1967), its body, including the sensory tissue, will vibrate (or oscillate) at nearly the same amplitude and phase as the surrounding water particles when a pressure wave passes through the body. However, the far denser (at least 3 times) otolith would lag behind the motion of the rest of the fish's body, and its motion would differ both in amplitude and phase from that of the sensory tissue. Since the ciliary bundles are coupled to the otolith by an otolithic

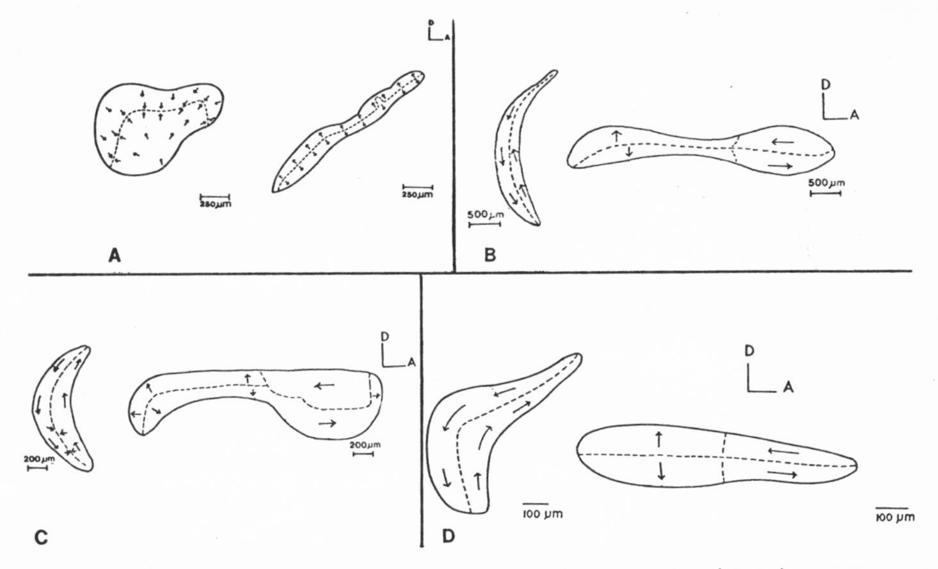

Figure 4. Hair cell orientation patterns on the saccular (right) and lagenar (left) maculae from several teleost species. In each figure the dashed lines indicate the approximate separation line between different hair cell orientation groups while the tips of the arrows indicate the side of the cells in each group upon which the kinocilium is found. A. goldfish, Carassius auratus (redrawn from Platt, 1977); B. lake whitefish, Coregonus (redrawn from Popper, 1976); C. a squirrelfish, Adioryx xantherythrus (redrawn from Popper, 1977); D. a clupeid, Sardenella marquensis (redrawn from Platt & Popper, in press).

membrane, the bundles would be bent by the relative movement created between otolith and sensory macula. Because the movement of the sensory tissue is essentially the same as surrounding water particles being displaced, the system can be thought of as displacement or velocity sensitive. It is important to point out, however, that the pattern of movement between otolith and macula may be quite complex and the shape of the otolith in combination with its suspension in the labyrinth and surrounding fluid as well as its coupling to the macula via the otolithic membrane may ultimately determine the pattern of movement.

Indirect stimulation, as opposed to direct stimulation described above, most likely results from the swimbladder being affected by changes in ambient pressure accompanying the sound pressure wave. Since the gas in the swimbladder is more compressible than the surrounding water, the swimbladder will expand and contract at a greater amplitude than the water in response to the pressure changes. The vibrating swimbladder will 're-transmit' the signal and act somewhat like a dipole sound source producing both pressure and particle displacement. Any displacement receptor, such as the ear, in the vicinity of the source will potentially be stimulated. The degree of stimulation of the ear will depend upon the proximity of the ear to the swimbladder, or, in other words, how far into the swimbladder's acoustic near-field the ear lies. Thus, the closer the ear is to the swimbladder, the higher will be the level of displacement stimulation. Although the end organs of the inner ear ultimately respond to displacements generated by the swimbladder, the system as a whole is responding in phase with changes in pressure and can be thought of as pressure sensitive.

Depending upon the way sound reaches the ear, the auditory system of fish may respond as either a velocity or a pressure detector. Behavioral and physiological evidence (Chapman & Sand, 1974; Fay & Olsho, 1979; Fay & Popper, 1974, 1975) indicate that (a) teleost inner ear systems most likely behave as velocity detectors for low (below 100 or 200 Hz) frequencies; (b) some species, such as Tilapia and several species of flatfish, respond as velocity detectors over their entire range of auditory sensitivity, which is generally narrower than the sensitivity ranges of pressure detecting species; (c) some species, particularly those with specialized connections between the swimbladder and the inner ear (e.g. Weberian ossicles) are pressure sensitive above several hundred Hz; and (d) systems that are pressure sensitive will respond as velocity detecting systems if the swimbladder is removed, while species normally without swimbladders, such as the flatfish, will show pressure sensitivity if an artificial swimbladder is provided.

Thus, it is becoming apparent that while the ultimate stimulus for the endorgan is displacement of the cilia, the way in which this occurs and hence the way in which the total system is responding (i.e. to pressure or velocity) may vary between fish and even within the same fish.

Because velocity sensitive systems respond to a vector quantity, the pattern of movement between the otolith and macula is going to be a function (but not necessarily a direct function) of the axial direction of wave propagation. In pressure sensitive systems, however, the direction of the shearing force between the otolith and macula will be independent of the direction of the incident sound wave if the swimbladder contracts symmetrically. Further, in data for the goldfish, a predominantly pressure-sensitive animal, the pattern of hair cell stimulation in the sacculus is a function of the phase (i.e. compression vs. rarefaction) of the pressure wave (Furukawa, 1978; Furukawa & Ishii, 1967). Thus, while velocity sensitive systems may get directional information, pressure sensitive systems are likely to get phase information on pressure changes. Since the relationship between pressure and velocity changes as a function of distance in the near-field, auditory systems sensitive to both dimensions may be getting information on distance from the sound source (Parvulescu, 1964). In addition, more precise directional information from comparing pressure and displacement has been proposed by Schuijf and his colleagues (see Schuijf, 1976a,b; Schuijf & Buwalda, 1975, in press).

In considering how sound may reach the ear, particularly in the direct vibration mode, it becomes apparent that the morphological similarity and close proximity of the three otolithic organs may result in all of them having an acoustic role. We might speculate that any otolithic organ preferentially located to receive sound channeled from the swimbladder or other similar pressure transducers might respond to acoustic pressure. This certainly seems to be the case for the sacculus in the goldfish, Carassius auratus, which is linked directly to the swimbladder via the Weberian ossicles. In general however, data delinating roles for each of the organs are quite limited. For the few teleost fish examined, most notably the goldfish, the sacculus has been implicated as the primary acoustic receptor responsive to pressure (Fay, 1974a, 1978a; Fay & Popper, 1975; Furukawa & Ishii, 1967; Sand, 1974a), while the lagena has been implicated only secondarily as a low frequency vibration or velocity/displacement detector (Fay & Olsho, 1979; Furukawa & Ishii, 1967). The utriculus, on the other hand, is generally thought to be a gravistatic organ (Lowenstein, 1971), although Denton, Gray and Blaxter (1979) suggest that the clupeid utricle, which is attached to an air bubble, is a pressure organ in this species, while the sacculus may serve as a d or u detector.

Several other aspects of the inner ear should be mentioned with regard to signal processing, but extensive studies will be needed before specific data can be presented regarding these points. It is now becoming apparent that there is substantial morphological variation in the ears of different teleost species. For example, the ciliary bundles in some regions of the saccular macula lie directly under the otolith while other regions (most often at the anterior end of the macula) are only in contact with the otolithic membrane (e.g. Popper, 1976, 1977, 1978a). The shapes of otoliths are also complex and the width of the sensory maculae vary along their lengths. Thus it is possible that different macula regions are stimulated by different signals, or in different ways, thereby providing some sort of basis for signal analysis along the sensory maculae, much like what has been described in some reptiles (Weiss et al., 1976). There is also substantial inter-specific variation in many of the features of the ear (see Fay & Popper, in press; Popper, 1978a; Popper, in press a, for a detailed discussion of the variation), and this has led to the suggestion that there may be significant inter-specific functional variation in the ear. However, data concerning such functional variation are essentially non-existent, although there are limited data showing differences in responses from different points along the saccular macula in several species (Enger et al., 1973; Fay & Olsho, 1979; Furukawa & Ishii, 1967; Sand, 1974b). Of particular interest are data indicating that the otolith moves in diferent patterns in response to different signals in the perch (Sand & Michelsen, 1978), suggesting that different macula regions may be responsive to different signals.

The Lateral-Line

Historically, the lateral-line and the inner ear have been considered a part of a single acoustico-lateralis system with implications for functional continuity or overlap between the two organs (see van Bergeijk, 1967). The basis for this supposition has been largely centered around three lines of anatomical evidence: the sensory tissue of both otolithic maculae and lateral-line neuromasts are composed of hair cells; the sensory tissue from each arises from the same tissue mass during embryological development; and the cranial nerves innervating the inner ear and the lateral-line project to the same location in the medulla. Recent experimental data, however, show that there is little if any overlap between central projections of the eighth and lateralis nerves (Maler et al., 1973a, b; McCormick, in press; Northcutt, in press). In addition, as Wever (1974) has pointed out, the lateral-line and the inner ear do not arise from precisely the same region of the ectoderm. Thus, considering that all special sense organs (including taste buds, olfactory organs, etc.) arise from closely related ectodermal regions (Balinsky, 1975), we must exlude this line of evidence (also see discussion in Northcutt, in press).

Yet the notion of an acoustic function for the lateral-line still persists and it is most often suggested that it may play a role in sound localization (van Bergeijk, 1964, 1967; Horch & Salmon, 1973; Tavolga, 1977b). It is now reasonably established that lateral-line organs do not respond to pressure changes (except possibly in clupeids, where the lateral-line is associated with an air bubble - Denton et al., 1979) but are sensitive to the displacement component of the propagated sound wave (Harris & van Bergeijk, 1962; Horch & Salmon, 1973; Strelioff & Honrubia, 1978), making it theoretically possible for the lateral-line to obtain directional information (e.g. van Bergeijk, 1964; Schwartz, 1967). Tavolga (1977b), for example, reports that the lateral-line system of the sea catfish, *Arius felis*, is capable of differential responses to different sound sources from 15 cm away in the frequency range of 50-150 Hz when the displacement levels were of sufficient magnitude.

In general, however, the lateral-line organs appear to respond to sound only under certain conditions favoring large water displacements (i.e. low frequency sounds close to the source) which are in the same direction as the canal axis (Harris & van Bergeijk, 1962). Hence, it is still not clear whether stimuli eliciting responses of the lateral-line observed by Tavolga (1977b) and Horch and Salmon (1973) are perceived by the fish as acoustic information or whether they fall into a more general class of stimuli associated with massed water movements, such as hydrodynamic swimming motions, possibly important in schooling behavior (e.g. Pitcher et al., 1976, Shaw, 1970). Although the argument is partly one of semantics, depending on one's definition of sound (see van Bergeijk, 1967), it should be pointed out that there is still no clear evidence comparable to that for the ear, designating the lateral-line as an acoustic receptor. While earlier data showing behavioral shifts in sensitivity were interpreted as a shift in sensory modalities from inner ear to lateral-line (Cahn, Siler & Wodinsky, 1969; Cahn, Siler & Auwarter, 1971; Tavolga & Wodinsky, 1963), it now appears possible that this shift could reflect the fish 'attending' to sound reaching the ear by different routes (see Inner Ear Stimulation).

Further complicating our knowledge of the teleost lateral-line is the fact that studies of lateral-line function in recent years have been directed at amphibians rather than fishes (Bauknight, Strelioff & Honrubia, 1976; Kroese et al., 1978; Strelioff & Honrubia, 1978), and we have no idea as to whether extrapolations regarding biological role from amphibian to fish is reasonable, even if the physiology of the systems is the same.

AUDITORY CAPABILITIES

Behavioral Sensitivity

The bulk of information we now have on auditory capabilities in fish is based upon studies using behavioral techniques to determine auditory sensitivity as a function of frequency. It is becoming increasingly apparent that there is substantial inter-specific variability in the sensitivity and frequency range of hearing. The most conspicuous differences appear when comparing fish with and without specialized connections between the swimbladder and inner ear. In general (although with some exceptions), fish belonging to the teleostean superorder Ostariophysi, such as goldfish, catfish and carp, have relatively good sensitivity (around -40 dB re: 1 μbar) and a wider frequency range (exceeding 2000 Hz) than non-ostariophysans such as *Tilapia* or the squirrelfish *Adioryx*, where best sensitivity is generally about -10 to -20 dB re: 1 μbar and where the frequency range does not extend beyond 1,000 Hz. The ostariophysans are distinguished by a series of bones, the Weberian ossicles, which mechanically link the anterior end of the swimbladder to a fluid-filled chamber in contact with the saccular vesicle. Since the swimbladder may function as both a pressure transducer and sound amplifier, the greater sensitivity in the Ostariophysi has been attributed to the mechanical linkage between swimbladder and inner ear.

Unfortunately, a good deal of the behavioral data showing variability in hearing capabilities among fish must remain somewhat suspect since differences in acoustic conditions and behavioral techniques have prevented legitimate comparisons across studies in many cases (see Popper & Fay, 1973). However, such comparisons are possible when animals are studied in the same experimental set-ups (see Popper, Chan & Clarke, 1973). When such comparisons are made (Fig. 5), striking differences in both hearing sensitivity and frequency range of hearing can be demonstrated. For the species shown in Fig. 5, all tested under identical conditions, best sensitivity (-50 dB) for the goldfish Carassius auratus (family: Cyprinidae) (Popper, 1971) was almost identical to that for a non-ostariophysan, the squirrelfish Myripristis kuntee (family: Holocentridae) (Coombs & Popper, 1979). Like the goldfish, Myripristis has a linkage between the swimbladder and inner ear consisting of anterior projections of the swimbladder which abut against a fenestra in the posterior portion of the skull enclosing the saccular endorgan (Nelson, 1955). In contrast, hearing sensitivity for another squirrelfish, Adioryx xantherythrus (family Holocentridae), whose

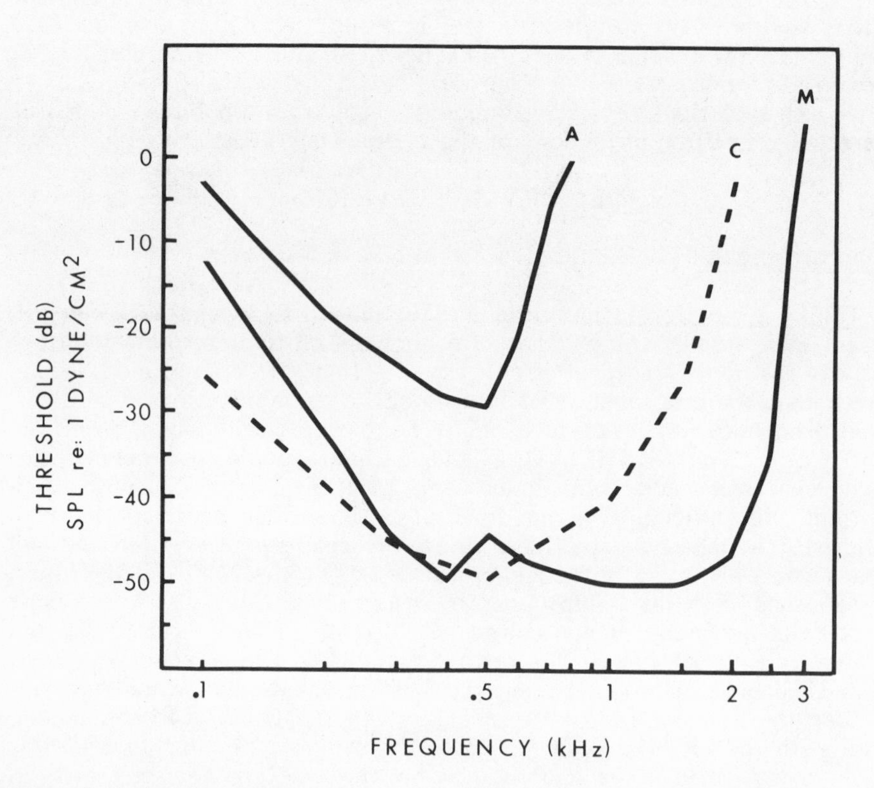

Figure 5. Behavioral audiograms for (C) goldfish, Carassius auratus, (A) Adioryx xantherythrus, and (M) Myripristis kuntee (Data from Popper, 1971 and from Coombs & Popper, 1979).

swimbladder is distant from the inner ear, is distinctly poorer than for either Myripristis or Carassius (Coombs & Popper, 1979).

While it is clear from the behavioral data that connections between the swimbladder and inner ear in Carassius and Myripristis are related to increased auditory sensitivity, it is not clear to what extent these linkages determine the frequency range of hearing. It is likely that the connections enhance high frequency sensitivity by enabling high frequency stimulation of the ear at greater amplitudes, thus effectively increasing the frequency range of hearing (Coombs & Popper, 1979). However, clearly there are other additional determinants of frequency responsiveness. For example, the clown knifefish, Notopterus chitala (family: Notopteridae) has a swimbladder arrangement very similar to that in Myripristis (Dehadrai, 1957; Greenwood, 1963, 1973), yet the frequency range of hearing for Notopterus is narrower than that of Myripristis (Coombs & Popper, in prep.). While subtle differences in swimbladder shape and stiffness may be involved in the variability, there is evidence that the properties of the inner ear, including inter-specific morphological variation, may determine frequency responsiveness (Fay & Popper, 1975). The presence of substantial differences in gross morphological and ultrastructural features of the saccular endorgans in Adioryx, Myripristis and Notopterus (Popper, 1977, 1979b), which are known to have distinct differences in frequency ranges of hearing, is suggestive of a functional relationship.

Discrimination

Although measures of auditory sensitivity provide basic information about what pure tones fish can and cannot hear, these data yield little information about other analyzing capabilities of the fish auditory system or about what capabilities are relevant to the fish's responses to biological signals. For example, we have no information regarding the behavioral significance of high frequency sensitivity to the goldfish or to Myripristis.

While behavioral data are limited, it is clear that fish can discrimin ate between frequencies and that discriminability may be better for ostariophysans than for non-ostariophysans (Dijkgraaf, 1952; Dijkgraaf & Verheijen, 1950; Fay, 1970; Jacobs & Tavolga, 1968; Wohlfahrt, 1939). There are also data indicating the ability of fish to discriminate between different intensities (Chapman & Johnstone, 1974; Jacobs & Tavolga, 1967), to detect amplitude modulated signals (Fay, 1972, 1977), and to detect specified sound signals in the presence of other sounds (Buerkle, 1969; Chapman, 1973; Chapman & Hawkins, 1973; Chapman & Johnstone, 1974; Fay, 1974b; Fay, Ahroon & Orawski, 1978; Hawkins & Chapman, 1975; Hawkins & Sand, 1977; Popper & Clarke, 1979; Tavolga, 1967b, 1974). The mechanisms for these abilities, however, have not been identified. These data are more informative than the simple hearing sensitivity data, because they begin to tell us what features of the sound signal the fish is capable of processing and hence which features may be potentially relevant to the bioacoustic behavior of the animal.

Sound Localization

Sound localization has frequently been cited as one of the most important auditory functions known for animals and one that may play a significant role in the evolution of auditory structures (Masterton, 1974; Masterton, Thompson, Bechtold and Robards, 1975). Recent studies demonstrate that several teleosts can orient to, and swim toward, a sound source (e.g. Chapman & Johnstone, 1974; Popper, Salmon & Parvulescu, 1973; Schuijf, Baretta & Wildschut, 1972; Schuijf & Buwalda, 1975, in press), even in the far-field, without the involvement of the lateral line (Schuijf, 1975; Schuijf & Siemelink, 1974).

Postulated mechanisms for sound localization in terrestrial animals are thought to operate under two general principles: (1) differential stimulation of more than one auditory receptor (such as two ears) with central analysis of peripheral differences providing directional information, and (2) directional sensitivity of the peripheral receptors. Differential stimulation of two receptors can result from differences in time of stimulus arrival, stimulus intensity or stimulus frequency characteristics. These differences depend on the separation of the auditory system into at least two effectively isolated receptor units; the magnitude of the difference will in part be determined by the distance between the two receptors (Erulkar, 1972; Mills, 1972).

The usefulness of binaural cues for fish was questioned by van Bergeijk (1964), who, in a pivotal paper, pointed out that due to longer wavelengths of sound in water and the closeness of the two ears in fishes, binaural information would be minimal, and it would be non-existent when both ears are stimulated by the single median swimbladder. Recent studies have shown, however, that in spite of the terrestrial-type of binaural cues being unavailable to fishes, localization can be accomplished through direct stimulation of the ears by the particle displacement component of the sound wave. Since the paired otolith maculae of fish are oriented in different body planes (Lowenstein, 1971; Popper, 1976), and since various hair cell polarization patterns have been shown to exist within one macula (Dale, 1976; Jorgensen, 1976; Popper, 1976, 1977, 1978b; 1979a), the inner ear of fish is ideally suited for extracting directional information. Thus, comparisons of outputs from different hair cell populations within an otolith macula, as well as between maculae, could provide information about the direction of the sound source. Of course, this would require the inner ear to behave as a velocity or displacement sensitive system, since it is this feature of the propagated sound wave that carries directional information (also see Dijkgraaf, 1960; Schuijf & Buwalda, in press).

Furthermore, since fish are as likely to be stimulated from signals from above and below as well as on the same plane, it has been suggested that their auditory systems have evolved mechanisms for localization of signals that come from virtually any direction (e.g. Fay & Popper, in press; Hawkins & Sand, 1977; Popper, 1978a, in press; Schuijf, 1976a, b; Schuijf & Buwalda, in press).

The actual direction from which the signal comes would be determined in the CNS using information from different receptor groups. Each point in space around the fish would have a unique ratio of response levels between the different receptor groups, with one inherent exception. The exception is that any two points exactly 180° from one another will have identical response ratios, thus making it theoretically impossible to discriminate between sounds coming from opposite directions (Schuijf & Buwalda, 1975, in press). Schuijf (e.g. 1976a) has proposed that fishes resolve this 180° ambiguity by using a phase reference which depends on sensitivity of the system to either the compression or rarefaction phase of the pressure wave. By simultaneously comparing the pressure phase and the direction of cilia displacement from the two ears, this ambiguity could be resolved. This mechanism requires substantial central integration, as well as separation of input from velocity and pressure detectors. At present, there are electrophysiological and anatomical data for preservation of phase information in different eighth nerve fibers innervating separate portions of the goldfish saccular macula (Furukawa, 1978; Furukawa & Ishii, 1967), and it is evident from behavioral demonstrations of the ability of fish to discriminate between compressions and rarefactions (Piddington, 1972; Schuijf, 1975) that the integrity of phase information is maintained in the sensory system.

It would be interesting to speculate if fish with specialized connec tions from the swimbladder to inner ear are better adapted for preserving pressure phase information than fish without such systems. Since a number of taxonomically unrelated groups of teleost fish seem to have evolved a swimbladder-inner ear connection functionally analogous to that found in ostariophysans, it would appear that this morphological arrangement has some significant advantages for these groups. To date, we have little information on what that advantage might be. It is possible that the wider frequency ranges noted for these fish are merely a "bi-product" of enhanced responsiveness to other features of the sound signal, such as pressure phase, which may be potentially more relevant to the fish than sensitivity to high frequency sounds.

SOUND PRODUCTION

Sounds produced by fishes generally consist of short pulses with a fairly wide spectral bandwidth but with the predominant energy at frequencies below 500 Hz (see Demski, Gerald & Popper, 1973; Tavolga, 1971, 1977a). However, a number of species, such as the toadfish (Opsanus, do produce longer duration sounds. The sound production mechanism varies in different species, but the basic mechanisms involve either the movement of body parts, like bones or teeth, against one another or the contraction of special muscles. Sound production often involves the swimbladder as an ancillary organ to match impedances between the sound producing structure and the water and to provide amplification of the sound (Salmon, Winn & Sorgente, 1968; Tavolga, 1962, 1964b).

Recent investigations on the behavioral role of fish sounds have been limited. However, a number of workers have now demonstrated that the temporal pattern of the sounds play a significant role in intra-specific communication and that small changes in the temporal pattern will affect the responsiveness of conspecifics in both damselfish (Pomocentridae) (Myrberg et al., 1978; Myrberg & Spires, 1972) and toadfish (Fine et al., 1977; Fish, 1972; Winn, 1967, 1972). Both Myrberg and Winn now argue that temporal patterning and not frequency spectrum is the major communicative component of a fish sound. This argument is especially compelling considering findings that the spectrum of a fish sound changes markedly with the substrate over which the fish swims (Schneider, 1967; Tavolga, 1962, 1964b). Consequently, whereas temporal patterns will not change, the timbre of a sound may vary and so the communicative value of information encoded in the spectrum of a sound would be 'lost' in species moving over different substrates. Thus, by primarily involving temporal characteristics in communication, fishes would be able to change the spectral composition of their sounds to meet particular environmental needs (see below) without altering the content of the 'message'.

AQUATIC ENVIRONMENT

Investigators studying acoustic communication in terrestrial vertebrates have suggested that the acoustic properties of an animal's habitat will produce selective forces favoring certain features of sound signals, depending on the communicative function of the signal (Michelsen, 1978; Morton, 1975; Wiley & Richards, 1978). In long distance communication, for example, selective pressures should favor the use of sounds that give the greatest broadcast area for a given output intensity. In the natural environment, there are a number of factors which will influence this broadcast area, including reduction of sound energy by absorption or redirection of energy by reflection or diffraction (Morton, 1975). These factors may also influence the spectral qualities of sounds detected. Certain open habitats, for example, such as grasslands, distort spectral qualities of a sound because higher frequency components attenuate more rapidly than lower frequencies (Morton, 1975). It might be expected that information transfer based on a temporal pattern of sounds (which, as long as the sounds are audible, would remain unchanged) in such a habitat would be more efficient than the use of frequency-coded information. Indeed, the rapidly modulated sounds produced by bird species in this habitat are consistent with such an hypothesis (Morton, 1975).

Similar types of selective factors may operate in the acoustic environment of fish. Several aspects of the underwater environment, for example, may be similar to the grasslands described by Morton, favoring temporal patterns of sounds over spectral content for transfer of information. Since the spectral content of fish sounds is known to change as a function of the environment (see sound production section), the spectral characteristics are an unreliable source of information. Parvulescu (1964) postulated other related reasons why the pulsed signal may be more suited for communication in the aquatic environment. Continuous 120 Hz sounds emitted by fish swimming near the surface should be almost completely cancelled out by reflections from the surface, rendering the sound inaudible to other fish in the same horizontal plane. However, by pulsing the sound, the cancellation effect would be slightly delayed, resulting in shorter but audible pulses of altered spectral composition. In addition, Banner (1971) found that in shallow water, pulsed broad band noises (80-640 Hz; 10-1200 msec) attenuate less with distance than do continuous noises.

A number of other features of the aquatic environment may influence the evolution of auditory receptor mechanisms. Probably one of the most significant environmental pressures is the nature of the ambient noise which affects the animal's ability to detect biologically relevant signals in the presence of irrelevant or masking noise. Studies measuring ambient noise levels in shallow ocean waters (Banner, 1968, 1971) and in coral reef habitats (Horch & Salmon, 1973) indicate that velocity levels of noise are higher than pressure levels, and for waters less than eight m deep, they are even higher than expected for the far-field (Horch & Salmon, 1973). Therefore, it would appear that in these environments, there are considerable velocity noise levels which could potentially interfere with sound detection by fish. Moreover, behavioral studies by Cahn, Siler and Auwarter (1971) indicate that sonic particle velocity noise did not interfere with the detection of pressure signals but that pressure signals did mask velocity signals. Thus, it has been suggested that in an environment where low frequency velocity levels are high, it would be advantageous to have a pressure sensitive system that was independent and unaffected by velocity noise. This same conclusion was reached by Horch and Salmon (1973), who also found that relative to background levels in the open environment, sounds produced by the squirrelfish (*Myripristis*) generally contained higher pressure levels than corresponding velocity levels at any given part of the spectrum. Thus, here is an example where pressure sensitivity may have distinct advantages for a fish which is known to have extraordinary pressure sensitivity.

Finally, it is likely that aquatic environments such as murky waters as well as nocturnal life styles which even further preclude the use of vision have led to the evolution of alternative sensory systems specialized to function under the given constraints. The highly specialized swimbladder connection to the ear in the squirrelfish, *Myripristis*, for example, may have in part evolved in response to selective pressures generated by the nocturnal feeding behavior of this species.

In this respect, it is also interesting to note that many of the other teleost fish which have evolved some type of swimbladder-inner ear connection (i.e. Mormyridae, Gymnarchidae, and Gymnotidae) have also evolved active electroreceptive systems (Bennett, 1971). The function of active electroreceptive systems is in many ways analogous to that of specialized auditory systems designed to echolocate. Thus, it is conceivable that some of the same selective pressures (such as the murky waters that mormyrids are known to inhabit (Norman & Greenwood, 1975)) have operated in the evolution of both auditory and electroreceptive sensory systems.

SUMMARY AND CONCLUSIONS

We have attempted, in this paper, to demonstrate that the teleost auditory system is substantially more complex, and more variable, than might have been imagined ten years ago. The variability in the system is most widely known, at the present, in the morphology of the system and, to a lesser degree, in behavioral data of auditory sensitivity. There have been fewer studies of the physiological responses of the inner ear and only two or three investigations of the auditory CNS. Thus, whether the variability apparent in the morphological and behavioral data is related to functional variability at different levels of the auditory system is still unknown. Regretfully, data on the anatomy of the ear are still limited enough for us not to be able to make many broad generalizations regarding taxonomic, and thus evolutionary, relationships of ears of different species (see Popper, 1978a, in press a, for a further discussion of this problems).

Certainly, one facet in the evolution of the teleost auditory system has been the selective pressures imposed upon the whole acoustic mechanism (including sound detection and production) by the environment. The evolution of sound communication per se no doubt resulted from the inability of other potential communicative channels to allow long distance, rapid, and directional transmission of information. Other factors, such as the background noise in the aquatic environment and the presence of coral reefs and rocks would affect specific aspects of the evolution of sound communication. It may be particularly relevant that we now have substantial data indicating that fishes can detect and use both pressure and displacement information. Such an ability may be related to extraction of information in the presence of masking signals, a significant problem that must be eliminated for there to be suitable acoustic communications.

It is also apparent that selective pressures have operated to ultimately produce a communications system, at least for two soniforous groups, the toadfish and damselfish, that circumvents the acoustic problems in the aquatic environment associated with attenuation over long distances and differences in substrates. It is also becoming clearer that fishes have evolved a highly efficient and complex mechanism for sound localization totally different from the system found in terrestrial vertebrates. While the system differs from that in terrestrial forms, it appears that fishes are able to extract a good deal of information about the position in space of sounds, including the position of sources on the vertical as well as horizontal axis. There are also limited data indicating that fish may be able to determine the distance of the emitter from them by comparing the phase of pressure and particle displacement information.

Finally, it is important to point out that we are just at the beginnings of understanding of the acoustic system of fishes. There has been tremendous growth of knowledge since the mid-1960's (e.g Tavolga, 1964a, 1967a), and it is more than likely that the next 10 years will see a further expansion of our understanding of fish audition, particularly with regard to the inter-specific variation in the auditory system and selective pressures operating in the evolution of these systems.

ACKNOWLEDGEMENTS

We would like to thank Dr. W. N. Tavolga for commenting upon an earlier version of this paper. Portions of the work discussed here were supported by grant NS-15090 from the National Institute of Neurological and Communicative Disorders and Stroke, grant BNS 78-22441 from the National Science Foundation, and NINCDS Research Career Development Award NS-00312 to A. N. P.; and by predoctoral fellowship MH-07441 from the National Institute of Menthal Health to S. C.

REFERENCES CITED

Albers, V.M. (Ed.) (1967) Underwater Acoustics. New York, Plenum Press.

Balinsky, B.I. (1975) An Introduction to Embryology, 4th edition. Philadelphia, Saunders.

Banner, A. (1968) Measurements of the particle velocity and pressure of the ambient noise in a shallow bay. J. Acoust. Soc. Am. 44(6): 1741-1742.

Banner, A. (1971) Propagation of sound in a shallow bay. J. Acoust. Soc. Am. 49(1): 373-376.

Bauknight, R. S., Strelioff, D. and Honrubia, V. (1976) Effective stimulus for Xenopus laevis lateral line hair cell system. Laryngoscope 86: 1836-1844.

Bennett, M. V. L. (1971) Electroreception. In: Fish Physiology, Vol. V. Eds. W. S. Hoar and D. J. Randall. New York, Academic, pp. 493-574.

Buerkle, U. (1969) Auditory masking and the critical band in Atlantic cod (Gadus morhua). J. Fish. Res. Bd. Canada 26: 1113-1119.

Cahn, P. H., Siler, W. and Auwarter, A. (1971) Acoustico-lateralis system of fishes: cross modal coupling of signal and noise in the grunt, Haemulon parrai. J. Acous. Soc. Am. 49: 591-594.

Cahn, P. H., Siler, W. and Wodinsky, J. (1969) Acoustico-lateralis system of fishes: Tests of pressure and particle-velocity sensitivity in grunts, Haemulon sciurus and Haemulon parrai. J. Acoust. Soc. Amer. 46: 1572-1578.

Chapman, C. J. (1973) Field studies of hearing in teleost fish. Helgolander wiss. Meeresunters 24: 371-390.

Chapman, C. J. and Hawkins, A. D. (1973) A field study of hearing in the cod, Gadus morhua L. J. Comp. Physiol. 85: 147-167.

Chapman, C. J. and Johnstone, A. D. F. (1974) Some auditory discrimination experiments on marine fish. J. Exp. Biol. 61: 521-528.

Chapman, C. J. and Sand, O. (1974) Field studies of hearing in two species of flatfish, Pleuronectes platessa (L.) and Limanda limanda (L.) (Family Pleuronectidae). Comp. Biochem. Physiol. 47A: 371-385.

Coombs, S. and Popper, A. N. (1979) Hearing differences among Hawaiian squirrelfish (family Holocentridae) related to differences in the peripheral auditory system. J. comp. Physiol. 132:203-207.

Cushing, D.H. (1967) The acoustic estimation of fish abundance. In: Marine Bio-Acoustics, Ed. W. N. Tavolga. Oxford, Pergamon, pp. 75-92.

Dale, T. (1976) The labyrinthine mechanoreceptor organs of the cod Gadus morhua L. (Teleostei: Gadidae). Norw. J. Zool. 24: 85-128.

Dehadrai, P.V. (1957) On the swimbladder and its relation with the internal ear in genus Notopterus (Lacepede). J. Zool. Soc. India 9: 50-61.

Demski, L. S., Gerald, J. W. and Popper, A. N. (1973) Central and peripheral mechanisms of teleost sound production. Amer. Zool. 13: 1141-1167.

Denton, E. J., Gray, J. A. B. and Blaxter, J. H. S. (1979) The mechanics of the clupeid acoustico-lateralis system: frequency responses. J. Mar. Biol. Assn. U. K. 59: 27-47.

Dijkgraaf, S. (1952) Uber die Schallwahrnehmugn bei Meeresfischen. Z. vergl. Physiol. 34: 104-122.

Dijkgraaf, S. (1960) Hearing in bony fishes. Proc. Roy. Soc. Lond. B. 152: 51-54.

Dijkgraaf, S. and Verheijen, F. (1950) Neue Versuche uber das Tonunterscheidungsvermogen der Elritze. Z. vergl. Physiol. 32: 248-256.

Enger, P. S. (1976) On the orientation of hair cells in the labyrinth of perch (Perca fluviatilis. In: Sound Reception in Fish. Eds. A. Schuijf and A. D. Hawkins. Amsterdam, Elsevier, pp. 396-411.

Enger, P. S., Hawkins, A. D., Sand, O. and Chapman, C. J. (1973) Directional sensitivity of saccular microphonic potentials in the haddock. J. Exp. Biol. 59: 425-433.

Erulkar, S. D. (1972) Comparative aspects of spatial localization of sound. Physiol. Rev. 52:237-360.

Fay, R. R. (1970) Auditory frequency discrimination in the goldfish (Carassius auratus. J. Comp. Physiol. Psychol. 73: 175-180.

Fay, R. (1972) Perception of amplitude-modulated signals in the goldfish. J. Acoust. Soc. Am. 52, 660-666.

Fay, R. R. (1974a) Sound reception and processing in the carp: Saccular potentials. Comp. Biochem. Physiol. 49(A): 29-42.

Fay, R. R. (1974b) The masking of tones by noise for the goldfish (Carassius auratus. J. Comp. Physiol. Psych. 87: 708-716.

Fay, R. R. (1977) Auditory temporal modulation transfer function for the goldfish. J. Acoust. Soc. Am. 62 (Suppl. 1): S88.

Fay, R. R. (1978a) Coding of information in single auditory nerve fibers of the goldfish. J. Acoust. Soc. Am. 63: 136-146.

Fay, R. R. (1978b) Sound detection and sensory coding by the auditory systems of fishes. In: The Behavior of Fish and Other Aquatic Animals. Ed. D. Mostofsky. New York, Academic Press, pp. 197-231.

Fay, R. R., Ahroon, W. and Orawski, A. (1978) Auditory masking patterns in the goldfish (Carassius auratus): Psychophysical tuning curves. J. Exp. Biol. 74: 83-100.

Fay, R. R. and Olsho, L. (1979) Response patterns of neurons in the lagenar branch of the goldfish auditory nerve. Comp. Biochem. Physiol. 62A: 377-386.

Fay, R. R. and Popper, A. N. (1974) Acoustic stimulation of the ear of the goldfish (Carassius auratus). J. Exp. Biol. 61: 243-260.

Fay, R. R. and Popper, A. N. (1975) Modes of stimulation of the teleost ear. J. Exp. Biol. 62: 379-388.

Fay, R. R. and Popper, A. N. (in press) Structure and function in teleost auditory systems. In: Comparative Studies of Hearing in Vertebrates. Eds. A. N. Popper and R. R. Fay. New York, Springer-Verlag.

Fine, M., Winn, H. and Olla, B. 1977. Communication in fishes. In: How Animals Communicate. Ed. T. A. Sebeok. Bloomington, Indiana University Press, pp. 472-518.

Fish, J. F. (1972) The effect of sound playback on the toadfish. In: Behavior of Marine Animals, Vol. 2. Eds. H. E. Winn and B. L. Olla. New York, Plenum Press, pp. 386-434.

Flock, A. (1965) Electron microscopic and electrophysiological studies on the lateral line canal organ. Acta Oto-laryngol. Suppl. 199: 1-90.

Flock, A. (1971) Sensory transduction in hair cells. In: Handbook of Sensory Physiology, Vol. I. Principles of Receptor Physiology. Ed. W. R. Lowenstein. Berlin, Springer-Verlag, pp. 396-411.

Furukawa, T. (1978) Sites of termination on the saccular macula of auditory nerve fibers in the goldfish as determined by intracellular injections of procion yellow. J. Comp. Neurol. 180: 807-814.

Furukawa, T. and Ishii, Y. (1967) Neurophysiological studies on hearing in goldfish. J. Neurophysiol. 30: 1377-1403.

Greenwood, P. H. (1963) The swimbladder in African Notopteridae (Pisces) and its bearing on the taxonomy of the family. Bull. Br. Mus. Nat. Hist. (Zool.). 11: 377-412.

Greenwood, P.H. (1973) Interrelationships of osteoglossomorphs. In: Interrelationships of Fishes. Eds. P.H. Greenwood, R.S. Miles and C. Patterson. London, Academic Press, pp. 307-332.

Hama, K. (1969) A study on the fine structure of the saccular macula of the goldfish. Z. Zellforsch. 94: 155-171.

Harris, G. G. (1964) Considerations on the physics of sound production by fishes. In: Marine Bio-Acoustics. Ed. W. N. Tavolga. Oxford, Pergamon Press, pp. 233-247.

Harris, G. G. and Milne, D. C. (1966) Input-output characteristics of the lateral-line sense organs of Xenopus laevis. J. Acoust. Soc. Am. 40(1): 32-42.

Harris, G. G. and van Bergeijk, W. A. (1962) Evidence that the lateral line organ responds to near-field displacements of sound sources in water. J. Acoust. Soc. Am. 34: 1831-1841.

Hawkins, A. D. and Chapman, C. J. (1975) Masked auditory thresholds in the cod Gadus morhua L. J. Comp. Physiol. 103A: 209-226.

Hawkins, A. D. and Sand, O. (1977) Directional hearing in the median vertical plane. J. Comp. Physiol. 122A: 1-8.

Horch, J. and Salmon, M. (1973) Adaptations to the acoustic environment by the squirrelfishes Myripristis violaceus and M. parlinus. Mar. Behav. Physiol. 2:121-139.

Hudspeth, A. J. and Corey, D. P. (1977) Sensitivity, polarity, and conductance change in the response of vertebrate hair cells to controlled mechanical stimuli. Proc. Natl. Acad. Sci. 74: 2407-2411.

Hudspeth, A.J. and Jacobs, R. (1979) Stereocilia mediate transduction in vertebrate hair cells. Proc. Natl. Acad. Sci. 76(3): 1506-1509.

Jacobs, D. W. and Tavolga, W. N. (1967) Acoustic intensity limens in the goldfish. Anim. Behav. 15: 324-335.

Jacobs, D. W. and Tavolga, W. N. (1968) Acoustic frequency discrimination in the goldfish. Anim. Behav. 16: 67-71.

Jenkins, D. B. (1979) A transmission and scanning electron microscopic study of the saccule in five species of catfishes. Am. J. Anat. 154: 81-101.

Jorgensen, J. M. (1976) Hair cell polarization in the flatfish ear. Acta Zool. 57: 37-39.

Kroese, A. B. A., Van der Zalm, J. M. and Van den Bercken, J. (1978) Frequency response of the lateral-line organ of Xenopus laevis. Archiv. 375: 167-175.

Kuiper, J. W. (1956) The microphonic effect of the lateral-line organ. Pub. Biophys. Group Naturk. Lab. Groningen Nethrl. 159 pp.

Kuiper, J. W. (1967) Frequency characteristics and functional significance of the lateral line organ. In: Lateral Line Detectors. Ed. P. H. Cahn. Bloomington, Indiana University Press, pp. 105-122.

Lowenstein, O. (1971) The labyrinth. In: Fish Physiology, Vol. 5. Eds. W. S. Hoar and D. S. Randell, New York, Academic Press, pp. 207-240.

Maler, L., Karten H. J. and Bennett, M. V. L. (1973a) The central connections of the posterior lateral line nerve of Gnathonemus petersii. J. Comp. Neurol. 151: 67-84.

Maler, L., Karten, H. J., Bennett, M. V. L. (1973b) The central connections of the anterior lateral line nerve of Gnathonemus petersii. J. Comp. Neur. 151, 67-84.

Masterton, R.B. (1974) Adaptation for sound localization in the ear and brainstem of mammals. Fed. Proc. 33: 1904-1907.

Masterton, B., Thompson, G.C., Bechtold, J.K., and Robards, M.J. (1975) Neuroanatomical basis of binaural phase-difference analysis for sound localization: a comparative study. J. Comp. Phys. Psych. 89(5): 379-386.

McCormick, C. (in press) Organization and evolution of the octavolateralis area in fishes. In: Fish Neurobiology and Behavior. Eds. R.G. Northcutt and R. E. Davis, Ann Arbor, Univ. of Michigan Press.

Michelsen, A. (1978) Sound reception in different environments. In: Sensory Ecology. Ed. M.A. Ali. New York, Plenum Press, pp. 345-373.

Mills, A.W. (1972) Auditory localization. In: Foundations of Modern Auditory Theory, Vol. 2. Ed. J. V. Tobias. New York, Academic Press, pp. 303-348.

Morton, E. (1975) Ecological sources of selection on avian sound. Am. Nat. 109: 965.

Myrberg, A. A., Jr., Spanier, E. and Ha. S. J. (1978) Temporal patterning in acoustical communication. In: Contrasts in Behavior. Eds. E. Reese and F. Lighter. New York, Wiley and Sons, pp. 138-179.

Myrberg, A. A., Jr. and Spires, J. Y. (1972) Sound discrimination by the bicolor damselfish, Eupomacentrus partitus. J. Exp. Biol. 57: 727-735.

Nelson, E. M. (1955) The morphology of the swimbladders and auditory bulla in Holocentridae. Fieldiana: Zool. 37: 121-137.

Norman, J.R. and Greenwood, P.H. (1975) A History of Fishes. New York, Halsted Press, John Wiley and Sons, Inc.

Northcutt, R. G. (in press) Central auditory pathways in anamniotic vertebrates. In: Comparative Studies of Hearing in Vertebrates. Eds. A. N. Popper and R. R. Fay. New York, Springer-Verlag.

Parvulescu, A. (1964) Problems of propagation and processing. In: Marine Bio-Acoustics. Ed. W. N. Tavolga. Oxford, Pergamon Press, pp. 87-100.

Piddington, R. W. (1972) Auditory discrimination between compressions and rarefactions by goldfish. J. Exp. Biol. 56: 403-419.

Pitcher, T.J., Partridge, B.L., and Wardle, C.S. (1976) A blind fish can school. Science 194: 963-965.

Platt, C. (1977) Hair cell distribution and orientation in goldfish otolith organs. J. Comp. Neurol. 172: 283-298.

Platt, C. and Popper, A. N. (in press) Otolith organ receptor morphology in herring-like fishes. In: Vestibular Function and Morphology. Ed. T. Gualterotti. New York, Springer-Verlag.

Popper, A. N. (1971) The effects of size on auditory capacities of the goldfish. J. Aud. Res. 11: 239-247.

Popper, A. N. (1972) Auditory threshold in the goldfish (Carassius auratus) as a function of signal duration. J. Acoust. Soc. Am. 52: 596-602.

Popper, A. N. (1976) Ultrastructure of the auditory regions in the inner ear of the lake whitefish. Science 192: 1020-1023.

Popper, A. N. (1977) A scanning electron microscopic study of the sacculus and lagena in the ears of fifteen species of teleost fishes. J. Morph. 153: 397-418.

Popper, A. N. (1978a) A comparative study of the otolithic organs in fishes. In: Scanning Electron Microscopy/1978. Eds. R. P. Becker and O. Johari. AMF O'Hare, Ill.,Scanning Electron Microscopy, Inc., Vol. 2, pp. 405-416.

Popper, A. N. (1978b) Scanning electron microscopic study of the otolithic organs in the bichir (Polypterus bichir) and shovel-nose sturgeon (Scaphirhynchus platorynchus). J. Comp. Neurol. 181: 117-128.

Popper, A. N. (1979a) The ultrastructure of the sacculus and lagena in a moray eel (Gymnothorax sp.). J. Morphol. 161:241-256.

Popper, A. N. (1979b) The ultrastructure of the ear in osteoglossomorph fishes. Soc. Neurosci. Absts. 5:29.

Popper, A. N. (in press a) Organization of the inner ear and auditory processing. In: Fish Neurobiology and Behavior. Eds. R.G. Northcutt and R. E. Davis, Ann Arbor, Univ. of Michigan Press.

Popper, A. N. (in press b) Scanning electron micrsocopic study of the sacculus and lagena in several deep sea fishes. Amer. J. Anat.

Popper, A. N., Chan, A. T. H. and Clarke, N. L. (1973) An evaluation of methods for behavioral investigations of teleost audition. Behav. Res. Meth. Instr. 5: 470-472.

Popper, A. N. and Clarke, N. L. (1979) Simultaneous and non-simultaneous auditory masking in the goldfish (Carassius auratus). J. Exp. Biol. 83:145-158.

Popper, A. N. and Fay, R. R. (1973) Sound detection and processing by teleost fishes: a critical review. J. Acoust. Soc. Am. 53: 1515-1529.

Popper, A. N. and Platt, C. (1979) The herring ear has a unique receptor pattern. Nature 280:832-833.

Popper, A. N., Salmon, M. and Parvulescu, A. (1973) Sound localization by the Hawaiian squirrelfishes, Myripristis berndti and M. argyromus. Anim. Behav. 21: 86-97.

Salmon, M., Winn, H. E. and Sorgente, N. (1968) Sound production and associated behavior in triggerfishes. Pac. Sci. 22: 11-20.

Sand, O. (1974a) Recordings of saccular microphonic potentials in the perch. Comp. Biochem. Physiol. 47A: 387-390.

Sand, O. (1974b) Directional sensitivity of microphonic potentials from the perch ear. J. Exp. Biol. 60: 881-899.

Sand, O. and Michelsen, A. (1978) Vibration measurements of the perch saccular otolith. J. Comp. Physiol. 123A: 85-89.

Schneider, H. (1967) Morphology and physiology of sound-producing mechanisms in teleost fishes. In: Marine Bio-Acoustics, Vol. II. Ed. W. N. Tavolga. Oxford, Pergamon Press, pp. 135-155.

Schuijf, A. (1975) Directional hearing of cod (Gadus morhua) under approximate free field conditions. J. Comp. Physiol. 98: 307-332.

Schuijf, A. (1976a) The phase model of directional hearing in fish. In: Sound Reception in Fish. Eds. A. Schuijf and A. D. Hawkins. Amsterdam, Elsevier, pp. 63-86.

Schuijf, A. (1976b) Timing analysis and directional hearing in fish. In: Sound Reception in Fish. Eds. A. Schuijf and A. D. Hawkins. Amsterdam, Elsevier, pp. 87-112.

Schuijf, A., Baretta, J. W. and Wildschut, J. T. (1972) A field investigation on the discrimination of sound direction in Labrus bergglyta (Pisces: Perciformes). Netherlands J. Zool. 22: 81-104.

Schuijf, A. and Buwalda, R. J. A. (1975) On the mechanism of directional hearing in cod (Gadus morhua L.). J. Comp. Physiol. 98: 333-343.

Schuijf, A. and Buwalda, R. J. A. (in press) Underwater localization - A major problem in fish acoustics. In: Comparative Studies of Hearing in Vertebrates. Eds. A. N. Popper and R. R. Fay. New York, Springer-Verlag.

Schuijf, A. and Siemelink, M. E. (1974) The ability of cod (Gadus morhua) to orient towards a sound source. Experientia 30: 773-774.

Schwartz, E. (1967) Analysis of surface-wave perception in some teleosts. In: Lateral Line Detectors. Ed. P. H. Cahn. Bloomington, Indiana University Press, pp. 123-134.

Shaw, E. (1970) Schooling in fishes: critique and review. In: Development and Evolution of Behavior. Eds. L. R. Aronson, E. Tobach, D. S. Lehrman and J. S. Rosenblatt, San Francisco, W.H. Freeman and Co., pp. 452-480.

Siler, W. (1969) Near and far fields in a marine environment. J. Acoust. Soc. Am. 46: 483-484.

Strelioff, D. and V. Honrubia. (1978) Neural transduction in Xenopus laevis lateral line system. J. Neurophysiol. 41: 432-444.

Tavolga, W. N. (1962) Mechanisms of sound production in the ariid catfishes Galeichthys and Bagre. Bull. Am. Mus. Nat. Hist. 124: 1-30.

Tavolga, W. N. (Ed.) (1964a) Marine Bio-Acoustics. Oxford, Pergamon Press, 413 pp.

Tavolga, W. N. (1964b) Sonic characteristics and mechanisms in marine fishes. In: Marine Bio-Acoustics. Ed. W. N. Tavolga. Pergamon Press, Oxford, pp. 195-211.

Tavolga, W. N. (Ed.) (1967a) Marine Bio-Acoustics, II. Oxford, Pergamon Press, 353 pp.

Tavolga, W. N. (1967b) Masked auditory thresholds in teleost fishes. In: Marine Bio-Acoustics, II. Ed. W. N. Tavolga. Oxford, Pergamon Press, pp. 233-245.

Tavolga, W. N. (1971) Sound production and detection. In: Fish Physiology, Vol. V. Eds. W. S. Hoar and D. J. Randall. New York, Academic Press, pp. 135-205.

Tavolga, W. N. (1974) Signal/noise ratio and the critical band in fishes. J. Acoust. Soc. Am. 55: 1323-1333.

Tavolga, W. N. (1976) Recent advances in the study of fish audition. In: Sound Reception in Fishes - Benchmark Papers in Animal Behavior, Vol. 7. Ed. W. N. Tavolga. Stroudsburg, Pa., Dowden, Hutchinson & Ross, pp. 37-52.

Tavolga, W. N. (1977a) Recent advances in the study of sound production in fishes. In: Sound Production in Fishes - Benchmark Papers in Animal Behavior, Vol. 9. Ed. W. N. Tavolga. Stroudsburg, Pa., Dowden, Hutchinson & Ross, pp. 47-53.

Tavolga, W. N. (1977b) Mechanisms for directional hearing in the sea catfish (Arius felis). J. Exp. Biol. 67: 97-115.

Tavolga, W. N. and Wodinsky, J. (1963) Auditory capacities in fishes. Pure tone thresholds in nine species of marine teleosts. Bull. Am. Mus. Nat. Hist. 126: 177-240.

van Bergeijk, W. A. (1964) Directional and nondirectional hearing in fish. In: Marine Bio-Acoustics. Ed. W. N. Tavolga, Oxford, Pergamon Press, pp. 281-350.

van Bergeijk, W. A. (1967) The evolution of vertebrate hearing. In: Contributions to Sensory Physiology. Ed. W. D. Neff. New York, Academic Press, pp. 1-49.

von Bekesy, G. (1960) Experiments in Hearing. New York, McGraw-Hill, 745 pp.

von Frisch, K. (1936) Uber den Gehorsinn der Fische. Biol. Rev. 11: 210-246.

Weiss, T. J., Mulroy, F., Turner, M. R. G. and Pike, C. L. (1976) Tuning of single fibers in the cochlear nerve of the alligator lizard: relation to receptor morphology. Brain Res. 115: 71-90.

Wever, E. G. (1974) The evolution of vertebrate hearing. In: Handbook of Sensory Physiology, Vol. V/1, Auditory System. Eds. W. D. Keidel and W. D. Neff. Berlin, Springer-Verlag, pp. 423-454.

Wiley, R. H. and Richards, D. G. (1978). Physical contraints on acoustic communication in the atmosphere: Implications for the evolution of animal vocalizations. Behav. Ecol. Sociobiol. 3:69-94.

Winn, H. (1964) The biological significance of fish sounds. In: Marine Bio-Acoustics. Ed. W. N. Tavolga. Oxford, Pergamon Press, pp. 213-231.

Winn, H. E. (1967) Vocal facilitation and biological significance of toadfish sounds. In: Marine Bio-Acoustics, II. Ed. W. N. Tavolga. Oxford, Pergamon Press, pp. 283-303.

Winn, H. E. 1972. Acoustic discrimination by the toadfish with comments on signal systems. In: Behavior of Marine Animals, Vol. 2. Eds. H. E. Winn and B. L. Olla. New York, Plenum Press, pp. 361-385.

Wohlfahrt, T. A. (1939) Untersuchungen uber das Tonunterscheidungsuermogen der Elritze (Phoxinus laevis) Agass.). Z. vergl. Physiol., 26: 570-604.

VISION IN FISHES:

ECOLOGICAL ADAPTATIONS

J.N. Lythgoe

University of Bristol
Department of Zoology
Woodland Road, Bristol BS8 1UG

GENERAL PROBLEMS OF VISION

The vertebrate eye is one of the most conservative of organs. At a casual glance a Coelacanth eye is not so very different from our own, and even a lamprey eye is recognisably the same organ as our own. The underwater light climate may scatter more light, may be more monochromatic and be darker in the daytime, but the basic laws of optics are the same underwater as on land. The most obvious adaptations that fishes show are to the dim light at depth, the monochromatic nature of the underwater light and the more directional distribution of the spacelight.

At root the problems of vision are statistical. It is necessary to sample enough photons to make reliable judgements about the nature of the image on the retina. As in any statistical problem, the more similar are two populations, the greater the sample required to distinguish them. In visual terms fine discriminations of contrast detail colour and movement can only be made if there is ample light. In dim light fine discriminations become impossible.

The brightness of the light can be measured in physical terms as the number of photons, per unit area, per unit time, per unit wavelength interval. These are units of radiance. The number of photons sampled can be increased by increasing the area of image that is sampled (with a resulting loss in spatial detail); or by increasing the time over which the sample is taken (with a resulting loss in the visibility of moving targets); or by increasing the spectral bandwidth that is sampled (with perhaps a loss of colour discrimination).

THE VISUAL ENVIRONMENT UNDERWATER

As any diver will testify it is more difficult to see underwater than on land. The three main problems are that it gets rapidly darker with depth, contrasts are much reduced with a resulting loss in visual range, and the spectral bandwidth available for vision is greatly reduced. In the very clearest ocean water at noon it may be possible for a man to see downwelling daylight at depths down to 800-1000 m (Clarke & Denton 1962). In clear coastal water 200 m would be considered good, and 100 m might be more typical, whilst in turbid inshore waters a lower limit for vision could be 10 m or less. It is not possible to see actual objects nearly so far through the water in any direction. Duntley (1960) has prepared nomograph tables to predict the visual range of objects of various size in waters of various light transmission characteristics. Duntley predicts that not even the most visible object will be seen at distances greater than 70 m or so. Lythgoe (1971) considers that the greatest visual range ever achieved in practice is 100 m. In the exceptionally clear waters of the Mediterranean a horizontal range of 40 m for large white targets would be good. If the water is turbid the visual range will reduce to a few metres and in very turbid water a diver would be fortunate to see his own outstretched hand.

Pure water is most transparent in the blue-green region of the spectrum between about 460 and 490 nm. See Smith and Tyler (1967), Jerlov (1968) and Lythgoe (1979) for reviews on which this section is based. This is molecular absorption and is the main reason why clear water appears blue. Another factor that contributes to the blue appearance is Rayleigh scatter from the water molecules and from those suspended particles less than a wavelength of light in diameter. Short wave blue light is scattered more than longer wave red light giving the blue colour to scattered light. The blue colour of pure water is a property of the water molecule itself, but natural water often appears green, yellow-green, or even red-brown in colour. The two most important agents that cause this colouration are the yellow products of vegetable decay, the so-called "yellow substances" or "Gelbstoffe", and chlorophyll itself. Obviously chlorophyll-rich water will have a strong green component; but of greater significance is the presence of "yellow substances" in the water. It has been known since the pioneer work of Kalle (1938, 1966) that yellow substances absorb most strongly in the blue, and least strongly in the red region of the spectrum. Because water absorbs red, and yellow substances absorb blue, only the mid-spectrum colours of green and yellow-green remain. Landlocked waters may be so heavily stained with yellow substances that they appear reddish brown rather than green.

The absorption of light by water is exponential with the result that as the depth increases the daylight becomes restricted to a narrow spectral band. Water thus acts as a monochromater (Tyler 1959) and the vision of fishes that live in different colours of water is markedly adapted to suit the colour of the ambient light, a subject that is pursued in later sections.

It is difficult to overstate the importance of scattered light in reducing visual range underwater. Large suspended particles of sand and silt and plankters as well as the minute particles responsible for Rayleigh scattering cause a reduction of visual contrast by interposing a bright veil of scattered light between underwater objects and the eye. Contrast is defined as (T-B)/B where T is the spectral radiance of a target object and B is the spectral radiance of the background. The addition of veiling light, V, to both T and B results in a reduced contrast of (T-B)/(B+V).

The angular distribution of light underwater is another factor that needs consideration. In shallow water the distribution of daylight is largely governed by the position of the sun and refractions at the air-water interface. An observer underwater on a calm day can only see out through the surface through a window, subtending a solid angle of 98°. This is called Snell's window because it is predictable from Snell's law of refraction. Peripheral to Snell's window total internal reflexion means that one can see the bottom reflected from the water surface. In real life the surface of the water is rarely absolutely flat and the waves and ripples passing over it act as a complex series of cylindrical lenses that fragment the edges of Snell's window and concentrate the sunlight into an ever-changing ripple pattern.

In deeper water Snell's window is no longer visible. Here the angular distribution of light is governed by the position of the sun and by the relative amount of scattering material in the water. The more the scattering the more uniform is the directional distribution of the light.

Because there are only a limited number of agents that have an important influence on water colour, Jerlov (1951, 1976) found it possible to classify oceanic and inshore (Baltic) waters on the basis of their spectral absorption of downwelling diffuse light. Jerlov's classification has been of great help to visual ecologists because it allows a quantitative comparison of the light environment underwater with the visual pigments of the fishes that live there. Studies based upon the relationship between water colour and visual pigments are outlined in later sections.

THE ENHANCEMENT OF SENSITIVITY

An eye needs to capture sufficient photons radiating from the visual scene to allow the brain to interpret what is going on there. This means that in dim light the greatest possible number of photons should be incorporated in the formation of the image on the retina, so that the neural analysis of the image can give the maximum amount of reliable information.

Geometric optics of the eye: The number of photons that are absorbed by a photoreceptor is related to the brightness of that part of the image that falls upon it. As in the camera, the brightness of the image of an extended light source is proportional to the f-number of the camera lens or eye. The f-number is defined as A/F where A is the diameter of the lens aperture and F is the focal length. Thus the brightness of the image depends upon the

Table 1

	f-number	
Fishes	0.78	Munk and Frederiksen, 1974
Cat	0.89	Vakkur and Bishop, 1963
Tawny Owl	1.3	Martin, 1977
Man	2.1	Martin, 1977
Man	3.3	Kirschfeld, 1974
Pigeon	4.0	Marshall et al, 1973

relative size of the aperture and the focal length of the lens, not upon the absolute size of the eye. (Rodieck 1973, Kirschfeld 1974).

In fishes there is a lower limit to the possible f-number. This is a result of the ineffectiveness of the cornea as a refractive surface that is due to the similar refractive index of aqueous humor, cornea and water. The focussing power of the eye relies entirely on the spherical lens. The refractive index of fish lens is about 1.65 and the focal length is 2.55 its radius (Matthiessen's ratio). In turn this means the lower theoretical limit for the f-number of fishes is 0.78 (Munk and Frederiksen 1974). Perhaps fortunately for fishes a lens aperture of 0.78 confers good sensitivity compared to other vertebrates.

When image-forming light enters the eye obliquely to the optic axis, the effective aperture is partly reduced by the pupil margin and the image brightness is correspondingly reduced. In some fishes, particularly predators, that need to see well in the forward direction, the pupil margin is pear-shaped leaving an area of pupil not occupied by the lens. This area is called the aphakic space. In some deep-living teleosts there is an aphakic space completely surrounding the lens. Munk and Frederiksen (1974) consider that this allows a bright image to be formed from all oblique directions. It is not clear how the "fogging" effect of unfocussed light entering through the aphakic space affects the quality of the retinal image.

In common with some other dim light vertebrates such as owls (Walls, 1942), the eyes of many deep-living marine teleosts are tubular in shape (for review see Marshall 1971, 1979; Locket 1977). Such eyes retain the advantages of size, but only for a restricted angle of view; the peripheral parts of the visual field are sacrificed to keep the volume of the eye down to that which can reasonably be accomodated in the head. The visual field of tubular eyes show a wide area of overlap. This may partly be to enhance stereoscopic vision (see for example Marshall 1979). Alternatively sensitivity is improved by 0.1 log units where two visual fields overlap (Pirenne

1943); and Campbell and Green (1965) have shown that binocular vision enhances contrast detection by $\sqrt{2}$ because 'noise' can be physiologically filtered out.

A feature of the underwater environment, particularly in the mesopelagic zone, is that downwelling light may be 2 log units brighter than upwelling light. Fishes are modified both in their camouflage and in their preferred direction of vision to allow for this (Lythgoe 1979 for a review). Because it is impossible to camouflage fishes seen in silhouette by mirror camouflage when they will always appear brighter than the water background, and because in the deeper part of the mesopelagic zone the only light bright enough to see comes directly from above, the visual axis of many mesopelagic fishes is directed upwards. As a counter strategy most bioluminescent species have ranks of downward directed photophores that raise the radiance of the silhouetted fish to equal that of the downwelling light (see Herring 1977). The penalty of restricted visual angle provided by tubular eye is mitigated by the 'lens pad' frequently present in these eyes (Locket 1977). These in fact act as wave guides, conducting light from beyond the periphery of the normal field of view onto an accessory retina in the wall of the tubular eye.

A method to enhance the brightness of the image at the photopigment layer is to place light-reflecting material behind the photoreceptors. These are tapeta and they act by reflecting light that has already passed through the visual cells back through the photoreceptors for a second chance of absorption. Tapeta are common in nocturnal and dim-light animals, both vertebrates and invertebrates. (There are many different types of reflecting layer in fishes (Nicol 1974, Nicol et al. 1973, Locket 1977). In common with other tapeta they probably enhance sensitivity, but at the expense of image sharpness and hence visual acuity.

Retinal morphology and 'grain': It is the function of the optical part of the eye to deliver a bright, good quality, image to the retinal photoreceptors. The retina has to transform the information contained in the optical image into the neural signals that the brain interprets as visual sensations.

There are two main kinds of morphological adaptations of the photoreceptor cells to dim light. The first is to increase the depth of the visual pigment layer, the second is to increase the effective capture area of each receptor unit. The increased depth of visual pigment in rods is often obtained by increasing the length of the outer segments or by arranging the rods several layers deep in the retina (Locket 1977). The effective capture area might be enlarged by increasing the cross-sectional area of single receptors as may be the case for the cover of crepuscular fishes (see Lythgoe 1979 for a review). Receptors may also be clumped together into what are optically single units. The best examples are found in several deep-sea fishes where groups of rods are bundled together, each bundle being contained in a cup of highly reflecting tapetum, (Locket, 1977).

Teleost cones are frequently arranged in a regular mosaic. The exact function, if any, of these mosaics is unclear. However double, and even triple cones are a feature of these retinas (Ahlbert 1975, Ali and Anctil 1976) and there is some suggestion that it is these cones that are adapted to the lower brightness limits of photopic vision. There is quite good reason to think that it is these cones whose visual pigments are matched in spectral absorption to the spectral maximum of the underwater spacelight.

More important than the enlarged cross-sectional area of individual receptors, is the way that receptor cells can pool their information together to form a functionally single unit sampling a large area of image. This is called retinal summation and the increased integration areas obtained improve sensitivity, but at the expense of acuity.

To get large enough photon samples for statistical reliability it is necessary to gather the sample over an extended time period (the integration time) as well as over an extended area (the integration area). The integration time, or retinal memory time, is measured experimentally by finding the time at which an increase in exposure time can no longer compensate for a reciprocal decrease in the intensity of the stimulus (see Ripps and Weale 1976). Related to the retinal integration time is the flicker fusion frequency. This is the frequency of a flashing light that cannot be distinguished from a continuous one. The flicker fusion frequency of a number of fishes has been published by Protasov (1970). The frequency in the species he examined ranged from 67 Hz in Atherine mochon to 10 Hz in the goldfish. This range is typical of vertebrates as a whole. For example man's flicker fusion frequency ranges from about 60 Hz in bright light to 10 Hz in dim light.

Just as an extended integration area carries the penalty of reduced spatial detail, so an increased integration time carries the penalty of reduced temporal detail. Dim-light animals with a long integration time may fail to detect images that move fast across the retina. The smaller the image or the lower the contrast it presents the greater will be the problem. Perhaps dim-light predators habitually ambush their prey so that their own movements do not add to the movement of their prey to produce images on the retina that move too fast to be seen. Elsewhere in this volume (Liem) it is shown that many fishes capture their prey by ambushing them and rapidly sucking them into the mouth. It would be interesting to know if such predators were more effective or more active in dim light.

Large eyes and small eyes: Small animals have relatively larger eyes than small ones (Kirschfeld 1976). Intuitively there is nothing surprising about this statement; but little of the foregoing discussion has actually offered an explanation. For if the brightness of the image from an extended source depends upon the ratio of pupil aperture to focal length rather than upon the actual value of either, what is to prevent the visual cells being reduced in size to enable a small bright image to be analysed just as finely as a larger image.

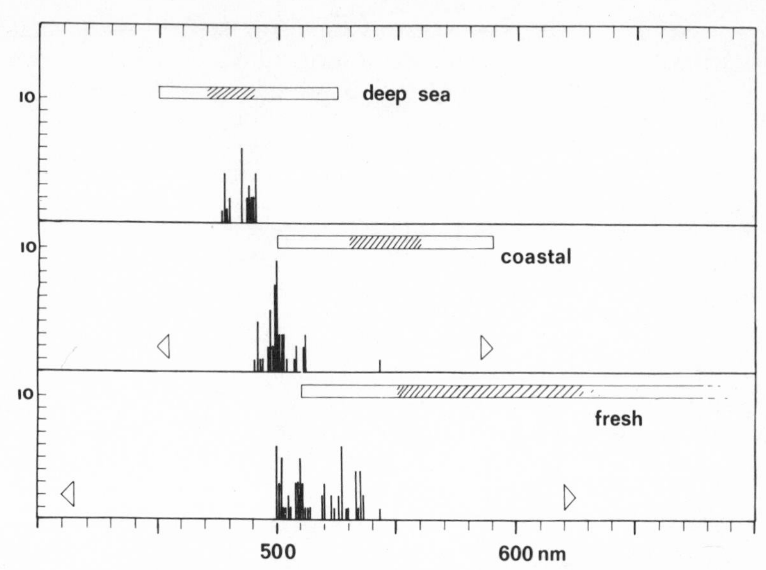

Fig. 1. Histograms of the numbers of species containing single rod visual pigments of known wavelength of maximum absorption (λmax). Deep-sea, coastal and freshwater species are shown. The horizontal bars show the possible range of visual pigment λmax that would give maximum sensitivity in each water type. The shaded areas show the most likely range. The open triangles show the λmax of the most blue sensitive and the most red sensitive visual pigments measured in the cones of fishes inhabiting that water type. (Data from Ali and Wagner 1975, Fernandez 1979).

The answer seems to lie in the wave nature of light rather than in its quantum nature that has been the basis of the discussion hitherto. Diffraction effects at the pupil aperture limit the sharpness of the image so that the smaller the pupil the poorer the acuity. This effect is different to the enhanced depth of field obtained with small apertures. At the visual cell level the wavelength of light sets a lower limit of 1 - 2 μm to the diameter of the visual cell outer segments. Anything smaller and the light energy is not guided down through the visual pigment but in a sheath outside. A situation that leads to a loss of sensitivity and optical "cross talk" between the receptor cells and thus to reduced acuity.

The visual pigments: It has been a question of interest for nearly half a century whether the absorptive properties of fish visual pigments are adapted to make them particularly sensitive to the ambient light underwater. (See Munz and MacFarland 1977, Lythgoe 1979, for reviews). For mesopelagic fishes there is an excellent match between the spectral band of available light in deep ocean water and the spectral absorption of the visual pigment that would absorb the most photons there (Fig. 1). There is also a tendency for the rods of fishes that live in freshwaters, which are most transparent at longer wavelengths, to be most sensitive to longer wavelengths. In coastal waters, that are intermediate in spectral transmission between fresh and oceanic water, the rod visual pigments are most sensitive to intermediate wavelengths.

There is, however, a difficulty: we find that it is only the deep sea fishes that have visual pigments matched to the ambient light to give maximum sensitivity. In coastal and freshwater, where the light can get every bit as dim, the visual pigments absorb most strongly at shorter wavelengths than one would expect if sensitivity was the only criterion. This is true despite the fact that visual pigments exist that would give the expected match.

There are various explanations for the mismatch, but none of them are very reliable. Common to most of them is the idea that the perception of contrast is as important as sensitivity (Lythgoe 1968). Rod visual pigments may be evolved as much to reduce visual 'noise' as to increase sensitivity. It has also been suggested that for bright objects underwater a visual pigment offset to shorter wavelengths will enhance visual contrasts and hence visibility. The relationship between visual pigment and contrast perception underwater is discussed in the next section.

Many land animals have visual pigments with absorption maxima near 500 nm (Lythgoe 1972b for a list) although there is no obvious reason why this should confer enhanced sensitivity at night (Dartnall 1975, Lythgoe 1979). It is possible that visual pigments of λ max near 500 nm are in some way easier to manufacture in rods or are in some way more stable.

Another explanation involves 'infra red noise' (Stiles 1948, Barlow 1957). Infra red quanta do not by themselves carry enough energy to isomerise the visual pigment molecule and hence initiate a visual signal. However it is possible that if the molecule is warm, the vibrational and rotational energy of the molecule can supplement the energy of the quantum sufficiently for isomerisation to occur. Visual pigments that absorb at longer wavelengths are the most likely to be excited by infra red quanta. Sensitivity to infra red would be visually undesirable because the eye itself and the animal's body radiate it. Images of the outside world would still be present, but its contrast would be degraded by non image-forming infra red "fogging".

CONTRAST AND COLOUR

The theme of this review is that it is necessary to sample sufficient photons to make reliable judgements about fine differences in radiance. In the previous discussions nearly all the emphasis has been placed on sensitivity, or, in other words, ways of increasing the number of photons sampled. The reciprocal approach is to concentrate on those features of the visual scene that show the greatest differences, and therefore require fewer photons to distinguish them.

In the brief discussion on the visual effects of infra red radiation, it was evident that it can be disadvantageous to see into the infra red. Short wave radiation is also a problem and it is noteworthy that most diurnal vertebrates have some kind of yellow (i.e. minus blue) filter in their optical path (Walls and Judd 1933, Walls 1942). The reasons for this may be to reduce Rayleigh-scattered light that is rich in the shorter wavelengths, or it may be to reduce chromatic aberration. In a comprehensive review of yellow filters in fishes Muntz (1975) shows that yellow carotenoid filters in the cornea or lens of teleost fishes are common but are confined to diurnal, bright light species. Muntz believes that the purpose of yellow filters in fishes, as in terrestrial vertebrates, is to cut out scattered light and to reduce chromatic aberration.

Whilst most deep sea fishes have a particularly transparent cornea, lens and humors (Denton 1956); Muntz (1976) has shown that some mesopelagic animals (2 squids and 4 fishes) have yellow lenses. These yellow filters will reduce sensitivity by up to two thirds. Muntz believes that the explanation may be that the spectral radiance of bioluminescence is richer in red light than the downwelling radiance in the mesopelagic zone. Animals that camouflage their silhouette with downward-directed photophores must adjust their brightness to match the downwelling daylight. A brightness match that is correct for most fishes not having yellow filters, will leave the photophores looking conspicuously brighter to fishes that possess yellow intra-ocular filters.

When it is present the yellow filters in the cornea are frequently concentrated in the dorsal area (Moreland and Lythgoe 1968) and thus selectively reduces the bright downwelling light that is characteristic of the light environment underwater. The unavoidable scattering of some of this very bright light within the eye reduces visual contrast and must be particularly troublesome when searching the dimly lit visual scene below. Another technique that may help to avoid this intra-ocular flare is used by many shallow living, chiefly marine and diurnal teleosts. These possess complex irridescent reflectors in the cornea that selectively reflect away the bright downwelling rays of the sun while allowing image-forming light from other directions to pass without significant loss (Lythgoe 1974 a, b, 1976).

On land a distant object takes on the colour and brightness of the sky; underwater the same process occurs, but within metres rather than kilometres. Fig. 2 shows the change in spectral radiance of a grey object as it recedes to merge with the background spacelight. The contrast of the object against its background which was defined earlier reduces exponentially as the distance between the eye and object increases. The exponent is $\propto$ the narrow beam attenuation co-efficient of light through water. This means that contrasts reduce less rapidly at wavelengths where the water is most transparent.

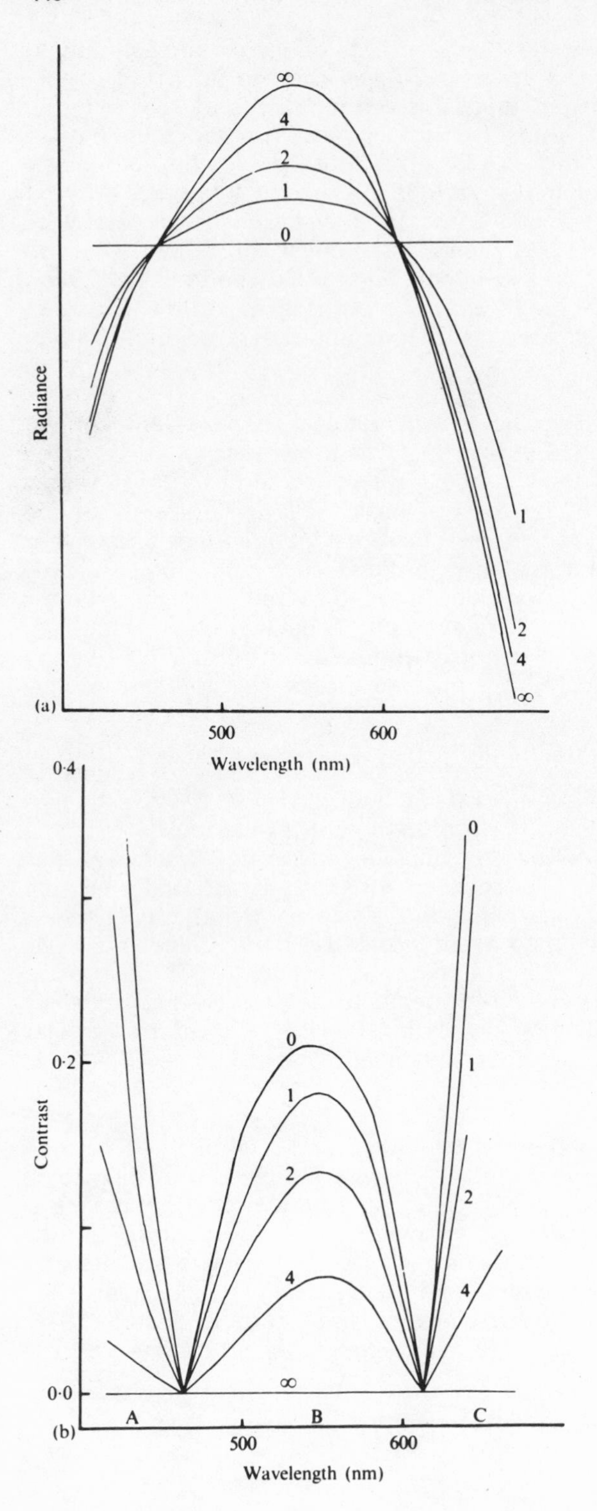

Fig. 2. (a) The spectral radiance of a grey target suspended in shallow water at zero range from the observer and at 1,2, and 4 m. At range ∞ the target is so far distant that it is indistinguishable from the background space-light. (b) As for (a) but this time it is the contrast between the object and the background space-light that is measured. At a large visual range (∞) there is no contrast and the object is invisible. A, B, C are the positions of the λ max of the hypothetical visual pigments mentioned in the text, A and C are offset; B is matched to the spectral radiance maximum of the water.

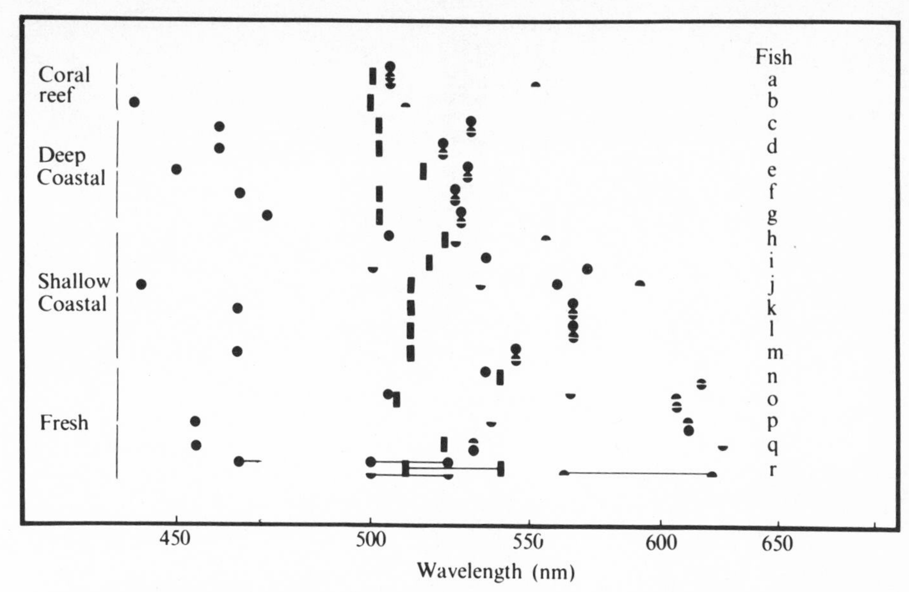

Fig. 3a. The visual pigments in the rods and cones of various teleost fishes. Rods, rectangles; single cones, filled circle; single member of twin or double cone, half-circle. The lines joining the symbols in (r) (rudd) represent the possible rhodopsin-porphyropsin shift in λ max. The fishes are: (a) damsel fish, Pomacentrus melanochir; (b) lion fish, Dendrochirus zebra; (c) saphirine gurnard, Trigla lucerna; (d) grey gurnard, Eutrigla gurnardus; (e) plaice, Pleuronectes platessa; (f) lemon sole, Microstomus kitt; (g) dab, Limanda limanda; (h) corkwing wrasse, Crenilabrus melops; (i) shanny, Blennius pholis; (j) 14-spined stickleback, Spinachia spinachia; (k) rock goby, Gobius paganellus; (l) shore rockling, Gaidropsarus mediterraneus; (m) flounder, Platichthys flesus; (n) perch, Perca fluviatilis; (o) blue acara, Aequidens pulcher; (p) tench, Tinca tinca; (q) goldfish, Carassius auratus; (r) rudd, Scardinius erythrophthalmus. (After Loew and Lythgoe 1978).

Objects suspended in the water are chiefly illuminated by the bright, broad spectral band, downwelling light; but are viewed against the less bright more monochromatic horizontal spacelight. At wavelengths offset from the wavelength of maximum water transparency the contrast presented by a close bright object should be large but is reduced faster as the viewing range increases. At wavelengths where the water is most transparent contrasts for bright close objects may be relatively low, but will reduce less rapidly with distance, until at the limits of visibility all (large) objects remain most easily seen at wavelengths where the water is most transparent.

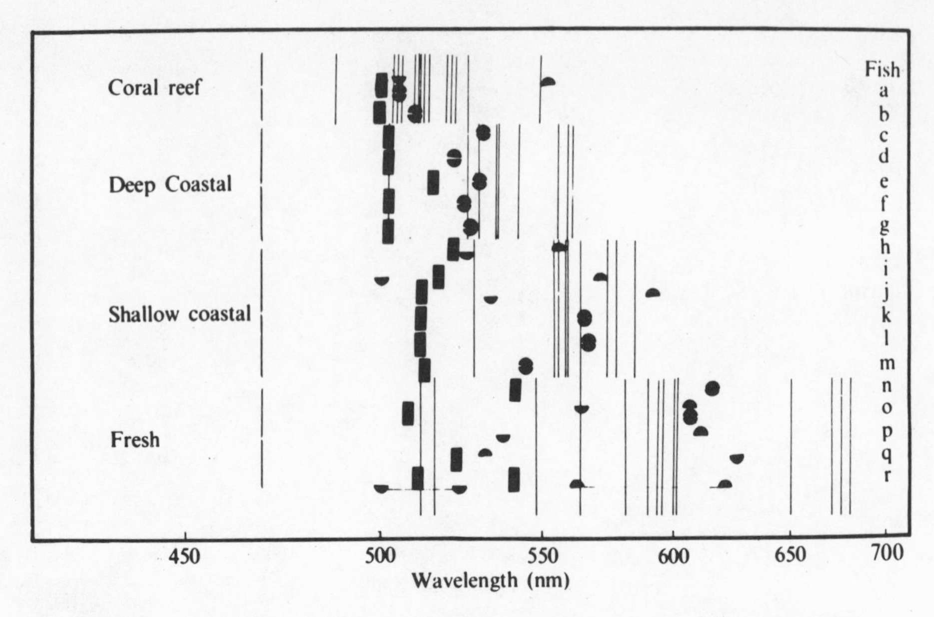

Fig. 3b. The visual pigments in the rods and paired cones of teleost fishes compared according to habitat. The vertical lines represent the λ max of the visual pigment that would have been most sensitive on particular occasions when spectral radiance measurements were made (see Loew and Lythgoe 1978 for details). The spectral bandwidth covered by the lines indicate the variation in the colour of water of each type that can be expected. Note that the paired cones appear to have the visual pigments that make them most sensitive, whereas the rods are generally less sensitive to red light than might be expected on sensitivity grounds alone. (After Loew and Lythgoe 1978).

In summary, therefore, an offset visual pigment will generally be best for detecting close objects that are brighter, (albeit marginally), than the water background. However, a matched visual pigment will be best for detecting all dark or distant objects. In deep water the selective absorption of the water will in any case confine vision to a narrow spectral waveband, for outside that band insufficient photons will be available for vision.

Arguments of this kind were used by Lythgoe (1968, 1972) to explain the mismatch between rod visual pigments and water transparency in all but extremely deep blue water. However, this approach may be better applied to cone vision where more light is available and the necessary reduction in sensitivity may be less of a disadvantage. Munz and McFarland (1977) have suggested that the possession of both matched and offset visual pigments to detect contrasts at different wavelengths might be the condition required for the evolution of colour vision. It is the ratio of signal strengths from two photoreceptors of different spectral sensitivity scanning the same area of image that the brain interprets as colour. For aquatic animals three pigment types might be useful, one matched to the wavelength of best water transparency, one offset to shorter wavelengths and one to longer.

In fishes there is now some evidence that the double cones are matched to the predominant waveband in the water spacelight (Loew and Lythgoe 1978) and Fig. 3. The visual pigments in the single cones are offset to shorter or longer wavelengths. Certainly this view is over-simple and recent work by Levine (this volume) has shown that within a body of water, fishes that share an ecological niche also possess remarkably similar visual pigments. It also seems to be true that fishes that live in deeper water, and are thus exposed to a more restricted waveband of daylight, correspondingly possess a set of visual pigments that span a narrower spectral range than their shallower counterparts. We are as yet only on the threshold of the study of the ecology of colour vision; but it promises to be a very fascinating one.

REFERENCES

Ahlbert, I. (1975). Organization of the cone cells in the retinae of some teleosts in relation to their feeding habits. Univ. of Stockholm. Thesis.

Ali, M.A. and Anctil, M. (1976). Retinas of Fishes: An Atlas. Heidelberg, Springer-Verlag, 284 p.

Ali, M.A. and Wagner, H.-J. (1974). Visual pigments: phylogeny and ecology. In: Vision in Fishes. Ed. M.A. Ali. N.Y. Plenum, p. 481-516.

Barlow, H.B. (1957). Purkinje shift and retinal noise. Nature (Lond.) 179: 255-256.

Campbell, F.W. and Green, D.G. (1965). Monocular versus binocular visual acuity. Nature (Lond.) 208: 191-192.

Clarke, G.L. and Denton, E.J. (1962). Light and animal life. In: The Sea, Vol. I. Ed. M.N. Hill.N.Y., London J. Wiley & Sons, p. 456-468.

Dartnall, H.J.A. (1975). Assessing the fitness of visual pigments for their photic environment. In: Vision in Fishes. Ed. M.A. Ali. N.Y., Plenum Press, p. 543-563.

Denton, E.J. (1956). Recherches sur l'absorption de la lumière par le cristallin des poissons. Bull. Inst. Oceanog., Monaco 1071: 1-10.

Duntley, S.Q. (1960). Improved nomographs for calculating visibility by swimmers (natural lights). Bureau of Ships Contract No. bs-72039, Rep. 5 - 3 Feb.

Fernandez, H.R.C. (1979). Visual pigments of bioluminescent and nonbioluminescent deep-sea fishes. Vision Res. 19: 589-592.

Herring, P.J. (1977). Bioluminescence of marine organisms. Nature (Lond.) 267: 788-793.

Jerlov, N.G. (1951). Optical studies of ocean water. Rep. Swedish Deep Sea Exped. 3: 1-59.

Jerlov, N.G. (1968). Optical Oceanography. Amsterdan, Elsevier.

Jerlov, N.G. (1976). Marine Optics. Amsterdam, Elsevier.

Kalle, K. (1938). Zum Problem der Meereswasserfarbe. Ann. Hydrol. Marine Mitt. 66: 1-13.

Kalle, K. (1966). The problem of Gelbstoffe in the sea. Oceanog. Marine Biol. Ann. Rev. 4: 91-104.

Kirschfeld, K. (1974). The absolute sensitivity of lens and compound eyes. Z. Naturfosch. 29c: 592-596.

Kirschfeld, K. (1976). The resolution of lens and compound eyes. In: Neural Principles of Vision. Eds. F. Zettler and R. Weiler. Berlin, Springer-Verlag, p. 354-369.

Locket, N.A. (1977). Adaptations to the deep-sea environment, in Handbook of Sensory Physiology, Vol. VII 5. Ed. F. Crescitelli. Berlin, Springer-Verlag, p. 67-192.

Loew, E.R. and Lythgoe, J.N. (1978). The ecology of cone pigments in teleost fishes. Vision Res. 18: 715-722.

Lythgoe, J.N. (1968). Visual pigments and visual range underwater. Vision Res. 8: 997-1012.

Lythgoe, J.N. (1971). Underwater Vision. In: Underwater Science. Eds. J.D. Woods and J.N. Lythgoe. Oxford University Press, p. 103-139.

Lythgoe, J.N. (1972a). The adaptation of visual pigments to their photic environment. In: Handbook of Sensory Physiology, Vol. 7/1. Ed. H.J.A. Dartnall. Berlin, Springer-Verlag, p. 566-603.

Lythgoe, J.N. (1972b). List of vertebrate visual pigments. In: Handbook of Sensory Physiology, Vol. VII/1. Ed. H.J.A. Dartnall. Berlin, Springer-Verlag, p. 604-624.

Lythgoe, J.N. (1974a). The structure and physiology of irridescent corneas in fishes. In: Vision in Fishes. Ed. M.A. Ali. N.Y., Plenum Press, p. 253-262.

Lythgoe, J.N. (1974b). The iridescent cornea of the sand goby, Pomatoschistus minutus (Pallas). In: Vision in Fishes. Ed. M.A. Ali. N.Y., Plenum Press, p. 263-278.

Lythgoe, J.N. (1976). The ecology function and phylogeny of irridescent multilayers in fish cones. In: Light as an Ecological Factor II. Eds. G.C. Evans, R. Bainbridge and O. Rackham. Oxford, Blackwell, p. 211-247.

Lythgoe, J.N. (1979). The Ecology of Vision. Oxford University Press.

Marshall, N.B. (1971). Explorations in the Life of Fishes. Cambridge, Mass., Harvard.

Marshall, N.B. (1979). Developments in Deep-Sea Biology. London, Blandford.

Marshall, J., Mellerio, J. and Palmer, D.A. (1973). A schematic eye for the pigeon. Vision Res. 13: 2449-2453.

Martin, G.R. (1977). Absolute visual threshold and scotopic spectral sensitivity in the tawny owl Strix aluco. Nature (Lond.) 268: 636-638.

Moreland, J.D. and Lythgoe, J.N. (1968). Yellow corneas in fishes. Vision Res. 8: 1377-1380.

Munk, O. and Frederiksen, R.D. (1974). On the function of aphakic apertures in teleosts. Videnskabelige Meddelelser fra Dansk Naturhistorisk forening 137: 65-94.

Muntz, W.R.A. (1975). The visual consequences of yellow filtering pigments in the eyes of fishes occupying different habitats. In: Light as an Ecological Factor II. Eds. G.C. Evans, R. Bainbridge and O. Rackham, Oxford, Blackwell, p. 271-287.

Muntz, W.R.A. (1976). On yellow lenses in mesopelagic animals. J. Mar. Biol. Ass. UK 56: 963-976.

Munz, F.W. and McFarland, W.N. (1977). Evolutionary adaptations of fishes to the photic environment. In: Handbook of Sensory Physiology, Vol. VII 5. Ed. F. Cresctelli. Berlin, Springer-Verlag, p. 193-274.

Nicol, J.A.C. (1974). Studies on the eyes of fishes: Structure and ultrastructure. In: Vision in Fishes. Ed. M.A. Ali. N.Y., Plenum Press, p. 579-607.

Nicol, J.A.C., Arnott, H.J. and Best, C.G. (1973). Tapeta lucida in bony fishes (Actinopterygii): a survey. Can. J. Zool. 51: 69-81.

Pirenne, M.H. (1943). Binocular and uniocular threshold for vision. Nature (Lond.) 152: 698-699.

Protasov, U.R. (1970). Vision and Near Orientation of Fish (translation from Russian 1968). Israel Programme for Scientific translation, Jerusalem.

Ripps, H. and Weale, R.A. (1976). The visual stimulus. In: The Eye, Vol. II a. Ed. H. Davson. N.Y., Academic Press, p. 43-99.

Rodieck, R.A. (1973). The Vertebrate Retina. San Francisco, W.H. Freeman & Co.

Smith, R.C. and Tyler, J.E. (1967). Optical properties of natural water. J. Opt. Soc. Am. 57: 589-601.

Stiles, W.S. (1948). The physical interpretation of the spectral sensitivity curve of the eye. In: Transactions of the Optical Convention of the Worshipful Company of Spectacle Makers. London, Spectacle Makers Co., p. 97-107.

Tyler, J.E. (1959). Natural water as a monochromator. Limnol. Oceanogr. 4: 102-105.

Vakkur, G.J. and Bishop, P.W. (1963). The schematic eye of the cat. Vision Res. 3: 357-382.

Walls, G.L. (1942). The Vertebrate Eye and Its Adaptive Radiation. N.Y., Hafner.

Walls, G.L. and Judd, H.D. (1933). The intra-ocular colour filters of vertebrates. Brit. J. Opthal. 17: 641-75, 705-25.

VISUAL COMMUNICATION IN FISHES*

Joseph S. Levine [1,2]
Phillip S. Lobel [1]
Edward F. MacNichol, Jr. [2,3]

[1] Museum of Comparative Zoology, Harvard University
Oxford Street
Cambridge, MA 02138

[2] Laboratory of Sensory Physiology
Marine Biological Laboratory
Woods Hole, MA 02543

[3] Department of Physiology
Boston University School of Medicine
Boston, MA

INTRODUCTION

The intensity and spectral distribution of the light encountered by an animal species set reasonably broad operational limits for the efficient generation and reception of visual signals. This phenomenon is of special importance to fishes because aquatic environments cause profound changes in the nature of light that penetrates to any depth (Blaxter, 1970; Brezonik, 1978; James and Birge, 1938; Jerlov, 1968; Kinney et al., 1967; Tyler, 1959). The complex and variable nature of the underwater photic environment offers a unique opportunity to examine the evolutionary responses of visual receptor systems and body colors to variations in the spectral quality of ambient light. It is of particular interest to determine the extent to which

* We gratefully acknowledge the generous support of the Rowland Foundation and other non-governmental donors. This work was also funded in part by an NSF Doctoral Dissertation Improvement Grant and a Grant-in-aid from Sigma-X' to J.S.L. and by a NIH Grant EY-02399 to E.F.M.

organisms have evolved in directions that either make maximum use of environmental conditions or make optimal compromises in the face of conflicting structural and functional requirements. In other words: how well is the visual communication system of the organism "engineered" to perform the tasks upon which the species depends for its survival?

Underwater illumination

The nature of the aquatic photic environment has been studied and discussed in great detail elsewhere (Lythgoe and Northmore, 1973; Lythgoe, 1975; Lythgoe, this volume; McFarland and Munz, 1975 a,b, 1977; Dartnall, 1975), and since we have no new data to add on this topic we will mention only points of special interest to the ensuing discussion on color communication.

The modal wavelength and spectral bandwidth of downwelling illumination in aquatic environments shift and narrow, respectively, with increasing path length. The direction of the shift (longwave or shortwave) depends on the presence, nature and concentration of dissolved organic compounds in the water. If no dissolved compounds are present in significant amounts, both fresh and salt water exhibit transmission maxima of 475-490nm (Jerlov, 1965, 1968; James and Birge, 1938; Tyler and Smith, 1970). The presence of chlorophylls and various other organic substances loosely termed "gelbstoff" (yellow substance) imparts to many lakes and some coastal marine waters a greenish color and a characteristic transmission maximum of 500-700nm. Tannins and lignins - along with large amounts of gelbstoff - push the transmission maximum in swamps, marshes and many tropical rivers and streams to well over 600nm (Brezonik, 1978; Kinney et al., 1967; Muntz, pers. com.).

The extent of the shifting and narrowing that occurs in each of these habitats depends on both the concentration of these organic compounds and on the total distance light has travelled through the medium to the point of measurement. This path-length-dependent effect is the cause of the well documented changes in downwelling illumination encountered at increasing depths in clear ocean water. Equally important from the standpoint of color vision is the influence of the path-length effect on the spectral nature of upwelling and horizontally scattered light. Munz and McFarland (1975, 1977) were the first to thoroughly document this effect in clear tropical seas. Their voluminous data are summarized in Figure 1a. Note the profound differences that can exist among the spectral characteristics of light encountered along different lines of sight radiating from the same point. A similar phenomenon will exist in organically-stained freshwater environments but the displacement of the modal wavelength will be in the opposite direction - towards longer wavelengths (Figure 1b). Although no data of this type have yet been gathered in tropical freshwaters, we can say with assurance that differences between downwelling and horizontally scattered and upwelling light will be even more striking in these environments than in clear marine water because of the heavily stained nature of the medium.

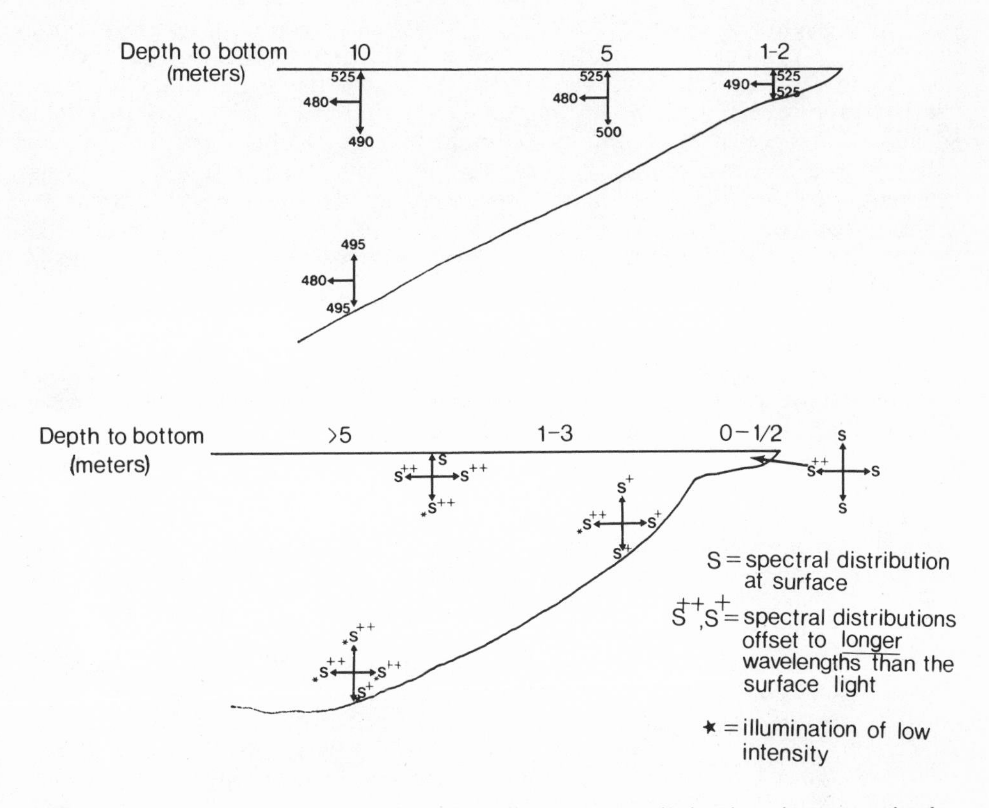

Figure 1a. Spectral properties of background light in clear tropical seas, adapted from Munz and McFarland, 1977). Numbers given represent that wavelength which equally divides the total number of available photons between 400 and 700 nm. With minor alterations, the measurements here would also represent clear bodies of fresh water like Crater Lake, Oregon and Lake Malawi at certain times of the year.

Figure 1b. Predictions regarding the differences in spectral properties among the microhabitats in aquatic environments containing substantial amounts of dissolved organic matter. Downwelling light at the surface (S) will show a broad, flat spectrum resembling the data collected near the surface of the sea by Munz and McFarland (1977) unless pre-filtered by over-hanging vegetation. The high shortwave absorptance of the medium shifts the dominant wavelength in the longwave direction with increasing path length. Further details are found in the text.

Receptor-systems

Vertebrate eyes owe their light sensitivity to one or more photolabile visual pigments, each of which absorbs a band of wavelengths covering a particular region of the spectrum. The presence of one visual pigment in an animal's eye provides it with the ability to discriminate between objects and backgrounds on the basis of brightness, endowing it with a monochromatic view of the world similar to the image on a black and white television screen. To serve as a maximally efficient photon catcher for a species - in other words, to maximize the animal's absolute sensitivity - the visual pigment of that species must absorb light at wavelengths that are transmitted well through the type of water in which the animal performs its visually-based tasks (Dartnall and Lythgoe, 1965; Munz and McFarland, 1977; Lythgoe, this volume).

The optimum peak spectral sensitivities of the visual pigments contained in the eyes of a species are not necessarily matched with the peak wavelength of available illumination unless the task to be performed is simply the detection of the shadows of predators or prey passing directly in front of the source of illumination. Contrast enhancement of several different sorts may be more important than absolute sensitivity in various underwater viewing situations. The maximization of contrast for light reflected from colored objects requires a visual pigment that is offset somewhat from the predominant wavelengths of the available illumination (Lythgoe, 1966, 1968; McFarland and Munz, 1975b). We have seen that the spectral nature of the background against which visual discriminations must be made varies with the line of sight. Different visual tasks may also, by their nature require different types of discriminations and would therefore be best served by visual pigments with different spectral locations. McFarland and Munz (1975b), and Munz and McFarland (1977) have extended this line of reasoning into a compelling argument to explain the origin of multi-pigment visual systems. With sufficient light available for vision, species could develop two or more offset visual pigment designed to maximize contrast in a variety of situations.

Once multi-pigment systems exist, the groundwork is laid for the development of color vision. True hue discrimination independent of relative brightness requires a minimum of two visual pigments of effectively different spectral sensitivities, which produce qualitatively different signals in the central nervous system (Cornsweet, 1970). This is accomplished in all eyes thus far studied by locating the pigments in separate receptor cells that respond differently to spectral stimuli containing different energy distributions. These neural signals interact in a number of ways at several levels in the retina and higher visual centers (MacNichol et al., 1973, for extensive review). Although it is common to refer to photoreceptor cells by assigning them color names on the basis of their λmax values, it is important that one not draw hasty conclusions about the relationship of receptor cells to perceived color stimuli. Only under rare circumstances is a visual perception based entirely on the response of one class of photoreceptors. In other words, stimulation of "red" (or more accurately "red-sensitive") cones is not the necessary and sufficient condition for producing a sensation of redness. Rather it is the neural comparison of the signals generated by all extant receptor classes in a given retina that results in the final perception. For this reason, the shape of a species' hue discrimination function will depend on both the individual spectral locations of its visual pigments and their positions relative to each other. Hue discrimination will be most acute at spectral positions where the absorption curves of a species' visual pigments overlap and where the rates of change of light absorption with wavelength are greatest and of opposite sign (Figure 2).

We now know that the visual pigment systems of fishes range from monochromatic, all rod systems - designed to maximize sensitivity in low-light situations - to complex, duplex retinas equipped with 4 (and occasionally 5) visual pigments that provide their owners with well-developed hue discrimination capability. Data on the photopic (cone) visual pigments that make color vision possible in fishes were first gathered by the technique of microspectrophotometry (MSP) (Hanaoka and Fujimoto, 1957; Marks and MacNichol, 1962; Liebman and Entine, 1964; Marks, 1965; Liebman, 1972; Ali et al., 1978; Loew and Lythgoe, 1978; Levine and MacNichol, 1979). The technique of extraction followed by partial bleaching analysis used by Wald (1937) to identify chicken cone pigments has also proved useful in certain fishes (Munz and McFarland, 1975). Recent advances both in MSP instrumentation and in techniques for on-line data analysis have streamlined that method sufficiently to permit the rapid accumulation of large amounts of information on cone pigments (MacNichol, 1978). As these data have become available for comparison with the substantial earlier studies by Munz and McFarland (1973, 1975), Loew and Lythgoe (1978) and Svaetichin et al. (1965), an increasingly complex, increasingly intriguing picture of aquatic visual systems has emerged. In addition, single cone action spectra recorded by micropipette electrodes have amply confirmed the microspectrophotometric data (Tomita et al; 1967; Burkhardt et al; in prep.).

Visual signal generation

The generation of visual signals by fishes is also subject to the pressure of natural selection. The color and texture of an animal's integument can serve either to convey information regarding the presence and identity of an individual or to conceal that information. Species endowed with the ability to change their colors over time can either hide themselves on different substrates like chameleons, or make themselves cryptic or conspicuous at will (seasonal breeding coloration). In this paper we will concern ourselves only with colors and patterns that we believe are designed to actively transmit information about the bearer.

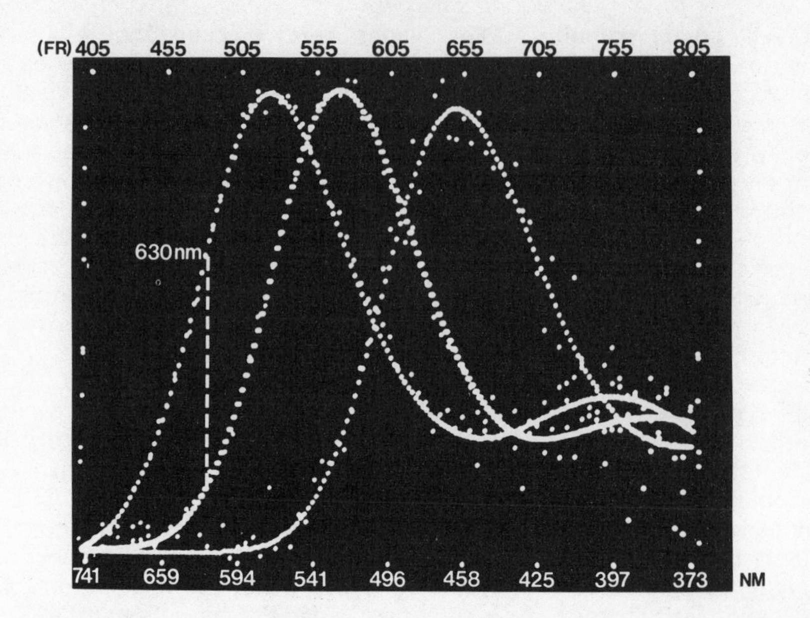

Figure 2. Visual pigments of the cichlid fish Cichlasoma longimanus as determined by transverse microspectrophotometry. Curves represent the averages of a number of records for each pigment type fitted with template curves. Absorption maxima for these spectra are 579 ± 5 nm, 532 ± 2 nm and 455 ± 5 nm. Curves are normalized and indicate relative optical density as a function of frequency. Actual (transverse) OD values ranged from 0.04 to 0.06. Dots at upper and lower margins indicate spectral location at 50 - fresnel intervals. Upper row is labelled in frequency units (fresnels), lower row in wavelength units (nanometers).

The extensive overlap of the visual pigment curves shown here is typical for multi-pigment visual systems. Neural interactions in the retina and higher visual centers compare the responses of cones containing different pigments in order to determine the final color perceived for any stimulus. The dashed line in this figure shows that a monochromatic stimulus of 630 nm will affect both the long- and middlewave sensitive pigments in this species. Monochromatic stimuli between 400 and 530 nm will affect all three cones pigments. Since natural stimuli are rarely monochromatic, normal viewing situations will usually result in the stimulation of at least two cone types.

By definition, a conspicuous visual signal must stand out or contrast from its surroundings. It can do this by differing from the areas adjacent to it in brightness, color, or a combination of both. Brightness contrast refers to a non color-dependent difference in apparent intensity. Color contrast, on the other hand, infers that the reflectance spectra sent from the object and background to the eye cause different sensations in the viewer that cannot be equated merely by altering their relative intensities.

In order for a pattern to be conspicuous, its components must differ from each other and the background in brightness, wavelength distribution or both. The most straightforward and universally conspicuous patterns consist of combinations of black and white. Black and white pigments absorb and reflect light uniformly across the visible spectrum and differ from each other only in the total amount of energy they reflect. For this reason black and white patterns will be highly visible regardless of the illumination that falls upon them.

Colored patterns, on the other hand, differ in appearance under illumination of different spectral composition, although not nearly as much as we would expect (McCann, et al; 1976; Land, 1977).

Colored objects are characterized by reflectance spectra that are non-uniform in some part of the visible spectrum. Where broad-band illumination is available, this region of non-uniform reflectance can be anywhere in the spectrum where the intended recipient of the signal has visual pigments that can detect the difference. In spectrally restricted aquatic environments, however the reflectance spectrum of the object must change within the envelope of available light.

For example, pigments whose reflectance spectra show major changes between 420 and 550 nm will appear colored in blue water, which transmits significant amounts of energy in this region of the spectrum. Pigments (and structural colors) that reflect large amounts of the energy incident upon them from 400 to around 450 nm and then very little elsewhere in the spectrum will appear dark blue. Pigments that reflect very little light between 400 and about 500 nm and whose reflectance rises sharply to a plateau shortly thereafter will appear yellow in this environment. A pigment characterized by low reflectance from 400 nm up to around 560 nm, followed by a rapid rise and plateau would be perceived as orange or red under broad-spectrum illumination but will apear black or grey (depending on its saturation) in the marine environment.

In habitats where the umbrella of available light is shifted in the longwave direction, the types of spectra that produce conspicuous colors also shift. Red becomes highly visible in green to reddish-brown water, as do greens, which simultaneously contrast strongly against the reds. Blues, in the meantime, slowly darken into chromatic obscurity. (For more detailed discussions of this phenomenon see Lythgoe and Northmore, 1973; Kinney et al; 1967).

The foregoing discussion has been based on certain implicit assumptions regarding the types of reflectance spectra generated by naturally colored objects. In order to justify these assumptions, it is necessary to discuss the nature of the structures in fish integuments that produce visual signals.

1. Achromatics: Black and white, the two colors that we noted above to be the most widely useful in signal generation are created in the integuments of fishes by two familiar compounds, melanin and guanine. Melanin - of which there are actually numerous forms - is ubiquitous in the animal kingdom. Most melanins, which are synthesized by vertebrates as required, have spectra characterized by low reflectance throughout the visible spectrum. Small, mobile, granular melanin organelles called melanosomes are typically contained in cells called melanophores that are characterized by extensive peripheral processes. In addition to the ability to either concentrate melanosomes centrally or disperse them into the peripheral cellular areas, many melanophores can actually change shape and position in the integument in order to effect lightening and darkening (Hawkes, 1974; Matsumoto, 1965; Fox, 1953).

Guanine, familiar to many as the white, reflecting component of the tapetum lucidum found in some vertebrate eyes (Rodieck, 1973), is also a prominent component of many integumentary patterns (Fries, 1958). Depending on the size, shape, and relative position of guanine crystals, they can be responsible for a number of different effects (Fox, 1953). When they are arranged to reflect light evenly across the spectrum, they can be responsible for either white or silvery effects.

2. Colored pigments: The most predominant pigments in fishes are the carotenoids, pterins and pteridines, which are responsible for most reds, yellows and oranges (Fox, 1953). Carotenoids cannot be synthesized *de novo* by most vertebrates; they must be obtained from the plants that manufacture them or from other animals that have collected them. Vertebrates can, however, slightly alter the carotenoids they have ingested, changing their optical properties in the process (Katayama and Tsuchiga, 1971). Carotenoids, pterines and pteridines are characterized by *in situ* spectra that take the form of middle-to-longwave bandpass filters (Figure 3; see also Lythgoe and Northmore, 1973; Lythgoe, 1975). These spectra - which differ considerably from the spectra exhibited by these compounds in dilute solution - are fortuitously amenable to the type of "conspicuousness analysis" with which we are concerned here.

The chromatophores that contain these pigments are usually intimately associated with melanophores and guanine platelet-containing structures known as irridiophores (Becher, 1924). Changes in the state of all three types of pigment-containing cells can effectively alter the lightness and saturation of the energy reflected by the integument as a whole. These changes have been elegantly documented in the treefrog *Hyla* by Nielson (1978) and Nielson and Dyck (1978). Short or long-term changes in the

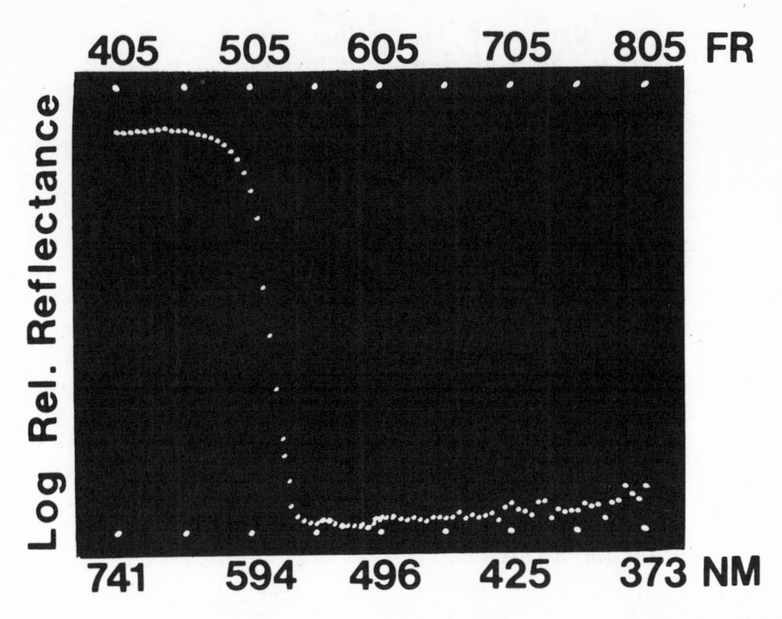

Figure 3. Log relative transmission spectrum of a lipid-soluble carotenoid pigment from a retinal cone oil droplet of the turtle Pseudemys. Dots and lables at upper and lower figure margins as in figure 2. This curve is virtually identical in shape to spectral reflectance curves obtained by Lythgoe and Northmore (1973) from the integuments of fishes. Note that this red carotenoid pigment effectively acts as a longwave bandpass filter. Concentrated solutions of yellow carotenoids have spectra that are of the same shape but are shifted 30-50 nm in the shortwave direction.

concentration of carotenoids within the chromatophores can also cause changes in the color of the organism. It is for this reason that we will consider as one "color" all areas in sampled species that match up to chips of different saturation and brightness on any one Munsell hue page. (See Materials and Methods).

3. Structural colors: Blue pigments are conspicuous by their absence from teleosts as a group. With remarkably few documented exceptions (Goodrich and Hedenburg, 1941), blue colors in fishes are produced by structural rather than by chemical pigments. For a detailed explanation of structural colors see Fox (1953). These structurally produced colors have

reflectance spectra that resemble either the transmission spectra of interference filters or those of short-middlewave bandpass filters. Often, colors like greens and purples are formed by the combination of structural blues with underlying pigmentary yellows and reds, respectively.

MATERIALS AND METHODS

Visual pigments

Specimens were obtained from aquarium dealers or through the courtesy of other researchers. Animals were maintained at a water temperature of 21-24 °C. Illumination of all specimens for at least a month prior to experiments was on a 12/12 hr light/dark cycle. Animals were fed either dried flake food (Tetra-Min) or "freeze dried plankton" according to their preference.

Dark adapted eyes were removed under infra-red illumination (Kodak Wratten filter 87C) with the aid of an infra-red image converter. Eyes were hemisected and the posterior portion was incubated in a low-calcium physiological saline for about one hour at room temperature. Small (circa $1 mm^2$) pieces of sensory retina were teased flat and mounted between two coverslips in a drop of saline ringed by silicone oil. Preparations on the stage of the computer-assisted, photon-counting microspectrophotometer (PMSP) were observed using infra-red illumination and an IR sensitive television monitoring system. Photoelectron counts taken with the measuring beam passing through photoreceptor outer segments (sample) were compared with those taken through a clear area of the preparation (blank) and optical density was calculated and plotted or printed out as a function of frequency at 5 fresnel* intervals from 400 to 800 fresnels (750-375 nm). Absorption maxima (λ max) and other relevant parameters were determined using an on-line computer system to generate and adjust template curves after the method of Harosi (1976). More detailed information on the instrumentation and techniques used in gathering these data are available elesewhere (MacNichol, 1978; Levine and MacNichol, 1979).

Biochromes and color patterns

1. Species selection: We are concerned in this paper only with colors and patterns that are designed to transmit information about the bearer. To this end we have examined the brightly colored fishes of the coral reef whose conspicuous coloration has been the subject of much attention for some time (Lorenz, 1962). Five families of reef fishes with both shallow and deep water representatives were selected and all of the members of those families found in Hawaii were examined. "Shallow reef" species are considered to be those found from the surface to a maximum depth of 20 meters. "Deep reef" species are defined as those found only at depths

* 1 fresnel (also known as terahertz) = 10^{12} hertz = 3×10^5 nm^{-1}
= 33 1/3 wave numbers

greater than 20 meters. Wide-ranging species were treated separately.

As freshwater subjects we have selected two groups of cichlid fishes - Mbuna from the relatively clear waters of the African rift lakes (Malawi, Tongonika), and several Cichlasoma species from the organically stained rivers and lakes of the New World tropics. Since cichlids often exhibit substantial changes in their body coloration over time, we have examined the breeding colors of these species in the belief that these are specifically designed to convey information regarding the bearer's species, sex and breeding condition. The availability of good quality color photographs (see below) was a significant limiting factor in the selection of the freshwater species.

2. Body colors: A sample of 103 color swatches selected from "Color-Aid" papers was matched to a set of Munsell chips under diffuse daylight illumination. This finite slection of color samples was then used to quantify the hues present in photographs that were known to have proper color rendition by matching under the same types of illumination. Further details about these color standards and the grouping of Munsell hues assigned to each of the simple color names shown in subsequent figures are given in appendix 1. Details regarding the precise hues assigned to each species are available from the senior author on request. For the purpose of simplicity, and because we have already stated the biological reason for doing so, we have treated all degress of lightness and saturation of each given Munsell hue as a single "color".

The organization of colors into patterns was quantified using a map-graphics digitizer interfaced with a PDP-11. A series of transects was taken across the body of each fish and the beginning and end of each occurrence of each color was translated by the digitizer into an internally consistant X - Y coordinate system. Simple software then calculated the total length of the transect line covered by each color while simultaneously keeping track of the number of separate occurrences of each color for each species. A detailed analysis of color and pattern in these fishes will be presented elsewhere; here we are concerned only with the "importance" values calculated for each color using standard ecological algorithms. (Further information on the software package is available from the senior author on request.)

3. Carotenoid spectra: A number of procedural difficulties makes it impractical to obtain meaningful reflectance curves from the intact integuments of many fishes, but it is a simple matter to obtain absorption/transmission spectra from naturally occurring carotenoids similar in structure and color to those found in fishes. Transmission spectra of naturally occurring carotenoids from the retinal cone oil droplets of various organisms were generated by inversion of the algorithm used to calculate absorptance from the raw photon counts stored in the data records. (Reference scans in these cases were made through clear areas of the preparation adjacent to the specimens.) Data were then graphed as log

relative transmission on a frequency axis (Figure 3).

RESULTS AND DISCUSSION

I. Visual pigments, habitat and behavior: The extent to which the number and spectral locations of a species' visual pigments are dominated by water color depends on both behavioral and environmental considerations. Non-migratory, mesopelagic fishes, for example, are truly at the mercy of the photic environment imposed upon them by the presence of hundreds of meters of water above them. Their retinas, designed to function in the almost total darkness of the deep oceans, usually contain only rod photoreceptors (Ali and Anctil, 1976). The solitary visual pigments contained in these rods were found long ago (Wald and Brown, 1957) to exhibit absorption maxima matched almost perfectly to the spectral distribution of the downwelling illumination. At the low light levels these animals continually encounter, their visual systems must maximize sensitivity at one particular spectral location determined entirely by the transmission characteristics of the water above them. The predominant wavelength in their environment is close to 485 nm, and that is precisely where their rod pigments absorb maximally. Since the dominant wavelength of most bacterial and invertebrate luminescence is also in this general region of the spectrum, such visual pigments are well-equipped to detect the majority of biological light sources they are apt to encounter as well.

Even among these deep-dwelling fishes, however, there are a few species known to have visual pigments that deviate substantially from the expected pattern. Fish from the genera *Pachystomias* and *Aristomias* have visual pigments with max values greater than 550 nm (Denton et al., 1971; O'Day and Fernandez, 1974), apparently grossly mismatched to local photic conditions. Both of these species possess unusual photophores (light-producing organs) that emit longwave light. The majority of deep-sea species are insensitive to longwave light. *Aristomias* and *Pachystomias*, then, can communicate visually on "secret" channels by producing signals of longwave light which only they have the proper visual pigments to detect. Thus, even in the otherwise visually monotonous deep sea, behavioral factors involved in intraspecies communication can have profound effects on the nature of the visual system.

The retinas of marine fishes from more moderate depths confront us with a more complicated situation. If we examine the group of temperate marine species whose visual pigments are shown in Figure 4, we see two distinct visual pigment/cell-type patterns. Some species have twin cones - paired cones in which both members contain the same pigment - with max values tightly clustered around 530 nm. Other fishes have double cones whose members contain different pigments with λ max values neatly straddling those of the sympatric species' twin cone pigments by about 20 nm on either side. The underwater spacelight in the coastal regions these fish inhabit is characterized by a dominant wavelength of 525-540 nm. The paired cones of the first group maximize sensitivity by matching this wavelength fairly closely, while the paired cones of the second group straddle the region of highest energy while providing better hue discrimination ability in middle and longwave areas. (Both sets of species possess blue-sensitive single cones with which signals from longwave receptors can interact.)

These pigment patterns are distributed across species that belong to widely separated taxonomic groups, can be found from extremely shallow water to reasonable depths and are either benthic and basically sedentary or active and pelagic (Bigelow and Schroeder, 1953). In the absence of any detailed behavioral observations on when and where these fishes are feeding and on the precise nature of other visual tasks they must perform, we must defer further discussion of these species until such information becomes available.

Among tropical freshwater species a still more complex assortment of visual pigments and receptor cell patterns exists. We have previously (Levine and MacNichol, 1979) divided the photopic pigment systems in freshwater species into 4 loosely-knit groups based on a combination of

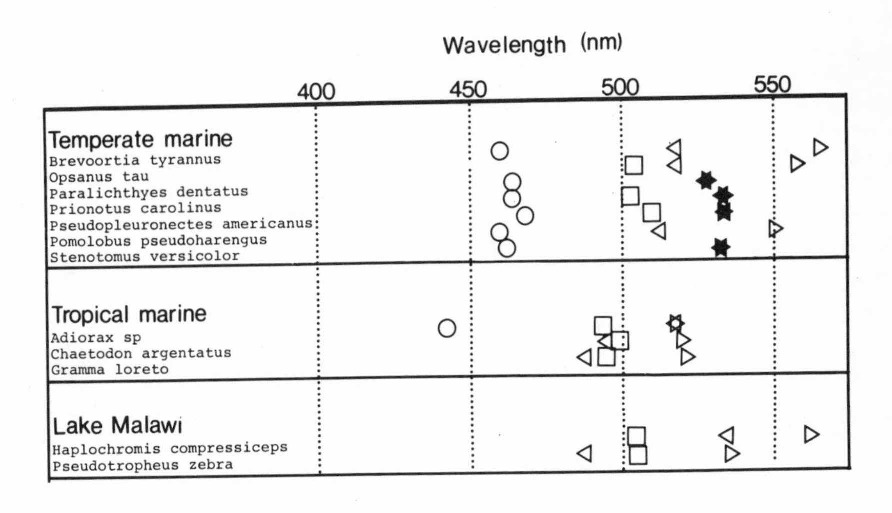

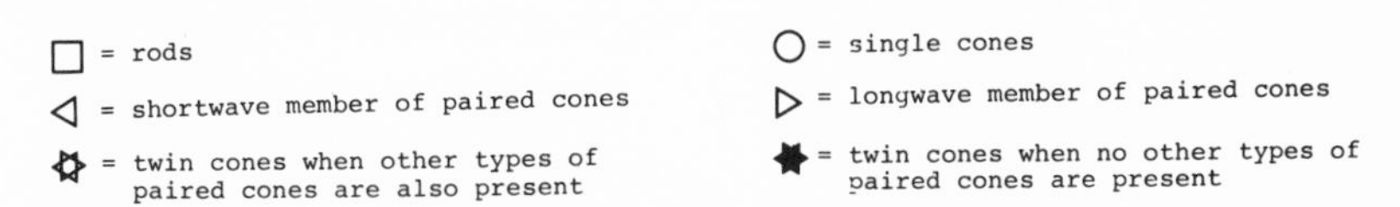

Figure 4. Absorption maxima of visual pigments from temperate marine, tropical marine and clear lacustrine species. Fishes have been grouped according to habitat in the broadest sense.

λ max values and photoreceptor patterns (Figure 5). Freshwater group I consists of species characterized by radically "shortwave-shifted" sets of visual pigments. These animals are strongly diurnal and seldom venture far from the surface of the water. Freshwater group II contains species with "goldfish-like" visual pigments spread across the spectrum. These fishes can best be called generalized midwater species. Freshwater group III, whose retinas are dominated by longwave-sensitive pigments contained in twin cones, are typically - though not exclusively - crepuscular and predaceous. Freshwater group IV is characterized by a total lack of shortwave - and often middlewave - sensitive cone pigments, although green-sensitive rods are found in all of them. These are primarily benthic species.

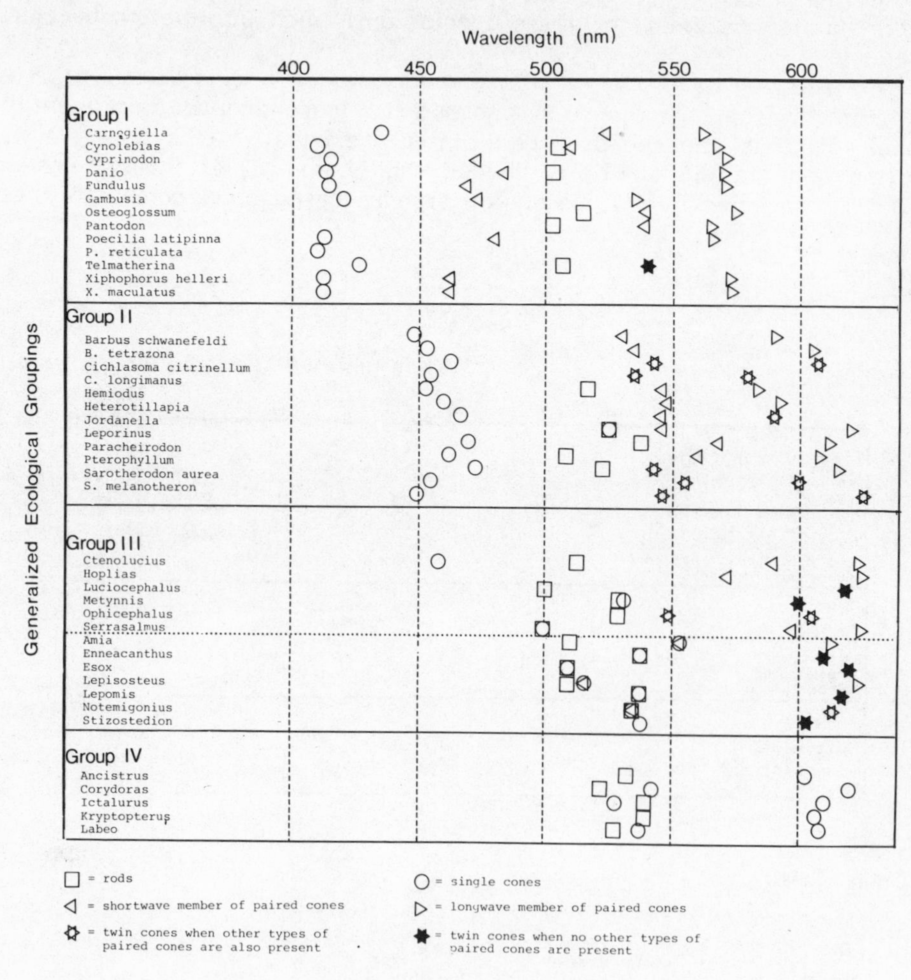

Figure 5. Absorption maxima of visual pigments from tropical freshwater fishes. Species have been grouped according to λ max values and assigned tentative ecological groupings in the text.

How can we explain these major differences in pigment/cell type patterns among species that are often sympatric? Despite the fact that many of these species can be found in the same streams, differences in their behavioral microhabitat selection and preferred feeding times can place them in visual worlds that are anything but similar. Many freshwater habitats are so strongly colored that differences in optical path length of less than a meter can cause more extensive changes in the spectral characteristics of light than are caused dy differences of 10 or 20 meters in clear tropical seas. Fishes active at the surface in this type of water would be exposed to intense, broad spectrum illumination at the same time that benthic species only a few meters beneath them would experience a spectrum of very low overall intensity almost totally lacking in shortwave light (Figure 1b). These dramatic differences in local photic conditions would be expected to result in changes among the visual systems of the fishes that inhabit the various microhabitats.

Differences in activity rhythm further exacerbate the situation. Mid-water species feeding crepuscularly at low overall light levels would encounter a preponderance of longwave illumination similar to that faced by benthic species during the day. Blue-sensitive cones would occupy valuable space in the receptor mosaic of these species and only rarely contribute useful neural signals, as most of the time too few shortwave photons would be available to activate such cones.

Only one photon is necessary to activate a rod (Hecht et al., 1942) however, and it should be noted that those species that lack both shortwave- and middlewave-sensitive cones invariably retain green-sensitive rods. Recent studies in goldfish have shown that rod-cone interaction can occur in at least one fish species at mesopic light levels (Powers and Easter, 1978 a, b), in a manner that permits rudimentary hue discrimination. If such rod-cone interaction exists in these benthic species, it would provide them with the luxury of hue discrimination under conditions too dim to insure adequate functioning of an all cone system. For unknown reasons, rod pigments with λ max values at wavelengths longer than around 540 nm do not appear to exist, so a middlewave-longwave two rod system - which would be functionally ideal for the situation under discussion - appears evolutionarily unfeasable. (Shortwave-middlewave two rod systems combined with longwave cones, however, do exist in certain amphibians (Witkowsky et al., in prep.)

Moving to the other extreme, the strictly diurnal animals active in the topmost portions of the water column will be exposed to illumination containing a great deal of energy in the shortwave end of the spectrum. Shortwave-sensitive cones are thus potentially useful to these species, offering as they do sensitivity to shortwave light, improved hue discrimination capabilities throughout the middle and lower ends of the spectrum and the high spatial and temporal acuity of cone vision. It is still unclear, however, why some of these species possess the most shortwave-sensitive (λ max 410-415 nm) visual pigments yet definitively identified in vertebrates. It seems likely that the precise nature of some crucial visual discrimination that these species need to make is evolutionarily responsible for this phenomenon. It is common knowledge that chlorophylls absorb strongly at both the longwave and shortwave ends of the spectrum. Many Poeciliids (Poecilia, Xiphophorus and their close relative) are largely herbivorous. Perhaps this pigment increases their ability to discriminate among algae and other plants that contain different chlorophylls and/or accessory pigments. Other members of the Poeciliidae (Belonesox, Gambusia) and the closely related Cyprinodontidae that have the same kinds of visual pigments, however, feed extensively on small invertebrates and even fishes. Perhaps - as in the cases of Aristomias and Pachystomias - the need to simultaneously advertise one's presence to conspecifics while hiding from potential predators has had an influence on these visual systems. Endler (1978, 1978) has shown the importance of conspicuous coloration in mating success among guppies. Haas (1976) made a similar demonstration for the Cyprinodontid Nothobranchius, and both researchers also demonstrated that the more conspicuously colored males were also significantly more susceptible to predation. Endler further indicated, through an unusually elegant series of experiments, that the spectral characteristics of the color patterns of male guppies are under direct pressure to minimize their conspicuousness to the visual systems of particular predators. If, in recent history, the major predators of the diurnal Poeciliids were crepuscular species endowed with visual systems relatively insensitive to shortwave light and simultaneously ill-equipped to discriminate among shortwave colors (Group III in Figure 4), the Poeciliids' shortwave-shifted visual systems might have enabled them to be conspicuous to each other while remaining cryptic in the eyes of their predators.

The lack of far red (circa 620 nm) sensitivity in these species is harder to explain. There is plenty of longwave light present in their microhabitats and several species in the family use significant amounts of red in their body coloration patterns. Once again we must fall back, regroup, and gather more data on the interactions of visual pigments in the hue discrimination process and on the precise nature of the visual discriminations that these species need to make before we can go any further.

II. Intraretinal variations in visual pigment/receptor cell distribution: Looking closely at the retinas of many fish species we find that more is happening than simple (or not so simple!) inter-species shifts in λ max values. The precise adjustment of the visual systems of aquatic animals to the underwater photic environment continues intra-retinally in the form of differential visual pigment/receptor cell type distribution. We have already discussed the data of Munz and McFarland (1975) which show that light approaching the eyes of marine fishes from different directions varies significantly in spectral composition. We have further stated that the same

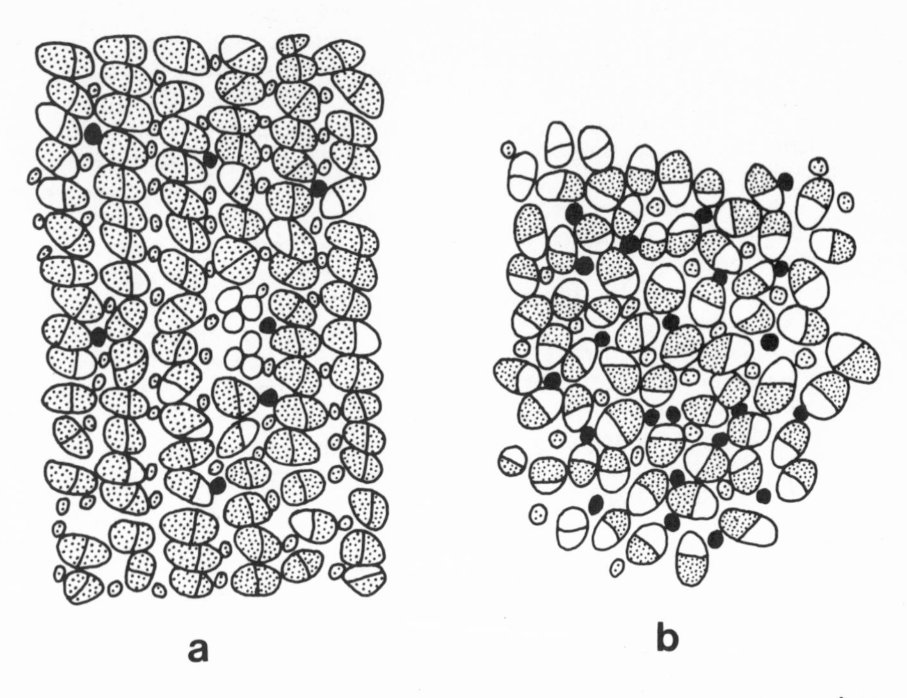

Figure 6 a: End-on view of photoreceptor mosaic from the ventral retina of the guppy (Poecilia reticulata). The visual pigments that can be assigned to these cells on morphological grounds are indicated as follows:
dotted cells: λ max. 545nm
clear cells: λ max. 468 nm
blackcells: λ max. 410nm

Figure 6 b: Photoreceptor mosaic from the dorsal retina of P. reticulata. Cellular pigment contents indicated as in (a).

type of phenomenon will occur in freshwater and that the effect in heavily stained freshwater will be even stronger. It is not too surprising, therefore, to discover that the eyes of at least some species of fishes have responded to this situation. We have discussed in detail elsewhere (Levine et al., 1979) the methods for studying this phenomenon; here we will merely describe the phenomenon and attempt to explain its relationship to the photic environment.

Figure 6 shows views of dorsal and ventral areas from the retina of the common guppy Poecilia (=Lebistes) reticulatus. The retinas of this species, studies in great detail by Müller (1952) and re-examined by MacNichol et al.

(1978), contain 4 morphologically distinct cone types: short (additional) single cones, long (central) single cones, identical twin cones and non-identical double cones. The visual pigments contained in each cell type was determined by MSP, and the chromatic organization of the retina were determined by studying the physical mosaic arrangement of the cells. The dorsal retina of this species contains all four cone types arranged into a reasonably regular mosaic pattern that contains all 3 photopic visual pigments. Large areas of the ventral retina, on the other hand, consist almost entirely of long rows of identical twin and long single cones, both of which contain the most longwave-sensitive pigment of the species (λmax 545 nm). We suspect that this upward-looking ventral area - which contains a visual pigment well-matched to the downwelling light near the surface - is designed to maximize contrast in the viewing situation that it most often encounters - objects silhouetted against the downwelling light. The upturned mouths of most Poeciliids and many Cyprinodontids are well engineered to facilitate feeding on floating objects. Although these fishes also feed off various substrates periodically, they make use of the ventral retina regularly in surface feeding behaviors.

The rest of the retina - which contains a thoroughly trichromatic mosaic arrangement - is apparently designed to permit crucial hue discrimination tasks in the bright, broad-spectrum illumination these species encounter in the shallow water in which they dwell. Exactly what most of these tasks are, we do not yet know. In light of our earlier discussion of male reproductive colors it is interesting to note that when the brightly colored male guppies display in an effort to woo females, they do so from positions ahead of and slightly below the object of their desires. The positions they take in the females' visual fields are projected squarely onto the trichromatically organized dorsal portion of the retina which contains numerous shortwave contrast-enhancing ellipsosomes (MacNichol et al., 1978).

Many animal species do not possess morphologically distinct cone types that uniquely contain one of the visual pigments present in the retina. The cichlid fish Cichlasoma longimanus has only 2 clearly different cone types; single cones and equal double cones. The single cones - which are located centrally in the cone mosaic pattern - contain the blue-sensitive pigment of the species. The equal double cones - whose members are physically indistinguishable from each other at the light microscope level - can contain either or both of the two remaining cone pigments in any of three possible combinations; Longwave-Longwave (L-L), LongwaveMiddlewave (L-M) or Middlewave-Middlewave (M-M).* The receptor mosaic pattern of this species does not vary significantly from one retinal area to another. Within

* Since the precise spectral positions of the paired cone pigments vary from species to species, color names such as "red" or "green" are not adequate to describe them as a class. We prefer to call them Longwave and Middlewave paired cone pigments (PCP's).

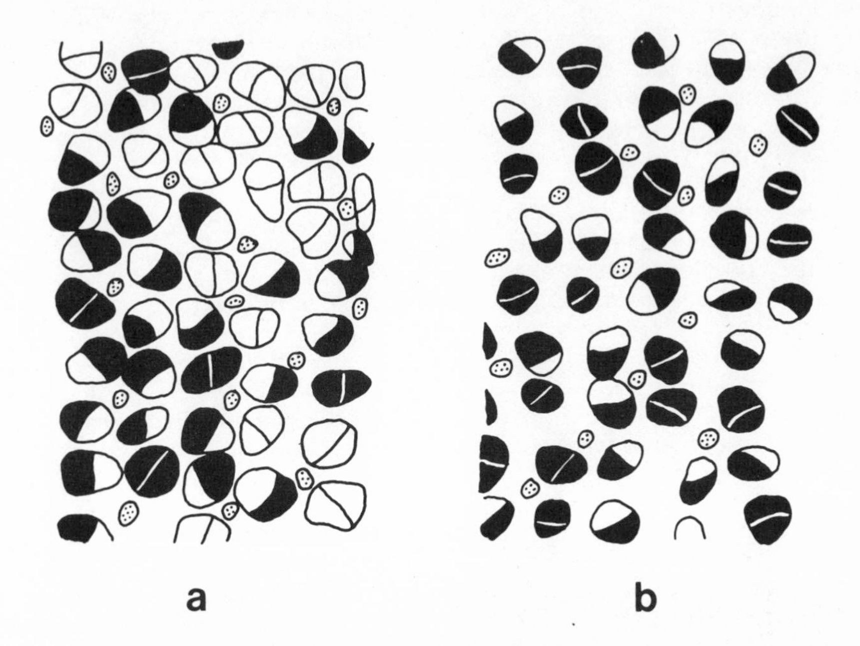

Figure 7 a: End-on view of photoreceptor mosaic from the ventral retina of Cichlasoma longimanus. Single cones (dotted) in this species always contain a visual pigment with a λ max of 455nm. Paired cones are morphological twins whose pigment contents were determined using the NBT vital staining technique. Black cells contain a visual pigment with λmax = 579nm. Clear cells contain a pigment with λmax = 531nm.

Figure 7 b: Photoreceptor mosaic from the dorsal retina of C. longimanus - key to pigments as in (a).

this physical mosaic structure, however, the realtive proportions of L-L, L-M and M-M pairs present (as determined by MSP and the Nitro-blue tetrazolium chloride reduction technique) change significantly as one moves from the dorsal to ventral retina (Levine et al., 1979). Figure 7 shows the chromatic makeup of typical dorsal and ventral areas from C. longimanus. The paired cone population in the ventral retina consists of a mixture of all three types, but the M-M and L-M pairs predominate. In the dorsal retina there is a much higher proportion of L-L pairs and the M-M pairs are almost completely absent.

Lacking detailed field observations on C. longimanus and similar species that share its regional specializations, we can only form hypotheses regarding interactions between the underwater spacelight and these intra-retinal pigment distribution patterns. The λmax values of New World cichlid middlewave PCP's fall between 530 and 540 nm, a range in which they are reasonably well-matched to the downwelling light in shallow areas of the rivers and lakes where they are found. There are proportionally more cells containing this pigment in the upward-looking ventral retina which could function similarly to the ventral area of the guppy eye.

The longwave PCP's of these species have maximum sensitivities that range from 580 to 625 nm. Paired with the middlewave PCP's in L-M double cones they provide a basis for hue discrimination in the middle and longwave portions of the spectrum. Concentrated in the dorsal part of the retina in L-L twins and L-M doubles, they would probably raise the local relative longwave sensitivity. Figure 1b predicts that light of longer wavelengths will predominate in the horizontally scattered and upwelling light that falls on this part of the eye. Obviously, a wide range of field observations, electrophysiological studies and psychophysical measurements are necessary to test this hypothesis (see Summary and Conclusions).

III. Distribution of integumentary colors among species from different habitats: The information presented in the introduction enables us to make certain predictions regarding the colors we would expect to find in "conspicuous" fishes in various environments. Black and white should be common everywhere. Blues and yellows should be major components of color patterns in clear water species, while greens and reds should predominate in organically stained environments. The range of wavelengths that can be seen as colors in a given habitat is controlled by the spectral bandwidth of the available illumination. For this reason, deep water species should exhibit a more restricted range of colors than shallow water species from the same environment. Our preliminary data in this area are encouraging but they also indicate the need for further study.

Figure 8a summarizes the relevant color data for the coral reef species while Figure 8b shows the data from the clearwater and blackwater cichlid species. The first obvious point is that the achromatic components - white, silver, grey and black - are heavily utilized by all groups, including the "poster colored" reef species. Shallow and deep-dwelling reef species share most of their colors which fall into two major groups centered on blue and yellow as predicted. Note that the shallow group does exhibit a somewhat broader range of colors than the deep-dwelling group. The comparison is complicated by a functional discrepancy regarding the use of red in certain deep-water forms. In several instances it is apparent that the red pigment present in the color patterns of deep-dwelling fishes is not used as a hue to be contrasted with another hue but rather as black to be contrasted with white. (Recall that red pigments illuminated with blue or green light appear black.) Oxycirrhites typus, for example, lives at such depths that its red-on-white body pattern is seen in nature strictly as black on white, even by divers who have been underwater long enough to adapt to the prevailing blue illumination. The use of red pigments - in place of melanin - to provide neutral, dark areas in body patterns is widespread among tropical reef fishes dwelling below about 20 m, occurring in widely separated families including the Serranidae and the Holocentridae. Whether this phenomenon is simply the result of random genetic drift - since red and black are visually identical in these cases - or whether it has arisen because it may be metabolically less demanding to concentrate carotenoids obtained from food items than it is to synthesize melanin is uncertain at this time.

Figure 8 b shows that the clearwater Mbuna possess biochromes that follow the same general pattern of importance as those of the tropical marine species with achromatics, blues and yellows predominating. The blackwater species - while maintaining their use of achromatic components - switch their signal colors to greens and reds as we have previously predicted.

During the course of this survey it became apparent that certain colors seemed to "run in the family". The more major taxa included in consideration, the more colors were counted in the group as a whole. New colors appeared mostly as new families - rather than just new species - were added to the list. Our saltwater sample included five families of higher teleosts, while the Mbuna are all members of a fairly restricted species flock and the Cichlasoma species are obviously closely related. The presence of a smaller total number of colors in the freshwater group (11) than in the marine group (17) can probably be attributed to the larger number of higher taxa in the latter sample. (Note that all the marine families studied had members in both shallow and deep-dwelling groups so that this phenomenon is not a source of error in that comparison.) We therefore attach no particular significance to the more monolithic appearance of the freshwater color histograms.

Another interesting observation was that many species exhibited different "colors" that upon examination turned out to be different degrees of lightness and saturation of the same hue. It should be recalled that major changes in lightness and saturation can be effected by changing the concentration of carotenoid pigments in chromatophores and altering the number and state of melanophores and irridiophores in the integument. Artists have long been aware that darkened, desaturated oranges and reds appear brown, while certain types of yellow appear green when mixed with neutral blacks. Biologists have recently become aware that many animals use this strategy to produce differently "colored" body areas through varying combinations of one or two carotenoids with melanin and guanine platelets (Dyck, 1966). Although the perceptual basis of this effect is not well understood, it means that even a species restricted to a very few carotenoids in its diet has at its disposal a color palette that is considerably more diversified than one might expect.

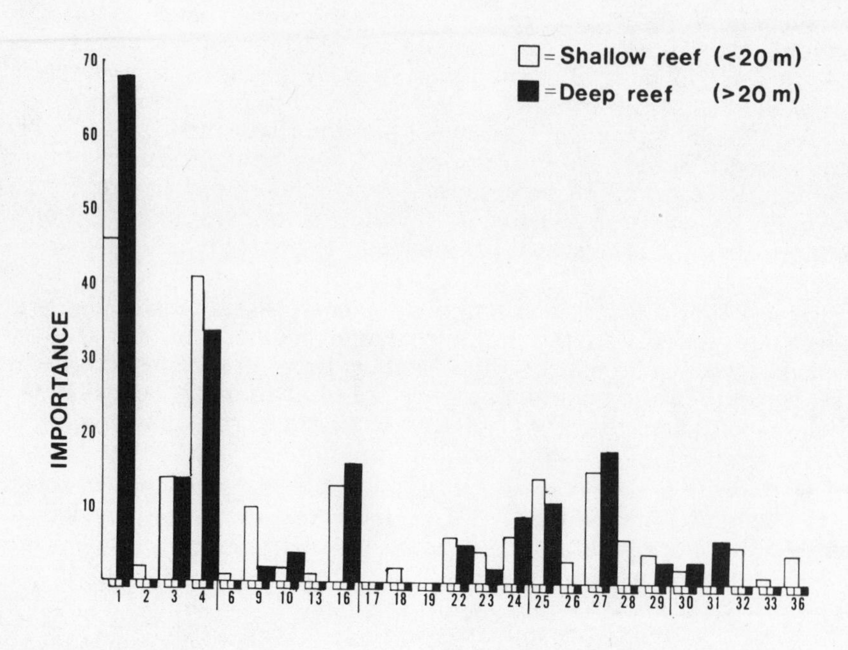

□ = Shallow reef (<20 m)
■ = Deep reef (>20 m)
IMPORTANCE
70
60
50
40
30
20
10
1 2 3 4 6 9 10 13 16 17 18 19 22 23 24 25 26 27 28 29 30 31 32 33 36

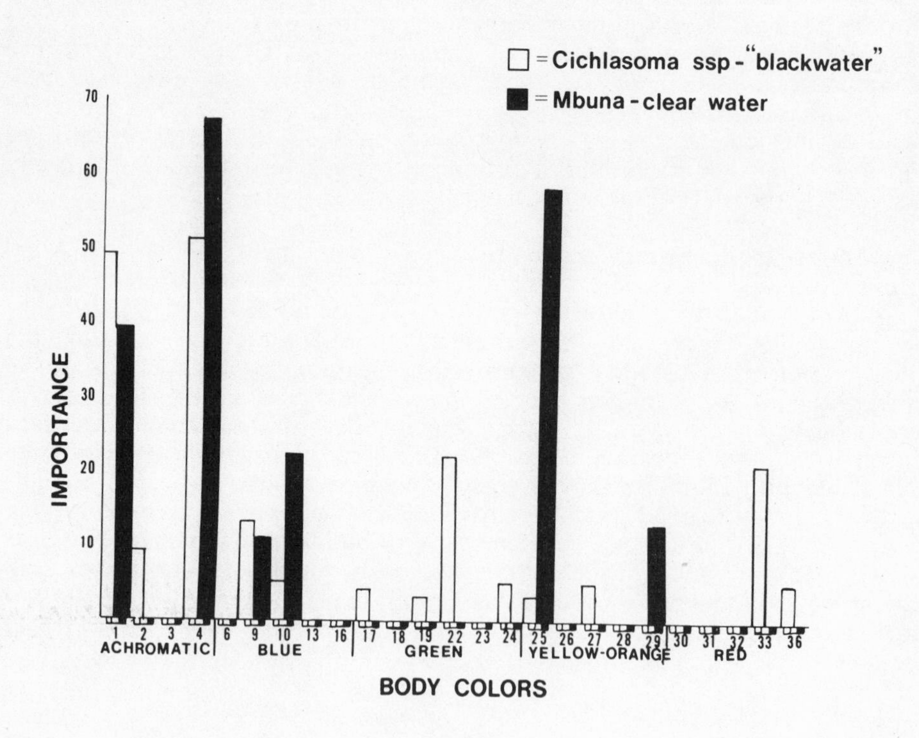

□ = Cichlasoma ssp - "blackwater"
■ = Mbuna - clear water
IMPORTANCE
70
60
50
40
30
20
10
1 2 3 4 6 9 10 13 16 17 18 19 22 23 24 25 26 27 28 29 30 31 32 33 36
ACHROMATIC
BLUE
GREEN
YELLOW-ORANGE
RED
BODY COLORS

SUMMARY

A diagrammatic view of the factors affecting aquatic visual communication systems is given in Figure 9. Despite a "bare bones" approach, the complexity of the interactions among components is as evident as is the lack of data in crucial places.

Possible visual receptor system functions can be arranged in an intensity-related hierarchy beginning with sensitivity and moving through simple contrast to full hue discrimination. Each succeeding function requires progressively higher levels of illumination for adequate functioning. Deep sea fish, with a single visual pigment matched to local photic conditions, maximize absolute sensitivity at certain wavelengths but sacrifice broad band sensitivity and hue discrimination to do so. Fishes that are active under high light levels often have three widely-spaced cone pigments (in addition to a rod pigment) that afford them with excellent hue discrimination capabilities across a broad spectral range. Receptor systems of intermediate design are found in species that must perform under intermediate levels of illumination. In all cases, behavioral factors - most of which are poorly understood - can effect the "fine points" of system design and in certain cases may be major determinants.

Body coloration patterns can allow animals to be cryptic, conspicuous or alternately cryptic and conspicuous. The most desireable strategy among these is directly determined by the toal biotic - as well as physical - environment of the species.

Figure 8a,b. Importance values for body colors in fishes from different habitats. Colors, assigned as described in Methods section are listed on the x-axis by their specific code numbers (see appendix for associated Munsell coordinates) and by commonly applied descriptive color names. "Importance" values are obtained from a combination of color coverage, relative color density and color frequency, each of which is calculated using standard ecological algorithms.

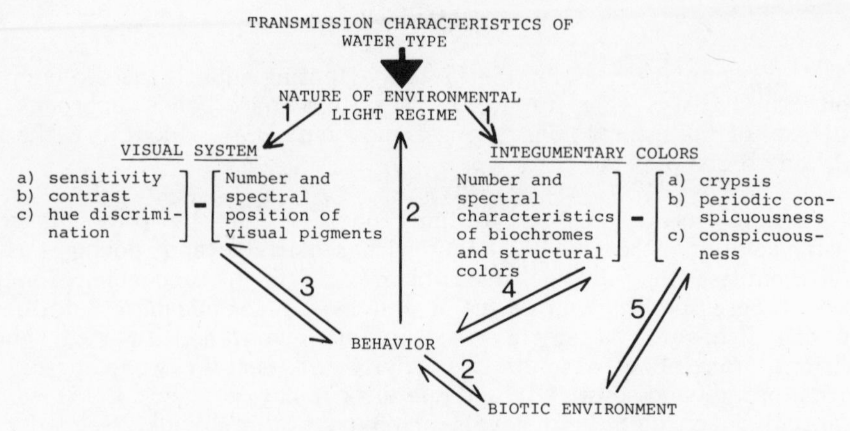

1) Intensity (photon flux) and spectral characteristics (distribution of photons) of available illumination

2) Position within habitat (microhabitat selection)
Activity period timing (crepuscular hunting vs high light feeding)

3) Nature of visual tasks crucial to survival
    identity of food items (plant/animal) and the nature of the background from which they must be distinguished
    need (or lack of it) for intra-specific communication; breeding behaviors, territoriality, juvenile/adult interaction

4) Nature of intra-specific interactions
    breeding systems, territoriality, juvenile/adult interaction
Predator - prey relationships
    species "X" as a predator
    species "X" as prey
        (nature of reaction to predators; "freezing, flight, etc.)

5) Nature of substrate
    visually
    physically (hiding places, etc.)
Nature of predation pressure
    visual systems of predators
    spatial and temporal hunting patterns of predators
Availability of various carotenoids in food items

Figure 9.

REFERENCES

Ali, M.A. and Anctil, M. (1976). Retinas of Fishes: An Atlas. New York, Springer Verlag.

Ali, M.A., Harosi, F.I. and Wagner, H.-J. (1978). Photoreceptors and visual pigments in a cichlid fish, *Nannacara annomala*. Sensory Processes 2: 130-145.

Bagnara, J.T. and Hadley, M.E. (1973). Chromatophores and color change. Englewood Cliffs, New Jersey, Prentice Hall.

Becher, H. (1926). Uber chromatische Organe und andere Chromatophorenzusammenlagerungen in der Haut einheimische Fische (*Esox lucius*, *Coregonus fera*, *Leuciscus rutilus*). Zool. Jarb. 45: 581-618.

Bigelow, H.B. and Schroeder, W.C. (1953). Fishes of the Gulf of Maine. Washington, D.C., U.S. Government Printing Office.

Blaxter, J. (1970). Fishes, Part 2, Light. In: Marine Ecology Vol. I, Environmental Factors. Ed. O. Kinne. New York, Wiley Interscience. p. 213-320.

Brezonick, P.L. (1978). Effect of organic color and turbidity on Secchi disk transparency. J. Fish. Res. Board Can. 35: 1010-1416.

Burkhardt, D.A., Levine, J.S., MacNichol, E.F. Jr. and Hassin, G. (In preparation). Electrical responses and photopigments of twin cones in the retina of the walleye.

Cornsweet, T.N. (1970). Visual Perception, New York, Academic Press.

Dartnall, H.J.A. (1975). Assessing the fitness of visual pigments for their photic environment. In: Vision in Fishes. Ed. M.A. Ali. New York, Plenum Press. p. 543-564.

Dartnall, H.J.A. and Lythgoe, J.N. (1965). The spectral clustering of visual pigments. Vision Res. 5: 81-100.

Denton, E.J., Gilpin-Brown, J.B. and Wright, P.G. (1971). On the "filters" in the photophores of a mesopelagic fish and on a fish emitting red light and especially sensitive to red light. J. Physiol. (Lond.) 208: 72-73.

Dyck,J. (1966). Determination of plumage colours, feather pigments and structures by means of reflection spectrophotometry. Dansk. Orn. Foren. Tidsskr. 60: 49-76.

Endler, J.A. (1978). A predator's view of animal colour patterns. In: Evolutionary Biology, Vol. II. Eds. M.K. Hecht, W.C. Steere and B. Wallace. New York, Plenum Press. p. 319-364.

Endler, J.A. (1979). Natural selection of color patterns in Poecilia reticulata. Evolution. (In press).

Fox, D.L. (1953). Animal biochromes and structural colours. Cambridge, Cambridge University Press.

Fries,E.F.B. (1958). Irridescent white reflecting chromatophores (Antauga-phores, Irridoleucophores) in certain teleost fishes, particularly in Bathygobius. J. Morph. 103: 203-253.

Fujii, R. (1969). Chromatophores and pigments. In: Fish Physiology, Volume III. Eds. W.S. Hoar and D.J. Randall. New York, Academic Press. p. 307-353.

Goodrich, H.B. and Hedenburg, M. (1941). The cellular basis of colors in some Bermuda parrot fish with special reference to blue pigment. J. Morph. 68: 493-505.

Haas, R. (1976). Sexual selection in Nothobranchius guntheri (Pisces: Cyprinodontidae). Evolution 20: 614-622.

Hanaoka, T. and Fujimoto, K. (1957). Absorption spectrum of a single cone in carp retina. Jap. J. Physiol. 7: 276-285.

Harosi, F.I. (1976). Spectral relations of cone pigment in goldfish. J. Gen. Physiol. 68: 65-80.

Hawkes, J.W. (1974). The structure of fish skin. II: The chromatophore unit. Cell Tiss. Res. 149: 159-172.

Hecht, S., Schlaer, C. and Pirenne, M.H. (1942). Energy, quanta and vision. J. Gen. Physiol. 25: 819-840.

James, H.R. and Birge, E.A. (1938). A laboratory study of the absorption of light by lake waters. Trans. Wisc. Acad. Sci. 31: 1-54.

Jerlov, N.J. (1965). The evolution of the instrumental technique in underwater optics. Progr. Oceanog. 3: 149-154.

Jerlov, N.J. (1968). Optical Oceanography. Amsterdam, Elsevier.

Katayama, T. and Tsuchiga, H. (1971). Mechanism of the interconversion of plant carotenoids into fish carotenoids. In: Proc. 7th Seaweed Symposium. Ed. K. Nisizawa. Tokyo, University of Tokyo Press.

Kinney, J.S., Lauria, S.M. and Weitzman, D.O. (1967). Visibility of colours underwater. J. Opt. Soc. Am. 57: 802-809.

Land, E.A. (1977). The retinex theory of color vision. Scient. Am. 237: 108-128.

Levine, J.S. and MacNichol, E.F. Jr. (1979). Visual pigments in teleost fishes: effects of habitat, microhabitat and behavior on visual system evolution. Sensory Processes. (In press).

Levine, J.S., MacNichol, E.F. Jr., Kraft, T. and Collins, B.A. (1979). Intraretinal distribution of cone pigments in certain teleost fishes. Science 204: 523-526.

Liebman, P.A. (1972). Microspectrophotometry of photoreceptors. In: Handbook of Sensory Physiology Vol VII/1. Photochemistry of Vision. Ed. H.J.A. Dartnall. New York, Springer Verlag. p. 482-528.

Liebman, P.A. and Entine, G. (1964). Sensitive low-light-level microspectrophotometric detection of photosensitive pigments of retinal cones. J. Opt. Soc. Am. 54: 1451-1459.

Loew, E.R. and Lythgoe, J.N. (1978). The ecology of cone pigments in teleost fishes. Vision Res. 18: 715-722.

Lorenz, K. (1962). The function of colour in coral reef fishes. Proc. Roy. Inst. Gr. Br. 139: 282-298.

Lythgoe, J.N. (1966).. Visual pigments and underwater vision. In: Light as an Ecological Factor. Eds. R. Bainbridge, G.C. Evans and O. Rackham. Oxford, Blackwell.

Lythgoe, J.N. (1968). Visual pigments and visual range underwater. Vision Res. 8: 997-1012.

Lythgoe, J.N. (1975). Problems of seeing colours underwater. In: Vision in Fishes. Ed. M.A. Ali. New York, Plenum Press. p. 619-634.

Lythgoe, J.N. and Northmore, D.P.N. (1973). Colours underwater. In: Colour '73. London, Adam Hilger. p. 77-98.

MacNichol, E.F. Jr. (1978). A photon-counting microspectrophotometer for the study of single vertebrate photoreceptor cells. In: Frontiers in Visual Science. Eds. S.J. Cool and E.L. Smith. New York, Springer Verlag. p. 194-208.

MacNichol, E.F. Jr., Feinberg, R. and Harosi, F.I. (1973). Colour discrimination processes in the retina. In: Colour '73. London, Adam Hilger. p. 191-251.

MacNichol, E.F. Jr., Kunz, Y.W., Levine, J.S., Harosi, F.I. and Collins, B.A. (1978). Ellipsosomes: Organelles containing a cytochromelike pigment in the retinal cones of certain fishes. Science 200: 549-552.

Marks, W.B. (1965). Visual pigments of single goldfish cones. J. Physiol. 178: 14-32.

Marks, W. and MacNichol, E.F. Jr. (1962). Bleaching spectra of single goldfish cones. Fed. Proc. 22: 519 (Abstract 2143).

Matsumoto, J. (1965). Studies on the fine structure and cytochemical properties of erythrophores in swordtail, Xiphophorus helleri, with special reference to their pigment granules (Pterinosomes). J. Cell Biol. 27: 493-504.

McCann, J.S., McKee, S.P. and Taylor, T.H. (1976). Quantitative studies in retinex theory. Vision Res. 16: 445-458.

McFarland, W.N. and Munz, F.W. (1975a). Part II: The photic environment of clear tropical seas during the day. Vision Res. 15: 1063-1070.

McFarland, W.N. and Munz, F.W. (1975b). Part III: The evolution of photopic visual pigments in fishes. Vision Res. 15: 1071-1080.

Muller, H. (1952). Bau und Wachstum der Nezhaut des Guppy (Lebistes reticulatus). Zool. Jarb. Abt. Fuer Allg. Zool. Physiol. der Tierre. 63: 275-324.

Muntz, W.R.A. (1975). Visual pigments and their environment. In: Vision in Fishes. Ed. M.A. Ali. New York, Plenum Pres.. p. 565578.

Muntz, F.W. and McFarland, W.N. (1975). Part I: Presumptive cone pigments extracted from tropical marine fishes. Vision Res. 15: 1045-1062.

Muntz, F.W. and McFarland, W.N. (1977). Evolutionary adaptations of fishes to the photic environment. In: Handbook of Sensory Physiology, Vol. VIII/5, The visual system in vertebrates. Ed. F. Crescitelli. New York, Springer Verlag. p. 193-274.

Nielson, H.I. (1978). Ultrastructural changes in the dermal chromatophore unit of Hyla arborea during color change. Cell Tiss. Res. 194: 405-418.

Nielson, H.I. and Dyck, J. (1978). Adaptation of the tree frog, Hyla cinerea, to colored backgrounds, and the role of the three chromatophore types. J. Exp. Zool. 205: 79-94.

O'Day, W.T. and Fernandez, H.R. (1974). Aristomias scintillans (Malacosteidae): A deep sea fish with visual pigments adapted to its own bioluminescence. Vison Res. 14: 545-550.

Powers, M.K. and Easter, S.S. (1978a). Wavelength discrimination by the goldfish near absolute threshold. Vision Res. 18: 1137-1147.

Powers, M.K. and Easter, S.S. (1978b). Absolute visual sensitivity of the goldfish. Vision Res. 18: 1149-1154.

Rodieck, R.W. (1973). The vertebrate retina. San Francisco, W.H. Freeman.

Svaetichin, G., Negishi, K. and Fatechand, R. (1965). Cellular mechanisms of a Young-Hering visual system. In: Ciba Foundation Symposium on Physiology and Experimental Psychology of Colour Vision. Ed. E.W. Wolstenhome and J. Knight. London, J.A. Churchill. p. 178203.

Tomita, T., Kaneko, A., Murakami, M. and Pauther, E.L. (1967). Spectral response curves of single cones in the carp. Vision Res. 7: 519531.

Tyler, J.E. (1959). Natural water as a monochromator. Limnol. Oceanogr. 4: 102-105.

Tyler, J.E. and Smith, R.C. (1970). Measurements of spectral irradiance underwater. New York, Gordon and Breach.

Witkovsky, P., Levine, J.S., Engbretson, G., Hassin, G. and MacNichol, E.F. Jr. (In prep.). A microspectrophotometric study of normal and artificially induced visual pigments in the retina of Xenopus laevis.

Wald, G. (1937). Photo-labile pigments of the chicken retina. Nature (Lond.) 140: 545.

Wald, G. and Brown, P.S. (1957). Visual pigments and depths of habitat of marine fishes. Nature (Lond.) 180: 969-971.

Zeigler-Gunder, I.Z. (1956). Untersuchungen uber die purin- und pterin-pigmente in der haut und in der augen der Weissfische. Zeit. Vergl. Physiol. 39: 163-189.

APPENDIX

The Munsell Book of Color is a standardized book of several hundred chips of carefully controlled color. In the Munsell system, each hue is given a code name (ex. 2.5 pb for 2.5 purple-blue), and all shades of lightness and saturation of that hue are placed in a two dimensional array on the same page of the book. By referencing back to this standard, we can allow for much more precise quantification of body colors than would be possible if we simply called them "red" or "green". Color Aid Papers (Geller Artist Materials, Inc. 116 E. 27th St., New York) are a set of 202 colored papers readily available in artist supply stores. The 40-fold difference in price from the Munsell standard - which is best left safely ensconced in a library - makes them attractive for use in day-to-day work and field studies.

LIST OF MUNSELL HUES AND ASSIGNED CODE NUMBERS

Hue	No.	
white	01	Achromatics
silver	02	
grey	03	
black	04	
2.5p	05	Purple
5p	06	
7.5p	07	
10p	08	
2.5pb	09	Purple-Blue
5pb	10	
7.5	11	
10pb	12	
2.5b	13	Blue
5b	14	
7.5b	15	
10b	16	
2.5bg	17	Blue-Green
2.5g	18	Green
5g	19	

7.5g	20	
5gy	21	Green-Yellow
7.5gy	22	
10gy	23	
2.5y	24	Yellow
7.5y	25	
2.5yr	26	Yellow-Red
5yr	27	
7.5yr	28	
10yr	29	
2.5r	30	Red
5r	31	
7.5r	32	
10r	33	
2.5rp	34	Red-Purple
5rp	35	
7.5rp	36	

PINEAL PHOTOSENSITIVITY IN FISHES

Tamotsu Tamura
Fisheries Laboratory, Faculty of Agriculture
Nagoya University
Furo-cho, Chikusa-ku
Nagoya, Japan

Isao Hanyu
Laboratory of Fish Physiology, Faculty of Agriculture
University of Tokyo
Bunkyo-ku
Tokyo, Japan

INTRODUCTION

The pineal organ or epiphysis cerebri develops embryonically as an evagination of the diencephalon in the same way as the retinas of the eyes. The pineal organ of fishes is located on the top of the brain, between the telencephalon and the optic tectum. In some species of fishes the head skin above the pineal is lacking in melanophores and translucent. The translucent region of the head skin was termed the "pineal window" by Rivas (1953).

Behavioral and morphological studies have been done since 1911 concerning the photosensitivity of the pineal organ in fishes, when von Frisch showed that the photoreceptors in the pineal region were involved in chromatophore responses of Phoxinus laevis. More direct evidence of such pineal photosensitivity came from electrophysiological studies (Dodt, 1963; Morita, 1966). The adaptive radiation of the vertebrate pineal system, including morphological, physiological, biochemical and functional aspects, has been reviewed by Hamaski and Eder (1977).

RECEPTOR POTENTIAL

Hanyu et al. (1969) have recorded a light-induced hyperpolarizing slow potential from the exposed pineal vesicle of the rainbow trout (Salmo irideus), the ayu (Plecoglossus altivelis) and the gigi (Pelteobagrus undiceps) by means of the glass capillary microelectrode method. At first, this potential was considered equivalent to the S-potential recorded intracellularly from the horizontal cell of the fish retina, but more recent evidence suggests that these potentials originate directly from the photoreceptor cells which are usually found lining lumina within the pineal epithelium (Tabata et al., 1975). The following description is mainly based upon the results obtained in our laboratory (Hanyu et al., 1969; Tabata et al., 1971; Tabata et al., 1975).

In order to locate the origin of the slow potential, Tabata et al. (1975) examined the effect of the size of the illuminated area on the potential (area effect) and marked the cell inserted by the electrode with a dye, procion yellow (marking with procion yellow).

AREA EFFECT

To stimulate only the very localized area around the tip of the electrode, a recording glass microelectrode filled with 3 M KCL solution and externally coated with black lacquer except at the tip was used (Fig. 1). The effective area of illumination throught the electrode was about ten microns in diameter. This was too small to allow direct measurements of the intensity of the illumination. The area effect, therefore, was examined and discussed only in comparison with that of the retinal S-potential which is known to be affected very much by the size of the illuminated area.

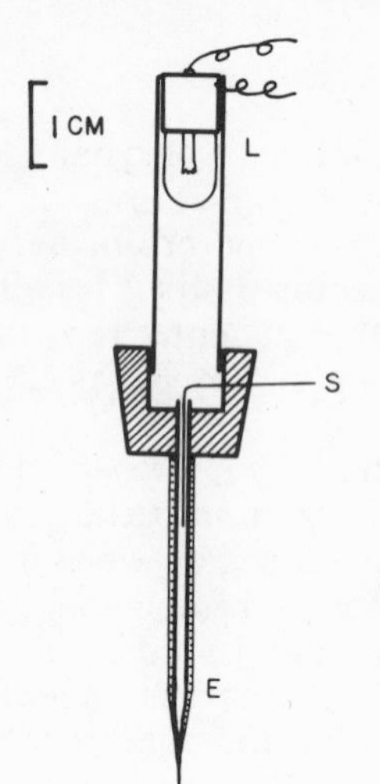

Fig. 1. Device for microillumination through the recording electrode. E, glass capillary electrode filled with 3 M KCl solution and externally coated with black lacquer except for its tip; L, miniature d.c. lamp; S, silver wire (0.1 mm dia) to the pre-amplifier. (Tabata et al., 1975).

Fig. 2 shows, for comparison, both S-potential and pineal responses obtained by both diffuse and local illumination. The response of the pineal organ to the diffuse illumination is a little smaller than the response to the local stimulus because of the low intensity of the former (Fig. 2 bottom). On the other hand, when the S-potential is recorded from the retina of the goldfish through the same electrode as that used for recording the pineal potentials, the response to the local illumination is much smaller than that due to the diffuse one (Fig. 2 upper). From these results, it can reasonably be concluded that the area effect is considerably smaller in pineal responses than in the S-potentials.

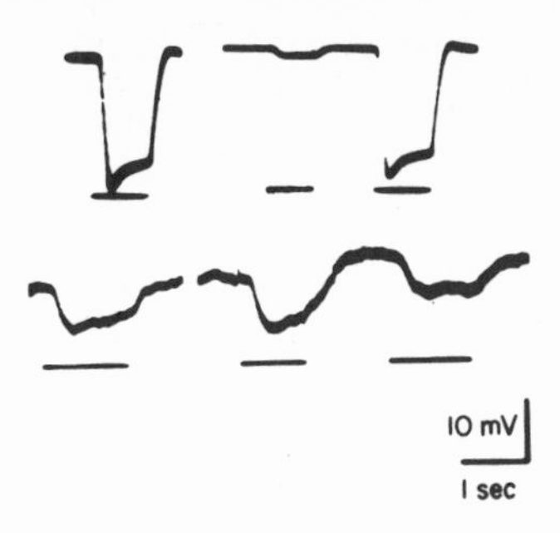

Fig. 2. Responses to diffuse and local illuminations, showing that the pineal response of ayu, in contrast to the S-potential of goldfish, has practically no area effect. Black bar under each response indicates the duration of stimulus. Upper, S-potential from the retina of goldfish; Lower, pineal slow potential of ayu (light adapted); Left, responses to diffuse illumination; Middle, responses to local illumination; Right, responses to diffuse illumination (repeat of left). (Tabata et al., 1975).

The absence of a distinct area effect in pineal responses may suggest that these potentials originate from first order neurons, the receptor cells. More conclusive evidence, however, may be obtained by identification of the impaled cell by means of dye injection experiments.

IDENTIFICATION OF THE RECORDING SITE

To identify the cell in which the tip of the electrode was located, we used glass microelectrodes filled with 6% water solution of procion yellow. After each recording of the stable response with the electrode, the dye was electrophoretically injected. More than five injections were made in each pineal preparation. The preparation was fixed in glutaraldehyde and then embedded in Epon, sectioned at 3-10 μm and was examined under the fluoresent microscope by the method of Kaneko (1970).

The cells marked by this procedure were usually found on the surface of infoldings of the lumen in the pineal vesicle. Each of the marks is composed of two parts; an oval body and a thin process. The oval body faces the lumen and the thin process starts from the basal part of the oval body and ends in the deeper layer.

In contrast to the cell arrangement within the retina, the arrangement of the pineal cells is more irregular, therefore it is not always easy to identify the cells under the fluorescent microscope. However, judging from the location of the marked cells facing the lumen, as well as from the general features of the preparation, it is concluded that these marks correspond to single photoreceptor cells. It is conceivable that the oval body is composed of the soma and the inner segment, with some part of the outer segment, whereas the long and thin process is the basal process of the photoreceptor cell. It has also been observed that the terminal of the basal process is slightly expanded. It is likely that synaptic contacts with the dendrites of the ganglion cell take place in this region. In a few cases, ramification into two branches has been detected in the basal process.

This experiment, together with the experiment of area effect, leads us to the conclusion that the slow potential picked up with the capillary microelectrode is the receptor potential of a single photoreceptor cell.

SOME PROPERTIES OF THE RECEPTOR POTENTIAL (RP)

1) Membrane potential of the receptor cell.

In order to record the RP, the electrode was advanced slowly into the pineal vesicle in darkness. The RP was always picked up following a sudden increase of the electrode negativity which was assumed to be approximately the level of the resting membrane potential of the receptor cell. The membrane potential ranged from 7 to 26 mV in the ayu and from 10 to 32 mV in the rainbow trout.

2) Effects of stimulus intensity.

The RP is a graded potential. The amplitude of the RP changes with the intensity of the light stimulus, but reaches its maximum at about 30 lux. Flashes of brighter light induced responses with shorter latency, longer duration and slower decay to the resting level.

3) Resistance change during stimulation.

Resistance changes in the cell membrane induced by the light stimulation were measured by a bridge circuit to pass constant current pulses across the cell membrane. The bridge was balanced at the darkness after penetration of the cell membrane so that any change in the membrane resistance had to result in an unbalance.

Fig. 3 shows the two records obtained from the same cell, but in different conditions of balancing of pulses. In Fig. 3 (a), pulses are balanced by adjusting the applied current in darkness, but are unbalanced and displaced downward by light stimulation. After the stimulation, pulses are gradually balanced again in the recovery phase (positive going phase). In Fig. 3 (b), pulses, unbalanced and displaced upward in darkness, are balanced by light stimulation, showing the gradual imbalance in the recovery phase.

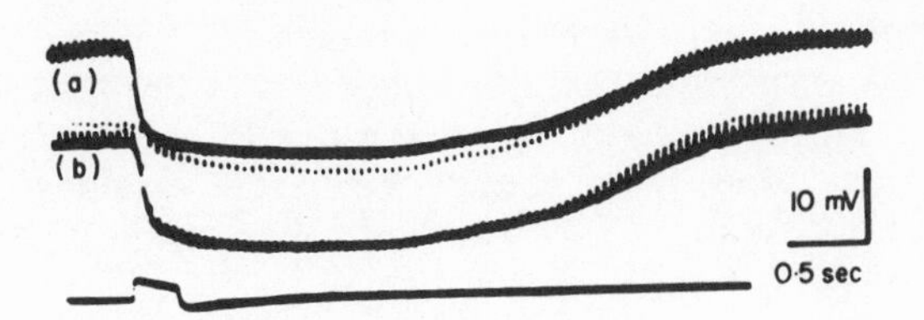

Fig. 3. Records of the resistance changes of the slow potential from the pineal organ of rainbow trout. Pulses are actually displaced downward by light stimulation in both records, see text in detail. Upward deflection of the lowest trace shows the light stimulation (20 lux. white light). (Tabata et al., 1975).

Both records show that the light stimulation displaces the pulses downward. This indicates that the membrane resistance decreases (conductance increases) during illumination. The greatest resistance change is observed at the peak of the response and this gradually falls off after the cessation of light. The stronger the light stimulus, the greater the resistance change. The maximum decrease in resistance due to the stimulation was about 5 M Ohm.

Retinal cone cells respond to the light they absorb with hyperpolarization (Tomita et al., 1967) associated with an increase in the membrane resistance (Toyoda et al., 1969; Baylor and Fuortes, 1970). An increase of the membrane resistance has been observed also in rods penetrated at the outer segment (Cervetto et al., 1977). By contrast, a decrease in the resistance was found when the electrode was inserted in the inner segment of the rod (Lasansky and Marchiafava, 1974; Werblin, 1975; Cervetto et al., 1977). The resistance changes of the pineal photoreceptor membrane resemble the changes observed at the cell body of rod. As in the rods one may conceive that the changes of the membrane resistance are the result of voltage dependent properties of the membrane rather than the direct effect of illumination.

ACTIVITY OF GANGLION CELLS

From the exposed pineal vesicle, Dodt (1963), first recorded spontaneous nerve discharges that were inhibited by illumination. Morita (1966) made a detailed study on the discharging pattern of individual pineal neurons in the rainbow trout. Features of the teleostean pineal as appearing in sensory nerve impulses have now been understood to a considerable degree. The following descriptions are mostly based on the results obtained in our laboratory (Hanyu and Niwa, 1970; Hanyu et al., 1978).

Spontaneous mass discharges of nerve impulse were recorded by a tungsten microelectrode at certain locations in the pineal vesicle, especially from its proximal part narrowing to the pineal stalk. This suggests that the mass discharges came from the nerve fibers composing the pineal tract contained in the stalk.

The spontaneous activity was inhibited, with a considerable latency, by a light stimulus. When the stimulus was of sufficient strength and duration, all the discharges disappeared temporarily. The stronger the stimuli, the shorter the latency of the inhibition and the more retarded the recovery of the nerve activity from the inhibition.

The threshold for the inhibitory response varied with the species and the state of adaptation. Fig. 4 shows the changes in the threshold, or the stimulus intensity for barely detectable inhibition, observed during dark adaptation following exposure to strong illumination for over 10 minutes.

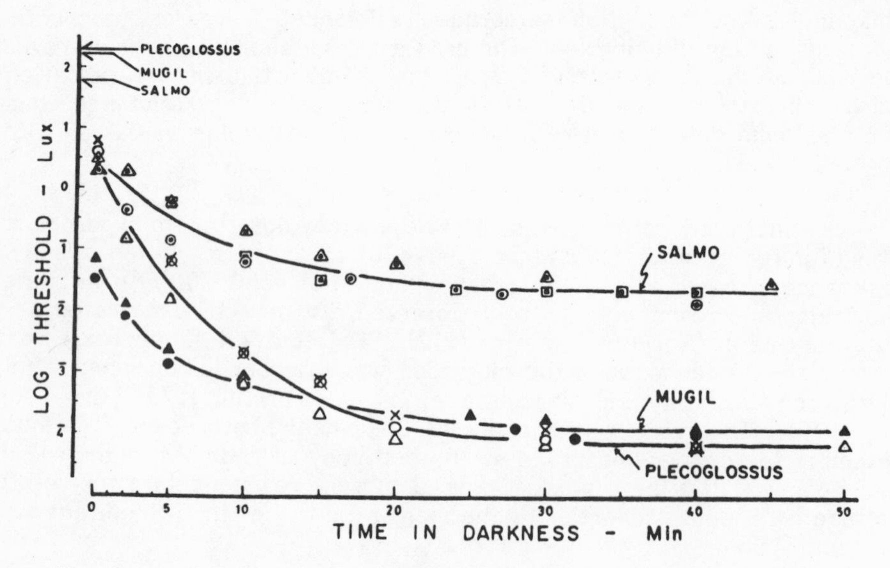

Fig. 4. Curves for dark adaptation of the pineal photosensitivity in three teleosts. Arrows at the top of ordinate indicate levels of preceding light adaptation. Symbols denote different individuals. (Hanyu and Niwa, 1970).

As shown in the figure, the pineal photosensitivity of the ayu and the mullet is a few hundred times higher than that of the rainbow trout, and the change in the sensitivity during dark adaptation is far greater in the ayu than in the other two species. The absolute threshold for dark-adapted rainbow trout obtained by Morita (1966) is 7.8×10^{-3} lux, on the average, a figure close to the values shown in Fig. 4.

The spectral aspect of the inhibitory response was examined with 12 colored lights of equal energy, ranging from 404 to 726 nm. Inhibition of the spontaneous discharge was affected by any strong light in the visible range except the red end. As the stimulus intensity was decreased, however, the response became confined to the middle portion of the spectrum, the last threshold response always occurring between blue green and green. The peak of the spectral sensitivity was estimated to lie near 495 nm in the mullet and near 525 nm in the rainbow trout and ayu. Aside from the sensitivity level, the pattern of spectral sensitivity in each species did not change in the light- and dark-adapted states. The so-called Purkinje shift in vision did not take place in the pineal response of these teleosts.

Most neurons or sensory units, which were recorded isolated, were found to discharge in quite regular rhythms. Fig. 5 shows the relationship between the illumination and the impulse frequency which changes with the adaptation time in the case of the ayu. This neuron was first adapted to complete darkness for about 40 minutes, and discharged steadily at a rate of 37 sec^{-1}. When illumination of 1.6×10^{-5} lux was given, the discharging frequency fell momentarily, but soon began an exponential rise, to attain a new steady rate of about 35 sec^{-1} (Fig. 5). Similar falls and rises of discharging frequency were repeated as the illumination was increased stepwise to 1.6×10^{-3}, 0.16, 16 and 1100 lux, the last being the maximal intensity used. When this illumination was adapted, the steady frequency was about 19 sec^{-1}. On the other hand, an opposite change occurred when the illumination was decreased, as seen in Fig. 5. The discharging impulses turned to a momentary burst in response to the decrease of illumination (off-response). Stronger off-responses were observed at higher levels of adapting illumination.

Plotting the steady frequency against the adapting illumination gave an elongated loop as shown in Fig. 6. The lower and upper courses of the loop represent adaptations to increasing and decreasing illuminations respectively. Either course is roughly divided into two segments, i.e. sloping and horizontal. The sloping segments represent the dependence of the steady frequency on the level of illumination in the low illumination range, while the horizontal ones correspond to a tendency to stabilization of the steady frequency in the range of bright illumination.

Fig. 7 shows several examples where the course of adaptation to increasing illumination started from darkness. Whereas considerable variation exists among individual courses, the above-mentioned general feature, i.e. division into sloping and horizontal segments may be easily recognized. Transition from the sloping to horizontal segments takes place largely between 10^{-1} and 10lux, and the steady frequency along the horizontal segment ranges from 50% to 70% of that in the full dark-adaptation.

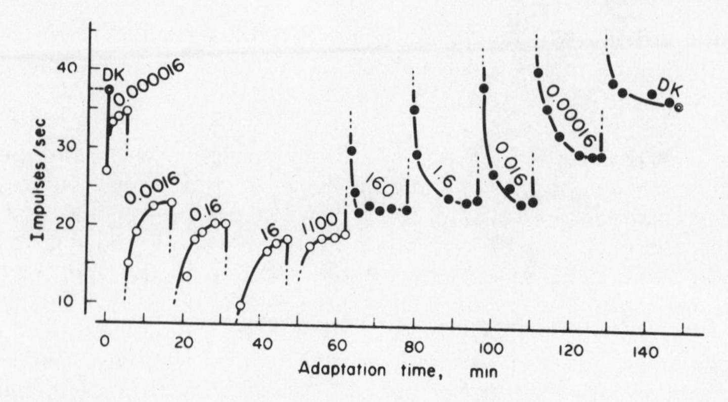

Fig. 5. Changes in the impulse discharging frequency of a pineal neuron of ayu, during stepping-up and -down adapting illumination. Open circle, stepping-up; filled circle, stepping-down; DK, darkness. Figures indicate illumination in lux. (Hanyu et al. 1978).

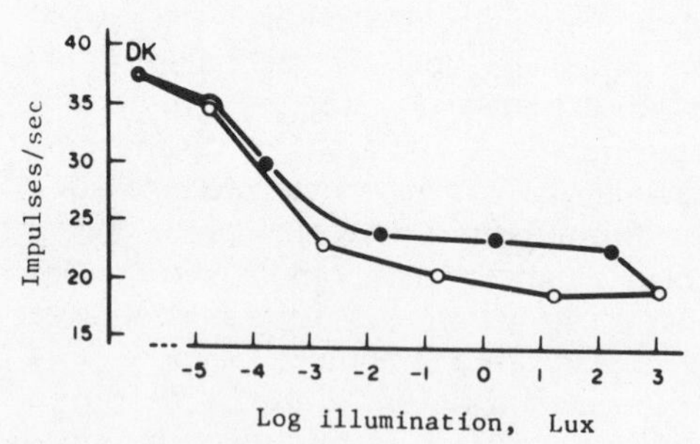

Fig. 6. Changes in the steady frequency of a pineal neuron of ayu during succesive adaptations to different levels of illumination. This figure was derived from Fig. 6. (Hanyu et al., 1978).

Exceptionally, one neuron was so sensitive that, beyond 10^{-1} lux, it did not keep discharging but only produced off-response.

In summary, the pineal neurons of the ayu mostly show the steady discharging frequency which decreases rather regularly as the level of adapting light increases within the low illumination range (Fig. 6 sloping segment). Conversely, the steady frequency increases with the ambient dimness. Similar results have been reported for the rainbow trout (Morita, 1966; Hanyu and Niwa, 1970). These facts point to the pineal function as a "dusk" detector to keep the fish's brain informed of how dim, not how bright, its environment is. On the other hand, the pineal neurons produce stronger off-responses in the high illumination range, where the steady freqeuency remains nearly stabilized (Fig. 6 horizontal segment). This probably means that detection of a "shadow" passing over the head is the main role for the pineal in the environment exposed to sunshine.

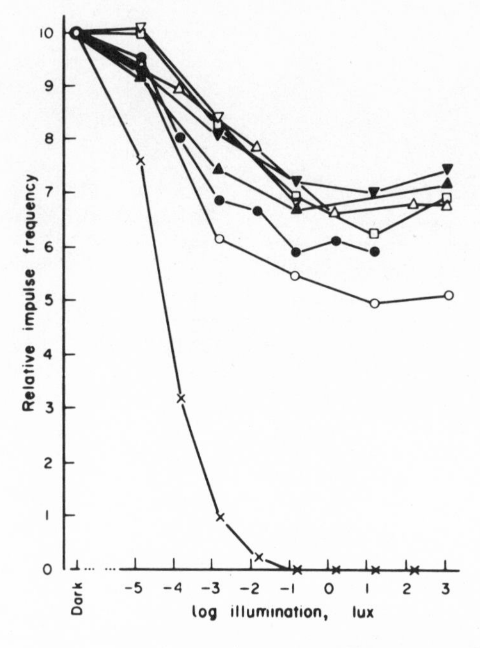

Fig. 7. Examples of decrease in the steady frequency of pineal neurons of ayu during stepping-up illumination from darkness. Symbols denote different neurons. (Hanyu et al., 1978).

ELECTROPINEALOGRAM

Like the electroretinogram (ERG) recorded from the retina, a light induced electrical response can be recorded from the isolated pineal organ by using a pair of electrodes placed on its surface. This response is referred to as the electropinealogram (EPG) (Morita and Dodt, 1973; Tabata, 1973, 1976).

The following description is mostly based on the results of Tabata (1976) who investigated the EPG with the rainbow trout, ayu, catfish (Parasilurus asotus), carp (Cyprinus carpio) and goldfish (Carassius auratus).

The EPG recorded is larger in the rainbow trout, ayu and catfish than in carp and goldfish. An example recorded from the rainbow trout is shown in Fig. 8. As shown in this figure, a small shift of the recording electrode produces a response with inverted polarity. This reversal may have deep relationship to the structure, i.e. infoldings of pineal epithelium to be described below. Fig. 9 shows the relation between the duration of the stimulation and the pattern of EPG.

It is well known that the sodium-aspartate blocks all the electrical activities of the retina (ERG) except a component produced by the receptor cells (e.g. Witkovsky et al., 1973). Application of the sodium-aspartate to the pineal suggests that the EPG may originate in the receptor cells, because EPG treated with sodium-aspartate did not differ from the untreated one (Fig. 10).

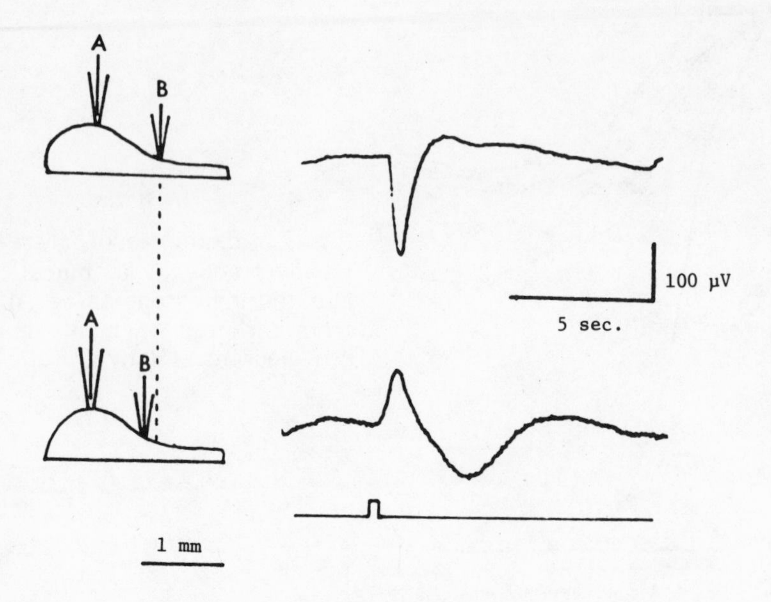

Fig. 8. Illustration showing the reversal of polarity of EPG caused by a small shift of the electrode (B). Rainbow trout. Time constant of the amplifier, 0.26 sec. (Tabata, 1976).

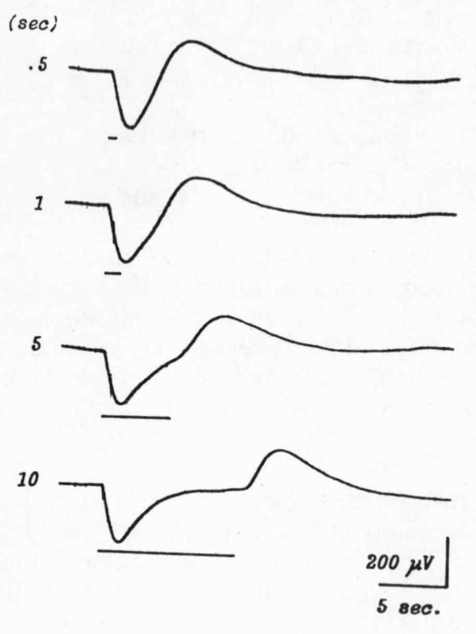

Fig. 9. EPG recordings in rainbow trout. Each figure on the left shows duration of the stimulus in sec. Stimulus intensity, 10 lux. Time constant of the amplifier, 1 sec. (Tabata, 1976).

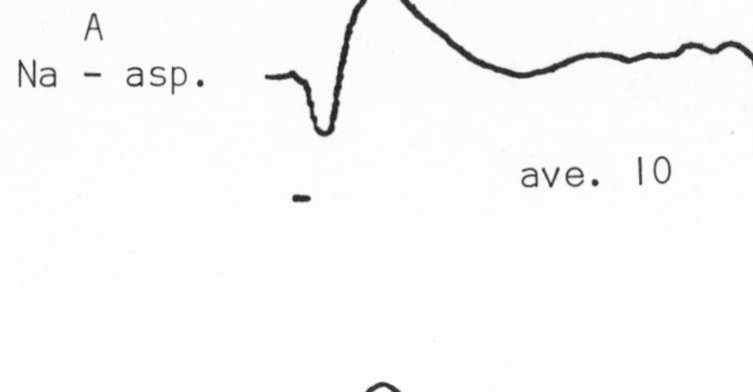

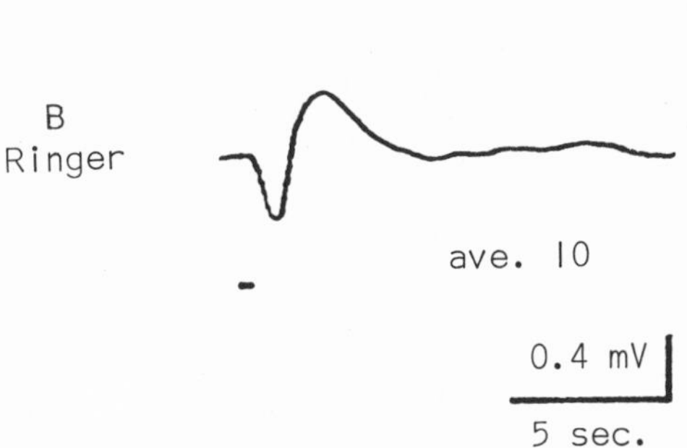

Fig. 10. EPG recording (A) after treated with Na-aspartate solution (120 mM, 15 minutes) does not differ from the recording (B) in the normal Ringer solution. Ave. 10 means ten times addition of recordings. White light (10 lux). Rainbow trout. (Tabata, 1976).

Using the amplitude of the EPG as an indicator, the process of dark adaptation (Fig. 11), the spectral sensitivity (Fig. 12) and the response to the flicker (Fig. 13) were obtained. Fig. 11 shows that 30-40 minutes are necessary for a complete dark adaptation similarly to the results obtained from threshold impulse frequencies (Hanyu and Niwa, 1970; Morita, 1966). Fig. 13 shows that the critical fusion frequency of the pineal is about 1 Hz which is much lower than that of the lateral eyes where it may reach values up to 20 Hz.

STRUCTURE FOR PHOTORECEPTION

Pineal photosensitivity mentioned in the previous sections is also supported by morphological studies of the pineal.

1) The reduction of light passing through the pineal window.

According to Hanyu et al. (1978), the pineal window of the ayu was found to have a marked selective absorption, increasingly higher toward shorter wavelengths within the visible range examined. Thus the calculated transparency increased almost linearly with the wavelength, being 0.26, 0.49 and 0.74 respectively at 400, 500 and 700 nm. The average value for all the wavelength was 0.54. Approximately 50% might be a reasonable estimate for overall transparency applicable to the underlying pineal organ whose maximal sensitivity is near 525 nm. This level of transparency is distinctly high when compared with other reported instances e.g. 5-10% *Phoxinus* (de la Motte, 1964), 1.43 - 10.2% for *Salmo* (Morita, 1966) and 0.89 - 3.55% for *Pterophyllum* (Morita and Bergmann, 1971).

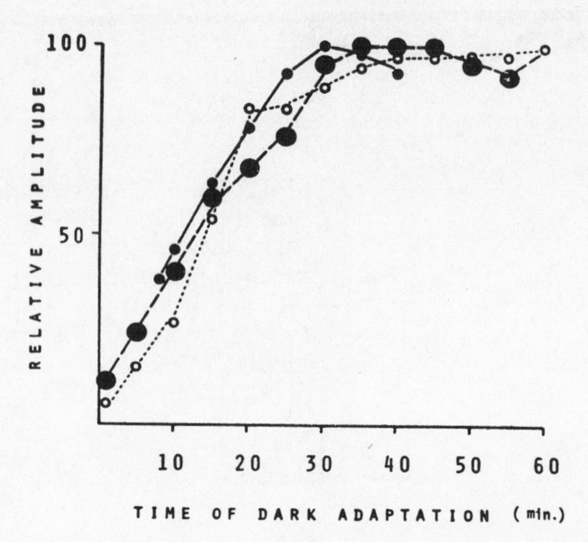

Fig. 11. Curves showing the time course of dark adaptation of the pineal photosensitivity indicated by the relative amplitude of EPG. At first the pineal was adapted to the illumination of 10 lux for 10 minutes, and then it was placed in darkness and tested by white light (30 lux) of 300 msec duration at every five minutes. Open circle, Salmo irideus; Small filled circle, Plecoglossus altivelis; Large filled circle, Parasilurus asotus. (Tabata, 1976).

2) Structure of the pineal organ.

The pineal organ of fish is composed of three kinds of cells - photoreceptor cells, ganglion cells and supporting cells. Oguri and Omura (1973) investigated the ultrastructure of pineal vesicle in five species of teleost - the ayu (Plecoglossus altivelis), the gigi (Pelteobagrus nudiceps), the guppy (Poecilia reticulata), the Japanese eel (Anguilla japonica) and the blind cave fish (Astyanax mexicanus) - and proposed a diagrammatic illustration of the pineal epithelium of teleost (Fig. 14).

The photoreceptor cell of the pineal is, like that in the retina, made up of an outer segment, an inner segment and a basal nucleated soma with a process. In a typical case of the ayu, the outer segment is composed of lamellae or stack of 30 to 70 flattened sacs, which are fewer in number than those in visual cells of the retina. The pineal photoreceptor cell of the blind cave fish was found to have a well developed outer segment which showed signs of degeneration when the fish had been placed in either constant light or dark (Omura, 1975; Herwig, 1976). In the pineal receptor cell of ayu, most of lamellae are continuous with the plasma membrane (Omura et al., 1969). This is just like the lamellae of the cone in the retina, whereas those

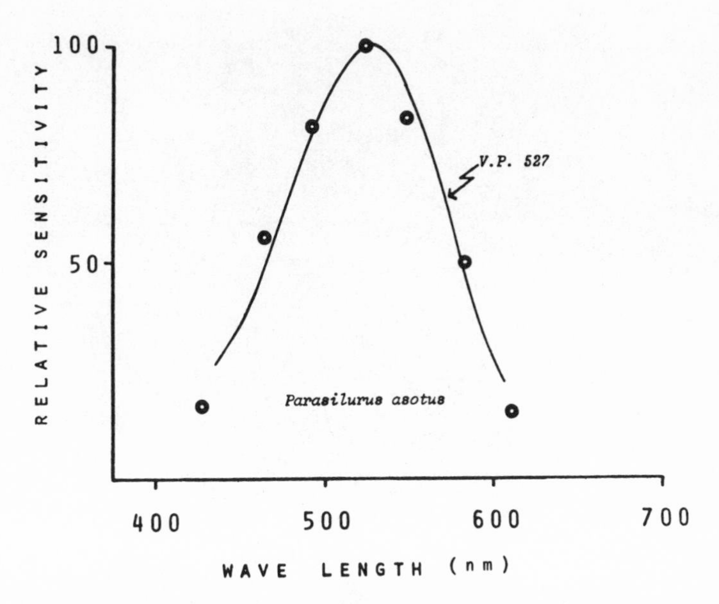

Fig. 12. Spectral sensitivity of the pineal organ of ayu measured by EPG. The continuous curve is Dartnall's nomogram (V.P. = 527 nm). (Tabata, 1976).

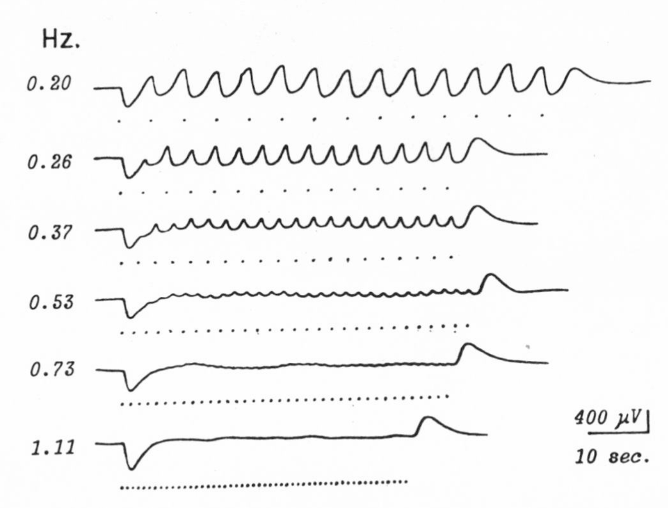

Fig. 13. Responses of EPG of rainbow trout to flicker stimuli. Stimulus light is 1 lux and 200 msec duration. Frequencies are shown on the left of the recordings. (Tabata, 1976).

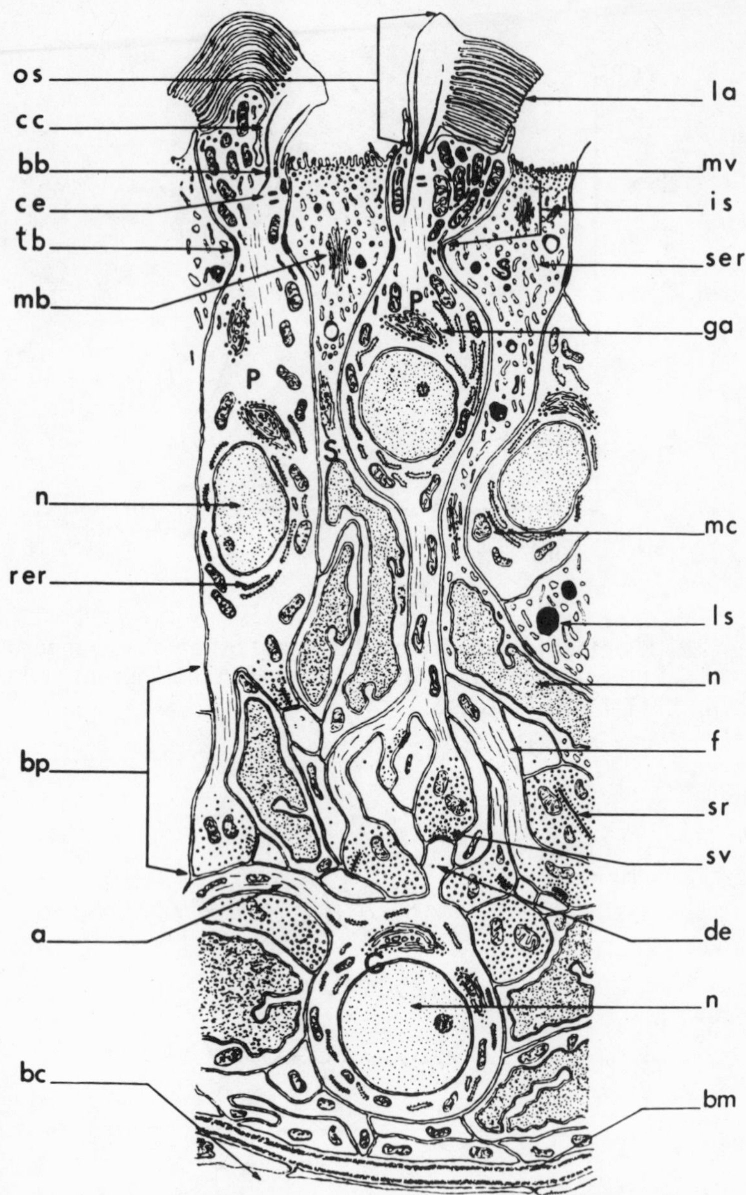

Fig. 14. Diagrammatic ultrastructure of the pineal epithelium of teleost. a, axon; bb, basal body; bm, basement membrane; bp, basal process; cc, connecting cilium; ce, centriole; de, dendrite; dj, desmosomal junction; f, bundle of filaments; G, ganglion cell; ga, Golgi apparatus; is, inner segment; la, lamellae; ls, lysosome; mb, myeloid body; mc, mitochondria; mv, microvilli; n, nucleus; no, nucleolus; os, outer segment; P, photoreceptor cell; rer, rough surfaced endoplasmic reticulum; S, supporting cell; ser, smooth surfaced endoplasmic reticulum; sr, synaptic ribbon; sv, synaptic vesicle. (Oguri and Omura, 1973).

of the rod are not continuous with the membrane but are free-floating disks (Young, 1970).

Ganglion cells were examined by means of electronmicroscopy in the pineal organ of the ayu, gigi and blind cave fish (Oguri and Omura, 1973). Every species with pineal photoreceptor cells is considered to have ganglion cells in the pineal organ, though their detection is not easy because of their sparse distribution. Near ganglion cells, neuropile zones are found where basal processes of photoreceptor cells and unmyelinated nerve fibers are gathered together to keep synaptic contact with each other (Takahashi, 1969; Omura et al., 1969). In the terminal region of basal process near the synapse, synaptic vesicles and synaptic ribbons were observed. These structures indicate that the information received by the photoreceptor cells are sent to the brain by way of the neuropiles, ganglion cells and their axons (pineal tract). According to the study of Wake (1973) in the goldfish, a group of 30-50 photoreceptor cells is considered to lie apposed to each neuropile formation in the rostral half of the pineal vesicle, and at the periphery of the neuropile an AChE-containing nerve cell (ganglion cell) is present. In the caudal half of the pineal vesicle as well as in the stalk, only 3 or 4 receptor cells appose to each ganglion cell to form a single neuropile. Such neuropiles are also found in the pineal epithelium of the rainbow trout. In this fish, the epithelium is thicker and has many infoldings, unlike that of the goldfish (Omura and Oguri, 1969).

The polarity reversal of the EPG (Fig. 8) caused by a small shift of the different electrode is considered to have deep relationship to the neuropiles and/or the infoldings of the epithelium (Fig. 15).

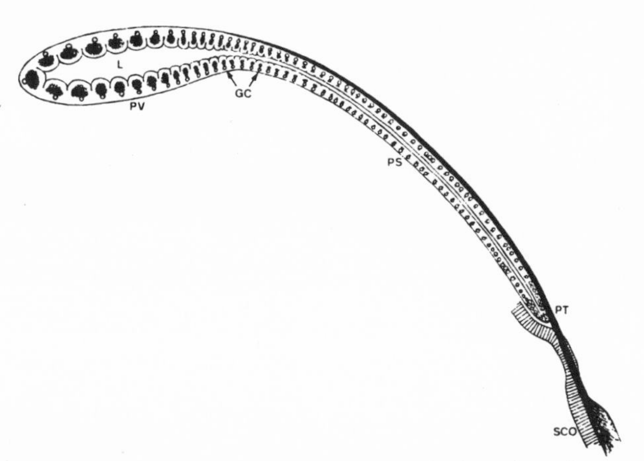

Fig. 15. Diagrammatic representation of the pineal organ of the goldfish to show its structural organization. PV, pineal vesicle; L, lumen; GC, nerve cell; PS, pineal stalk; PT, pineal tract; SCO, subcommisural organ. (Wake, 1973).

BIOLOGICAL SIGNIFICANCE OF PINEAL FUNCTION

1) Body color and phototaxis.

Hafeez and Quay (1970) demonstrated that pinealectomy abolished day-night change in the body color of the blinded rainbow trout, probably by suppressing blanching at night. Pinealectomy is also known to cause darkening of the body color in other Salmonids (Hoar, 1955). These effects may be taken to reflect the loss of tonic impulses from the pineal organ as a dusk-detector.

In constrast, observations on the contribution of the pineal to the phototactic movements in the teleosts are contradictory. Some are for (Breder and Rasquin, 1947; Hoar, 1955; Pang, 1965), while others are against (Fenwick, 1970; Hafeez and Quay, 1970) the pineal involvement in the phototaxis. Further investigations should be undertaken with proper consideration to the pineal serving as a shadow detector and, being unpaired, not as a suitable receptor for mediating oriented behavior of fishes (Hanyu et al., 1978).

2) Gonadal maturation.

The pineal organ, as in mammals, has been indicated to play a role in controlling the gonadal maturation in fishes. Thus in Notemigonus crysoleucas during the early spawning season, de Vlaming (1975) has observed that pinealectomy either caused gonadal regression or retarded maturation if the fishes are maintained on a long photoperiod-warm temperature regime. On the other hand, pinealectomy stimulated gonadal activity in fish kept on a short photoperiod-warm temperature. Therefore, it appears that the pineal organ stimulates gonadal development in fish exposed to long photoperiod, turning into an inhibitory action on a short photoperiod.

Similarly in Oryzias latipes (Urasaki, 1972) and in the goldfish (de Vlaming and Vodicnik, 1978), the pineal may stimulate or inhibit the sexual maturation depending on photoperiod. It is likely, at least in some teleosts, that the pineal organ acts in the photo-sexual responses as a tonic darkness receptor by measuring the dark phase of daily light-dark cycle. Whether or not melatonin, known as a gonad inhibitor, mediates the photo-sexual responses is an interesting question yet to be solved (Fenwick, 1970; Oguri et al., 1968).

The ayu, a Salmonid species, is a short-day fish and spawns in autumn; Both male and female ayu usually die after spawning. For ayu-culturists, prolongation of the life span and acceleration of gonadal maturation are two big problems. The former problem was solved by the extention of the daily photoperiod with illumination over ponds. Illumination is effective because it prevents the gonadal development. This is now popularly used among ayu-culturists. Regarding the latter problem, shortening the day length in early summer is known to be effective in accelerating the gonadal maturation. This kind of experiment was carried out in trout by Hoover (1937) and by Hazard and Eddy (1951), but proved very laborious since it is difficult to cover the outdoor ponds completely with light-proof sheets (a few lux of illumination is reported to be critical for the gonadal maturation of the ayu).

Fushiki (1978), however, found that shortening the day's length is not always necessary for accelerating maturation, provided that fish are exposed to longer photoperiod regime in spring and return to the natural day length in early summer. For instance, the ayu kept under 20 hr light a day (20 L - 4 D) for five weeks (from Apr. 23 to May 28) and returned to the natural day length (about 14.5 L - 9.5 D) have matured their gonads earlier than the control fish by 40 days. When the period of the longer light regime was changed from 5 weeks to 4, 3, 2 and 1 week, the gonadal maturation became more and more retarded thus approaching the control. From these results one can conclude that the gonadal maturation is induced by some amount of decrement, rather than by an absolute length of the light period in a day. Preliminary investigations have, however, indicated that neither pinealectomy nor enucleation of eyes affects significantly gonadal maturation on a short photoperiod regime and gonadal suppression on a long photoperiod regime. It is quite obvious that more detailed studies are needed before a conclusion is made regarding this matter.

ACKNOWLEDGEMENTS

We wish to thank Dr. Luigi Cervetto for his helpful criticisms of the manuscript. Fig. 1, 2, 3, 5, 6 and 7 were reprinted with permission of Pergamon Press, Ltd., Fig. 4 with permission of Les Presses de l'Université de Montréal and Fig. 15 with permission of Springer-Verlag.

REFERENCES

Baylor, D.A. and Fuortes, M.G.F. (1970). Electrical responses of single cones in the retina of the turtle. J. Physiol. (Lond.). 207: 77-92.

Breder, C.M., Jr. and Rasquin, P. (1947). Comparative studies in the light sensitivity of blind characins from a series of Mexican caves. Bull. Am. Mus. Nat. Hist. 89: 325-351.

Cervetto, L., Pasino, E. and Torre, V. (1977). Electrical responses of rods in the retina of Bufo marinus. J. Physiol. (Lond.) 267: 17-51.

Dodt, E. (1963). Photosensitivity of the pineal organ in the teleost Salmo irideus (Gibbons). Experientia (Basel) 19: 642-643.

Fenwick, J.C. (1970a). Demonstration and effect of melatonin in fish. Gen. Comp. Endocrinol. 14: 86-97.

Fenwick, J.C. (1970b). Effects of pinealectomy and bilateral enucleation on the phototactic response and on the conditioned response to light of the goldfish, Carassius auratus L. Can. J. Zool. 48: 175-183.

Frisch, K. von (1911). Veber das Parietalorgan der Fische als funktionierendes Organ. Sitzsber. Ges. Morphol. Physiol. (München) 27: 16-18.

Fushiki, S. (1978). On acceleration of gonadal maturation of ayu-fish

(Plecoglossus altivelis T. & S.) by means of controlling illumination. II The effect of term and length of the photoperiod in spring season on the gonadal maturation (In Japanese). Reports of the Fisheries Experimental Station of Shiga prefecture No. 30: 1-7.

Fushiki, S. (1979). Effects of vernal long photoperiod regimes on maturation of the ayu (in Japanese). Doctorate Thesis, the Univ. of Tokyo.

Hafeez, M.A. and Quay, W.B. (1970). The role of the pineal organ in the control of phototaxis and body coloration in rainbow trout (Salmo gairdneri Richerdson). Z. Vergl. Physiol. 68: 403-416.

Hamasaki, D.I. and Eder, D.J. (1977). Adaptive radiation of the pineal system. In: Handbook of Sensory Physiology Vol. VII/5. Ed. F. Crescitelli. Berlin, Springer-Verlag, p. 497-548.

Hanyu, I., Niwa, H. and Tamura, T. (1969). A slow potential from the epiphysis cerebri of fishes. Vision Res. 9: 621-623.

Hanyu, I. and Niwa, H. (1970). Pineal photosensitivity in three teleosts, Salmo idideus, Plecoglossus altivelis and Mugil cephalus. Rev. Can. Biol. 29: 133-140.

Hanyu, I., Niwa, H. and Tamura, T. (1978). Salient features in photosensory function of teleostean pineal organ. Comp. Biochem. Physiol. 61A: 49-54.

Hazard, T.P. and Eddy, R.E. (1951). Modification of the sexual cycle in brook trout (Salvelinus fontinalis) by control of light. Trans. Am. Fish. Soc. 80: 158-162.

Henderson, N.E. (1963). Influence of light and temperature on the reproductive cycle of the eastern brook trout Salvelinus fontinalis (Mitchell). J. Fish. Res. Board Can. 20: 859-897.

Herwig, H.J. (1976). Comparative ultrastructural investigations of the pineal organ of the blind cave fish, Anoptichthys jordani, and its ancestor, the eyed river fish, Astyanax mexicanus. Cell Tiss. Res. 167: 297-324.

Hoar, W.S. (1955). Phototactic and pigmentary responses of sockeye salmon smolts following injury to the pineal organ. J. Fish. Res. Board Can. 12: 178-185.

Hoover, E.E. (1937). Experimental modification of the normal sexual cycle of trout. Science 86: 425-426.

Kaneko, A. (1970). Physiological and morphological identification of horizontal, bipolar and amacrine cells in goldfish retina. J. Physiol. (Lond.) 207: 623-633.

Lasansky, A. and Marchiafava, P.L. (1974). Light-induced resistance changes in retinal rods and cones of the tiger salamander. J. Physiol. (Lond.) 236: 171-191.

Morita, Y. (1966). Entladungsmuster pinealer Neurone der Regenbogenforelle (Salmo irideus) bei Belichtung des Zwischenhirns. Pflügers. Archiv. 289: 155-167.

Morita, Y. and Bergmann, G. (1971). Physiologische Untersuchungen und weitere Bemerkungen zur Struktur des lichtempfindlichen Pinealorgans von Pterophyllum scalare Cuv. et Val. (Cichlidae, Teleostei). Z. Zellforsch. 119: 289-294.

Morita, Y. and Dodt, E. (1973). Slow photic responses of the isolated pineal organ of lamprey. Nova Acta Leopoldina 211: 331-339.

Motte, I. de la (1964). Untersuchungen zur vergleichenden Physiologie der Lichtempfindlichkeit geblendeter Fische. Z. Vergl. Physiol. 49: 58-90.

Oguri, M., Omura, Y. and Hibiya, T. (1968). Uptake of ^{14}C-labelled 5-hydroxytryptophan into the pineal organ of rainbow trout. Bull. Jap. Soc. Sci. Fish. 34: 687-690.

Oguri, M. and Omura, Y. (1973). Ultrastructure and functional significance of the pineal organ of teleost. In: Responses of Fish to Environmental Changes. Ed. W. Chavin. Springfield, Charles C. Thomas, p. 412-432.

Omura, Y. (1975). Influence of light and darkness on the ultrastructure of the pineal organ in the blind cave fish, Astyanax mexicanus. Cell Tiss. Res. 160: 99-112.

Omura, Y., Kitoh, J. and Oguri, M. (1969). The photoreceptor cell of the pineal organ of ayu, Plecoglossus altivelis. Bull. Jap. Soc. Sci. Fish. 35: 1067-1071.

Omura, Y., Oguri, M. (1969). Histological studies on the pineal organ of 15 species of teleosts. Bull. Jap. Soc. Sci. Fish. 35: 991-1000.

Pang, P.K.T. (1965). Light sensitivity of the pineal gland in blinded Fundulus heteroclitus. Am. Zool. 5: 682.

Rivas, L.R. (1953). The pineal apparatus of tunas and related scombrid fishes as a possible light receptor controlling phototactic movements. Bull. Mar. Sci. Gulf and Caribbean 3: 168-180.

Tabata, M. (1973). Electropinealogram (EPG) elicited by light stimulation in fish (in Japanese). Dobutsugaku Zasshi 82: 294.

Tabata, M. (1976). Electrophysiological studies on the photosensitivity of epiphysis cerebri in fishes (in Japanese). Doctorate Thesis, Nagoya Uni.

Tabata, M., Niwa, H. and Tamura, T. (1971). On a slow potential from the epiphysis cerebri of fishes. Bull. Jap. Soc. Sci. Fish. 37: 487-490.

Tabata, M., Tamura, T. and Niwa, H. (1975). Origin of the slow potential in the pineal organ of the rainbow trout. Vision Res. 15: 737-740.

Takahashi, H. (1969). Light and electron microscopic studies on the pineal organ of the goldfish, Carassius auratus L. Bull. Fac. Fish. Hokkaido Univ. 20: 143-157.

Tomita, T., Murakami, M. and Pautler, E.L. (1967). Spectral response curves of single cones in the carp. Vision Res. 7: 519-531.

Toyoda, J., Nosaki, H. and Tomita, T. (1969). Light-induced resistance changes in single photoreceptors of Necturus and Gekko. Vision Res. 9: 453-463.

Urasaki, H. (1972). Role of the pineal gland in gonadal development in the fish, Oryzias latipes. Annot. Zool. Japan. 45: 152-158.

Vlaming, V.L. de (1975). Effects of pinealectomy on gonadal activity in the cyprinid teleost, Notemigonus crysoleucas. Gen. Comp. Endocrinol. 26: 36-49.

Vlaming, V.L. de and Vodicnik, M.J. (1978). Seasonal effects of pinealectomy on gonadal activity in the goldfish, Carassius auratus. Biol. Reprod. 19: 57-63.

Wake, K. (1973). Acetylcholinesterase-containing nerve cells and their distribution in the pineal organ of the goldfish, Carassius auratus. Z. Zellforsch. 145: 287-298.
Werblin, F.S. (1975). Regenerative hyperpolarization in rods. J. Physiol. (Lond.) 244: 53-81.
Witkovsky, P., Nelson, J. and Ripps, H. (1973). Action spectra and adaptation properties of carp photoreceptors. J. Gen. Physiol. 61: 401-423.
Young, R.W. (1970). Visual cells. Sci. Am. 223: 80-91.

THE ROLE OF TEMPERATURE IN THE ENVIRONMENTAL PHYSIOLOGY OF FISHES

William W. Reynolds & Martha E. Casterlin

Marine Biology Program, Center for Life Sciences
University of New England,
Biddeford, Maine 04005, U.S.A.

INTRODUCTION

Temperature is among the most pervasive and important physical factors in the environment of an organism. It is a measure of the average rate of random motions of atoms and molecules: the higher the temperature the faster the motion. Properties such as viscosity or fluidity, and changes in state from solid to liquid to gas, depend upon temperature. Diffusion rates increase as temperature increases, because the particles are moving faster. Only at absolute zero (0°K, or -273°C) does the motion virtually cease. Temperature also affects the rates of chemical reactions, since in order to react with one another to form new molecular combinations, two atoms or molecules must collide or come into close proximity to one another. The higher the temperature, the faster the random motion, and thus the more frequent will be the collisions. The life processes of living organisms, which are physicochemical in nature, are therefore profoundly affected by temperature. In general, higher temperatures tend to speed up these processes, but also tend to disrupt the structural integrity of the organism. As temperatures change, the rates of various processes must be balanced and coordinated. The organism must either compensate for the rate changes induced by changes in temperature (acclimation or acclimatization), or it must try to prevent or minimize changes in its body temperature (thermoregulation). A combination of these strategies can also be employed.

CHANGES IN THE PHYSICAL STATE OF WATER

A lower temperature limit for aquatic life, including fishes, is the freezing point of water. Pure water freezes at 0°C, while increasing concentrations of solutes progressively lower the freezing point of aqueous solutions such as salt water or the blood and tissue fluids of organisms. Some fishes inhabiting polar seas live at temperatures slightly below 0°C, aided by natural antifreeze in their bodies.

The boiling point of water, 100°C for pure water at sea level, would likewise place an upper limit on the thermokinetic zone of organisms based on aqueous chemistry. However, the proteins of most organisms denature far below the boiling point of water, and only a few thermophilic prokaryotes withstand temperatures much above 50°C. Few fishes can withstand temperatures above 40°C, and the upper lethal limits of some species fall near 35°C, 25°C, 15°C or even 5°C. On the other hand, the lower lethal temperatures of some fishes are considerably above the freezing point of water, even exceeding 10°C for some tropical species.

ACCLIMATION AND ACCLIMATIZATION

Acclimation is a term applied to the compensatory responses made by an organism to constant temperatures in the laboratory, whereas acclimatization refers to compensatory responses to a complex array of varying temperatures and to nonthermal factors such as photoperiod in the natural environment (cf. Reynolds and Casterlin, 1979a). For many purposes it is unnecessary to distinguish between these, so we will routinely use the compound form "acclim(atiz)ation" wherever the distinction seems unnecessary.

ENZYME THERMAL OPTIMA AND ACCLIM(ATIZ)ATION

The chemical reactions in living organisms, collectively referred to as metabolism, are catalyzed by enzymes. There is a specific enzyme for each of the thousands of kinds of chemical reactions in the organism. Enzymes are proteins, which have a complex three-dimensional tertiary structure (cf. Reynolds, 1975, pp. 24-25) that gives them their specificity (ability to catalyze a specific reaction). This tertiary structure is thermolabile, i.e., easily denatured by excessive temperature which renders the structure unstable. Thus, as temperature increases, the progressive denaturation of the protein tends to decrease the reaction rate as the concentration of undenatured enzyme decreases. The reaction rate increases with thermomolecular motion up to the temperature at which the effect of enzyme denaturation causes the reaction rate to decrease again (cf. Reynolds, 1975, p. 59). The result of these counteracting influences is that each enzymatically catalyzed reaction has a specific thermal optimum at which the rate is maximal. Different enzymes have differing thermal optima, depending on their relative thermolability.

There are alternative forms of enzymes, with differing thermal optima, that catalyze the same reaction. The alternative forms are known as isozymes or allozymes (Shaklee et al., 1977; Johnson, 1977; Somero, 1978; Kuz'mina and Morozova, 1977; Niemierko et al., 1977). Natural selection induces changes in gene frequencies in species populations that adapt the population to its thermal habitat (Johnson, 1977). The genes control the forms of enzymes that are produced by the organism. However, an individual organism often has the capability of producing alternative enzyme forms in response to seasonal or other changes in temperature. These biochemical changes at the organismic level constitute acclim(atiz)ation, permitting metabolic reaction rates to be adjusted to compensate for changes in temperature. Genetically based changes at the population level constitute evolutionary adaptation, including the ability to acclimat(iz)e. In addition to genetically based changes and changes resulting from gene regulation, there are changes in enzyme conformation directly induced by temperature at the post-transcriptional level (Somero, 1978).

THERMAL ADAPTATIONS IN STRUCTURAL PROTEINS

Structural proteins, as well as enzyme proteins, denature above a certain temperature. Thus, genetic adaptation and acclim(atiz)ation also involve production of different structural proteins with differing thermal properties. The thermal stability of muscle proteins (Vitvitskii, 1977) and collagen proteins (Rigby, 1977) correlate with the preferred habitat temperatures of different species.

CHANGES IN MEMBRANE STRUCTURE AND FUNCTION

Thermal acclim(atiz)ation involves changes in membrane structure and function. Changes in the lipid composition of cellular membranes affect the fluidity of the membranes (Cossins and Prosser, 1978; Wodtke, 1978). Changes in the fluidity of mitochondrial membranes affect the rate of oxidative metabolism (Wodtke, 1978), presumably by altering rates of diffusion through the membrane. Rahmann (1978) observed changes in ganglioside compounds in synaptic membranes in the brain during thermal acclim(atiz)ation, which correlated with changes in electrophysiological parameters and in motor activity. Hulbert (1978) attributes the changes in membrane structure to thyroid hormonal influences (cf. also Reynolds and Casterlin, 1980).

ACID-BASE REGULATION

Following changes in temperature, fishes must make compensatory changes in blood pH (Crawshaw, 1977, 1979) which are intimately correlated with physicochemical requirements for the maintenance of cellular metabolism (Withers, 1978).

OSMOTIC AND IONIC BALANCE

The gills of fishes are a major site of ionic exchange (both active and passive) with the water. Changes in respiratory rate (cf. Reynolds, 1977a) that occur during changes in temperature can cause changes in osmotic and ionic balance (Crawshaw, 1979). Active transport mechanisms can be incapacitated at low temperatures, and at high temperatures rates of passive ion flux may be excessive (Crawshaw, 1979). Acclim(atiz)ation processes must include compensatory adjustments for thermal effects on osmotic and ionic balance in addition to other homeostatic adjustments.

GAS EXCHANGE ADJUSTMENTS

Metabolic rate changes in response to changing temperatures alter oxygen requirements and carbon dioxide production (the latter in turn affecting acid-base balance). Therefore gill ventilation rates, as well as cardiac output, respond to changes in temperature (Reynolds, 1977a; Reynolds and Casterlin, 1978a). Through peripheral thermoreceptors which quickly sense changes in ambient temperatures, fishes are able to make anticipatory adjustments in gas exchange rates which precede changes in core temperatures (Crawshaw, 1977, 1979).

CLASSIFICATIONS OF ORGANISMS BASED ON ACCLIM(ATIZ)ATION ABILITY

Organisms which have very little ability to acclimat(iz)e to temperature changes are characterized as being stenothermal. Organisms which can acclimat(iz)e over a wide range of temperatures are eurythermal. These terms are, of course, relative. Stenotherms can be further characterized as being "cold" or "warm" stenotherms, but again these are relative terms. Generally, one would expect polar species to be cold stenotherms, tropical species to be warm stenotherms, and temperate species to be eurytherms.

ALTERNATIVE THERMAL ADAPTIVE MECHANISMS

An organism has two alternative (but not mutually exclusive) adaptive mechanisms for coping with changes in environmental temperature (Reynolds, 1978): it can restructure itself biochemically as discussed above (acclimatization), or it can attempt to minimize changes in its body temperature (thermoregulation). Thermoregulation can be achieved by either behavioral or physiological means, or by some combination of these. Fishes tend to utilize a combination of acclim(atiz)ation and thermoregulation to achieve a suitable degree of internal homeostasis. Both processes incur certain costs to the organism, which must be weighed against the benefits derived therefrom. At the species population level, the cost-benefit ratio is adjusted by natural selection, tending to optimize the species' overall strategy for coping with temperature changes. The relative degree of eurythermality or stenothermality will generally tend to correlate with the thermal stability of the environment and with the ability of the species to thermoregulate within that environment. In a very constant thermal environment, such as the deep sea or polar regions, or some tropical regions, there may be little opportunity or necessity to either thermoregulate or acclimat(iz)e. By contrast, temperate freshwater environments typically experience wide seasonal fluctuations in temperature which demand highly developed abilities for coping with such changes.

THERMOREGULATORY CLASSIFICATIONS

Ability to thermoregulate perfectly results in a condition referred to as homeothermy. Conversely, poikilothermy is a term applied to a very labile body temperature. Heterothermy is an intermediate condition, in which body temperatures change from time to time, as in seasonal (hibernation, estivation) or daily torpor, but some degree of control is retained, including the capacity for spontaneous arousal from torpor. Many smaller species of mammals and birds are heterothermic (Reynolds, 1979).

High rates of metabolic heat production and relatively low rates of heat loss, usually combined with some degree of control over both, constitute a condition known as endothermy, in which the body core temperature is relatively independent of ambient environmental temperatures. This condition is best exemplified by the homeoendothermic mammals and birds. Some insects, a snake, and even certain plants are facultative endotherms, able to raise their body temperatures significantly above ambient temperatures for limited periods of time through elevated metabolic rates (Reynolds, 1979). Some fishes, such as tunas (Dizon and Brill, 1979) and lamnid sharks, achieve a degree of regional endothermy by means of a vascular specialization (a *rete mirabile*) that reduces heat loss from the active lateral swimming muscles.

Most fishes can be considered ectotherms, in that their metabolic heat production makes no significant contribution to their body temperatures. Not only are their metabolic rates low, being only a fraction of the metabolic rate of an endotherm of the same size and at the same temperature, but their rates of heat loss are very high: they lack effective thermal insulation. Heat is lost to the ambient water, which is a very effective heat sink because of its high thermal conductance and specific heat, through the gills and integument. The relative amounts of heat lost through each route varies among species (Crawshaw, 1976, 1977, 1979; Erskine and Spotila, 1977) and depends upon the activity of the fish (Mueller, 1976).

Internally, the flow of heat from the core to the periphery occurs by conduction and by convection. In live fish, the convective component is attributable to blood flow. The relative magnitude of the convective component is revealed by comparing heating and cooling rates of live fish and of dead fish. Dead fish heat and cool at equal rates, and much more slowly than live fish. Live fish heat and cool at different rates, due to differences in blood flow (Crawshaw, 1976; Reynolds, 1977a; Reynolds and Casterlin, 1978a). Blood flow accounts for about 10-30% of the internal heat flow of a live fish (Reynolds and Casterlin, 1978a), presaging the vasomotor refinements of endotherms. Vasomotor influences do not (except in tunas - cf. Dizon and Brill, 1979), however, significantly affect the steady-state core temperatures of most fishes, which differ from a few tenths to a few degrees Celsius above ambient water temperatures depending on body mass (Stevens and Fry, 1970, 1974; Dean, 1976; Mueller,

1976; Reynolds et al., 1976). Allometric relationships between the mass of metabolizing tissue and the surface area through which heat is lost account for the size differences in excess core temperatures.

Most studies on thermoregulation in fishes have concentrated on temperate freshwater Osteichthyes (Coutant, 1977), with relatively little attention having been paid to tropical marine fishes (Reynolds and Casterlin, in prep.), sharks (Crawshaw and Hammel, 1973; Reynolds and Casterlin, 1978b; Casterlin and Reynolds, 1979 a), or lampreys (McCauley et al., 1977; Lemons and Crawshaw, 1978; Reynolds and Casterlin, 1978c).

Thermoregulatory behavior serves both physiological (homeostatic) and ecological (habitat selection) functions (Reynolds and Thomson, 1974a; Matthews and Hill, 1979). Temperature serves as a proximate (Reynolds, 1977b) or directive (Fry, 1947) factor to guide the animal into, and keep it in, an environment conducive to its survival, growth and reproduction. The thermal responses of an animal therefore comprise an important component of the species' ecological niche.

Thermoregulatory behavior in fishes appears to be controlled by central nervous system mechanisms similar and probably homologous to those of higher vertebrates (Crawshaw and Hammel, 1973, 1974; Crawshaw, 1977, 1979), with the major differences between ectotherms and endotherms arising on the effector side of the circuit (Kluger, 1979). Fishes exhibit thermoregulatory responses to various neuropharmacological agents which are similar to the responses of mammals (Green and Lomax, 1976; Reynolds and Casterlin, 1980). Fishes, like other vertebrates (Casterlin and Reynolds, 1977; Kluger, 1979; Reynolds and Casterlin, 1980), respond to infection with a febrile response (Reynolds, Casterlin and Covert, 1976, 1978; Reynolds, 1977c; Reynolds, Covert and Casterlin, 1978) which enhances survival from the infection (Covert and Reynolds, 1977). Behavioral fever (Reynolds, Casterlin and Covert, 1976) is expressed in ectotherms as an increase in preferred temperature. As in mammals, the response is blocked by antipyretics such as acetaminophen (Reynolds, 1977 c).

BEHAVIORAL THERMOREGULATION

While fishes (vertebrate classes Agnatha, Chondrichthyes and Osteichthyes) are, by and large, ectotherms, they are nevertheless able to thermoregulate behaviorally, by selecting appropriate water temperatures and avoiding those which are harmful. Behavioral thermoregulation, which is also utilized extensively by endotherms, can provide a considerable degree of homeothermy if the thermal structure of the environment is suitable and other factors do not interfere. Thermoregulatory behavior of fishes involves locomotion (secondary orientations sensu Fraenkel and Gunn, 1961), since postural (primary) orientations are ineffective for thermoregulation in water (Reynolds and Casterlin, 1979a).

Given a free choice in a thermal gradient, a fish will occupy a limited range of temperatures. Some measure of central tendency (cf. Reynolds

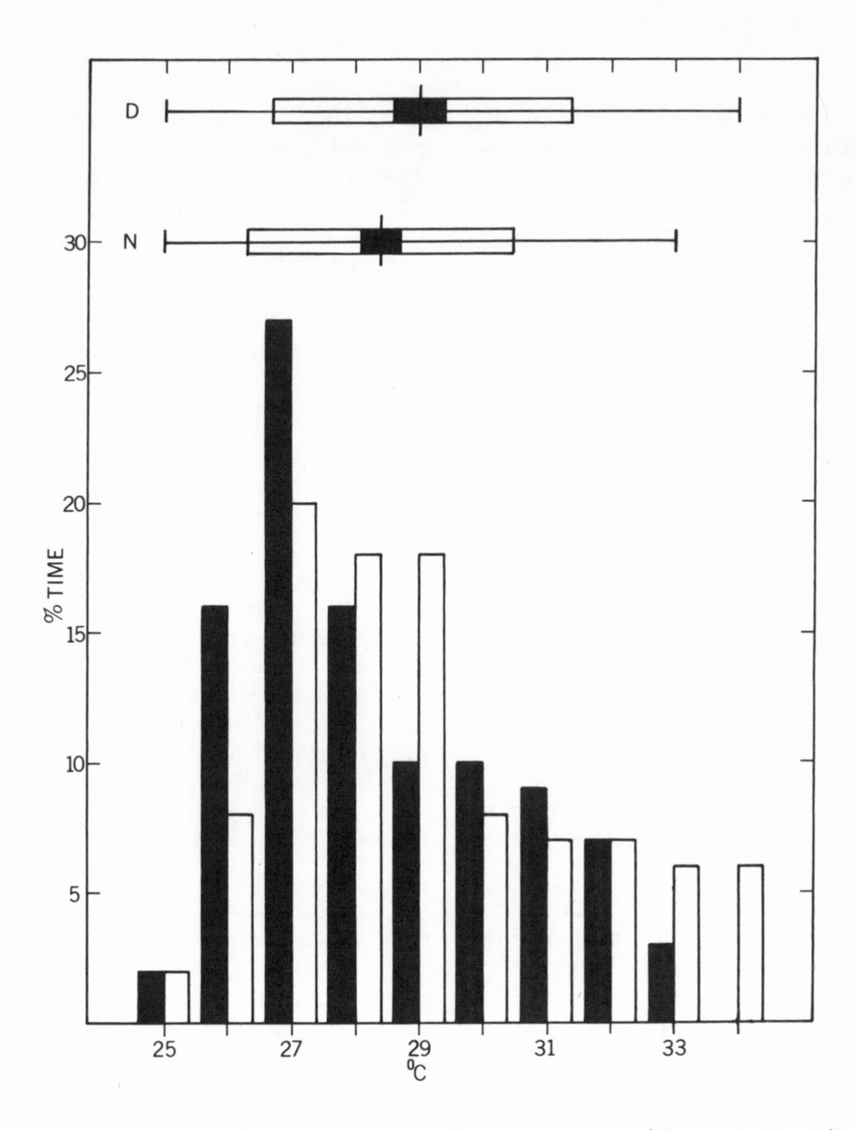

Fig. 1. Diurnal (D, unshaded bars) and nocturnal (N, shaded bars) relative frequency distributions of temperatures voluntarily occupied by bluespotted sunfish, Enneacanthus gloriosus, thermoregulating in an Ichthyotron electronic shuttlebox. Diurnal and nocturnal frequency distributions are positively skewed and have identical modal preferenda of 27°C. Modified Dice-Leraas diagrams (cf. Simpson et al., 1960) above show ranges (horizontal lines), means (vertical lines), ± 1 S.D. (unshaded horizontal bars), and ± 2 S.E. (shaded horizontal bars) for day (D) and night (N).

and Casterlin, 1976a, 1979a; Reynolds, 1977b) in this range is called the preferred or selected temperature, or thermal preferendum (Fry, 1947; Reynolds, 1977b). These terms are used empirically, with no implication of the mechanisms involved, in contrast to such terms as thermokinesis or thermotaxis which bear precise mechanistic definitions (cf. Fry, 1958; Fraenkel and Gunn, 1961). Avoided temperatures are those which are not voluntarily occupied, i.e., the limits of the occupied range (Reynolds, 1977b). Since the frequency distribution of occupied temperatures is often skewed (Fig. 1; DeWitt and Friedman, 1979), various measures of central tendency may fail to coincide (Reynolds and Casterlin, 1976a; Reynolds, 1977b). DeWitt and Friedman (1979) offer a hypothesis that accounts for negative skewness, but not for positive skewness as exemplified in Fig. 1. As in mammals, the response is blocked by antipyretics such as acetaminophen (Reynolds, 1977c).

ACUTE AND FINAL THERMAL PREFERENDA

In discussing temperature preference, it is useful to distinguish between acute or short-term thermal preferenda (Reynolds and Casterlin, 1979a) and the final thermal preferendum (Fry, 1947), which is defined as that temperature at which preference and acclimation are equal, and to which an animal will finally gravitate in a thermal gradient (Richards et al., 1977). While acute preferenda (measured within 2 hours or less in a gradient) are influenced by prior thermal experience or acclim(atiz)ation state (Fig. 2; Reynolds and Casterlin, 1979a), the final preferendum is not (Fry, 1947). Final preferenda are usually attained within 24 hours or less in a thermal gradient (Richards et al., 1977; Reynolds and Casterlin, 1979a). Fishes usually do not simply move to a single temperature and then remain there, however, but rather continue to shuttle between temperatures above

Fig. 2. Acute thermal parameters relative to acclimation temperature for brown bullhead (Ictalurus nebulosus), bluegill (Lepomis macrochirus), smallmouth bass (Micropterus dolomieui), and rainbow trout (Salmo gairdneri). E = line of equality where response temperature equals acclimation temperature; UL = upper incipient lethal temperature; LL = lower incipient lethal temperature; UA = upper avoidance temperature; LA = lower avoidance temperature; P = preferred temperature; CTM = critical thermal maximum. Incipient lethal temperatures delimit thermal tolerance polygons. Final preferenda are intersections of P and E lines. Rainbow trout avoidance and upper lethal temperatures are for juveniles. Brown bullhead data are from Brett (1956), Crawshaw (1975), and Richards and Ibara (1978). For sources of bluegill data, see Reynolds and Casterlin (1979a). Bass data are from Fry (unpublished), Horning and Pearson (1973), Barans and Tubb (1973), Cherry et al. (1975, 1977), Reynolds and Casterlin (1976a), and Stauffer et al. (1976). Trout data are from Garside and Tait (1958), Javaid and Anderson (1967), McCauley and Pond (1971), Cherry et al. (1975, 1977), McCauley and Huggins (1976, 1979), McCauley et al. (1977), Hokanson et al. (1977), and Kwain and McCauley (1978).

and below the final preferendum (Reynolds, 1978), so that the final preferendum is simply a statistical measure of central tendency, for example, the mean of a diel rhythm (Reynolds, Casterlin, Matthey, Millington and Ostrowski, 1978) of preferred temperature.

ACCLIMATION INFLUENCES

Various acute physiological and behavioral parameters respond differently to acclimation temperature (Fig. 2) for different fish species. Available data indicate that the upper (UL) and lower (LL) acute incipient lethal temperatures, as well as acute upper (UA) and lower (LA) avoidance temperatures, and acute thermal preferenda (P), of the brown bullhead (Ictalurus nebulosus) vary approximately linearly, but with differing positive slopes, with acclimation temperature (Fig. 2). Other species, such as the bluegill sunfish (Lepomis macrochirus) and the smallmouth bass (Micropterus

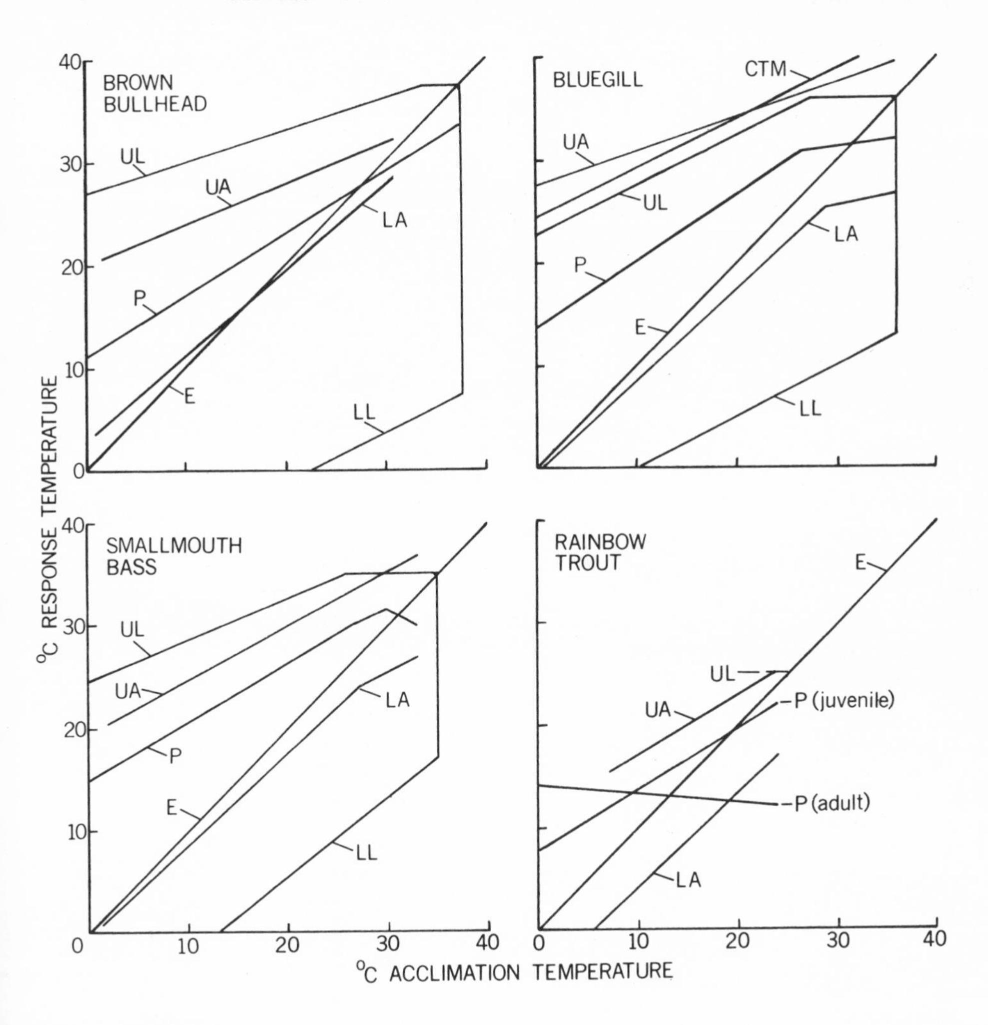

dolomieui) exhibit considerable nonlinearity in the region of the final preferendum (Fig. 2): the upper incipient lethal temperature reaches a horizontal plateau (the "ultimate upper incipient lethal temperature" - Fry, 1947), and the preference and lower avoidance temperatures exhibit varying degrees of slope inflection (cf. Zahn, 1962). This nonlinearity in the critical final preferendum region leads to the necessity for caution in extrapolating linearly from acute preferenda to obtain the final preferendum (Richards et al., 1977; Reynolds and Casterlin, 1979a).

The most extreme example is the rainbow trout Salmo gairdneri (Fig. 2), which exhibits a marked ontogenetic change not only in the final preferendum, but in the sign of the slope of the acute-preference line. While juveniles are typical in that acute preferenda increase with acclimation temperature, the adults are very unusual in that acute preferenda actually decrease with increasing acclimation temperature (McCauley and Huggins, 1979).

In most species and under most conditions, the upper and lower acute avoidance temperatures keep a fish well within its "zone of tolerance" (Fry, 1947; Reynolds and Casterlin, 1979a), bounded by the upper and lower incipient lethal temperatures for each acclimation temperature (Fig. 2). Some anomalies occur, however. For example, in the bluegill the upper avoidance (UA) temperature exceeds the upper incipient lethal (UL) temperature (Fig. 2). At higher acclimation temperatures, the UA falls between the UL and critical thermal maximum (CTM) lines, i.e., within the "zone of resistance" (Fry, 1947; Reynolds and Casterlin, 1979a). Within the resistance zone, survival is limited by a combination of time and temperature; brief incursions can be made without ill effect (cf. Reynolds and Thomson, 1974a). However, at lower acclimation temperatures, the UA of bluegills exceeds the CTM and so falls outside the resistance zone, with lethal result. This phenomenon is referred to as "low thermal responsiveness" (Meldrim and Gift, 1971; Beitinger and Magnuson, 1975; Reynolds and Casterlin, 1979a), or LTR.

ACCLIMATIZATION INFLUENCES

The chief value of the final preferendum (Fry, 1947) is that it yields a single species-specific preferred temperature value which is independent of acclimation temperature (Reynolds and Casterlin, 1979a). However, it does not take into account various non-thermal influences on preferred temperature which might collectively be referred to as "acclimatization influences" (Reynolds and Casterlin, 1979a). These include photoperiod or other seasonal factors, reproductive status, ontogenetic stage, nutritional status, health (pyrogenic effects of pathogens, etc.), neuropharmacological agents (pesticides, heavy metals, antipyretics, neurotransmitters, hormones, prostaglandins, etc.), biotic interactions (predators, competitors, prey, dominance hierarchies, territoriality, etc.), and temporal variations (circadian or other rhythms, etc.) (Reynolds, 1977b; Reynolds and Casterlin, 1979a, 1980).

Thermal tolerances and optima are also affected by acclimatization factors (Alderdice, 1972). Interactive variables include salinity, dissolved oxygen, nutritional factors, temporal factors, stresses (biotic or abiotic), and ontogenetic stage. Tolerances and optima of gametes, eggs, larvae, juveniles and adults may differ (cf. Reynolds and Thomson, 1974b).

TEMPORAL CHANGES IN TOLERANCE AND PREFERENCE

Various temporal changes have been found to occur in the thermal physiology and behavior of fishes. Ontogenetic changes occur in thermal tolerance and preference. For example, Reynolds and Thomson (1974b) found that larval Gulf grunion (*Leuresthes sardina*) are more eurythermal than are conspecific juveniles. The same is true for California grunion, *L. tenuis* (Reynolds, Thomson and Casterlin, 1976).

Similarly, ontogenetic changes occur in thermoregulatory behavior. Larval *L. sardina* do not thermoregulate before 3 weeks post-hatching (Reynolds and Thomson, 1974a). Juvenile yellow bullheads (*Ictalurus nebulosus*) prefer a mean temperature 1.2°C higher than do conspecific adults (Reynolds and Casterlin, 1978d). Rainbow trout (*Salmo gairdneri*) exhibit a similar change of even greater magnitude (Fig. 2; McCauley and Huggins, 1979). Ferguson (1958) suggested that such a change might be a general phenomenon in fishes, but Beitinger (personal communication) noted no such change in the bluegill.

McCauley and Huggins (1979) recently reviewed the evidence for non-thermally mediated seasonal changes in preferred temperatures of fishes. Such effects, reported by Barrans and Tubb (1973) and by Reutter and Herdendorf (1974) among others, might be mediated by photoperiod. There is scanty evidence as yet for endogenous cirannual rhythms in fishes.

Recent work has revealed the existence of diel or circadian thermoregulatory rhythms in some fish species. Diel rhythms of preferred temperature have been found in *Cyprinodon macularius* (Barlow, 1958; Lowe and Heath, 1969), *Oncorhynchus nerka* (Brett, 1971), *Carassius auratus* (Reynolds, Casterlin, Matthey, Millington and Ostrowski, 1978; Reynolds, 1977d), *Amia calva* (Reynolds, Casterlin and Millington, 1978), *Micropterus salmoides* and *M. dolomieui* (Reynolds and Casterlin, 1976a, 1978e; Reynolds, 1977e), *Esox masquinongy* (Reynolds and Casterlin, 1979b), *Salmo trutta* (Reynolds and Casterlin, 1979c), *Perca flavescens* (Reynolds and Casterlin, 1979f), *Zebrasoma flavescens* and *Forcipiger longirostris* (Reynolds and Casterlin, in prep.), and *Enneacanthus gloriosus* (Fig. 3). Winkler (1979) reported that *Gambusia affinis* thermoregulates only during daylight hours.

These thermoregulatory rhythms do not necessarily correspond with diel activity patterns, which may be out of phase or bimodal (Fig. 3). The functions of thermoregulatory rhythms are not fully understood. Barlow (1958) and Lowe and Heath (1969) suggest a relationship to diel changes in the thermal environment, while Brett (1971) presents evidence for energetic

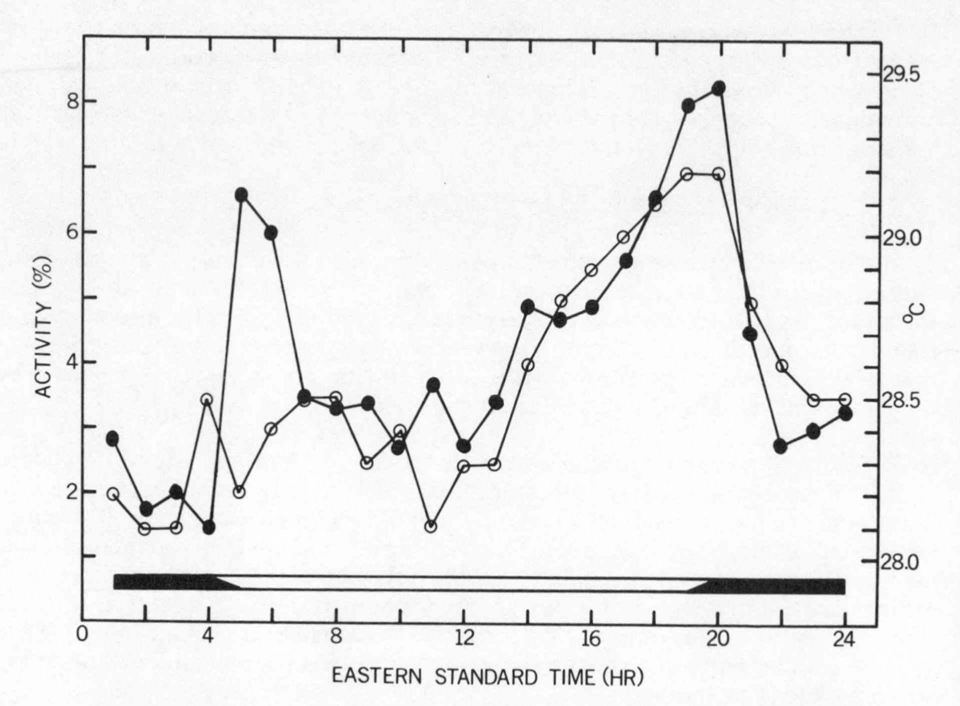

Fig. 3. Mean hourly locomotor activity (quantified as photocell-monitored light-beam interruptions per hour; solid symbols) expressed as a percentage of total 24-hour activity, and mean hourly preferred temperatures (open symbols) of bluespotted sunfish (Enneacanthus gloriosus), relative to time of day and photoperiod. Unshaded portion of horizontal bar at bottom indicates day, shaded portion night, with tapered crespuscular transitions at dawn and dusk. Previously unpublished figure from data of Casterlin and Reynolds (1979b, 1980).

advantages and a relationship to the feeding cycle. Different thermoregulatory rhythms in sympatric congeners may serve a niche differentiation function (Reynolds and Casterlin, 1978e). The thermoregulatory rhythm of Carassius auratus (Reynolds, Casterlin, Matthey, Millington and Ostrowski, 1978) has been shown to maximize growth and reproductive capacity (Spieler et al., 1977a; Reynolds, 1977d). Behaviorally cycled temperatures may also maximize thermal resistance and tolerance (Heath, 1963; Bacon et al., 1967; Feldmeth et al., 1974; Otto, 1974; Reynolds, 1977d; Hutchison and Maness, 1979). Rhythms of thermal tolerance exist in at least some species (Spieler et al., 1977b), possibly related to hormonal rhythms. Ward (1976) suggests that very constant thermal environments are unusually rigorous for organisms adapted to varying temperatures. Because of these cyclic effects, preferenda or tolerances in such rhythmic species must either be reported as 24-hour averages, or testing must be conducted at a standardized time relative to the photoperiod. Some fish species, of course, do not exhibit diel rhythms of preferred temperature - e.g., the bluegill L. macrochirus (Reynolds and Casterlin, 1976a) - although they may exhibit rhythms of locomotor activity (Reynolds and Casterlin, 1976b).

GENETIC ADAPTATIONS TO DIFFERING THERMAL ENVIRONMENTS

Species populations inhabiting different thermal environments often exhibit differences in thermal physiology and behavior which presumably have a genetic basis influenced by natural selection. For example, Johnson (1977) found evidence of heterozygosity and allozyme polymorphism correlated with temporal variability of the thermal environment of crested blennies (Anoplarchus purpurescens) in the Puget Sound area.

Some studies have indicated the existence of interpopulational differences in thermal resistance (CTM) within a species (Holland et al., 1974), while other data (Brown and Feldmeth, 1971) show that thermal tolerance can be remarkably conservative phylogenetically among closely related species inhabiting very different thermal environments. Most species show some ability to acclimat(iz)e to different thermal regimes. A way of expressing both acclimatory ability and the range of tolerance for a given acclimation temperature on a single graph is by means of a thermal tolerance polygon (Fig. 2), the area of which yields a measure of eurythermality (Brett, 1956; Reynolds and Thomson, 1974b) useful for interspecific comparisons.

Winkler (1979) found that populations of Gambusia affinis affinis from very different thermal environments exhibit identical final thermal preferenda. Reynolds and Casterlin (1976a) compared final preferendum data from geographically diverse populations of bluegills and other centrarchids, and found remarkably good agreement, indicating little or no geographic variation (cf. also Reynolds and Casterlin, 1979a). Hall et al. (1978), however, reported different thermal preferenda among different geographic populations of Morone americana. Two closely related grunion species (L. sardina and L. tenuis) inhabiting different thermal environments differ by 7°C in final thermal preferenda (Reynolds and Thomson, 1974b; Reynolds, Thomson and Casterlin, 1976). Thermal responses differ greatly between San Francisco Bay and Gulf of California populations of the extremely eurythermal goby Gillichthys mirabilis (DeVlaming, 1971; Winkler, 1979). Little experimental work has been done on the behavioral genetics of thermal preference (Goddard and Tait, 1976). This is a fruitful area for further research.

INTERACTIONS OF TEMPERATURE, THERMOREGULATION, ACTIVITY AND METABOLISM

The fabric of causes and effects regarding temperature, activity and metabolism is intricately woven (Reynolds and Casterlin, 1979d). Metabolic rate is influenced by temperature and by locomotor activity. Locomotor activity is also influenced by temperature. Metabolic heat production is influenced by activity, but so is heat loss to the environment. Directed locomotor activity in the form of behavioral thermoregulation influences body temperature. Therefore, temperature affects, and is affected by, activity and metabolism. Furthermore, thermal acclimation can affect, and be affected by, thermoregulatory behavior.

Fry (1964) reviewed earlier attempts to determine causality between temperature and locomotor activity, noting that differing results were dependent upon whether temperature was constant or changing, and upon whether the test animal was acclimated to the test temperature. Peterson and Anderson (1969) found that, in Salmo salar, an acute increase or decrease in temperature stimulated locomotor activity. They reported that the rate of temperature change was significant, and that there was typically an overshoot in acutely measured activity following a temperature change, followed by a stabilized phase as the fish became reacclimated. A plot of activity during the stabilized phase versus temperature revealed a plateau or dip in the general region of the final preferendum (preferendum data from Coutant, 1977). Similar activity and metabolic minima have been found for other species as well (Power and Todd, 1976; Bartholomew, 1977; Reynolds and Casterlin, 1979d, 1979e).

The bluespotted sunfish Enneacanthus gloriosus exhibits a crepuscular pattern of locomotor activity (Fig. 3), with a major activity peak at dusk and a smaller one at dawn (data from Casterlin and Reynolds, 1980). The 24-hour mean preferred temperature of this species is 28.7°C (Casterlin and Reynolds, 1979b), with little difference between diurnal (29.0°C) and nocturnal (28.4°C) means (Fig. 1). A diel rhythm of preferred temperature is evident, however (Fig. 3), with a peak at dusk coinciding with the major activity peak. During the dawn activity peak, however, the preferred temperature is quite low (Fig. 3). A plot of mean hourly activity and preferred temperature values (Fig. 4) reveals the presence of an anomalous region near the final preferendum, including a minimum or "activity well" (Reynolds and Casterlin, 1979d). A similar pattern occurs in the goldfish and in the bluegill (Reynolds and Casterlin, 1979d), in the lobster Homarus americanus (Reynolds and Casterlin, 1979e), and in the yellow perch Perca flavescens (Reynolds and Casterlin, 1979f). This seems to be a general phenomenon, suggesting that the unusual relationship between temperature, activity and metabolism in the final preferendum region may provide an explanation for the adaptive value of the final preferendum as a thermal region in which metabolic rate and activity are little affected by increasing temperature (cf. Fry, 1958b), conferring a degree of homeostasis greater than that which would otherwise obtain.

Neill (1979) has presented a computer simulation model for fish thermal preference based on a thermokinetic model. However, the activity of bluegills, for example, differs by a factor of 2 or more between night and day at the same temperature (Reynolds and Casterlin, 1976a, 1976b), so it is obvious that activity of fishes is not mechanistically controlled by

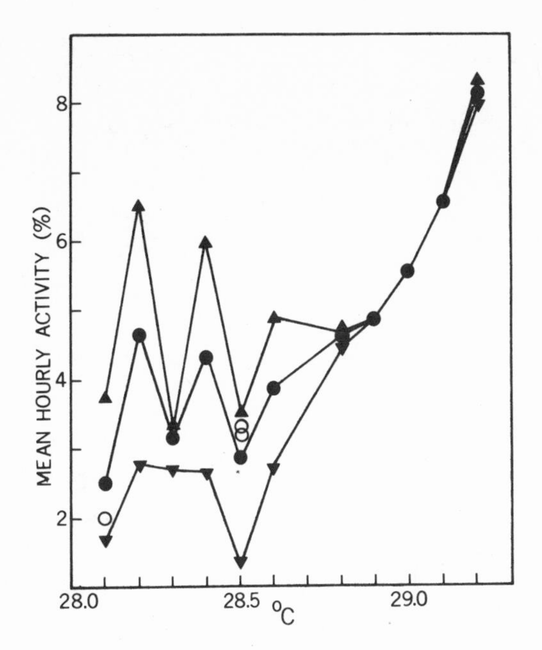

Fig. 4. Mean hourly activity (as in Fig. 3) plotted against mean hourly preferred temperatures of bluespotted sunfish, E. gloriosus. Upright triangles indicate maximum activity values, solid circles are mean values, inverted triangles are minimum values, open circles are additional values for each temperature.

temperature as has been suggested by Ellgaard et al. (1975). Rozin and Mayer (1961) demonstrated that fish will thermoregulate by means of a completely arbitrary conditioned response (see also Frank, 1971). Sullivan (as reported by Fry, 1964, p. 720) pointed out that thermoregulatory behavior of fish cannot be properly termed "thermokinesis" sensu Fraenkel and Gunn (1961), since "the thermal experiences of the fish from moment to moment may be continuously modifying its response in the gradient, a situation by no means within the scope of their definition." This would seem to rule out any simple thermokinetic argument to explain fish thermoregulatory behavior. Indeed, a considerable body of work (e.g., Hammel et al., 1969, 1973; Crawshaw and Hammel, 1973, 1974; Reynolds, Casterlin and Covert, 1976, 1978; Green and Lomax, 1976; Reynolds and Casterlin, 1980) suggests that the central nervous system mechanisms controlling thermoregulation are similar and homologous among vertebrate classes, including fishes.

REFERENCES

Alderdice, D.F. (1972). Responses of marine poikilotherms to environmental factors acting in concert. In: Marine Ecology, Vol. 1. Ed. O. Kinne. New York, Wiley-Interscience, p. 1659-1722.

Bacon, E.J. Jr., Neill, W.H. Jr. and Kilambi, R.V. (1967). Temperature selection and heat resistance of the mosquitofish, Gambusia affinis. Proc. Ann. Conf. Southeast. Assoc. Game Fish Comm. 21: 411-416.

Barans, B.A. and Tubb, R.A. (1973). Temperatures selected seasonally by four fishes from western Lake Erie. J. Fish. Res. Board Can. 30: 1697-1703.

Barlow, G.W. (1958). Daily movements of the desert pupfish Cyprinodon macularius in shore pools of the Salton Sea, California. Ecology 39: 580-587.

Bartholomew, G.A. (1977). Body temperature and energy metabolism. In: Animal Physiology, Principles and Adaptations. Ed. M.S. Gordon. New York, MacMillan, p. 364-449.

Beitinger, T.L. and Magnuson, J.J. (1976). Low thermal responsiveness in the bluegill, Lepomis macrochirus. J. Fish. Res. Board Can. 33: 293-295.

Brett, J.R. (1956). Some principles in the thermal requirements of fishes. Quart Rev. Biol. 31: 75-87.

Brett, J.R. (1971). Energetic responses of salmon to temperature. A study of some thermal relations in the physiology and freshwater ecology of sockeye salmon (Oncorhynchus nerka). Am. Zool. 11: 99-113.

Brown, J.H. and Feldmeth, C.R. (1971). Evolution in constant and fluctuating environments: thermal tolerance of the desert pupfish (Cyprinodon). Evolution 25: 390-398.

Casterlin, M.E. and Reynolds, W.W. (1977). Behavioral fever in anuran amphibian larvae. Life Sci. 20: 593-596.

Casterlin, M.E. and Reynolds, W.W. (1979a). Shark thermoregulation. Comp. Biochem. Physiol., 64A: 451-453.

Casterlin, M.E. and Reynolds, W.W. (1979b). Thermoregulatory behavior of the bluespotted sunfish, Enneacanthus gloriosus. Hydrobiologia 64:3- 4.

Casterlin, M.E. and Reynolds, W.W. (1980). Diel activity of the bluespotted sunfish, Enneacanthus gloriosus. Copeia 1980: 344-345.

Cherry, D.S., Dickson, K.L. and Cairns, J. Jr. (1975). Temperatures selected and avoided by fish at various acclimation temperatures. J. Fish. Res. Board Can. 32: 485-491.

Cherry, D.S., Dickson, K.L. and Cairns, J. Jr. (1977). Preferred, avoided and lethal temperatures of fish during rising temperature conditions. J. Fish. Res. Board Can. 34: 239-246.

Cossins, A.R. and Prosser, C.L. (1978). Evolutionary adaptation of membranes to temperature. Proc. Natl. Acad. Sci. U.S.A. 75: 2040-2043.

Coutant, C.C. (1977). Compilation of temperature preference data. J. Fish. Res. Board Can. 34: 739-745.

Covert, J.B. and Reynolds, W.W. (1977). Survival value of fever in fish. Nature (Lond.) 267: 43-45.

Crawshaw, L.I. (1975). Attainment of the final thermal preferendum in brown bullheads acclimated to different temperatures. Comp. Biochem. Physiol. 52A: 171-173.

Crawshaw, L.I. (1976). Effect of rapid temperature change on mean body temperature and gill ventilation in carp. Am. J. Physiol. 231: 837-841.

Crawshaw, L.I. (1977). Physiological and behavioral reactions of fishes to temperature change. J. Fish. Res. Board Can. 34: 730-734.

Crawshaw, L.I. (1979). Responses to rapid temperature change in lower vertebrate ectotherms. Am. Zool. 19: 225-237.

Crawshaw, L.I. and Hammel, H.T. (1973). Behavioral temperature regulation in the California horn shark, Heterodontus francisci. Brain Behav. Evol. 7: 447-452.

Crawshaw, L.I. and Hammel, H.T. (1974). Behavioral regulation of internal temperature in the brown bullhead, Ictalurus nebulosus. Comp. Biochem. Physiol. 47A: 51-60.

Dean, J.M. (1976). Temperatures of tissues in freshwater fishes. Trans. Am. Fish. Soc. 105: 709-711.

DeVlaming, V.L. (1971). Thermal selection behavior in the estuarine goby Gillichthys mirabilis Cooper. J. Fish. Biol. 3: 277-286.

DeWitt, C.B. and Friedman, R.M. (1979). Significance of skewness in ectotherm thermoregulation. Am. Zool. 19: 195-209.

Dizon, A.E. and Brill, R.W. (1979). Thermoregulation in tunas. Am. Zool. 19: 249-265.

Ellgaard, E.G., Bloom, K.S., Malizia, A.A. Jr. and Gunning, G.E. (1975). The locomotor activity of fish: an anology to the kinetics of an opposed first-order chemical reaction. Trans. Am. Fish. Soc. 104: 752-754.

Erskine, D.J. and Spotila, J.R. (1977). Heat-energy-budget analysis and heat transfer in the largemouth blackbass (Micropterus salmoides). Physiol. Zool. 50: 157-169.

Feldmeth, C.R., Stone, E.A. and Brown, J.H. (1974). An increased scope for thermal tolerance upon acclimating pupfish (Cyprinodon) to cycling temperatures. J. Comp. Physiol. 89: 39-44.

Ferguson, R.G. (1958). The preferred temperatures of fish and their midsummer distribution in temperate lakes and streams. J. Fish. Res. Board Can. 15: 607-624.

Fraenkel, G.S. and Gunn, D.L. (1961). The Orientation of Animals. New York, Dover. 376 p.

Frank, L.H. (1971). A technique for measuring thermoregulatory behavior in the fish. Behav. Res. Meth. Instrum. 3: 250.

Fry, F.E.J. (1947). Effects of the environment on animal activity. Univ. Toronto Stud. Biol. Ser. 55, Publ. Ontario Fish. Res. Lab. 68: 1-62.

Fry, F.E.J. (1958a). The experimental study of behavior in fish. Proc. Indo-Pac. Fish. Counc. 3: 37-42.

Fry, F.E.J. (1958b). Temperature compensation. Ann. Rev. Physiol. 20: 207-224.

Fry, F.E.J. (1964). Animals in aquatic environments: fishes. In: Handbook of physiology. Ed. D.B. Dill, E.F. Adolph and G.C. Wilber. Washington, D.C., Am. Physiol. Soc., p. 715-728.

Garside, E.T. and Tait, J.S. (1958). Preferred temperature of rainbow trout and its unusual relationship to acclimation temperature. Can. J. Zool. 36: 563-567.

Goddard, C.I. and Tait, J.S. (1976). Preferred temperatures of F_3 to F_5 hybrids of Salvelinus fontinalis x S. namaycush. J. Fish. Res. Board Can. 33: 197-202.

Green, M.D. and Lomax, P. (1976). Behavioural thermoregulation and neuroamines in fish (Chromus chromus). J. Thermal Biol. 1: 237-240.

Hall, L.W. Jr., Hocutt, C.H. and Stauffer, J.R. Jr. (1978). Implication of geographic location on temperature preference of white perch, Morone americana. J. Fish. Res. Board Can. 35: 1464-1468.

Hammel, H.T., Crawshaw, L.I. and Cabanac, H.P. (1973). The activation of behavioral responses in the regulation of body temperature in vertebrates. In: The Pharmacology of Thermoregulation. Ed. Basel, Karger, p.

Hammel, H.T., Stromme, S.B. and Myhre, K. (1969). Forebrain temperature activates behavioral thermoregulatory response in arctic sculpins. Science 165: 83-85.

Heath, W.G. (1963). Thermoperiodism in sea-run cutthroat trout (Salmo clarki clarki). Science 142: 486-488.

Hokanson, K.E.F., Kleiner, C.F. and Thorslund, T.W. (1977). Effects of constant temperatures and diel temperature fluctuations on specific growth and mortality rates and yield of juvenile rainbow trout, Salmo gairdneri. J. Fish. Res. Board Can. 34: 639-648.

Holland, W.E., Smith, M.H., Gibbons, J.W. and Brown, D.H. (1974). Thermal tolerances of fish from a reservoir receiving effluent from a nuclear reactor. Physiol. Zool. 47: 110-118.

Horning, W.B. II and Pearson, R.E. (1973). Growth temperature requirements and lower lethal temperatures for juvenile smallmouth bass (Micropterus dolomieui). J. Fish. Res. Board Can. 30: 1226-1230.

Hulbert, A.J. (1978). The thyroid hormones; a thesis concerning their action. J. Theor. Biol. 73: 81-100.

Hutchison, V.H. and Maness, J.D. (1979). The role of behavior in temperature acclimation and tolerance in ectotherms. Am. Zool. 19: 367-384.

Javaid, M.Y. and Anderson, J.M. (1967). Thermal acclimation and temperature selection in Atlantic salmon, Salmo salar, and rainbow trout, S. gairdneri. J. Fish. Res. Board Can. 24: 1507-1513.

Johnson, M.S. (1977). Association of allozymes and temperature in the crested blenny Anoplarchus purpurescens. Marine Biol. 41: 147-152.

Kluger, M.J. (1979). Fever in ectotherms: evolutionary implications. Am. Zool. 19: 295-304.

Kuz'mina, V.V. and Morozova, E.N. (1977). Effects of temperature on - amylase activity of freshwater bony fishes. Vopr. Ikhtiol. 17: 922-929.

Kwain, W. and McCauley, R.W. (1978). Effects of age and overhead illumination on temperatures preferred by underyearling rainbow trout, Salmo gairdneri, in a vertical temperature gradient. J. Fish. Res. Board Can. 35: 1430-1433.

Lemons, D.E. and Crawshaw, L.I. (1978). Temperature regulation in the Pacific lamprey (Lampetra tridentata). Fed. Proc. 37: 929.

Lowe, C.H. and Heath, W.G. (1969). Behavioral and physiological responses to temperature in the desert pupfish (Cyprinodon macularius). Physiol. Zool. 42: 53-59.

Matthews, W.J. and Hill, L.G. (1979). Influence of physico-chemical factors on habitat selection by red shiners, Notropis lutrensis (Pisces: Cyprinidae). Copeia 1979: 70-81.

McCauley, R.W., Elliott, J.R. and Read, L.A.A. (1977). Influence of acclimation temperature on preferred temperature in the rainbow trout Salmo gairdneri. Trans. Am. Fish. Soc. 106: 362-365.

McCauley, R.W. and Huggins, N. (1976). Behavioral thermal regulation by rainbow trout in a temperature gradient. In: Thermal Ecology II. Ed. G.W. Esch and R.W. McFarlane. Springfield, Va., Technical Information Service, p. 171-175.

McCauley, R.W. and Huggins, N. (1979). Ontogenetic and non-thermal seasonal effects on thermal preferenda of fish. Am. Zool. 19: 267-271.

McCauley, R.W. and Pond, W.L. (1971). Temperature selection of rainbow trout fingerlings in vertical and horizontal gradients. J. Fish. Res. Board Can. 28: 1801-1804.

McCauley, R.W., Reynolds, W.W. and Huggins, N. (1977). Photokinesis and behavioral thermoregulation in adult sea lampreys (Petromyzon marinus). J. Exp. Zool. 202: 431-437.

Meldrim, J.W. and Gift, J.J. (1971). Temperature preference, avoidance and shock experiments with estuarine fishes. Ichthyol. Assoc. Bull. 7: 1-76.

Mueller, R. (1976). Investigations on the body temperature of freshwater fishes. Arch. Fischereiwiss. 27: 1-28.

Neill, W.H. (1979). Mechanisms of fish distribution in heterothermal environments. Am. Zool. 19: 305-317.

Niemierko, S., Kramska, J.S., Mleczko, M. and Suszczewski, S.R. (1977). The effect of the assay temperature on brain acetylcholinesterase activity of two Antarctic fish species. Bull. Acad. Pol. Sci. Biol. 25: 821-826.

Otto, R.G. (1974). The effects of acclimation to cyclic thermal regimes on heat tolerance of the western mosquitofish. Trans. Am. Fish. Soc. 103: 331-335.

Peterson, R.H. and Anderson, J.M. (1969). Influence of temperature change on spontaneous locomotor activity and oxygen consumption of Atlantic salmon, Salmo salar, acclimated to two temperatures. J. Fish. Res. Board Can. 26: 93-109.

Power, M.E. and Todd, J.H. (1976). Effects of increasing temperature on social behavior in territorial groups of pumpkinseed sunfish, Lepomis gibbosus. Environm. Pollut. 10: 217-223.

Rahmann, H. (1978). Gangliosides and thermal adaptation in vertebrates. Jap. J. Exp. Med. 48: 85-96.

Reutter, J.M. and Herdendorf, C.E. (1974). Laboratory estimates of the seasonal final temperature preferenda of some Lake Erie fish. Proc. Conf. Great Lakes Res. 17: 59-67.

Reynolds, W.W. (1975). Laboratory Manual for Man, Nature and Society. Dubuque, Iowa, Wm. C. Brown Co., 253 p.

Reynolds, W.W. (1977a). Thermal equilibration rates in relation to heartbeat and ventilatory frequencies in largemouth blackbass, Micropterus salmoides. Comp. Biochem. Physiol. 56A: 195-201.

Reynolds, W.W. (1977b). Temperature as a proximate factor in orientation behavior. J. Fish. Res. Board Can. 34: 734-739.

Reynolds, W.W. (1977c). Fever and antipyresis in the bluegill sunfish, Lepomis macrochirus. Comp. Biochem. Physiol. 57C: 165-167.

Reynolds, W.W. (1977d). Circadian rhythms in the goldfish Carassius auratus L.: preliminary observations and possible implications. Rev. Can. Biol. 36: 355-356.

Reynolds, W.W. (1977e). Fish orientation behavior: an electronic device for studying simultaneous responses to two variables. J. Fish. Res. Board Can. 34: 300-304.

Reynolds, W.W. (1978). The final thermal preferendum of fishes: shuttling behavior and acclimation overshoot. Hydrobiologia 57: 123-124.

Reynolds, W.W. (1979). Perspective and introduction to the symposium: Thermoregulation in ectotherms. Amer. Zool. 19: 193-194.

Reynolds, W.W. and Casterlin, M.E. (1976a). Thermal preferenda and behavioral thermoregulation in three centrarchid fishes. In: Thermal Ecology II. Ed. G.W. Esch and R.W. McFarlane. Sprinfield, Va., Technical Information Service, p. 185-190.

Reynolds, W.W. and Casterlin, M.E. (1976b). Locomotor activity rhythms in the bluegill sunfish, Lepomis macrochirus. Am. Midl. Nat. 96: 221-225.

Reynolds, W.W. and Casterlin, M.E. (1978a). Estimation of cardiac output and stroke volume from thermal equilibration and heartbeat rates in fish. Hydrobiologia 57: 49-52.

Reynolds, W.W. and Casterlin, M.E. (1978b). Thermoregulatory behavior in the smooth dogfish shark, Mustelus canis. Fed. Proc. 37: 427.

Reynolds, W.W. and Casterlin, M.E. (1978c). Behavioral thermoregulation by ammocoete larvae of the sea lamprey (Petromyzon marinus) in an electronic shuttlebox. Hydrobiologia 61: 145-147.

Reynolds, W.W. and Casterlin, M.E. (1978d). Ontogenetic change in preferred temperature and diel activity of the yellow bullhead, Ictalurus natalis. Comp. Biochem. Physiol. 59A: 409-411.

Reynolds, W.W. and Casterlin, M.E. (1978e). Complementarity of thermoregulatory rhythms in Micropterus salmoides in M. dolomieui. Hydrobiologia 60: 263-264.

Reynolds, W.W. and Casterlin, M.E. (1979a). Behavioral thermoregulation and the "final preferendum" paradigm. Am. Zool. 19: 211-224.

Reynolds, W.W. and Casterlin, M.E. (1979b). Thermoregulatory rhythm in juvenile muskellunge (Esox masquinongy): evidence of a diel shift in the lower set-point. Comp. Biochem. Physiol. 63A: 523-525.

Reynolds, W.W. and Casterlin, M.E. (1979c). Thermoregulatory behavior of brown trout, Salmo trutta. Hydrobiologia 62: 79-80.

Reynolds, W.W. and Casterlin, M.E. (1979d). Effect of temperature on locomotor activity in the goldfish (Carassius auratus) and the bluegill (Lepomis macrochirus): presence of an "activity well" in the region of the final preferendum. Hydrobiologia 63: 3-5.

Reynolds, W.W. and Casterlin, M.E. (1979e). Behavioral thermoregulation and locomotor activity in the lobster Homarus americanus. Comp. Biochem. Physiol. 64A: 25-28.

Reynolds, W.W. and Casterlin, M.E. (1979f). Behavioral thermoregulation and locomotor activity of Perca flavescens. Can. J. Zool. 57: 2239-2242.

Reynolds, W.W. and Casterlin, M.E. (1980). The pyrogenic responses of non-mammalian vertebrates. In: Handbook of Experimental Pharmacology. Pyretics and Antipyretics. Ed. A.S. Milton. Berlin, Springer-Verlag, Chapter XIX.

Reynolds, W.W. and Casterlin, M.E. (in press). Thermoregulatory behavior of a tropical reef fish: Zebrasoma flavescens. Oikos 34: in press.

Reynolds, W.W., Casterlin, M.E. and Covert, J.B. (1976). Behavioural fever in teleost fishes. Nature (Lond.) 259: 41-42.

Reynolds, W.W., Casterlin, M.E. and Covert, J.B. (1978). Febrile responses of bluegill (Lepomis macrochirus) to bacterial pyrogens. J. Thermal Biol. 3: 129-130.

Reynolds, W.W., Casterlin, M.E., Matthey, J.K., Millington, S.T. and Ostrowski, A.C. (1978). Diel patterns of preferred temperature and locomotor activity in the goldfish Carassius auratus. Comp. Biochem. Physiol. 59A: 225-227.

Reynolds, W.W., Casterlin, M.E. and Millington, S.T. (1978). Circadian rhythm of preferred temperature in the bowfin Amia calva, a primitive holostean fish. Comp. Biochem. Physiol. 60A: 107-109.

Reynolds, W.W., Covert, J.B. and Casterlin, M.E. (1978). Febrile responses of goldfish (Carassius auratus) to Aeromonas hydrophila and to Escherichia coli endotoxin. J. Fish Diseases 1: 271-273.

Reynolds, W.W., McCauley, R.W., Casterlin, M.E. and Crawshaw, L.I. (1976). Body temperatures of behaviorally thermoregulating largemouth blackbass (Micropterus salmoides). Comp. Biochem. Physiol. 54A: 461-463.

Reynolds, W.W. and Thomson, D.A. (1974a). Responses of young Gulf grunion, Leuresthes sardina, to gradients of temperature, light, turbulence and oxygen. Copeia 1974: 747-758.

Reynolds, W.W. and Thomson, D.A. (1974b). Temperature and salinity tolerances of young Gulf grunion, Leuresthes sardina (Atheriniformes: Atherinidae). J. Marine Res. 32: 37-45.

Reynolds, W.W., Thomson, D.A. and Casterlin, M.E. (1976). Temperature and salinity tolerances of larval Californian grunion, Leuresthes tenuis (Ayres): a comparison with Gulf grunion, L. sardina (Jenkins & Evermann). J. Exp. Mar. Biol. Ecol. 24: 73-82.

Reynolds, W.W., Thomson, D.A. and Casterlin, M.E. (1977). Responses of young California grunion, Leuresthes tenuis, to gradients of temperature and light. Copeia 1977: 144-149.

Richards, F.P. and Ibara, R.M. (1978). The preferred temperatures of the brown bullhead, Ictalurus nebulosus, with reference to its orientation to the discharge canal of a nuclear power plant. Trans. Am. Fish. Soc. 107: 288-294.

Richards, F.P., Reynolds, W.W., McCauley, R.W., Crawshaw, L.I., Coutant, C.C. and Gift, J.J. (1977). Temperature preference studies in environmental impact assessments: an overview with procedural recommendations. J. Fish. Res. Board Can. 34: 728-761.

Rigby, B.J. (1977). Thermal transitions in the collagenous tissues of poikilothermic animals. J. Thermal Biol. 2: 89-93.

Rozin, P.N. and Mayer, J. (1961). Thermal reinforcement and thermoregulatory behavior in the goldfish, Carassius auratus. Science 134: 942-943.

Shaklee, J.B., Christiansen, J.A., Sidell, B.D., Prosser, C.L. and Whitt, G.S. (1977). Molecular aspects of temperature acclimation in fish: contributions of changes in enzyme activities and isozyme patterns to metabolic reorganization in the green sunfish. J. Exp. Zool. 201: 1-20.

Simpson, G.G., Roe, A. and Lewontin, R.C. (1960). Quantitative Zoology. New York, Harcourt Brace and World, 440 p.

Somero, G.N. (1978). Temperature adaptation of enzymes: biological optimization through structure-function compromises. Ann. Rev. Ecol. Syst. 9: 1-29.

Spieler, R.E., Noeske, T.A., DeVlaming, V. and Meier, A.H. (1977a). Effects of thermocycles on body weight gain and gonadal growth in the goldfish, Carassius auratus. Trans. Am. Fish. Soc. 106: 440-444.

Spieler, R.E., Noeske, T.A. and Seegert, G.L. (1977b). Diel variations in sensitivity of fishes to potentially lethal stimuli. Progr. Fish-Cult. 39: 144-147.

Stauffer, J.R. Jr., Dickson, K.L., Cairns, J. Jr. and Cherry, D.S. (1976). The potential and realized influences of temperature on the distribution of fishes in the New River, Glen Lyn, Virginia, Wildl. Monogr. 50: 1-40.

Stevens, E.D. and Fry, F.E.J. (1970). The rate of thermal exchange in a teleost, Tilapia mossambica. Can. J. Zool. 48: 221-226.

Stevens, E.D. and Fry, F.E.J. (1974). Heat transfer and body temperatures in non-thermoregulatory teleosts. Can. J. Zool. 52: 1137-1145.

Vitvitskii, V.N. (1977). A comparative analysis of heat stability and electrophoretic migration of muscle proteins of fishes living at different depths. Ekologiya 6: 88-92.

Ward, J.V. (1976). Effects of thermal constancy and seasonal temperature displacement on community structure of stream macroinvertebrates. In: Thermal Ecology II. Eds. G.W. Esch and R.W. McFarlane. Springfield, Va., Technical Information Service, p. 302-307.

Winkler, P. (1979). Thermal preference of Gambusia affinis affinis as determined under field and laboratory conditions. Copeia 1979: 60-64.

Withers, P.C. (1978). Acid-base regulation as a function of body temperature in ectothermic toads, a heliothermic lizard and a heterothermic mammal. J. Thermal. Biol. 3: 163-172.

Wodtke, E. (1978). Lipid adaptation in liver mitochondrial membranes of carp acclimated to different environmental temperatures: phospholipid composition, fatty acid pattern, and cholesterol content. Biochim. Biophys. Acta 529: 280-291.

Zahn, M. (1962). Die Vorzugstemperaturen zweier Cypriniden und eines Cyprinodonten und die Adaptationstypen der Vorzugstemperatur bei Fischen. Zool. Beitr., n.s., 7: 15-25.

EFFECTS OF TEMPERATURE ON THE MAXIMUM SWIMMING SPEED OF FISHES

C.S. Wardle

DAFS Marine Laboratory

PO Box 101, Aberdeen, UK

INTRODUCTION

Teleost fish inhabit a great variety of ecological niches throughout the world where temperature may range from below the freezing point of water to as high as 40°C. There are probably few or none of the 20,000 species that can survive even a slow change through this full range of temperature, but some species have evolved with quite wide temperature tolerance in order to survive in particular temperate or extreme climate habitats. The temperature of the body tissues of most teleost species is generally considered to equilibrate rapidly with any change in water temperature. Changes in the temperature of the tissues cause complicated changes in the rates of metabolic processes and enzymes and other components of the tissue biochemistry can become more or less active. Maximum swimming speed is severely reduced in cold water probably as a result of the effects of low temperature on the biochemical and physiological processes involved in muscle contraction. Whatever the cause the observation is that the twitch contraction of the anaerobic fast white swimming muscle of the fish becomes longer as the muscle is cooled. This paper defines the effect of changing the temperature of the swimming muscle on the muscle contraction time and examines the consequent effect on the maximum swimming speed. Some adaptation and compensation tactics that might help regain the speed lost in cold water are also examined.

THE RELEVANCE OF MAXIMUM SPEED SWIMMING

High speed swimming is always accompanied by short endurance in teleost fish. In towed fishing gears such as trawls and seine nets, all sizes of fish can be observed keeping station with the moving netting patterns swimming at the speed of the net (Hemmings, 1973). Small fish swimming in the net are seen to be moving with very rapid tail beats whereas large fish are making easy slow tail sweeps at the same speed. The small fish are moving at close to their maximum speed and swim for only a short period and drop back, the larger fish can swim for long periods and have scope for much greater speed. If the observing diver reaches out to grab at a large fish easily swimming in this position the startled fish dashes away and clears the area of the net. The significance to the fish of these fast bursts of swimming when made even for one tenth of a second can be the difference between life and death. Recent findings have shown that maximum swimming speed of most fish species are similar for the same fish length and surprisingly high for short periods. TV recordings have been made showing a 10cm fish swimming at 26 lengths/second, that is $2.6ms^{-1}$ at 12°C and showing a tail beat frequency of more than 30Hz. However, the larger the fish the slower the tail beat and the fewer lengths it can swim in one second. A 100cm cod can manage only 5-6 lengths per second at about 9Hz, however, this is $5\text{-}6ms^{-1}$ twice the speed of the 10cm fish (Wardle, 1975). Anyone who has watched or hunted fish underwater will be aware of the apparent speed with which they can dart away. Fish in their natural environment are often close to shelter. The shelter can have many different forms, coral reef fish dart into a hole in the reef or can gain protection among the tentacles of a tolerant sea anemone, the shelter is always within a metre or so and with a well tested escape route. In the open sea a school of identical individuals can be reached by a quick dash. Visibility through water is often short and a fish can disappear from view by dashing only a few metres. Other species disappear in a cloud of sand thrown up while they bury themselves in the sea bed. A pelagic flying fish passes through the surface and glides through the air above the water. High speed for brief periods is important for escape it is also essential for a hunting fish to place its mouth close to its preys. To gain high speed in a short time a very high acceleration rate is achieved by fish in standing starts (Webb, 1978). A few fish species like salmon use high speed bursts to swim up or leap otherwise impassible water falls, allowing the fish to migrate up a river to reach a spawning ground. Typically, brown trout shelter in quiet water below the fast moving water of the river and are able to dash rapidly to and from the surface, to pick up food particles drifting past, without loosing position in the river. Our conclusion must be that high speed swimming for short periods of time is important for the survival of teleost fish.

THE REASONS FOR LOW ENDURANCE AT HIGH SPEED

It is reasonable to expect that at the maximum swimming speed the contraction of the whole of the fast muscle in each myotome is required to generate sufficient force to overcome the drag of the water on the body (Wardle and Videler, 1980). The fish has no way of acquiring oxygen from the water at a sufficient rate to generate the energy required for maximum speed swimming. The fast muscle is anaerobic and contracts using energy derived from the fuel stored within the contractile cells as glycogen. The energy is released from the glycogen in the muscle cells by anaerobic pathways leading to the product lactic acid. This means that the muscle can only contract with its maximum force while there is sufficient of this limited store remaining in the muscle cells. It is well known that when a rested fish, that is a fish with a high level of glycogen in its white muscle, is made to swim fast the first bursts of swimming are the most rapid and each burst gets weaker. A behaviour change is associated with this exhaustion and the initial fast movements give way to a seek shelter type of behaviour when the fish is reluctant to swim (Black et al., 1961; Black, 1958; Wardle, 1978). Recovery of the swimming muscle can take up to 24 hours following its complete exhaustion. During the recovery period most of the lactic acid can remain in the fish muscle (Wardle, 1978) and may be converted back to glycogen without leaving the muscle cells (Batty and Wardle, 1979). As has been described in this volume the hydrogen ions associated with lactic acid production probably pass immediately from the exhausting muscle cells to the circulation (see chapters by Holeton and Heisler).

MUSCLE CONTRACTION TIME, LIMITS MAXIMUM SPEED

It is now well established that during constant speed swimming the swimming speed of fish is closely related to the tail beat frequency (Bainbridge, 1958; Hudson, 1973; Hunter and Zwiefel, 1971; Pyatetski, 1970). All these authors have found in a number of species that when swimming at constant speed the fish moves forwards one stride equal to about 0.7 lengths of the fish for each complete tail beat. Note that when slowing the stride can be much longer and when accelerating much shorter than 0.7L. In a paper examining the maximum swimming speed of fish, Wardle (1975) showed that the maximum tail beat or stride frequency was determined by the twitch contraction time of the muscle. The twitch contraction time of the lateral swimming muscle and thus the stride completion time was short for small fish and became much longer as the fish became larger. The muscle contraction time was also found to be sensitive to the temperature of the fish, cold muscle contracted slowly and the same muscle made warm contracted more rapidly. A general formula was established where the maximum speed of swimming equals the stride length divided by the time to complete the stride or $u\ max = 0.7L/2t$, where L = fish length and t is muscle contraction time. The stride length $0.7L$ holds good for a wide range of speeds in the tail beat frequency studies quoted above. The contraction time (t) is easily measured in a small isotonic frame designed to record the movement of the muscle when it is stimulated electrically (20 volts, .002s square pulse). Examples of the contraction of pieces of lateral swimming muscle from a sprat (7cm) and a cod (75cm) are shown in Figs. 1 and 2. These isotonic contraction records are used to measure the time from the electrical stimulation of the muscle to the time for the completion of shortening. Notice the application of different loads to the contracting muscle of the cod (Fig. 2) does not affect the time of shortening, addition of loads alter only the velocity of shortening, that is the distance moved (vertical scale Fig. 2) by the free end of the muscle (in this case attached to an optical wedge transducer) during the shortening time.

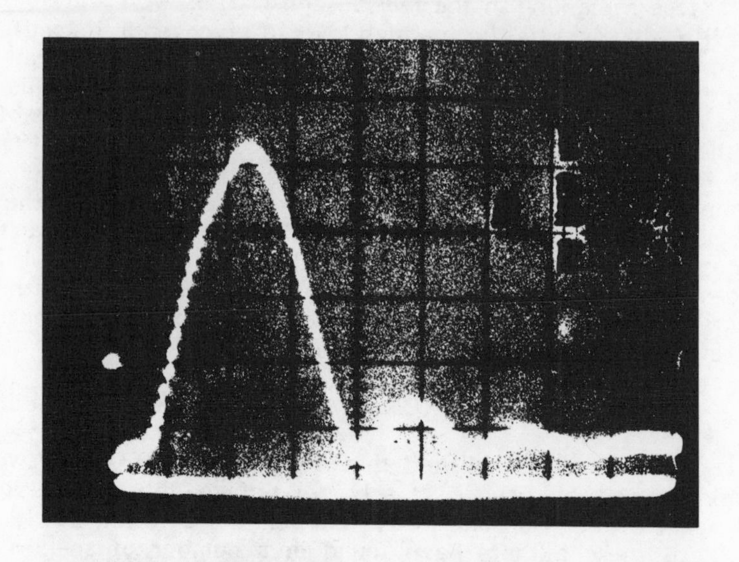

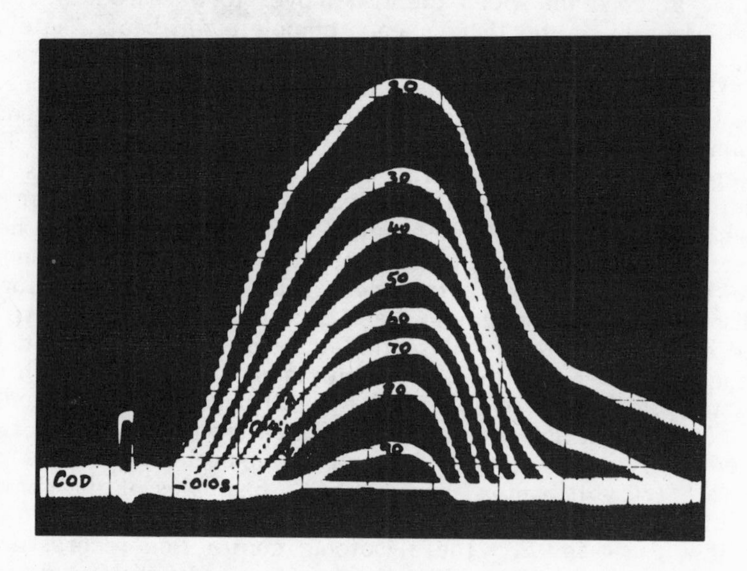

Figs. 1 & 2. The oscilloscope traces show records of twitch contraction times for a 7cm sprat (Fig. 1) and a 75 cm cod (Fig. 2). Horizontal scale time 0.01 s per div., vertical scale is distance moved. Both at 14°C. In Fig. 2 the muscle load was varied from 20 to 90g.

THE EFFECT OF TEMPERATURE ON CONTRACTION TIME

Preliminary experiments (Wardle, 1975) indicated that increasing the temperature of the swimming muscle decreases the contraction time by about 1ms per 1°C. A further study by Wardle and Mojsiewicz (in preparation) extended these experiments by examining in detail the effect of a range from lowest to highest viable temperatures. The procedure was to measure the contraction time of isolated pieces of anaerobic swimming muscle equilibrated for several minutes at the chosen temperature. The measurement was repeated with the same muscle sample at a number of temperatures in a series between 2°C and up to 18°C. Spot-checks were made by returning the muscle samples to a previous temperature and repeating the measurement. In this way a single muscle sample gave a repeatable series of contraction time measurements. Examples of the effect of temperature on contraction time of muscle sampled for four species of fish of about the same sizes are shown in Fig. 3. These results show that the four species had similar properties and that a change of 1°C had different effects on contraction time at different temperatures. At temperatures between freezing and 5°C the contraction time changed by as much as 8 milliseconds (ms) per °C (see Fig. 3 dashed line). Between 5 and 10°C this rate was halved to 4ms per 1°C and between 10 and 15°C to 2ms per 1°C. Measurements of the contraction time of muscle from slightly larger skipjack tuna by Brill and Dizon (1979) at three warmer temperatures indicate that the trend continues to shorter contraction times at a rate between 1 and 2ms per 1°C up to 30°C (Fig. 3). Measurements from muscle of larger cod are included in Fig. 3 to demonstrate the smaller but important effect of size on contraction time. It is significant that fish of quite different mode of life have such similar muscle properties. The plaice, *Pleuronectes platessa* L., and lemon sole, *Microstomus kitt* (Walbaum), are bottom living flat fish, the cod, *Gadus callarias* L., is a demersal preditor and mackerel, *Scomber scombrus* L., and skipjack tuna, *Katsuwonus pelamis* L., ocean cruising pelagic fish with very streamlined bodies.

TEMPERATURE AND MAXIMUM SWIMMING SPEED

We have seen how increased temperature shortens the contraction time of the swimming muscle (Fig. 3). The tail can beat and complete one stride in a time equal to twice this contraction time. The effect of temperature on stride frequency can be plotted for the same examples in Fig. 4. Each completed stride during steady swimming moves the fish forwards 0.7 fish lengths and the derived maximum swimming speed is plotted against temperature for the same examples in Fig. 5. The large cod shows a lower maximum stride frequency (Fig. 4) but a higher maximum swimming speed (Fig. 5) for further discussion on the effect of size on swimming speed see Wardle (1975, 1977). From the maximum speed predictions shown in Fig. 5 it is clear that if fish do remain at the same temperature as the water in which they swim their speed is severely reduced in cold water.

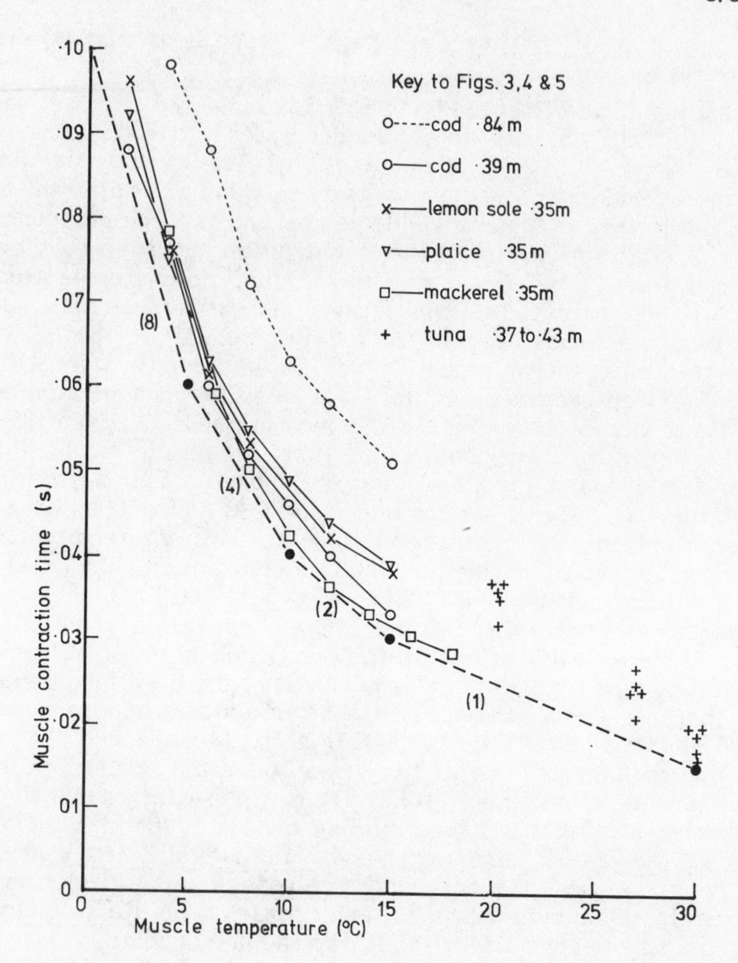

Fig. 3. The effect of temperature on the contraction time (t) of one sample of anaerobic swimming muscle from each of the fish of the size and species indicated. Skipjack tuna from Brill and Dizon (1979). A reference slope (filled circles and dashed line) is indicated by the figures in brackets in milliseconds per 1°C.

It is thought that with the continuous exposure of the fish's blood to the water in the gills, in order to take up oxygen, the body is obliged to remain at the water temperature. Recent studies have confirmed that small teleosts show body temperatures close to water temperature and that as white muscle is made to work it can show an excess temperature of up to 0.6°C above ambient in Pacific mackerel (Roberts and Graham, 1979). In five different teleost species, resting temperatures were between 0.1 and 1.1°C higher than the water temperature (Dean, 1976). In rainbow trout

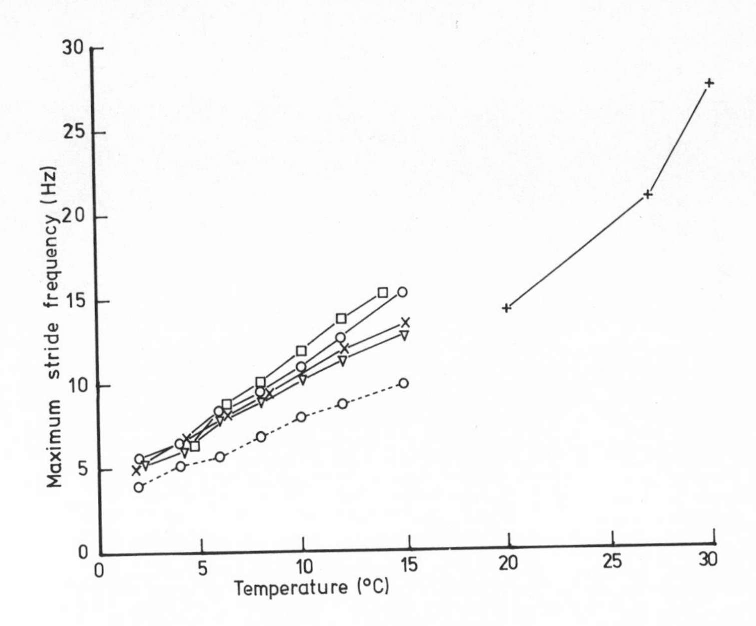

Fig. 4. The effect of temperature on the maximum tail beat frequency or stride rate calculated from the measurements shown in Fig. 3 (stride rate = reciprocal 2t).

Dean found that when muscle was already warmed up to 1.4°C above ambient water temperature, the cooling rate was greater when the trout was forced to swim actively, indicating the possibility of an increased blood flow bringing oxygen to the more active tissue but also aiding equilibration of the muscle temperature as activity increased. In both these studies the water temperatures were relatively high and it would be interesting to know what happens in fish swimming in colder water. For example can the circulation to the white anaerobic muscle be closed with consequent warm up of the white muscle and thus increase the maximum swimming speed. White muscle is anaerobic and does not need oxygen while working. In the large blue fin tuna a counter current blood supply in the form of a rete allows maintenance of a high temperature in the aerobic cruising red muscle while preserving its blood circulation. The red muscle in these big fish (up to 400kg) is positioned usually deep in the body (Carey and Teal, 1969). The white muscle of the same fish does warm up to 5-10°C above the water temperature when these fish are exhausted while played on a fishing line or removed from fish traps (Carey and Teal, 1969).

DOES MUSCLE CONTRACTION TIME ADAPT TO TEMPERATURE CHANGE?

The natural seasonal fluctuations in water temperature often exceed 10°C and even in temperate climates can swing from near zero to 20°C in the sea surface. A change of 10°C alters maximum swimming speed by $2ms^{-1}$

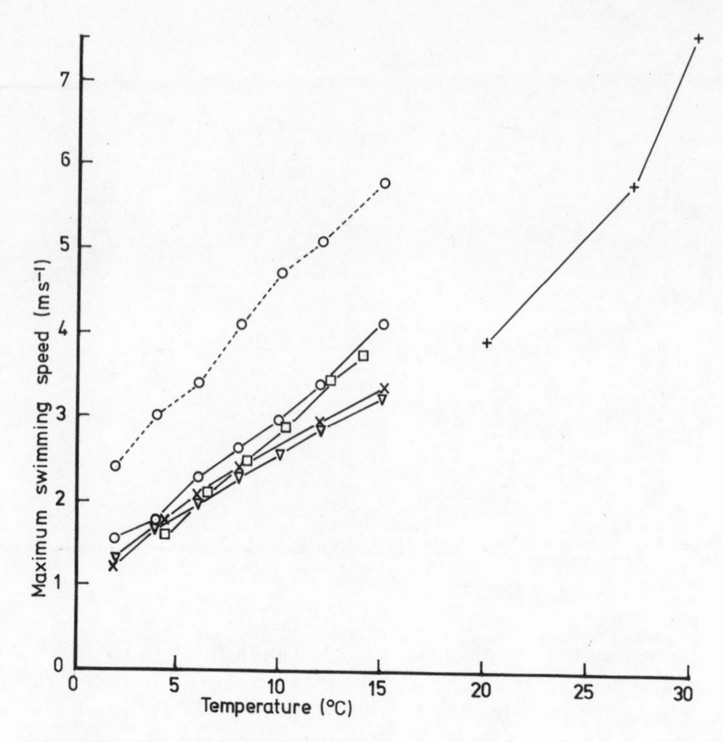

Fig. 5. The effect of temperature on the maximum swimming speed calculated from measurements in Figs. 3 and 4 and a stride length of 0.7 x fish length.

in a 35-40cm fish and one might expect alterations in contraction time to compensate for this obligate change in swimming performance. No adaptation could be detected in plaice held in aquaria at 8°C and 18°C for 8 weeks (Wardle and Mojsiewicz, in preparation) and the results are summarised in Fig. 6. The graph shows the muscle contraction time of the muscle from five plaice sampled when first caught at 8°C compared with two groups sampled eight weeks later having been held in aquaria at 8°C (five fish) and at 18°C (three fish). The change in contraction time to be expected if adaptation had compensated for the 10°C change in performance is indicated by the dashed line and arrow (Fig. 6).

IS IT POSSIBLE TO BREAK THE SPEED LIMIT?

The poikilothermic teleost is clearly at a disadvantage when hunted by the homoiothermic mammal when both are in cold water. As well as warming cold muscle to shorten the fishes contraction time other tricks might be used to get better speed from the handicapped cold muscle. A discussion by Wardle and Videler (1980) points out that some high speed swimming has been measured at just about twice the speed predicted by

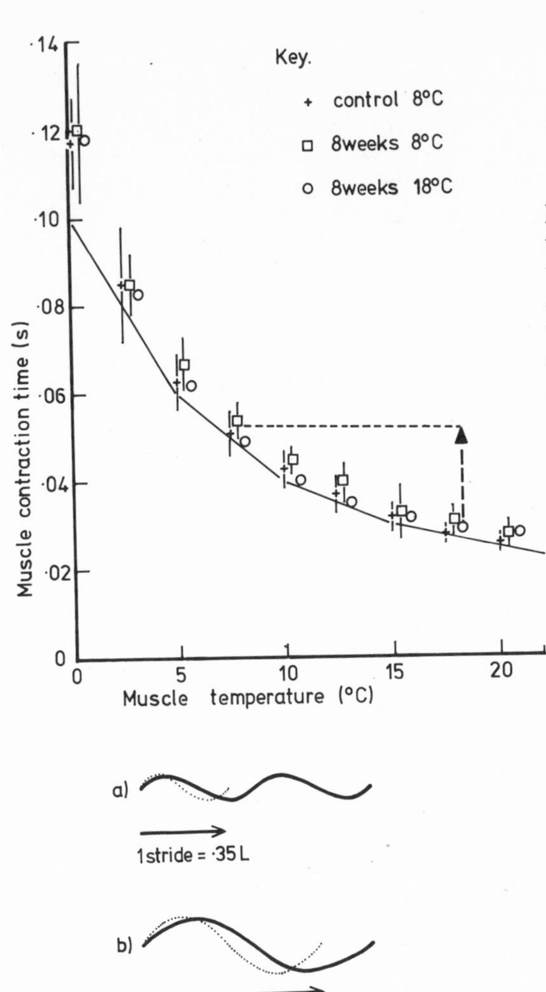

Fig. 6. The effect of temperature on the contraction time of the muscle of 13 plaice (length .30 to .36m) previously kept at the temperatures for the time shown (mean and SD for the 3 groups). The firm line is the same reference slope seen in Fig. 3.

Fig. 7. The diagrams show the body wave (firm line) and the track of the tail tip after one complete tail cycle (dotted line). The horizontal axis is distance moved. For explanation see text. (Adapted from Wardle and Videler, 1980).

Wardle (1975). These authors suggested that if the power could be made available to double the speed then this speed might be achieved by halving the wave length of the wave of contraction on the body of the fish. This in effect doubles the distance moved forwards for each stride from 0.7 x length to 1.4 x length. Three contrasting examples are shown in the diagrams (Fig. 7) to illustrate this point. The first example (Fig. 7a) shows a fish with two complete wave lengths on its body (firm line) and indicates the track of the tail tip traced out during one complete tail cycle (dotted line). In this case the distance moved forwards (1 stride) following completion of one tail cycle is only 0.35 x length. The more normal fish is shown in the second diagram (Fig. 7b) where one completed tail cycle moves the fish forwards 0.7 x length. The third diagram (Fig. 7c) shows the proposed method of swimming giving double the speed for the same contraction time. In this last case the body of the fish contains half a wave length and on completion of one complete tail cycle the fish moves forwards 1.4 x length. One problem with this last style of swimming is how the power available from the swimming muscle might be increased. Normal fish swimming at their maximum speed are probably contracting the whole of their fast white anaerobic muscle in each myotome in order to develop sufficient force in the wave of contraction to counteract the drag of the water on the fast moving body. This force (F) increases with the swimming speed to the power 1.8 where $F = 1/2\rho\ Au^2\ 1.2Cf$ and power Fu increases with $u^{2.8}$ (Wardle and Reid, 1977). The speed remains variable but the other components are relatively constant being the water density (ρ), the body surface area (A) and the coefficient of drag (Cf). To double the speed, 7 x the power is required, equivalent to 7 x the volume of swimming muscle, a fish swimming with half a wave of contraction within its body length (Fig. 7c) might reasonable be expected to have a body diameter 3 to 6 times greater than a fish of the same length but with a complete wave of contraction within its body length (Fig. 7b). However, twice the length of each lateral muscle will be involved in developing the movement when the body is swimming using 1/2 wave length motion (Fig. 7c) so this estimate might be halved to 1 ½ to 3 times the body diameter. In fish swimming in cold water with cold muscle and consequent increased contraction time there is a big reduction in maximum power output at the obligate slower tail beat frequencies. By switching to a 1/2 wave swimming mode double the speed might be achieved using this cold handicapped muscle. Plotting the swimming speed against tail beat frequency for these different modes of swimming (Fig.8) demonstrates the potential of changing the swimming mode from a stride length of 0.7 x length to a stride length of 1.4 x length. A similar graph published by Magnusson (1978), summarises the relationship between tail beat frequency and swimming speed in various tuna species (his Fig. 16). He summarises a series of measurements where at slower swimming speeds the distance moved forwards by the tuna on completing one tail beat <u>is</u> greater than one fish length.

CONCLUSIONS

The quite different teleost species examined from flat fish, like plaice and lemon sole, through cod and mackerel to the skipjack tuna, appear to have very similar contraction time characteristics in their swimming muscle. All these species show a similar change in swimming muscle performance as temperature is changed giving very similar predictions of maximum swimming speeds. Fig. 5 shows that fish between .35 and .40 m in length can swim no faster than $1ms^{-1}$ at 0°C and gain in speed by $2ms^{-1}$ for

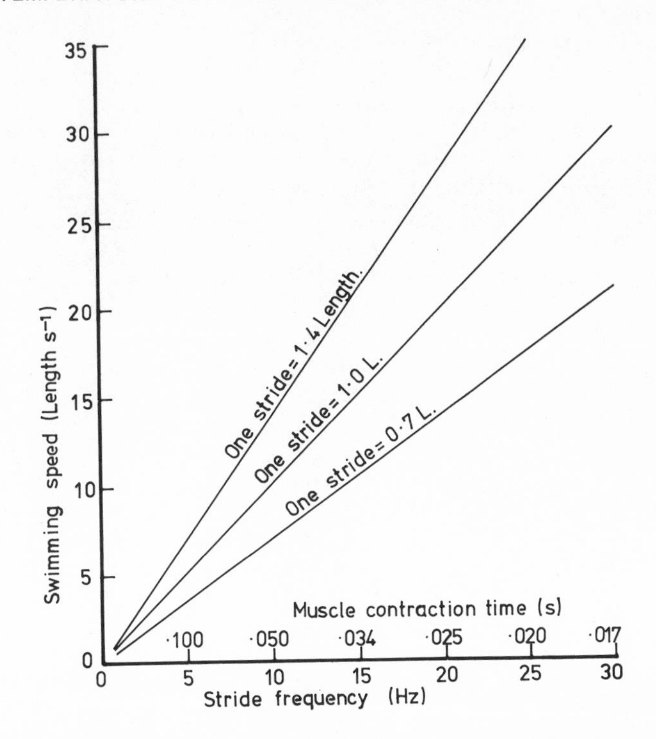

Fig. 8. The diagram relates the muscle contraction time and maximum stride frequency to swimming speed, showing the effect of altering the stride length relative to fish length, see text and Fig. 7.

each 10°C rise in temperature. It appears that muscle contraction time may not adapt to change in temperature and other methods of compensation or protection may be used. The warming of the anaerobic muscle has been observed in various species of large tuna and might also be expected in smaller fish made to swim rapidly in cold water. There is some evidence that swimming with a half wave movement of the fish's body may exist and that it might double the speed predicted from the same muscle contraction time. Further evidence for doubling the stride length will be obtained most easily by noticing the tail beat frequency in relation to distance moved during steady swimming.

REFERENCES

Bainbridge, R. (1958) The speed of swimming of fish as related to size and to the frequency and amplitude of tail beat. J. Exp. Biol. 35: 109-133.

Batty, R.S., and Wardle, C.S. (1979) Restoration of glycogen from lactic acid in the anaerobic swimming muscle of plaice Pleuronectes platessa L. J. Fish. Biol. 15: 509-519.

Black, E.C. (1958) Energy stores and metabolism in relation to muscular activity in fishes. In: The Investigation of Power Problems. Ed. P.A. Larkin. Univ. British Columbia, p. 51-67.

Black, E.C., Robertson, A.C. and Parker, R.R. (1961) Some aspects of carbohydrate metabolism in fish. In: Comparative Physiology of Carbohydrate Metabolism in Heterothermic Animals. Ed. A.W. Martin. Seattle, Univ. Washington Press, p. 89-122.

Brill, R.W. and Dizon, A.E. (1979) Effect of temperature on isotonic twitch of white muscle and predicted maximum swimming speeds of skipjack tuna Katsuwonus pelamis. Env. Biol. Fish 4: 199-205.

Carey, F.G. and Teal, J.M. (1969) Regulation of body temperature by the blue fin tuna. Comp. Biochem. Physiol. 28: 205-213.

Dean, J.M. (1976) Temperature of tissues in freshwater fishes. Trans. Am. Fish. Soc. 1976: 709-711.

Hemmings, C.C. (1973) Direct observation of the behaviour of fish in relation to fishing gear. Helgolander wiss. Meeresunters. 24: 348-360.

Hudson, R.C.L. (1973) On the function of the white muscles in teleosts at intermediate swimming speeds. J. Exp. Biol. 58: 509-522.

Hunter, J.R. and Zweifel, J.R. (1971) Swimming speed, tail beat frequency, tail beat amplitude and size in the jack mackerel, Trachurus, symmetricus, and other fishes. U.S. Fish. Bull. 69: 253-266.

Magnusson, J.J. (1978) Locomotion by scombrid fishes: Hydromechanics, morphology and behaviour. In: Fish Physiology. Ed. W.S. Hoar and D.J. Randall. London, Academic Press. Vol. VII, p. 239-313.

Pyatetski, V. Ye. (1970) Kinematic swimming characteristics of some fast marine fish. In: Hydrodynamic Problems of Bionics. Bionica No. 4 1970 translated from Russian, J.P.R.S. 52605, Natl. Tech. Inf. Serv. Washington, D.C. (1971) p. 12-23.

Roberts, J.L. and Graham, J.B. (1979) Effects of swimming speed on the excess temperatures and activities of heart and red and white muscles in the mackerel Scomber japonicus. U.S. Fish. Bull. 76: 861-867.

Wardle, C.S. (1971) New observations on the lymph system of the plaice Pleuronectes platessa and other teleosts. J. Mar. Biol. Ass. U.K. 51: 977-990.

Wardle, C.S. (1975) Limit of fish swimming speed. Nature (Lond.) 255: 725-727.

Wardle, C.S. (1977) Effect of size on swimming speeds of fish. In: Scale Effects in Animal Locomotion. Ed. T.J. Pedley. New York Academic Press, p. 299-313.

Wardle, C.S. (1978) Non-release of lactic acid from anaerobic swimming muscle of plaice Pleuronectes platessa L.: A stress reaction. J. Exp. Biol. 77: 141-155.

Wardle, C.S. and Mojsiewicz, W.M. (in preparation) The effect of size and temperature on muscle contraction time and the maximum swimming speed of fish.

Wardle, C.S. and Reid, A. (1977) The application of large amplitude elongated body theory to measure swimming power in fish. In: Fisheries Mathematics. Ed. J.H. Steele. London, Academic Press, p. 171-191.

Wardle, C.S. and Videler, J.J. (1980) Fish swimming. In: Aspects of Animal Movement. Ed. H.Y. Elder and E.R. Trueman. Society of Experimental Biology Seminar Series No. 5. Cambridge Univ. Press, p. 125-150.

Wardle, C.S. and Videler, J.J. (1980) How do fish break the speed limit? Nature (Lond.) 284: 445-447.

Webb, P.W. (1978) Fast start performance and body form in seven species of teleost fish. J. Exp. Biol. 74: 211-226.

PHOTOPERIODIC AND ENDOGENOUS CONTROL OF THE ANNUAL REPRODUCTIVE CYCLE IN TELEOST FISHES

Bertha Baggerman

Zoological Laboratory,University of Groningen

9750 AA Haren (Gr.), The Netherlands

I. INTRODUCTION

Most organisms, particularly those living at higher latitudes, produce their offspring only in that season of the year in which food is most abundantly present for prosperous growth. Survival of the species therefore depends on a proper timing of the onset as well as termination of the breeding period. It has been believed for a long time, that breeding was controlled solely by external factors, such as photoperiod and temperature. However, it is now known that the correct timing of the breeding season often depends on a co-operation between external factors and endogenous rhythms, circadian and/or circannual.

As is presently known, all eukaryotic organisms have developed endogenous rhythms (or biological clocks), with periods matching those of a day, while in addition, several animal species also possess clocks with a period of about one year. Biological clocks, by their nature, do not have the accuracy of geophysical rhythms and their periods always deviate to some extent from that of a 24-hour day and a 365-day year. Therefore, these rhythms are known as circadian and circannual rhythms respectively (from circa = about). To ensure that biological rhythms do not get out of phase with the corresponding geophysical rhythms, the organism uses certain external signals as "Zeitgeber", by which the former become synchronised with the environmental rhythms, i.e. acquire periods of more exactly 24 hours, or one year. For the synchronisation of circadian rhythms, signals from the light-dark cycle are very strong Zeitgebers, while circannual rhythms may be entrained by the annual cycle of the daylength. Biological rhythms can have several functions. In the first place they provide the organisms with a means to anticipate coming events. In the second place they serve as time-measuring mechanisms (clocks), for instance in photoperiodic responses, in compass-orientation of migrating species and in the time-sense of bees.

In many animal species gonadal recrudescence and regression are to a large extent controlled by the length of the photoperiod. However, in order to be able to give a differentiated response to photoperiods of different lengths, they must possess some kind of time-measuring mechanism or clock. It has been shown that in many species this mechanism operates on the basis of a circadian rhythm of sensitivity to light (Palmer, 1976; Rensing, 1973; Saunders, 1977). The existence of circannual rhythms has become established only during the last decade. In some species of birds there are strong indications that annual reproduction is based on a circannual gondadal rhythm (Pengelley, 1975), while this may also be the case in the catfish Heteropneustes fossilis (Sundararaj and Sehgal, 1970b; Sundararaj and Vasal, 1973, 1976).

The aim of the present paper is to show how in certain species of teleost fishes the regulation of the annual reproductive cycle is achieved by a co-operation between external factors - particularly the photoperiod - and endogenous rhythms. The main focus of attention will be on the ways in which the animals are able to use photoperiodic information in their timing of the onset and end of the breeding period. It will be shown that the time-measuring mechanism involved in the photoperiodic responses is based on a daily, most probably circadian rhythm of sensitivity to light. Further, evidence will be presented indicating that (in addition) an annual, possibly circannual rhythm, may also play a role in the regulation of the reproductive cycle. Finally, some attention will be paid to the components of the reproductive neuroendocrine system which may be involved in the co-operation between environmental factors and biological rhythms. In dealing with these subjects I will mainly use my knowledge on the regulation of the gonadal cycle in the stickleback (Gasterosteus aculeatus L.) as a framework, into which findings in other species will be fitted in at the appropriate places. The paper is not intended, however, to review the literature on the influence of the photoperiod on gonadal development in fishes. The reader interested in this information is referred to a critical review by de Vlaming (1972) and a paper by Sundararaj and Vasal (1976).

II. DESCRIPTION OF ANNUAL GONADAL CYCLES

Investigations on photoperiodic control of breeding require knowledge on the gonadal cycle under natural conditions. It should be known at which time of the year recrudescence begins, whether it proceeds continually or shows an interval of relative quiescence and finally, at which time of the year breeding begins and how long it lasts. Only when this information is available will it be possible to set up the right kind of experiments, to give the best possible interpretation of the findings and to properly appreciate their significance in the light of what happens under natural conditions. It is not possible, however, to make generalisations concerning the annual gonadal cycles of the different species, as almost every species has its own specific cycle, which guarantees the best possible reproductive success in the ecological niche it occupies. Since this paper will be mainly concerned with the regulation of the gonadal cycle of the stickleback, only this cycle will be described in some detail to enable better understanding of the findings. Readers interested in gonadal cycles of other species are referred to e.g. Barr (1968) and Lofts (1968).

The gonadal cycle of the three-spined stickleback Gasterosteus aculeatus L.

The male. Sticklebacks in Holland breed between approximately mid-April and mid-July. Already in late summer the first signs of recrudescence can be observed and this process proceeds in two phases. Phase 1 consists of spermatogenesis and leads to the formation of viable spermatozoa as early as November/December. During winter a period of quiescence occurs in which little development if any takes place until January/February, when phase 2 begins to develop. This phase consists of an increase in Leydig-cell activity, which leads to the gradual appearance of nuptial colouration and eventually to the performance of reproductive behaviour. Male sticklebacks build a nest of plant fragments glued together by a mucous substance secreted by the kidneys. To this end the kidneys are transformed (under the influence of androgen secreted by the Leydig cells) into a mucus producing organ, in which the tubule cells have greatly increased in height. This increase in height has proved to be an excellent parameter for the amount of androgen produced by the Leydig cells: the higher the androgen level, the greater the height of the kidney-cells (Mourrier, 1972). Figure 1 summarises the sequence of events in the course of the year.

The female. Oögenesis begins in late summer and (like in the males) develops in two phases. Phase 1 is completed before the onset of winter and consists of an increase in oöcyt diameter to about 350 μ. In the largest eggs some vacuolisation occurs, which is considered to be a very first sign of yolk formation. After the completion of phase 1, a period of relative quiescence occurs, after which the development of phase 2 begins around January/February. This phase mainly consists of vitellogenesis, which expresses itself in the deposition of large eosinophilic yolk granules, resulting in an increase in egg diameter to about 900 μ. At this time the eggs are fully mature and await ovulation and oviposition. The sequence of events is summarised in fig. 1.

In studies on the influence of photoperiod and temperature on gonadal development many different criteria have been used. Baggerman (1957, 1972) seems to be the only one using the occurrence of reproductive behaviour as criterion, which has the advantage that the performance of this behaviour is definite proof that the external conditions induced full maturation, including the formation of viable gametes and the necessary endocrine secretions. In most other studies, data on the gonadosomatic index (weight gonads/100 g body weight) and/or histological data are taken as criterion (see e.g. review by de Vlaming, 1972), while in only a very few studies pituitary and plasma levels of gonadotropin are determined (e.g. Breton and Billard, 1977). The use of different parameters sometimes makes it difficult to properly evaluate the significance of the findings.

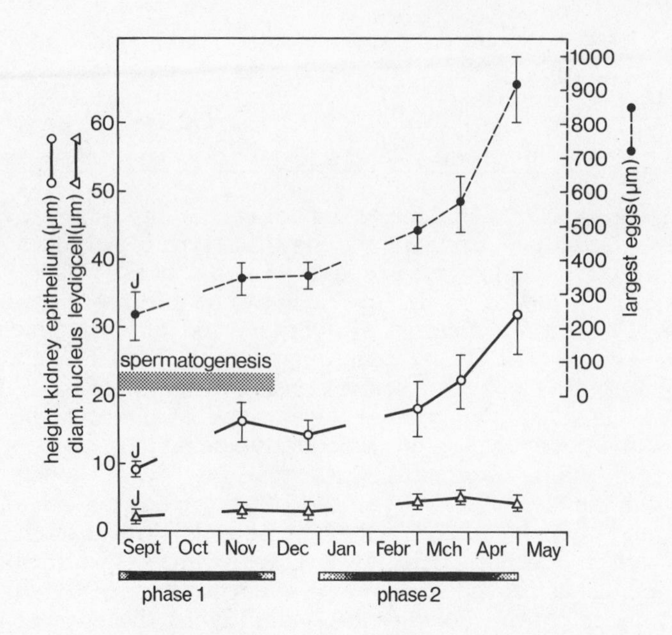

Fig. 1. Gonadal recrudescence in sticklebacks living under natural conditions. The height of the kidney epithelium cells is a measure for androgen production by the Leydig cells. J = data from underyearlings; all other data from yearlings. Further explanation in the text.

III. GONADAL RECRUDESCENCE

III. I. Influence of the photoperiod

Gasterosteus aculeatus. In her studies Baggerman (1957, 1972) used the performance of reproductive behaviour as criterion to assess the influence of the photoperiod. To this end she always exposed groups of 15-20 fish (males and females together) in 60 L all glass aquaria to various photoperiods, usually at a temperature of 20°C. A male was considered to be sexually mature when he built a nest, for such males are known to be willing to court ripe females and to fertilise eggs. As soon as a male had built a nest he was removed from the group. A female was considered to be mature when she was willing to respond to a courting male by creeping into his nest. This response is only shown after ovulation has taken place. In such females the eggs can easily be stripped by exerting a slight pressure on her abdomen, while stripping is impossible before ovulation. Therefore, stripping was used as a criterion for the attainment of complete maturity. Ripe females were also removed from the group. This kind of experiments always yielded two kinds of data: in the first place data on the number of days of exposure to a given external condition until maturity was attained (= response-time) and in the second place data on the number (percentage) of animals within a given group, which had attained maturity by the time the experiment was discontinued (response-percentage). Experience has learned that 65 days is sufficiently long to obtain conclusive results. Therefore, all experiments lasted at least 65 days. This paper will be mainly based on data concerning the percentage of fish attaining maturity under the various conditions. The data on males and females have been taken together since in the vast majority of cases there were no differences between their responses. Although part of the results have already been published (Baggerman, 1957, 1972), it is thought to be necessary for a good understanding of the many, as yet unpublished data presented in this paper, to repeat some of the earlier published findings.

Figure 2A gives data obtained in an experiment started in November (1972) with underyearling animals caught in nature in a large outside pond, where they had lived since hatching in June. They were caught a few days prior to the beginning of the experiment for acclimation to the temperature of 20°C used in the experiment. On November 16 the group was divided into three groups of 19-20 animals each. One group was exposed to a long photoperiod, 16 hours of light followed by 8 hours of darkness, = 16L-8D. The second group was exposed to 12L-12D and the third to 8L-16D. Figure 2A shows that exposure to 16L-8D induced maturity in all animals within 40 days, i.e. about 4 months before breeding would start in nature. This precocious maturation was not caused by the high temperature, for fig. 2A shows that none of the fish exposed to 8L-16D at the same temperature had responded by the time the experiment was discontinued after 100 days. Autopsy at this time showed that the gonads of the fish were still in phase 1. The group of fish exposed to 12L-12D responded in a remarkable way. Figure 2A shows that only two of them attained maturity within 65 days, while none of the remaining 17 animals responded during the next 35 days. Like in the 8L-16D animals, their gonads were still in phase 1 when the experiment was discontinued after 100 days. I shall return to this finding after the description of an exactly similar experiment, which started January 25 (1971). Again fish were used which had been caught in the same outside pond, where they had lived since hatching in June.

Figure 2B shows that all fish exposed to 16L-8D matured within 45 days, while also all fish exposed to 12L-12D matured in about the same time (50 days). This time also 6 of the 19 fish exposed to 8L-16D attained maturity and again within about the same period of time (35 days). However, of the 13 remaining fish none had matured when the experiment was discontinued after 100 days; like in the previous experiment the gonads of the latter fish at that time were still in phase 1.

In view of the main subject of this paper, attention will be focussed mainly on the experiments in which only part of the animals attained maturity, i.e. the 12L-12D group in fig. 2A and the 8L-16D group in fig. 2B. In both cases the fish which did mature responded within a relatively short time (usually within 45 days), which did not differ significantly from the response-times of the 16L-8D animals. Therefore, it is very remarkable that

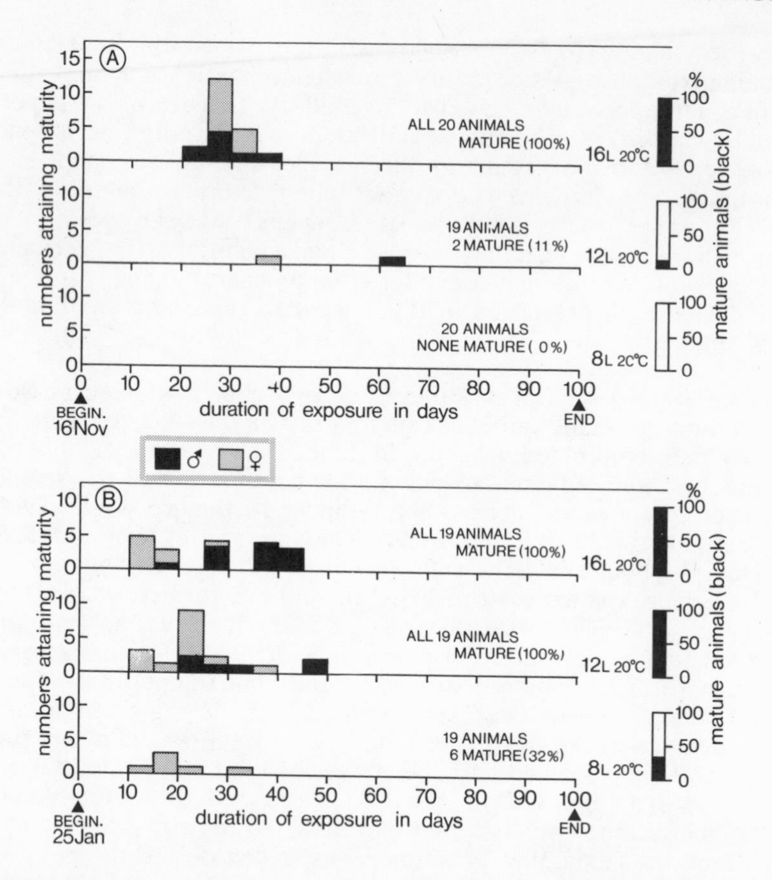

Fig. 2. Results of two experiments, started November 16 (A) and January 25 (B) in which groups of 19-20 underyearling fish (which had been living under natural conditions) were exposed to three different photoperiods, (16L-8D, 12L-12D and 8L-16D) at a temperature of 20°C. Represented are the number of animals attaining maturity in the course of the experiments. The narrow columns on the right represent the percentage of animals in each group which had attained maturity (black) when the experiments were discontinued after 100 days. Further explanation in the text.

none of the remaining animals responded, not even when the experiments were continued for a much longer period of time. Moreover, it should be recalled that when the experiments were discontinued, the gonads of none of these fish had developed beyond phase 1.

Baggerman (1972) explained these results by postulating the existence of a photo-sensitivity threshold, which determines whether or not the animal will be able to respond to a given photoperiod. As I have since discovered that this term is often misunderstood, I have replaced it in the present paper by photo-reactivity threshold. This hypothesis assumes that the higher this photo-reactivity threshold, the longer the photoperiod has to be to overcome this threshold and to induce maturity. In addition, it is assumed that the threshold has an "all-or-none" effect: the fish either respond within a relatively short period of time (45 days), or they do not respond, not even when the experiment is continued for at least 65 days. In the latter case the gonads will not develop beyond phase 1.

Since data of the kind as given in fig. 2 are the basis of all ideas developed in the following part of this paper, it is necessary to deal with them somewhat further. It should be recalled that the data were obtained with samples from a population of (underyearling) fish, which had lived under natural conditions until the experiments began in November and January. Therefore, the reponses give an indication of the height of the photo-reactivity threshold in the same population at two different times of the year. Figure 2A shows that in November the threshold was rather high, for it could be overcome only by 16L-8D in all animals and in but a few also by 12L-12D. Therefore, it is concluded that at this time of the year the average height of the threshold of the population must have been close to 12L-12D. Figure 2B shows that in January the threshold of the population had apparently declined, for now all animals were also able to respond to 12L-12D, while a few of them even responded to 8L-16D. Thus, the average threshold level of the population at this time of the year was much lower and close to that of 8L-16D. The final conclusion from figs. 2A and 2B is that in a population of sticklebacks living under natural conditions the photo-reactivity threshold declines from high in November to relatively low in January.

This conclusion is supported by findings represented in fig. 3. They come from all available experiments, similar to those of fig. 2, in which fish had been exposed to different photoperiods (ranging from 16L-8D to 8L-16D, at 20°C) at different times of the year. The animals used were sampled from a population of underyearling fish, which had lived under natural conditions (outside pond), since the time of their hatching in June until the experiments started. It can be seen very clearly that in the autumn most animals only respond to photoperiods of about 16L-8D, whereas in the course of autumn, winter and early spring they are able to respond increasingly better also to the shorter photoperiods. This finally leads to the situation that in February/March all animals are able to attain maturity even under the very short photoperiod of 8L-16D (at 20°C).

In fig. 3 a line has been drawn connecting those results indicating a response of only part of the fish exposed to a given photoperiod. This has been done because (as was explained above) such data roughly indicate the height of the threshold in the population at that particular time of the year. The drawn line in fig. 3, therefore, indicates the course of threshold decline between autumn and spring in fish living under natural conditions. Looking at the course of this threshold decline it will be clear that the onset of breeding in the spring is to a large extent determined by the rate of

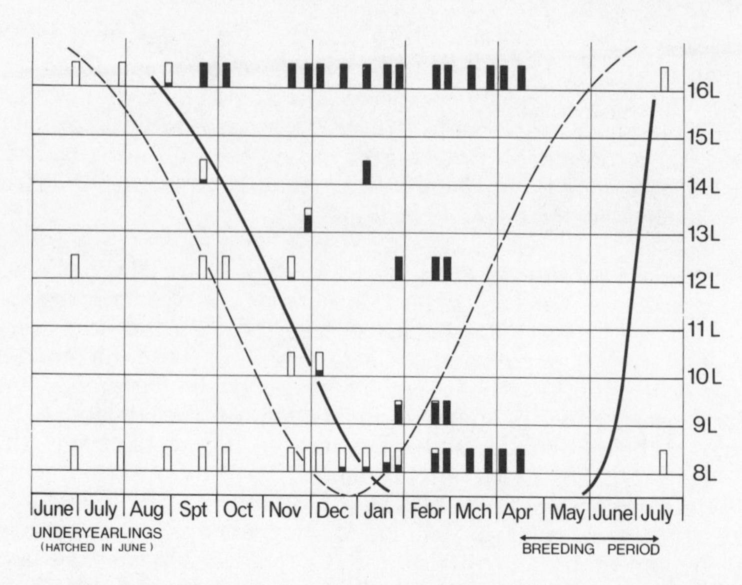

Fig. 3. Percentages of sticklebacks attaining maturity (black) within 65 days after being caught in nature at different times of the year and subsequently exposed to different photoperiods at a temperature of 20°C. Example: at the end of September animals were caught and divided into four groups of 20 animals each. These groups were exposed to 8L-16D, 12L-12D, 14L-10D and 16L-8D (horizontal lines) and the columns show that none of the animals of the first two groups matured (0%), while 10% and 100% matured in the 14L and 16L groups respectively. The heavy drawn line represents the annual cycle of the photoreactivity threshold and the dotted line the number of hours of light per day at the latitude of Holland, varying from more than 16L in the summer to less than 8L in the winter. Further explanations in the text.

threshold-decline: the faster the decline the earlier breeding would start; or, the slower the decline the later the onset of breeding. It is important, therefore, to analyse the factors which determine the rate of this decline. We shall return to this subject in section III.4.

Figure 3 further shows that it is not correct to conclude (as I did in my 1957 paper) that breeding of sticklebacks in the spring is determined by the animals responding to the already rather long day lengths in March/April (see broken line in fig. 3). Rather, it should be concluded that breeding becomes possible, because the fish at this time of the year (due to the low threshold), have become able to respond also to the still very short photoperiods in January and February. The result is, that at that time gonadal development can be resumed (fig. 1 and section II) and is completed by mid-April, the onset of the natural breeding season (fig. 3). Finally, this figure also explains why sticklebacks in the autumn do not respond to the still rather long day lengths at that time of the year: evidently the threshold is still too high to be overcome even by these rather long photoperiods (compare the drawn line in fig. 3 indicating the height of the threshold, with the broken line indicating the length of the photoperiod). Thus, breeding at this time of the year (which would be highly unfavourable for the growth and survival of the young) is prevented. Perhaps in temperate zone birds a similar explanation may hold for the absence of breeding in the autumn, which for the same reason is often considered to be rather puzzling (see e.g. Farner, 1967).

At this point some attention should be paid to the relation between the photo-reactivity threshold and the onset of development of the gonadal phases 1 and 2. When dealing with fig. 2, it has been mentioned that the gonads of animals not responding to the different photoperiods had not developed beyond phase 1. Therefore, phase 2 apparently can only start its development when the threshold is overcome by the photoperiod. This agrees very well with the findings that under natural conditions phase 2 only begins to develop in January/February (fig. 1), i.e. at the time of the year when the threshold can be overcome by the prevailing photoperiods (fig. 3). In contrast, phase 1 seems to be able to develop under any photoperiod, for I have data showing that this phase develops completely in underyearling fish exposed in September to 8L-16D or 16L-8D, both at 20°C. Since it also develops under natural conditions between late summer and November/December (fig. 1), when the daylengths and temperatures are decreasing, the development of phase 1 seems to be to a large extent independent of photoperiod and temperature. The onset of its development may thus be under the control of some endogenous mechanism (section V), although the rate of its development is enhanced by higher temperatures (see below).

Influence of the temperature

Although this paper is mainly concerned with the effects of the photoperiod, some attention has to be paid also to the possible effects of temperature, since these two factors invariably are part of the environment. In his review de Vlaming (1972) rightly stresses the fact that in most investigations not enough combinations of photoperiod and temperature have been studied and that this may have led to erroneous conclusions in a number of cases.

This criticism also holds to a certain extent for my stickleback investigations, which have been mainly concerned with the effects of different photoperiods, usually at only one temperature, viz. 20°C. However, I did run a few experiments like those of fig. 2, in which fish were exposed to a given photoperiod at two temperatures, 20° and 15°C. In those cases no differences were observed with respect to the number of animals responding. Whether the same would hold at lower temperatures is not known.

However, I have evidence that temperature is involved in at least two ways in the regulation of the annual gonadal cycle. In the first place, I have found that once the threshold is overcome by a given photoperiod, the response-times (see beginning of this section) are 2.5 times longer under 9°C than under 20°C. This means that phase 2 develops much faster when the temperatures are higher. Thus, breeding under natural conditions may start somewhat earlier or later, depending on the temperature during the early months of the year. As to phase 1, Craigh-Bennett (1931) mentions that under natural conditions this phase develops faster when the autumn temperatures are higher. In the second place I have evidence that temperature is also involved in the regulation of threshold decline between late summer and early spring; this aspect will be dealt with in section III.4.2.

Other species

As mentioned before, this paper is not intended to give a review of the literature on photo-thermal regulation of annual breeding in fishes. Nevertheless, it is considered to be useful to give at least a general outline of the findings; details can be found in a review by de Vlaming (1972) and in Sundararaj and Vasal (1976). From this literature the picture emerges that the regulation of the annual reproductive cycle is in some species more photoperiod dependent, in others more temperature dependent, while in again others the cycle seems to be equally dependent on both. This is not surprising, since both factors are an integral part of the natural environment. It is not difficult to imagine that, depending on the geographical location, habitat and ecological niche, each species will have adjusted its gonadal cycle to those external factors which ensure the production of young at the most favourable time of the year.

For the present paper it suffices to know that there are teleosts which use the length of the photoperiod for the timing of annual breeding. Apart from the stickleback this has been shown also in, for instance, some species of trout and salmon like *Salmo gairdnerii* (Breton and Billard, 1977), *Oncorhynchus nerka* (Combs et al, 1959), *Salvelinus fontinalis* (Henderson, 1963), in the bridled shiner *Notropis bifrenatus* (Harrington, 1959) and the asian catfish *Heteropneustes fossilis* (Sundararaj and Vasal, 1976). With respect to the occurrence of a photo-reactivity threshold in other species, it seems that the only information on this matter comes from literature concerning the so-called refractory period. The latter is defined as that period of the year in which the gonads cannot be stimulated by long photoperiods, which at other times are found to be highly stimulatory. Harrington (1959) working with *Notropis bifrenatus* and Sehgal and Sundararaj (1970) working with *Heteropneustes fossilis* found that these species were not responsive to long photoperiods in the autumn, whereas they readily responded to the same photoperiods a few months later. In terms of my threshold theory these findings indicate that in the autumn the photo-reactivity threshold was too high to be overcome even by these long photoperiods. As we shall see in section V, the same occurs in sticklebacks at the end of the breeding season in late summer. It is not known, however, whether in other species the refractory period ends "abruptly", or as the consequence of a gradually declining photo-reactivity threshold, like in the stickleback.

III.2. A daily rhythm of light-sensitivity as basis of photoperiodic responses

From studies on photoperiodism in widely divergent animal groups like insects and birds, the picture has emerged that there are in principle two different time-measuring mechanisms by which organisms are able to distinguish between photoperiods of different lengths. One is known as the hour-glass mechanism and this has been particularly well demonstrated in some insects (see e.g. Lees, 1966). The other mechanism is based on a daily rhythm in sensitivity to light, which has been amply demonstrated in birds (see e.g. Farner, 1970) and which is most likely also used in at least two teleost species, as will be shown in this section.

The hour-glass, or interval timer measures the length of the dark period in the 24-hr cycle, which is of central importance in this concept of time-measurement. It lies outside the scope of this paper to go into further detail and the reader is referred to Lees (1966) for more information. For the present paper it is of interest to point out the most essential difference between the hour-glass timer and that based on a circadian rhythm of light-sensitivity. The hour-glass timer operates like its name implies; it is set in motion by some external signal, e.g. dusk or dawn and then runs its course. Under the natural light/dark cycle this timer is thus set into motion every day. However, when the signal is absent, like under continuous light (LL) or darkness (DD), the mechanism is unable to operate. This is in contrast to the time-measuring mechanism which is based on an endogenous daily (= circadian) rhythm of light-sensitivity, which continues to oscillate even in the absence of any external signal.

Time-measuring based on a daily rhythm in sensitivity to light can best be explained as follows. At the beginning of each day, i.e. when it becomes light at zero hour, the system underlying the photoperiodic response is not, or very little sensitive to stimulation by light and this remains so during the first 6-8 hours. However, during the next hours its sensitivity increases until a maximum is reached around 16 hours, after which it declines again to zero at the end of the 24-hour day. This light-sensitivity cycle is repeated every day and it has been shown that it continues to occur even under LL and DD, be it with a period deviating somewhat from 24 hours (circadian rhythm, section I). This means, that this daily cycle is not set in motion by external signals like the hourglass timer, but that it has an endogenous origin. This daily rhythm is thought to be involved in photoperiodic gonadal stimulation as follows. Under a short photoperiod, 8L-16D, the neuro-endocrine gonadal system is not able to respond, since the animal is exposed to light only in the period of the day when its gonadal system is still insensitive to it. Under photoperiods longer than 8 hours increasingly better responses will occur, with a maximum under 16L-8D, since in these cases the organism is exposed to light also during the period of the day when the gonadal system becomes increasingly more sensitive to light.

Different types of experiments can be used to demonstrate a daily light-sensitivity rhythm. In one of them, the dark component of a "short day" photoperiod is interrupted systematically at various times by a supplementary light pulse of relatively short duration. They are known as night-interruption experiments and the photoperiods as "skeleton-photoperiods". The aim of this type of experiment is to scan the animals' responsiveness to the light pulse in the course of a 24-hour day. This can best be further explained by taking fig. 4 as an example. In this figure the horizontal bars indicate five different skeleton-photoperiods, the numbers II-VI, together with one "normal" photoperiod, no. I = 8L-16D. In all these light regimens the animals are exposed to a total of 8 hours of light per 24 hours, but in all of them (except I) the photoperiod is broken up into a main light-period of 6 hours, plus a two hour light pulse given at various times in the dark period. It is important to add that the total duration of light given per 24 hours (i.e. 8 hours) is always chosen on the basis of previous findings, that the animals used will not be responsive to it. The expectation of experiments like this is that in case the animals would be equally responsive to light in the course of the day none of them would respond to any of the skeleton-photoperiods. On the other hand, if there would be a daily light-sensitivity rhythm, responses should occur to at least some of the skeleton-photoperiods.

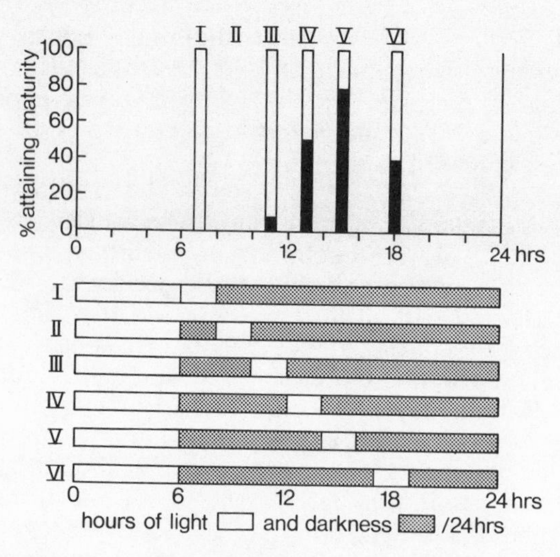

Fig. 4. The percentage of animals attaining maturity (black) within 65 days when caught in nature in the second half of November and subsequently exposed to different photoperiods at 20°C. One group was exposed to 8L-16D (photoperiod I) and the other four to the skeleton-photoperiods III-VI. All five groups were daily exposed to the same total amount (8 hours) of light. Further explanation in the text.

The experiments of which the data are given in fig. 4 were carried out with underyearling sticklebacks (hatched in June in the outside pond), which were exposed to this series of skeleton-photoperiods in the second half of November. The data are the means of four experiments carried out in two different years. It can be seen that none of the fish responded to 8L-16D, which (as was mentioned) is a prerequisite for this kind of experiment. Figure 4 further shows that 8% responded to photoperiod III (6L-4D-2L-12D), 52% to IV (6L-6D-2L-10D), 78% to V (6L-8D-2L-8D) and 41% to VI (6L-11D-2L-5D). These responses together form a curve, with a response-maximum between 14 and 16 hours after the onset of the main light-period.

It is important to add that sticklebacks respond to skeleton-photoperiods in the same way as to "normal" photoperiods: they either attain maturity within a relatively short period of time, or they do not mature at all, not even when exposed for 65 days or longer. Moreover, in almost none of the unresponsive animals of fig. 4 had the gonads developed beyond phase 1 when they were autopsied at the end of the experiment. All this, therefore, is similar to what is found in experiments with normal photoperiods described in the previous section.

In addition, it has been found that very similar responses are obtained when sticklebacks are exposed to a given skeleton-photoperiod, or to the "corresponding" normal (= uninterrupted) photoperiod. This can best be explained in fig. 4. The corresponding normal photoperiod of skeleton-photoperiod II, 6L-2D-2L-14D has been found to be 10L-14D, because the dark period in between the main light-period and the two-hour light pulse is apparently regarded as being part of a total light-period of 6 + 2 + 2 = 10L. In the same way the responses to skeleton photoperiods III-VI of fig. 4 are comparable to those obtained by exposure to the normal photoperiods of 12L-12D, 14L-10D, 16L-8D and 19L-5D respectively. Examples can be found in fig. 5, in which response-percentages to some of the skeleton-photoperiods can be directly compared to those obtained in parallel experiments with the corresponding normal photoperiods (narrow columns marked with an asterisk). The reason for the great similarity in the responses will be clear: it is not so much the exposure to light during the first 8-10 hours of the daily light period which counts most, but rather the exposure during the later period of the day up to 16 hours (fig. 4).

With all this in mind, the results of the experiment carried out in November with normal photoperiods (fig. 2A) can be easily explained. At this time of the year the fish are not sensitive to light during the first 8 hours of the day (fig. 4); therefore, they are unable to respond to 8L-16D. When exposed to 16L-8D the situation is quite different, for they are now exposed to light also in that period of the day in which they are most sensitive to it, viz. between 14 and 16 hours (fig. 4). The intermediate response to 12L-12D can be explained along a similar reasoning: fig. 4 shows that only a small percentage of the fish at that time is able to react to light impinging upon them between 10 and 12 hours.

Experiments like that of fig. 4 show that photoperiodic gonadal responses can be explained on the basis of a daily rhythm of light-sensitivity. However, they do not give conclusive evidence for a possible endogenous (= circadian) nature of the underlying rhythm (although this is very likely), because light/dark signals are present in every 24-hour cycle. To obtain conclusive evidence on this matter night-interruption experiments like that of fig. 4 should be carried out not only in 24-hour cycles, but also in, for instance, 72-hour cycles with a very long dark period, like 6L 66D. Using such cycles the very long dark period could be scanned by 2-hour light pulses in the same way as in fig. 4. If the light-sensitivity rhythm would have a circadian nature, good responses are expected to skeleton-photoperiods in which the light pulse falls between 14 and 16 hours on the first "day" (thus 6L-8D-2L-56D), but also to those in which this light pulse falls 24 hours later (6L-32D-2L-32D), or 48 hours later (6L-56D-2L-8D). On the other hand, no responses would be expected with the light pulse given between 8 and 10 hours on the first "day"(6L-2D-2L-62D) and also not when it is given 24 hours later (6L-26D-2L-38D), or 48 hours later (6L-50D-2L-14D). I intend to carry out such experiments in the near future in order to more conclusively demonstrate the endogenous nature of the daily light-sensitivity rhythm in the stickleback.

Sundararaj and Vasal (1973, 1976) reported that also in the asian catfish Heteropneustes fossilis the ovarian responses to a long photoperiod are based on a daily rhythm of light-sensitivity. Since they also used only skeleton-photoperiods within a 24-hour cycle, there is no conclusive evidence as yet (like in the stickleback) that this rhythm has an endogenous (circadian) origin, altough this seems very likely. It is interesting to note that in both species maximal light-sensitivity occurs in about the same period of the day, i.e. around 16 hours. The same has been found in birds (see e.g. Farner, 1970).

Finally, it is important to point out that light from the daily light/dark cycle has two different functions. In the first place, as shown in this and the preceding section, light is the actual stimulator of the neuroendocrine gonadal system, when the photoperiod is long enough to overcome the photo-reactivity threshold. In the second place, as mentioned in section I, signals from the daily light/dark cycle (like "lights-on" and "lights-off") serve as Zeitgeber for the circadian light-sensitivity rhythm, so that it remains in phase with the 24-hour daily rhythm. Only in this way is this circadian rhythm able to function as a reliable time-measuring mechanism in photoperiodic responses.

III.3. A seasonal change in the daily light-sensitivity rhythm

At the beginning of this section it should be recalled that fig. 4, based on experiments carried out in November, could easily explain the results of the experiment of fig. 2A, likewise carried out in November. However, the high response-percentages to the shorter photoperiods (8L-16D and 12L-12D) given in fig. 2B and fig. 3 (early months of the year) cannot be explained on

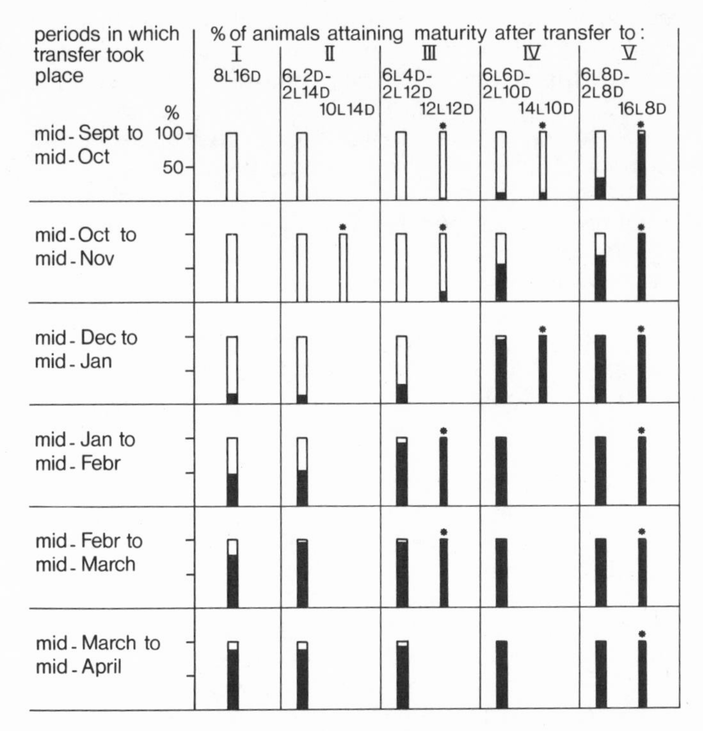

Fig. 5. Percentages of animals attaining maturity (black) within 65 days when caught in nature at various times between September and April and subsequently exposed to 8L-16D (photoperiod I) and the skeleton-photoperiods II-V at 20°C. The narrow columns marked with an asterisk give the response-percentages of animals exposed in parallel experiments to uninterrupted photoperiods of 10L-14D, 12L-12D, 14L-10D and 16L-8D. Further explanation in the text.

the basis of fig. 4, for according to the latter the fish should have been only little or even not responsive at all to these photoperiods. This discrepancy was already noted by Baggerman (1972) and she postulated that perhaps the "shape" of the daily light-sensitivity curve would change in the course of the year, so that the animals would become sensitive also to light to which they are exposed earlier in the day. In terms of fig. 4 this would mean that increasingly higher responses would be obtained in the course of the year, first to photoperiod IV, then to III, a.s.o.

To test this idea, it was decided to repeat the experiment of fig. 4 with regular intervals between September and April. Only photoperiod VI was excluded, since it was considered to be of little importance in this context. The animals used were again underyearling sticklebacks (hatched in June in the outside pond) and caught a few days prior to the beginning of the experiments for thermal acclimation to 20°C. This series was repeated in two successive years. As the results of this experimental series have never been published before, full details will be given in a separate paper. Figure 5 only gives the means of the response-percentages found.

To begin with, fig. 5 reveals two things. As was expected, exposure to skeleton-photoperiods in which the two-hour light pulse was given later in the day, always resulted in higher response-percentages: horizontal comparison of the columns in each period of the year. In the second place, fig. 5 shows that in the course of the year increasingly higher responses were obtained to each of the photoperiods: vertical comparison of the columns for each photoperiod. This means that in the months between September and April the daily light-sensitivity cycle changed, in that the sticklebacks also became responsive to light to which they were exposed earlier in the day. This therefore, confirms the hypothesis put forward by Baggerman (1972).

The data of fig. 5 have been represented in a different way in fig. 6, so as to show more clearly the way in which the "shape" of the daily sensitivity cycle changes between autumn and spring. It should be noted that in fig. 6 only that part of the daily curve of fig. 4 is drawn, which pertains to the reactions to photoperiods I-V. The curves of fig. 6 have been obtained by connecting points which represent the response percentages taken from fig. 5. However, since the responses to photoperiod V obtained between September and November were somewhat lower than expected on the basis of the high response-percentages to the corresponding normal photoperiod of 16L-8D (fig. 5), the latter percentages have been used. Since the number of points in fig. 6 is very small, the courses of the different curves can only be approximations.

When trying to read the meaning of fig. 6, it should be recalled that similar response-percentages as found by exposure to the skeleton-photoperiods II-V would have been obtained, had the animals been exposed to the corresponding normal photoperiods, thus to 10L-14D, 12L-12D, 14L-10D and 16L-8D respectively (section III.2.). Taking this into account, the curve for September/October indicates that responses of 50% or higher at this time of the year could only be obtained with photoperiods longer than 15 hours; looking at fig. 3 we see that this is exactly what was found. Again, the curve for January/February shows that responses of 50% or more could already be obtained by photoperiods between 9 and 10 hours and fig. 3 shows that this is also what was actually found.

The main conclusion from figs. 5 and 6 is that the shape of the daily light-sensitivity curve changes between late summer and spring. This change is such that sticklebacks become increasingly more responsive to light to which they are exposed in the earlier hours of the day.

Comparing the data of figs. 5 and 6 with those of fig. 3 the following important conclusion can be drawn. The decline of the photo-reactivity threshold between late summer and spring (fig. 3) is based on the change

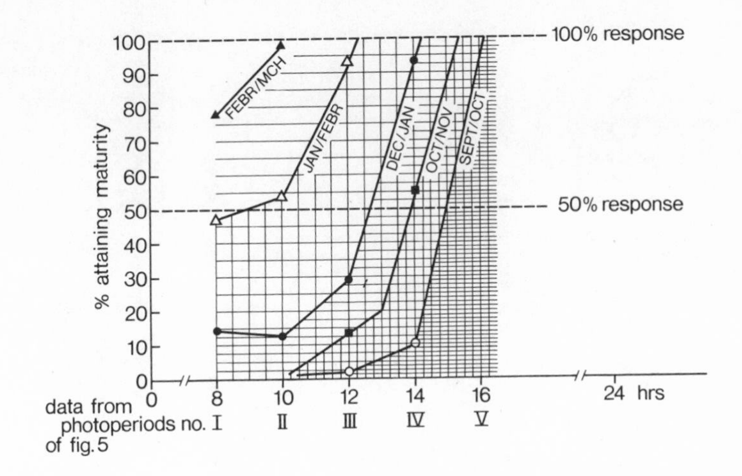

Fig. 6. The same response-percentages as given in fig. 5, but represented in a way which shows how between September and March the animals become increasingly more responsive also to the (shorter) photoperiods IV, III, II and I respectively. When compared to fig. 4, the curves show that in the course of the year a change takes place in the 'shape' of the daily light-sensitivity rhythm. Further explanation in the text.

taking place in the shape of the daily light-sensitivity curve (figs. 5 and 6).

The occurence of a seasonal change in the daily light-sensitivity rhythm has (to my knowledge) not been demonstrated before. However, Farner (1967) has postulated the existence of such a change in birds in order to explain "... the puzzling observations that there is a range of photoperiods that are photostimulatory in spring, but not in fall after the birds have become photo-sensitive". He suggested that this difference in response might be caused by a phase-shift of the daily light-sensitivity cycle, in that the whole curve (see fig. 4) might be shifted to the left between autumn and spring. As can be seen, such a shift would also result in the animals becoming more sensitive to shorter photoperiods. It seems possible that the changes depicted in fig. 6 are due to such a phase-shift, although it seems to me that, in addition, the shape of the rhythm also changes, because its upward slope becomes less steep.

III.4. Factors controlling the seasonal change in the daily light-sensitivity rhythm

In section III.1. it was mentioned that the onset of breeding apparently is determined by the rate of decline of the photo-reactivity threshold (fig. 3) and that it is, therefore, very important to analyse the factors involved in its control. In the previous section we have seen that, underlying this threshold decline is the change which takes place in the daily light-sensitivity cycle (figs. 5 and 6). Therefore, if we would know the factors controlling

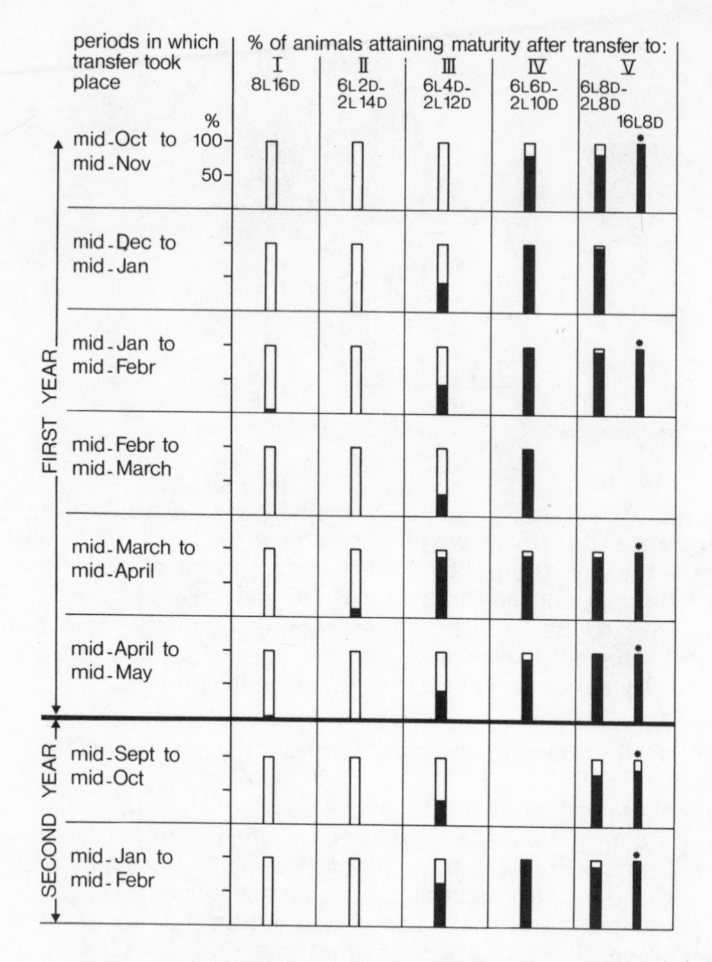

Fig. 7. Response-percentages obtained in an experiment similar to that of fig. 5, but carried out with animals maintained continuously under 8L-16D, 20° from the beginning of August, i.e. from about two months after hatching in nature. They were exposed to the photoperiods I-V at various times during the next one year and a half. Further explanation in the text.

the latter change, we would also know those controlling the rate of threshold decline. It does not seem unlikely that endogenous and/or external factors (like photoperiod and temperature) may be involved. In the next two sections evidence will be presented supporting this idea.

III.4.1. Factors having an endogenous origin

To prove the existence of endogenous factors it is necessary to keep animals for long periods of time under unchanging external conditions. In the experiments described in this section this was done by exposing underyearling sticklebacks continuously to 8L-16D at 20°C for over one year and a half. The fish had hatched in June in the outside pond and exposure to the constant condition started at the beginning of August. From September onwards the responses of these animals to the same skeleton-photoperiods as those of fig. 5 were studied with regular intervals. Only the mean results of this as yet unpublished experiment are represented in fig. 7 and more details will be given in a separate paper.

It can be seen that (like in fig. 5) the response-percentages were higher when the two-hour lightpulse was given later in the day (horizontal comparison of the columns in each period of the year). Again (like in fig. 5) increasingly higher responses to each of the photoperiods were obtained between September and May (vertical comparison of the columns for each photoperiod). However, unlike in fig. 5, only two animals (one in January/February and one in April/May) ever responded to photoperiod I (8L-16D). Likewise only two animals = 14% responded to photoperiod II (6L-2D-2L-14D), viz. in March/April. Thus, there is a large discrepancy between the responses to these two photoperiods of animals which had been living under natural conditions (fig. 5) and of those which had been living under the unchanging conditions of 8L-16D, 20° since August (fig. 7). The differences with respect to the other photoperiods were much smaller or even negligible.

Looking at the responses of the animals maintained continuously under 8L-16D, 20° for longer than one year (fig. 7), we see that hardly any further changes occurred. Neither in September/October, nor in January/February of the second year were there any responses to photoperiods I and II, while those to photoperiod III became somewhat lower in April/May and from then on remained rather low. Like in the previous year, the responses to photoperiods IV and V remained high. The data are insufficient to show that even under constant external conditions there is annual rhythm in light-sensitivity, although looking at the data of fig. 7 one may perhaps see an optimum between March and April of the first year.

The conclusion from fig. 7 is that even in animals maintained under unchanging external conditions (8L-16D, 20°), a change occurs in their daily light-sensitivity cycle, as happens in animals living under natural conditions. However, in the animals living under constant conditions this change was less far-reaching than in those living under natural conditions, for hardly any responses occurred to the shorter photoperiods I and II. This means that in animals living continuously under 8L-16D, 20° the threshold does not decline as low as it does in animals living under natural conditions (fig. 3). Instead, the decline "stops" at about the levels of skeleton-photoperiod III or perhaps II, which are comparable to the normal photoperiods of 12L-12D and 10L-14D respectively (section III.2.).

These findings lead to at least two conclusions. In the first place, since a change occurs in the daily light-sensitivity of animals living under constant external conditions, there can be little double that this change has an endogenous origin. The bit of doubt expressed here stems from the fact that the fishes used for this experiment had been exposed to natural conditions (long photoperiod and high temperature) between hatching in June and August, when they were transferred to 8L-16D, 20°. This change in external conditions, due to the transfer, might have triggered the onset of the change in the daily cycle. To rule out this possibility, a similar experiment should be carried out with animals having been exposed to 8L-16D, 20° from the time of hatching. In 1978/1979 I ran a pilot experiment with underyearling fish which had hatched in June from eggs that were laid and fertilised by parent animals already living under 8L-16D, 20°. The next year, beginning March 12, nine animals were exposed to photoperiod II and nine to photoperiod III. The result was that none of the fish exposed to photoperiod II did attain maturity within 65 days, whereas two = 20% matured when exposed to photoperiod III. These responses are, therefore, very similar to those of the animals of fig. 7 in the same period of the year (February/March). Although this experiment should be repeated with larger numbers of fish and at different times of the year, there can be no doubt in my opinion that the change in the daily light-sensitivity curve also occurs in fish hatched and maintained continously under the constant external conditions of 8L-16D, 20°.

The second conclusion from these experiments is that, apart from endogenous factors, external factors are also involved in the control of the change in the daily rhythm. This can be concluded from the already discussed difference between the data of figs. 5 and 7, particularly with respect to the responses to photoperiods I and II. Apparently, the extent of the change is greater in fish living under natural conditions between late summer and spring than in those living constantly under 8L-16D, 20°. It is difficult, however, to make a suggestion as to which of the external factors this difference should be attributed. Under natural conditions between summer and January, both photoperiod and temperature decline, whereas they increase again from January onward. This means that a change in photoperiod, or in temperature, or in both may be involved in the regulation of the change in the daily light-sensitivity rhythm. So far I have only run an experiment to test the influence of temperature; the results are dealt with in the next section.

III.4.2. Factors originating in the environment

In 1977/1978 an experiment was carried out similar to that of fig. 7, with underyearling sticklebacks, which had hatched in the outside pond and had been transferred to the laboratory at the beginning of August. From that time onward they were exposed constantly to 8L-16D, but at two different temperatures, viz. 20° and 15°C. The mean results of this as yet unpublished experiment are given in fig. 8 and full details will be presented in a separate paper. To facilitate comparison, the data of figs. 5 and 7 have been once again represented in fig. 8.

Comparison of the responses of the 8L-16D animals maintained at 20° and 15° shows very clearly that the latter exhibited the largest range of

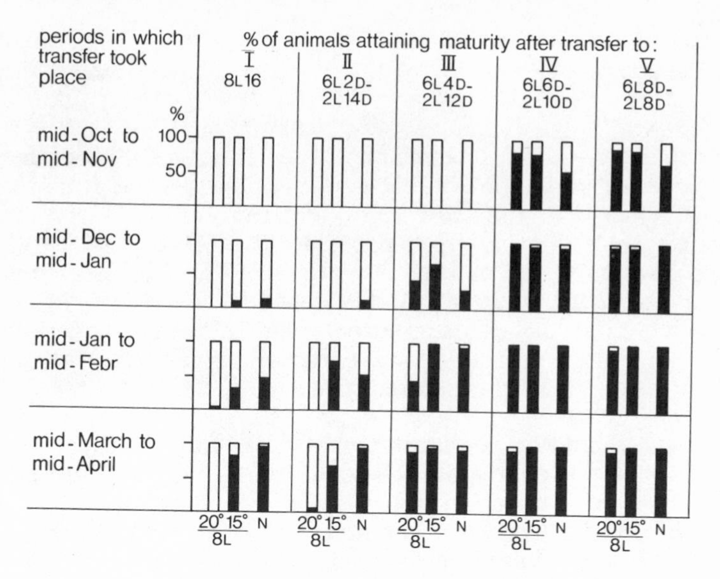

Fig. 8. Comparison of the response-percentages of animals which had been living under different external conditions from about two months after hatching in nature. One group continued to live under natural conditions (N) until exposed to photoperiods I-V. The two other groups were both maintained under 8L-16D, but one group at a temperature of 20°C (8L, 20°) and the other at 15°C (8L, 15°), until the times they were exposed to photoperiods I-V. Further explanation in the text.

change in the daily light-sensitivity cycle. This is particularly evident when the responses to photoperiods I and II are compared in the period between mid-March and mid-April. For instance, whereas none of the animals of the 20° group responded to photoperiod I in March/April, 83% of the 15°- group responded. Again, whereas only 14% of the 20°group reacted to photoperiod II in March/April, 68% of the 15°-group attained maturity. Comparable differences in the responses to photoperiods I and II also occurred at other times of the year. With respect to photoperiods III, IV and V the differences were (much) less outspoken, or absent.

In addition, fig. 8 shows that the seasonal change of the daily rhythm took place faster under 15° than under 20°. This is most evident with respect to photoperiod II. For instance, the 20°-group responded for the first time (with 14%) to this photoperiod in March/April, whereas already 73% of the 15°-group had responded to this condition as early as January/February. The responses to photoperiod I and III show a similar trend. All this means that the rate of change of the daily cycle was faster under the lower than under the higher temperature.

Comparing the responses of the 15°-group with those of the fish which had been living under natural conditions during the same period of time (N in fig. 8), it can be seen, that the responses of these two groups are very much the same. This similarity holds for all photoperiods and for all times of the year.

The conclusions of this experimental series with the lower temperature confirms those arrived at in the previous section. In the first place, since the seasonal change in the daily light-sensitivity rhythm also occurred in animals which were continually exposed to 8L-16D at 15°, this is further evidence in favour of the endogenous origin of this change. In the second place, since the range over which this change takes place, as well as the speed with which it occurs is greater under 15° than under 20°, this is further evidence that at least one external factor, viz. temperature, is involved in the control of this seasonal change.

Finally, it should be considered what these findings mean with respect to the decline of the photo-reactivity threshold between late summer and spring (fig. 3). which underlies the onset of breeding in the spring (section III.1.). In section III.3. it was concluded that the decline of the photo-reactivity threshold is based on the change taking place in the same period of the year in the daily light-sensitivity rhythm, making the animals more responsive to the shorter photoperiods. Therefore, on the basis of the findings of fig. 8 it can now be concluded that the onset of breeding in the spring is controlled by endogenous as well as external factors because 1) endogenous factors initiate the thresholddecline in late summer, while 2) the rate of its decline, as well as the level to which it declines is controlled by endogenous factors in co-operation with external factors, among them the ambient temperature. This also means that the onset of breeding in spring is to a large extent controlled by physiological processes taking place well in advance of it, viz. between late summer and early spring.

IV. GONADAL REGRESSION

Considering that the annual gonadal cycle consists of alternating periods of recrudescence and regression, the timing of breeding is determined not only by factors controlling its onset, but as well as by those governing its termination. However, much less research has been carried out on the latter subject. This may have been cuased by difficulties in bringing and maintaining fishes in breeding condition in the laboratory.

The stickleback Gasterosteus aculeatus is one of the teleosts that can be most easily induced to breed under laboratory conditions. Baggerman (1957) studied the duration of the breeding condition of sticklebacks maintained under different constant external conditions. To determine the duration of breeding the following criteria were used. Males in breeding condition will rebuild their nests within one or two days after they have been destroyed. Therefore, nests were destroyed with weekly intervals and breeding was considered to be ended, when an animal had not rebuilt its nest within a week after its destruction. The duration of the breeding condition could thus be expressed as the number of days between the building of the male's first nest of the season and the day its last nest was destroyed.

During the breeding season female sticklebacks produce clutches of eggs with intervals of 5-7 days (when food conditions are optimal). After ovulation the females can deposit their eggs in the nest of a male, or at any place in the aquarium when no male with a nest is present. As mentioned in section II, eggs can also be stripped easily once ovulation has taken place. The duration of breeding in the female was expressed as the number of days between the production of her first and last clutch of the season.

Under natural conditions sticklebacks breed in Holland between about mid-April and mid-July, i.e. at a time of the year when the photoperiod is relatively long (13-16 hrs; fig. 3) and the temperature relatively high (15-20°). Baggerman (1957) determined the duration of breeding under two different constant external conditions viz. 16L-8D and 8L-16D, both at 20°. The animals used had been caught in their natural habitat in March and from that time onward exposed to the conditions mentioned. As fig. 3 shows, the fish readily mature under both conditions at this time of the year.

When exposed to 16L-8D, 20° the average duration of breeding in the males was 91 days (extremes 23-198 days) and in the females 60 days (extremes 31-77 days). In this period the females laid on the average 11 clutches of eggs. The shorter duration of breeding in the females may have been due to sex differences, and/or to the occasional lack of sufficient amounts of proper food (live Daphnia). The duration of breeding in males kept at 16L-8D, 20° (± 3 months) agrees very well with that of males living under natural conditions (mid-April - mid-July).

When exposed to 8L-16D, 20° breeding in the males was much shorter, viz. 46 days on the average (extremes 5-112 days). The females under this condition produced no more than three clutches at the most and the mean duration of their breeding period was 7 days on the average (extremes 4-13 days).

From these data Baggerman (1957) has drawn two conclusions. In the first place that breeding is ended by factors of endogenous origin, since it occurred under constant external conditions. In the second place she concluded that, in addition, external factors are involved, since breeding lasted so much longer under 16L-8D, 20° than under 8L-16D, 20°.

Baggerman (1972) has tried to fit these data into the threshold theory. She postulated that breeding ends because some endogenous process induces the photo reactivity threshold to start rising shortly after having reached its lowest level: this hypothetical rise has been indicated in fig. 3. Due to this threshold increase, breeding stops earlier when the animals are exposed to short than to long photoperiods. This can be deducted from fig. 3, which shows that the rising threshold first passes the 8L-16D level, which means that the photoperiod is then no longer able to overcome this threshold, with the result that breeding stops. The rising threshold reaches the 16L-8D level later, with the result that breeding under this condition lasts longer than under 8L-16D. In this hypothesis, therefore, the termination of breeding is caused by some endogenous mechanism inducing an increase of the photo-reactivity threshold. As yet nothing is known concerning factors controlling the rate of threshold increase, but it is not unlikely that endogenous as well as external factors may be involved.

As to the possible causes underlying this threshold increase, the following suggestion can be made. In view of the finding that a change in the daily light-sensitivity cycle is involved in the control of threshold-decline, it seems possible that a similar change, but now in the opposite direction (compare fig. 6), may be involved in the control of the threshold-increase. To stretch speculations even further, it could be postulated that the regulation of the breeding season depends on an annual cycle of the photo-reactivity threshold, which in its turn is based on an annual cycle in the shape, or phase of the daily light-sensitivity curve (section III.3.). We shall return to this point in section V.

The asian catfish Heteropneustes fossilis, seems to be the only other teleost in which factors affecting gonadal regression have been studied. Sundararaj and Sehgal (1970b) and Sundararaj and Vasal (1976) found that in this species regression occurs spontaneously under various constant external conditions, including continuous darkness (DD) and continuous light (LL), as well as short and long photoperiods. Thus, also in the catfish breeding is ended by some endogenous mechanism. In addition, the same authors found that external conditions are also involved in the control of the duration of breeding, since it lasted longer under long than under short photoperiods (like in the stickleback), while moreover, it could be prolonged by a high temperature (30°), regardless of the photoperiod.

V. DO FISHES POSSESS AN ENDOGENOUS ANNUAL REPRODUCTIVE RHYTHM?

In the preceding sections evidence has been presented that endogenous factors (mechanisms) are involved in the onset, termination and regulation of different events of the annual gonadal cycle. Therefore, the question arises as to whether or not these endogenous factors come into play at the respective times because their appearance would be based on an underlying circannual rhythm.

Such a rhythm can only be established in longlasting experiments carried out under constant external conditions and covering at least two (circ) annual cycles. Some authors believe that the animals should be kept in continuous darkness (DD), or continuous light (LL) in order to obtain conclusive evidence on this matter. Others maintain that such conditions might endanger the animals' health and hold that circannual rhythms can also be demonstrated when animals are exposed to different constant external conditions with respect to photoperiod and temperature. When a cycle would appear under each of these constant conditions with a period

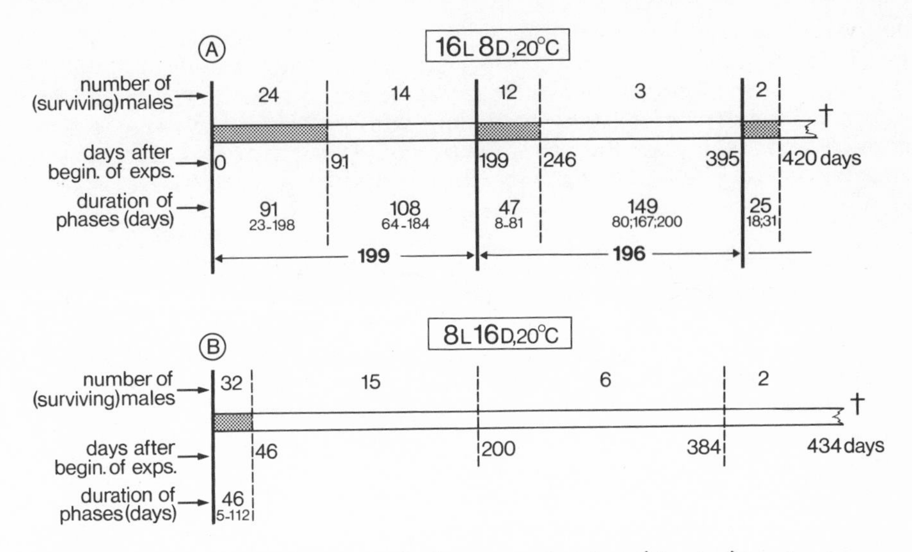

Fig. 9. Duration in days of the reproductive (shaded) and non-reproductive phases (unshaded) of male sticklebacks which (since early spring) had been maintained continuously under 16L-8D, 20° (A) and 8L-16D, 20° (B) for 420 and 434 days respectively. Further explanation in the text.

deviating significantly from that of one year, this could also be considered as conclusive evidence in favour of a circannual cycle.

Baggerman (1957) exposed sticklebacks, from about one month prior to the onset of breeding in nature, for more than 400 days to constant photoperiods of 8L-16D and 16L-8D, both at 20°. These are the same animals of which data on the duration of breeding were presented in the previous section. During this whole period they were supplied with ample nesting material, while it was tried to feed them *ad libitum* with high quality food (live *Daphnia*), although this was not always available. In the males, the day on which they built a new nest after a period of reproductive inactivity, was regarded as the onset of a new breeding period, while in the females this was the day on which they produced a new clutch of eggs after a period of ovarian inactivity. The duration of the second and third reproductive periods was determined in the same way as described in the previous section. The results of these long lasting experiments are given in fig. 9. Attention should be drawn to the fact that in the course of time many animals died, so that the average durations of the reproductive and non-reproductive periods are based on increasingly smaller numbers of animals (these numbers are also given in fig. 9).

Figure 9 shows that under 16L-8D, 20° the males showed alternating reproductive and non-reproductive phases or periods. The average duration of the first breeding period was 91 days (see previous section) and this was followed by a non-reproductive phase lasting 108 days. Of the original 24 males 12 started a second breeding period which lasted 47 days and was followed by a non-reproductive period lasting 149 days. Of the three surviving males two started a third breeding phase, which only lasted 25 days and ended shortly before they died. In addition fig. 9 shows that the period of each gonadal cycle, i.e. the number of days between the onset of two consecutive breeding periods, was 199 and 196 days respectively, thus about 200 days.

In females continuously maintained under 16L-8D, 20° the duration of the first breeding phase was 60 days on the average (see previous section). This was followed by a non-reproductive period lasting 168 days, after which three of the four surviving females started a second breeding period. Due to the absence of sufficient amounts of live food no reliable data were obtained on the duration of the second breeding period. However, it is clear that a breeding cycle also occurs in females continuously exposed to 16L-8D and 20°C. The finding that the period of the cycle of the males (about 200 days) and that of the females (228 days) both deviate significantly from the period of one year occurring under natural conditions, is an important indication for the endogenous (circannual) nature of this breeding rhythm.

However, this result is in complete contrast to that obtained when animals were continuously exposed to a short photoperiod, 8L-16D at 20°; (fig. 9). In the males the average duration of the first breeding period was 46 days (see previous section), after which none of the surviving animals ever started a second breeding period. The females showed a similar response: the duration of the first breeding period was on the average 7 days (see previous section) and none of the surviving fish ever started a second breeding period. The conclusion, therefore, is that under a constant short photoperiod of 8L-16D at 20° there are no overt signs of a (circannual) breeding rhythm. As mentioned, this conclusion pertains to the occurrence of actual breeding, viz. nestbuilding in the male and the production of ripe eggs by the female, since these were my criteria. The absence of actual breeding, however, does not imply that the gonads may not have shown some covert signs of development and regression. This possibility should be investigated in the future on the basis of histological studies of the gonads. If this were found to be true, then the conclusion would have to be, that even under continuous exposure to 8L-16D and 20°C a (circannual) gonadal cycle occurs, even though this cycle does not express itself in overt reproductive behaviour (see further below).

It is interesting to note that the most likely explanation for the failure of a breeding rhythm to express itself under 8L-16D at 20° can be read from figs. 7 and 8, which show that at this high temperature the photo-reactivity threshold does not decline low enough to be overcome by 8L-16D. Instead, as was argued in section III.4.1., it declines only to a level which can be overcome by photoperiods longer than about 10L-14D. If this explanation is correct, a circannual breeding cycle should be able to express itself even at the high temperature of 20°C, but only when the photoperiods are longer than 10L-14D. In addition, fig. 8 has shown that even under the short photoperiod of 8L-16D the photo-reactivity threshold is able to decline to that level, but only when the temperature is 15°C. This would mean that a circannual breeding cycle should be able to express itself even under 8L-16D, but only when the temperature is 15°C. These two predictions will be tested in future research. If they would hold, then it would be justified to conclude that in the stickleback annual breeding is based on an endogenous annual rhythm of the photo-reactivity threshold, although certain external conditions will have to be present to allow this rhythm to obtain its maximal "amplitude" (permissive action).

There is only one other teleost in which the problem of a circannual rhythm has been studied. Sundararaj and Sehgal (1970b) and Sundararaj and Vasal (1973, 1976) studied the ovaries of the catfish Heteropneustes fossilis, when the animals were maintained for 34 months under continuous darkness (DD) or continuous light (LL) both at 25°C. By using changes in ovarian weight/100 g bodyweight as main criterion, they found that under both conditions ovarian cycles occurred with periods deviating considerably from that of one year (although this difference was significant only under LL). Particularly the observation that the ovaries of fish maintained under DD were able to produce eggs in which apparently considerable amounts of yolk had been deposited, is significant and supports the idea of an underlying circannual rhythm. It should be added, however, that the maximal weight the ovaries obtained under DD was about half the weight of fully mature ovaries of fish caught in nature. The circannual rhythm, therefore, only seems to pertain to this first, apparently light-independent phase of recrudescence. This may mean that the second phase of recrudescence can only develop when the animals are exposed to external conditions having a daily light/dark cycle. This would tie in with the finding that the catfish is a truly photoperiodic species, using a daily light-sensitivity rhythm as clock (section III.2). The whole situation in the catfish would then be very similat to that in the stickleback, in which the photo-reactivity threshold also only pertains to the second phase of gonadal development (section III.1.). However, I have no data on whether or not phase 1 in the stickleback would be able to develop in DD. The only thing I know concerning this phase is, that it is able to develop under widely different external conditions (section III.1.) and thus perhaps also in DD.

The refractory period

The literature on photoperiodic regulation of breeding often deals with the existence of a refractory period. As mentioned before (section III.1.) this is a period in the annual cycle in which the gonads are unresponsive to a long photoperiod, which at other times in the year is very stimulatory. My own experiments with sticklebacks show that also in this species a refractory condition occurs. We have seen (fig. 9) that even under optimal conditions (16L-8D, 20°) breeding eventually comes to an end, due to some endogenous mechanism (section IV). By definition, the fish have thus entered the refractory period. My data are not suited, however, to draw conclusions as to the duration of this period. I can only say that this period is included in the non-reproductive phase represented in fig. 9, together with the period needed for complete gonadal recrudescence up to maturity (which was my criterion). The occurrence of alternating breeding and non-breeding periods in fish continuously maintained under 16L-8D, 20° further shows that the refractory period is not only induced by some endogenous mechanism, but apparently also endogenously terminated, since no change took place in the external conditions.

In terms of my threshold theory, the refractory period begins as soon as the rising photo-reactivity threshold (fig. 3) has surpassed the level which can be overcome by 16L-8D, 20° (section IV). From the same point of view, refractoriness ends when the endogenous factors (either or not in co-operation with external factors), induce the threshold to decline again with the result that at a given moment its level can be overcome once again by 16L-8D, 20°.

Not much information is available on the refractory period in other teleosts. Harrington (1959) working with *Notropis bifrenatus* and Sehgal and Sundararaj (1970) working with *Heteropneustes fossilis* have given evidence for the existence in both species of a refractory period. They found that their animals were unresponsive to a relatively long photoperiod (with a high temperature) in the autumn when the fish are in their postspawning period, while they readily responded to the same condition a few months later.

In the literature on photo-refractoriness in birds it is often reported (e.g. Wolfson, 1959) that in several species the refractory period does not end, unless the birds have first been exposed to a short photoperiod. The findings concerning gonadal cycling under 16L-8D, 20° (fig. 9) show that in sticklebacks exposure to a short photoperiod is not necessary to end the refractory period. However, in the catfish Sundararaj and Sehgal (1970a) reported that it was possible to obtain fishes with responsive ovaries also during the earlier part of the post-spawning period, by first exposing them to a decreasing photoperiod. This means that the refractory period could be made to end earlier by prior exposure of the animals to this condition. However, prior exposure to a decreasing photoperiod is apparently not essential for the termination of refractoriness when the temperature is 25°C or higher (Sundararaj and Vasal, 1976). This also explains how a circannual ovarian cycle can occur in catfish continuously exposed to LL and DD at 25°C, as we have seen above.

To end this section I would like to point out that there seems to be a difference between the nature of the annual (presumably circannual) rhythms of stickleback and catfish. In the stickleback there is evidence for a circannual rhythm of the photo-reactivity threshold which, by definition, will only be able to express itself under photoperiodic conditions. In the catfish, on the other hand, a circannual gonadal rhythm is able to express itself even in continuous darkness (DD) which, therefore, pertains to a light-independent process. It seems possible that this difference is due to species specific properties. However, it is equally well possible that the difference is only seeming, since the rhythms evidently pertain to two different phases of recrudescence, viz. the second phase in the stickleback (phase 2) and the first phase in the catfish. Therefore, it seems possible that both types of circannual rhythms may occur in both species, each pertaining to a different phase. The circadian light-sensitivity rhythm, which has been demonstrated in both species (section III.2.) would then function as time-measuring mechanism involved in the photoperiodically controlled (second) phase of gonadal development.

The question which external factor may be used as Zeitgeber for the circannual rhythms will not be dealt with in this paper, because it lies outside its scope and the reader is referred to the general statement made in section I. The very interesting problem of how circadian and circannual rhythms may be "coupled" in their combined control of annual reproduction, likewise will not be dealt with, again because this would lead too far from the main topic of this paper. I only would like to point out that the finding in the stickleback, that breeding is based on an annual (probably circannual) change in the shape, or phase of the circadian light sensitivity rhythm merits being mentioned once again, because as far as I know such a coupling of rhythms has not been shown before in any other animal species.

VI. THE NEURO-ENDOCRINE GONADAL SYSTEM AND THE WAYS IN WHICH IT IS AFFECTED BY EXOGENOUS AS WELL AS ENDOGENOUS FACTORS

The neuro-endocrine system involved in the gonadal cycle of teleost fishes is to a large extent comparable to that of the higher vertebrates. Specialised neurons (neuro-secretory cells) in the hypothalamus produce a secretion, which is called a regulating factor or hormone (RF = RH) and which is transported through a short bloodvessel system, or through perivascular spaces to the adenohypophysis. In this gland the RF stimulates certain cells to produce gonadotropic hormones, which in their turn are carried by the systemic bloodflow to the gonads. It is still not clear whether the pituitary gland of teleost fishes produce only one, or two different gonadotropins like in mammals (FSH and LH). There is good evidence, however, that the stickleback has two different gonadotropic cell types, one of them being involved in the regulation of phase 1 and the other in phase 2 (Slijkhuis, 1978). Under the influence of gonadotropins the gonads are stimulated to develop to complete maturity with respect to gametogenesis and the production of gonadal hormones. More information can be found e.g. in Ball and Baker (1969) and Bentley (1976).

It has been shown in the previous sections that gonadal development is affected by both photoperiod and temperature. Therefore, the neuro-endocrine system must be connected to photo- and thermoreceptors. As we have seen there is also good evidence that this system is to a large extent controlled by two endogenous rhythms, the "pacemakers" of which must be located either inside or outside the above described neuro-endocrine system. A detailed discussion of these points lies outside the scope of this paper, but a brief survey is thought to be necessary to complete the picture. The information dealt with comes mainly from studies in mammals and birds, as comparatively little research has been carried out in fishes. References will be made only to papers which will give good starting points for acquiring further information.

The location of receptors involved in the control of annual breeding

Although there is conclusive evidence that certain gonadal processes in fishes are affected by the temperature, very little seems to be known about the location of the temperature receptors involved.

In mammals, the photoreceptors involved in the regulation of the gonadal cycle seem to be exclusively located in the retina of the eyes (Reiter, 1974). In birds, it has been shown that photoreceptors mediating photoperiodic responses are located not only in the eyes, but also in the brain; the latter are known as encephalic receptors, but their precise location within the brain is not known (Menaker and Keatts, 1968). There is not much evidence available on neural pathways leading from retinal and encephalic receptors to the RF-producing cells in the hypothalamus.

In fishes, it seems likely that the eyes will be involved in the photoperiodic responses, although little research has been carried out on this subject. It seems not unlikely that fishes may also possess encephalic receptors. In addition, numerous photoreceptors are located in the pineal organ, which is situated on top of the brain directly under the skull. In birds and mammals this organ (no longer) has photoreceptor cells, but instead has cells which synthesise and secrete melatonin, which has a pronounced inhibitory influence on the gonadal cycle (see below). In recent years melatonin has also been demonstrated in fishes (see below), and therefore in these animals the pineal organ may be involved in the mediation of photoperiodic responses in two different ways. In the first place, it may serve as a photoreceptor, the information of which is transmitted to the hypothalamus along neural pathways. In the second place, the pineal may relay the information from its own, or from retinal photoreceptors to the melatonin producing cells and thus affect the neuro-endocrine system along a humoral pathway.

The pineal organ (gland)

The pineal gland synthesizes melatonin from seratonin, by means of the successive action of two enzymes. Synthesis of melatonin is almost completely restricted to the dark hours of the day. Apart from melatonin the pineal also secretes a number of polypeptides, but their function is as yet unknown.

Particularly in mammals much research is being carried out on the role of the pineal gland and melatonin in reproduction. It has been found that melatonin has an inhibitory influence on gonadal function, either directly, or indirectly by affecting the gonadotropic cells. In addition, it is known that the gonads of mammals become highly stimulated in continuous light, which ties in with the finding that hardly any melatonin is produced during the hours of light. There is some evidence that the pineal receives its information on the daily light/dark cycle by way of retinal photoreceptors. More information on the role of the pineal in mammalian reproduction can be found in Reiter (1972 and 1974).

In birds much research is being done on the role of the pineal in reproduction, but many conflicting data have been reported, indicating stimulatory as well as inhibitory effects (Takahashi et al, 1978).

Although in lower vertebrates much information is available on the structure of the gland, its function is still enigmatic. Fenwick (1970a) has given a review of all aspects of pineal morphology and physiology in fishes and Fenwick (1970b), Vasal and Sundararaj (1975), and Urasaki (1976) give some additional information. In summary, the following main points can be mentioned. In fishes, the pineal organ is able to respond to local illumination, while there is some evidence, that it may also receive information by way of the eyes. Melatonin has been demonstrated in the pineal of a number of fishes. In salmon, the pineal melatonin store is lower in spawning animals than in immature ones. In several species, injections with melatonin have been shown to have an inhibitory effect on gonadal function. In view of these data it is, therefore, very surprising that quite a few studies have reported no effect of pinealectomy on gonadal function. Fenwick (1970b), however, has suggested that this may have been due to the experiments having been carried out in the wrong periods of the year. For in his own studies he found that pinealectomy was only effective in the goldfish (*Carassius auratus*) between January and May, i.e. the season in which the gonads are maximally responsive to stimulation by long photoperiods. Urasaki (1976) working with the medaka (*Oryzias latipes*) concluded that the pineal gland may exert two opposite functions, depending on the length of the photoperiod. Under a long photoperiod it would exert a stimulatory and under a short photoperiod an inhibitory influence.

Very little information is available on the ways in which the pineal gland may affect gonadal activity in fishes: Fenwick (1970b) has made some suggestions.

Endogenous rhythms (biological clocks)

Circadian rhythms have been demonstrated at all levels of organization, from single cells to tissues and organs, and from the level of the individual animal to that of whole populations. Many examples of the countless processes and activities showing circadian rhythms can be found in introductory books on biological rhythms, such as those of Palmer (1976), Rensing (1973) and Saunders (1977). Data on circannual rhythms, the occurrence of which has only been demonstrated during the last decade, can also be found in those books, as well as in Pengelley (1975).

It is generally assumed that in the body of multicellular organisms not one master-clock is present regulating all circadian processes, but many clocks each controlling one, or perhaps a group of processes. All these pacemakers are thought to have varying degrees of mutual connections, while synchronisation with the environment may be achieved via Zeitgebers (section I).

Ever since the existence of biological clocks became known, much attention has been paid to the nature and location of the circadian pacemakers. It is generally assumed that they are based on fundamental properties of cellular organization, but so far only hypothetical explanations have been put forward (Palmer, 1976; Saunders, 1977). More information is available on the location of these pacemakers, particularly in insects and mammals (see review by Kawamura and Ibuka, 1978). In mammals, evidence is accumulating that the suprachiasmatic nucleus may be the site of an important circadian pacemaker. In addition, reports are accumulating indicating the existence of a nervous pathway between the eyes and this hypothalamic nucleus. This could be the pathway along which signals from the light/dark cycle could serve as Zeitgeber for the synchronisation of the rhythm of the pacemaker with that of the environment.

It is remarkable that in birds many data seem to point to the pineal organ as the site of a biological clock, which is particularly involved in the control of locomotory activities, whereas in mammals reports on such a role of the pineal seem to be lacking.

In fishes very little research if any seems to have been carried out as yet on the location of biological clocks. Therefore, with respect to the evidence reported in this paper that two endogenous rhythms, a circadian and a circannual one are involved in the control of the annual reproductive cycle, only the following very general remarks can be made. The pacemakers of these rhythms could be located either "outside", or "inside" the neuro-endocrine system underlying gonadal maturation. If they would be situated outside this system, a location somewhere in the brain seems to be the most likely place. If they would be situated inside the neuro-endocrine system, they could be located at any place, in the RF-cells of the hypothalamus, in the hypophysis or even in the gonads themselves. The latter, however, seems unlikely since Sundararaj et al (1972) have reported that the ovaries of the catfish Heteropneustes fossilis are responsive to stimulation by (salmon) gonadotropins, even during their refractory period (see section V), when they are unresponsive to photo-stimulation.

SUMMARY

The aim of the paper is to analyse the ways in which photoperiodic teleost fishes are able to control annual breeding by means of an integrated action between photoperiod and endogenous rhythms. The evidence presented mainly comes from investigations on the stickleback (Gasterosteus aculeatus L.), but findings in other species are discussed at appropriate places. Evidence is given that the stickleback becomes increasingly more responsive to shorter photoperiods between late summer and early spring. On the basis of these findings the theory is put forward that this increased responsiveness is due to a decline of the photo-reactivity threshold. It will further be shown that a daily, most probably circadian rhythm in sensitivity to light is used as a time-measuring mechanism, enabling the differentiated responses to photoperiods of different lengths. Evidence will be presented that this daily light-sensitivity rhythm changes its shape, or phase between late summer and early spring, which most likely is the basis of the decline of the photo-reactivity threshold. Subsequently, evidence will be presented showing that this change in the daily light-sensitivity cycle is controlled by endogenous as well as external factors, among them temperature. It will further be shown that gonadal cycling may occur even when animals are maintained continuously under constant external conditions for 34 months or longer. The data and considerations presented have led to the hypothesis that in the stickleback a circannual rhythm in the circadian light-sensitivity rhythm controls the course of the photo-reactivity threshold and thus underlies annual breeding under natural conditions. Finally, some attention is paid to the neuro-endocrine system underlying breeding and the ways in which it may be affected by external and endogenous factors.

REFERENCES

Baggerman, B. (1957). An experimental study on the timing of breeding and migration in the three-spined stickleback (Gasterosteus aculeatus L.). Arch. Néerl. Zool. 12: 105-318.

Baggerman, B. (1972). Photoperiodic responses in the stickleback and their control by a daily rhythm of photosensitivity. Gen. Comp. Endocrinol. Suppl. III: 466-476.

Ball, J.N. and Baker, B.I. (1969). The pituitary gland: anatomy and histophysiology. In: Fish Physiology II. Eds. W.S. Hoar and D.J. Randall. New York, Academic Press, p. 1-110.

Barr, W.A. (1968). Patterns of ovarian activity. In: Perspectives in Endocrinology. Eds. E.J.W. Barrington and C. Barker Jørgensen. New York, Academic Press, p. 164-237.

Bentley, P.J. (1976). Comparative vertebrate endocrinology. Cambridge Univ. Press.

Breton, B. and Billard, R. (1977). Effects of photoperiod and temperature on plasma gonadotropin and spermatogenesis in the rainbow trout Salmo gairdnerii Richardson. Ann. Biol. Anim. Biochem. Biophys. 17 (3A): 331-340.

Combs, B.D., Burrows, R.E. and Bigej, R.G. (1959). The effect of controlled light on the maturation of adult blueback salmon. Progr. Fish. Cult. 21: 63-69.

Craig-Bennet, A. (1931). The reproductive cycle of the three-spined stickleback, Gasterosteus aculeatus L. Phil. Trans. Roy. Soc. Lond. B 219: 197-279.

Farner, D.S. (1967). The control of avian reproductive cycles. In: Proc. XIV Int. Ornithol. Congr. Ed. D.W. Snow. Oxford, Blackwell, p. 107-133.

Farner, D.S. (1970). Predictive functions in the control of annual cycles. Environ. Res. 3: 119-131.

Fenwick, J.C. (1970a). The pineal organ. In: Fish Physiology IV. Eds. W.S. Hoar and D.J. Randall. New York, Academic Press. p. 91-108.

Fenwick, J.C. (1970b). The pineal organ: photoperiod and reproductive cycles in the goldfish, Carassius auratus L. J. Endocr. 46: 101-111.

Harrington, R.W. (1959). Photoperiodism in fishes in relation to the annual cycle. In: Photoperiodism and Related Phenomena in Plants and Animals. Ed. R.W. Withrow. Washington, D.C., Publ. 55 of the A.A.A.S. p. 651-667.

Henderson, N.E. (1963). Influence of light and temperature on the reproductive cycle of the eastern brook trout, Salvelinus fontinalis Mitchill. J. Fish. Res. Board Can. 20: 859-897.

Kawamura, H. and Ibuka, N. (1978). The search for circadian rhythm pacemakers in the light of lesion experiments. Chronobiologia 5: 69-88.

Lees, A.D. (1966). Photoperiodic timing mechanisms in insects. Nature (Lond.) 210: 986-989.

Lofts, B. (1968). Patterns of testicular activity. In: Perspectives in Endocrinology. Eds. E.J.W. Barrington and C. Barker Jørgensen. New York, Academic Press. p. 239-304.

Menaker, M. and Keatts, H. (1968). Extra retinal light perception in the sparrow. II. Photoperiodic stimulation of testis growth. Proc. Natn. Acad, Sci. U.S.A. 60: 146-151.

Mourrier, J.P. (1972). Etude de la cytodifferation du rein de l'épinoche (Gasterosteus aculeatus L.) au cours de sa transformation muqueuse. Z. Zellforsch. 123: 96-111.

Palmer, J.D. (1976). An Introduction to Biological Rhythms. New York, Academic Press.

Pengelley, E.T. (ed.) (1975). Circannual Clocks: Annual Biological Rhythms. New York, Academic Press.

Reiter, R.J. (1972). The role of the pineal in reproduction. In: Reproductive Biology. Eds. H. Balin and S. Glasser. Amsterdam, Excerpta Med. p. 71-114.

Reiter, R.J. (1974). Circannual reproductive rhythms in mammals related to photoperiod and pineal function: a review. Chronobiologia 1: 365-395.

Rensing, L. (1973). Biologische Rhythmen und Regulation. Stuttgart, Fischer Verlag.

Saunders, D. (1977). Biological Rhythms. Glasgow and London, Blackie.

Sehgal, A. and Sundararaj, B.I. (1970). Effects of various photoperiodic regimens on the ovary of the catfish, Heteropneustes fossilis (Bloch), during the spawning and postspawning periods. Biol. Reprod. 2: 425-434.

Slijkhuis, H. (1978). Ultrastructural evidence for two types of gonadotropic cells in the pituitary gland of the male three-spined stickleback, Gasterosteus aculeatus. Gen. Comp. Endocrinol. 36: 639-641.

Sundararaj, B.I., Anand, T.C. and Donaldson, E.M. (1972). Effects of partially purified salmon pituitary gonadotropin on ovarian maintenance, ovulation and vitellogenesis in the hypophysectomized catfish, Heteropneustes fossilis (Bloch). Gen. Comp. Endocrinol. 18: 102-114.

Sundararaj, B.I., Nath, P. and Jeet, V. (1978). Role of circadian and circannual rhythms in the regulation of ovarian cycles in fishes: a catfish model. In: Comparative Endocrinology. Eds. P.J. Gaillard and H.H. Boer. Amsterdam, Eslevier/North Holland, Biomedical Press, p. 141-144.

Sundararaj, B.I. and Sehgal, A. (1970a). Responses of the pituitary and ovary of the catfish Heteropneustes fossilis (Bloch) to accelerated light regimen of a decreasing followed by an increasing photoperiod during the post-spawning period. Biol. Reprod. 2: 435-443.

Sundararaj, B.I. and Sehgal, A. (1970b). Short- and long-term effects of imposition of total darkness on the annual ovarian cycle of the catfish Heteropneustes fossilis (Bloch). J. Interdiscipl. Cycle Res. 1: 291-301.

Sundararaj, B.I. and Vasal, S. (1973). Photoperiodic regulation of reproductive cycle in the catfish Heteropneustes fossilis (Bloch). In: Endocrinology. Proc. IV Int. Congr. Endocrinol. Washington, D.C. (1972). Int. Congr. Ser. 273, Amsterdam, Excerpta Medica. p. 180-184.

Sundararaj, B.I. and Vasal, S. (1976). Photoperiod and temperature control in the regulation of reproduction in the female catfish Heteropneustes fossilis. J. Fish. Res. Board Can. 33: 959-973.

Takahashi, J.S., Norris, C. and Menaker, M. (1978). Circadian photoperiodic regulation of testis growth in the house sparrow: is the pineal gland involved? In: Comparative Endocrinology. Eds. P.J. Gaillard and H.H. Boer. Amsterdam, Elsevier/North Holland Biomedical Press. p. 153-156.

Urasaki, H. (1976). The role of the pineal and eyes in the photoperiodic effect on the gonad of the madaka, Oryzias latipes. Chronobiologia 3: 228-234.

Vasal, S. and Sundararaj, B.I. (1975). Responses of the regressed ovary of the catfish Heteropneustes fossilis (Bloch) to interrupted-night photoperiods. Chronobiologia 2: 224-239.

Vlaming, V.L. de (1972). Environmental control of teleost reproductive cycles: a brief review. J. Fish. Biol. 4: 131-140.

Wolfson, A. (1959). The role of light and darkness in the regulation of spring migration and reproductive cycles in birds. In: Photoperiodism and Related Phenomena in Plants and Animals. Ed. R.W. Withrow. Washington, D.C., Publ. 55 A.A.A.S. p. 679-716.

FISH REPRODUCTION AND STRESS*

Shelby D. Gerking

Department of Zoology, Arizona State University

Tempe, AZ 85281, USA

REPRODUCTION AND STRESS

This chapter will draw attention to the importance of performing more research on the effects of stress on the reproductive performance of fishes. The subject will be approached from the point of view that reproduction may have a narrower tolerance to stress than any other life function, while enjoying at the same time a pivotal role in the success or failure of a population in nature. The sensitivity of reproduction to stress has emerged during the last 10 years. Mount (1968) announced this point of view for the first time in his studies of the effect of copper on the fathead minnow (Pimephales promelas). He said, "It appears from these data that there is a range of concentration of copper in water in which indefinite survival is possible but reproduction is inhibited and that this effect is not detected by measurements of growth or histopathological examination." This statement has been echoed many times since, and it is now taken for granted that the bioassay of any stressor must take into account the whole life cycle with the strong possibility that reproduction will be the most sensitive part. Reproductive bioassays are as much as 200-500 times more sensitive than acute survival bioassays (Eaton 1973).

Reproduction might be expected to be a critical period in the life history. The life functions - coordination, excretion, respiration, nutrition, growth and reproduction are all affected to some extent by sublethal stresses. Of these, coordination, excretion, respiration and nutrition (using these functions in a broad sense) must have broad tolerance and well developed compensating mechanisms between upper and lower lethal limits

* This research was performed under contract No. EY-76-5-02-2498 sponsored by the U.S. Department of Energy.

or the fish would not remain alive. On the other hand, an individual fish need not grow or reproduce to maintain its integrity, and these functions can be expected to have narrower tolerance limits to stress.

In reviewing the U.S. Water Quality Criteria 1972, Sprague (1976) concluded that growth studies have not been productive in providing support for water quality criteria. He called this surprising in view of the fact that the growth process integrates many physiological functions and might be expected to express whatever stress the organism might be experiencing. Having performed many growth experiments in the laboratory, I would agree. I note, however, that toxicity studies often measure growth by feeding a group of fish ad libitum (see Brungs 1969, Spehar et al. 1978 as examples). Fish are notorious for forming size hierarchies where the larger individuals consume the greater share of food. Thus, the individuals of the group exhibit a tremendous variability in size when measured at the end of the experiment. Size differential mortality under stress has also been observed (Brungs 1969). Both of these sources of variability may conspire to produce growth rates which may not be related to the stress at all. It might be prudent to give growth another chance by reassessing the methods employed in its measurement. With a substantial history of laboratory-based bioenergetic studies behind us (Webb 1978, Brett and Groves 1979), we may be in a better position to critically measure the growth response to stress than we think.

The reproductive studies that are needed to appraise pollutants are in their infancy. Much is known about reproductive behavior (Breder and Rosen 1966), reproductive endocrinology (Hoar 1969, Pickford and Atz 1957), fecundity (Bagenal 1978), and egg development (Oppenheimer 1947). An extensive literature on egg metabolism also exists, as well as some factors that influence egg survival, such as temperature, salinity, electrolyte balance and dissolved oxygen (Blaxter 1969). It is fair to say, however, that we could not piece together from this literature how sublethal stress on adults might affect egg laying and egg survival, and yet these parts of the reproductive process are key elements of a self-sustaining fish population.

DEFINITION OF REPRODUCTION

Reproduction is defined here as the events associated with egg maturation in the ovary, egg laying, and egg viability. Post-hatching events are certainly important to overall reproductive success as are courtship and mating, but they are not included here for arbitrary reasons only. Following the reasoning of Rosenthal and Alderdice (1976), reproductive success should be considered in the light of the total life cycle, since each stage in the life history is dependent on the one preceding it. Thus, eggs may hatch successfully but give rise to abnormal larvae, or larvae might survive a stressful environment but develop into fry with a low survival potential, and so on. The concept has considerable merit and is one to which I subscribe, but from a practical standpoint some lines must be drawn in this early phase of the research, and I prefer to limit our studies to those we can handle with confidence.

Note the heavy emphasis on the female in the above definition of reproduction. Even though the male is a partner in the process, only modest efforts have been made to measure effects on the testes or on the sperm. Eggs are easy to see and count; sperm are not. Right away we have identified a gap in our research that surely deserves more attention than it has received.

METHODS USED FOR REPRODUCTIVE BIOASSAYS

Reproductive bioassays are relatively new, and a standard method has not yet been adopted. The problems revolve around the species selected for the test, numbers of fish in a test, group testing, individual variation and statistical reliability. These elements are all interrelated and they cannot be entirely separated from one another.

Species. In terms of the species used for reproductive bioassays, investigators have been torn between using sport or commercial species which are not well adapted to the laboratory and smaller laboratory adapted species with short generation times which have no economic importance. The rearing of bluegills (_Lepomis macrochirus_) and brook trout (_Salvelinus fontinalis_) to maturity in the laboratory (Allison and Hermanutz 1977, see Appendix) is an accomplishment to be proud of inasmuch as they are familiar, important species and they represent both cold and warm water forms. The disadvantages are also obvious. Considerable time is required to rear them to maturity (2 years) with the accompanying possibility of nutritional and disease problems. The size of the fish imposes a holding space problem that only specialized laboratories have at their disposal. This may result in experiments performed on few individuals. For example, the effects of copper, mercury, toxaphene, cadmium and lead on the reproduction of brook trout were determined with as few as one and a maximum of six females in each replicate treatment (McKim and Benoit 1971, McKim et al. 1976, Mayer et al. 1975, Benoit et al. 1976, Holcombe et al. 1976). An even greater drawback is the sporadic nature of spawning with intervals of months between egg-laying bouts.

The first species to enter the scene with more desirable characteristics was the fathead minnow, mentioned previously. Immense progress has been made in bioassay work with this fish since its introduction to laboratory culture. It can be easily reared in the laboratory; it has a relatively short life cycle, and its eggs can be separated and reared individually in quantity.

Other investigators have turned to members of the killifish family (Cyprinodontidae). The rapid life cycle of 2-4 months from egg to maturity, the nearly year round spawning, and the habit of laying eggs singly yield the characteristics needed for experiments on egg production and egg viability. These fish are also extraordinarily adaptable to laboratory culture. For example, no disease has attacked our pupfish colony from its inception five years ago to the present. The most frequently used species are the flagfish, Jordanella floridae, and the sheepshead minnow, Cyprinodon variegatus. About the same time these species were brought into the laboratory, we began to work with the desert pupfish, Cyprinodon n. nevadensis when its potential for a quantitative measure of reproduction was recognized.

Group spawning. The fathead minnow and flagfish are spawned in small groups. The fathead groups range from two to nine males and three to nine females with females usually predominating (Carlson 1971, Nebeker et al. 1974). Two males and five females are recommended for the flagfish (Smith 1973). The fathead minnow glues its eggs on the under side of half tiles while the flagfish deposits its eggs singly on two yarn mats strung on a metal frame. The half tiles and yarn mats are removed every 24 hours when the eggs are counted and placed in nylon baskets suspended in water for further observation.

Group spawning is very productive for experiments on egg survival, but it is limited for an analysis of reproductive performance because individual variation is not measurable. As Hermanutz (1978) states the average number of eggs laid per female is the only datum obtainable, whereas the number of spawns per female and number of eggs per spawn would be preferable. This information on individual females is essential for statistical analysis and reliability.

Group spawning also results in egg predation. Although the adult fathead minnow presumably does not feed on the eggs, the flagfish and sheepshead minnow have this habit (Hermanutz op cit., Hansen, Schimmel and Forester 1977). The females attempt to eat the eggs, and the defending male is not always able to ward off two or more intruding females. Egg predation may not be the same in each tank, thereby affecting the average number of eggs laid per female. Hansen op cit. have offered a solution for the sheepshead minnow. Small spawning chambers were constructed for individual pairs which can be immersed in the test solution. The chambers have a bottom of nylon screen through which the eggs can fall into a nylon-lined drawer below. Eggs are then collected periodically from the drawer. This procedure also provides a measure of individual female performance in addition to preventing egg predation.

Spawning periodicity. The fathead minnow spawns from April to August (Brungs 1969). This 4-5 month interval is sufficient to perform a considerable amount of research, but year-long egg-laying would be preferable. The flagfish and sheepshead minnow can be spawned at any time of year. The flagfish females have a 14-day reproductive cycle in which egg production reaches a peak about midway through the period and then tapers off (Smith 1973). The sheepshead minnow has been induced to spawn by injecting human chorionic gonatotropin (Hansen, Goodman and Wilson 1977, Schimmel et al. 1974). Both events introduce additional sources of variability unless the procedures are standardized.

Individual variation. A large amount of individual variation in spawning among fathead and flagfish females can be deduced from experimental results even though they spawn in groups. In three studies on the fathead minnow (Nebeker et al., 1974, Pickering and Gast, 1972, Carlson, 1971) the average ratio between eight duplicate measurements of eggs female^{-1} was 2.55, excluding one extraordinarily high value of 70. These studies were testing the effects of Aroclor 1254 (PCB), carbaryl (Sevin), and cadmium on reproductive performance. Variability in the flagfish is considerably less. Averaging the ratios between replicates in two studies (Spehar 1976, Spehar et al. 1978) gave 1.76 for 15 pairs of observations. No measure of variability could be obtained for the sheepshead minnow. Since the fathead and flagfish are spawned in groups, the individual variability must be even greater than the replicates suggest. Spawning individual pairs of these species may become possible in the future to provide the measure of variability required in experimental design. Otherwise all except major differences may lie undiscovered.

Pupfish system. I do not wish to leave the impression that the above three species are not valuable additions to the bioassay arsenal or that the data on reproduction obtained by spawning them is unreliable. Far from it. Each in its own way has added a new dimension to progress in obtaining pollutant effects over the whole life cycle. My only plea here is for species that provide an even more sensitive measure of reproductive performance. Assuming that we agree that such an experimental animal is desirable, I offer the pupfish as another step toward developing reliable measurements. The following procedure has been designed to take full advantage of its reproductive features.

Pairs of these small fish are spawned in separate 20 l tanks. Male and female are separated by a plastic divider except for 30 minutes each day. During that short period the male courts the female by chasing and vibrating his body against hers (Barlow 1961). If she is prepared to spawn, she dives into a black yarn mop accompanied by the male. Eggs are lightly attached to the mop and immediately fertilized. At the end of the 30-minute period the divider is replaced; the mop is removed; eggs are removed with tweezers, counted and either placed in nylon mesh baskets suspended in the spawning tank or transferred to a test tank. The procedure can be followed at nearly any time of year, although late winter months are less productive than the rest of the year. A standard test lasts three weeks. Results are expressed as percentage of days eggs are laid, eggs laid per spawning, and eggs laid per gram body weight per day. Egg viability is followed and is expressed as percentage hatch. Egg diameter and gonosomatic index have yielded useful information as well.

The pupfish system avoids two pitfalls mentioned above with respect to the flagfish and sheepshead minnow. Egg predation is nullified by the short exposure of male and female and the immediate removal of the spawning mop after egg-laying is completed. Individual variability in egg production is incorporated in the design by spawning several pairs (usually 10) separately. We probably were sensitive to the latter element in the design because the objective from the start was to measure reproductive performance whereas the flagfish and sheepshead minnow were first used mainly in survival tests of several organic chemicals and heavy metals.

The pupfish also has its shortcomings for reproduction experiments. A few eggs are found on the bottom of the tank unattached to the mop. This amounts to about five percent or less of the total. Adult male and female are sexually dimorphic but dominant males repress the subordinates in a large group, and the coloration of the latter is indistinguishable from females. Sexing mistakes are made, usually by placing two males in the same tank. We can partially overcome this difficulty by removing the dominant males from the group, thereby allowing the distinctive coloration to develop in the subordinates. This circumstance often delays us in setting up an experiment which requires a considerable number of males.

The last source of difficulty is more serious. The variation in day to day egg production is much greater than is desirable. We have not been able to lower this variability by inbreeding; the females of the seventh inbred generation seem to be just as variable as the first. A typical 21-day record is: 16, 0, 0, 3, 1, 3, 6, 0, 9, 8, 1, 14, 2, 0, 0, 0, 16, 31, 20, 5, 7 (Gerking et al. 1979). The mean of 7.48 ± 8.36 gives a coefficient of variation of 112 percent. The variability among females is also large, giving coefficients of variation in the neighborhood of 50-70 percent. Under the circumstances, it is not surprising that a group of females may be selected by chance whose egg production departs from the average performance by a startling amount. Recently two groups were tested under the same conditions a few weeks apart. One group averaged 14 and the other 4 eggs g^{-1} day^{-1}. Our only weapons against such events are to use as many pairs as possible in the control and treatments and to exercise judgement in recognizing great departures from expectation.

PUPFISH AS A MODEL

Thus, the pupfish, with its disadvantages and advantages, is offered as a model, not as a final selection of a species to be used as a standard in reproductive tolerance tests. On the one hand, the pupfish is useful to point out certain principles involved in selecting criteria to be used for successful reproduction and in putting these criteria to use in evaluating various stresses on reproductive performance. On the other hand, we realize that the pupfish is adapted to live in the desert, and for this reason its thermal tolerances are shifted up the temperature scale and its tolerance to salinity is broad (Gerking and Lee 1980). Nevertheless, the species provides an opportunity to determine whether reproductive tolerance limits to stresses are narrow or broad regardless of where they might fall in the spectrum of many species; how the limits for growth, for example, relate to those for reproduction; how the limits of growth and reproduction relate to those of survival; and finally how the limits might be related to events of practical significance in fishery biology. The next section of the paper reports some of the results that we have obtained in the past few years.

THERMAL EFFECTS ON GROWTH AND REPRODUCTION IN THE PUPFISH.

Reproduction. The method described above has provided quantitative measures of reproduction for each pair of fish: (1) percentage of days on which eggs are laid, (2) eggs per spawning, which eliminates the zero entries when no eggs are laid, and (3) eggs day^{-1} or eggs g^{-1} day^{-1}. Since the response pattern has been the same for all three measures of egg production, only the results expressed as eggs g^{-1} day^{-1} will be used here. Egg production is coupled with egg viability to yield a measure of reproductive success, and egg viability is expressed as percentage of eggs that hatch. A standard 21-day test using 7-14 pairs has been done at constant temperatures of 18, 20, 22, 24, 26, 28, 30, 32, and 36 C and fluctuating temperatures of 16-24, 28-32, 28-36, 28-38, 30-39 C Acclimation of reproductive performance was tested by exposing a group of pupfish for an entire generation from oocyte maturation to adult maturity at a given temperature and then testing egg production and egg viability at that temperature or another. Two previously determined marginal conditions were used, 24 and 32 C, along with the optimum, 28 C. The mature progeny were divided into three groups; one was exposed to the rearing temperature and the other two were transferred to temperatures which they had never before experienced. Details of the experiments have been presented by Shrode and Gerking (1977) and Gerking et al. (1979).

An analysis of variance on log transformed data demonstrated clearly that significant differences in egg production occurred at the different constant temperatures ($F = 14.8$ with 14 and 134 d.f.; $P < 0.01$). Eggs g^{-1} day^{-1} averaged less than one at the extreme temperatures of 18-22 and 36 C in contrast to about eight at the peak egg production at 30 C (Fig. 1). The broad range of optimality was 24-30 C. Due to the large amount of individual variability, a more precise optimum cannot be established. For example, no statistical distinction can be made between 28 and 30 C, the two temperatures at which egg production was the highest. Egg viability was also highest in the range of 24-30 C, averaging about 78 percent and declining sharply above and below these temperatures to values of 10 percent or less. Fluctuating temperatures had only a slight effect on egg production. In general egg production corresponded to that obtained at the mean of the minimum and maximum. Performance was reduced if the extremes fell below or above a range of 24 to 32 C where egg production was typically high.

An interesting contrast between egg production and egg viability appeared at 32 C. Egg production was high, about 5 g^{-1} day^{-1}, but only about 10 percent of the eggs hatched. The eggs had soft chorionic membranes and many had no yolk. The detrimental effect of 32 C (and above) was also expressed in a smaller yolk diameter and lower gonosomatic index than at the optimum temperature range. This event taught us that egg

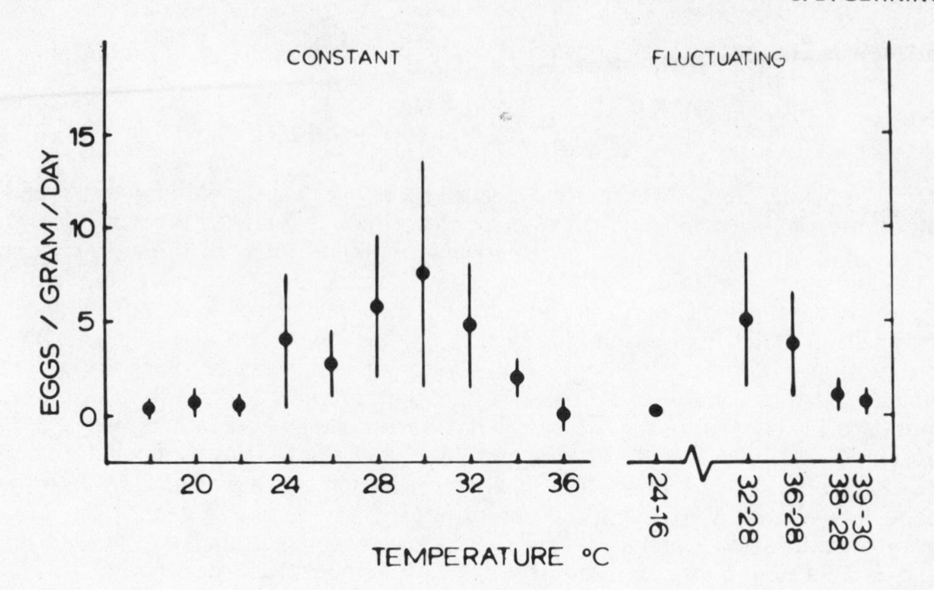

Fig. 1. Mean (± SD) number of eggs g^{-1} day^{-1} produced by the desert pupfish, Cyprinodon n. nevadensis. The fish were acclimated to each test temperature and then bred for 21 days.

production alone could not be relied upon as a measure of reproductive success. We finally adopted a dual criterion of reproductive success as the range of optimal egg production with at least 50 percent hatchability. Reference to the reproductive temperature tolerance limits will be made later in the paper.

The results of acclimation of reproductive performance to temperature yielded a simple conclusion. No temperature compensation took place. Generation-long exposure to temperature extremes did not shift the optimum away from 28 C. Also, when the fish were transferred from the optimum to the extremes, reproductive performance was typical of the new temperature to which they had never before been exposed. A suggestion that temperature compensation was lacking appeared in the reproductive temperature tolerance experiments. Egg production was consistent from the first to the third week at most of the temperatures, and no predictable pattern emerged at three temperatures where statistical differences were demonstrated. Thus, egg production showed no adaptation to temperature during the course of the three week experiments. We have no explanation for the lack of temperature compensation. Apparently the reproductive function is regulated in such a complex manner that the responses of multiple physiological mechanisms -- neural, endocrinological, genetic - are not synchronized to produce a compensatory shift away from a typical egg-laying performance even in the face of generation-long exposure to an unaccustomed temperature.

Growth. Growth experiments on the pupfish have recently been completed (Gerking and Lee MS). They were performed in 600 ml of water in 1000 ml beakers at temperatures of 12, 16, 20, 24, 28 and 32 C (photoperiod 16L 8D as in the reproduction tests). Rations in the form of Tetra-min, a balanced food for aquarium fishes, were fed at the rates of 1.5, 3.0 and 4.5 percent of body weight day^{-1}, except at 36 C where the 1.5 percent ration led to severe undernourishment. The 36 C rations were 3.0, 4.5 and 6.0 percent body weight day^{-1}. Each feeding trial lasted seven weeks except for one at 20 C which lasted 5 weeks. Proximate composition of the food and fish flesh was determined before and after the trial for dry weight, ash, caloric content (Parr Adiabatic Calorimeter), protein (Lowry technique) and lipid (Folch wash). Feces were collected by filtration.

Assimilation was constant (mean = 75.3 percent) at all temperature-ration combinations except 12 and 16 C. At these low temperatures the fish were torpid, ate little and assimilated only about 25 percent of the daily ration. The body composition measurements were instrumental in revealing once again that lipid metabolism is primarily responsible for the energy the fish relies on during periods when they are assimilating less than a maintenance amount.

Growth was measured as specific growth rate (percentage increase or decrease of initial dry weight wk^{-1}) and as calories wk^{-1}. Specific growth curves rose sharply between 16 to 20 C and decreased just as abruptly between 32 and 36 C. Otherwise the mode was between 20 and 24 C (Fig. 2). The curves for each ration are skewed to the left. Significance among the three rations appeared in the statistical analysis and also among the four temperatures. Growth measured as caloric gain or loss had the same spectrum of temperature response as specific growth rate, with positive responses between 20 and 32 C and an inability to grow at 12, 16 and 36 C (Fig. 3). The failure to grow at 12 and 16 C was traced to a low degree of responsiveness, lack of appetite and poor assimilation. Behavior was just the opposite at 36 C where the fish were hyperactive, voracious and had normal assimilation. Their metabolism was so great at this temperature that they would not have been able to meet maintenance requirements if they had been fed to satiation. At 36 C the fish are just below their upper incipient lethal temperature of 38 C, as determined on *C. milleri*, a close relative (Otto and Gerking 1973).

Comparison of thermal limits for growth and reproduction. In order to compare the thermal limits for growth and reproduction both functions are expressed as a percentage of peak performance. Using the results that have been presented earlier (Figs. 1, 2, 3) we have summarized them as smooth curves of percentage of peak performance, using the optimum response as 100 percent (Fig. 4). The temperature limits for successful reproduction and growth are established arbitrarily as 50 percent of peak performance. The choice is based solely on an opinion that less than 50 percent performance in reproduction, growth, or both will not lead to a self-sustaining population over the long run. Since the curves are steep near the 50 percent levels, the temperature limits are not materially broadened at 25 or even 15 percent of

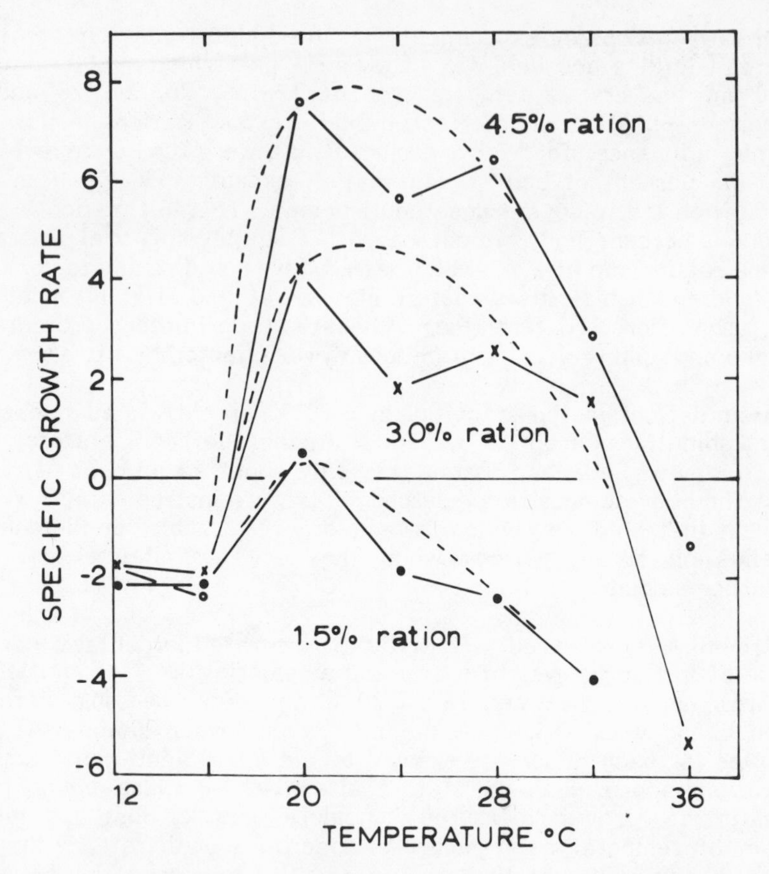

Fig. 2. Specific growth rate (% change in dry weight wk^{-1}) of the desert pupfish, Cyprinodon n. nevadensis, in relation to temperature and ration. Rations are expressed as percentage body weight day^{-1}.

peak performance.

The thermal limits for growth at 50 percent peak performance are 17 to 29 C using specific growth rate as a measure, and 18 to 32 C using energy gain or loss. Reproductive limits are more difficult to assign because both criteria of egg production and egg viability must be employed. The 50 percent level of peak performance lies, therefore, between 26C (egg production) and 29 (egg viability). The conclusion is reached that the thermal limits for growth are broader than those for reproduction and that while the upper limits of both functions are about the same (29 C), growth is possible at a much lower temperature (17-18 C) than that which allows the fish to reproduce successfully (26 C).

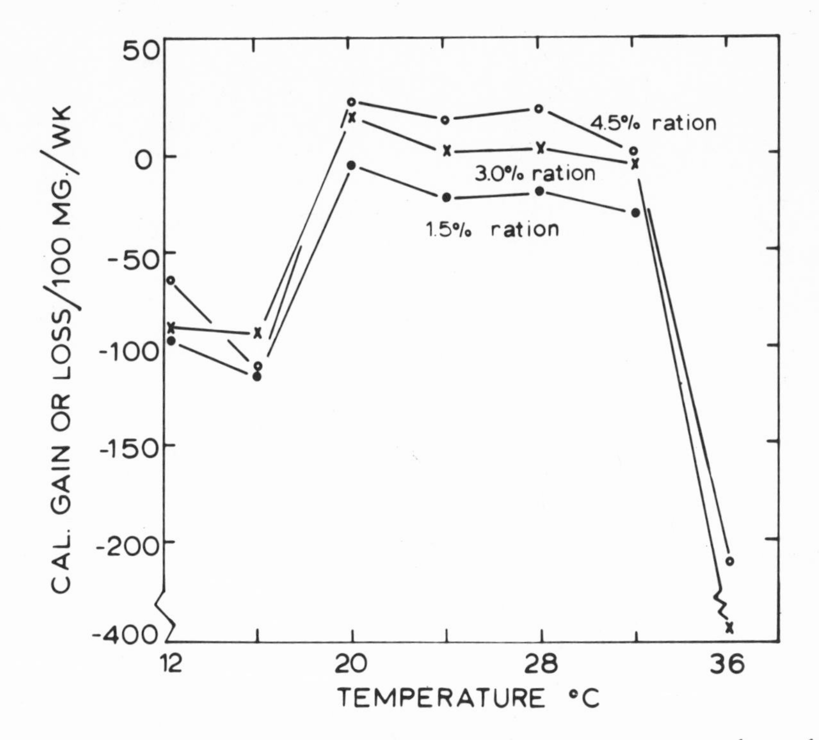

Fig. 3. Change in caloric content (cal 100mg dry weight^{-1} wk^{-1}) of desert pupfish, Cyprinodon n. nevadensis, in relation to temperature and ration. Rations are expressed as percentage body weight per day.

These limits can be compared with others which have been established during the course of our work (Fig. 5). Here the thermal limits for reproduction and growth are portrayed as being considerably narrower than the critical thermal limits. Thus, reproductive success can be expected over 10 percent of the survival spectrum, and growth 29 percent. Reproductive limits are known to be considerably narrower than survival limits in detailed studies of salinity and pH stress as well. The aggressive sexual behavior and coloration of the male along with other elements of courtship are displayed over a much broader range (12-38 C) than is egg production and egg viability (26-30 C). We have learned not to rely on sexual behavior as an indication that egg laying and hatching are in the offing. Developmental temperature tolerance is discussed in the next section.

OOGENESIS AS A SENSITIVE STAGE IN THE LIFE CYCLE

Note the broad-domed curve (dashed line) on the egg viability graph (Fig. 4). This was obtained by spawning the pupfish at 28 C and testing egg viability at a variety of other temperatures (Shrode 1975). From this curve

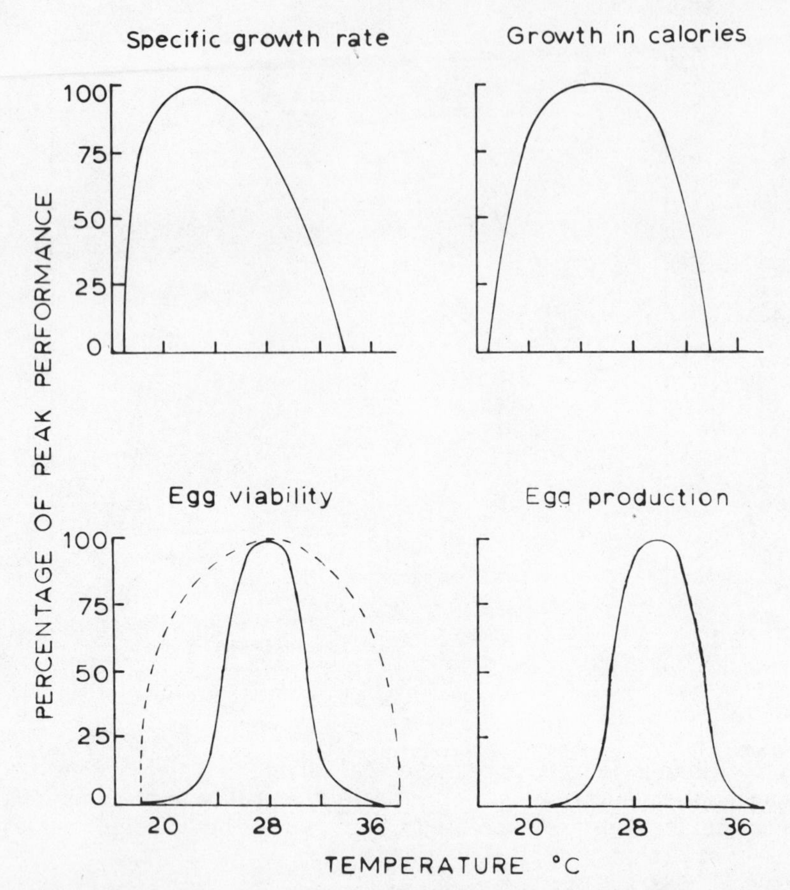

Fig. 4 Diagrammatic representation of specific growth rate, growth in calories, egg viability and egg production of the desert pupfish, Cyprinodon n. nevadensis, in relation to temperature. Each measurement is expressed as a percentage of peak performance.

the developmental tolerance limits, mentioned above, were derived. The curve beneath it (solid line) represents egg viability in a situation more closely resembling that which might be encountered in nature. Egg viability was tested at the same temperature to which the female was exposed. The difference between the curves suggests that the egg was somehow affected by temperature during a stage in its maturation. We call it the "oogenesis effect" and have used it to suggest that oogenesis might be the most sensitive stage in the life cycle, at least to temperature.

In addition to the temperature effect on oogenesis, just described, the most direct evidence of stress effects on developing ova comes from a study of egg maturation in flagfish exposed to varying levels of acidity (Ruby et al. 1977). The ovaries of females exposed to pH 6.7 (control), 6.0, 5.5, 5.0,

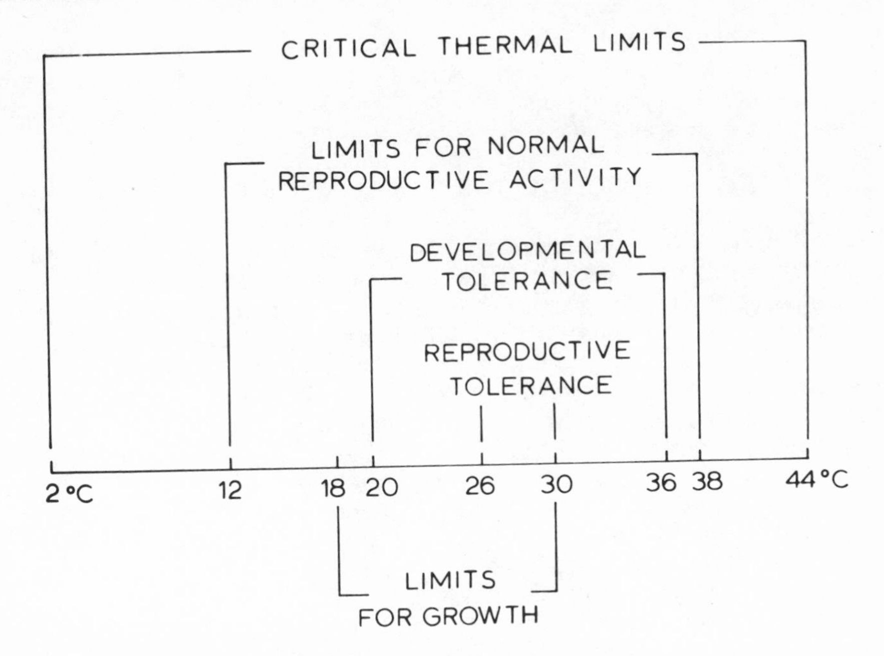

Fig. 5. A comparison of the thermal limits for growth and reproduction with the critical thermal limits for survival.

and 4.5 were examined histologically. The acid stress caused the failure of oogenesis and oocytes to progress to more advanced stages. Secondary yolk deposition and protein synthesis were both disrupted, and the later stages of oogenesis were nearly absent. Suggestive evidence was also presented that atresia (oocyte resorption and degeneration) was accelerated by acid levels higher than the control. With less precise methods we have corroborated these observations in the pupfish. A lower proportion of oocytes were classified as mature at pH 5.5 and 5.0 than at the control of 8.3 (Lee and Gerking 1981). With such evidence before us, it is clear that more attention should be focused on possible alterations in the egg while it is in the ovary.

The literature contains some suggestive information on the oogenesis effect. At this point a distinction should be made between non-specific, systemic stresses, such as temperature and pH, and specific, chemical poisons. The latter can affect the egg by being transmitted through the body of the exposed female to the ovary. Sprague (1971) reviewed DDT contamination by describing cases where the chemical was transferred from exposed females to the yolk lipids with correspondingly poor hatching success. A PCB (polychlorinated biphenyl, trade name Arochlor 1254) has been added to aquaria containing sheepshead minnows. The fish bioaccumulated the PCB and produced contaminated eggs (Hansen et al. 1973). Fertilization of the eggs was unaffected, but the hatching rate and the survival rate of fry was less than that from the eggs of control, unexposed fish. Several other chemicals apparently cause poor hatchability of eggs or poor survival of fry which were derived from exposed parents and tested in pollutant-free water: diazinon (Allison and Hermanutz 1977), kepone (Hansen, Goodman and Wilson 1977) and endrin (Hermanutz 1978, Hansen, Schimmel and Forester 1977).

RAPID CHANGES IN THE CHARACTER OF THE EGG

The permeability of the egg changes rapidly after being laid and affects its survival. These changes were brought to our attention after pondering Craig and Baksi's (1977) results on flagfish egg survival at high levels of acidity. Eggs laid at pH 6.7 (control) and transferred within 24 hours to pH 4.5 and 5.0 exhibited a high survival of 85 percent. Similar experiments on pupfish eggs yielded low survival (Lee and Gerking 1980). It struck us that the elapsed time between egg laying and transfer to the pH stress might provide an explanation for the difference in results. Our transfers were made within an hour after the eggs were laid. This transfer time was compared with one of 12 hours where the eggs were taken from pH 8.3 (control) and placed in solutions of pH 6.5, 6.0 and 5.5. Survival percentages for the one hour transfer were 32.3 and 21.1 and 12.5; for the 12-hour transfer they were 55.9, 32.2 and 30.0. Then the one-hour transfer was compared with no transfer; that is, eggs that were laid directly in acid solutions by females that were exposed to the same solutions. Again the differences are obvious. Control hatching percentages at pH 8.3 were virtually the same (50.6 and 57.0 percent), but significant differences showed up at pH 7.0 (1-hour = 44.0; direct exposure = 32.0), pH 6.5 (21.2 and 8.7) and pH 6.0 (15.4 and 3.3). No eggs hatched at pH 5.5. Therefore, Craig and Baksi's high egg survival rates in acid solutions can be attributed at least in part to delayed transfer.

The egg rapidly changes in character in the first few hours. Only a few from a multitude of references are offered to support the statement. Permeability of the chorion decreases rapidly in the first hour (Kao et al. 1954), and chorion hardening is virtually complete by 12 hours (Cykowska and Winniki 1972). After hardening, the chorion and plasma membranes form a nearly impenetrable barrier to ion exchange until the eyed stage when permeability again increases (Terner 1968).

Thus, experiments to test the survival of eggs under stress conditions must be designed carefully with the age of the egg foremost in mind. To mimic nature more closely, future investigators should bear in mind that both the female and the eggs she produces are subjected to the same stress at the same time.

CONCLUSIONS

The following general conclusions can be drawn if the reproductive function of other fishes responds to stress like that of the pupfish. Some of the conclusions have been extracted from the text of this paper and others from unpublished work. No effort has been made to supply supporting evidence in the latter case. The conclusions are divided into two sections, those of more basic biological interest and those of practical significance to fishery biology.

BASIC BIOLOGICAL IMPORTANCE

1. Reproduction is the most sensitive of all the life functions to stress. For example, reproductive success can be expected only over about 10 percent of the range of temperatures over which the fish can survive. Limits for reproductive success have been established for salinity and pH as well.

2. The criteria of reproductive success must include both egg production and egg viability. Egg laying alone cannot be used as a criterion of success, because eggs are laid under conditions that mitigate against hatching. The act of courtship and sexual coloration are evident over a much broader range of stress conditions than that which allows 50 percent of peak performance in egg laying and hatching.

3. Oogenesis may be the most sensitive stage in the reproductive process. This has been shown to be true for systemic stresses, such as temperature and pH, and for chemical poisons, such as insecticides.

4. The egg changes its character rapidly after being laid. The decrease in permeability which accompanies chorion hardening imposes restrictions on experimental design of egg survival tests which may not have been fully appreciated.

5. Immediately after being laid, eggs are more sensitive to stress than larvae, and larvae are more sensitive than adults.

6. Reproductive performance shows no temperature compensation even after generation-long exposure to conditions which are considered to be marginal. No physiological adaptation of reproductive performance to pH was detected as well.

7. Reproductive performance is more sensitive to temperature stress than is growth. The upper limits of these two functions is about the same, but growth is possible at much lower temperatures than is reproduction. Using specific growth rate as a measure of growth, the temperature optimum for growth is considerably below that for reproduction.

PRACTICAL SIGNIFICANCE

1. Year class strength may be determined in part by the temperature experience of the female during egg maturation, since oogenesis is the most sensitive stage of reproduction.

2. Fish confined in water of unaccustomed temperature or pH will not accommodate their reproductive function to the new regime. Because physiological adaptation of the reproductive function to stress is lacking, fish populations which cannot escape thermal effluents or acidified lakes whose conditions lie beyond their reproductive performance limits are doomed even though they live in these conditions for long periods of time.

3. Survival tests on eggs, larvae and adults may not yield critical information on the ability of populations to survive stressful regimes, since these tests may define survival limits which are not suitable for egg production and egg viability.

4. Survival tests on eggs are more easily extrapolated to nature when the females and the egg are exposed to the same stress at the same time.

5. Water drawn from hydroelectric dams is often cooler than the original stream water. Soon after the dam is installed, it is sometimes noticed that a native fauna may survive, grow large and finally disappear. This response may be an illustration of our finding that growth, especially at lower temperatures, is quite satisfactory in the face of environmental temperature that mitigates against successful reproduction. A similar observation has been made in the acidified, LaCloche Mountain Lakes (Beamish and Harvey 1972) where the fish population may live a considerable period of time beyond the time when young are absent.

ACKNOWLEDGEMENT

I wish to thank Mr. M. Lee, research associate on the project, for his loyalty and devotion to the work.

REFERENCES

Allison, D.T. and Hermanutz, R.O. (1977). Toxicity of diazinon to brook trout and fathead minnow. Ecol. Res. Ser. EPA 600/3-77-060. SFA 23. 69 pp.

Bagenal, T.B. (1978). Aspects of fish fecundity. In: Ecology of Freshwater Fish Production, Ed. S.D. Gerking. Oxford, Blackwell Sci. Publ., p. 75-101.

Barlow, G.W. (1961). Social behavior of the desert pupfish, Cyprinodon macularius, in the field and in the aquarium. Am. Midl. Nat. 65: 339-359.

Beamish, R.J. and Harvey, H.H. (1972). Acidification of the LaCloche Mountain Lakes, Ontario, and the resulting fish mortalities. J. Fish. Res. Board Can. 29: 1131-1143.

Benoit, D.A., Leonard, E.N., Christensen, G.M. and Fiandt, J.T. (1976). Toxic effects of cadmium of three generations of brook trout (Salvelinus fontinalis). Trans. Am. Fish. Soc. 105: 550-560.

Blaxter, J.H.S. (1969). Development: eggs and larvae. In: Fish Physiology. Vol. II. Eds. W.S. Hoar and D.J. Randall. New York, Academic Press,

p. 177-252.

Breder, C.M.J. and Rosen, D.E. (1966). Modes of reproduction in Fishes. Garden City, N.Y., Natural History Press.

Brett, J.R. and Groves, T.D.D. (1979). Physiological energetics. In Fish Physiology. Vol. VII. Eds. W.S. Hoar, D.J. Randall and J.R. Brett. New York, Academic press, p. 280-352.

Brungs, W.A. (1969). Chronic toxicity of zinc to the fathead minnow, Pimephales promelas Rafinesque. Trans. Am. Fish. Soc. 98: 272-279.

Carlson, A.R. (1971). Effects of long term exposure to carbaryl (Sevin) on survival, growth and reproduction of the fathead minnow (Pimephales promelas). J. Fish. Res. Board Can. 29: 583-587.

Craig, G.R. and Baksi, W.F. (1977). The effects of depressed pH on flagfish reproduction, growth and survival. Water Res. ll: 621-626.

Cykowska, C. and Winnicki, A. (1972). Embryonic development of the Baltic seatrout (Salmo trutta) in buffer solutions. Acta Ichthyologica et Piscatoria (Poland) 2: 3-12.

Eaton, J.G. (1973). Recent developments in the use of laboratory bioassays to determine "safe" levels of toxicants for fish. In: Bioassay Techniques and Environmental Chemistry, Ed. G.E. Glass, Ann Arbor, Michigan; Ann Arbor Science Publ. Inc., p. 107-115.

Gerking, S.D. and Lee, R.M. (1980). Reproductive performance of the desert pupfish (Cyprinodon n. nevadensis) in relation to salinity. Environ. Biol. Fishes (in press).

Gerking, S.D. and Lee, R.M. (MS). A comparison of the thermal limits for growth and reproduction in the desert pupfish, Cyprinodon n. nevadensis.

Gerking, S.D., Lee, R. and Shrode, J.B. (1979). Effects of generation-long temperature acclimation on reproductive performance of the desert pupfish, Cyprinodon n. nevadensis. Physiol. Zoöl. 52: 113-121.

Hansen, D.J., Goodman, L.R. and Wilson, A.J. (1977). Kepone: chronic effects on embryos, fry, juvenile and adult sheepshead minnows. Chesapeake Sci. 18: 227-232.

Hansen, D.J., Schimmel, S.C. and Forester, J. (1973). Arochlor 1254 in eggs of sheepshead minnows; effect on fertilization success and survival of embryos and fry. 27th Annual Conf. S.E. Assn. Game and Fish Commissioners. p. 420-426.

Hansen, D.J., Schimmel, S.C. and Forester, J. (1977). Endrin: effects on the entire life cycle of a saltwater fish, Cyprinodon variegatus. J. Toxicol. and Environ. Health. 3: 721-733.

Hermanutz, R.O. (1978). Endrin and malathion toxicity to flagfish (Jordanella floridae). Arch. Environ. Contam. Toxicol. 7: 159-168.

Hoar, W.S. (1969). Reproduction. In: Fish Physiology, Vol. III. Eds. W.S. Hoar and D.J. Randall, New York, Academic Press. p. 1-72.

Holcombe, G.W., Benoit, D.A., Leonard, E.N. and McKim, J.M. (1976). Long-term effects of lead exposure on three generations of brook trout. J. Fish. Res. Board Can. 33: 1731-1741.

Kao, C.Y., Chambers, P. and Chambers, E.L. (1954). Internal hydrostatic pressure of the Fundulus egg. II. Permeability of the chorion. J. Cell. Comp. Physiol. 44: 447-461.

Lee, R.M. and Gerking, S.D. (1980). Sensitivity of fish eggs to acid stress. Water Res. (in press).

Lee, R.M. and Gerking, S.D. (1981). Survival and reproductive performance of the desert pupfish, Cyprinodon n. nevadensis, in acid waters. J. Fish Biol. (in press).

Mayer, F.L. Jr., Nehrle, P.M. and Droyer, W.P. (1975). Toxaphene effects on reproduction, growth and mortality of brook trout. EPA. Ecol. Res. Ser. 600/3-75-013. p. vii + 41.

McKim, J.E. and Benoit, D.A. (1971). Effects of long-term exposures to copper on survival, reproduction and growth of brook trout (Salvelinus fontinalis Mitchill). J. Fish Res. Board Can. 28: 655-662.

McKim, J.M., Olson, J.F., Holcombe, J.W. and Hunt, E.P. (1976). Long term effects of methylmercuric chloride on three generations of brook trout (Salvelinus fontinalis): toxicity, accumulation, distribution and elimination. J. Fish Res. Board Can. 33: 2726-2739.

Mount, D.I. (1968). Chronic toxicity of copper to fathead minnows (Pimephales promelas Rafinesque). Water Res. 2: 215-223.

Nebeker, A.V., Puglisi, F.A. and DeFoe, D.L. (1974). Effect of polychlorinated biphenyl compounds on survival and growth of fathead minnows and flagfish. Trans. Am. Fish. Soc. 103: 562-568.

Oppenheimer, J. (1947). Organization of the teleost blastoderm. Quart. Rev. Biol. 22: 105-118.

Otto, R.G. and Gerking, S.D. (1973). Heat tolerance of a Death Valley pupfish (Genus Cyprinodon). Physiol. Zoöl. 46: 43-49.

Pickering, Q.H. and Gast, M.H. (1972). Acute and chronic toxicity of cadmium to the fathead minnow (Pimephales promelas). J. Fish Res. Board Can. 29: 1099-1106.

Pickford, G.E. and Atz, J.W. (1957). The Physiology of the Pituitary Gland. New York, N.Y. Zoological Soc.

Rosenthal, H. and Alderice, D.F. (1976). Sublethal effects of environmental stressors, natural and pollutional, on marine fish eggs and larvae. J. Fish. Res. Board Can. 33: 2047-2065.

Ruby,S.M., Aczel, J. and Craig, G.R. (1977). The effects of depressed pH on oogenesis in flagfish, Jordanella floridae. Water Res. 11: 757-762.

Schimmel, S.C., Hansen, D.J. and Forester, J. (1974). Effects of Arochlor 1254 on laboratory reared embryos and fry of sheepshead minnows (Cyprinodon variegatus). Trans. Am. Fish. Soc. 103: 582-586.

Shrode, J.B. (1975). Developmental temperature tolerance of a Death Valley pupfish (Cyprinodon nevadensis). Physiol. Zoöl. 48: 378-389.

Shrode, J.B. and Gerking, S.D. (1977). Effects of constant and fluctuating temperatures on reproductive performance of a desert pupfish, Cyprinodon n. nevadensis. Physiol. Zoöl. 59: 1-10.

Smith, W.E. (1973). A cyprinodontid fish, Jordanella floridae, as a laboratory animal for rapid chronic bioessays. J. Fish. Res. Board Can. 30: 329-330.

Spehar, R.L. (1976). Cadmium and zinc toxicity to flagfish, Jordanella floridae. J. Fish. Res. Board Can. 33: 1939-1945.

Spehar, R.L., Leonard, E.N. and DeFoe, D.L. (1978). Chronic effects of cadmium and zinc mixtures on flagfish (Jordanella floridae). Trans.

Am. Fish. Soc. 107: 354-360.
Sprague, J.B. (1971). Measurement of pollutant toxicity of fish III. Sublethal effects and "safe" concentrations. Water Res. 5: 245-266.
Sprague, J.B. (1976). Current status of sublethal tests of pollutants on aquatic organisms. J. Fish. Res. Board Can. 33: 1988-1992.
Terner, C. (1968). Studies of metabolism in embryonic development. I. The oxidative metabolism of unfertilized and embryonated eggs of the rainbow trout. Comp. Biochem. Physiol. 24: 933-940.
Webb, P.W. (1978). Partitioning of energy into metabolism and growth. In: Ecology of Freshwater Fish Production. Ed. S.D. Gerking. Oxford, Blackwell Sci. Publ. p. 184-214.

MIGRATORY PERFORMANCE AND

ENVIRONMENTAL EVIDENCE OF ORIENTATION

F.-W. Tesch

Biologische Anstalt Helgoland (Zentrale)

Hamburg 50, Palmaille 9, FRG

INTRODUCTION

Three major types of migratory and orientation performance are presented: (1) rheotactic orientation, (2) migration along or perpendicular to a shoreline, (3) orientation across large water bodies. The significance of different senses and environmental factors as stimuli in order to initiate and to maintain the migration route is discussed. Of great interest is the nature of the chemical stimulant responsible for the orientation especially during the spawning run of salmonids, and geomagnetism as a possible cue for long range compass course movements. Examples of organisms exhibiting transoceanic migration and orientation include tuna, eel, salmon, and shark.

A. Type of orientation performance. If the degree of difficulty of orientation performance is considered three main types of active migration may occur.

1. Rheotactic orientation. - The fish is forced to swim in a channelled body of water. Banks or shallow water on both sides of this route limit the direction of movement. In most cases water flow in one direction or the other allows rheotactic orientation. Activity is released by, for example, a rise or decrease in temperature, an increase of current, an odour stimulus, light or an internal factor. Many freshwater fishes and diadromous fishes during their movement in freshwater or in tidal streams use this type of orientation. Much experimental work has been performed to prove it.

2. Shoreline orientation. - The route of migration is not limited by two banks, but follows the shoreline of a lake or the open sea. Orientation on the basis of rheotaxis is mostly not possible. Stimulation other than

the current must cause the fish to move in one of two directions along the shoreline or away from or towards the shore. The simplest mechanism would be that the fish reacts to a change in some specific factor (e.g. warmer or colder water in one direction or different salinities); it has to follow a gradient. Movement towards the coast is mechanically limited by too shallow water or the sense of pressure indicates that the water becomes shallower. The sense of pressure may also be involved in movement towards or away from the shore to shallower or deeper water. The shore may be detected by either sound (of the breakers) or vision (of the bank). Finally the sun can provide cues to encounter or to avoid the bank. Probably many anadromous fish species orient in this way prior to their arrival at the home river estuary.

3. Orientation across large water bodies. - When oceans or great lakes are to be crossed, no coastline or bottom structure is available to provide the fish with cues for orientation. Possibly the course is determined by the sun compass or by the earth's magnetic field directly or indirectly through electric currents induced by the flow of oceanic currents or by movement of the fish itself in the geomagnetic field. Other cues such as temperature, salinity and odour gradients are less likely because the gradients over thousands of kilometers are too small; furthermore they are interrupted by gradients to the opposite side. Even less likely is migration along temperature and salinity boundaries or isolines although this may occur during movement over short distances. Tunas, sharks, salmon, eels, herring gadids and other fishes exhibit this type of orientational performance.

B. Examples of and mechanisms for the different types of orientation.

1. Rheotactic orientation. - There are ecological examples and much experimental evidence to show that this category of orientation is often used by the fish. It includes short range as well as distant migration. The long range movements are of major interest here. The fundamental mechanism of this type of orientation and migration is rheotaxis (see review by Arnold, 1974). A prerequisite for positive or negative rheotactic behaviour is the fact that the water moves. This movement has to be detected by reference to the background which is mainly possible by optical or tactical stimuli. Electrical stimulation is also possible but may not be considered as a rheotactic response during up or downstream migration of the first type (see third type).

Rheotaxis may be affected by a number of environmental factors (Arnold, 1974), such as temperature and chemical stimuli as well as internal (endocrine) factors, which can determine whether the taxis is positive or negative or alter the intensity of the kinetic response.

The most studied and familiar example of this type of orientation is that of river migration of salmonids (Hasler, 1966). Before juvenile salmon migrate to the sea they become imprinted with the distinctive odour of their native tributary, and adult salmon use this odour as a stimulant for rheotactic movement and as a cue for homing when they migrate through the home-stream network to the home-tributary.

The nature of such an odour is still unresolved. It is possible to imprint different salmonids (Oncorhynchus kisutch, Salmo gairdneri and S. trutta with the odour of an artificial substance (morpholine) so that the fish homes to a river enriched with morpholine (Cooper et al., 1976; Hasler, 1978; Scholz et al., 1978). The value of morpholine as an effective olfactory stimulant has been doubted (Hara, 1974). The morpholine-effect caused by a 1% solution in rainbow trout (S. gairdneri) differs in many respects from the normal olfactory bulb response.

Possibly pheromones released in the home water by conspecifics could be responsible for the attractive odour. Electrophysiological tests were not found to be conclusive (Dizon et al., 1973) although they have yielded positive results (Oshima, 1969). In the presence of ovarian fluid obtained from spawning female rainbow trout (Salmo gairdneri), male rainbow trout in a stream tank exhibited an increase in the frequency of upstream orientation and movement (Hara and McDonald, 1976). Field observations of Nordeng (1971) give weight to the pheromone hypothesis. He indicated that the homing of a migratory population of the char (Salvelinus alpinus) in Norway was influenced by pheromones and the attractant might be secreted in the mucus. Support for the results came from observation of a salmon population (Salmo salar) in South-West England (Solomon, 1973).

The discrepancy between results of different workers concerning the nature of stimulating odours is probably caused by neglecting the concentration of calcium ions (Bodznick, 1978). Sockeye salmon (Oncorhynchus nerka) are very sensitive to different calcium concentrations of natural water. The ions are discriminated by the sense of smell and probably very often accentuate the effect of other odorants.

The hypothesis for such activation of rheotaxis is supported by the results of investigations on other migratory fish. Silver eels (Anguilla anguilla) exhibited positive rheotactic responses when they were exposed to seawater. Cauterization of the olfactory capsule resulted in loss of this capacity. In seawater which contained conspecifics the silver eels showed no specific response. Freshwater, on the contrary, released negative rheotaxis in silver eels caught during their seaward migration in autumn (Hain, 1975). That this catadromous migration is no passive drift downstream has recently been shown by tracking experiments in a river (Tesch, unpublished). Glass eels, which have shown positive rheotaxis when exposed to unfiltered, natural freshwater (Creutzberg, 1961) also seemed to react differently when exposed to natural water of different electrolyte content (Miles, 1968). But it should be added that the attractiveness of the water increased when it contained adult eels and decreased when elvers were present. Hence, in some salmonids as well as in eels ions seem to evoke rheotaxis and a combind effect together with pheromone is likely.

That rheotactic behaviour is widely used for homing is evident by the fact that other groups of fishes orient by means of this mechanism. Alewifes (Alosa pseudoharengus) tested in experimental chambers oriented strongly to currents regardless whether fresh- or seawater was used (Richkus, 1975). Field experiments with American shad (Alosa sapidissima) on the basis of conventional tagging as well as tracking transmitter-tagged and sensory-impaired fish showed that an olfactory-rheotactic mechanism forms the basis for homing in a freshwater stream and in tidal currents (Dodson and Leggett, 1974).

A modification of the first type of orientation and migration might be the capability of fish to use the tidal stream as a transport mechanism. Arnold (1974) calls it "modulated drift". Such migratory behaviour is performed by elvers (Anguilla anguilla) during their invasion from tidal area to continental waters as mentioned above (Creutzberg, 1961) and described by Deelder (1958). No banks are present but a coastward directed flood tide. No permanent rheotactic movement is performed but a drifting after the perception of the beginning of a flood tide. Similar behaviourial patterns may be used by the sole (Solea solea) (De Veen, 1967), the plaice (Pleuronectes platessa) (Greer Walker et al., 1978), the yellow tail flounder (Limanda ferruginea) (Smith et al., 1978) and during the inshore migration of the invertebrate (Penaeus duorarum) (Hughes, 1969). Except for the elvers, which receive the freshwater odour as a stimulus to hide at the bottom and perhaps to swim against the current, the tidal steering mechanism for other animals is not known. The author (Tesch, 1965) proposed that an internal clock could determine the rhythm of the change from drifting to rheotactic behaviour on the bottom. Changing temperature and salinity (perhaps by odour perception!) may also produce a change of behaviour.

If in addition, a heading during drifting takes place (Greer Walker et al., 1978) the third type of orientation (see below) is used or an inertial system (as well as geodetic orientation, Merkel, 1978) enables the fish to maintain a direction for a certain time after leaving the bottom.

2. Shoreline orientation. - Migration along a shoreline is reported for many fish species which have their feeding ground in the sea or in a lake but their spawning ground in a river. The upstream migration in the river is a type one movement (rheotactic). In order to reach the estuary of the river they have to orient along the coast or from the feeding grounds perpendicular to the coast. This type of orientation is very likely to be performed by the American shad (Alosa sapidissima) (Fig. 1) which migrate to and from the spawning river parallel to the North American Atlantic coast or from offshore to the coast (Leggett,

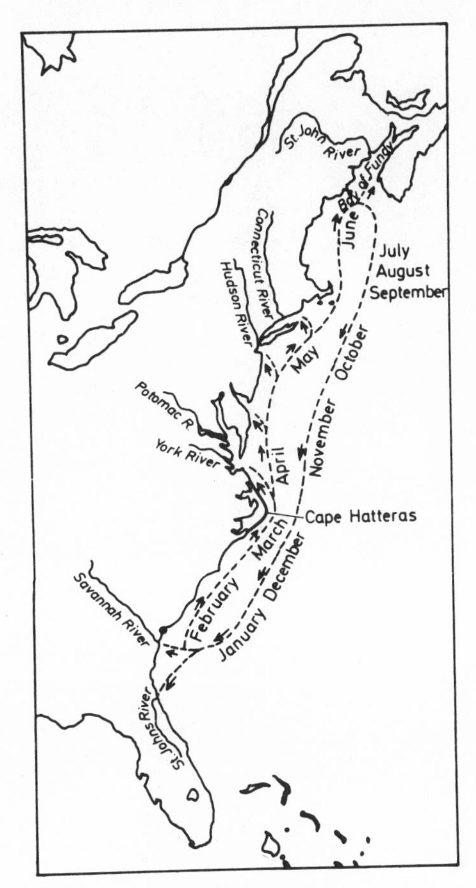

Fig. 1. The migration route of the American shad (Alosa sapidissima) in the course of the year and some of its major spawning rivers (Leggett, 1973).

1973; Neves and Depres, 1979). The shad starts out North in January from Georgia and northern Florida when the temperatures begin to rise (Leggett, 1973). The northernmost occurrence and entrance into the rivers is the Bay of Fundy (lowest temperatures). The return migration is driven by decreasing temperature from midsummer to winter. The movements are probably directed by the coastline or, if perpendicular to the coast, by increasing or decreasing depth. When the estuary is reached temperature or salinity gradients may provide directional stimuli. The decision whether the migration along the coast is North or South may be provided by decreasing or increasing temperature but this is doubtful because the gradient is probably too small over such a wide range.

The Baltic whitefish (Coregonus lavaretus) is also known to migrate along the coast (e.g. Lind and Kaukoranta, 1974). The reproduction of this fish takes place in rivers flowing into the Gulf of Bothnia and it feeds predominantly in the southern part of this Baltic Bay (Fig. 2).

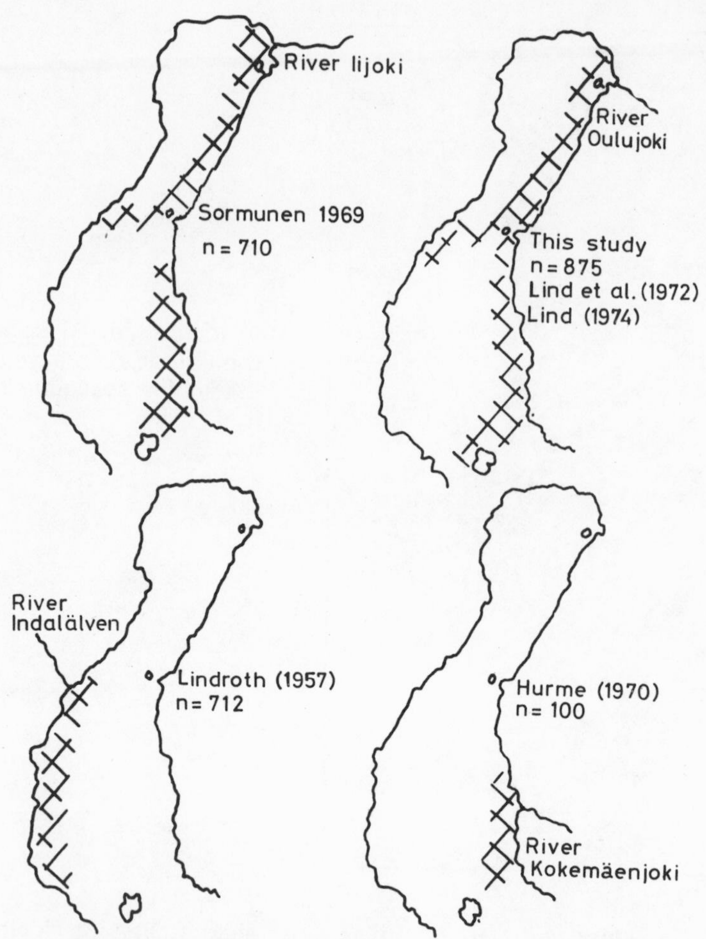

Fig. 2. Migration areas (shaded) of whitefish (Coregonus lavaretus) between different spawning rivers flowing into the Baltic Gulf of Bothnia and feeding grounds as determined by tagging (in the spawning river) and recapture experiments (after Lind and Kaukoranta, 1974).

The farther north the river the farther is the migration from and to the spawning river. Some populations have to move over a distance up to 650 km. Arriving at the mouth of the river this fish may also be activated for rheotaxis by the sense of smell. But the movements in the sea cannot be guided by the cue of odour; the currents during the spawning of the Oulujoki River whitefish are in the direction from the feeding grounds to the home rivers. Lind and Kaukoranta (1974) hypothesise that orientation is effected by "underwater guide posts" or by a certain type of sun compass mechanism. It seems likely that the

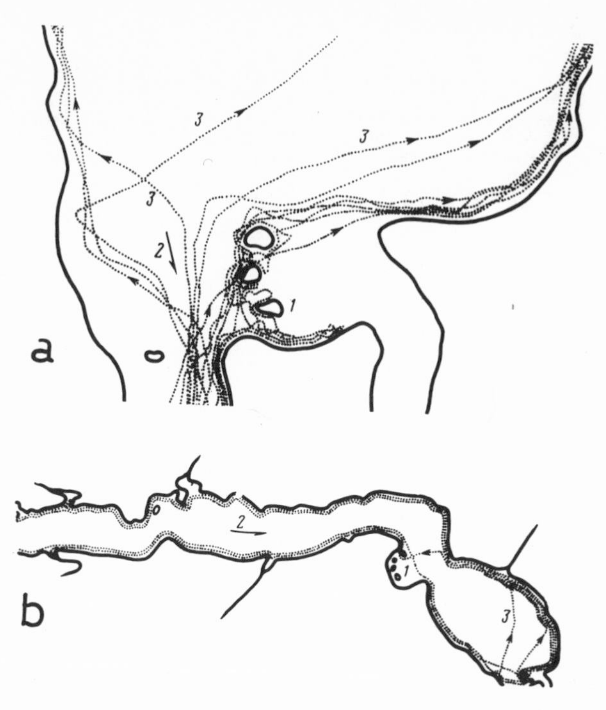

Fig. 3. Tracks of salmon (3) in an artificial lake a) entering the lake and swimming along shoreline of isles (1); b) moving upstream (2) along shorelines of the lake (after Malinin et al., 1974).

fish is guided by the coast line which includes underwater guide posts like barriers and shallow water.

The second type of orientation seems to be more common than originally thought. It was proposed by McCleave (1967) and McCleave and LaBar (1972) when they analysed the movements of cutthroat trout (*Salmo clarki*) transplanted during spawning migration in a lake. After release the fish move along the coast; this course could more likely guide them to the river estuary than random movements. Even sunapee trout (*Salvelinus alpinus*) tracked in a familiar environment (flood ponds in Maine) partly followed the shoreline (McCleave and LaBar, 1977). Tracing of Atlantic salmon through storage lakes of the Kola peninsula demonstrated that in order to reach their spawning tributaries the fish moved along the shoreline (Fig. 3-5; Malinin et al., 1974). Migration occurred mainly during daylight which suggests visual orientation. Similarly McCleave and LaBar (1977) postulated visual features as orientational cues for the above mentioned sunapee trout.

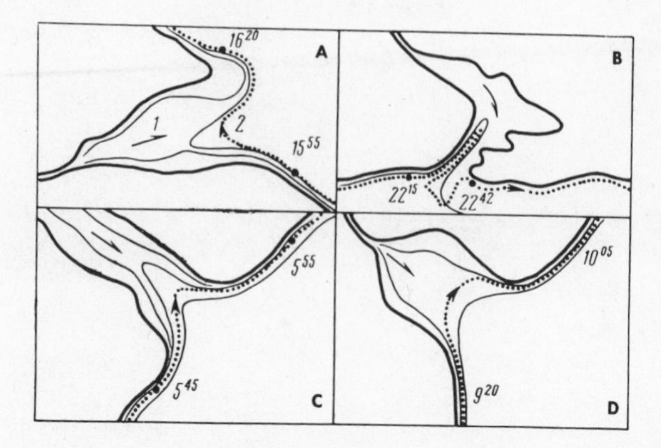

Fig. 4. Tracks of salmon (2) passing the inflow (1) of non-natal rivers with time on positions before and after the passage (after Malinin et al., 1974).

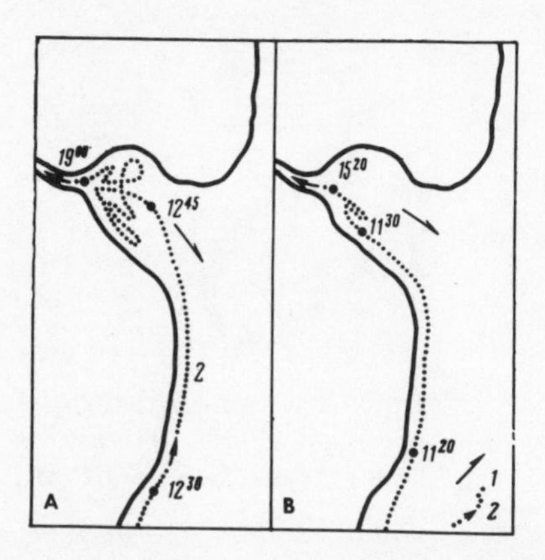

Fig. 5. Tracks of two different salmon (2) encountering the inflow of their natal river with time on positions before and during entering the river and with current direction (1) of non-natal water (after Malinin et al., 1974).

In other fish species the tendency to meet and to follow a shoreline is likewise evident. Stationary yellow eels which have a strong tendency to home after transplantation (e.g. Tesch, 1967; Vladikov, 1971) mostly orient in this way. Long distant transplantations (over 100 km) of eels in the North Sea (Deelder and Tesch, 1970) showed poorer homing success than short range transplantations. The long distant transplanted eels were mainly recaptured during attempted homing movements which pointed straight to the continental coast; tracking experiments with transplanted yellow eels confirmed these observations (Tesch, 1974). The direction of their swimming to the coast was quite different from the direction to the home area and successful homing would have required inadequate long distant travelling along the coast which explains their lack of success. That eels during their seaward migration move in a similar way as described for the Kola salmon (see above Fig. 3) follows from the experience of the commercial fyke net fishery. The eels move along inlets and outlets of lakes passed. In order to put the fyke nets in the most favourable position, the bank near the outlet or even the whole shoreline up to the inlet stream is chosen rather than the central area of the lake. This means that the eels orient themselves along the shore until the outlet is reached. They are then guided by negative rheotaxis (Tesch, unpublished). Fish such as cyprinids are captured by the fyke net fishery during their spring spawning migration which gives further evidence that the shoreline oriented migration is widely used by the fish.

The orientation during the coastward directed movement of the transplanted eels mentioned above is unknown and probably falls under type three orientation. In other cases(e.g. Tesch, 1979) decreasing depth and therefore decreasing pressure could have been the cue for the course to the coast; the sense of pressure in some fishes is extremely sensitive (Tsvetkov, 1969, 1972; Kuhne and Strodtkoetter, 1967). If the coast is not too distant, its direction may be indicated by the noise of breakers. Directional hearing in fishes has been experimentally proven (e.g. Schuijf, 1974, 1975; Banner, 1968, 1972).

3. Orientation across large water bodies. - The crossing of great lakes and transoceanic migrations has been reported from tagging and recapture experiments on some fish species. For other species, breeding and feeding areas are known to be situated far distant from one another and orientation during migration between these areas is not guided by coastlines. How is orientation during crossing of these great water bodies pelagically possible?

Sun compass orientation is known to play a role in the migration of birds, and there is evidence that it may be important for fishes (Tesch, 1975). Recent field experiments confirm that some fishes are able to use the sun as a directional cue. Released in a circular pool young largemouth bass (Micropterus palmoides) only choose the predicted compass direction on clear days (Loyacano and Chappell, 1977). Field experiments with some cyprinids in an artificial lake showed that homing ability was most reduced when vision combined with the sense of lateral line organ was impaired (Abrasimova, 1976). - From the results of earlier laboratory experiments it is evident that not only directional finding on the basis of the azimuth is possible but also compensation of the movement of the sun (see review by Tesch, 1975).

The biological significance of this is indicated by field experiments of Winn et al. (1964; Fig. 6), who have shown that parrot fishes use the sun to find their feeding places near the shore of Bermuda after resting at night and leaving offshore caves. The ranges of movement are not more than a few hundred metres in shallow and mostly very clear water. Orientational performance by celestial cues over longer ranges, during day and night and under variable hydrographical conditions are not known to exist or are speculative.

There is even less evidence that fishes use the sense of smell in order to orient over long ranges. Speculation arose when Teichmann (1959) found that the eel (Anguilla anguilla) has such a powerful sense of

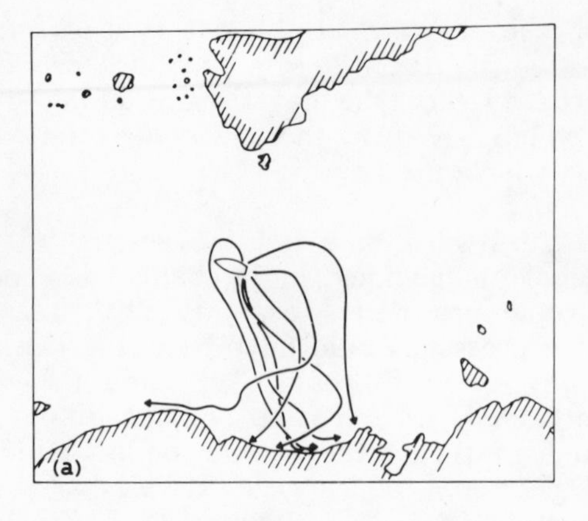

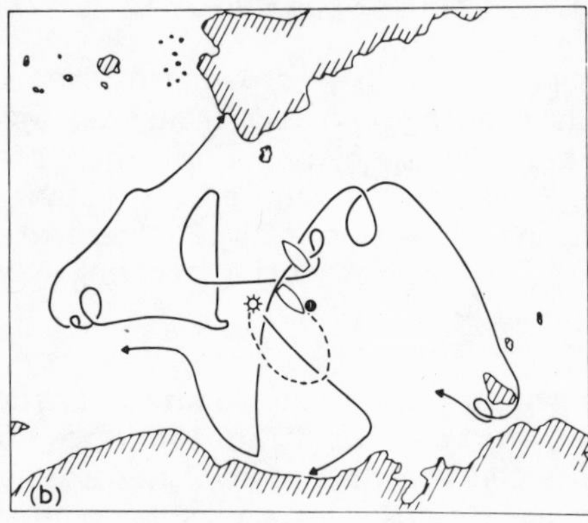

Fig. 6. Paths of adult parrot fishes (Scarus guacamaia and S. coelestinus) released during the afternoon (July/August) in Balley's Bay (Bermuda). a) Seven individuals during bright sunshine. b) Three individuals (upper release point) under complete cloud cover, one individual (lower release point; dotted line) under cloud cover during release. As the circled point was attained the sun came into full view and the fish oriented immediately southeasterly (after Winn et al., 1964).

smell that it can perceive two molecules of an odorous substance entering its nasal cavity. This perhaps could be useful for orientation, when it has attained the boundaries of spawning places. But even in homing experiments with yellow eels displaced over far shorter distances no proof could be given that the sense of smell is important for homing success (Tesch, 1970; see also Deelder and Tesch, 1970). It seems impossible that the eel is in a position to locate the direction of its 6000 km distant spawning places on the basis of increasing odour concentration in the North-Atlantic current and in the Gulf Stream. Even less possible is the tracing of these currents along their edge by trial and error. The edge of the huge current is of such irregular shape that tracing would result in left and right and back and forth swimming without progress in the direction of the Sargasso Sea. Similarily doubtful is the use of temperature or salinity if long distance courses are involved.

Another factor involved in finding far distant locations may be perception of the geomagnetic field directly (Leask, 1978; Tesch, 1977; Becker, 1974; Kundtson and Stimers, 1977) or by means of electrical induction (Rommel and McCleave, 1973; Kundtson and Stimers, 1977; Brown and Ilyinsky, 1978). Other animals especially birds (e.g. Walcott and Green, 1974; review of Schmidt-Koenig, 1975) have proven to be sensitive to magnetism or to use the geomagnetic field in homing experiments. Whether fishes use geomagnetism to steer a compass course during migration is unknown. But as some fishes are affected by magnetism and as other senses are not very likely to be involved, migratory performances of some fishes are discussed below with respect to geomagnetism as a cue.

a) Tuna: The most striking and fastest transoceanic migrations are known from the tunas (*Thunnus thynnus*); the population of western North Atlantic waters is related to the population of the coast of France as shown by electrophoretic studies on enzyme polymorphism. The tunas are reported to cross the Atlantic as evidenced by different tagging and recapture experiments (Mather, 1962; Tiews, 1963; Mather et al., 1967; Int. Mar. Angler 1975; after Fisch v. Fang 16: 78, 1978). The speed calculated by connection of the points of capture and recapture (distance 5.500 to 8.000 km) was 3 km h^{-1}, probably more; this based on the assumption of straight line course. Similar long range migrations are known from the Pacific Skipjack tunas (*Katsuwonus pelamis*). Travelling from the North American coast to the area of Hawaii (5,000 - 6,000 km) lasted 9 months and resulted in a speed of at least 1 km h^{-1} (Matsumoto, 1975). Tracking experiments with this species (Yuen, 1970) and the albacore (*Thunnus alalunga*) (Laurs et al., 1977) could be performed during local movements of these fishes. The albacore demonstrated that oceanographic conditions play an important role in the local concentrations and movements. The fishes were attracted to upwelling areas but avoided water of low temperature. The upwelling fronts within which tunas reduced their speed has a width of 4 to 14 km with a change in temperature of at least 0.5 °C (0.003 °C m^{-1}). Steffel et al., (1976) tested temperature discrimination of captive free-swimming tuna and found a discrimination threshold of temperature of 0.10 - 0.15 °C. Thus in order to sense this difference the fishes have to swim at least 30 m which requires a certain "temperature memory". Laurs et al (1977) speculate that this memory could enable tunas to sense temperature gradients as small as 0.0001 °C m^{-1} which could qualify tuna to sense 0.1 °C after 1 km of swimming. In the North Atlantic, which is crossed by *Thunnus thynnus*, far lower gradients occur (see e.g. Worthington, 1976). In the most favourable case on a North-South section (about 45 °N; 35 °W) the temperature increased 1 °C per 100 km (0.00001 °C m^{-1}). Another case (NE-SW section about 40 °N 40 °W) shows no gradient at all on a range of 500 km). Other sections exhibit an increase in temperature which is then followed by a decrease. As indicated for olfactory orientation (see above) orientation along the Gulf Stream by means of temperature seems similarily impossible for the same reason.

Compass course swimming is therefore more likely. This is also

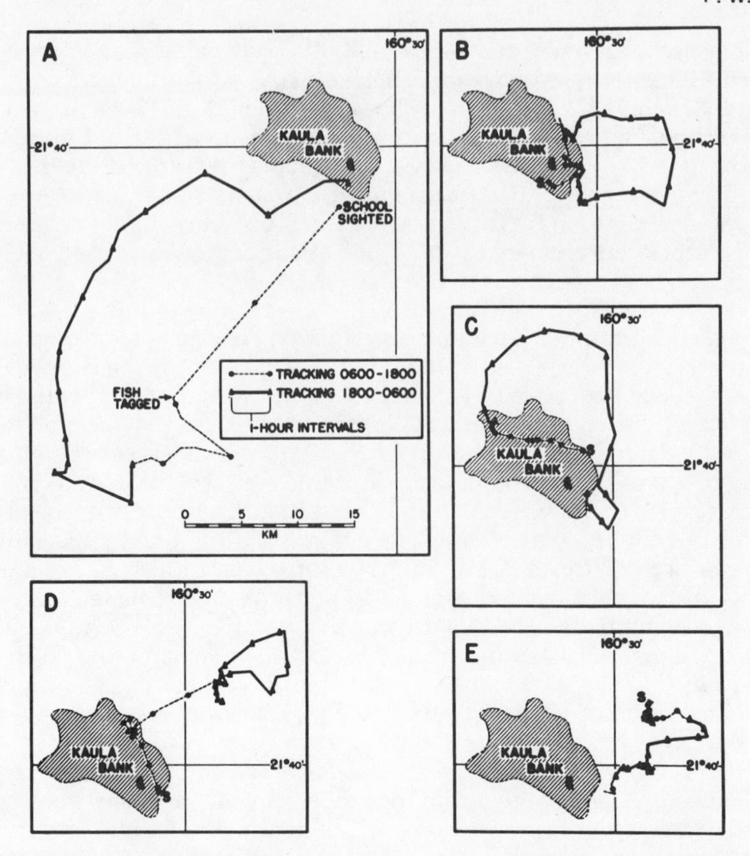

Fig. 7. Track of a tagged skipjack tuna (length: 44cm) at kaula Bank, Hawaii. (A) from 14^{52}, August 30, to 6^{00}, August 31; (B) from 6^{00} August 31, to 6^{00}, September 1; (C) from 7^{38}, September 3, to 6^{00}, September 4; (D) from 6^{00} September 4, to 6^{00} September; (E) from 6^{00}, September 5, to 7^{30}, September 6 (after Yuen, 1970).

assumed by Yuen (1970), who tracked skipjack tunas during local movements but over comparatively long ranges. These fish performed nightly journeys of 25 to 106 km away from a bank near Hawaii (Fig. 7), and returned every morning. During the night the tunas swam close to the surface which might have facilitated celestial orientation in this area characterised by good weather conditions. But to my knowledge no experimental evidence has been presented to show that fish can orient by vision of stars. There is a possibility that magnetic orientation may play a part although, as the fish vary their direction from night to night, another mechanism of orientation could be involved as well: the geodetic orientation mentioned above (Merkel, 1978). The fish must have a memory for the time and the direction of its night excursions. Generally at $2^{\underline{00}}$ the fish returned reversing their swimming directions by 180° and arrived at the bank by sunrise.

For the Atlantic Ocean migration of Thunnus thynnus visual orientation seems possible during the southwestern part of the journey which is generally favoured by good weather and clear water conditions. But the northeast Atlantic Ocean section rarely offers sufficiently good weather conditions for a menotactic sun or star course. Skipjack tuna migration in the Pacific (Matsumoto, 1975) coincides with the circulation of the major ocean currents which could be interpreted as passive drift (see also Williams, 1972). But the author states that the movement of the skipjack tuna in some populations involves more than just being transported by the surface currents. Hence the guiding mechanism can be interpreted as induction produced by the current flow through the geomagnetic field and received by the fish. It has been shown that salmon, eel and elasmobranchs (see below) can sense electrical currents of such magnitude provided the hydrographic current is strong enough. For the Atlantic tuna (Thunnus thynnus) such a strong current is only available in the area of the Gulf Stream. Most of the North Atlantic current is too slow to produce adequate electrical currents. In some areas it reverses its direction and the fish would be misled. As celestial orientation in the often overcast North Atlantic is also unlikely, direct sensing of the geomagnetic field and active electro-orientation by swimming in the geomagnetic field - as evident up to now for elasmobranchs only (Kalmijn, 1978) - are the only remaining possibilities for finding the right compass direction in the open ocean. The relationships for eel and salmon migration in the North Atlantic (see below) are similar.

b) Eel: One of the fishes with oceanic migration which is increasingly preferred for orientation experiments in both laboratory and field is the eel (Anguilla spp.). It crosses thousands of kilometres of Atlantic as well as Pacific and Indian Ocean areas both as adult and larva. For the North Atlantic Ocean the same problems exist with respect to optical and current dependent orientation (Westerberg, 1975) as with the tuna. Optical orientation is even less probable because as adult it becomes a mesopelagic organism which prefers nearly complete darkness (Tesch, 1978 a, b, c). Westin and Nyman (1977) conclude from their laboratory studies and field experiments in the Baltic that temperature is the stimulating and guiding agent for the westward migration of silver eels (A. anguilla). This may be true as far as stimulation or releasing migratory activity is concerned (see discussion of Tesch, 1978 b), but with respect to orientation the gradient of the temperature in the North Atlantic is much too weak and sometimes even reverses direction (see tuna: above). Orientation by current induced electricity is doubtful as well because in depths of 300 to 700 m (Tesch unpubl.), preferred by the eels, currents are mostly too slow (Richardson and Knauss, 1971) to produce electricity of the order which has been found to stimulate eel (Anguilla rostrata) and salmon

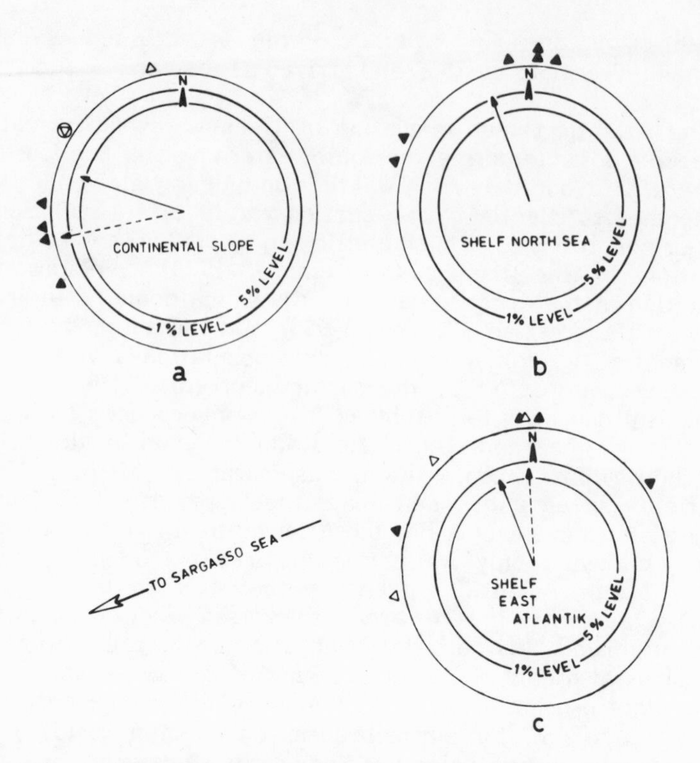

Fig. 8. Circular distribution of directions (triangles) of tracked silver eels (A. anguilla) and mean vectors with indications of significance levels (not valid for dashed arrows). (a) Continental slope; dashed arrow includes the mean of all black triangles; white triangles indicate specimens with part of their movements on the continental shelf. (b, c) Different areas on the East Atlantic shelf (for further explanations see Tesch, 1978b).

(Salmo salar) (McCleave et al., 1971). But it is conceivable that the eel senses the electricity produced by its own movement through the geomagnetic field as proposed by Branovov et al. (1971) and Gleiser and Khordorkovskii (1971) and as shown for example in the elasmobranchs by Kalmijn (1978). Direct perception of geomagnetism seems also possible as shown by investigations on A. anguilla by Tesch (1974) on Carassius auratus by Becker (1974), as discussed by Tesch (1975) and explained theoretically by Leask (1978). Tracking experiments in the Baltic (Tesch, 1979; Westin and Nyman, 1977), in the North Sea (Tesch, 1974) and in the area of the European continental slope (Tesch, 1978 a, b, unpublished data) always exhibited a certain compass direction preference: in the continental shelf areas NW and over the deep sea (continental slope) W to S (Fig. 8). A hypothetical migratory route of the European eel is illustrated in Fig. 9. If the eel encounters barriers

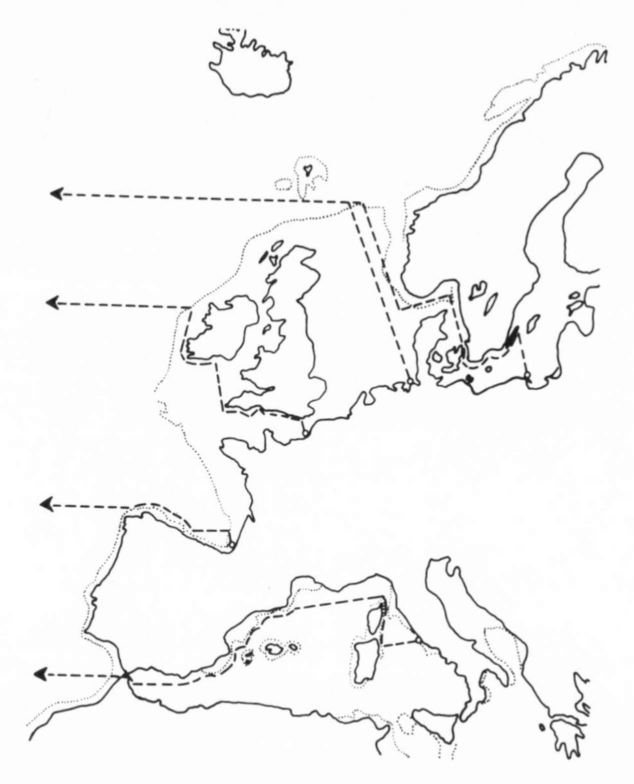

Fig. 9. Hypothesis of the eels' (A. anguilla) spawning migration in different East Atlantic shelf and offshelf areas. On the shelf: NNW compass course; when coastlines are encountered: a Type 2 migration; offshelf: WSW compass course (after Tesch, 1978d).

and coastlines it changes to directions more South or more North and the movement changes to a type 2 migration. In the Eastern and Central Baltic, for example, it is deviated by the Swedish coast to southwestern and by the Danish isles to a northern course as shown by tagging (see review by Tesch, 1975) and tracking experiments (Tesch, 1979). The change from the north western to the south-eastern direction obviously takes place at the continental slope (Tesch, 1978 b). The greater depth (pressure) could act as releaser for the change in direction. Both Atlantic eel species exhibited frequent diving excursions during tracking experiments (Stasko and Rommel, 1974; Tesch 1978 a, b, 1979); this may have been to estimate depth. The southwestern compass course obviously continues until the spawning area in the Sargasso Sea is reached. Tracking experiments in the northern to central Sargasso Sea showed a mean horizontal progress of 247^{0} of three silver eels treated with gonadotropic hormones; as controls have shown the gonads of these eels exhibited an advanced stage of ripeness (Tesch et al., 1979).

c) Salmon: The transoceanic migration patterns of the salmon Oncorhynchus sp. in the North Pacific have been presented as a model by Royce et al. (1968) and possible orientation by magneto-hydrodynamical effects dependent on the prevailing ocean currents has been postulated by McCleave et al. (1971). According to the model several species of Oncorhynchus in their oceanic migration present a type three migration; they follow the major ocean currents, i.e. a circular instead of a linear route. A later model published by French and Bakkala (1974) is described on the basis of the Oncorhynchus nerka migratory pattern (see also Fujii, 1975). They write: "It could not be demonstrated that defined oceanographic features of the North Pacific Ocean had any direct influence on the North-South movements and distribution of sockeye salmon. Their movements and distribution may be governed by other environmental conditions such as water temperature or food abundance". Hence an orientation on the currents by passive electric magneto-hydrodynamical effects seems unlikely. The search for food (small fish) may influence part of the salmon migration and the temperature preferences of the small fish can also influence the salmon movements. But as discussed for tuna and eel (see above) long range orientation e.g. to the spawning area on the basis of temperature or odour gradients is not likely. It has experimentally shown that pink salmon can use the sun compass (Churmasov and Stepanov, 1977); sockeye salmon can perceive polarised light (Dill, 1971; see discussion by Tesch, 1975). But a menotactic course depending on visual stimuli, especially in the North Pacific Ocean characterised by bad weather (Royce et al., 1968), seems highly impossible. More likely is the active electro-location (induction by the fish's own movement in the geomagnetic field) or direct sensing of the geomagnetism as discussed for other fishes (see above). But experimental evidence for the salmon is not available at present.

Transoceanic migrations are also performed by the Atlantic salmon (Salmo salar) (May, 1973). Substantial proportions of the smolt from many North Atlantic countries migrate to Greenland and adjacent waters in the Labrador Sea. Their orientation from and to North America and European rivers is similarly problematical as with Pacific salmon when it is postulated that orientation is based on currents, odour, temperature and visual cues. These factors may be involved locally. Future field and laboratory experiments on salmon should mainly consider the earth's magnetic field as a possible cue for long range orientation (see also discussion of Tesch, 1975).

d) Elasmobranchs: The high electric sensitivity of many elasmobranchs provides them with the capability to sense the geomagnetic field. Several experiments have shown that their threshold sensitivity is close to the magnetic strength present in their natural environment.

Assuming that the fish, the water or the magnetic field is moving or changing. The ampullae of Lorenzini, as studied on skates (e.g. Brown and Ilynski, 1978) and sharks (Kalmijn, 1978), are responsible for the perception of the geomagnetic field.

Hence these animals must perform a type three migration for crossing large oceanic areas. Indeed, several trans-Atlantic migrations are reported by tagging and recapture experiments with two species of shark. Figure 10 shows spiny dogfish (Squalus acanthias) records which are described and reviewed by Templeman (1976). Recapture took place several years after tagging except for one specimen which was tagged north of Scotland and recaptured 5 months later west of Iceland. Its speed must have been at least 0.5 km h^{-1}. For the relatively slow moving sharks this must have been a migration without detours; possibly a menotactic compass course was involved and geomagnetism could have provided the directional stimuli. Figure 11 presents long and short range migrations of blue sharks (Prionace glauca) across the North Atlantic and North-South migrations in the North-East Atlantic (Stevens, 1976). Other long range migrations are reported from the North-West Atlantic (Casey and Stillwell, 1970, after Stevens, 1976). The author suggests a migration route "along the

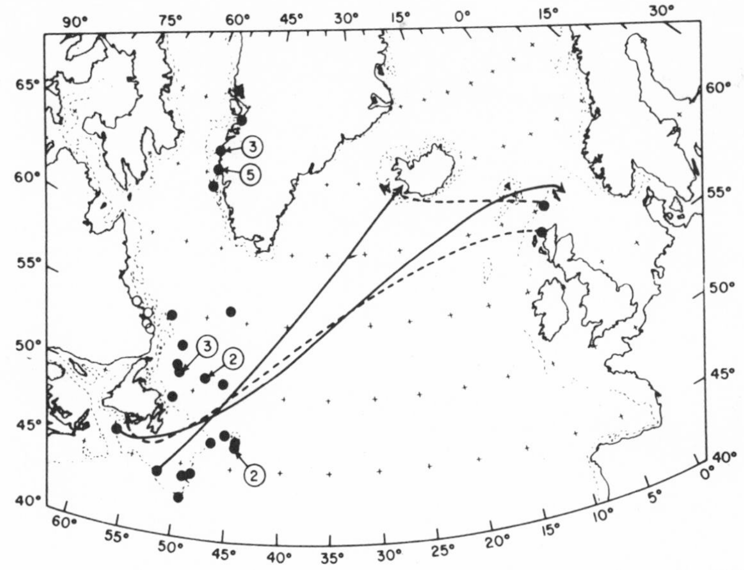

Fig. 10. Transatlantic migrations of tagged spiny dogfish (Squalus acanthias) and the most northward and eastward records of spiny dogfish (black dots, partly with numbers of recorded fish). Solid lines show eastward transatlantic migrations of tagged dogfish from points of tagging to points of recapture (arrows). Dashed lines represent the longest westward migrations (after Templeman, 1976).

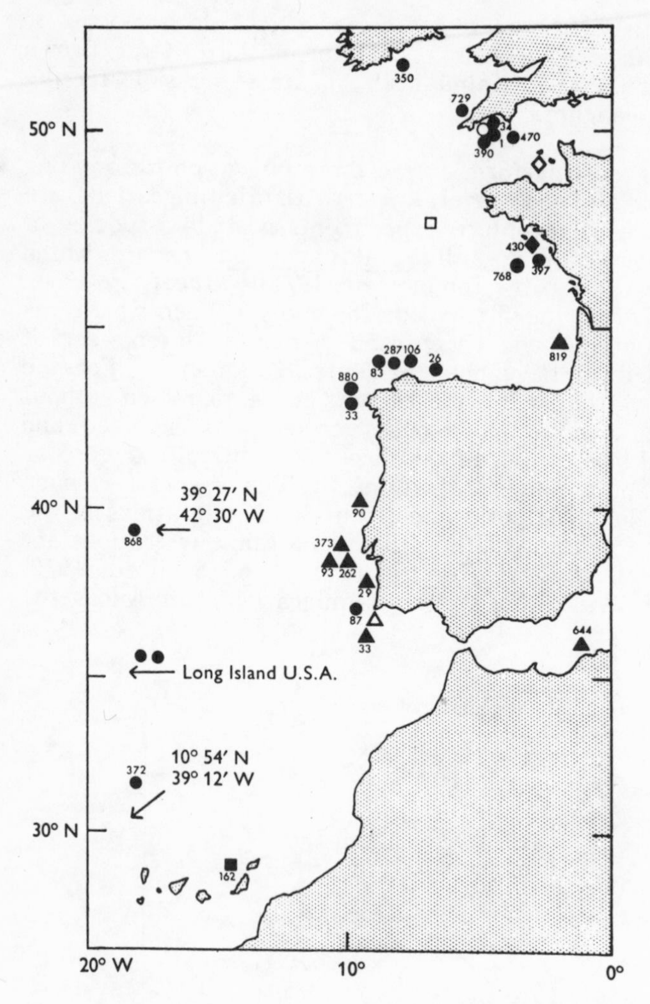

Fig. 11. Long and short range migrations of blue sharks (Prionace glauca) as indicated by tagging (tagging and release sites: open symbols) and recapture (solid) symbols) experiments. Numbers indicate days at liberty. An additional shark was recaptured after 2036 days at 05° 05'N, 20° 50'W (after Stevens, 1976).

North Atlantic gyre". The fastest mean rate of travel based on the shortest distance between tagging and recapture was 0.73 miles h^{-1} (nautical miles?) or 0.2. body lengths - sec^{-1}; others exhibited a speed half as fast. These migratory performances must have been performed by a comparatively straight line compass course; blue sharks are slow or moderate swimmers compared, for example, with Scombridae and Salmonidae. Both the gill area and the amount of red muscle are relatively small (Stevens, 1976). It is suggested that blue sharks used geomagnetism for directional guidance although sun-compass steering cannot be excluded; during a recent research cruise the author himself sighted a shark (obviously a blue shark) during daylight in the open East Atlantic between the Azores and Spain swimming at the surface, which may have been assisted by celestial orientation.

In order to examine whether a type three migration is performed sharks are good objects for field experiments for telemetric studies; because of their size they offer no difficulties. Standora and Nelson (1977) obtained good results by telemetric studies of local small range movements of Pacific angel sharks (Squatina californica) using ultrasonic transmitters as large as 3.5 x 15 cm provided for transmission of up to eight channels of information. These channels included sensors for measuring swimming speed, depth, ambient light, and water temperature. Possibly the impulse activity of nerve fibres supplying the ampullae of Lorenzini is transmitted as well as the compass direction of the shark. Shallow or surface swimming species can be tagged by radio transmitters which are technically easier than ultrasonic transmission and reception. In addition to the transmitters, sharks more readily than smaller fish species, can sustain magnets or even Helmholtz coils, to disturb the magnetic field.

ACKNOWLEDGEMENTS

I wish to thank M. Blake for assistance preparing the English text and C. Schuster for typing the manuscript.

REFERENCES

Abrasimova, A.M. (1976). Role of sense organs in the orientation of the bream, blue bream, roach and white bream in the Kiev reservoir (russ.). Vestn. Zool. 3: 40-44.

Arnold, G.P. (1974). Rheotropism in fishes. Biol. Rev. 49: 515-576.

Banner, A. (1968). Attraction of young lemon sharks, Negaprion brevirostris (Poey) by sound. Copeia: 871-872.

Banner, A. (1972). Use of sound in predation by young lemon sharks, Negaprion brevirostris. Bull. Mar. Sci. 22: 251-283.

Becker, G. (1974). Einflub des Magnetfeldes auf das Richtungsverhalten von Goldfischen. Naturwissenschaften 61: 220-221.

Bodznik, D. (1978). Calcium ion: an odorant for water discriminations and the migratory behavior of sockeye salmon. J. Comp. Physiol., A. 127: 157-166.

Branover, G.G., Vasiliev, A.S., Gleiser, S.I. and Tsinober, A.B. (1971). A study of the behaviour of eel in artificial and natural magnetic fields and the analysis of their mechanism of reception. (russ.) Vop. Ikhtiol. 11: 720-727.

Brown, H.R. and Illyinsky, O.B. (1978). The ampullae of Lorenzini in the magnetic field. J. Comp. Physiol. A. 126: 333-341.

Casey, J.G. and Stillwell, C. (1970). Ecology of ocean game fish. In: Progress in Sport Fishery Research. Washington, Bureau of Sport Fisheries Research, 1969. p. 192-194.

Churmasov, A.V. and Stepanov, A.S. (1977). Sun-compass and other types of orientation in pink salmon. Mar. Biol. (Vladivostok) 5: 55-62.

Cooper, J.C. and Scholz, A.T. (1976). Homing of artificially imprinted steelhead (rainbow) trout, Salmo gairdneri. J. Fish. Res. Board Can. 33: 826-829.

Cooper, J.C., Scholz, A.T., Horral, R.M., Hasler, A.D. and Madison, D.M. (1976). Experimental confirmation of the olfactory hypothesis with homing, artificially imprinted coho salmon. J. Fish. Res. Board Can. 33: 703-710.

Creutzberg, F. (1961). On orientation of migrating elvers (Anguilla vulgaris Turt.) in a tidal area. Neth. J. Sea Res. 1: 257-338.

Deelder, C.L. (1958). On the behaviour of elvers (Anguilla vulgaris Turt.) migrating from the sea into fresh water. J. Cons. perm. int. Explor. Mer 23: 83-88.

Deelder, C.L. and Tesch, F.-W. (1970). Heimfindevermögen von Aalen (Anguilla anguilla), die über grobe Entfernungen verfplanzt worden waren. Mar. Biol. 6: 81-92.

Dill, P.A. (1971). Perception of polarized light by yearling sockeye salmon (Oncorhynchus nerka). J. Fish. Res. Board Can. 28: 1319-1322.

Dizon, A.E., Mooral, R.M. and Hasler, A.D. (1973). Olfactory electroencephalographic responses of homing coho salmon, Oncorhynchus kisutch, to water conditioned by conspecifics. Fish. Bull. 71: 603-910.

Dodson, J.J. and Leggett, W.C. (1974). Role of olfaction and vision in the behavior of American shad (Alosa sapidissima) homing to the Connecticut River from Long Island Sound. J. Fish. Res. Board Can. 31: 1607-1619.

French, R.R. and Bakkala, R.G. (1974). A new model of oceanic migrations of Bristol Bay sockeye salmon. Fish. Bull. 72: 589-614.

Fujii, T. (1975). On the relation between the homing migration of the western Alaska sockeye salmon Oncorhynchus nerka (Walbaum) and oceanic conditions in the eastern Bering Sea. Mem. Fac. Fish. Hokkaida Univ. 22: 100-191.

Gleiser, S.T. and Khordorkovskii, V.A. (1971). Experimental determination of geomagnetic reception in Anguilla anguilla (Russ.) Dokl. Akad. Nank SSSR 201: 964-967.

GreerWalker, M., Harden Jones, F.K. and Arnold, G.P. (1978). The movements of plaice (Pleuronectes platessa L.) tracked in the open sea. J. Cons. Int. Mer. 38: 58-86.

Hain, J.H.W. (1974). The behaviour of migratory eels, Anguilla rostrata, in response to current, salinity and lunar period. Helgoländer wiss. Meeresunters. 27: 211-233.

Hara, T.I. (1974). Is morpholine an effective olfactory stimulant in fish? J. Fish. Res. Board Can. 31: 1547-1550.

Hara, T. and McDonald, S. (1976). Olfactory responses to skin mucous substances in rainbow trout (Salmo gairdneri). Comp. Biochem. Physiol. A. Comp. Physiol. 54: 41-44.

Hasler, A.D. (1966). Underwater Guideposts. Madison, Univ. of Wisconsin Press.

Hasler, A.D. (1978). Olfactory imprinting in coho salmon (Oncorhynchus kusutch). In: Eds. K. Schmidt-Koenig and W.T. Keeton: Animal Migration, Navigation and Homing. Sympos. in Tübingen, Berlin, Springer, p. 357-369.

Hughes, D.A. (1969). Responses to salinity change as a tidal transport mechanism of pink shrimp, Penaeas duorarum. Biol. Bull. Mar. Biol.

Lab., Woods Hole 136: 43-53.

Kalmijn, A.J. (1978). Experimental evidence of geomagnetic orientation in elasmobranch fishes. In: Eds. K. Schmidt-Koenig and W.T. Keeton: Animal Migration, Navigation and Homing; Sympos. in Tübingen, Berlin, Springer, p. 347-363.

Kuhne, O. and Strotkoetter, E. (1967). Untersuchungen von Drucken auf den tierischen Körper. I. Druckrezeption bei Fischen und ihre Mitwirkung bei der Orientierung im Raum. Forsch. Ber. Land N. Rhein. Westf. 1857: 1-57.

Kundtson, B.K. and Stimers, J.R. (1977). Notes on the behaviour of elasmobranch fishes exposed to magnetic fields. Bull. South Calif. Acad. Sci. 76: 202-204.

Laurs, R.M., Yuen, S.H.Y. and Johnson, J.H. (1977). Small-scale movements of albacore, Thunnus alalunga, in relation to ocean features as indicated by ultrasonic tracking and oceanographic sampling. Fish. Bull. 75: 347-355.

Leask, M.J.M. (1978). Primitive models of magnetoreception. In: Eds. K. Schmidt-Koenig and W.T. Keeton: Animal Migration, Navigation and Homing; Sympos. in Tübingen. Berlin, Springer, p. 318-322.

Leggett, W.C. (1973). The migration of the shad. Sci. Am. 228: 92-100.

Lind, E.A. and Kaukoranta, E. (1974). Characteristics, population structure and migration of the white fish, Coregonus lavaretus (L.) in the Onlukoki river. Ichthyol. Fenn. Borealis 4: 160-217.

Loyacano, H.A. and Chappell, J.H. (1977). Sun compass orientation in juvenile large-mouth bass, Micropterus salmoides. Trans. Am. Fish. Soc. 106: 77-79.

Malinin, L.K., Poddubnyi, A.J. and Swirskyi, A.M. (1974). Behaviour of salmon (Salmo salar L.) in the course of spawning migration through the waterbody. Zhurn. Obshthl. Biol. 25: 645-649.

Mather, F.J. (1962). Transatlantic migration of two large bluefin tuna. J. Cons. perm. int. Expl. Mer. 27: 325-327.

Mather, F.J., Bartlett, M.R. and Becket, J.S. (1967). Transatlantic migrations of young bluefin tuna. J. Fish. Res. Board Can. 24: 1991-1997.

Matsumoto, W. (1975). Distribution, relative abundance and movement of skipjack tuna, Katsuwonus pelamis in the Pacific Ocean based on Japanese tuna longline catches, 1964-67. NOAA Techn. Rep. NMFS SSRF 695: 1-30.

May, A.-W. (1973). Distribution and migration of salmon in the Northwest Atlantic. Spec. Publ. Ser. Int. Salmon Found. 4: 373-382.

McCleave, J.D. (1967). Homing and orientation of cutthroat trout (Salmo clarki) in Yellowstone Lake, with special reference to olfaction and vision. J. Fish. Res. Board Can. 24: 2011-2044.

McCleave, J.D. and LaBar, G.W. (1972). Further ultrasonic tracking and tagging studies of homing cutthroat trout (Salmo clarki) in Yellowstone Lake. Trans. Am. Fish. Soc. 101: 44-54.

McCleave, J.D. and LaBar, G.W. (1977). Within-season homing movements of displaced, mature sunapee trout (Salvelinus alpinus) in flood ponds, Main. Trans. Am. Fish. Soc. 106: 156-162.

McCleave, J.D., Rommel, S.A. and Cathcart, C.L. (1971). Weak electric and magnetic fields in fish orientation. Ann. N.Y. Acad. Sci. 188: 270-282.

Merkel, F.W. (1978). Angle sense of painted quails: A parameter of geodetic orientation. In: Eds. K. Schmidt-Koenig and W.T. Keeton: Animal Migration, Navigation and Homing; Sympos. in Tübingen. Berlin, Springer, p. 269-274.

Miles, S.G. (1968). Rheotaxis of elvers of the American eel (Anguilla rostrata) in the laboratory to water from different streams in Nova Scotia. J. Fish. Res. Board Can. 25: 1591-1602.

Neves, R.J. and Depres, L. (1979). The oceanic migration of American Shad, Alosa sapidissima, along the Atlantic coast. Fish. Bull. 77: 199-212.

Nordeng, H. (1971). Is the local orientation of anadromous fishes determined by pheromones? Nature (Lond.) 233: 411-413.

Oshima, K., Hahn, W. and Gorbman, A. (1969). Olfactory discrimination of natural waters of salmon. J. Fish. Res. Board Can. 26: 2111-2121.

Richardson-P.L. and Knauss, J.A. (1971). Gulf Stream and western boundary undercurrent observations at Cape Hatterras. Deep Sea Res. 19: 108-109.

Richkus, W.A. (1975). The response of juvenile alewifes to water currents in an experimental chamber. Trans. Am. Fish. Soc. 100: 494-498.

Rommel, S.A. and McCleave, J.D. (1973). Sensitivity of American eels (Anguilla rostrata) and Atlantic salmon (Salmo salar) to weak electric and magnetic fields. J. Fish. Res. Board Can. 30: 657-663.

Royce, F.W., Smith, S.L. and Hart, A.C. (1968). Models of oceanic migrations of Pacific salmon and comments on guidance mechanisms. Fish. Bull. U.S. Fish. Wild. Serv. 66: 441-462.

Schmidt-Koenig, K. (1975). Migration and Homing in Animals. Berlin, Springer-Verlag, 99 pp.

Scholz, A.T., Cooper, J.C., Horrall, R.M. and Hasler, A.D. (1978). Homing of morpholine imprinted brown trout, Salmo trutta. Fish. Bull. 76: 293-295.

Schuijf, D.A. (1974). Field Studies of Directional Hearing in Marine Teleosts. Utrecht, Eliukwijk.

Schuijf, D.A. (1975). Directional hearing of cod (Gadus morhua) under approximate free field conditions. J. Comp. Physiol. 98: 307-332.

Solomon, D.J. (1973). Evidence for pheromone-influenced homing by migrating Atlantic salmon, Salmo salar (L.). Nature (Lond.) 244: 231-232.

Standora, E. and Nelson, D.R. (1977). A telemetric study of the behaviour of free-swimming Pacific angel shark. Bull. South Calif. Acad. Sci. 76: 193-201.

Stasko, A.B. and Rommel, S.A. (1974). Swimming depth of adult American eels (Anguilla rostrata) in a salt water bay as determined by ultrasonic telemetry. J. Fish. Res. Board Can. 31: 1184-1150.

Steffel, S., Dizon, A.E., Magnuson, J.J. and Neill, W.H. (1976). Temperature discrimination by captive free-swimming tuna, Euthynnus affinis. Trans. Am. Fish. Soc. 105: 588-591.

Stevens, J.D. (1976). First results of shark tagging in the north-east Atlantic, 1972-1975. J. Mar. Biol. Assoc. U.K. 56:

Stillwell, C. and Casey, J.G. (1968). Migratory habits of sharks. In: Progress in Sport Fisheries Research 1967, Washington, Bureau of Sport Fish. and Wildlife, Div. of Fish. Res. p. 184-185.

Teichmann, H. (1959). Ober die Leistung des Gerichssinnes beim Aal (Anguilla anguilla L.). Z. Vergl. Physiol. 42: 206-254.

Templeman, W. (1976). Transatlantic migration of spiny dogfish (Squalus acanthias). J. Fish. Res. Board Can. 33: 2605-2609.

Tesch, F.-W. (1965). Verhalten der Glasaale (Anguilla anguilla) bei ihrer Wanderung in den Astuarien deutscher Nordseeflüsse. Helgoländer wiss. Meeresunters. 12: 404-419.

Tesch, F.-W. (1967). Homing of eels (Anguilla anguilla) in the Southern North Sea. Mar. Biol. 1: 2-9.

Tesch, F.-W. (1970). Heimfindevermögen von Aalen Anguilla anguilla nach Beeinträchtigung des Geruchssinnes, nach Adaption oder nach Verpflanzung in ein Nachbarästuar. Mar. Biol. 6: 148-157.

Tesch, F.-W. (1974). Influence of geomagnetism and salinity on the directional choice of eels. Helgoländer Wiss. Meeresunters. 26: 382-395.

Tesch, F.-W. (1974). Migratory behaviour of displaced homing yellow eels (Anguilla anguilla) in the North Sea. Helgoländer Wiss. Meeresunters. 27: 190-198.

Tesch, F.-W. (1975). Fishes: Orientation in space: animals. In: Marine Ecology 2 (2) Ed. O. Kinne, London, Wiley, p. 657-707.

Tesch, F.-W. (1977). The Eel, Biology and Management of Anguillid Eels. London, Chapman and Hall, 434 pp.

Tesch, F.-W. (1978a). Telemetric observations on the spawning migration of the eel (Anguilla anguilla) west of the European continental shelf. Env. Biol. Fish. 3: 203-209.

Tesch, F.-W. (1978b). Horizontal and vertical swimming of eels during the spawning migration at the edge of the continental shelf. In: Eds. K. Schmidt-Koenig and W.T. Keeton: Animal Migration, Navigation and Homing; Sympos. in Tübingen, Berlin, Springer, p. 378-391.

Tesch, F.-W. (1978c). Occurrence of eel larvae (Anguilla anguilla) west of the European continental shelf. 1971-1977. Envir. Biol. Fish (in preparation).

Tesch, F.-W. (1978d). Blankaalwanderung im Meer. Fisch u. Fang 19: 180.

Tesch, F.-W. (1979). Tracking of silver eels (Anguilla anguilla) in different shelf areas of the Northeast Atlantic. Rapp. Proc. Verb. Reun. Cons. Int. Explor. Mer. (in press).

Tesch, F.-W., Kracht. R., Schoth, M., Smith,D.G. and Wegner, G. (1979). Report on the eel expedition of FRV "Anton Dohrn" and R.K. "Friedrich Heincke" to the Sargasso Sea. ICES, Anadromous and Catadromous Fish Commitee C.M. 1979 M: 6.

Tiews, K. (1963). Der Thunbestand (Thunnus thynnus Linnaeus) in der Nordsee, seine Wanderungen, seine transatlantischen Beziehungen und seine Nutzung durch die deutsche Fischerei. Arch. FischWiss. 14: 105-108.

Tsvetkov, V.I. (1969). On the threshold sensibility of some freshwater fishes to the rapid change of pressure (Russ.). Vopr. Ikhtiol. 9: 928-935.

Tsvetkov, V.I. (1972). The sensitivity of load (Misgarnus fossilis L.) to changes in pressure (Russ.). Vopr. Ikhtiol. 12: 950-953.

Veen, J.F. De (1967). On the phenomenon of soles (Solea solea L.) swimming at the surface. J. Cons. Serm. Int. Explor. Mer. 21: 207-236.

Vladykov, V.D. (1971). Homing of the American eel, Anguilla rostrata, as evidenced by returns of transplanted tagged eels in New Brunswick. The Canadian Field-Naturalist 85: 241-248.

Walcott, C. and Green, R. (1974). Orientation of homing pigeons altered by a change in the direction of an applied magnetic field. Science 184: 180-182.

Westerberg, H. (1975). Counter-current orientation in the migration of the European eel (Anguilla anguilla L.). Göteborg Univ. Oceanogr. Inst. Rep. 9: 1-18.

Westin, L. and Nyman, L. (1977). Activity, orientation and migration of Baltic eel (Anguilla anguilla (L.)). Rap. Proc. Verb. Reun. Cons. Int. Explor. Mer. (in press).

Williams, F. (1972). Consideration of three proposed models of migration of young skipjack tuna (Katsuwonus pelamis) into the eastern Pacific Ocean. Fish. Bull. 70: 741-762.

Winn, H.E., Salmon, M. and Roberts, N. (1964). Sun-compass orientation by parrot fishes. Z. Tierpsychol. 21: 798-821.

Worthington, L.V. (1976). On the North Atlantic Circulation. Baltimore, Johns Hopkins Univ. Pr., 110 pp.

Yuen, H.S.H. (1970). Behaviour of skipjack tuna, Katsuwonus pelamis, as determined by tracking with ultrasonic devices. J. Fish. Res. Board Can. 27: 2071-2079.

BIOLOGICAL RHYTHMS: THEIR ADAPTIVE SIGNIFICANCE

Horst O. Schwassmann
Department of Zoology, University of Florida
Gainesville, FL 32611, USA

For the essence of beauty lies in rhythm and harmony
(Cloudsley-Thompson, 1961, p. 201)

INTRODUCTION

Environmental physiology concerns itself with the functional characteristics of organisms that have evolved as adaptations to particular environmental situations. Such an approach towards understanding a most important aspect of biology would remain incomplete and would prove unsuccessful if it did not also encompass details of temporal organization, the precise order according to which all the different physiological changes and behavioral patterns are scheduled to occur in anticipation of, rather than in reaction to, the many cyclic environmental events.

An earlier extensive review of information concerning biorhythms in fish was prepared about a decade ago (Schwassmann, 1971a). At that time it seemed expedient, even necessary, to consider rhythmic behavior of fishes within the framework of recently accumulated knowledge accrued from studies of other organisms, mainly birds and mammals. These two classes of homeothermic vertebrates, regulating their body temperature by physiologic means, provided a vast amount of laboratory data of a precision sufficient to permit qualitative predictions and generalizations concerning the mechanisms underlying biological rhythmicity. Comparable quantitative data for fishes had been achieved by studying the activity pattern of gymnotid electric fish in nature and in the laboratory (Lissmann and Schwassmann, 1965; Schwassmann, 1971b). These studies demonstrated not only the endogenous nature of the activity rhythm in fishes but also uncovered circadian features such as the dependence of endogenous period length on the intensity of constant light, known an Aschoff's circadian rule, and the phase determining effect of brief light exposures recognized as the basis for entrainment by the light-dark transitions of the natural environment. These findings, together with the recognition of time-compensated sun-compass orientation in many animals, including fishes, as endogenous circadian rhythms (Schwassmann, 1960), mandated a treatment of biological rhythmicity in fishes as an expression of innate timing mechanisms that are inherent characteristics of living systems. The formerly assumed "triggering" action of certain environmental periodic factors was now considered a mere zeitgeber or synchronizer-action (Schwassmann, 1971a).

During the last ten years considerable additional evidence on biorhythms in fishes has appeared and one symposium on the subject has been held in 1977 (Thorpe, ed., 1978). This chapter will not provide an all inclusive list of more recent publications but will instead emphasize a few new findings that appear to contribute significantly to our understanding of the adaptive value of rhythmic behavior in fishes and will in this way present a general overview of biochronometry in fishes seen from an ecological perspective.

PERIODICITIES OF THE EARTH'S ENVIRONMENT

Life on our planet has evolved under the constant influence of, and in adaptive response to, three predominant environmental periodicities. These are the daily cycle, the annual cycle, and the lunar monthly cycle. The daily cycle is caused by the rotation of our planet around its polar axis resulting in the regular alternation of night and day. The annual cycle lasts about 365 solar days, the time it takes for the earth to complete one revolution around the sun. Due to the progressive changes of the angle of inclination of the planet's polar axis in relation to the sun we observe the well-known phenomenon of the seasons. The third periodicity, the lunar monthly cycle, has a pronounced effect on the changing amplitudes of the tides. Tides are movements of the ocean water caused by the attraction exerted by the mass of the moon and, to a lesser extent, by the sun. There are approximately two high and two low tides occurring on each solar day which is due to the earth's daily rotation around its axis and the resulting relative positions of moon and sun. The moon takes about 29.53 solar days for one revolution around the earth and, since its movement is towards east, we notice a delay of about 50 minutes for each day of the meridian passage of the moon. This changes the relative positions of sun and moon. When moon and sun are in opposition (full moon), or in apposition (new moon), the gravitational forces of both bodies add and high "spring tides" result which occur in intervals of about 14.77 solar days. Relatively low "neap tides" are observed during the quarter and three-quarter moon. At higher latitudes a certain inequality in the two highs and two lows per day is present, a consequence of the changing declination of the moon's arc. When the moon's arc passes near the equator, very little diurnal inequality results, but when it approaches the two tropics the two tides become unequal. This inequality can go as fas as eliminating one of the tidal peaks altogether, as is the case on the Mississippi coast in the Gulf of Texas. In addition, there exists considerable variability in the tidal pattern due to geographic location and geomorphology of the continental slopes.

EVOLUTION OF CHRONOBIOLOGICAL ADAPTATIONS TO PERIODIC CHANGES IN THE AQUATIC MEDIUM

Organisms, plants as well as animals, have evolved in such a way that their activities are timed to correspond with the periodicities of the earth's environment. The resultant endogenous, or innate, adaptive features of organisms are the circadian, the circannual, and circatidal rhythms which regulate in time physiological and behavioral processes in such perfect manner that their high selective value seems immediately obvious. This rhythmic organization has become part of the "phylogenetic memory store" of the majority, possibly of all, existing species, at least to the degree to which such genetically fixed anticipation of, or preparedness for, specific periodic events of the environment proved to be of advantage to the species. Rhythms are, therefore, the result of evolution by natural selection; and the physiological basis of these biorhythms is as much inherited as are the stereotyped innate behaviors, or all the species-specific functional and morphological features.

In the aquatic habitat there are daily periodic oscillations in certain physical factors of which light intensity and temperature are probably the most conspicuous abiotic variables. Their amplitude changes are pronounced in shallow water, and especially in small bodies of water. Light level changes are greatly attenuated with increasing depth while daily cyclic temperature changes are present in small streams that are continually mixed but are negligible in the ocean and in larger lakes where they are greatly surpassed by the existing temperature differences of different layers due to summer stratification. In some tropical and subtropical areas daily periodic rains occurring in the early afternoon during at least part of the year can cause cyclic changes in water volume, current velocity, and turbity, as well as chemical composition due to run-off of rain water from the watershed. Other physical factors that are found to cycle in a daily pattern are caused by the periodically changing action of aquatic organisms. These are the hydrogen ion concentration, dissolved oxygen and carbon dioxide, and certain nutrients and waste products. Their levels depend on the photosynthetic cycle and nutrient uptake of aquatic forms of life. The daily variations of these factors in relation to the drift pattern of algae have been demonstrated in small streams by Müller-Haeckel (1965, 1966, 1976) and they are an essential part of the synecological analysis of running water biotopes in the subartic by Müller (1978) who clearly recognized the impact of temporal niche dimensions as well as their interdependencies and changes throughout the year.

The annual cycle is characterized by gradually changing daylength and temperature variation. These two climatic factors are most conspicuous at higher latitudes, away from the equator, while they are quite insignificant in regions near the equator. In the tropics one can observe an alternation between rainy and dry seasons with subsequent effects on water level, discharge, and physico-chemical composition of aquatic habitats. Outside of the tropics the shifting inclination of the sun's arc causes the progressive changes in duration and in intensity of insolation resulting in the familiar seasons of winter and summer with the intervening transitional periods of spring and autumn. The temperate zone summer is the principal growing season; its relatively high temperatures and long lasting strong insolation provide the necessary conditions for high rates of production, thereby resulting in an abundance of food for all other organisms. This is also the case for aquatic life forms, especially fishes. While those populations of the very mobile birds that spend the summer at highest latitudes in subarctic zones, can escape the harshness of winter by migrating towards the tropics, some of the less mobile mammals have evolved physiological adaptations for passing the unfavorable season in a state of reduced energy expenditure, or hibernation. Hibernation is also the rule for those poikilothermic, better termed ectothermic, vertebrates which live at very high latitudes. At near arctic latitudes, the aquatic environment provides suitable conditions for surviving the cold winter months. Temperature fluctuations are levelled out due to the high heat capacity of water and, while the freezing point lies at zero degrees, its highest density is reached at four degrees, causing the formation of a protective ice cover on the lake surface only, keeping liquid water below. Fishes and aquatic invertebrates are known to have invented other very efficient means of coping with changing, especially low temperatures in form of enzymatic seasonal adaptations; specific sets of iso-enzymes have rate constants adjusted for different temperatures. Conditions in tropical freshwaters seem less favorable for fish, at least at first sight, especially when considering the coincidence of low oxygen content and high metabolic requirements at the high temperature. In addition, there are many areas where a long-lasting dry season causes smaller bodies of water to disappear. In these areas, the tropical cerrado, or llanos, we encounter the "annual" fishes which have a stage of true developmental arrest, obligatory in most investigated species, which permits these populations to estivate as partially developed embryos inside a dessication-resistant egg.

EVIDENCE FOR CIRCADIAN RHYTHMS IN FISHES

Endogenous rhythms of approximately 24-hour periodicity have been reported for several species of fish. Though almost all species display cyclic patterns of alternating activity and rest in their natural habitat, demonstration of the endogenous nature of the underlying timing mechanism requires continuation of the cyclic pattern for many periods in constant conditions, mostly constant temperature and continuous light at a certain intensity, and the appearance of a "free-running" period which is different from exactly 24 hours. This could be convincingly shown in experiments with certain species of gymnotoid electric fishes from Brazil (Lissmann and Schwassmann, 1965; Schwassmann, 1971b). Earlier indications of free-running, non-24 hour, periods in constant light were found in sun-compass orientation studies with a centrarchid, *Lepomis cyanellus* (Schwassmann, 1960), and be recording the activity pattern in constant conditions of larval and adult *Petromyzon* (Kleerekoper et al., 1961). Further evidence for free-running activity rhythms was obtained by Müller (1968) who also noted non-synchronized free-running in brown trout during mid-summer near the arctic circle (Müller, 1969).

Additional extensive work by Müller and his group in Sweden demonstrated the seasonal changes of the circadian activity pattern in several species, and illustrated the adaptive significance of a certain built-in flexibility of the circadian system with respect to dualistic phasing and naturally occuring phase inversions in response to the annual changes of the light environment. These results have been summarized recently by Müller (1978a,b) and Eriksson (1978). While Salmo trutta had formerly been considered a day-, or a night-active species, or both, depending on season and latitude (Müller, 1978b), its basic pattern now appears to be crepuscular, showing activity peaks at dawn and dusk, according to Eriksson (1978). Similar to findings on Atlantic salmon (Richardson and McCleave, 1974), Eriksson's trout are predominantly day-active in summer, night-active in winter, and crepuscular during autumn and spring (Eriksson, 1978). Two night-active species, burbot (Lota lota) and sculpin (Cottus poecilopus), show phase inversions to a day-active pattern during the winter near the arctic circle (Müller, 1973, 1978a). The normally night-active brown bullhead (Ictalurus nebulosus) can shift gradually to a day-active pattern when exposed to very dim light intensities in the laboratory (Eriksson, 1978). Data from recording the swimming speed, or activity, of bluefish (Pomatomus saltatrix) suggest an endogenous component although no constant light conditions were employed (Olla and Studholme, 1972, 1978). The study by Kavaliers (1978) on the circadian activity pattern and its annual changes of the lake chub (Couesius plumbeus) is in line with, and in extension of, the findings of Müller (1978b), and suggests a possible mechanism underlying the annual changes of the circadian activity pattern. Recording swimming activity in conditions of constant dark of individual freshly-caught lake chub, Kavaliers (1978) found the activity pattern to persist for up to 20 days. Remarkably, the fish exhibited relatively short free-running periods, about 25 hours, when tested during summer, and they had much longer spontaneous periods, near 28 hours, in winter. This seasonal change of the circadian period is interpreted as an after effect of the natural photoperiod to which these fish had been exposed immediately preceding the experiments. That no spontaneous periods of less than 24 hours were recorded is in accord with Aschoff's circadian rule in the case of a day-active species in continuous darkness. A comparable effect of seasonally changing photoperiod had been noted in experiments investigating the circadian rhythm of quantitative compensation of the sun's daily movement, sun-compass orientation, in centrarchid fish in Wisconsin (Schwassmann and Braemer, 1961). In these experiments groups of the sunfish were exposed for several weeks to "long" and "short" days respectively. Those fish conditioned to long photoperiod were found to alter the trained angle to the sun faster during the hours around noon than the fish that had been conditioned to the short winter days. Although monitoring two different expressions of circadian organization, Kavaliers' (1978) and our data appear to document the same effect of photoperiod: the clock seems to run faster in summer when the days are longer than during short winter days. This is shown in case of the lake chub in summer by a shorter free-running period, and in case of the sun-orientation rhythm by the faster change of the orientation angle during the hours around midday only, because these latter fish were not in free-running but in entrained conditions of the experimental light-dark cycle. The data obtained by Kavaliers (1978) with lake chub in constant dark show endogenous free-running patterns for up to 20 days, becoming arhythmic from then on. Similar breaking up of the circadian pattern in constant conditions was noted in our studies with electric fish and was found to be due to the non-natural laboratory conditions. This discrepancy between the pattern observed in nature and in the laboratory led to the generalized statement that "the rhythm finds full expression only under suitable environmental conditions" (Lissmann and Schwassmann, 1965).

Among other recent attempts to record circadian rhythms in fish one could mention those which combined the monitoring of locomotor activity with temperature preference, or selection, in bluegill (Beitinger, 1975; Reynolds, 1976) and in bowfin, Amia calva (Reynolds et al., 1978). Edel (1976) found evidence for endogenous timing in the locomotor activity pattern of female American eels. Other experiments combined swimming activity with heart rate recording (Kneis and Siegmund, 1976), and heart rate telemetry seems to be a novel and useful technique for monitoring the circadian pattern of unrestrained fish in their normal environment (Priede, 1978).

APPARENT ADAPTIVE VALUE OF CIRCADIAN ORGANIZATION IN FISH

It can be assumed that all fish, including those living in caves and in the deep ocean, go through periodically changing states of high activity and decreased activity, or rest. In some species one might be tempted to use the term "sleep" for the less active phase, as in many wrasses, and the electric sandfish, and several of the parrot fishes; the first ones burying themselves into the bottom substrate, and the last ones secreting an elaborate mucous envelope for the night.

In addition to the literature reviewed earlier (Schwassmann, 1971a) which documents daily periodic activities in fish in nature, mention may be made of the report on daily feeding patterns of many species from Hawaiian waters (Clarke, 1978), and on daily periodic spawning activity near sunset in bluefish (Norcross, 1974), and in a pomacanthid fish from Hawaii (Lobel, 1978). With respect to schooling behavior, a report by Müller (1978c) emphasized the importance of near-natural conditions for an arctic coregonid species. While the circadian activity pattern persisted for up to ten days in conditions of constant dark within a group of these fish, apparently facilitating school formation during daylight hours alternating with dispersal and reduced swimming at night in natural conditions, individual fish showed no such cyclic pattern in constant dark, suggesting abnormal behavior due to the absence of some social, or group-related, factor. The same paper also reports data on normally occuring "arythmic" activity during the height of the arctic summer when the sun remains above the horizon continually, the non-rhythmic pattern being more pronounced in solitary fish than in groups (Müller, 1978c).

A definite advantage of a built-in circadian timing mechanism lies in the resultant capability to be ready in advance of, and prepared for, predictable environmental opportunities. This often involves active movements to certain feeding stations prior to the time of food presence or of optimal feeding conditions. Figuratively speaking, the circadian mechanism warms up the organism, often hours before actual emergence from the resting place in the sand (Lissmann and Schwassmann, 1965), or initiates and controls other physiological mechanisms. To cite another example, retinomotor changes in the eye begin long before light levels are sufficient for seeing (John and Haut, 1964; Davis, 1962). These photomechanical changes in the eyes of fishes involve movements of molecules of shielding pigment as well as contraction and relaxation of the myoids of rods and cones. They are basically not in direct dependence on altered light levels but occur spontaneously in a true circadian pattern which persists in constant darkness. The advantage of such a mechanism preparing their eyes for immediate optimal vision at rapidly increasing light levels of dawn without necessitating a long adjustment phase as a direct response to light seems clearly indicated for predator and prey alike. This interpretation is supported by the observation that very many species go through their most active feeding states during times of changing light levels at dawn and dusk.

One should also concede, on the other hand, that there are many instances where an advance timing of certain activities seems not as vitally important as might be implied by an endogenous timing capability of such precision that day-to-day variations were found to be of the order of a few minutes only (Lissmann and Schwassmann, 1965). Vertical migrations of pelagic marine fishes are believed to occur in response to light level changes or, perhaps, in pursuit of planktonic food organisms where these daily vertical movements may have an endogenous basis (Enright and Hamner, 1967). The daily cyclic pattern of schooling behavior in fishes seems very dependent on light levels, but the tendency for swimming in tight schools could be shown to follow a circadian pattern in constant conditions (Hunter, 1966, Müller, 1978c).

We do not always recognize an adaptive advantage in all of the overt circadian rhythms. This might be due to our state of relative ignorance of natural history and behavior in fishes, but it should also be recognized that all measurable physiological and behavioral rhythms are manifestations, or outward signs, of a very basic feature of living beings on this planet, which is circadian organization. Many of the overt periodicities are of functional significance, others many not be critically involved in survival, but all are circadian since this is the way organisms are made and are functioning. Probably, the most important adaptive value lies in the fundamental role of the circadian timing system in photoperiodic induction, and actual spawning. Photoperiodic induction occurs when light coincides with a certain responsive, highly sensitive, phase; and the normally entrained circadian system renders this reponse mechanism to be very precise. Such involvement of the circadian system in photoperiodic time measurement has been demonstrated in several species of plants, insects, birds, and mammals, and recently also in fishes. Using the "interrupted night" technique, Baggerman (1972) could achieve this in Gasterosteus. Chan (1976) induced ovarian growth in the medaka, Oryzias latipes, when one hour of light fell near the center of the 17-hour long dark phase. Similar results were obtained with the same technique in female catfish, Heteropneustes fossilis (Sundararaj and Vasal, 1976). The subject of photoperiodism in fish is treated in a separate chapter (Baggerman, this volume).

CIRCANNUAL RHYTHMS IN FISH

The choice of the term "circannual" implies the presence of an internal, or endogenous, component of approximately one-year period. These circannual rhythms are the basic substrate of periodic reproduction and related activities, as preceding migratory movements, and they could easily be envisaged as simply the physiological changes involved in gonadal maturation controlled by the many intricate endocrine mechanisms. These rhythmic changes can be accelerated within certain limits by interfering with normally acting environmental factors, like photofraction and temperature change, or by injection of pituitary hormones. The reproductive cycle may in some fish last only one year and completion of spawning may be followed by death of the parents, as in the case of annual cyprinodonts where the partially developed embryo passes the unfavorable season in diapause protected inside the egg against dessication, or, as in some small tetras in Northern Brazil which apparently die after their first and only complete spawning. Other species may also spawn only once in their lifetime but require several years before reaching maturity and they need to undertake a long migration to the distant spawning grounds, as is the case with eels of the genus Anguilla. Although there seem to exist a few exceptions to the rule, a few species of marine teleosts are reported to spawn throughout the year (Munro et al., 1973), reproduction in fish in general is annually periodic. Within the oviparous fishes we can distinguish two basic types. The first are the total spawners where a usually great number of eggs are maturing and are released simultaneously, and where the timing of the one spawning effort a year is precise, and synchronized within a population, often preceded by a migration, or spawning run. Eggs are small, and embryonic development is fast, parental care is absent. The other type consists of those species which go through several successive egg-laying bouts during a prolonged spawning season which may last for several months. These are the intermittent, multiple, or partial spawners which produce a relatively small number of large eggs. Many of these species exhibit elaborate courtship behaviors, nest-building, often parental care; even mouth-brooding is known in some. Successive spawning bouts can follow in regular intervals, eight to nine days in populations of green sunfish (Hunter, 1963) where mutual behavioral synchronization within local populations takes place, quite similar to the electric sandfish, Gymnorhamphichthys hypostomus where populations of different, but adjacent streams may be out of phase with each other (Schwassmann, 1978). The timing of repeated spawnings can be in strict coincidence with some environmental periodicity, for example, the 15 day intervals of the semi lunar-monthly high spring tide cycle in the California grunion (Walker, 1949), or in Fundulus heteroclitus (Taylor, et al., 1979). In contrast, the intervals between repeated nesting sessions of tropical cichlids seem more variable, and subject to environmental factors and mutual synchronization by behavioral interactions seems the rule rather than entrainment by environmental cycles.

Systematically changing daylength and temperature have been found to play an important role in the advance timing of reproduction-related events in fish at higher latitudes and photoperiodic control of annual breeding is the subject of another chapter (Baggerman, this volume). The important role of endocrines and their interactions has been reviewed recently by Billard and Breton (1978). In this chapter I should like to bring a few references that deal with the intriguing problem of how advance timing of migration and reproduction could possibly be effected in the tropics where annual daylength and temperature changes are insignificant, or non-existing. That such a problem exists and that reproduction in the tropics is, indeed, periodic has been recognized only recently. The older viewpoint, supporting aperiodic reproduction throughout the year, can be found expressed in a standard zoology text (Hesse and Doflein, 1943), apparently based on examples of rather anecdotal nature in a previously published text (Hesse, 1924). This view translates as follows: "Because of even temperatures in many tropical areas - the coolest month being only two to three degrees cooler than the warmest - annual changes seem restricted to an alternation of a more rainy with a less rainy season which would be almost unnoticeable for animals. One can therefore observe a no longer rhythmical sequence, or order, in the other areas of periodical events of animal life. In the tropics reproduction is not coupled to season, instead, one finds at all times young animals at various stages of development." Some examples are cited in support, lepidoptera and other insects, and tadpoles of frogs that can be found year-round, as well as the continuous chirping of crickets, locusts, katydids, and the mating calls of frogs. Reproduction in birds, one can read in the same paragraph, is not bound to a particular time of year (Hesse and Doflein, 1943). This opinion, probably very common, may also lie at the basis of Bünning's (1967) statement that: "Information about seasonal changes would be entirely superfluous in tropical areas which are continually moist." The assumption of uniform, year-round reproduction was contradicted by the many cases of strictly seasonal reproduction in bats and birds reported by the members of the Oxford University Expedition to the New Hebrides in the 1930ies (Baker, 1939, 1947; Baker and Baker, 1936; Baker and Bird, 1936; Baker et al., 1940). Unfortunately, the location of the islands so far away from the equator, Espirito Santo at about 15 degrees south, makes an evaluation of possible timing factors difficult. At that high a latitude, there occurs an annual change in the photofraction of more than one hour, and possible photoperiodic entrainment cannot be ruled out. Problems concerning advance timing of spawning in tropical fishes were emphasized in an earlier review and the seasonal changes resulting from different amounts of rainfall were listed for the different aquatic biotopes (Schwassmann, 1971a). At that time the most urgent need seemed to be for more information regarding the natural times of spawning of tropical species in relation to local meteorological conditions. Since then, a wealth of knowledge about natural history of tropical freshwater fishes and about seasonally changing conditions of their natural habitats, including data and observations on reproductive activities, has been compiled and presented in ecological perspective by Lowe-McConnell (1975). Some additional information on the aquatic biotopes of Amazonian rainforest and on reproductive timing of some species of local teleosts is contained in Schwassmann (1978). With respect to the degree of noticeably changing seasonal conditions, two basic types of aquatic habitats can be singled out. The floodplain lakes, the first type, show considerable annual variation in surface area, depth of water, and related factors, due to their open connection with the main rivers which undergo great variations in water discharge and level because of the different amounts of rain water runoff from their catchment basins during the course of a year. At the other extreme are small streams of the firm ground which are at sufficiently high elevation not to be affected by water level changes of the large rivers, they are often separated from them by rapids and falls. These small streams, especially those in Central Amazonia where there is little variation in rainfall, exhibit few changes in hydrological and physicochemical factors throughout the year.

There seems to exist a significant correlation with respect to the reproductive methods employed by fish populations inhabiting these two different aquatic biotopes. In the "Varzea" lakes one finds mostly total spawners which discharge their many eggs within a very brief time span after the first filling of the floodplain, often after undertaking a short migration, while the fish populations in the small upland streams are partial spawners. It may be that in this last mentioned biotope of so very little seasonality we could eventually succeed in demonstrating free-running spawning rhythms as suspected by Bünning (1967). Several of my own data obtained during he last few years on gymnotoid streamdwelling fishes seem to favor such a possibility (Schwassmann, 1978). The two different reproductive methods, sometimes referred to as -r- and -K- strategies, appear well adapted to the two situations. The considerable augmentation in water volume, amounting to an eventually increased water level of two or more meters, provides for a rather sudden and enormous enlargement of surface area due to the low elevation and flatland character of the "Varzea". Subsequent sedimentation of the now no longer agitated and relatively nutrient-rich river water permits sunlight penetration which results in greatly increased primary production. Immediate release of a large number of eggs and fast embryonic development, as it is done by the total spawners, takes advantage of this so suddenly created opportunity of an ample food supply for the many hatching and growing fish. On the other hand, the intermittent spawners in the much more uniform environment of the small streams introduce a small number of larger eggs into their community at more or less regularly spaced intervals. Thus, they are not likely to waste their entire reproductive complement at one time. This mode of interval spawning also seems to facilitate mutual synchronization among the population members with the result that both sexes can be ready for spawning at the same time. In the electric sandfish such intrapopulation synchronization is present and populations inhabiting adjacent, often even interconnecting, streams can be found at a different developmental stage and in a different phase of the spawning cycle (Schwassmann, 1978).

With respect to potential environmental factors that regulate periodic breeding in the intermittent spawners, Kirschbaum (1975) succeeded in inducing gonadal recrudescence, maturation, and spawning, as well as in rearing of the young, in Eigenmannia, one of the gymnotid electric fishes. Among the factors that could be systematically varied in the laboratory were water level change, conductance and acidity, and rain simulation while daylength was continually kept at 13 hours of light and 11 hours of darkness. After completing a four-year study, the same author presents a full account of this important work (Kirschbaum, 1979). Not only did he breed a gymnotoid species in captivity, to my knowledge not having been accomplished previously, he could also utilize for his experiments the aquarium-bred first, second and third generations of his original imported stock. Starting out with fish having regressed gonads, gonadal recrudescence was induced by lowering the conductance and pH, together with water level increases and rain simulation, up to four times within one year's period; no state of post-breeding refractoriness seemed to occur. Of the four factors, the hydrogen ion concentration did not appear to be important while the decrease in conductivity was the most potent stimulus for achieving spawning readiness. In this, specific ions did not seem to matter, in fact, total and carbonate hardness could be increased while conductance lowering still proved effective. Once started, in response to these changed factors, spawning would continue while these factors were kept constant. Intervals between successive spawning bouts were three to four days under optimal conditions in the laboratory-reared fish, while the specimens of the imported stock, although reaching maturity in the aquaria, could be induced to spawn only very irregularly. A female released from 100 to 200 eggs during each of the successive spawnings. Kirschbaum's result must be considered a significant break-through in our understanding of the timing of reproduction in tropical freshwater fishes. Although several of these factors had been suspected earlier, experimental documentation about their respective roles had been lacking. Kirschbaum also mentions that he tested five other species of gymnotoids and that they showed a similar reactiveness, at least the same gonadal responses, as Eigenmannia virescens. Partial intermittent spawning is common in the gymnotoids and was first reported for Gymnorhamphichthys hypostomus where longer interbout intervals than in Eigenmannia were suspected because of the distinct size classes of ova, each batch containing only half the number found by Kirschbaum in Eigenmannia (Schwassmann, 1976).

In another study on reproductive seasonality in tropical streams Kramer (1978) found a spacing of spawning times, the temporal niche parameter, and also a separation of spawning sites, the spatial niche aspect, making for minimal niche overlap, among six different sympatric populations of characoid fish in a small stream in Panama.

ULTIMATE AND PROXIMATE CAUSES, AND THE ZEITGEBER CONCEPT

In the older literature a useful distinction was made between ultimate and proximate causes, or factors, of rhythmicity. For example, in annually periodic breeding the ultimate cause is the adaptive value which lies in the necessity of producing offspring at a time, usually the higher latitude summer, when the lake provides sufficient food for the growing young. Proximate causes are those environmental factors that were assumed to result in spawning, like increasing temperature and daylength, or tropical rains in equatorial zones. Since there is now good evidence for the existence of endogenous circannual rhythms as the basis for gonadal, endocrine-mediated, development, it seems no longer appropriate to apply the term "proximate causes". Environmental periodic factors act as mere synchronizers of the inherent rhythm, and the new term "zeitgeber" is properly used. This new term reflects our present interpretation of biochronological adaptation, and is even more appropriate in the case of circadian rhythms where the changes between light and dark, dawn and dusk in nature, do not directly cause, or trigger, activity or inactivity, but where they merely tend to synchronize,or maintain in proper phase position, the endogenous rhythm. Similar to the action of light in phasing circadian rhythms, annual cycles of behavioral and physiological function are under the adjusting control, advancing or delaying, of specific environmental variables which act on the endogenous circannual rhythm. This was, perhaps, first recognized by Baker and Ranson (1938), but the relative mutual interaction of the inherent rhythm with the environmental zeitgeber was conceptualized by Aschoff (1958).

TIDAL AND LUNAR RHYTHMS

This topic has been reviewed recently by Gibson (1978) who could bring together a great number of new data on this interesting aspect of the life habits of marine fishes, most of this information having become available during the last ten years only. After having been the first to demonstrate a tidal rhythm in an intertidal fish which persisted in the laboratory, getting out of phase with the natural tidal cycle, he succeeded in adding several more species to the list of those marine teleosts already known for tidal rhythmicity (Gibson, 1967, 1978). He also succeeded in demonstrating the apparently easy interconvertibility of tidal and bimodal circadian patterns of activity (Gibson, 1973a,b; 1976; Gibson et al., 1978). Several flatfish, plaice and flounders, are known to possess a circadian pattern of nocturnal activity (Verheijen and DeGroot, 1967; DeGroot, 1971). With the exception of the dab, the other flatfish species are found on sandy beaches while young, when they exhibit a tidal rhythm of activity which turns into a circadian nocturnal pattern after about one week in "non-tidal" laboratory conditions. Such strong relationship between the two rhythms and their interconvertibility could possibly be accomplished simply by strong action of the respective zeitgeber. The zeitgeber for tidal rhythms in fish could be the actual pressure changes in fish that rest on the bottom substrate (Gibson, 1976, 1978). Gibson distinguishes three types of intertidal species, ranked in order of degree and timing of intertidal residence. The true residents spend most of their life in the littoral, partial residents are there only as juveniles, and tidal visitors are those that come into the littoral with the high tide only. True residents are members of the goby, blenny, and clingfish families (Gibson, 1978). The longest lasting record known of an endogenous circatidal rhythm, persisting in constant laboratory conditions up to 50 days, was found in a mudskipper species from Japan (Nishikawa and Ishibashi, 1975).

Tide-related movements into and out of brackish water lagoons were reported for a population of Anableps, the four-eyed fish, from the Amazon estuary. The lagoons served as breeding grounds and also as feeding sites for the young fish which remained there while the adults were merely visiting with the high tide (Schwassmann, 1967). A tidal rhythm is also shown in the timing of the three, or four, successive spawnings during each of the semi-lunar monthly spawning runs of the California grunion, Leuresthes tenuis. Only the higher nightly tides are used for spawning (Walker, 1949). A similar pattern of tidal spawning was reported recently for Fundulus heteroclitus in Delaware Bay by Taylor et al. (1979) who find evidence for the action of a superimposed circadian rhythm of ovulation responsible for the restriction of actual spawning to the hour of high tide at night. Quite similar to the reproductive pattern of the California grunion, the mummichog also shows a semi-lunar monthly periodicity in phase with the high spring tides of new and full moon of late spring and summer. As in the grunion, eight or more successive spawning runs are observed, each of which can last for five days (Taylor, et al., 1979). To my knowledge, we have no idea of the actual zeitgeber that entrains the semi-lunar spawning runs of grunion or of Fundulus. The precise timing and the duration of actual egg-laying and fertilization in the grunion is well adapted to the particular wave pattern of their spawning beaches. Leuresthes tenuis of the Southern California Pacific beaches requires about 20 seconds for successful spawning, in which the female buries herself deep into the sand, several males usually curled around her, releasing sperm. As every local surfer knows, the surf pattern at those beaches is typical by having about every fith wave of higher amplitude. The fish do not come up on the beach with every single wave, nor do the surfers ride on the intervening smaller waves, but both utilize the very big ones. For the fish there remains sufficient time for completion of spawning before another very high wave facilitates their return to the ocean.

While rhythms of semi-lunar monthly period seem common in the marine environment, Gibson remarks on the apparent paucity of full lunar monthly rhythms (Gibson, 1978). These latter are well documented only for the European eel (Jens, 1953, Deelder, 1970; Tesch, 1977), and in the seaward movement of coho salmon fry (Mason, 1975).

IN RETROSPECT

Looking back and considering how our early hopes and excitement carried us through the past 20 years, and how our views changed and matured in the field of biological rhythms, we may be able to agree on a few general truths and insights. The early enthusiasm of systems analysis and model building which began in the late fifties has been helpful for clarifying concepts and developing ideas. Hopes for universality and general mechanisms common to all different organisms came partially true, as in the case of the light response as basis of entrainment of circadian rhythms, but did not in other areas, as in the detailed mechanism underlying photoperiodic entrainment of circannual rhythms. It seems more probable that the physiological details by which different organisms have invented the ability for proper orientation in a circadian, circannual, or circatidal time pattern must have come about separately and in different ways; organisms must have found slightly diverse avenues of achieving the same end. All organisms are inherently periodic, not because we as observers are able to recognize, with different degrees of success, an adaptive value in such rhythmic organization, but because all life forms evolved under the continuous impact of an environment that was by no means uniform, or stable, but was and is, instead, fundamentally periodic. Being pre-programmed to perform certain functions in a cyclic fashion, and being responsive to entrainment by certain reliable periodic variables of the environment, must be of paramount importance. We must accept the existence of a temporal aspect to the ecological niche, in addition to the reality of a process of evolution by natural selection that favors those who will be able to do the right thing at the proper time of their environment. Our reward from realizing such relationship is a feeling of satisfaction that comes from understanding an important aspect of life, the beauty of which lies in rhythm and harmony.

REFERENCES

Aschoff, J. (1958). Tierische Periodik unter dem Einfluss von Zeitgebern. Z. Tierpsychol. 15: 1-30.

Baggerman, B. (1972). Photoperiodic responses in the stickleback and their control by a daily rhythm of photosensitivity. Gen. Comp. Endocrinol. (Suppl.) 3: 466-476.

Baker, J.R. (1939). The relation between latitude and breeding season in birds. Proc. Zool. Soc. London A. 108: 557-582.

Baker, J.R. (1947). The seasons in a tropical rain-forest (New Hebrides), final part: Summary and general conclusions. J. Linn. Soc. (Zool.) 41: 248-258.

Baker, J.R. and Baker, Z. (1936). The seasons in a tropical rain-forest (New Hebrides), part 3: Fruit bats (Pteropidae). J. Linn. Soc. (Zool.) 39: 123-141.

Baker, J.R. and Bird, T.F. (1936). The seasons in a tropical rain-forest (New Hebrides), part 4: Insectivorous bats (Vespertilionidae and Rhinolophidae). J. Linn. Soc. (Zool.) 40: 143-161.

Baker, J.R. and Ranson, R.M. (1938). The breeding seasons of southern hemisphere birds in the northern hemisphere. Proc. Zool. Soc. London A. 153: 101-141.

Baker, J.R., Marshall, A.J. and Harrison, T.H. (1940). The seasons in a tropical rain-forest (New Hebrides), part 5: Birds (Pachycephala). J. Linn. Soc. (Zool.) 41: 50-70.

Beitinger, T.L. (1975). Diel activity rhythms and thermoregulatory behavior of bluegill in response to unnatural photoperiods. Biol. Bull. 149: 96-108.

Billard, R. and Breton, B. (1978). Rhythms of reproduction in teleost fish. In: Rhythmic Activity of Fishes. Ed. J.E. Thorpe. London, Academic Press, p. 31-53.

Bünning, E. (1967). The Physiological Clock. 2nd ed., Berlin, Springer.

Chan,K. K-S. (1976). A photosensitive daily rhythm in the female medaka, Oryzias latipes. Can. J. Zool. 54: 852-856.

Clarke, T.A. (1978). Diel feeding patterns of 16 species of mesopelagic fishes from Hawaiian waters. Fish. Bull. 76: 495-514.

Cloudsley-Thompson, J.L. (1961). Rhythmic Activity in Animal Physiology and Behaviour. New York and London, Academic Press.

Davis, R.E. (1962). Daily rhythm in the reaction of fish to light. Science 137: 430-432.

Deelder, C.L. (1954). Factors affecting the migration of the silver eel in Dutch inland waters. J. Cons. Perm. Int. Expl. Mer. 20: 117-185.

DeGroot, S.J. (1971). On the interrelationships between morphology of the alimentary tract, food and feeding behaviour in flatfish (Pisces, Pleuronectiformes). Neth. J. Sea Res. 5: 121-196.

Edel, R.K. (1976). Activity rhythms of maturing American eels (Anguilla rostrata) Mar. Biol. 36: 283-289.

Enright, J.T. and Hamner, W.M. (1967). Vertical diurnal migration and endogenous rhythmicity. Science 157: 937-941.

Eriksson, L.O. (1978). Nocturnalism versus diurnalism; dualism within fish individuals. In: Rhythmic Activity of Fishes. Ed. J.E. Thrope. London, Academic Press, p. 69-89.

Gibson, R.N. (1967). Experiments on the tidal rhythm of Blennius pholis. J. Mar. Biol. Ass. U.K. 47: 97-111.

Gibson, R.N. (1973a). The intertidal movements and distribution of young fish on a sandy beach with special reference to the plaice (Pleuronectes platessa L.). J. Exp. Mar. Biol. Ecol. 12: 79-102.

Gibson, R.N. (1973b). Tidal and circadian activity rhythms in juvenile plaice, Pleuronectes platessa. Mar. Biol. 22: 379-386.

Gibson, R.N. (1976). Comparative studies on the rhythms of juvenile flatfish. In: Biological Rhythms in the Marine Environment. Ed. P.J. DeCoursey. Columbia, S.C., University of South Carolina Press, p. 199-213.

Gibson, R.N. (1978). Lunar and tidal rhythms in fish. In: Rhythmic Activity of Fishes. Ed. J.E. Thorpe. London, Academic Press, p. 201-213.

Gibson, R.N., Blaxter, J.H.S. and DeGroot, S.J. (1978). Developmental changes in the activity rhythms of the plaice (Pleuronectes platessa L.). In: Rhythmic Activity of Fishes. Ed. J.E. Thorpe. London,

Academic Press, p. 169-186.

Hesse, R. (1924). Tiergeographie auf Økologischer Grundlage. Jena, G. Fischer, p. 416.

Hesse, R. and Doflein, F. (1943). Tierbau und Tierleben, in ihrem Zusammenhang betrachtet. vol. 2: Das Tier als Glied des Naturganzen. 2nd ed., Jena, G. Fischer, p. 195.

Hunter, J.R. (1963). The reproductive behavior of the green sunfish, Lepomis cyanellus. Zoologica 48: 13-24.

Hunter, J.R. (1966). Procedure for analysis of schooling behaviour. J. Fish. Res. Board Can. 23: 547-562.

Jens, G. (1953). Uber den lunaren Rhythmus der Blankaalwanderung. Arch. Fisch. Wiss. 4: 94-110.

John, K.R. and Haut, M. (1964). Retinomotor cycles and correlated behaviour in the teleost Astyanax mexicanus (Fillipi). J. Fish. Res. Board Can. 21: 591-595.

Kavaliers, M. (1978). Seasonal changes in the circadian period of the lake chub, Couesius plumbeus. Can. J. Zoology 56: 2591-2596.

Kirschbaum, F. (1975). Environmental factors control the periodical reproduction of tropical electric fish. Experientia 31: 1159-1160.

Kirschbaum, F. (1979). Reproduction of the weakly electric fish Eigenmannia virescens (Rhamphichthyidae, Teleostei) in captivity. Behav. Ecol. Sociobiol. 4: 331-355.

Kleerekoper,H., Taylor, G. and Wilson, R. (1961). Diurnal periodicity in the activity of Petromyzon marinus and the effects of chemical stimulation. Trans. Am. Fisheries Soc. 90: 73-78.

Kneis, P. and Siegmund, R. (1976). Heart rate and locomotor activity in fish; correlation and circadian and circannual differences in Cyprinus carpio L. Experientia 32: 474-475.

Kramer, D.L. (1978). Reproductive seasonality in the fishes of a tropical stream. Ecology 59: 976-986.

Lissmann, H.W. and Schwassmann, H.O. (1965). Activity rhythm of an electric fish, Gymnorhamphichthys hypostomus, Ellis. Z. Vergl. Physiol. 51: 153-171.

Lobel, P.S. (1978). Diel, lunar, and seasonal periodicities in the reproductive behavior of the pomacanthid fish, Centropyge potteri, and some other reef fishes in Hawaii. Pacific Science 32: 193-208.

Lowe-McConnell, R.H. (1975). Fish Communities in Tropical Freshwaters. London, Longman.

Mason, J.C. (1975). Seaward movement of juvenile fishes, including lunar periodicity in the movement of coho salmon (Oncorhynchus kisutch) fry. J. Fish. Res. Board Can. 32: 2542-2547.

Müller, K. (1968). Freilaufende circadiane Periodik von Ellritzen am Polarkreis. Naturwissenschaften 55: 140.

Müller, K. (1969). Jahreszeitlicher Wechsel der 24 h Periodik bei der Bachforelle (Salmo trutta L.) am Polarkreis. Oikos 20: 166-170.

Müller, K. (1973). Seasonal phase shift and the duration of activity time in the burbot Lota lota (L.) (Pisces, Gadidae), J. Comp. Physiol. 84: 357-359.

Müller, K. (1976). Chronological studies on whitefish (Coregonus lavaretus

L.) at the Arctic circle. Arch. Fisch. Wiss. 27: 121-132.
Müller, K. (1978a). Locomotor activity of fish and environmental oscillations, In: Rhythmic Activity of Fishes. Ed. J.E. Thorpe. London, Academic Press, p. 1-19.
Müller, K. (1978b). The flexibility of the circadian system of fish at different latitudes. In: Rhythmic Activity of Fishes. Ed. J.E. Thorpe. London, Academic Press, p. 91-104.
Müller, K. (1978c). Locomotor activity in whitefish-shoals (Coregonus lavaretus). In: Rhythmic Activity of Fishes. Ed. J.E. Thorpe. London, Academic Press, p. 225-233.
Müller-Haeckel, A. (1965). Tagesperiodik des Siliziumgehaltes in einem Fliessgewässer. Oikos 16: 232-233.
Müller-Haeckel, A. (1966). Diatomeendrift in Fliessgewässern. Hydrobiologia (Den Haag) 28: 73-87.
Müller-Haeckel, A (1976). Migrationsperiodik einzelliger Algen in Flieccgewässern. Vaxtekol. stud. (Uppsala), 10: 1-36.
Munro, J.L., Gaut, V.C., Thompson, R. and Reeson, P.H. (1973). The spawning season of Caribbean reef fishes. J. Fish. Biol. 5: 69-84.
Nishikawa, M. and Ishibashi, T. (1975). Entrainment of the activity rhythm by the cycle of feeding in the mud-skipper, Periophthalmus cantonensis (Osbeck). Zool. Mag. (Tokyo) 84: 184-189.
Norcross, J.J., Richardson, S.L., Massmann, W.H. and Joseph, E.B. (1974). Development of young bluefish (Pomatomus saltatrix) and distribution of eggs and young in Virginian coastal waters. Trans. Am. Fish. Soc. 103: 477-497.
Olla, B.L. and Studholme, A.L. (1972). Daily and seasonal rhythms of activity in the bluefish Pomatomus saltatrix. In: Behavior of Marine Animals: Current Perspectives in Research, Vol. 2, Eds. H.E. Winn and B.L. Olla, New York, Plenum Press, p. 303-326.
Olla, B.L. and Studholme, A.L. (1978). Comparative aspects of the activity rhythm of tautog, Tautoga onitis, bluefish, Pomatomus saltatrix, and Atlantic mackerel, Scomber scombrus, as related to their life habits. In: Rhythmic Activity of Fishes. Ed. J.E. Thorpe. London, Academic Press, p. 131-151.
Priede, I.G. (1978). Behavioural and physiological rhythms of fish in their natural environment as indicated by ultrasonic telemetry of heart rate. In: Rhythmic Activity of Fishes. Ed. J.E.Thorpe. London, Academic Press, p. 153-168.
Reynolds, W.W. (1976). Locomotor activity rhythms in the bluegill sunfish, Lepomis macrochirus. Am. Midland Natur. 96: 221-224.
Richardson, N.E. and McCleave, J. (1974). Locomotor activity rhythms of juvenile Atlantic salmon (Salmo salar) in various light conditions. Biol. Bull. 147: 422-432.
Schwassmann, H.O. (1960). Environmental cues in the orientation rhythm of fish. Cold Spring Harbor Symp. Quant. Biol. 25: 443-449.
Schwassmann, H.O. (1967). Orientation of Amazonian fishes to the equatorial sun. In: Atas do Simposio sôbre a Biota Amazônica. Vol. 3. Ed. H. Lent. Rio de Janeiro, p. 201-220.
Schwassmann, H.O. (1971a). Biological rhythms. In: Fish Physiology, Vol. 6,

Eds. W.S. Hoar and D.J. Randall. New York, Academic Press, p. 371-428.

Schwassmann, H.O. (1971b). Circadian activity patterns in gymnotid electric fish. In: Biochronometry. Ed. M. Menaker. Washington, D.C., Natl. Acad. Sciences, p. 186-199.

Schwassmann, H.O. (1976). Ecology and taxonomic status of different geographic populations of Gymnorhamphichthys hypostomus Ellis (Pisces, Cypriniformes, Gymnotoidei). Biotropica 8: 25-40.

Schwassmann, H.O. (1978). Times of annual spawning and reproductive strategies in Amazonian fishes. In: Rhythmic Activity of Fishes. Ed. J.E. Thorpe. London, Academic Press, p. 187-200.

Sundararaj, B.I. and Vasal, S. (1976). Photoperiod and temperature control in the regulation of reproduction in the female catfish Heteropneustes fossilis. J. Fish. Res. Board Can. 33: 959-973.

Taylor, M.H., Leach, G.J., DiMichele, L., Levitan, W.M. and Jacob, W.F. (1979). Lunar spawning cycle in the mummichog, Fundulus heteroclitus (Pisces: Cyprinodontidae). Copeia 1979 (2): 291-297.

Tesch, F.-W. (1977). The Eel. Biology and Management of Anguillid Eels. London, Chapman and Hall.

Verheijen, F.J. and DeGroot, S.J. (1967). Diurnal activity of plaice and flounder in aquaria. Neth. J. Sea Res. 3: 383-390.

Walker, B.W. (1949). Periodicity of spawning in the grunion, Leuresthes tenuis. Ph.D. thesis, University of California, Los Angeles, California.

THE PINEAL ORGAN AND CIRCADIAN RHYTHMS OF FISHES

Martin Kavaliers
Department of Zoology and Entomology
Colorado State University
Fort Collins, Colorado 80523, U.S.A.

INTRODUCTION

Most fishes living under natural conditions express daily rhythms in their behaviour, physiology and biochemistry. Much of what fishes do is temporally organized with respect to the environmental day-night cycle. Investigations of the timing of diel rhythms on either a daily or seasonal basis should consider four major classes of criteria: (i) endogenous (circadian) rhythms; (ii) natural photoperiod; (iii) photoreceptive physiology and; (iv) specific ecological and behavioural characteristics of the species being investigated. The latter category includes responses to the various chemical and physical factors that are discussed in this volume. An appreciation of circadian rhythms and their controls is crucial to fully understanding diel rhythms and their evolutionary history, ontogeny, physiology, and biochemistry.

Circadian rhythms are inherent endogenous, genetically determined functions of the organism (Bünning, 1973). In constant laboratory conditions from which periodic fluctuations have been eliminated circadian rhythms continue with a free running period (τ) that is approximately but not exactly the environmental period (T) of 24 hours. Under natural conditions circadian rhythms are synchronized (entrained) to a period of exactly 24 hours in an adaptive temporal phase relationship with the day-night cycle. Light, temperature, tidal, and feeding cycles are the dominant synchronizers for fishes.

Circadian rhythms are used by fishes as clocks for; (i) measurement of the lapse of time e.g. photoperiodic time measurement and reproductive control (Baggerman, this volume); (ii) recognition of local time e.g. sun-compass orientation, migration (Schwassmann, 1971, this volume); (iii) providing internal temporal co-ordination e.g. determination of hormonal temporal coordination and phase relations (Meier, 1975; Spieler, 1979). Circadian rhythms coordinate the activity of the internal environment of the organism with its external environment to achieve maximally effective daily and seasonal adaptation.

Until relatively recently there were few studies dealing with the physiology of circadian rhythms in fishes. Most of those dealt with systematizing and analyzing the effects of environmental variables on circadian rhythms (Schwassmann, 1971, this volume; Thorpe, 1978). Studies concerned with the anatomical location, biochemical mechanisms and physiological relationships to other systems were neglected. There is now evidence that the pineal organ plays an important role in the control of circadian rhythms of many lower vertebrates including teleost fishes (Menaker et al., 1978; Kavaliers, 1979 a, b, 1980; Rusak and Zucker, 1979). The present review addresses the role of the pineal organ in the circadian systems of teleost fish and its role in determining environmental adaptations.

GENERAL CHARACTERISTICS OF THE PINEAL

The pineal organ arises as a neuroectodermal process and forms part of the diencephalic brain roof of most vertebrates. The adult teleost pineal consists of a well-vascularized end vesicle (epiphysis cerebi) connected by an epiphyseal stalk to the posterior diencephalon (Fenwick, 1969). Pinealofugal fibers extend to the lateral habenular nucleus, several thalamic nuclei and various portions of the mesencephalon and diencephalon (Hafeez and Zerihum, 1974). There also is a close anatomical relationship with and neural connections to the ependymal regions of the third ventricle and epithalamic diverticulum and probably the suprachiasmatic nuclei of the hypothalamus (Oksche and Hartwig, 1975); suggesting the existence of intimate communications and interactions between the pineal and other brain regions.

There appears to have been a gradual shift from a direct photosensory function of the pineal organ in the more primitive vertebrates to a neuroendocrine role in the advanced forms (Käppers, 1971). The latter receive photic input indirectly via sympathetic neural pathways. There is evident, a concurrent evolution of the principal cell type of the pineal, the photosensory pinealocyte giving rise to an endocrine cell (Collin, 1971). Consequently, the pineal organ of fishes functions both as a photosensory and secretory (neuroendocrine) organ (Fenwick, 1969, 1970 a,b).

Morphologically identifiable photoreceptors, along with nerve and "supporting" cells, have been demonstrated by electron microscopy in the pineal organs of several species of teleosts (Fenwick, 1969; Tamura, this volume), while behavioural studies have shown a direct photosensitivity (Fenwick, 1970a). Interpretations from behavioural observations must be tempered by the responses of other encephalic photosensors in fish (Oksche and Hartwig, 1975; van Veen et al., 1976). Electrophysiological studies indicate that the pineal organ is directly photosensitive. There is apparently a two-fold division in photosensory responses (Hanyu et al., 1978). At low light intensities (10^{-5}-10^{-1} lux) one type of pineal neurons show a steady discharging frequency which decreases regularly as the light level increases. This tonic sensitivity, which is similar to that of retinal rod cells, permits the teleost pineal organ to be an effective photoreceptor at low light levels such as are present during twilights. At higher illuminations there is a steady discharge of other neuron types in the light-adapted pineal. The steady discharge frequency increases only when there is a significant decrease in light intensity. This type of response could function in shadow detection and the mediation of phototactic responses. Indirect photic input via the lateral eyes can apparently affect the "supporting" cells in the pineal organ. Hafeez et al. (1978) suggest that these cells may be involved in a photocomparator system which integrates the direct photoresponses of the pineal with photic inputs from the lateral eyes.

The teleost pineal organ is the principal, though not the sole site of the synthesis and release of the indoleamide 5-methoxy-N-acetyltryptamine, or melatonin, the putative pineal hormone (Fenwick, 1969, 1970b; Gern et al., 1978a, b; Gern and Ralph, 1979). The levels of melatonin vary rhythmically within the plasma of rainbow trout, Salmo gairdneri, maximal levels occurring during the dark phase (scotophase) of a light-dark cycle (Gern et al., 1978a). Light depresses the levels and activities of a number of enzymes, including N-acetyltransferase which mediates the conversion of serotonin to N-acetylserotonin, and Hydroxyindole-0-methyltransferase (HIOMT) that converts N-acetylserotonin to melatonin. Investigations with birds suggest that N-acetyltransferase actually undergoes a circadian rhythm in activity that is normally entrained by the light-dark cycle (Binkley et al. 1978). Studies in optic tract-sectioned trout have shown persistent nychthemeral rhythms of plasma melatonin, indicating a direct photosensory input to the melatonin-secreting source(s) in the teleost pineal (Gern et al., 1978b).

Melatonin is believed to be a chemical transducer of the photic environment and has been implicated in a number of functions in vertebrates including reproduction (Reiter, 1978), thermoregulation (Ralph et al., 1979) and circadian activity rhythms (Turek et al., 1976; Gwinner and Benzinger, 1978). Melatonin has been suggested as a key compound for circadian integration (Underwood, 1979). This multiplicity of functions may arise through melatonin's actions on the central nervous system. Although at present evidence for direct effects of melatonin in fishes is fragmentary, similar actions appear likely.

The pineal gland also contains and releases a large number of peptides; a neurohypophysical hormone arginine vasotocin, hypothalmic releasing factors including lutropin releasing hormone and other unidentified substances that may be anti- or pro-gonadotrophic in nature (Benson, 1977). The actions of these products have to be determined before a complete analysis of the pineal's functions is possible.

Until recently (Kavaliers, 1979 a, b, 1980) there was little direct experimental evidence that the pineal organ had any definite role in the circadian system of fishes. Earlier investigations had shown that the pineal organ and possibly other extraretinal receptors could directly mediate the entrainment of locomotor activity rhythms by light-dark cycles (Eriksson, 1972; Siegmund and Wolff, 1972, van Veen et al., 1976). There were numerous reports of pinealectomy altering hormonal and metabolic cycles as well as reproductive development of a variety of species of fish (Fenwick, 1970b; de Vlaming, 1975; Delahunty et al., 1978; Vodicnik and de Vlaming, 1978). At present the available evidence supports a role for the pineal in mediating photosexual development of temperate region fish, though the extent of involvement may vary from species to species (Urasaki, 1976; Vodicnik and de Vlaming, 1978).

EXPERIMENTAL EVIDENCE FOR PINEAL INVOLVEMENT IN CIRCADIAN ORGANIZATION

FREE RUNNING CIRCADIAN ACTIVITY

Investigations were made of the effects of pineal removal on the circadian locomotor activity of a number of species of north temperate teleost fishes. Diurnally and nocturnally active species of fish displayed circadian rhythms of locomotor activity under constant darkness (DD) or constant illumination (LL) (Figs. 1, 2). Circadian rhythms of fish are generally more stable under DD than LL. This is a characteristic feature of most circadian rhythms, constant illumination being a highly stressful state (Bünning, 1973). Free running circadian rhythms of locomotor activity persisted in most pinealectomized fishes under DD or LL. This was true regardless of the state of sexual development, the adult sexual condition, or the time of year experiments were performed at. Responses of different stages of ontogenetic development have so far not been examined. Caution, however, must also be taken in making interpretations of functions solely from lesion and ablation studies, especially when neural tissues are involved (Schoenfeld and Hamilton, 1977). Even though control manipulations and sham pinealectomies had no effects on circadian parameters, these procedures may not always have consistently duplicated the peripheral damages.

Although a circadian rhythmicity persisted in most of the pinealectomized fishes, its pattern was affected. The most obvious effect was that of a general loss of stability and an increased "sloppiness" of the rhythm. For example, when a lake chub, Couesius plumbeus (Agassiz), free running in DD is pinealectomized, its activity becomes more variable, This change is indicated by the increased variability of the slope of the free running period as evidenced by the increased scatter in onset and end of the main activity period in the activity records (Fig. 2 and Kavaliers 1979a).

Pinealectomy also significantly altered the length of the circadian period. This is shown by the significant change in the slope of the onset of activity. The direction of change, that is whether τ was lengthened or shortened (Figs. 2,3), was influenced by the initial value of the period and

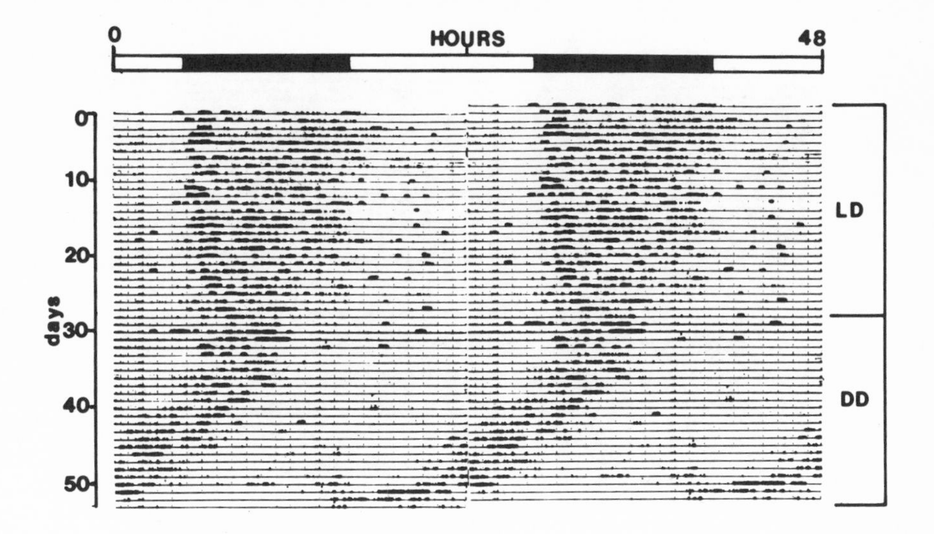

Fig. 1. Activity record of a single white sucker Catostomus commersoni, under a light-dark cycle (12:12 LD) followed by constant darkness (DD). The records have been duplicated on a 48 h time base to aid in interpretation. Successive days are plotted from top to bottom. Under LD the fish was nocturnally (dark) active. With the imposition of DD it displayed a free running circadian period less than 24 as shown by the slope to the left in the onsets.

therefore the initial circadian organization. The exact value of the free running period depends on species, genetic history and constitution, ontogenetic state, previous history of the individual and the particular set of constant conditions (see discussion in Kavaliers, 1978). The free running circadian period of lake chub and burbot, Lota lota, captured at different times of the year at 54°N. latitude and measured under DD show significant seasonal change. (Kavaliers, 1978, 1980). These seasonal changes in period length were considered to be history dependent "after-effects" (Kavaliers, 1978, 1979b) of the natural light-dark cycle the animals had been entrained to, though the possibility of circannual rhythms cannot be entirely discounted (Schwassman, this volume). Pinealectomy decreased τ values of lake chub during the winter and increased period values during the summer (Fig. 3); conversely in nocturnally active burbot pinealectomy lengthened τ in winter and shortened τ in summer (Kavaliers, 1980). These annual variations in the effects on circadian period could explain some of the seasonal discrepancies reported in the effects of pinealectomy on hormonal, metabolic, and especially reproductive functions (previously cited).

The entrained and free running circadian activity of fish is made up of non-randomly distributed shorter activity "bouts" that are composed of a

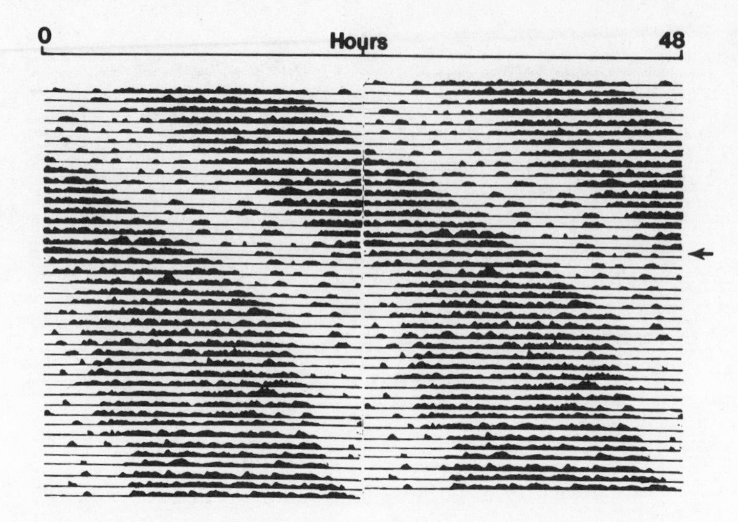

Fig. 2. The effect of pinealectomy on the free running rhythm of a diurnally active lake chub under DD. The pineal was removed at the time indicated by the arrow. The records have been duplicated on a 48 h time base to aid in interpretation. Successive days are plotted from top to bottom. Fish were in DD except for brief exposure to light during surgery (From Kavaliers 1979b).

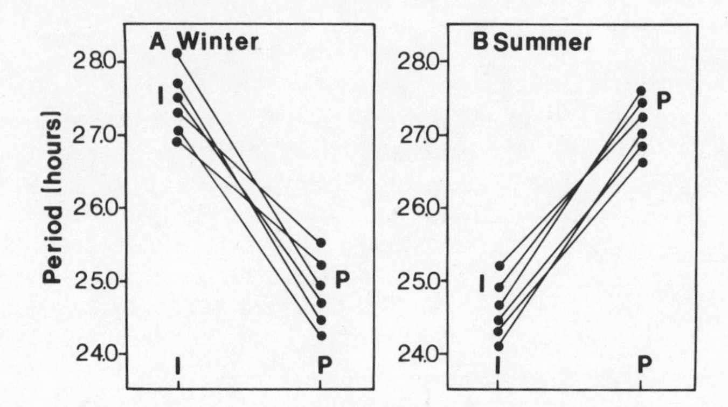

Fig. 3. Seasonal differences in the effects of pinealectomy on the circadian period length of lake chub. Initial (I) period values and the period after pinealectomy (P) of fish captured in winter (A) and summer (B) are shown. Period values determined before and after pinealectomy are connected. Initial summer period values are significantly shorter ($p < 0.01$ t-test) than winter values. Pinealectomy significantly ($p < 0.01$ t-test) altered the period length at both times of the year. (From Kavaliers, 1979b).

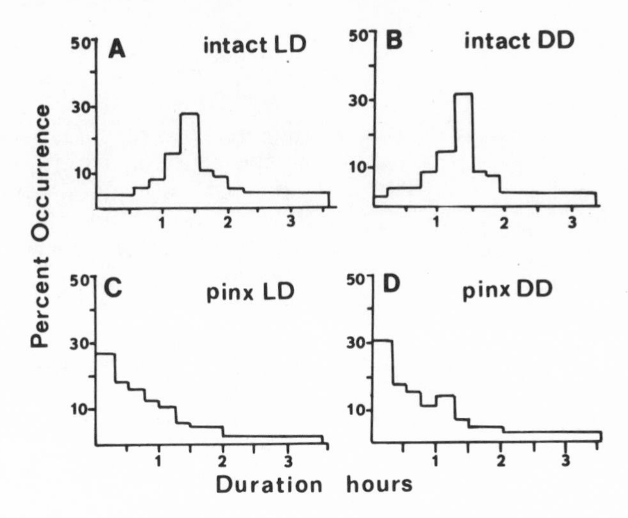

Fig. 4. Frequency duration histograms of the short-term activity ('bouts') components of 10 lake chub over 5 days. A. Entrained (LD) activity of intact fish. B. Circadian (DD) activity of intact fish. C. Entrained (LD) activity of pinealectomized fish (pinx). D. Circadian (DD) activity of pinealectomized fish. Pinealectomy results in activity components being distributed in a random (exponential) form. Activity components of intact fish are non-randomly distributed according to a skewed normal distribution.

variety of appetitive and consummatory behaviours (0.5 - 2.0 h duration). Frequency-duration and frequency interval analyses (Kavaliers, submitted) showed that pinealectomy eliminated this behavioural organization and led to a random distribution of the activity components of lake chub (Fig. 4). Surprisingly, there was no change in the total amount of locomotor activity only in the distribution of activity.

ENTRAINMENT TO LIGHT-DARK CYCLES

Circadian rhythms of locomotor activity can be readily entrained to light-dark cycles. During entrainment τ takes on the period T (24h) of the entraining stimulus. For a circadian rhythm to be entrained there must be a daily phase shift equal to the difference between τ and T. Pinealectomized lake chub were entrained by laboratory LD cycles. However, phase relations were more variable in pinealectomized fish than in intact ones. Longer times were required to attain stable entrainment in the pinealectomized fish (Fig. 5). Strikingly different entrainment patterns were evident between intact and pinealectomized fish when examined under light-dark cycles that included duplicated twilights and were changing at their naturally occurring rate of change. Intact fish showed stable phase relations to the light-dark cycles while the pinealectomized fish displayed erratic patterns and failed to establish any consistent or stable relations with the light-dark cycle (Fig. 6).

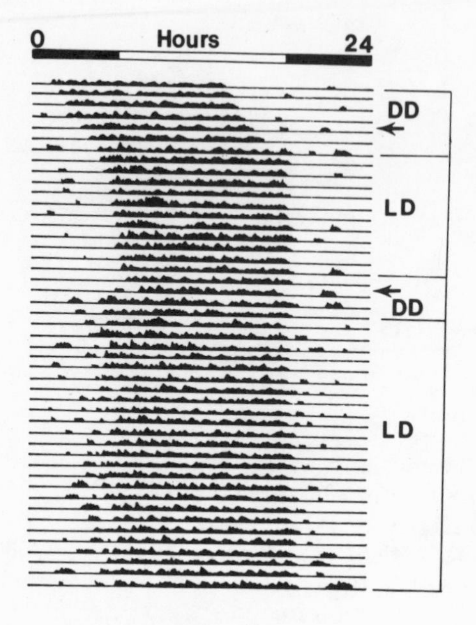

Fig. 5. The effect of pinealectomy on the phase relations of a lake chub from DD entrained to a 12 hour light-12 hour dark light-dark cycle (12:12 LD; L= 180 μw cm^{-2}; 270-775 nm). The light and dark bars at the top of the record indicate the LD cycle. The upper arrow indicates the day at which a control procedure was performed (the fish was anesthetized, removed from the tank exposed to brief, 1 min light pulse and returned to the tank); the lower arrow marks pinealectomy. After pinealectomy it takes longer to achieve entrainment as well as a change in the phase relationship between the onset of activity and light.

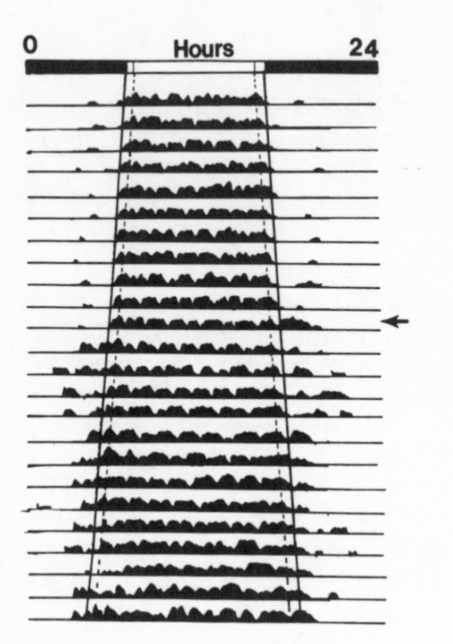

Fig. 6. The effect of pinealectomy on the phase relations of a lake chub entrained to a light-dark cycle that included simulated twilights, and was occurring at its natural rate of daily change. The light and dark bars at the top of the record indicate the initial LD cycle. The solid lines in the record indicate the exact times of sunrise and sunset while dotted lines enclose civil twilights (sun 6° below the horizon). The upper arrow indicates control procedures (described in Figure 5) and lower arrow pinealectomy. Note the immediate disruption of stable phase relationships between the onset of light in twilight and activity after pinealectomy.

DISCUSSION

These observations indicate that the pineal organ plays an important role in the control of the circadian organization of teleost fishes. The persistence of circadian rhythmicity after pinealectomy demonstrates that the pineal organ is not the entire source of circadian organization in teleosts. It cannot be considered as a driving or master oscillator that has either direct action on a central pacemaker or directly controls overt rhythms.

The entire circadian system consists of much more that the pacemaker. There are at least two kinds of inputs to the pacemaker(s). Those carrying information about the external environment and those feeding back information about the internal state of the organism. Since the temporal organization of a great many physiological processes is being concurrently determined the pacemaker(s) of necessity have extensive outputs. In addition, since different rhythms can have different phase relationships to a single entraining cycle the pacemaking structure(s) must also function to adjust the phase of particular physiological rhythms.

In a simplified model circadian rhythms have been described as arising from an interacting multi-oscillator system (Pittendrigh, 1974). Splitting of the circadian activity of a number of animals, including Arctic fishes (Eriksson, 1973, 1978), has provided evidence for a labile circadian organization that can be described by a coupled oscillator system. Moreover, entrained circadian activity rhythms behave in many respects like systems of at least two oscillators one of which is coupled to dawn and the other to dusk (Pittendrigh 1974). The latitudinal flexibility and lability of circadian organization in fishes (Eriksson, 1978; Müller, 1978) provided additional support for a coupled oscillator model of circadian organization in teleosts. Seasonal changes in the circadian period of fishes can be interpreted as arising from alterations in oscillator coupling and temporal integration resulting from seasonal shifts in entrainment (Kavaliers, 1978, 1980).

Changes in the absolute length and greater instability of the circadian period following pinealectomy, point to alterations in circadian organization and coupling. The pineal organ may function (i) either as a direct internal or an indirect hypothalamically mediated synchronizer or coupler of two or more oscillators which are weakly self-sustaining and coupled and, (ii) as part of the effector system. Indeed, in support of (i) pinealectomy of a white sucker, *Catostomus commersoni*, resulted in the splitting of circadian activity into a number of free running components (Fig. 7), indicating a coupling or integrative function for the teleost pineal in circadian organization. This may be similar to that proposed by Underwood (1977) for lizards. Seasonal variance in the effects of pinealectomy on τ reflect differences in initial circadian organization or coupling and consequent variations in the resultant coupling and period after ablation. These seasonal differences in effect suggest that the pineal organ may function to

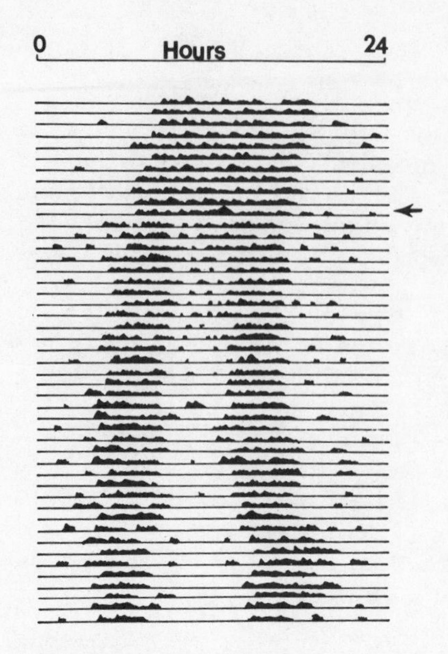

Fig. 7. The effect of pinealectomy on the free-running circadian activity of a nocturnally active white sucker under LL. The pineal was removed at the indicated day. Pinealectomy resulted in the "splitting" of activity into two components that free ran separately. Successive days are plotted from top to bottom.

synchronize or integrate circadian organization with the seasonally changing photoperiod. This would also partially explain the seasonal variations in the effects of pinealectomy on gonadal development of teleosts. The reproductive functions of the pineal may be related to its temporal organization of physiological events through a central nervous system modulation.

Elimination of short-term rhythmicities and changes in the distribution of behavioural activity resulting from pinealectomy can also be explained on the basis of alterations in oscillator coupling and temporal organization of hormonal and physiological cycles. If ultradian rhythms arise either endogenously or as sub-harmonics of circadian frequencies pinealectomy could limit ultradian structure and short-term behavioural organization by altering the coupling or integration of circadian oscillators. Alternatively, organized activity may arise through circadian modulation of stochastic behaviour sequences (Lehmann, 1976). Pinealectomy could disrupt the integrated temporal patterning of a variety of hormones, in particular that of prolactin, cortisol and thyroid (Spieler, 1979). Support for such a role is provided by the disruption of liver and plasma metabolites that occurs in goldfish following pinealectomy (Delahunty et al., 1978). This action would alter temporal hormonal coordination and hence circadian modulation of short-term activity and behavioural components, leading to a relatively unorganized behavioural structure. Additional evidence for the pineal being involved in behavioural and circadian organization is provided from the findings that pinealectomy of white sucker disrupts their diel rhythms of thermoregulatory behaviour and entrainment to temperature cycles (Kavaliers and Ralph, in press).

It has been suggested that melatonin may transfer circadian modulations to subordinate systems (Turek et al., 1976; Gwinner and Benzinger, 1978; Underwood, 1979). Melatonin injections have been shown to decrease the locomotor activity of trout (Hafeez, 1970). The doses used, however, were at a pharmacological rather than at a physiological level. However, treatment of white suckers with P-chlorophenylalanine, an inhibitor of melatonin synthesis, does alter their circadian activity and temperature responses (Kavaliers, unpublished). Rainbow trout show a day-night change in plasma melatonin (Gern et al., 1978a). Pinealectomy does not completely eliminate plasma melatonin or the nychthemeral rhythms of its secretion (Gern et al., 1978b), the trout retina being an additional major site of melatonin synthesis (Gern and Ralph, 1979). Whether these variations in plasma melatonin persist under constant conditions and are involved in the determination of circadian organization of fishes is at present not known. Preliminary results have shown that a circadian rhythm of locomotor activity persists in blinded and pinealectomized lake chub (Kavaliers, unpublished). This does not, however, exclude melatonin from other sources such as the retina, optic tract and gut (Bubenick et al. 1977). Relations of the pineal to other possible sites of circadian organization, such as the suprachiasmatic nuclei of the hypothalmus, the putative circadian oscillators in mammals (Rusak and Zucker, 1979) as well as other integrative regions of the nervous system need to be investigated.

The decreased precision of entrainment to light-dark as well as temperature cycles occurring after pinealectomy can also result from alterations in circadian integration. Entrainment to light-dark cycles is suggested to entail a selective coupling of oscillator populations (dawn and dusk populations) with the external 24 h cycle. Pinealectomy would decrease the precision of this coupling and lead to longer times being required for the establishment of relatively stable phase relations. This instability was most clearly evident in fish examined under photoperiods that were proceeding at their natural rates of daily change. As discussed earlier electrophysiological and behavioural studies have shown that the teleost pineal is an effective mediator of photic information. The teleost pineal organ may: (i) expand spectral sensitivity, (ii) affect the threshold for entrainment, (iii) detect different stimulus parameters of light than the lateral eyes or (iv) reflect an association between photoperiod and circadian pacemakers. Certainly, at the light intensities present in twilights, the pineal organ can function as an effective detector of ambient illumination (Hanyu et al., 1978). Photic input from the pineal organ along with that from the lateral eyes and other extraretinal photosensors may be integrated in a photo-neuro-endocrine complex (Scharrer, 1964) with the circadian system to effect stable entrainment. Direct actions on circadian integration are indicated by the disruptions of entrainment by temperature cycles following pinealectomy (Kavaliers and Ralph, in press). Pineal involvement in temperature responses of fishes is not entirely unexpected temperature cycles being potent synchronizers of circadian based hormonal cycles in fish (Spieler, 1979; Spieler et al., 1978).

Although much is known about the formal mechanisms of entrainment the physiological mechanisms are unknown. Neither the precise photoreceptor nor the neural pathways by which signals from photopigments are transmitted are known. Seasonal variations in the effects of pinealectomy may, therefore, be influenced by the extent of pineal involvement in photoperiodic entrainment, its mediation of photic input in circadian coupling, and its chronometric organization of the central nervous system (CNS). The teleost pineal organ through a direct photosensory action and determination of circadian organization can be considered to be directly involved in effecting hormonal, physiological and behavioural adjustments to a seasonally changing light-dark and temperature cycles.

In summary there is substantial evidence that the pineal is involved in determining circadian and seasonal organization in teleost fish. These effects may arise from the pineal's photoreceptive functions, its roles in circadian integration, and roles in temporal organization of hormonal physiological and behavioural events through CNS modulation.

ACKNOWLEDGEMENTS

The research described here was supported by a National Research Council of Canada grant to D. M. Ross (NRC-1455), the Department of Zoology, University of Alberta, and in part by an NIH grant to C.L. Ralph (NIH 12257).

REFERENCES

Benson, B. (1977). Current status of pineal peptides. Neuroendocrinology 24, 241-258.

Binkley, S.A., Reibman, J.B. & Reilly, K.B. (1978). The pineal gland: A biological clock in vitro. Science 202: 1198-1201.

Bubenik, G.A., Brown, G.M. & Grota, L.J. (1977). Immunohistochemical localization of melatonin in the rat digestive tract. Experientia 33: 662-663.

Bünning, E. (1973). The Physiological Clock. New York, Springer Verlag, 258 p.

Collin, J.P. (1971). Differentiation and regression of the cells of the sensory line in the epiphysis cerebri. In: The Pineal Gland, Eds. G.E. Wolsteholme and J. Knight. Edinburgh, Churchill Livingstone, p. 79-126.

Delahunty, G., Bauer, G., Prack, M. & de Valming, V. (1978). Effects of pinealectomy and melatonin treatment on liver and plasma metabolites in the goldfish, Carassius auratus. Gen. Comp. Endocrinol. 35: 99-109.

de Vlaming, V.L. (1975). Effects of pinealectomy on gonadal activity in the cyprinid teleost, Notemigonus crysoleucas. Gen. Comp. Endocrinol. 26: 36-49.

Eriksson, L.O. (1972). Die Jahresperiodik augen-und Pinealorganlosen Bachsaiblinge Salvenius fontinalis Mitchell. Aquilo Ser. Zool. 13: 8-12.

Eriksson, L.O. (1973). Spring inversion of the diel rhythm of locomotor activity in young sea-going trout, Salmo trutta trutta L. and Atlantic salmon, Salmo salar. L. Aquilo Ser. Zool. 14: 68-79.
Eriksson, L.O. (1978). Nocturnalism versus diurnalism; Dualism within fish individuals. In: Rhythmic Activity of Fishes, Ed. J.E. Thorpe. New York, Academic Press, p. 69-90.
Fenwick, J.C. (1969). The pineal organ In: Fish Physiology, Vol. 4. Eds. W.S. Hoar and D.J. Randall. New York, Academic Press, p. 91-108.
Fenwick, J.C. (1970a). Effects of pinealectomy and bilateral enucleation on the phototactic response and the conditioned response to light of the goldfish, Carassius auratus. L. Can. J. Zool. 48: 175-182.
Fenwick, J.C. (1970b). Demonstration and effect of melatonin in fish. Gen. Comp. Endocrinol. 14: 86-97.
Gern, W.A., Owens, D.W. & Ralph, C.L. (1978a). Plasma melatonin in the trout: day-night change demonstrated by radioimmunoassay. Gen. Comp. Endocrinol. 34: 453-458.
Gern, W.A., Owens, D.W. & Ralph, C.L. (1978b). Persistence of the nychthemeral rhythm of melatonin secretion in pinealectomized or optic tract-sectioned trout (Salmo gairdneri). J. Exp. Zool. 205: 371-376.
Gern, W.A. & Ralph, C.L. (1979). Melatonin synthesis by the retina. Science, 204: 183-184.
Gwinner, E. (1978). Effects of pinealectomy on circadian locomotor activity rhythms in european starlings, Sturnus vulgaris. J. Comp. Physiol. 126: 123-129.
Gwinner, E. & Benzinger, I. (1978). Synchronization of a circadian rhythm in pinealectomized european starlings by daily injections of melatonin. J. Comp. Physiol. 127: 209-213.
Hafeez, M.A. (1970). Effect of melatonin on body coloration and spontaneous swimming activity in rainbow trout, Salmo gairdneri. Comp. Biochem. Physiol. 36: 639-656.
Hafeez, M.A., Wagner, H.H. & Quay, W.B. (1978). Mediation of light-induced changes in pineal receptor and supporting cell nuclei and nucleoli in steelhead trout (Salmo gairdneri). Photochem. & Photobiol. 28: 213-218.
Hafeez, M.A. & Zerihun, L. (1974). Studies on the central projections of the pineal nerve tract in rainbow trout, Salmo gairdneri Richardson, using cobalt chloride iontophoresis. Cell Tiss. Res. 154: d485-510.
Hanyu, I., Niwa, H. & Tamura, T. (1978). Salient features in photosensory function of teleostean pineal organ. Comp. Biochem. Physiol. 61A: 49-54.
Käppers, J.A. (1971). The pineal organ: an introduction. In: The Pineal Gland. Eds. G.E.W. Wolstenholme and J. Knight. Edinburgh, Churchill Livingstone, p. 3-34.
Kavaliers, M. (1978). Seasonal changes in the circadian period of the lake chub, Couesius plumbeus. Can. J. Zool. 56: 2591-2596.
Kavaliers, M. (1979a). Pineal involvement in the control of circadian rhythmicity in the lake chub, Couesius plumbeus. J. Exp. Zool., 209: 33-40.

Kavaliers, M. (1979b). The pineal organ and circadian organization of teleost fish. Rev. Can. Biol. (in press).

Kavaliers, M. (1980). Circadian locomotor activity rhythms of the burbot, Lota lota: Seasonal differences in period length and the effect of pinealectomy. J. Comp. Physiol. (in press).

Lehmann, U. (1976). Stochastic principles in the temporal control of activity behaviour. Int. J. Chronobiol., 4: 223-236.

Meier, A.H. (1975). Chronoendocrinology of vertebrates. In: Hormonal Correlates of Behavior, Vol. 2. Eds. B.E. Eleftheriou and R.L. Sprott. New York, Plenum Press, p. 49-149.

Menaker, M., Takahashi, J.S. & Eskin, A. (1978). The physiology of circadian pacemakers. Ann. Rev. Physiol. 40: 501-526.

Müller, K. (1978). The flexibility of the circadian system of fish at different latitudes. In: Rhythmic Activity of Fishes. Ed. J.E. Thorpe. New York, Academic Press,p. 91-104.

Oksche, A. & Hartwig, H.G. (1975). Photoneuroendocrine systems and the third ventricle. In: Brain Endocrine Interaction II. The Ventricular System. Eds. K.M. Knigge, D.E. Scott, K. Kobayashi & S. Ishii. Basel, Krager, p. 40-53.

Pittendrigh, C.S. (1974). Circadian oscillations in cells and the circadian organization of multicellular systems. In: The Neurosciences: Third Study Program. Eds. F.O. Schmitt and F.G. Worden. Cambridge Mass., M.I.T. Press, p. 437-458.

Ralph, C.L., Firth, B.T., Gern, W.A. & Owens, D.W. (1979). The pineal complex and thermoregulation. Biol. Rev., 54: 41-72.

Reiter, R. (1978). The Pineal. Montreal, Eden press, 174 p.

Rusak, B. & Zucker, I. (1979). Neural regulation of circadian rhythms. Physiol. Rev. 59: 449-526.

Scharrer, E. (1964). Photo-neuro-endocrine systems: general concepts. Ann. N.Y. Acad. Sci. 117: 13-22.

Schoenfeld, T.A. & Hamilton, L.W. (1977). Secondary brain changes following lesions: a new paradigm for lesion experimentation. Physiol. Behav. 18: 951-967.

Schwassmann, H.O. (1971). Biological Rhythms. In: Fish Physiology, Vol. 6. Eds. W.S. Hoar and D.J. Randall. New York, Academic Press, p. 371-428.

Siegmund, R. & Wolff, D.L. (1972). Die Aktivitatsperiodik von Fischen (Leucaspius delineatus und Carassius carassius) unter Berucksichtigung der extraretinale, Lichtwahrnehmung. Forma et Functio. 5: 273-298.

Spieler, R.E. (1979). Diel rhythms of circulating prolactin, cortisol and thyroid hormones in fishes: A review. Rev. Can. Biol. (in press).

Spieler, R.E., Meier, A.H. & Noeske, T.A. (1978). Temperature-induced phase shift of daily rhythm of serum prolactin in gulf killifish. Nature 271: 469-470.

Thorpe, J.E. (1978). The Rhythmic Activity of Fishes. New York, Academic Press, 263 p.

Turek, F.W., McMillan, J.P. & Menaker, M. (1976). Melatonin: Effects on the circadian locomotor rhythm of sparrows. Science, 194: 1441-1443.

Underwood, H. (1977). Circadian organization in lizards: The role of the pineal organ. Science, 195: 587-589.

Underwood, H. (1979). Melatonin affects circadian rhythmicity in lizards. J. Comp. Physiol. 130: 317-323.

Urasaki, H. (1976). The role of the pineal and eyes in the photoperiodic effect on the gonad of the medaka, Oryzias latipes. Chronobiologia 3: 228-234.

Van Veen T., Hartwig, H.G. & Müller, K. (1976). Light-dependent motor activity and photonegative behavior in the eel (Anguilla anguilla L.). J. Comp. Physiol. 111: 209-219.

Vodicnik, M.J. & de Vlaming, V.L. (1978). The effects of pinealectomy on pituitary prolactin levels in Carassius auratus exposed to various photoperiod-temperature regimes. Endocr. Res. Comm. 5: 199-210.

ON PITFALLS IN QUANTITATIVE MEASUREMENTS IN PHYSIOLOGY

Sigfús Björnsson

Laboratory for Signal Processing and Bioengineering

University of Iceland, Reykjavik 107, Iceland

INTRODUCTION

The following analysis centers around selected topics in measurement and data acquisition technology which tend to be sources of irrevocable errors in experimental physiology.

The approach here is not one of cataloging rules of dos and don'ts. Such an approach has the tendency to backfire on us sooner or later, if it is not supported by some background. Emphasis is rather on the fundamental aspects and on elementary understanding of the problems involved, to which the non-specialist can apply his reasoning and come to his own conclusions.

Approximately the first one-third of the material, i.e. on preamplifiers, impedance matching, grounding and shielding, deals with subjects usually covered in works of this kind addressed to the physiologist. Here, this is followed by a relatively extensive discussion on signal filters, primarily for two reasons:

Firstly, it should, in my opinion, be given a more serious consideration by the physiologist, especially for quantitative work. Secondly, it serves here as an intermediary to sampling, digital filtering and digital data acquisition, i.e. subjects of future importance to the physiologist.

The material of the chapter is intended to be self-contained and does not need references to outside sources. In striving for cohesiveness, all examples demonstrated were made especially for the purpose of this presentation and staged on the same experiment (with the exception of two actual cases at the end).

The material should be a reasonable reading even to the least technically trained, provided that it be considered as a complete unit. Concepts and technical vocabulary are defined as they appear and should suffice to alleviate the obstacle of semantic problems in further communication with consulting experts.

Figure 1, in framing the subject, defines the scope of this analysis. There we divide for our purpose the experimental process into 3 stages, i.e. the aspects concerning the physiological preparation, those of doing the measurements on it and finally the processing and evaluation of the data.

It is the middle stage, that of measuring, which will particularily concern us, as it is the prime source of irrevocable pitfalling. In contrast to data processing, which is also a subject prone to errors where at worst we can do the data analysis over again, pitfalls in measuring can cost us the experiment.

Further, let us subdivide the process of measuring into 3 stages, i.e.

1. Transduction of the physiological variables into a convenient electrical signal (a subject usually specific to the particular case).
2. Pretreatment of signals.
3. Registration of signals (non-specific).

It is the 2nd stage we will pay most attention to, since it is primarily the improper treatment of the physiological response prior to its registration which is critical.

I. MATCHING TRANSDUCERS: CHOICE OF PREAMPLIFIERS

Loading errors: A signal source should have an output signal independent of the load placed across it; otherwise we have a loading error. In that sense, the preamplifier should not load the responding preparation we work with.

In physiological measurements we can subdivide preamplifier loading into resistive- (i.e. dc-loads) and capacitive loads.

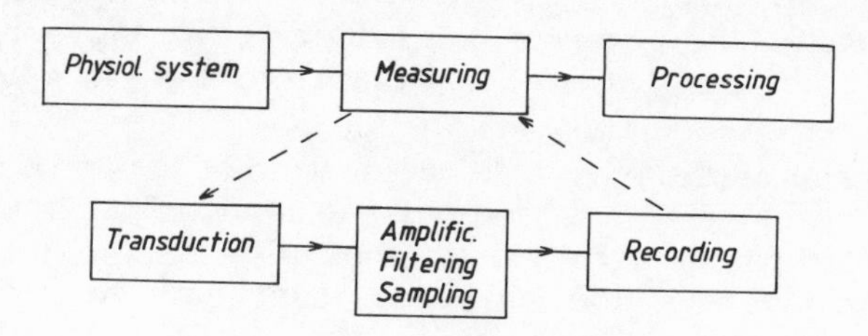

Fig. 1. In framing the subject.

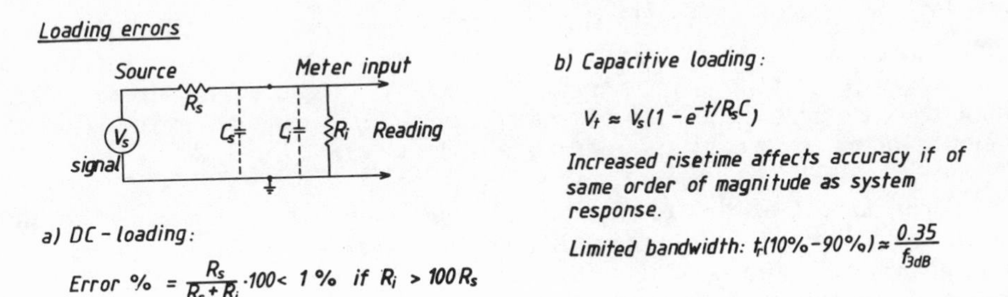

Single ended vs. differential amplifiers

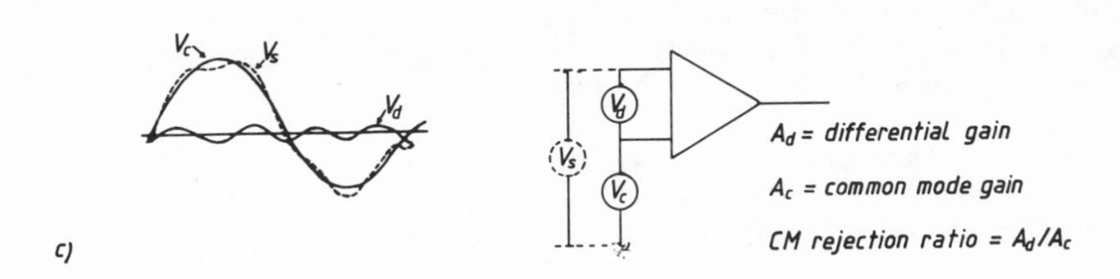

Fig. 2. On matching preamplifiers to the signal source (see text).

Resistive loading: If we look at the equivalent circuit in Fig. 2, and ignore to begin with the shunt capacitances, the meter would measure the source signal V_s attenuated by a voltage divider consisting of the source output resistance R_s and the input resistance R_i of the meter. Simple calculations of the difference of the input potential of the meter and V_s lead to an expression for the dc-loading error given in that figure.

In general a voltage source tends to have its signal independent of the load if its output impedance is much smaller than the input impedance of the preamplifier; this difference should be at least two orders of magnitude.

It is sometimes desirable to have the current from a signal source independent of the load placed across it; a current source is for instance the piezoelectric transducer. This can be satisfied by applying a preamplifier with very low input impedance compared to that of the output of the source (e.g. an ammeter).

By replacing the voltage source in Fig. 2 a, with a current source and doing the error calculations again would lead to the condition that the input impedance of the meter must now be 2 orders of magnitude less than the output impedance of the signal source, if we are to keep loading errors within 1%.

Capacitive loading: If we include in our discussion capacitors as in Fig. 2 a, they correspond to the shunt capacity of the electrode and that of the amplifier input. Considering a voltage measurement from a high impedance source like that of micropipette electrodes, these capacitors must be charged through the electrode resistance R_s.

The effect of this charging, which adds up (i.e. $C=C_s+C_i$), is to slow down the response of the preamplifier. This is shown by the expression V_t in Fig. 2b, which is approximately the response of the meter to a step function input, often called the "step response". We see that the exponential rise of V_t depends on the parameter $\tau = R\,C$, called the "time constant". For a given source impedance R_s, the capacitive loading of the preamplifier input (including the shield of the signal conductor) and the shunt capacity of the transducer is to increase the time it takes the meter to fully respond ("response time") or, as we will see, is equivalent to limiting the response of the meter to lower frequencies (i.e. limiting bandwidth).

Limited space prohibits me to go into details of methods to alleviate capacitive loading, but with two sources C_s and C_i responsible, they are in general efforts in one way or another to cancel them, i.e.:

1. To select carefully the input wiring components and preamplifiers for low shunt capacitances to ground. In critical circumstances the cable shield can be actively driven to follow the input signal at any time and thereby neutralize the mutual capacitance it has with the signal conductor.

2. With help of an equalization circuit built into the preamplifier one tries to neutralizethe capacitance the electrode has with the conducting medium in which it is inserted. What it does in principle is to boost the response of the preamplifier at higher frequencies where the electrode capacitance starts to attenuate the signal.

The often celebrated "negative capacitance" feedback in preamplifiers is one version of this neutralization technique. It usually involves adjusting for an improvement in the combined step response of a particular electrode and the preamplifier. I hope that people using it realize, that they are at best eyeballing a correction to capacitive loading with it; that they really would have to readjust it each time the impedance of the microelectrode changes, which often occurs during the experiment, otherwise it can markedly affect the shape of the signal being measured.

But the "state of the art" in digital equalizers would allow us to cope with this problem properly, which is a point I will touch upon later.

Time constant and bandwidth: It is intuitive that the limited response time caused by capacitive loading would affect the accuracy of a measurement if it reaches the same order of magnitude as the response time of the preparation we are investigating. But in order to arrive at a safe upper limit to the response time in a given situation, we need a better criterium than our intuition.

For that a practical relationship between the time constant and the bandwidth is needed, based on following definitions: The "rise time" t_r is often defined as the time it takes V_t in Fig. 2 b, to rise from 10% to 90% of a step function input. As an example, if we during an experiment would wait only one "rise time" between readings of the meter the error due to capacitive loading would be at least 10%. Now the expression V_t in Fig. 2 b, tells us, that the "rise time" stands in a logarithmic relationship to the error it causes. So if we wait two "rise times" the error has decreased by an order of magnitude or is of the order of 1%, etc.

In terms of the popular "time constant" τ , the expression for V_t shows that:

$$t_r = t_{90\%} - t_{10\%} = \ln \frac{1}{0,1} - \ln \frac{1}{0,9} = 2,2 \cdot \tau \; .$$

That is, it takes the duration of 4 to 5 "time constants" of the preamplifier to meet the criterion for an accurate reading (= 1%).

It should be obvious, that we could not keep on increasing the time constant and consequently the time span between measurements at will. Sooner or later the time spans will have increased so that the signal will be changing considerably in between consecutive samples and some information in the signal is lost in our sparse observations of it. This is just another point of view in stating the intuitive argument we made at the outset regarding the time constant of the preamplifier versus that of the preparation.

This viewpoint, which we will take up later when discussing signals bandwidths, invites the phrasing of the problem in terms of a minimum sampling rate needed to carry the full information capacity of a signal in a series of discrete measurements. I mention this now, to prevent us from looking at sampling and filtering of signals later as concepts unrelated to the preamplifier topics discussed here.

For our purpose now we only need to know that the information content of a signal is closely related to its frequency content, which in turn sets its bandwidth. The preamplifier has at all times to pass the full signal bandwidth and any setting of its time constant may not interfere with that.

Now, people often increase an adjustable time constant on a recording instrument quite liberally to cut down noise, i.e. to clean up the record as we say, but unknowingly in doing so might bring its magnitude too close to that of the signal source and thereby cut off some of the signal information.

With regard to the frequency point of view, we mentioned briefly before, that increasing the time constant of the instrument inversely affects its bandwidth, i.e. it cuts down the range of frequencies over which the instrument responds. It is therefore not surprising that the above action effectively cuts down the noise, i.e. its higher frequency components, but it might just do the same to our signal.

In order to arrive at guidelines to prevent what we need to quantify the relationship between the time constant and the bandwidth: Let us think of the signal source driving the meter with a sinusoid of constant amplitude V_i but with an increasing frequency. The 3 dB bandwidth (or the half power bandwidth) of the meter is defined as the range from dc up to that frequency, let us denote if f_{3dB}, at which the amplitude V_o of the preamplifier output has decreased to 70% of V_i due to bandlimiting. The amplitude squared, which is proportional to the signal power, would then have diminished by one half:

$$\Delta P = 10 \log V_o^2/V_i^2 = -10 \log 2 \simeq -3dB.$$

Based on this definition, a practical approximation to the relationship between bandwidth and the rise time is:

$$f_{3dB} \simeq 0{,}35/t_r.$$

This relationship gives the bandwidth limitation of an instrument due to a given setting of its time constant and vice versa. The question of bandwidth compatability of the limited preamplifier with the experiment has to do with estimating the frequency content of the physiological signal, which is a subject we will look at in due time.

Single-ended vs. differential amplifiers: Before we leave the subject of amplifiers to turn to effective measures against artefacts and noise, there is a distinction we should stress between single-ended and differential preamplifiers.

In a single-ended preamplifier the signal input and output have a common zero-signal reference conductor (like in Fig. 2 a), connected to the negative terminal of the power supply. It is important to notice that there is a low resistance path for signal currents leading from the lower input terminal to other parts of the amplifier, to the power supply ground and to the lower output-terminal.

On the other hand, a differential amplifier provides two high impedance input terminals in addition to a common reference terminal and an output-terminal giving a signal proportional to the difference of the signals at its two inputs, called the differential signal (see Fig. 2 c).

This arrangement has two very important properties: Effectively no current is drawn from either input terminal. Thus the preparation is isolated from effects caused by currents flowing in the power supply return leads (which is a frequent cause of artefacts).

Secondly, the differential property, if properly implemented, can be effective in picking small signal variations of interest out of a much larger common background-potential. This is illustrated in Fig. 2 c: If a signal V_s with respect to the instrument ground is composite of a small V_d, called a "differential signal" and a large component V_c common to both inputs, called a "common mode signal", the differential amplifier amplifies V_d but effectively rejects V_c, provided that the ratio of the corresponding amplifications, called the "common mode rejection ratio" or CMRR, is very high. A value of 10^6 is in order in physiological measurements.

The CMRR is an important "figure of merit". It is the degree to which the amplifier, while amplifying the minute differential signal, is able to reject large noise signals like a 50 Hz pick-up from the power line (ex. in Fig. 2 c) or drift in the background. In designing the differential amplifier a high CMRR is achieved partly by keeping an utmost balance or symmetry in the impedances of the two inputs.

But even the highest rejection ratio serves no purpose if the user throws this balance off with imbalanced electrodes, when he hooks them up as extensions to the inputs. Unavoidably, high impedance electrodes will in general be quite different. So in using differential amplifiers it is more important then ever, if a high CMRR and noise immunity is to be preserved, that the electrode impedance be kept a negligibly small part of the total input impedance to the amplifier, quite in excess of the "no loading" rule in Fig. 2 a.

In conclusion: When the conditions prevail of a common reference for input and output and the reference points can be grounded externally at a single point, we can apply single-ended amplifiers. This is often an advantage in the case of piezo-electric transducers and resistance bridges, to name some.

Differential amplifiers must be used when the signal source and output reference points must be kept ohmically disconnected; floating signal sources are the most notable case.

For other reasons differential amplifiers should be employed almost exclusively in electrophysiology; this results in an isolation of a signal source from ground, reduction of various artefacts, particularily from stimulation and reduction in AC-pickup, as exemples will show.

II. PROPER GROUNDING AND SHIELDING

When stimulating an electrophysiological system while recording, the stimulus current is often 3 to 4 orders of magnitude larger than the response

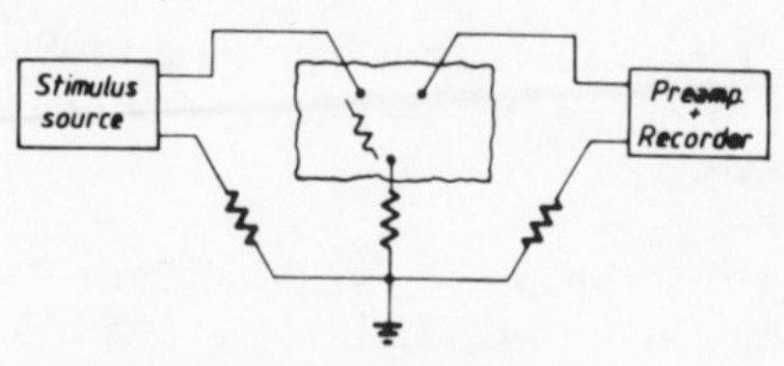

a) Improper arrangement.

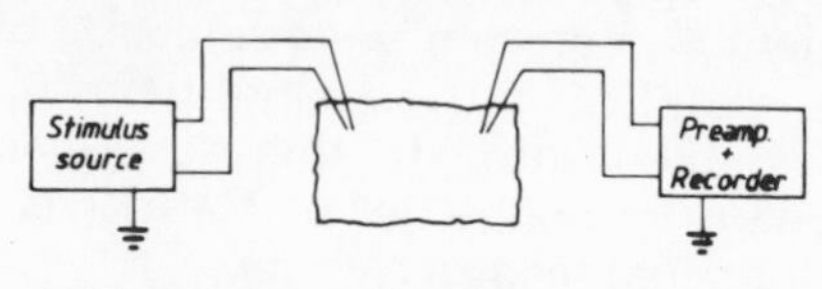

b) Differential recording.

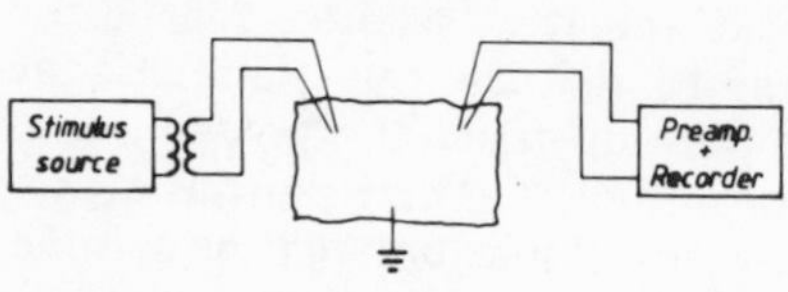

c) Stimulus isolation.

Fig. 3. Means of eliviating stimulus artefacts.

and is prone to reach the recording electrode and obscure the response with a stimulus artefact.

Artefacts: A familiar arrangement is given in Fig. 3; there finite resistances in the preparation, in leads and connections are lumped together symbolically. Although grounded at a single point (in a "star", see later), this arrangement is likely to give us stimulus artefacts for at least two reasons:

1. The electric field from the stimulus current could intercept the recording electrode.

2. The relatively large stimulus current on its return to ground through the finite resistance of the preparation can elevate passively the potential around the electrodes. This common mode potential would show up in all its force in a single-ended preamplifier.

Consequently we try to improve this in Fig. 3, by applying a differential amplifier. In order to keep the recording arrangement as unresponsive as possible to fields generated by the stimulation, the stimulus electrodes are brought as close together as possible and as far away from the recording electrodes as the experiment allows.

None the less, there will be a large common mode voltage for the differential preamplifier to cope with. If its CMRR is not sufficient to reject it, the stimulus electrodes must be better isolated from the ground path of the stimulator (i.e. floated). Fig. 3 shows that this is accomplished with an isolation transformer.

Grounding and Shielding: We have seen that floating inputs and differential preamplifiers can provide relief from noise and artefacts. Besides removing the source of interference if obvious, we know that it helps to ground instruments and build them into shielded cabinets.

Grounding to a single earthed point (as in Fig. 3) is to equalize potential differences between them.

Shielding is based on the ideal case, that if a conductor with a charge Q is fully enclosed by an electrical screen, the screen takes on the charge -Q and their net external field becomes zero. Consequently, the conductor should not influence nor be influenced by an external object changing its charge. Shielding should thus prevent pickup of extraneous electric fields.

Things are not quite that simple when it comes to practice. The extraneous field patterns in the laboratory can be extremely complex, originating in wiring in walls, ceiling lighting and various machinery and terminating in any bare connectors and leaks in shielding of the experimental set-up. Besides that, a plainly improper grounding and shielding can in itself be a major cause of noise interference, as the following cases will show.

Regarding grounding, conductors and their connections have unavoidably some resistances, shunt capacitances and inductances to ground, that can carry ground-loop currents and induce cross-coupling. These currents consequently inject unwanted voltage drops over the finite resistances of the conductors which superimposes as noise on the measured signal.

Such ground loops are bound to occur in multiple instrument set-ups when we ground two or more instruments independently with short heavy leads to a single earthed point as prescribed (i.e. in a nodal tree), if the instruments are additionally grounded through their power cords and 3-prong safety plugs.

The same thing happens often with shielding cabinets, which are more commonly what is grounded through the power cord. If we interconnect these instruments with shielded coaxial cables we have ground loops on our hands through the shieldings, which is not any better than through the instrument ground-leads.

Thus in short, to minimize ground loops in multiple instrument set-ups:

1. Use heavy short cables to ground instruments independently to a single earthed point, usually choose the ground terminal of the instrument with the heaviest current demand. Connect them to this common point in a"nodal tree" or in parallel, if current demands are low, to

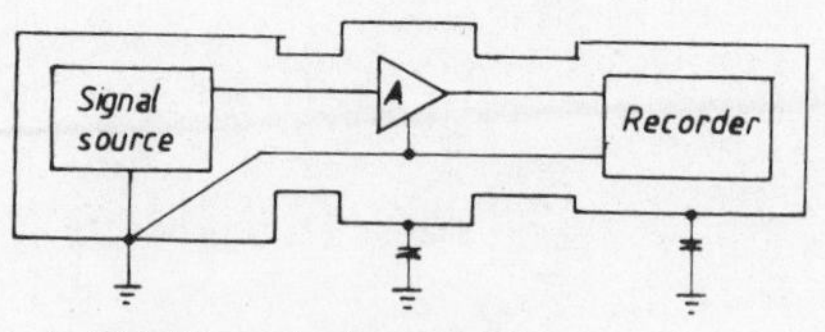

a) Shield and signal returns to a common point.

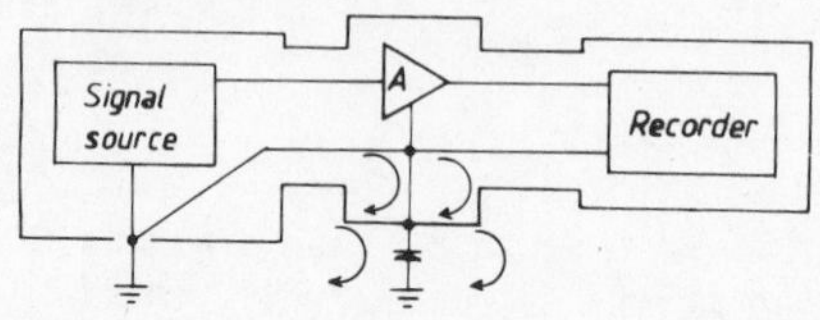

b) Incorrect return.

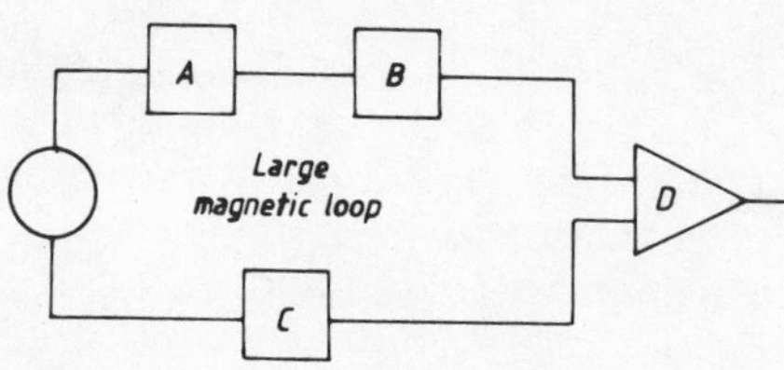

c) Inviting electromagnetic interference.

Fig. 4. On reducing external interference.

equalize potential differences between them, (Fig. 4 a). In the case of extremely sensitive measurements, ground loops can occur through unshielded transformers in power supplies of instruments. Battery operated preamplifiers can be useful to eliminate that problem.

2. Ground the chassis independently to the same nodal point as ground connectors (see Fig. 4 a). Never use chassies and shields like those in coaxial cables as a signal return path to ground (Fig. 4 b); this means that at least two coaxial cables must go to a "floated" instrument. Disconnect the shield on one end in coaxial cables between instruments if they would interconnect chassies.

3. To minimize pick-up from magnetic fields reduce the loop area formed by the signal conductors and the ground return path, by running them as close together as possible (Fig. 4 c). Twisting them together reduces mutual electromagnetic interference with other signal lines. In stringent cases, encase the set-up in a magnetic shielding.

Proper equipment grounding and shielding is only the start, what remains as noise at the preamplifier input must be filtered.

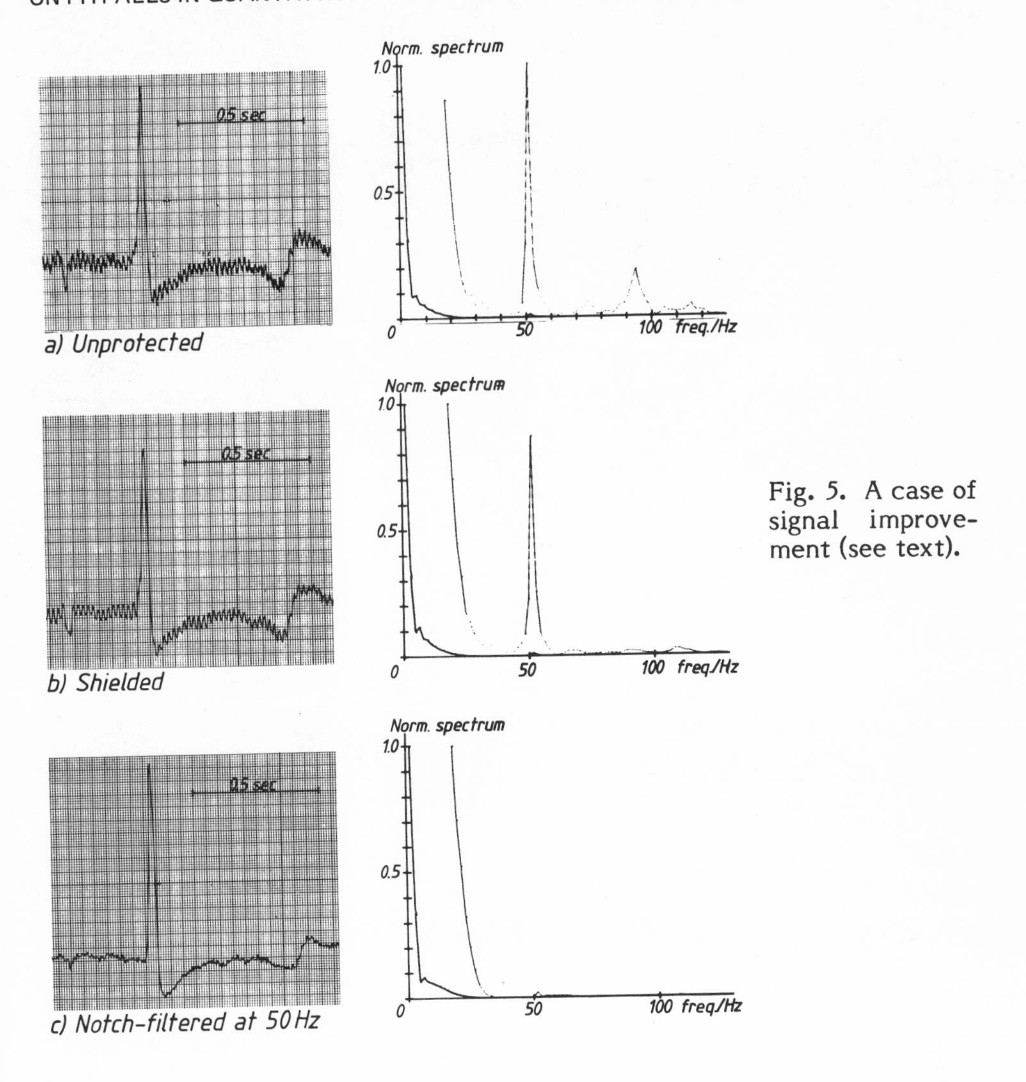

Fig. 5. A case of signal improvement (see text).

III. FILTERING FOR SIGNAL IMPROVEMENT

If we apply the guidelines set so far to a real case, figure 5 demonstrates the results. The left hand column shows records of responses; the corresponding power spectral distributions are given for two different magnifications in the right hand column.

We may recall that measured signals can be thought of as being composite of sinusoids of different amplitudes and frequencies (i.e. a Fourier representation); the spectral distribution plots give the relative weights of the amplitudes squared (power) of those sinusoids as a function of their frequency.

The example chosen in Fig. 5 is an electrocardiogram of a trout, which we will use for demonstration throughout this work.

Fig. 5 a shows a record from a properly grounded set-up employing a differential amplifier and enclosed in a Faraday-cage but no particular attention is paid to the finer details of shielding. The result is not too bad and someone might say he could live with that. But again this is purposely not a severe case;this noise level, which is totally unnecessary, could be detrimental if we were dealing with minute signals.

The normed spectrum in Fig. 5 a shows that besides the dominating interference from the power line at 50 Hz there are components spread at higher frequency, notably around 100 Hz, which definitely do not belong to

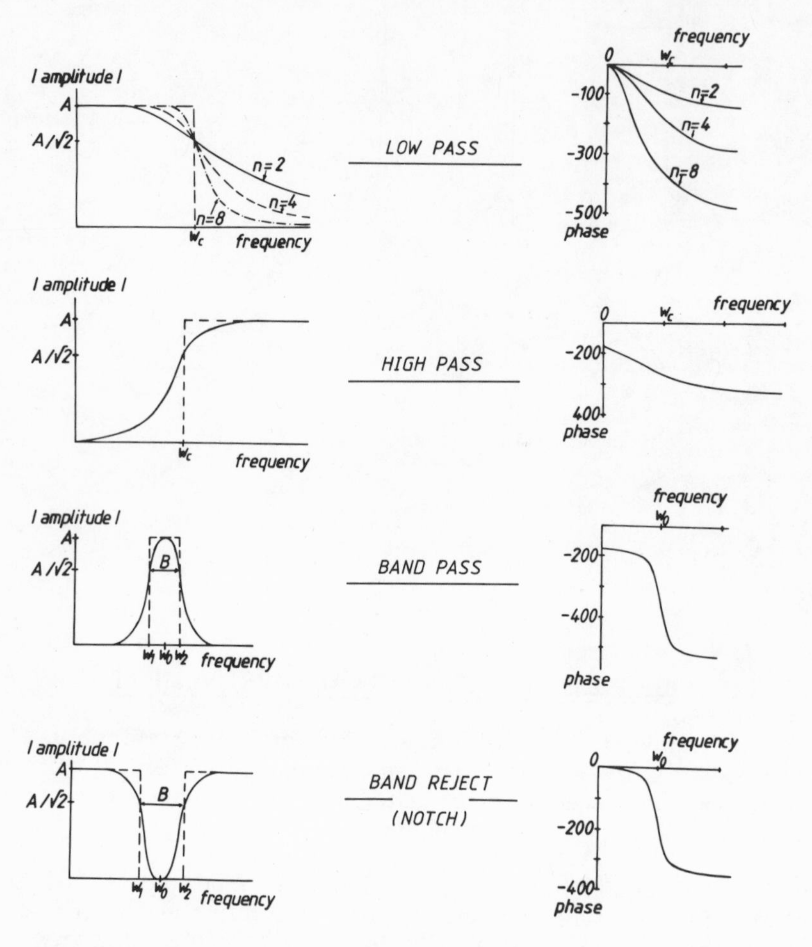

Fig. 6. Amplitude and phase characteristics of common filter categories; note nonlinear phases.

the ECG. The ECG signal has most of its power below 30 Hz and does therefore not coincide in frequency with the bulk of the noise spectrum. This makes our example an easy case, as we will better appreciate in the next section.

In Fig. 5 b, we have the set-up properly shielded and except for the 50 Hz peak, most of the noise spectrum in the measurement is gone.

To get rid of the 50 Hz power line interference, which most likely came through the AC power supplyof the preamplifier, we resort to a special filter which rejects components in a narrow frequency region around 50 Hz, but passes everything else in our range (called a band reject or notch filter).

This brings us to the subject of filtering in general.

Categories of signal filters: Figure 6 compiles the four categories of filters we need to concern us with. The figure is largely self-explanatory, the left hand column gives the pass-band and the stop-band regions respectively, the right hand column gives the relative phase relationship the filter exerts on the frequency components of the signal as it passes.

When the bulk of the noise spectrum lies outside the frequency range of the signal of interest, we can improve the signal-to-noise ratio (SNR) by applying low-pass and/or high-pass filters appropriately. When either the signal itself of the noise is of narrow bandwidth (like the 50 Hz interference in previous section) but their spectra overlap, we can apply beneficially the band pass or the band reject filters respectively.

We see from the "low pass" case in Fig. 6 that filters get sharper with increase in their order n (i.e. number of network poles), but most often also more nonlinear in their phases.

As a following example will show, the inherently nonlinear phase characteristics of filters are a potential source of error becoming more pronounced in these days of increasing quantification in physiological research. In order to put some judgement on these characteristics and select filters wisely for a specific purpose, we need to have some idea about the criteria and compromises taken in their design.

On distortion by filters: Even the best looking and respectable filter boxes picked off the shelf do distort, the question is where and to what degree.

Fig. 7 demonstrates how a filter can distort only with its phase characteristics. In the first row we add sinusoids of 3 different frequencies and amplitudes together and the result, the solid line, is already resembling a Fourier synthesized square wave. We would finally have a perfect one, if we would keep on adding sinusoidal components appropriately. The spectrum consisting of the 3 frequency components we used (a line spectrum) is shown to the right.

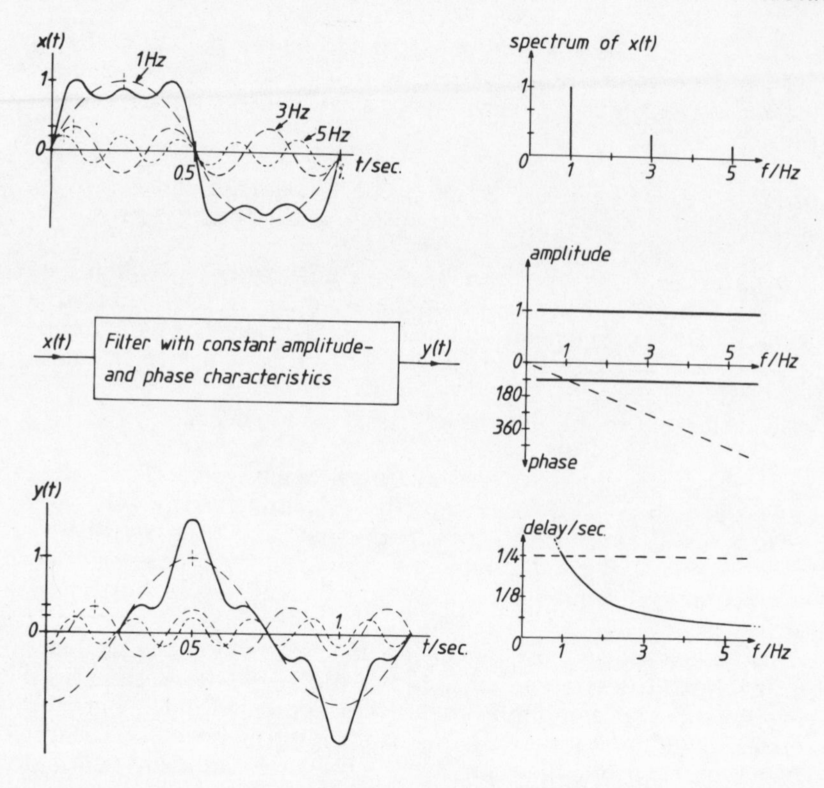

Fig. 7. A case of signal distortion by a nonlinear phase characteristics.

If we would choose a filter with constant amplitude and constant phase characteristics (which doesn't sound bad!), as shown by the solid line in the spectral plot of the 2nd row of Fig. 7, the filter should pass all amplitudes untouched and shift all phases the same amount of degrees, here 90°, independent of their frequency. Phase shifts in an output relative to the input are unavoidable since it always takes some time for a signal to pass through a physical element.

In the 3rd. row of Fig. 7, we have the 3 sinousoids again with unchanged amplitudes but phase shifted 90° each. If we add them together (the full drawn line) it becomes apparent that by dispersing the components in time the filter is distorting our square wave almost in to a triangular wave. Constant 90° phase shift does not mean the same transmission delay in the filter for the 3 frequency components; the 90° phase shift of the 5 Hz appears at the output in 1/20 sec, that of the 3 Hz in 1/12 sec and that of 1 Hz in 1/4 sec (in general $\frac{1}{4f}$ sec; f being the frequency).

This is admittedly an extreme case, but no real filter is free of phase distortion. What we prefer is that the signal would appear at the filter output with the same inherent phase relationship between its components as it has at the input, independent of frequency; in other words, it should have the same transmission delay for all frequencies (see dotted line in the frequency plot of 3rd row in Fig. 7).

Simple calculations of what a constant delay, e.g. $\frac{1}{4}$ sec., means in phase shifts to our 3 sinusoids, results in 3 points which fall on a straight line (see dotted phase plot of the 2nd row of Fig. 7). That is, the truly nondispersive signal filter should have a strictly linear phase characteristics, which is an ideal case we never can reach in practice. The same is true of amplitude characteristics of filters; a finite constant in the passband and zero elsewhere is a criterium we can never quite reach. In realizing filters we can at best approximate the ideal characteristics, the more we stress one point (e.g. distortion free amplitudes) the more we have to give off elsewhere (e.g. in phase distortion), and vice versa.

We discussed a little the different categories of filters (low-pass, high-pass, etc.). Each category is subdivided into types according to the approximation to the ideal case chosen each time in their design, i.e. a

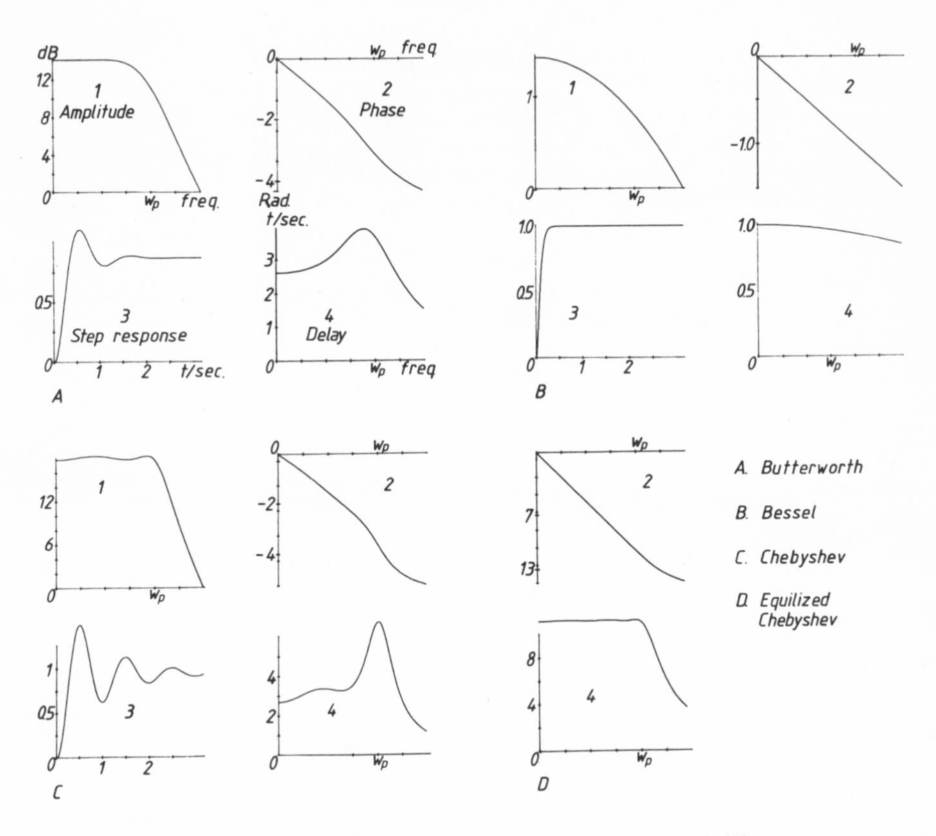

Fig. 8. Some essentials for a critical selection of filters.

subdivision based on where a certain filter type comes close to the optimum and where it distorts excessively. These questions are obviously fundamental to the user who needs to select filters critically.

Types of filters: Figure 8 comprehends the essentials on the types of filters I think we need to know for physiological measurements.

The first type, called the Butterworth filter, is an approximation which has an optimal amplitude characteristics at low frequencies (maximally flat at dc, see A1, Fig. 8). This filter type pays for this property in its rather nonlinear phase characteristics A2, which results in ringing in the step response of the filter A3 and in delay characteristics which is far from being flat A4. Because of this frequency dependent delay, this filter type in general should obviously not be applied in "transient" analysis, i.e. when the rapid time information in the signal is of prime importance, e.g. when measuring pulse- and step responses, or in quantitative work on neuronal spikes.

A Bessel filter (Fig. 8 B) is a type optimized with respect to the phase (see B2), which gives it an almost faultless step response B3 and delay characteristics B4, but is also responsible for its sloppy amplitude characteristics B1, i.e. instead of being constant within the passband region (i.e. below the "cut off" frequency w_p) it has there a continuous increase in attenuation of amplitudes as the frequency increases. Consequently the Bessel type should in general not be used in measurements where the "steady state" of a response is of prime concern, like that with sinusoidal responses (Bode analysis), but it is properly applied in pulse work.

The 3rd type we consider, named the Chebyshev filter, is designed for a minimum deviation within the whole passband region, which results in so called equiripple characteristics, and an increased sharpness around w_p. This improvement in the amplitude characteristics (see C1) compared to the Butterworth filter is paid for with a deterioration in the phase linearity (C2 as compared to A2), which shows up in an excessive ringing in the step response C3 and a gigantic peak in the delay characteristics around w_p. It is worth noting in this example how very sensitive the delay is to relative small changes in the phase characteristics at higher frequencies (i.e. around w_p).

This leads the subject to an important filter element called an equalizer, which is a specialized circuits designed to correct the amplitude and/or phase characteristics of filters. Figure 8D shows the phase and delay characteristics of a Chebyshev filter again, but this time from one supplemented with a phase equilizer, which linearizes the phase without touching the heights of the amplitudes of Fourier components. The result is a near perfectly flat delay- and amplitude characteristics within w_p, as seen in D4 and C1.

This is an example of a high quality filter, as one must specify for highly quantitative work like in pre-treatment of data (i.e. data filtering) for computer analysis. It is often advantageous to do data filtering after the measurement record has been converted into numbers (i.e. digitized or sampled, as we discuss in next section). Then the filter itself, hithertofore an electronic circuit accepting a voltage signal, must now also be converted to its digital analogon, to accept the time dependent signal in form of a number series (often called a "time series"). Everything said so far on categories and types of filters carries into the digital world, with the added convenience, that with a mere change in a program we can alter our equalized filter and tailor it to our specific needs at any time. We will have more on that in the section on future trends.

On digital processing filters: Let me leave the subject of signal filters by posing the question what to do if the signal bandwidth is not any less than that of the disturbing noise.

Obviously, applying conventional frequency selective filters would not do us any good; on the contrary it could be disastrous.

But is this then the end of filtering? By no means. We have not touched upon an extensive subject we could call "signal processing filters", where improvement in the condition of the signal is achieved by computational procedures. This subject classified under the last box in our introductory scheme in Fig. 1, i.e. "Processing of physiological data", would take us a whole new chapter to do any justice.

I would though like to stress, that as in the case of any well planned effective processing of data in general the power of applying signal processing filters lies in taking them into full account at the early design of the experiment.

For example: If an experiment would permit a repeated application of the stimulus, we could extract a response signal out of the camouflaging noise, even though the signal- and the noise spectra would fully coincide, by applying a signal averaging filter. If the basic shape which a response signal embedded in noise should have is known apriori, one can design a more effective processing filter, a "matched filter", to detect its presence or not.

In dealing with multiple responses overlapping in time, like we often encounter with compound signals in physiology neither of the above signal procedures would be optimal. If a proper design of the experiment and fundamental knowledge of physiology, will allow us to assume the basic shape each component in the composite response should have, a family called "deconvolution filters" would optimally decompose the compound signal. And even if physiological reasoning could not lead to the basic shape a whole new class, that of nonlinear processing filters ("homomorphic filters", long-pass and short-pass filtering), might do. This goes to show that the subject of filtering is far from being closed, though we leave it here.

But the application of these recent methods is only feasible in a digitized form, which stresses the future importance of digital sampling in physiology.

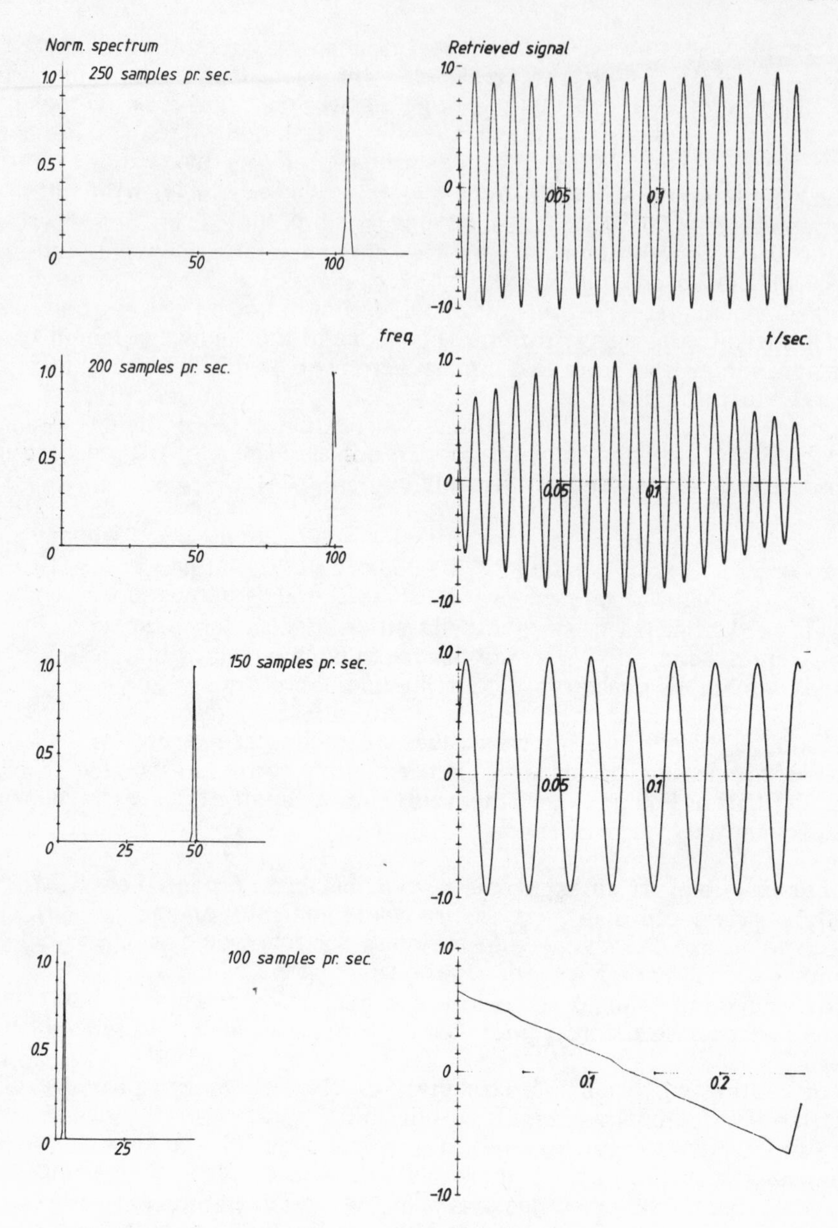

Fig. 9. Testing a sampling rate by retrieval.

IV. SAMPLING

Taking discrete values from a continuously varying signal, i.e. sampling, is a process prone to errors if we do it unaware of basic laws of signal processing governing it. Errors due to improper sampling (in general called aliasing errors) show up often as puzzling peculiarities along the way of the analysis procedure and can be very hard to spot and to trace from the analyzed data back to its origin, if possible at all. Preventive measures are therefore of eminent importance.

Proper sampling rate: To start with, Fig. 9 demonstrates sampling errors on a signal simple enough to disclose them without further ado to the naked eye. There we have a sinusoid slightly in excess of 100 Hz and we sample it with 4 different sampling frequencies. Sampling is done with an input unit, called analog-to-digital converter (A/D) to be found on many laboratory computers and digital instruments. After sampling, the sinusoid is in the form of a number series in the computer; if we read it out through another unit a digital-to-analog converter (D/A), we would retrieve the sinusoid as a time varying voltage signal identical to the original one, if the sampling was properly done. Fig. 9 demonstrates such a retrieval test. Having sampled at different rates, we compute the spectrum of what we have got (left hand column) and try then to retrieve the sinusoid for each sampling rate tested (right hand column).

The rate of 250 samples per sec (sps) preserves the frequency and the shape of the sinusoid, except for small amplitude variations, which is an artefact we will turn to later. At 200 sps, which is slightly less than twice the signal frequency, obviously something is already wrong. We retrieve a good sinusoid, but its frequency is slightly less than it should be (it amounts to the "difference between the sample- and the signal frequency") and has a peculiar slow-varying amplitude variation ("beating"). By sampling at 150 sps, retrieving gives us a cleaner sinusoid than ever! but the computed spectrum shows that the sampled version has unfortunately its frequency

Rule of thumb: Have at least 2 or 3 samples cover the most rapidly varying portion of the signal.

Example:

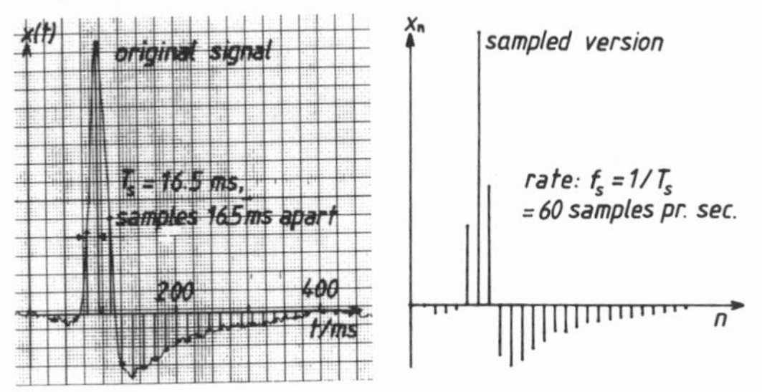

Fig. 10. A graphical estimation of signal bandwidth and proper sampling rate.

Resulting bandwidth: $B = 1/2T = 30$ Hz

Sampling theorem:
The signal must be bandlimeted. It must be sampled with a rate at least twice as high as its highest frequency component. Or else it cannot be retrieved.

less than one-half of what it should be (it obeys the "difference rule" again"). Finally at 100 sps, which corresponds to less than one sample per period, we get our spectral needle again obeying the "difference rule" and retrieving gives us something I don't suppose would pass anyone's criteria for a sinusoid.

But how can we predict the proper sampling rate in the case of a complex more or less unknown signal, which we need to do if sampling is done while measuring (i.e. before recording) as today's "real time" digital laboratory instruments often require. Undersampling will have cost us the experiment; oversampling will not only cost us dearly in computer time, but projects can literally suffocate under today's computerized data acquisition if the evaluation cannot keep pace. It is the more tragic if the datapile accumulated is largely redundant due to precautionary oversampling.

Here some very practical results of information theory can come to the rescue. The rule we broke in all but the first case of the sinusoid example of Fig. 9 is Shannon's sampling theorem, a part of which we formally state in Fig. 10. Its usage requires of us knowing the bandwidth of the signal to be sampled. Fig. 10 gives also a practical "rule of thumb" to estimate bandwidth from a graphical record; its usage is demonstrated there on the ECG signal previously introduced and should be largely self-explanatory.

The result of the example in Fig. 10 is that the ECG-signal has approximately 30 Hz bandwidth and, if sharply bandlimited at that (with a filter), could be properly sampled with a rate as low as 60 sps. We will put this result of the graphical "thumb rule" under some scrutiny by the retrieval method in the section on aliasing, stating here that our experience with it in physiological work (slow potentials, compound signals) shows that it tends to be on the safe side of the exact bandwidth (i.e. tends to overestimate it slightly). Neuronal spikes are an example where we however should think before applying the rule. If individual spikes (i.e. action potentials) are of interest the rule applies if we can expand the time scale of the record sufficiently to make graphical estimations from it. If the signal message carried by the spiking neuron is what we are after, it is not contained in the shape of the spike but in its rate of occurance and the graphical rule in the form applied above would vastly overestimate the minimum sampling rate; if we could run the signal first through some appropriate rate-to-analog converter the rule might apply.

While on the subject of bandwidth estimation, especially regarding sensory physiology, I would like to mention, that in a case like spiking neurons, where it is conceivable to estimate the maximum rate of information flow, i.e. the number of "all or none" responses per unit time, the wide ranging results of information and communication theory, which quite universally relate this to channel bandwidth, might become practical.

Aliasing: In order to get some insight into sampling and a little confidence in the graphical bandwidth estimation method in Fig. 10, we

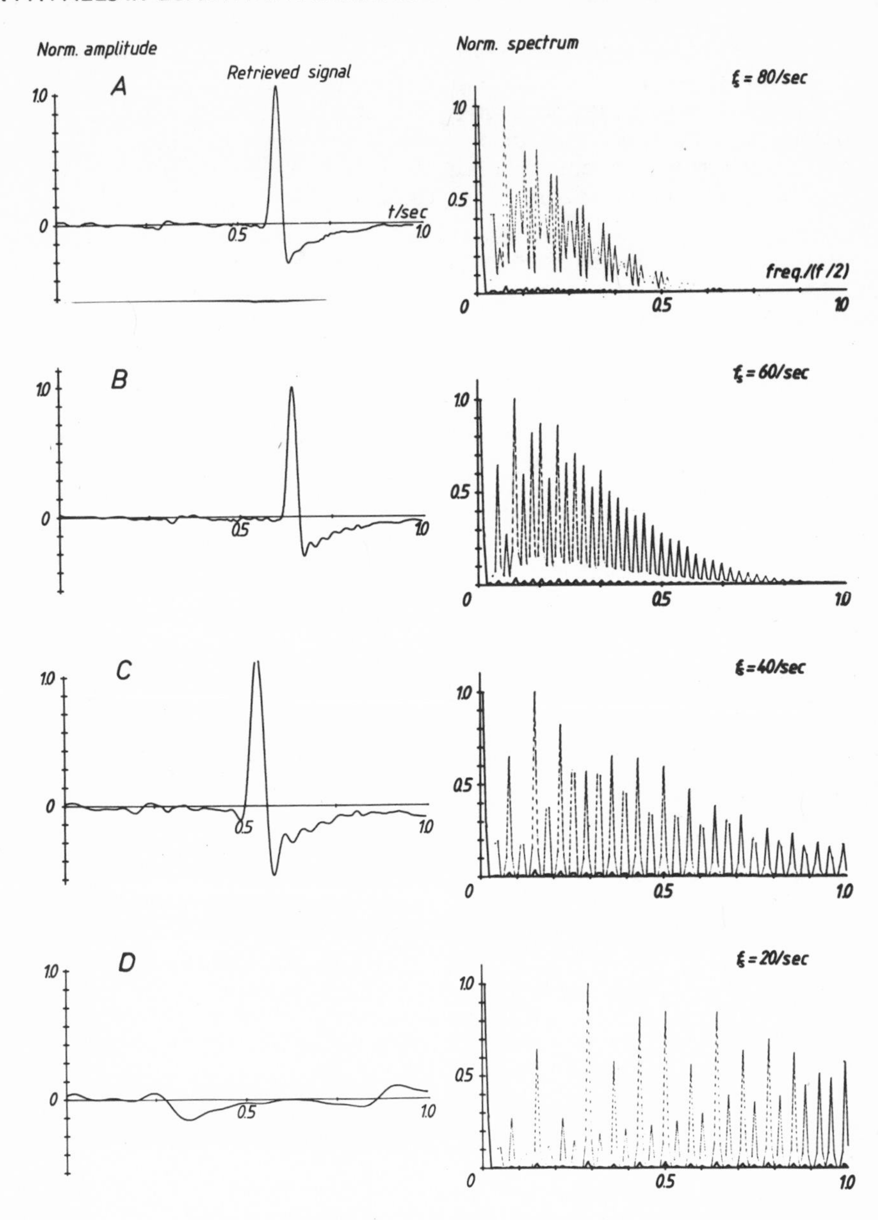

Fig. 11. Testing the "thumb rule" of Fig. 10 with a real signal.

digitized the same ECG signal four times with different sampling rates. Prior to sampling we secured the bandwidth limiting required of the sampling theorem by filtering the signal prior to sampling with a 4th order Butterworth low pass filter with the "cut off" frequency set at 30 Hz. The result is given in Fig. 11. In example A we have sampled with 80 sps, i.e.

about 2.6 times the 30 Hz upper limit of the signal bandwidth, as estimated by the graphical rule. If we compare the retrieved signal with the analog one in Fig. 10, we see that sampling has not done it any harm and filtering the analog signal above 30 Hz prior to sampling has not either. This shows that the "thumb rule" at least has not under-estimated the bandwidth; there is no or negligible information in this ECG-signal above 30 Hz.

Therefore, we should get away with sampling at the theoretical minimum rate set by the sampling theorem, as we have attemped in example B. None the less the spectrum in B has markedly changed compared to A and extra oscillations can be seen in the retrieved signal. This will to some degree always be the case when sampling at the absolute minimum rate, due to the fact that we never can with real filters sharply limit bandwidth at a given frequency. In this case with a filter not sharper than the Butterworth type, we definitely have in the measured signal always some traces of the noise spectrum passing through the filter above 30 Hz which is in violation of the condition set by sampling at 60 Hz. These traces interfere with the sampling process (aliasing) and cause smearing of the signal power along the frequency scale as seen in Fig. 11. The practical minimum sampling rate is 2,5-3 times the maximum frequency of the signal, depending on how good we are at filtering.

We might note that in case C where we are with the signal itself clearly in violation of the sampling theorem and we get consequently a heavily distorted spectrum, the retrieved signal still resembles the ECG and as an only criterium might not be suspected of an aliasing error by the qualitative eye. This goes to show that aliasing is not always easy to spot.

Antialiasing filters: In order to stress the importance of bandlimiting filters being always put immediately before the A/D-converter, no matter what guaranties we have of a limited bandwidth of the signal at its source, we repeat the situation in Fig. 11A but this time with no such precautionary data filter.

We have shown that the ECG has no spectral components above 30 Hz and we know that sampling at 80 sps suffices.

The result is given in Fig. 12, first row. The traces of noise seen in the original signal are enough to cause aliasing with the sampling process with disastrous results.

Reinstating the antialiasing filter restores the previous results.

V. ON AVOIDING SYSTEM COMPLEXITES

From the viewpoint of a signal processing, i.e. usually in a late stage of a research project, one is often faced with complications which could have been circumvented by forethought at the design stage of an experiment.

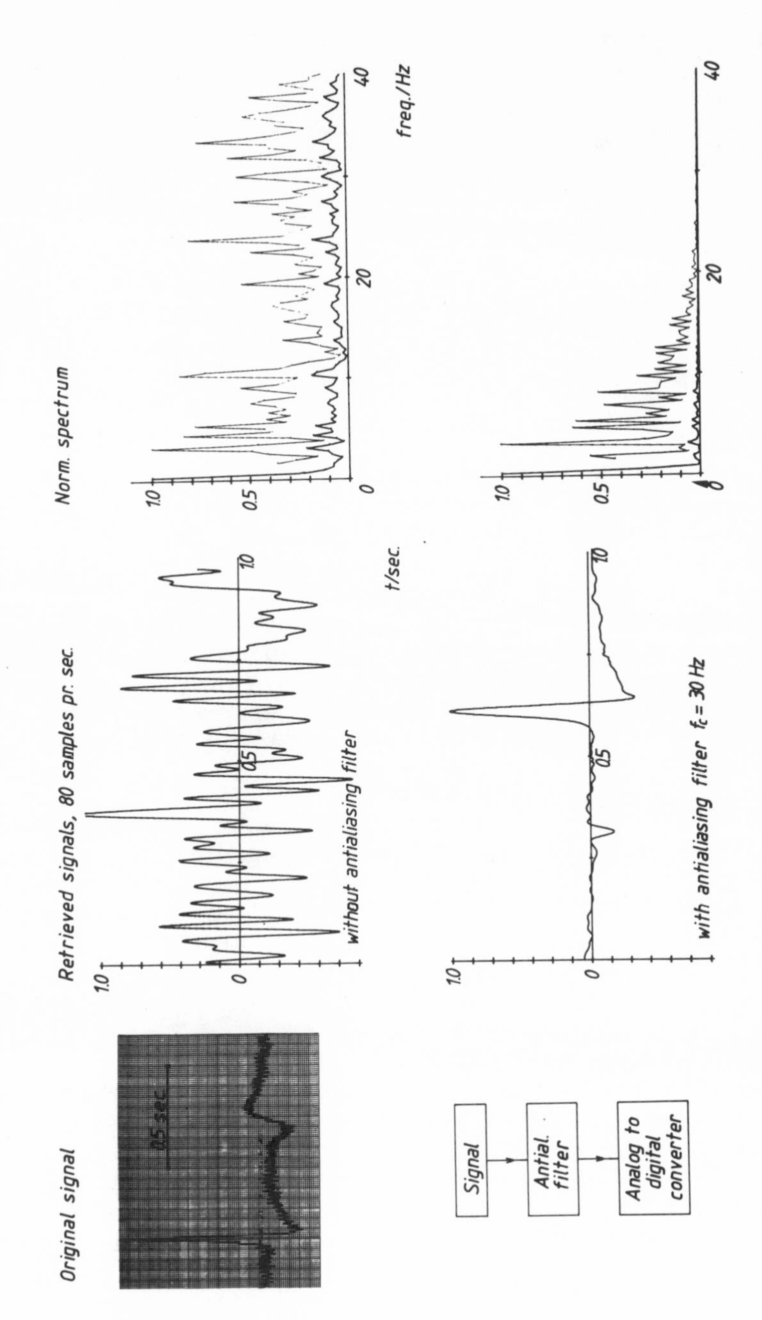

Fig. 12. The effect of bandwidth limiting (data filtering) prior to sampling.

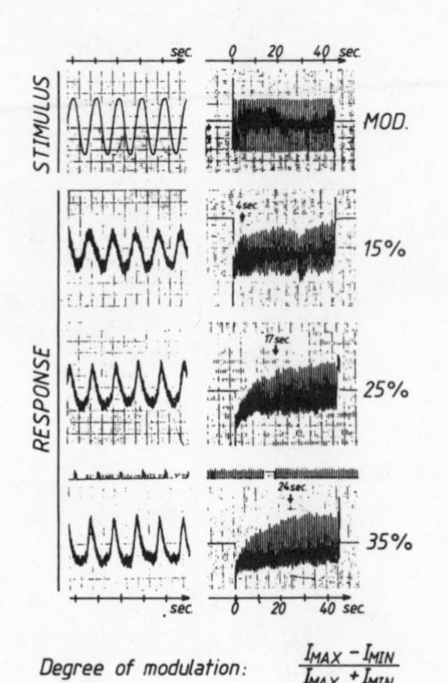

Fig. 13. On preventive measures regarding nonlinearity and nonstationarity.

It is easy to be wise in hindsight, I admit; but I think more concern in advance about the conditions and limitations of the particular analysis techniques we decide to use, would make us more aware of the constraints which necessarily have to be put on the experiment to achieve the proper analysable data. To practice data evaluation concurrent to measuring as much as possible, is of course vital to get the necessary feedback.

The subject of system analysis, which I am about to touch upon regarding two fundamental concepts, is wide ranging and hard to generalize on in physiology because of its relatively recent entry in the field. Just to raise the issue, I want to stress the concepts of nonlinearity and nonstationarity which we deal with more or less in all physiological systems. A little insight here might be a help on our way to simpler and better analysable quantitative experiments.

The example in Fig. 13 is taken from a pilot study on a heavily integrated system (intracellular measurements in the vertebrate retina), done for a concurrent work on system methods[1]. As the degree of stimulus modulation increases, as seen in the left hand column of Fig. 13, the heavy non-linearity of the system becomes more apparent. That is, at 35% modulation the response resembles the logarithm of the sinusoidal stimulus (not surprisingly); on the other hand at 15% or less, the response can be assumed to be reasonably linear.

If we look at the time behavior by watching longer records as in the right hand column of Fig. 13, we see that the system is heavily adaptive (i.e. nonstationary) and this is coupled to the degree to which we modulate.

ms, i.e. a good order of magnitude less time than over which we see a definate adaptive change take place. This we would define as the transient region.

In short: In a "simplistic" approach, like applying linear stationary until we get into the analysis phase (i.e. leave it to some theory to cope with), would we stand a chance to analyse such an experiment at all. I should remark that there is no workable nonlinear instationary system theory in general existing and the partial results there are, often lead to complexities very hard to sort out and verify experimentally.

I am not on general principals against a global analysis approach (as the above would be), on the contrary I believe strongly in it as a general framework to comprehend current status and for deductive work. In experimental work, it is obviously safer (besides we have a little choice) to approach our goal from the opposite direction, i.e. from the special ("simple") progressively to the more general ("complex"). In a majority of cases, which serve special purposes, the simplistic approach does perfectly well. Even in general modelling it is conceivable for instance in the above example, that one could reach just as far piecewise armed with well founded methods of linear and quasistationary analysis as with a global approach, with less likelihood of loosing touch with reality.

From this last point of view, let us look again at the retinal example in Fig. 13.:

Regarding the nonlinearity it would be conceivable to piece the steady state behavior of this system together from linear increments (i.e. small signal linear analysis at 15% modulation), spaced evenly along a nonlinear operating path (representing different states of adaption). It could be tedious, but in principle it is simple.

Regarding instationarity, we would much simplify the analysis and make it less error prone, if we could separate the experiment into 3 time domains, that of a transient-, a transitional and a steady state region, according to the following: In the 35% case, a transition region takes approximately 24 sec from the onset of the stimulus until steady state sets in. An analysis of it would tell us about the rate and range of adaption under given circumstances. Were we doing a "steady state" analysis under these circumstances, we would have to discard the first 24 sec of each record in order to avoid ambiguity in our results.

At 15% modulation, where we could well apply linear methods (small signal dynamic analysis), the steady state region sets in after about 4 sec. We would have to discard that part in each record.

If a measurement is to be done on the system in the particular state it is in, at the onset of the stimulus, we must resort to transient analysis. In this case we had better be done with each test within a time span of 100-200

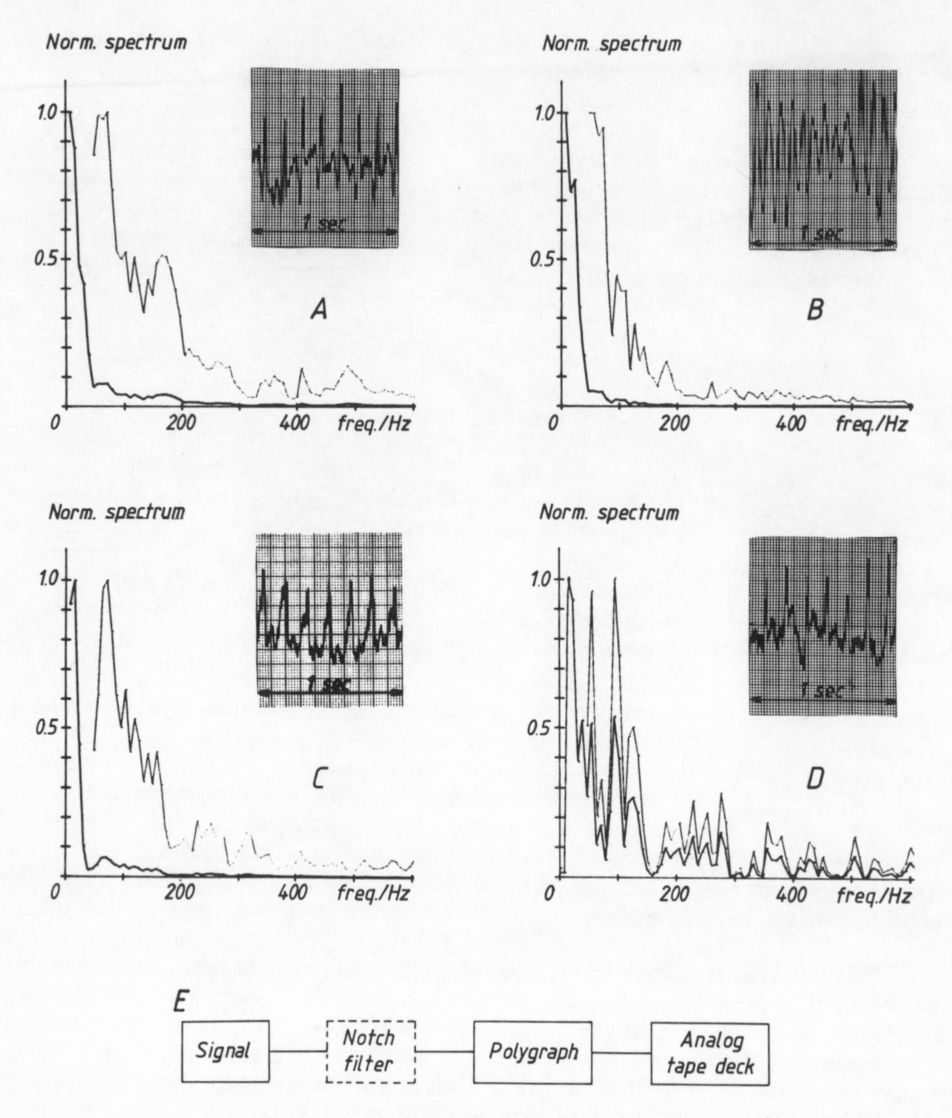

Fig. 14. An actual case of "pitfalling".

Now if we would plunge into an experiment with a system as integraded as this one, armed with a repertoire of stimuli (flashes, white noise; spots and annulli) and leave all worries about instationary and nonlinear transitions, which might occur during the stimulus application, methods to analyse physiological systems, one must be on guard for heavy constraints on the experiments. It leads to erroneus result, would we use Bode analysis mechanically on the 35% case in Fig. 13 or would we do stationary spectral analysis on the first 24 sec of that record.

To exemplify such constraints and appropriately conclude the chapter, our last example will be an actual case of pitfalling (a recent one of ours!). It is taken from a preliminary computer analysis[2] of experiments done with rats, which had been fed with different diets and during the experiment driven with stress hormones into cardiac arithmia.

Fig. 14 A, B and C demonstrate that situation before, during and after such a cardiac attack respectively; the insets give ECGs, the graphs give the corresponding power spectra.

A further analysis of this as a system was not asked for, but if so linear methods like the power spectrum estimator would not apply unless extended to complex higher order spectra (nonlinear analysis with polyspectra). The small signal linear analysis mentioned above does not apply either because the response is not a direct cause of an external stimulus, which modulation degree we could adjust.

The system is nonstationary over the period in which the attack sets in, but we circumvent that by taking short quasistationary excerpts periodically from the record and analysing them with a rapidly converging spectrum estimator.

If we look at Fig. 14 A, B and C we see, in spite of a drastic change in heart beat in B (see inset), that the bulk of the power spectrum (i.e. below 100 Hz) is not much different in all 3 cases. It does not resolve what we want.

In case D we computed again the ECG prior to the attack with a much higher resolution. This reveals that two dominating peaks, suspiciously at 50 Hz and 100 Hz, are disguising the most delicate part of the ECG spectra, actually ruining the measurement.

The 50 Hz signal obviously originates with the power lines and we discussed earlier in detail how to alleviate that. Its first harmonic at 100 Hz is a secondary effect from the power line caused by some nonlinearity in the set up or its surrounding, through which the 50 Hz pass.

REFERENCES

Björnsson, S. (1979). Comparative System Analysis in Sensory Physiology. Research Report. Laboratory for Signal Processing and Bioengineering, University of Iceland.

Björnsson, S., Axelsson, J.,Vilhjalmsson, J. (1979). Computer Evaluation of Electrocardiographic Data. Research Rpt., Laboratory for Physiology, University of Iceland.

CONCLUDING COMMENTS

M.A. Ali, J.H.S. Blaxter, George F. Holeton
B. Lahlou, Karel F. Liem, and Gunnar Sundnes

> Nay Gentlemen, prepare not to be gone;
> We have a trifling foolish banquet towards.
>
> Romeo and Juliet, Act 1, Scene 5

INTRODUCTION

The last day of the ASI was devoted to a general view of the meeting and an informal discussion. Seven of the participants, chosen a week before the end of the ASI, were each assigned certain topics for discussion. Each speaker gave a brief exposé which was followed by discussion. These discussions were long or short depending on the topic and the interest elicited by the speaker. In spite of interruptions to present spades (shovels) to Karel Liem and a friend of his and an award to George Holeton for his ability to handle five steaks, and other miscellaneous awards, the discussion proceeded well into the afternoon. The following is a brief compte rendu of the points discussed. Each speaker was asked to give a written account of what he said and what followed. The discussions had also been recorded on tapes. One of us compiled all this together, and edited it. Obviously, a presentation of this sort is bound to lack continuity but it was felt that an effort should be made anyway to present the spirit of the day's proceedings.

SOME RETROSPECTIVES AND PERSPECTIVES

In this brief overview we will deviate from the traditional approach of summarising the highlights of a meeting. Instead, we will deal with the question whether or not broad conceptual and theoretical progress has been made in this field. The purpose of this review is to provide some background for the appreciation of the chapters in this book and to focus on some especially promising research challenges that lie ahead in the environmental physiology of fishes.

General background

With a few exceptions the pervasive approach in this series of presentations has been distinctly mechanistic. At first the components of the fish are systematically disassembled for reductive analysis. This approach is of course a necessary phase of the development of environmental physiology. Subsequently most authors have successfully reassembled the physiological bits and pieces into the regulated systems and organisms in which they function. Several authors have even gone one step further by discussing their findings on regulated systems in terms of the whole organism's relation to its environment. In our opinion such a pansynthesis is the major intellectual goal for environmental physiology, although we are aware that this opinion is not universally acknowledged among those rallied to the banner of environmental physiology in this ASI.

Most lectures and seminars in this ASI dealt with strictly proximate causes by asking "how a particular component works", and "how it interacts with the prevailing natural environment". By isolating the studied phenomenon sufficiently many variables have been either eliminated or controlled. The corollary of most of these approaches is that the ideal of a purely or chemical experiment has been achieved. However by the deliberate disregard of the historical factors or genetic programmes of the organisms studied the goal of many of the contributors has shifted to the search for optima of biological functions. Accordingly, the common research strategy is to expose an isolated physiological component of the fish to an extreme chemical or physical factor, even though such "hyper-stimuli" may never occur in the fish's natural habitat. While the optimal rate of a physiological process as a function of, for example, pH may be fairly easy to determine and understand, the simultaneous optima for a multiplicity of relevant interacting factors, the metabolic pathways, the respiratory processes, the ionic regulation, the whole organism and the numerous factors in its environment become a very complex strategy exceedingly difficult to decipher.

To simplify their conceptual models environmental physiologists have not taken into consideration the historical factors (i.e. phylogenetic or evolutionary or genetic). Yet these factors govern the range of adaptive expressions of the fish to environmental perturbations. Another omission is the failure to appreciate that the various components do not function in isolation. The interrelated parts exert constraints on each other preventing the simple one-to-one optimal solutions advocated by many of the contributors. The study of internal constraints and redundant systems in relation to continuously fluctuating environments has not yet been undertaken by environmental physiologists working on fishes, although it represents an area of considerable promise.

Some Challenges for the Future

If we define evolution as changes in the diversity and adaptation of populations of living organisms, environmental physiology can play a key role in developing an understanding of the whole organism's relation not only to its spatial but also to its temporal environment. Even though most authors have not implicitly stated the intellectual targets of their comparative methods, one can recognise a common fabric throughout the proceedings of this ASI i.e. that superimposed on the basic similarities in physiological mechanisms by all living fishes there appeared to be a significant array of differences related to habit, environment or evolutionary history. Because evolutionary history has been ignored the current environmental physiological landscape can be likened to seeing a few mountain tops above a heavy cloud cover. Evolution proceeds by the simultaneous change of many closely interrelated attributes of the phenotype as these respond to the environment and adjust to each other. The future of environmental physiology lies in the experimental determination of not only the typological optima of components, but especially the correlations and built-in constraints of physiological parts, and the assesment of the influence of variation in each group of correlated traits on evolutionary fitness. To achieve this goal, environmental physiologists must analyse the environment much more thoroughly than hitherto done, taking in consideration seasonal fluctuations etc. The extensive multiplicity and polymorphism of many physiological mechanisms in fishes is now well established. It appears possible that functionally different multiple components and polymorphism may play a key role in (1) enlarging the habitable environment, (2) adjusting to unpredictably fluctuating environments, and (3) in serving the changing needs during growth and development. Modulatory multiplicity, polymorphisms and redundant systems are areas holding great promise for future research in environmental physiology of fishes.

Conclusion.

This meeting has generated an extensive and coherent array of facts on physiological variations correlated with environmental factors in fishes. The discovery of an intellectually satisfying "theme" on which the variations depend has not emerged. While this may seem discouraging, the way ahead obviously lies in the direction of systematically gathering comprehensive comparative data on the interrelationships of physiological parts, environmental fluctuations and, especially, evolutionary constraints. The art is to find and follow the best course between premature meaningless generalisation and endless gathering of new random facts. Although this ASI has perhaps failed to make giant steps forward, it has made stimulating progress by the many small advances and by exposing the big gaps and multitude of omissions.

RESPIRATION AND GAS EXCHANGE

One factor that comes across very clearly is the obvious convergence of interests between workers studying respiration and ionoregulation. The meeting ground is in the area of pH regulation and acid-base balance. Norbert Heisler's contribution, in particular, should be of interest to physiologists in a wide range of disciplines.

A second observation is the remarkable range of approaches there are to the study of environmental physiology. We have seen presentations of work on fish living undisturbed in their normal habitat and, at the other extreme, presentations of work on pieces of fish in the laboratory dish. It is certain in some cases that it must be, at times, difficult for workers at one of these extremes to appreciate the value of what those at the other extreme are doing. Nonetheless we feel that such a diversity of approaches is an indication of health and strength, and that we are on the proper course of action. Certainly there is little value in concerning ourselves with a tight definition of what, exactly, constitutes environmental physiology.

An outstanding problem area of environmental physiology which relates to a number of areas is the difficult jump in being able to say what an organism will do in its normal environment. At most, we can just give some idea of what an organism can do. We suspect much of the continued progress in this area will be due to improved technology in monitoring intact unrestrained fish.

A major stumbling block in studying respiration and ionoregulation of fish is due to what George Holeton calls the "Flashbulb syndrome". Fish muscle is mostly white muscle which usually functions anaerobically. That muscle, which looks so tasty on the dinner table, is used by the fish mainly for sprinting; to escape predation or to catch prey. This muscle can anaerobically generate a very high work output for a short time while lactic acid accumulates as a by-product. However this process is short-lived and once it happens it takes several hours or a day before it can be restored to its original status again. Hence the analogy with the flashbulb, you usually only see it go off once.

As a result of a few seconds of furious activity, a fish suffers a build up of large amounts of lactic acid in its white muscle (which constitutes half of the fish). The lactic acid can grossly distort the normal internal milieu of the fish with regard to a number of important physiological variables. Let us give a practical example.

The blood is a particularly convenient indicator of physiological status of a fish. However it is crucially important that the blood is sampled without disturbing the fish. Disturbed fish struggle thereby setting off the white muscle "flashbulb". All too often the method for sampling blood, particularly in the field and with small fish, is to use the technique of "grab and stab" whereby a fish is chased, seized, subdued and violent access made to the circulatory system to collect blood. By sampling blood in a less traumatic manner via chronically implanted catheters we can get some insight on the magnitude of the problems inherent in "grab and stab" sampling which arise from the "flashbulb syndrome".

Heisler, Neumann and Holeton have looked at what happens to the blood of rainbow trout as a result of a brief period of violent struggling. The fish, weighing about 1 kg, was cannulated in the dorsal aorta and allowed to recover undisturbed for 24 hours. After a series of control blood samples were withdrawn, the fish was induced to struggle strenuously. Then blood samples were taken immediately and at intervals after the struggles ceased. The following table will give an impression of what they found.

Table I: Changes in arterial blood of Rainbow Trout as a result of 4 minutes struggling at 15°C.

Variable	Control	Time after struggling (minutes) 0	4	30	1320
pH	7.789	7.204	7.325	7.369	7.834
PCO_2 (torr)	1.84	3.29	-	3.23	1.93
Hematocrit %	25.6	30.0	34.2	37.2	23.5
Σ of change in Plasma ions $mM \cdot l^{-1}$	0	+14.5	+26.6	+31.0	-3.7
(lactate) $mM \cdot l^{-1}$	1.2	4.22	6.52	10.7	1.45
(Na^+) $mM \cdot l^{-1}$	147.6	154.7	162.9	163.1	145.4
(HCO_3^-) $mM \cdot l^{-1}$	4.09	2.26	-	2.65	4.12

The very first samples taken after a short spell of exercise were drastically displaced from the control values. Values continued to change and most variables remained changed for several hours though they were more or less back to normal within 24 hours.

We think that consideration, even of this one example, of such changes should serve as a reminder that where CO_2 ions and pH are to be measured in fish the experimentor should try to use cannulated fish. At the very least, conclusions based upon "grab and stab" sampling should be made with a great deal of caution.

OSMOREGULATION

The relevent talks presented at this meeting dealt with plasma concentrations, gill exchanges, hormones, mechanisms of transport.

Some fishes have been found to live in media of unusually high salt concentration: up to 80% salinity, i.e. more than twice sea-water. This extends the range of natural biotopes in which fishes may be studied.

Salt adaptation is currently presented as resulting from changes in "permeability". This term should be used with caution however for all substances except water. Permeability is best expressed by the "permeability coefficient" which is drawn from Fick's law of diffusion and has the dimension of speed (ignoring the presence of electrical potentials). However ion exchanges comprise several components which should be determined separately: simple diffusion, passive facilitated diffusion, active transport, before a statement can be made safely about permeability changes.

A striking feature in ionic exchanges concerns couplings between ions of different species. "Neutral" transport of NaCl taking place in intestine, nephrons, urinary bladder, etc., seems to be associated with water absorption or excretion. Independent transport of Na^+ (exchanges with NH_4^+) and Cl^- (exchanged with HCO_3^-) at the gill level appears to provide flexibility for efficient acid-base regulation, in addition to maintenance of blood ionic concentrations functions.

Ion exchanges seem to be modified when the animals are faced with high hydrostatic pressures from the environment. Enzymatic activities associated with ion transport (such as Na/K-ATPase) then undergo large variations. This implies reorganisation of cellular membranes at the molecular level.

It seems well established that a number of hormones participate in osmoregulation. Much remains to be done about their mechanism of action in fishes. At least some of these substances display rhythmic variations which are not paralleled by similar changes in the ion fluxes they have been shown to control.

Thus, osmoregulation of fishes remains an important ecophysiological problem. At the present time, the mechanisms it involves should be looked at, at the cellular level, using a physiological and biochemical approach.

SOUND

The subject of mechanoreception has been covered by Dale's and Popper's chapters with some reference to sound by Blaxter. The relative role of near and far field effects was not fully brought out. A sound source in water will produce a sinusoidal change of sound pressure, detectable by hydrophones, which will fall off as $1/r$ from the source where r is the distance from the source. A back-and-forth motion of the water particles will also be produced which can be detected by velocity metres suspended in the water. This particle motion or particle displacement will fall off as $1/r^2$ for a pulsating source (like a resonating swimbladder) or as $1/r^3$ for a vibrating source (like a fish tail). Thus sound pressure will be relatively more important than particle displacement at a great distance from the source (the far field); particle displacement will be more important near the source (the near field). The near field will depend on the frequency of the sound, being longer at low frequencies. Usually it is defined as $\lambda/2\pi$ where λ is the wavelength of the sound.

Fish may have the potential to respond to both types of stimuli, to particle displacement by the neuromast or otolith organs and to sound pressure via the swimbladder. As shown by Dale the neuromast and otolith organs have a common structure with hair cells usually coupled in some way to a gelatinous cupula. The cupula is partially calcified in the otolith organs of the labyrinth. The hair cells have a polarity in that the relative position of the kinocilium and stereocilia cause depolarisation or hyperpolarisation depending on whether the kinocilia are bent towards or away from the stereocilia. The hair cells thus have the ability to respond to the direction of the stimulus source. As a fish is likely to be small in relation to the wavelength of the sounds it responds to, and being a rigid structure (with an axial vertebral column) the whole fish will tend to vibrate back-and-forth in the sound field. The walls of the lateral line being rigid will also vibrate in a similar way but the water within the lateral line will vibrate as $1/r^3$. Thus there will be differential reactions between the water and lateral line organs at different distances from the stimulus source which will give the fish a measure of range. Directional information will be given where the lateral line has branches orientated at different angles to the axis of the fish. Only canals with openings orientated towards the stimulus source will tend to have particle motion transmitted down the length of the canal. In the labyrinth there will tend to be an inertial effect with a calcified otolith, the macula vibrating with the body of the fish and with the otolith lagging. The hair cells of the maculae are aligned with axes of sensitivity in different directions. Depending on the orientation of the axis of the fish to the stimulus source, different groups of hair cells will be maximally stimulated so giving the fish further directional information.

The swimbladder acts as a resonator, transducing sound pressures to displacements which are re-radiated to the labyrinth. This is neatly shown by the work of Chapman and Sand who found that the sensitivity and frequency range of the dab (a flatfish without a swimbladder) was enhanced by placing a small resonating balloon near its head. In many fishes the swimbladder is closely apposed to the labyrinth as shown by Popper. In the Ostariophysi the Weberian ossicles provide a very close coupling indeed and this is refelcted in their audiograms which show high sensitivity and response to frequencies of 10 KHz or more. Usually the coupling is to the sacculus but the clupeids have a gas filled otic bulla which is close to the utriculus and they also have a higher frequency response than less specialised fishes. The clupeids also have a unique coupling between the labyrinth and lateral line. Beside the bulla a lateral membrane in the wall of the skull vibrates in sympathy with the bulla membrane. This lateral membrane is at the focal point of the lateral line system which is confined to the head. The lateral line is thus stimulated from a central position by sound pressures transduced to particle displacements by the bulla and also stimulated by particle displacements received directly from the stimulus source. This may give them the ability to measure range since the ratio of sound pressure to particle displacement varies with the distance of the source.

A resonating swimbladder may to some extent reduce the directional hearing of fish since the displacements produced within the fish's body may mask the directional displacements received directly from the source. The swimbladder will, however, tend to increase the sensitivity of the fish, especially in the far field, where sound pressure is the dominant stimulus.

A number of interesting and unresolved questions have arisen in the meeting. It is not clear whether all hair cells in the mechanoreception have a cupula. If the cupula has a density similar to water, it may have very little inertial effect. Its role may be to reduce excessive bending of the hair cells. Alternatively it may act to restore the hair cells quickly to some type of "resting" position.

Popper mentioned that the pulse pattern of sound produced by fish may be more important than actual frequency perception. It seems quite likely that the high frequency responses of fish may be a "trick" to enable them to respond to transient noises of very short duration, i.e., to stimuli with a very high rise line at their onset.

PRESSURE-BUOYANCY-SWIMBLADDER

Hydrostatic pressure is a specific parameter for the aquatic environment. The over all effect of this factor has only been looked into in the last decades. In earlier days the hydrostatic effect was related to higher animals like vertebrates. Today we know that the hydrostatic pressure does influence the physiology of all living organisms from bacteria to mammals.

The direct hyperbaric effect on biochemical reactions would make a chapter of its own and will not be dealt with here. Here we will discuss flotation in fish and the papers dealt with in this field, in other words; swimbladder physiology.

There are more functions served by the swimbladder. There is evidence of the swimbladder being an acoustical organ for sound production of love or sexsongs between fish (Freytag 1978). The drumming muscle of the haddock, Gadus aeglefinus, is a good example of this function.

Its function as a receptor organ of acoustical waves may also be discussed. In herring there is no doubt that there is anatomical evidence that the swimbladder is in direct contact with the central acoustical receptor organ (Blaxter and Denton 1976).

As mentioned by John Blaxter in the literature swimbladders are often divided into two major groups, viz, the physoclist swimbladder and the physostome swimbladder. The first group is known to be a closed cavity equipped with a gas gland and an oval. The physostome swimbladder is a cavity with open connexion to the alimentary system of the fish and has no gas gland or oval. The connexion between the physostome swimbladder and tne alimentary system is the so called pneumatic duct. By this major grouping of the swimbladders one may also divide them in two major functional groups, but it is not that simple. The physoclist and the physostome swimbladders as described here do represent the two outer ends of a slide rule where all types of swimbladders are the figures with connexions to each other.

First of all, the swimbladders are designed in different ways with one or more chambers and the geometry varies to a great extent (Fig. 1).

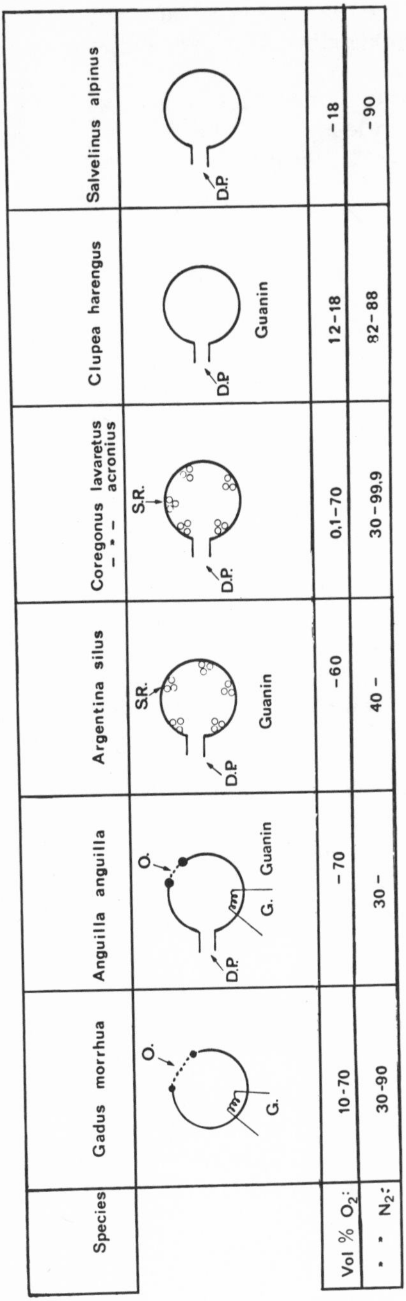

Species	Gadus morrhua	Anguilla anguilla	Argentina silus	Coregonus lavaretus – " – acronius	Clupea harengus	Salvelinus alpinus
Structures	O., G.	O., G., D.P., Guanin	S.R., D.P., Guanin	S.R., D.P.	D.P., Guanin	D.P.
Vol % O_2:	10-70	-70	-60	0,1-70	12-18	-18
" " N_2:	30-90	30 -	40 -	30-99,9	82-88	-90

Fig. 1. Different swimbladders schematically outlined. G = gas gland, DP = pneumatic duct, O = oval, SR = single retes. Guanin means inclusions of crystals which make the swimbladder wall less penetrable for gases.

Before going further with what's going on in the swimbladder another major question has also to be answered, that is: how much does the swimbladder act as a hydrostatic organ as such? Are the fishes capable of being buoyant over the whole depth range of their normal habitats? If they happen to be buoyant only in the upper regions of their habitats we can forget the whole problem of their ability to compensate increasing hydrostatic pressure. The total gas volume in a fish caught at a certain depth indicates whether the fish was near buoyancy or not at the depth of catch. The gas mixture indicates if the fish has been at the depth of catch for some time (Sundnes, 1969). In a steady state situation at depth, the diffusion loss gives an increase of the nitrogen content. On the other hand, while gas pressure is increased to compensate for increasing depth, a high percentage of oxygen is to be found in the swimbladder.

There are also other indications - that many fishes are buoyant at depth, but the most problematic fishes in this case are the Clupeiformes like the char, Salvelinus alpinus, and the herring, Clupea harengus. There is no doubt that they swallow air at the surface but this swallowing of air cannot make them buoyant at the depth where they are to be found. When caught, herring or char, always expel gas on being taken to a lower pressure. This expelling of air has also been observed in the field. In fact, before the echosounder became current the upward migration of herring was always observed by the expelled gas. In the char gas is expelled also due to stress. Cooperation of the char in swimbladder experiments in aquaria is mostly non-existent. They never get tame like the cod and they always expel air as soon as the human part of the experimental team shows up.

On the other hand, we have observations of herring migrating upwards from 200-300 metres depth where they release gas (Sundnes & Bratland 1972). We think that this gas release takes place to avoid overflotation. That means that the herring must have been buoyant at this depth. Swallowing air at the surface to be buoyant at 200 metres depth seems impractical. No organic gases have been found in the swimbladder under normal conditions. The question of gas filling mechanism in herring is still open. Buoyant char have been caught in gillnets but not deeper than 60 metres. Also here the gas filling mechanism is unknown. All these fishes caught in nature have been in steady state situation or on their way to lower pressure. It has so far not been possible to catch either char or herring in a situation where they are compensating for increasing pressure. The most certain way to get more knowledge about the real physostomes is to perform aquarium experiments on herring since the char is so uncooperative.

But all this discussion about the physostomes does not mean that the gas filling mechanism in the physoclists is understood. Is we recall Fig. 1 the gas analyses from both physoclist and physostomes indicates that the gas filling mechanism for those which are buoyant may be the same.

The most favourable situation to get data from the fish is down in the open water. Most of the swimbladder data found in the literature from fish in nature are based on captive fishes. These fishes have been forced to the surface before gas sampling. A few years ago Gunnar Sundnes had the opportunity to use the German submersible laboratory Helgoland. This was located at the seabed outside Kiel at a depth of approx. 30 metres (Sundnes et al. 1977). By sampling gas from the fish at depth of catch and also sampling gas from fish after being brought to the surface, it was shown that the gas mixture can change when the fish are taken to the surface before gas sampling.

To get relevant data from fish with regard to swimbladder physiology one has either to resort to diving or simulating deep water situation in aquaria. The first method has limitation to depth and the survival of divers. Simulation of deep water in aquaria has been performed in several ways as for example sending aquaria with fish inside, down to the actual depth for the experiment or by the use of pressure aquaria connected to high pressure pumps (Sundnes et al. 1962). The deep location of aquaria has all the disadvantages mentioned earlier of catching fish at a depth and sampling gas at the surface. The pressure aquarium gives the investigator the possibility to telemeter the measurements over short distances and keep the fish under the desired pressure while measuring. However, pumping system give undesired side effects such as cavitation in the pump whereby gas is trapped (Sundnes 1979). The system suggested is a simple one in which depth can be

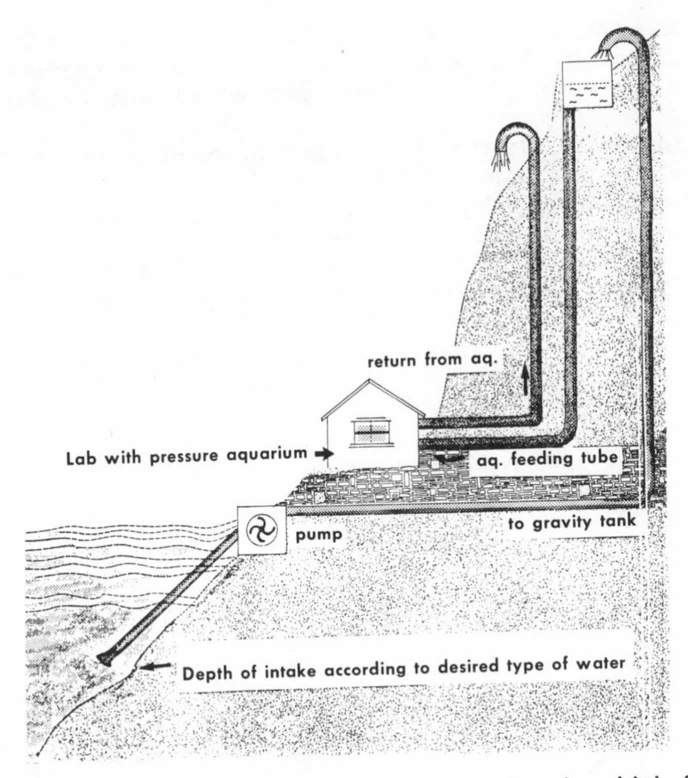

Fig. 2. Diagrammatic presentation of an aquarium in which fishes could be subjected to pressures of 200-300 metres and the gas content of their swimbladders sampled.

can be simulated and in which physical and chemical parameters are correct (fig. 2). The most important part of it is a mountain of wanted height. (In Norway there are some still left!) By using a pressure pump (if necessary more pumps and more steps) water can be pumped up to any height to a gravity tank from which the pressure aquarium is fed. By a pressure variable valve or a simple cock on a return tube from the aquarium up the mountain a nice and normal deep water environment could be obtained in the aquarium with a flow of water for keeping fish alive for the time required. By acoustic telemetry there is a lot to be done. Still there are problems of stress on the fish to be overcome under any experimental conditions (Sundnes loc. cit.).

REFERENCES

Blaxter, J.H.S. & Denton, E.J. (1976) Function of the swimbladder-inner ear-lateral line system of herring in the young stages. J. Mar. Biol. Ass. U.K. 56: 487-502.

Freytag, G. (1968) Ergebnisse zur marinen Bioakustik. Protok. Fischereitech. 52 (11): 252-352.

Sundnes, G. (1979) Forskningsakvariet - behov og virkelighet. Univ. of Trondheim - Trondhjem Biologiske Stasjon. Akvarie-Tanken 4: 1-7.

Sundnes, G. & Bratland, P. (1972) Notes on the gas content and neutral buoyancy in physostome fish. FiskDir. Skr. Ser. HavUnders. 16: 89-97.

Sundnes, G., Bratland, P. & Strand, E. (1969) The gas content in the coregonid swimbladder. FiskDir. Skr. Ser. HavUnders. 15: 274-278.

Sundnes, G., Gulliksen, B. & Mork, J. (1977) Notes on the swimbladder physiology of cod (*Gadus morhua*) investigated from the Underwater Laboratory "Helgoland". Helgoländer wiss. Meeresunters. 29: 460-463.

SUBJECT INDEX

SYSTEMATIC INDEX